Biofortification of Food Crops

Ummed Singh • C S Praharaj • S S Singh
N P Singh
Editors

Biofortification of Food Crops

Editors
Ummed Singh
Division of Crop Production
ICAR-Indian Institute of Pulses Research,
Kanpur, UP, India

C S Praharaj
Division of Crop Production
ICAR-Indian Institute of Pulses Research,
Kanpur, UP, India

S S Singh
Division of Crop Production
ICAR-Indian Institute of Pulses Research,
Kanpur, UP, India

N P Singh
ICAR-Indian Institute of Pulses Research,
Kanpur, UP, India

ISBN 978-81-322-2714-4 ISBN 978-81-322-2716-8 (eBook)
DOI 10.1007/978-81-322-2716-8

Library of Congress Control Number: 2015960950

Springer New Delhi Heidelberg New York Dordrecht London

Printed on acid-free paper

Springer (India) Pvt. Ltd. is part of Springer Science+Business Media (www.springer.com)

Foreword

Micronutrient malnutrition or *hidden hunger* is an alarming public health issue in most parts of the world including India as more than half of the world's population is iron deficient, while around one-third is zinc deficient. These deficiencies cause enormous loss in both quantity and quality in human life and its endeavors to progress and prosper. One of its probable remedies is in realizing the potential of biofortified crops or plants for nourishing nutrient-depleted soils, raising crop productivity, and providing nutritional benefits to the components of our own ecosystem – the plants, humans, and livestock. As biofortification usually refers to producing staple foods whose edible portions are denser in bioavailable minerals and vitamins, these could revolutionize our efforts in rendering more people above undernourished or malnourished.

Transforming India from green to protein (and micronutrients) revolution begins with the eradication of hunger to eradication of malnutrition. In this endeavor, our role in enhancing food production is immense and needs to be more reinforced over space and time. Despite the inherent constraints associated with cultivation of quality output from crops through production sustainability, our country is marching gradually toward realizing a target of 25 million tons of pulses by 2020 for elimination of protein malnutrition. Although it is a mammoth task to achieve keeping in view the harsh realities on ground, we would strive hard for it racing against time and space. In addition to targeted production, as pulses are known to be rich in protein, carbohydrates, fiber, Fe, and Zn, enhancing bioavailability of micronutrients through biofortification is a possible way to further strengthen or enrich their micronutrient contents. Strategically increasing mineral bioavailability with reduced antinutrient levels could have repercussions in appropriately enhancing mineral bioavailability. In this direction, improved agricultural tools, techniques, and approaches could be handy in enabling desired quality enrichment in food crops through reduction in antinutrient levels. In this context, this book on *Biofortification of Food Crops* is timely and handy as it would provide necessary impetus among the different stakeholders in putting in their efforts for a successful mission toward elimination of malnutrition from our country.

I would like to congratulate ICAR-Indian Institute of Pulses Research, Kanpur, and the editors of the book for their efforts in bringing out this publication. I am sure that this publication will be quite useful to all those involved with the larger issues related to malnutrition especially in our clientele and the poor. I urge the readers to be united in our efforts to bolster our support for a strong, healthy, and optimally nourished civil society enabling the nation to be self- and nutritionally sufficient through individual and collective contribution to solution of malnutrition through biofortification of food crops. I wish the editors for successful and timely completion of this valued publication.

Secretary, DARE and DG,
Indian Council of Agricultural Research,
New Delhi, India
23 February, 2015

S. Ayyappan

Foreword

Micronutrient deficiencies are public health issues in most parts of the world including India. Among these the deficiencies due to iron and zinc are major concern to all of us. Exploiting the potential of biofortified crops could be an important approach by nourishing the nutrient-deficient soils, upscaling crop productivity, and supplying adequate nutritional benefits to the plants, humans, and livestock. Therefore, biofortification has an immense potential in correcting these imbalances by providing good quality nutrition.

Pulses are known to be rich in protein, carbohydrates, fiber, Fe, and Zn. In addition to targeted production of 25 million tons of pulses by 2020, bioavailability of micronutrients through biofortification is a possible way to further strengthen the micronutrient contents in pulses. In this context, improved agronomy, transgenics, and quality breeding strategies could be handy in enabling desired quality enrichment in food crops. I am confident that this book *Biofortification of Food Crops* would provide necessary impetus among the different stakeholders in putting in their best efforts for elimination of micronutrient malnutrition or hidden hunger from our country.

I would like to congratulate ICAR-Indian Institute of Pulses Research, Kanpur, for their work in this area and place on record my appreciation for the efforts of editors of the book in bringing out this publication. I am highly optimistic that this publication will be quite useful to all those involved in the efforts to correct micronutrient malnutrition. I hope that a wide range of readers will find this publication to be of immense value and many of them will use these approaches for biofortification of important crops. I expect that with the inputs of readers, many editions of this book will come out in the future. I am very happy to write this foreword for this important book and convey my best wishes to all stakeholders.

Formerly-Secretary, Department of Health Research and DG, Indian Council of Medical Research, New Delhi, India
September11, 2015

V.M. Katoch

Preface

ICAR-Indian Institute of Pulses Research is a premier institute in the crop science division of the Indian Council of Agricultural Research (ICAR). The institute is mandated with the basic, strategic, and applied research on major pulse crops. With the key role in developing technologies and materials toward pulses improvement, production, and protection and giving its fruits to our clientele, *the farmers*, its activities also revolve round generating basic knowledge and information including human resource development by adequate training and education through tactical linkages and strategic coordination with the network on pulses research programs across the country and the globe.

Biofortification refers to producing staple foods whose edible portions are more dense in bioavailable minerals and vitamins. It is more relevant in the context of micronutrient malnutrition or *hidden hunger* which is an alarming public health issue in most parts of the world including India. More than half of the world's population is iron deficient, while around one-third is that of zinc. About 40 % of women and preschool children in Asia and Africa have low hemoglobin levels mainly due to Fe deficiency. These micronutrient deficiencies could cause stunting, respiratory tract infections, malaria, diarrhea, and others in human; and the solution to these is probably possible through crops that naturally reduce anemia, cognitive impairment, and other nutritionally related health problems. Here, biofortified plants could come to our rescue with the proven potential to nourish nutrient-depleted soils, help increase crop productivity, and provide nutritional benefits to plants, humans, and livestock.

Since pulses and cereals have the lion share in major dietary components of the food chain, their bioavailability through biofortification (with desired micronutrients) provides a truly feasible means of reaching the poor and undernourished. The pulses (chickpea, lentil, pigeon pea, mung bean, and urd bean), for example, are rich sources of protein, complex carbohydrates, dietary fiber, and micronutrients, viz., Fe and Zn. Biofortification of micronutrients in food crops is in fact dependant on seed type, growing environment, inherent varietal difference, agronomic constraints, and several key antinutritional factors. As an antidote, suitable strategies with increased mineral bioavailability promoter profiles with reduced antinutrient levels could appropriately enhance bioavailability of minerals. Thus, biofortification strategies involve increasing the mineral and vitamin content in food plants through modern breeding approaches, improved agronomy, and physiological, microbiological, and biotechnological tools enabling reduction in antinutrients to safe levels in food staples and thereby promoting bioavailability of nutrients. In this context, the book entitled *Biofortification of Food Crops* is handy as it could provide a suitable platform in our collective efforts for an appropriate dialogue among the scientists, researchers, entrepreneurs, policy makers, and farmers in reducing the budding issues of malnutrition through different means.

We are thankful to the contributors of the book chapters for their efforts in bringing out this publication. We are confident that this edited publication prepared out of experts' opinions and suggestions will be quite useful to all those directly or indirectly concerned with the emerging issues related to malnutrition and a possible solution to it through biofortification.

Kanpur, UP, India	Ummed Singh
Kanpur, UP, India	C S Praharaj
Kanpur, UP, India	S S Singh
Kanpur, UP, India	N P Singh

Contents

Part I

Biofortification for Health and Nutrition

Biofortification: Introduction, Approaches, Limitations, and Challenges

1

Ummed Singh, C S Praharaj, S.K. Chaturvedi, and Abhishek Bohra

Abstract

Micronutrient malnutrition is known to affect more than half of the world's population and considered to be among the most serious global challenges to humankind. Modern plant breeding has been historically oriented toward achieving high agronomic yields rather than nutritional quality, and other efforts related to alleviating the problem have been primarily through industrial fortification or pharmaceutical supplementation. Micronutrient malnutrition or the hidden hunger is very common among women and preschool children caused mainly by low dietary intake of micronutrients, especially Zn and Fe. Biofortification, the process of increasing the bioavailable concentrations of essential elements in edible portions of crop plants through agronomic intervention or genetic selection, may be the solution to malnutrition or hidden hunger mitigation. The Consultative Group on International Agricultural Research has been investigating the genetic potential to increase bioavailable Fe and Zn in staple food crops such as rice, wheat, maize, common beans, and cassava.

Keywords

Biofortification • Breeding • Genetic engineering • Limitations • Strategy

1.1 Introduction

Biofortification, the process of breeding nutrients into food crops, provides a comparatively cost-effective, sustainable, and long-term means of delivering more micronutrients. This approach not only will lower the number of severely malnourished people who require treatment by complementary interventions but also will help them maintain improved nutritional status. Moreover,

U. Singh (✉) • C S Praharaj
Division of Crop Production, ICAR-Indian Institute of Pulses Research, Kanpur, Uttar Pradesh 208 024, India
e-mail: singhummed@yahoo.co.in; cspraharaj@hotmail.com

S.K. Chaturvedi • A. Bohra
Division of Crop Improvement, ICAR-Indian Institute of Pulses Research, Kanpur, Uttar Pradesh 208 024, India
e-mail: sushilk.chaturvedi@gmail.com; abhi.omics@gmail.com

U. Singh et al. (eds.), *Biofortification of Food Crops*, DOI 10.1007/978-81-322-2716-8_1

biofortification provides a feasible means of reaching malnourished rural populations who may have limited access to commercially marketed fortified foods and supplements.

The biofortification strategy seeks to put the micronutrient-dense trait in those varieties that already have preferred agronomic and consumption traits, such as high yield. Marketed surpluses of these crops may make their way into retail outlets, reaching consumers in first rural and then urban areas, in contrast to complementary interventions, such as fortification and supplementation, that begin in urban centers. Biofortified staple foods cannot deliver as high a level of minerals and vitamins per day as supplements or industrially fortified foods, but they can help by increasing the daily adequacy of micronutrient intakes among individuals throughout the life cycle (Bouis et al. 2011).

1.2 Minerals and Vitamins

Minerals, in the context of the human diet, are inorganic chemical elements (or more properly their dissociated ions) that are required for biological or biochemical processes including the accumulation of electrolytes. Carbon, hydrogen, nitrogen, and oxygen are excluded from the list as these are found in common organic molecules. There are 16 essential minerals, but 11 of them are required in such small amounts and/or are so abundant in food and drinking water that deficiency arises only in very unusual circumstances. The remaining five are present in limiting amounts in many foods, so a monotonous diet can easily result in deficiency. These minerals are iodine (I), iron (Fe), zinc (Zn), calcium (Ca), and selenium (Se). Deficiency diseases arise when diets are based predominantly on staple foods, such as milled cereals, which have a low bioavailable mineral content (Christou and Twyman 2004). Mineral deficiency is therefore most prevalent in developing countries, where there is poor access to fresh foods, although Ca deficiency is also widespread in the industrialized world (Galera et al. 2010).

1.2.1 Iodine

Iodine is an essential component of the thyroid hormones thyroxine and triiodothyronine, which regulate growth and development and maintain the basal metabolic rate. However, only 30 % of the body's iodine is stored in the thyroid gland, and the precise role of the 70 % distributed in other tissues is unknown. It may overlap with the function of other minerals such as Se or Fe and Zn (Lyons et al. 2004). Goiter is another important symptom of iodine deficiency and results from the lack of thyroxine inducing the production of thyroid-stimulating hormone, which in turn causes the thyroid gland to swell (Dunn 2003). India is one of the worst affected countries in the world, with more than 50 million cases of goiter and more than two million of cretinism.

1.2.2 Vitamin A

Deficiency associated with blindness and increased risk of disease and death for small children and pregnant women can be addressed through supplements, which are now estimated to reach children at least once a year in 40 countries. The UN Standing Committee on Nutrition (UN/SCN) estimates that 140 million children and 7 million pregnant women are VA deficient, primarily in Africa and South/Southeast Asia. In 1998, WHO, UNICEF, Canadian International Development Agency, USAID, and the Micronutrient Initiative launched the VA Global Initiative. This provides support to countries in delivering VA supplements.

1.2.3 Iron

Iron has numerous important functions in the human body, reflecting its ability to act as both an electron donor and acceptor. In this role, it forms the functional core of the heme complex, which is found in the oxygen-binding molecules hemoglobin and myoglobin, and the catalytic center of cytochromes, which carry out redox reactions. Iron is therefore required for oxygen transport in the body and for energy metabolism, also contributing to the catalytic activity of a range of nonheme enzymes such as ribonuclease reductase (WHO/FAO 1998). The immediate outcome of Fe deficiency is iron deficiency anemia (IDA), which is thought to affect at least two billion people worldwide. More than half of these cases could be addressed by increasing the amount of Fe in the diet, but as for iodine this is difficult in developing countries where the population relies on staples, because cereal grains contain very low levels of Fe and also contain antinutritional compounds such as phytate that inhibit Fe uptake (Zimmermann et al. 2004).

1.2.4 Zinc

Zinc is an essential functional component of thousands of proteins. Many contain zinc prosthetic groups (e.g., zinc finger, zinc twist) and approximately 100 enzymes require Zn as a cofactor. Some olfactory receptors cannot function without Zn. Many cells in the body also secrete Zn as a signaling molecule, including cells in the immune and nervous systems (WHO/FAO 1998). Nearly two billion people are at risk of Zn deficiency, predominantly children and pregnant women. Signs of severe zinc deficiency include hair loss, skin lesions, wasting, and persistent diarrhea. The mineral appears to be particularly important during periods of rapid growth, and insufficient intake during childhood and adolescence can delay growth, sexual development, and psychomotor development (WHO/FAO 1998).

1.2.5 Calcium

Calcium is the most abundant mineral in the human body, accounting for 1–2 % of an adult's body mass. Over 99 % of Ca is stored in the teeth and bones, where it plays an important structural role (WHO/FAO 1998). However, Ca, like Zn, is also an enzyme cofactor and an important signaling molecule (a secondary messenger). It plays a pivotal role in the blood clotting cascade. Calcium deficiency has a profound impact on bone health, resulting in rickets if deficiency occurs in the young and osteoporosis if it persists into old age.

1.2.6 Selenium

Selenium is found in two unusual amino acids—selenocysteine and selenomethionine—which are the principal functional components of selenoenzymes. It is an essential cofactor in approximately 50 enzymes, including those whose function is to reduce antioxidant enzymes (such as glutathione peroxidase) and those whose function is to remove mineral ions from other proteins (such as thyroid hormone deiodinases) (Lyons et al. 2004). Se is an antioxidant with health benefits including the prevention of cancer and heart disease (WHO/FAO 1998).

1.2.7 Folate

Deficiency associated with increased risk of maternal death and complications in birth and also associated with neural tube defects in infants and with an estimated 200,000 severe birth defects every year can be addressed through fortification of wheat products (Haddad et al. 2004).

1.3 Alleviating Hidden Hunger: Interventions

The term "hidden hunger" has been used to describe the micronutrient malnutrition inherent in human diets that are adequate in calories but lack vitamins and/or mineral elements. The diets of a large proportion of the world's population are deficient in Fe, Zn, Ca, Mg, Cu, Se, or I, which affects human health and longevity and therefore national economies. Mineral malnutrition can be addressed by increasing the amount of fish and animal products in diets, mineral supplementation, and food fortification and/or increasing the bioavailability of mineral elements in edible crops. However, strategies to increase dietary diversification, mineral supplementation, and food fortification have not always proved successful. For this reason, the biofortification of crops through the application of mineral fertilizers, combined with breeding varieties with an increased ability to acquire mineral elements, has been advocated (White and Broadley 2009).

Food fortification and supplementation are currently the most cost-effective strategies to address global mineral malnutrition. The most successful strategy has been salt iodization (fortification with iodine) which has reduced the incidence of goiter and other IDD symptoms markedly where the scheme has been introduced (Galera et al. 2010). Most strategies to improve mineral nutrition have been less successful because of political, socioeconomic, infrastructure-related, and technical constraints that are apparent in most developing countries.

1.3.1 Food Fortification

Food fortification is one of the most cost-effective long-term strategies for mineral nutrition (Horton 2006). Fortification of dairy products such as bread and milk with different minerals (and vitamins) has been successful in industrialized countries However, this strategy is difficult to implement in developing countries because it relies on a strong food processing and distribution infrastructure. Fortification takes place during food processing and increases the product price. These factors make fortified products unaffordable to the most impoverished people living in remote rural areas. Since many parts of the world suffer from multiple deficiencies, strategies must also be developed to fortify foods simultaneously with several micronutrients without adverse interactions among them (Zimmermann et al. 2004). The addition of a single micronutrient would have more or less the same cost implications as the addition of several (Alavi et al. 2008).

Zinc fortification has been implemented in the industrial world but rarely in developing countries. One exception is Zn-fortified wheat and maize flours in Mexico, which are used to make bread and tortillas, the two principal staples (IZINCG 2007). Organizations such as the Zinc Task Force (ZTF) and the International Zinc Nutrition Consultative Group (IZiNCG) are fighting Zn malnutrition by promoting diverse strategies to eliminate it. As Zn and Fe deficiencies tend to go hand in hand, it has been suggested that double fortification would be effective with little additional cost, particularly if Fe fortification were already in place.

1.3.2 Industrial Fortification

The marketed supply of a widely consumed staple food can be fortified by adding micronutrients at the processing stage, and historically this is how micronutrient deficiencies have been addressed in the developed world. Concentration in the food industry also tends to strengthen compliance and quality assurance. Consumption of wheat flour products is growing around the world, even where wheat is not a traditional food staple, opening new fortification opportunities at the milling stage. Public support for traditional fortification has recently been enhanced by new promotion and coordination efforts: Micronutrient Initiative (based in Canada), Flour Fortification Initiative (based in Emory University), Mid Day Meal Scheme (India), and the Global Alliance for Improved Nutrition (GAIN, based in Geneva). Other important global actors include the Network for Sustained Elimination of Iodine Deficiency and the International Zinc Nutrition Consultative Group. In addition, efforts are underway to set regional standards for fortification. The Flour Fortification Initiative provides an assessment of global progress, and they report that 26 % of the global wheat market is fortified, benefiting 1.8 billion people. Most wheat fortification efforts in the developing world are still preliminary or on a pilot scale; they are primarily in the Western Hemisphere, with little sustained activity in Asia and Africa, where most of the micronutrient deficient populations live. Certain kinds of fortification may be impractical for some important food staples (e.g., VA fortification of milled rice) or may introduce off colors or flavors (e.g., VA fortification of white maize).

Industrial fortification will only apply to marketed supplies and therefore may not reach those among the poor who obtain food outside of commercialized channels. Given these limitations, it is clear that industrial fortification of food cannot provide a complete solution to the problem of micronutrient deficiencies in the medium term. It is in this context that a role emerges for biofortification as a complementary strategy.

1.3.3 Promotion of Dietary Diversification

Education is an important element in ensuring that improvements in income result in better maternal and child health. However, dietary diversification is constrained by resource availability for poor households and seasonal availability of fruits and vegetables. Promotion of home gardens is often touted, but the poor have a high opportunity cost for their labor and often limited land. Increased production of fruits and vegetables for household use reduces resources available for other income-earning or food production activities. This type of effort is also relatively expensive and difficult to sustain on any large scale.

1.3.4 Food Supplementation

Supplementation is the best short-term intervention to improve nutritional health, involving the distribution of pills or mineral solutions for immediate consumption. This helps to alleviate acute mineral shortages but is unsustainable for large populations and should be replaced with fortification at the earliest opportunity (Shrimpton and Schultink 2002). In industrialized countries, with few mineral malnutrition problems, supplementation is focused on a small subset of the population with specific deficiencies resulting from medical conditions. In developing countries, where acute and chronic deficiencies are common, supplementation is highly recommended to complement the diet (fortified or otherwise) of the entire population (Nantel and Tontisirin 2002).

Periodic provision of supplements (often in the form of tablets) can address deficiencies of micronutrients that are stored in the body, such as vitamin A and iron. Supplementation can be cheap compared to the large public health benefit. The total annual cost of iron tablet supplementation in India to reach 27 million women and 128 million children at risk is only $5.2 million. Yet even small budgets can be difficult to sustain year after year when they are dedicated to the welfare of politically weak or socially marginal beneficiaries. Furthermore, while certain populations are easy to reach through existing institutions (e.g., schoolchildren through schools), it is often difficult to accomplish full coverage of those most at risk—poor women and very young children. Thus, supplementation has often been most effective when delivered together with other maternal and child health interventions.

The distribution of vitamin A supplements has been one of the most cost effective and successful acute intervention programs in the developing world (Shrimpton and Schultink 2002), but this is a rarity. Like fortification, successful supplementation strategies require a robust infrastructure and a government determined to improve the nutritional health of its population (Shrimpton and Schultink 2002). Even more than fortification, supplementation requires compliance monitoring because more people will neglect to take regular supplements at prescribed intervals than fortified staple foods. Mineral supplements are prescribed for acute deficiency diseases in industrialized countries as well as serving a niche health market. Zn supplements rank very highly according to the Copenhagen Consensus report cost–benefit analysis.

1.3.5 Biofortification

Conventional interventions have a limited impact, so biofortification has been proposed as an alternative long-term approach for improving mineral nutrition (Zhu et al. 2007). Biofortification focuses on enhancing the mineral nutritional qualities of crops at source, which encompasses processes that increase both mineral levels and their bioavailability in the edible part of staple crops. The former can be achieved by agronomic intervention, plant breeding, or genetic engineering, whereas only plant breeding and genetic engineering can influence mineral bioavailability. Plant breeding and genetic engineering are often compared because, in contrast to agronomic interventions, both involve changing the genotype of a target crop. The two processes are similar in aim, albeit different in scope. Both attempt to create plant lines carrying genes that favor the most efficient accumulation of bioavailable minerals—plant breeding achieves this by crossing the best performing plants and selecting those with favorable traits over many generations—whereas genetic engineering accesses genes from any source and introduces them directly into the crop. Plant breeding is limited to genes that can be sourced from sexually compatible plants, whereas genetic engineering has no taxonomic constraints and even artificial genes can be used.

The main advantage of genetic engineering and plant breeding approaches for mineral enhancement is that investment is only required at the research and development stage, and thereafter the nutritionally enhanced crops are entirely sustainable. Furthermore, as stated above, mineral-rich plants tend to be more vigorous and more tolerant of biotic stress, which means yields are likely to improve in line with mineral content (Frossard et al. 2000; Nestel et al. 2006). Unlike conventional intervention strategies, genetic engineering and plant breeding are both economically and environmentally sustainable (Stein et al. 2008). Although there are no commercial nutritionally enhanced plants derived from either method at the current time, this approach has the greatest long-term cost-effectiveness overall and is likely to have an important impact over the next few decades. Biofortification is also likely to be more accessible than conventional interventions in the long term because it removes hurdles such as the reliance on infrastructure and compliance. Moreover, plants assimilate minerals into organic forms that are naturally bioavailable and contribute to the natural taste and texture of the food. Economic studies have shown many potential health benefits of biofortification strategies, especially in combination with conventional strategies (Buois 2002; Stein et al. 2008).

1.3.5.1 Plant Breeding

Plant breeding programs focus on improving the level and bioavailability of minerals in staple crops using their natural genetic variation (Welch and Graham 2005). Breeding approaches include the discovery of genetic variation affecting heritable mineral traits, checking their stability under different conditions, and the feasibility of breeding for increasing mineral content in edible tissues without affecting yields or other quality traits. Breeding for increased mineral levels has several advantages over conventional interventions (e.g., sustainability); no high-mineral varieties produced by this method have been introduced onto the market thus far. This reflects long development times, particularly if the mineral trait needs to be introgressed from a wild relative. Breeders utilize molecular biology techniques such as quantitative trait locus (QTL) maps and marker-assisted selection (MAS) to accelerate the identification of high-mineral varieties, but they have to take into account differences in soil properties (e.g., pH, organic composition) that may interfere with mineral uptake and accumulation. For example, the mineral pool available to plant roots may be extremely low in dry, alkaline soils with a low organic content (Cakmak 2008).

1.3.5.2 Conventional Plant Breeding

This allows crop scientists to make significant improvement in the nutritional, eating quality, and agronomic traits of major subsistence food crops. Conventional breeding is limited, however, because it can only use the genetic variability already available and observable in the crop being improved or occasionally in the wild varieties that can cross with the crop. Furthermore, conventional breeders usually have to trade away yield and sometimes grain quality to obtain higher levels of nutrition. One example is quality protein maize (QPM), which has taken decades of conventional plant breeding work to develop into varieties acceptable to farmers. However, multiple gains are at times possible, as with iron and zinc in rice and wheat, where the characteristics that lead to more iron and zinc in the plant can also lead, by some accounts, to higher yield. Other biofortified crops, such as the orange-fleshed sweet potatoes (OFSP) promoted through the HarvestPlus program in Africa, have been successfully selected and developed for both nutrient and (at least rainy season) yield traits (Unnevehr et al. 2007).

1.3.5.3 Mutation Breeding

Mutation breeding has been used extensively in developed and developing countries to develop grain varieties with improved grain quality and in some cases higher yield and other traits. This technique makes use of the greater genetic variability that can be created by inducing mutations with chemical treatments or irradiation. The FAO/International Atomic Energy Agency (IAEA) website contains more than 2500 varieties that have been developed through mutation breeding (Mutant varieties database). Of these, 1568 are in Asia, 695 in Europe, and 165 in the United States. Most of the European and US mutants are flowers, but most in Asia are basic food crops such as wheat, rice, maize, and soybeans. According to their website, FAO/IAEA include biofortification as one of the objectives of their mutagenesis program, but there do not seem to be any applicable results yet. Varieties produced using mutagenesis can be grown and certified as organic crops in the United States, whereas transgenic crops developed using recombinant DNA (rDNA) technology cannot.

1.3.5.4 Molecular Breeding

Also called marker-assisted breeding, this is a powerful tool of modern biotechnology that encounters little cultural or regulatory resistance and has been embraced so far even by organic growers because it relies on biological breeding processes rather than engineered gene insertions to change the DNA of plants. This technique is expanding rapidly with the development of genomics, which is the study of the location and function of genes, and with the rapid decline in costs of screening plant tissue. Once scientists have identified the location of a gene for a desirable trait, they build a probe that attaches itself only to a DNA fragment, a so-called marker, unique to that gene. They then can use this marker as a way to monitor and speed up their efforts to move this trait into relatives of the plant using conventional breeding. For example, since the marker can be detected in the tissue of new seedlings, the presence or absence of the desired trait can be determined without having to wait for a plant to mature, often reducing by years the length of a typical crop development process. If molecular breeding reduces the number of generations required to develop a pure line variety by three generations, this can save 3 years of research time. The use of molecular breeding has increased dramatically both by private seed companies and government plant breeders in developed countries, and it is gradually spreading to developing countries (Pray 2006). Using this technique, plant breeders also can stack into one variety several different genes that code for different traits, for example, QPM, disease resistance, and drought tolerance in maize (Pray 2006). This technique has also been used to find recessive traits in plants that cannot be located by conventional breeding or other techniques.

1.3.5.5 Genetic Engineering

Genetic engineering is the latest weapon in the armory against mineral deficiency and uses advanced biotechnology techniques to introduce genes directly into breeding varieties. The genes can come from any source (including animals and microbes) and are designed to achieve one or more of the following goals (Zhu et al. 2007): (a) improve the efficiency with which minerals are mobilized in the soil, (b) reduce the level of antinutritional compounds, and (c) increase the level of nutritional enhancer compounds such as inulin. Genetic engineering, or rDNA, is a technique that offers still greater speed and reach because it moves specific genes with desired traits from a source organism—one which does not have to be a related organism—directly into the living DNA of a target organism. The *transgenic* trait is added without normal biological reproduction, but once in the plant it becomes inheritable through normal reproduction. Scientists first developed this technique in the laboratory in 1973 and have been using it to transform agricultural crop plants since the 1980s. Once a useful gene has been identified (which can require a major research project and many years), it is attached to both marker and promoter genes and then inserted into a plant, usually using a nonviable virus called *Agrobacterium* as a carrier. GE produces plants that are known as transgenics or less precisely as GMOs. GE has great reach because it can add valuable characteristics that are not currently found in the seeds of individual plant species. GE was necessary for the development of golden rice, which contains the precursor to VA from a daffodil plant. This was a trait missing from rice plants, and it could not be introduced conventionally since daffodils cannot be crossed with rice plants. In addition, GE can take much less time to incorporate desired traits into a crop plant than either traditional or molecular breeding. The choice of which technology to use when biofortifying crops comes down to a calculation by breeders of how to get the best results most quickly, given their budget constraint. Conventional plant breeding requires less investment in labs or highly trained human resources (molecular biologists) than either marker-assisted selection or genetic engineering, and it faces lower and less costly regulatory hurdles. However, if there are no genes for the VA precursors in the genome of a crop (as one example), no amount of conventional plant breeding can put them there, and scientists must turn to GE. Molecular breeding and GE also have advantages over traditional breeding because they make it easier to *develop* crops with multiple desired nutritional traits, maintain agronomic viability of biofortified crops, adapt agriculture improvements arising in the United States for obscure crops in developing countries, etc.

1.3.5.6 Tissue Cultures

Modern tissue culture techniques can allow scientists to reproduce plants from a single cell. These techniques are now used extensively to produce disease-free planting material of clonally propagated crops such as bananas. When tissue culture is combined with embryo rescue techniques, plant breeders can use the genes from wild and weedy relatives of a crop, which would normally not cross with the cultivated crop. This allows breeders to increase genetic variability of the cultivated crop and then bring in valuable traits of the wild and weedy relatives. These techniques have allowed scientists to cross Asian and African rice varieties and develop Nerica rice varieties with agronomic traits, such as higher yield and resistance to water stresses, that have met with growing success in Africa. Tissue culture is an important tool for propagation of roots and tubers, such as potatoes and cassava, and both of these crops are part of current biofortification research.

1.3.6 Microbiological Interventions

1.3.6.1 *Plant Growth Promoting Rhizobacteria* (PGPR)

These include beneficial bacteria that colonize plant roots and enhance plant growth by a wide variety of mechanisms. The use of PGPR is steadily increasing in agriculture, as it offers an attractive way to reduce the use of chemical fertilizers, pesticides, and related agrochemicals (Rana et al. 2012). Interventions using PGPR or other biological agents are limited. Secretion of phytosiderophores by microorganisms and plants in restricted spatial and temporal windows represents an efficient strategy for uptake of iron and other micronutrients by plants from the rhizosphere. Analysis of the complex interactions between soils, plants, and microbes in relation with micronutrient dynamics represents a unique opportunity to enhance our knowledge of the rhizosphere ecology. Such progress can provide information and tools enabling us to develop strategies to improve plant nutrition and health with decrease in the application of chemical inputs. Microorganisms are known to differ significantly in competing with higher plants for micronutrients (Steven 1991). Among bacteria, a lot of attention has been dedicated to the siderophore-mediated iron uptake by fluorescent pseudomonads. A number of other mechanisms are also involved in the sequestration and transformation by microorganisms in soil such as production of acids, alkalis, etc. PGPR constitute a significant part of the protective flora that benefit plants by enhancing root function, suppressing disease, and accelerating growth and development (Glick 1995). Microorganisms differ in competing with higher plants for micronutrients. Species of *Azotobacter* differed in their competitiveness with wheat plants in extracting Fe and Zn (Shivay et al. 2010). Biofortification of crops through application of PGPR can be therefore considered as a possible supplementary measure, which along with breeding varieties can lead to increased micronutrient concentrations in wheat crop, besides improving yield and soil fertility.

1.3.6.2 AM Fungi

Most plants, including all major grain crops and almost all vegetables and fruits, are associated with mycorrhizal fungi that improve the uptake of essential mineral elements from soils and, therefore, enhance plant growth and productivity. These symbiotic fungi, therefore, change, directly or indirectly, the mineral nutrition of plant products that are also essential for humans. However, the role of mycorrhizas on element biofortification may be piloted through agricultural practices. Mycorrhizas can potentially offer a more effective and sustainable element biofortification to curb global human malnutrition. Approximately 90 % of land plants form mycorrhizas (literally "fungus roots"): they exist everywhere, from tiny home gardens to large ecosystems (Smith and Read 1997). Six types of mycorrhizas (arbuscular, arbutoid, ecto-, ericoid, monotropoid, and orchid) are categorized by their distinct morphological characteristics (Wang and Qiu 2006). Of them, arbuscular mycorrhiza (AM) is the most common and predominant type. Arbuscules, specific "little-tree-shaped" fungal structures inside root cortical cells, serve as the main sites of nutrient exchange between the plant and the fungus. AM also has external hyphae that provide an extensive surface area or network for nutrient uptake from soils. Thus, AM is the most important mycorrhiza in agriculture and closely relates to human nutrition. Obligately depending on plant photosynthates as energy sources, the extensive AM mycelial systems (the vegetative parts of the fungus) effectively explore soil substrates and acquire soil inorganic nutrients, including major macronutrients N, P, and K and micronutrients Cu, Fe, and Zn (Caris et al. 1998), with some capacity for acquiring organic N and P (Koide and Kabir 2000). These soil-derived nutrients are not only essential for AM development but are also partly transferred to the host plant. It is believed that many plants that usually form this symbiotic relationship would be unable to survive without the mycorrhiza. Mycorrhizal mycelia and their exudates also constitute a large carbon source (20–30 % total soil microbial biomass) for the functioning of other belowground microorganisms that ameliorate soil nutrient availability through decomposition of organic compounds and weathering of inorganic materials. So, not only do the activities of AM fungi have multiple functions that enhance plant performance but they also play crucial roles in the development of soil properties and the health of the entire ecosystem.

1.3.7 Agronomic Intervention

Farmers have applied mineral fertilizers to soil for hundreds of years in order to improve the health of their plants, but within certain limits the same strategy can also be used to increase mineral accumulation within cereal grains for nutritional purposes (Rengel et al. 1999). This strategy only works if the mineral deficiency in the grain reflects the absence of that mineral in the soil and if the mineral fertilizer contains minerals that are rapidly and easily mobilizable. Also, even if plants can absorb minerals efficiently from the soil, they may store the mineral in leaves but not fruits or seeds, or they may accumulate the mineral in a form that is not bioavailable, thus having no impact on nutrition (Frossard et al. 2000). Like supplements and fortification, agronomic intervention is probably best applied in niche situations or in combination with other strategies (Cakmak 2008). One drawback of agronomic intervention is the cost and impact of the fertilizers. Fertilizer use is likely to increase the cost of food, thus reducing its availability to the most impoverished people. The expensive fertilizers must be applied regularly, with no direct yield incentive to farmers in developing countries, so the intervention would likely be omitted to save costs even though seeds produced under rich mineral conditions germinate more vigorously than those in poor soils (Cakmak 2008). There is also concern about the impact of increased fertilizer use on the environment (Graham 2003).

Agronomic strategies to increase the concentrations of mineral elements in edible tissues generally rely on the application of mineral fertilizers and/or improvement of the solubilization and mobilization of mineral elements in the soil (White and Broadley 2009). When crops are grown where mineral elements become immediately unavailable in the soil, targeted application of soluble inorganic fertilizers to roots or to leaves is practiced. In situations where mineral elements are not readily translocated to edible tissues, foliar applications of soluble inorganic fertilizers are made. It has been observed that the human population of the world has exceeded the carrying capacity of low-input agriculture, and modern inorganic fertilizers are necessary to obtain the crop yields required to prevent starvation (Graham et al. 2007). Essential plant nutrients are mainly applied to soil and plant foliage for achieving maximum economic yields. Soil application method is more common and most effective for nutrients, which are required in higher amounts. However, under certain circumstances, foliar fertilization is more economic and effective. Foliar symptoms, soil and plant tissue tests, and crop growth responses are principal nutrient disorder diagnostic techniques. Soil applications of fertilizers are mainly done on the basis of soil tests, whereas foliar nutrient applications are mainly done on the basis of visual foliar symptoms or plant tissue tests. Hence, correct diagnosis of nutrient deficiency is fundamental for successful foliar fertilization. In addition, there are some more requirements for successful foliar fertilization. Foliar fertilization requires higher leaf area index for absorbing applied nutrient solution in sufficient amount; it may be necessary to have more than one application depending on severity of nutrient deficiency.

1.3.7.1 Parboiled Rice

Fe fortification in parboiled rice is a rapid and cost-effective solution to Fe deficiency anemia in economically disadvantaged populations with rice as the major staple food and poor access to animal proteins. It focused initially on the feasibility of this innovative approach, by examining the effectiveness of Fe fortification and retention, the solubility of Fe in the grain in response to fortification treatments, and the likely pathway of Fe movement into the endosperm. NaFeEDTA has been recently approved as an ingredient to be used in supervised food fortification programs (Hurrell 2003) and is the most promising Fe fortification compound for food additives and is used intensively to prevent oxidation and color changes in food and promote its bioavailability in the human diet. The effectiveness of enhancing Fe density in white rice through the Fe-fortified parboiling process is far greater than that achieved from conventional and transgenic rice breeding. For example, Fe concentration in milled rice of IR68144-2B-2-2-3, an improved rice cultivar by conventional breeding, is 7–13 mg Fe kg^{-1} (Graham et al. 1999) and 37 mg Fe kg^{-1} in rice containing transferred soybean ferritin gene (Vasconcelos et al. 2003). In contrast, Fe concentration in the milled rice grains Fe fortified in the parboiling had 70–144 and 30–110 mg Fe kg^{-1} in 60 and 120 s milled grains, respectively.

Comparatively, a substantial loss of Fe sprayed on raw rice surface occurs if the rice is rinsed before cooking. Although a polymer coating technique has been advocated for painting Fe on the rice grain surface to minimize Fe loss from washing and/or cooking, these techniques are expensive and not practical in rice mills of developing countries. In comparison, parboiled rice is a rapid and cost-effective vehicle to deliver Fe nutrition benefits through the already established parboiling infrastructure, market network, and consumers' acceptance in Asia (particularly in the subcontinent) and Africa, where the high risk of Fe malnutrition-induced anemia is present (Graham et al. 1999). Most of the fortified Fe in the milled rice grains remained in the dilute acid-soluble pool, which is considered as potentially bioavailable in the human diet in all cultivars.

One of the significant advantages of parboiled rice is that parboiling resulted in a significant inward movement of the fortified Fe into the endosperm, countering milling-induced Fe loss in raw rice grains due to restricted distribution of Fe in the aleurone and embryo (the bran fraction) of brown rice (Bhattacharya 2004). Parboiling itself may cause inward migration of some mineral nutrients present in the surface layers of rice grain, resulting in a higher retention rate of these nutrients when being milled to produce white rice (Ali and Bhattacharya 1980; Palipane and Swarnasiri 1985), such as P, Ca, Fe, Mn, Mo, and Cr in milled parboiled rice.

1.3.8 Biofortification of Feed for Livestock

For many decades plant breeding primarily focused on yield, and little attention was given to the nutritional value of cereal residues (bran and straw) that were not used for human consumption. However, bran and straw are among the most important feed for ruminant livestock in many parts of the world, where sorghum and millet are important staple cereals in addition to rice and wheat. In sorghum and pearl millet, various crop management interventions (Reddy et al. 2003) and plant breeding (Zerbini and Thomas 2003) strategies were shown to influence yield and improve quality of the straw for animal feed. These efforts are now being complemented with the introduction of molecular markers (QTL mapping and MAS) as tools to increase breeding efficiency (Hash et al. 2003). Progress has also been reported for a completely different approach to improving cereal nutrient availability. In transgenic animals like the Enviropig, phytase is produced in the salivary glands and the active enzyme is secreted into the saliva (Forsberg et al. 2003). These pigs show improved phosphorus uptake. Such a transgenic pig might also show improved iron uptake in the intestine, owing to lower content of the antinutritional phytate.

1.4 Hindrances or Limiting Factors

1.4.1 Antinutrients

Phytate and tannins are the limiting factors in the absorption of Fe, Zn, and Ca by the gut (Mendoza 2002). Phytate occurs widely in plant tissues but is concentrated in seeds or grain. There is considerable intraspecific variation in phytate concentration in edible portions (Glahn et al. 2002; Coelho et al. 2005) that is independent of variation in Fe and Zn concentrations. In addition, several low phytic acid (lpa) mutants have been produced by non-transgenic techniques in rice, maize, wheat, barley, and soybean (Banziger and Long 2000). Fortuitously, plants with lpa mutations often show raised levels of seed Fe, Zn, and Mg (or similar levels to those found in wild type), although they do have reduced concentrations of seed Ca. Tannin concentration in edible tissues also varies greatly between varieties (Lin et al. 2005). Hence, breeding for reduced concentrations of these antinutrients appears feasible.

Phytic acid, as well as other metabolites produced by plants, such as PP, is considered an "antinutrient" because by chelating iron, it can reduce its absorption in the human gut (Jin et al. 2009). In plants, however, phytic acid fulfills essential biological functions (Murgia et al. 2012). Phytate, a mixed cation salt of D-myo-inositol, hexakisphosphate, commonly known as PA, InsP6, or IP6, constitutes up to 1–8 % of mature seed dry weight and accounts for up to 90 % of phosphorus content in cereal grains, legumes, nuts, and oil seeds. Phytate represents an important metal cation reserve (magnesium, potassium, calcium, manganese, barium, and iron) in seeds, either in the aleurone cell layer or in the seed embryo, depending on the plant species. Degradation of phytate occurs during seed germination, by means of phytases, a class of phosphatases capable of releasing at least one phosphate from phytic acid (IP6) (Bohn et al. 2008). The consequent release of phosphorus and mineral nutrients supports growth and development of the seedling. Besides its well-known role in mineral storage in seeds, IP6 also acts in the leaves in the signaling cascade triggered by drought/osmotic stress leading to stomatal closure. In guard cells ABA produces rapid changes in IP6 which trigger release of Ca2+ from endomembrane stores and inhibition of K+ inward rectifying channels. Some low IP6 cereals are less tolerant to stress and possess undesirable agronomical traits, such as reduced seed yield and lower seed viability, as observed in bread wheat (*Triticum aestivum*), rice (*Oryza sativa*), and barley (*Hordeum vulgare*) (Zhao et al. 2008), suggesting that IP6 is indeed involved in essential biological functions in the whole plant. More recently, a key role for IP6 in the maintenance of basal resistance against a wide range of pathogens has been demonstrated transgenic potatoes with compromised synthesis of IP6 (through IPS antisense RNA) are less resistant to virus infection. Disruption of IP6 biosynthesis in *Arabidopsis* is also associated with increased susceptibility to viruses and to bacterial and fungal pathogens (Murphy et al. 2008).

1.4.2 Promoters

Some organic compounds stimulate absorption of essential mineral elements by humans (Table 1.1). These include ascorbate (vitamin C), b-carotene (provitamin A), protein cysteine, and various organic and amino acids. There is considerable intraspecific variation in both ascorbate and b-carotene concentrations in fruit and vegetables (Frossard et al. 2000). For example, ascorbate concentration in cassava varied by 250-fold in leaves and 40-fold in roots among the 530 accessions of the CIAT core collection, whereas b-carotene concentration varied by 3.7-fold in leaves and 10-fold in roots. Similarly, ascorbate concentration varied almost 20-fold among *Dioscorea alata* accessions. There is also appreciable intraspecific variation in amino acid concentrations in edible tissues (Guzman-Maldonado et al. 2000). However, the complement of amino acids present in different foodstuffs is constrained by evolutionary heritage such that cereal and vegetable crops contribute complimentary amino acids to the diet (White and Broadley 2005).

Table 1.1 Mineral nutritional enhancers and antinutrients

Nutritional enhancers	Antinutrients
b-Carotene (provitamin A)	Oxalic acid (oxalate)
Inulin	Phytic acid (phytate)
Long-chain fatty acids	Polyphenols
Certain amino acids (cysteine, lysine, etc.)	Tannins
Certain organic acids (ascorbic acid, citrate, etc.)	Others
Vitamin D	Others

Adapted from White and Broadley (2005); Welch and Graham (2005)

1.5 Assessing Iron Bioavailability from Biofortified Foods

Iron bioavailability from biofortified food can be assessed through four different approaches: (i) algorithmic approximations, (ii) in vitro digestion and Caco-2 intestinal cell uptake assay, (iii) animal studies, or (iv) human studies (Cockell 2007; Fairweather-Tait 2001). The algorithmic method is the least suited to predict the effects of new circumstances, such as the nutritional impact of a new biofortified crop. By contrast, the choice between the remaining methods should take into account experimental costs, short- and long-term responses, and differences in iron absorption between laboratory animals and humans (Cockell 2007; Walter et al. 2003). It is however important to note that at least two types of human bioassays for bioavailability have qualified as necessary: efficacy and effectiveness trials. Efficacy trials assess the beneficial effect of iron biofortification under ideal conditions and thus depend only on biological factors (such as in clinical trials); effectiveness trials take into account behavioral factors, because trials are performed under "real-life situations" (Davidsson and Nestel 2004).

1.6 Biofortification: Strategic Advantages

The biofortification strategy seeks to take advantage of the consistent daily consumption of large amounts of food staples by all family members, including women and children who are most at risk for micronutrient malnutrition. As a consequence of the predominance of food staples in the diets of the poor, this strategy implicitly targets low-income households.

After the one-time investment is made to develop seeds that fortify themselves, recurrent costs are low and germplasm may be shared internationally. It is this multiplier aspect of plant breeding across time and distance that makes it so cost-effective. Once in place, the biofortified crop system is highly sustainable. Nutritionally improved varieties will continue to be grown and consumed year after year, even if government attention and international funding for micronutrient issues fade. Moreover, biofortification provides a truly feasible means of reaching malnourished populations in relatively remote rural areas, delivering naturally fortified foods to people with limited access to commercially marketed fortified foods, which are more readily available in urban areas.

Biofortification and commercial fortification, therefore, are highly complementary. Breeding for higher trace mineral density in seeds will not incur a yield penalty. In fact, biofortification may have important spin-off effects for increasing farm productivity in developing countries in an environmentally beneficial way. Mineral-packed seeds sell themselves to farmers because, as recent research developments proved that seeds rich in trace elements are stronger to resist against biotic and abiotic stresses including diseases and environmental stresses (Bouis 2003). Further, fortified or enriched seeds also have more plant vigour, seedling survival, faster initial emergence and grain yield.

1.7 Future Challenges

- Produce crops for human nutrition with increased iron concentration. Biofortification strategies alternative to reduction in concentration of phytic acid or polyphenols should be explored further, in order to increase iron absorption without loss of their beneficial effects. When overexpressing ferritin, such crops should be tested for concentration of various heavy metals, in laboratory as in open-field trials, before releasing to the public. Detailed knowledge on mechanisms regulating iron compartmentalization in various plant organs will offer a major contribution for reaching such goal.
- Expand research on prebiotics and iron absorption. Crops biofortified with prebiotics have the potential to partially circumvent the "iron paradox" caused by host–pathogen competition for iron, by favoring amelioration of gut health and gut-associated immune defense.
- Promote initiatives supporting large-scale prospective studies on the effects of iron biofortified crops on effectiveness of the adopted biofortification strategy in relieving iron deficiency anemia and in improving general health.
- Improve the efficiency with which minerals are mobilized in the soil.
- Improve the efficiency with which minerals are taken up from the soil into the roots of the plant.
- Improve the transport of minerals from the roots to storage tissues, such as grain.
- Increase the capacity of storage tissues to accumulate minerals in a form that does not impair plant vegetative growth and development, but remains bioavailable for humans.
- Reduce the level of antinutritional compounds such as phytic acid, which inhibit the absorption of minerals in the gut.

Glossary

Anemia condition in which the number of red blood cells or their oxygen-carrying capacity is insufficient to meet physiological needs. The number of red blood cells is dependent on age, gender, and altitude and is altered by smoking, or during pregnancy the hemoglobin threshold used to define anemia is <11 g/dL

Antinutrient a substance that impairs the absorption of an essential element by the gut

Bioavailability measure of fractional utilization of orally ingested nutrient and also defined as the proportion of a particular nutrient that can be used by the body to provide its associated biological function. In simple words it is the amount of an element in a food constituent or a meal that can be absorbed and used by a person eating the meal. Bioavailability also refers to the portion of an ingested nutrient that can be absorbed in the human gut. The bioavailability of minerals can be reduced or enhanced by the consumption of food rich in antinutrients (inhibitors of absorption) or nutritional enhancers, respectively

Biofortification process for improving the nutritional value of the edible parts of the plants, through mineral fertilization, conventional breeding, or transgenic approaches. It can also be defined as the process of increasing the bioavailable concentrations of an element in edible portions of crop plants through agronomic intervention or genetic approaches

Disability-adjusted life year (DALY) a time-based parameter for assessing global burden of disease that combines years of life lost due to premature mortality (years of life lost, YLL) and years of life lost due to time lived in states of less than full health (years lived with disability, YLD): DALY=YLL+YLD

Hidden hunger the term "hidden hunger" has been used to describe the micronutrient malnutrition inherent in human diets that are adequate in calories but lack vitamins and/or mineral elements

Hunger the physical sensation of desiring food. When politicians, relief workers, and social scientists talk about people suffering from hunger, they usually refer to those who, for sustained periods, are unable to eat sufficient food to meet basic nutritional needs

Food fortification with iron a way to increase iron concentration in food by adding an iron compound (e.g., ferrous sulfate heptahydrate, ferrous gluconate, and sodium FeEDTA among others) to processed food (e.g., infant formula or cereals, wheat flour products, and corn meal among others)

Fortification the addition of an ingredient to food to increase the concentration of a particular element

Malnutrition the condition that results from eating a diet in which certain nutrients are lacking, in excess (too high in intake), or in the wrong proportions. The verb form is "malnourish"; "malnourishment" is sometimes used instead of "malnutrition." A number of different nutrition disorders may arise, depending on which nutrients are under- or overabundant in the diet. In most of the world, malnutrition is present in the form of undernutrition, which is caused by a diet lacking adequate calories and protein and not enough food, and of poor quality. Extreme undernourishment is starvation, and its symptoms and effects are inanition. While malnutrition is more common in less developed countries, it is also present in industrialized countries. In wealthier nations it is more likely to be caused by unhealthy diets with excess energy, fats, and refined carbohydrates (WHO 2001)

Promoter a substance that stimulates the absorption of an essential element by the gut

Reference nutrient intake (RNI) the amount of an element that is enough, or more than enough, for most people in a group (usually at least 97 %). If the average intake of a group is at the RNI level, then the risk of deficiency in the group is small

Regulators organic compounds, either natural or synthetic, are able to modify or control plant growth and/or development. Also, plant growth regulators (PGRs)

Supplementation the addition of an element to the diet to make up for an insufficiency

References

Alavi S, Bugusu B, Cramer G, Dary O, Lee TC, Martin L, McEntire J, Wailes E (2008) Rice fortification in developing countries: a critical review of the technical and economic feasibility. Institute of Food Technologists, Washington, DC

Ali SZ, Bhattacharya KR (1980) Pasting behaviour of parboiled rice. J Texture Stud 11:239–245

Banziger M, Long J (2000) The potential for increasing the iron and zinc density of maize through plant-breeding. Food Nutr Bull 21:397–400

Bhattacharya KR (2004) Parboiling of rice. In: Elaine TC (eds) Rice chemistry and technology, 3rd edn. Am Assoc Cereal Chemists, Inc, St. Paul, pp 329–404

Bohn L, Meyer AS, Rasmussen SK (2008) Phytate: impact on environment and human nutrition. A challenge for molecular breeding. J Zhejiang Univ Sci B 9 (3):165–191

Bouis HE, Hotz C, McClafferty B, Meenakshi JV, Pfeiffer WH (2011) Biofortification: a new tool to reduce micronutrient malnutrition. Food Nutr Bull 32(Supplement 1):31S–40S

Buois HE (2002) Plant breeding: a new tool for fighting micronutrient malnutrition. J Nutr 132:491S–494S

Bouis HE (2003) Micronutrient fortification of plants through plant breeding: can it improve nutrition in man at low cost? Proc Nutr Soc 62:403–411

Cakmak I (2008) Enrichment of cereal grains with zinc: agronomic or genetic biofortification? Plant Soil 302:1–17

Caris C, Hordt W, Hawkins HJ, Romheld V, George E (1998) Studies of iron transport by arbuscular mycorrhizal hyphae from soil to peanut and sorghum plants. Mycorrhiza 8(1):35–39

Christou P, Twyman RM (2004) The potential of genetically enhanced plants to address food insecurity. Nutr Res Rev 17:23–42

Cockell KA (2007) An overview of methods for assessment of iron availability from foods nutritionally enhanced through biotechnology. J AOAC Int 90:1480–1491

Coelho CM, Tsai SM, Vitorello VA (2005) Dynamics of inositol phosphate pools (tris-, tetrakis- and pentakisphosphate) in relation to the rate of phytate synthesis during seed development in common bean (*Phaseolus vulgaris*). J Plant Physiol 162(1):1–9

Davidsson L, Nestel P (2004) Efficacy and effectiveness of intervention to control iron deficiency and iron deficiency anemia. International Nutritional Anemia Consultative Group. International Life Sciences Institute, Washington, DC, URL: http://www.popline.org/node/238859. Accessed 12 Feb 2015

Dunn JT (2003) Iodine should be routinely added to complementary foods. J Nutr 133:3008S–3010S

Fairweather-Tait SJ (2001) Iron. J Nutr 131:1383S–1386S

Forsberg CW, Phillips JP, Golovan SP, Fan MZ, Meidinger RG, Ajakaye A, Hilborn D, Hacker RR (2003) The Enviropig physiology, performance, and contribution to management advances in a regulated environment: the leading edge of change in the pork industry. J Anim Sci 81(S2):E68–E77

Frossard E, Bucher M, Machler F, Mozafar A, Hurrell R (2000) Potential for increasing the content and bioavailability of Fe, Zn and Ca in plants for human nutrition. J Sci Food Agric 80:861–879

Galera SG, Rojas E, Sudhakar D, Zhu C, Pelacho AM, Capell T, Christou P (2010) Critical evaluation of strategies for mineral fortification of staple food crops. Transgenic Res 19:165–180

Glahn RP, Cheng Z, Welch RM (2002) Comparison of iron bioavailability from 15 rice genotypes: studies using an in vitro digestion/Caco-2 culture model. J Agric Food Chem 50:3586–3591

Glick BR (1995) The enhancement of plant growth by free-living bacteria. Can J Microbiol 41:109–117

Graham RD (2003) Biofortification: a global challenge program. Int Rice Res Notes 28(1):4–8

Graham R, Senadhira D, Beebe S, Iglesias C, Monasterio I (1999) Breeding for micronutrient density in edible portions of staple food crops: conventional approaches. Field Crop Res 60:57–80

Graham RD, Welch RM, Saunders DA, Ortiz-Monasterio I, Bouis HE, Bonierbale M, Haan DE, Burgos G, Thiele G, Liria R, Meisner CA, Beebe SE, Potts MJ, Kadian M, Hobbs PR, Gupta RK, Twomlow S (2007) Nutritious subsistence food systems. Adv Agron 92:1–74

Guzman-Maldonado SH, Gallegos-Acosta J, Parades-Lopez O (2000) Protein and mineral content of a novel collection of wild and weedy common bean (*Phaseolus vulgaris* L). J Sci Food Agric 80 (13):1874–1881

Haddad L, Ross J, Oshaug A, Torheim LE, Cogill B (2004) 5th report on the world nutrition situation: nutrition for improved development outcomes. United Nations, Standing Committee on Nutrition, Geneva, pp 143. URL: http://www.popline.org/node/236111#sthash.zssi67QS.dpuf

Hash CT, Raj AGB, Lindup S, Sharma A, Beniwal CR, Folkertsma RT, Mahalakshmi V, Zerbini E, Bluemmel M (2003) Opportunities for marker-assisted selection (MAS) to improve the feed quality of crop residues in pearl millet and sorghum. Field Crop Res 84:79–88

Horton S (2006) The economics of food fortification. J Nutr 136:1068–1071

Hurrell RF (2003) Influence of vegetable protein sources on trace element and mineral bioavailability. J Nutr 133:2973S–2977S

IZINCG (2007) Technical brief no 4: zinc fortification. URL: http://www.izincg.org/index.php. Accessed 11 Jan 2015

Jin F, Frohman C, Thannhauser TW, Welch RM, Glahn RP (2009) Effects on ascorbic acid, phytic acid and tannic acid on iron availability from reconstituted ferritin measured by an in vitro digestion-Caco-2 cell model. Br J Nutr 101:972–981

Koide RT, Kabir Z (2000) Extraradical hyphae of the mycorrhizal fungus Glomus intraradices can hydrolyse organic phosphate. New Phytol 148:511–517

Lin L, Ockenden I, Lott JNA (2005) The concentrations and distribution of phytic acid phosphorus and other mineral nutrients in wild-type and low phytic acid1-1 (lpa1-1) corn (*Zea mays* L.) grains and grain parts. Can J Bot 83(1):131–141

Lyons GH, Stangoulis JCR, Graham RD (2004) Exploiting micronutrient interaction to optimize biofortification programs: the case for inclusion of selenium and iodine in the HarvestPlus program. Nutr Rev 62:247–252

Mendoza C (2002) Effect of genetically modified low phytic acid plants on mineral absorption. Int J Food Sci Technol 37:759–767

Murgia I, Arosio P, Tarantino D, Soave C (2012) Biofortification for combating hidden hunger for iron. Trends Plant Sci 17(1):47–55

Murphy AM, Otto B, Brearley CA, Carr JP, Hanke DA (2008) A role for inositol hexakisphosphate in the maintenance of basal resistance to plant pathogens. Plant J 56(4):638–652

Nantel G, Tontisirin K (2002) Policy and sustainability issues. J Nutr 132:S839–S844

Nestel P, Buois HE, Meenakshi JV, Pfeiffer W (2006) Biofortification of staple food crops. J Nutr 136:1064–1067

Palipane KB, Swarnasiri CDP (1985) Composition of raw and parboiled rice bran from common Sri Lankan varieties and from different types of rice mills. J Agric Food Chem 33:732–734

Pray C (2006) The Asian Maize Biotechnology Network (AMBIONET): a model for strengthening national agricultural research systems. CIMMYT, Mexico

Rana A, Joshi M, Prasanna R, Shivay YS, Nain L (2012) Biofortification of wheat through inoculation of plant growth promoting rhizobacteria and cyanobacteria. Eur J Soil Biol 50:118–126

Reddy BVS, Reddy SP, Bindiger F, Bluemmel M (2003) Crop management factors influencing yield and quality of crop residues. Field Crop Res 84:57–77

Rengel Z, Batten GD, Crowley DE (1999) Agronomic approaches for improving the micronutrient density in edible portions of field crops. Field Crop Res 60:27–40

Shivay YS, Prasad R, Rahal A (2010) Studies on some nutritional quality parameters of organically or conventionally grown wheat. Cereal Res Commun 38 (3):345–352

Shrimpton R, Schultink W (2002) Can supplements help meet the micronutrient needs of the developing world? Proc Nutr Soc 61:223–229

Smith SE, Read DJ (1997) Mycorrhizal symbiosis, 2nd edn. Academic, London

Stein AJ, Meenakshi JV, Qaim M, Nestel P, Sachdev HPS, Bhutta ZA (2008) Potential impacts of iron biofortification in India. Soc Sci Med 66 (8):1797–1808

Steven FJ (1991) Organic matter-micronutrient reactions in soil. In: Mortvedt JJ, Cox FR, Shuman LM, Welch RM (eds) Micronutrient in agriculture. Soil Sci Soc Am, Madison, pp 145–186

Unnevehr L, Pray C, Paarlberg R (2007) Addressing micronutrient deficiencies: alternative interventions and technologies. AgBioforum 10(3):124–134

Vasconcelos M, Datta K, Oliva N, Khalekuzzaman M, Torrizo L, Krishnan S, Oliveira M, Goto F, Datta SK (2003) Enhanced iron and zinc accumulation in transgenic rice with the ferritin gene. Plant Sci 164 (3):371–378

Walter T, Pizarro F, Olivares M (2003) Iron bioavailability in corn-masa tortillas is improved by the addition of disodium EDTA. J Nutr 133:3158–3161

Wang B, Qiu YL (2006) Phylogenetic distribution and evolution of mycorrhizas in land plants. Mycorrhiza 16:299–363

Welch RM, Graham RD (2005) Agriculture: the real nexus for enhancing bioavailable micronutrients in food crops. J Trace Elem Med Biol 18:299–307

White PJ, Broadley MR (2005) Biofortifying crops with essential mineral elements. Trends Plant Sci 10:586–593

White PJ, Broadley MR (2009) Biofortification of crops with seven mineral elements often lacking in human diets – iron, zinc, copper, calcium, magnesium, selenium and iodine. New Phytol 182:49–84

WHO/FAO (1998) Vitamin and mineral requirements in human nutrition: report of a joint FAO/WHO expert consultation, 2nd edn. Bangkok, 21–30 Sept 1998. URL: http://apps.who.int/iris/handle/10665/42716#sthash.niRWmJ9o.dpuf

World Health Organization (2001) Water-related diseases: malnutrition. World Health Organization – Water Sanitation and Health (WSH). URL: www.who.int/water_sanitation_health/disease/malnutrition/en/. Accessed 22 Feb 2015

Zerbini E, Thomas D (2003) Opportunities for improvement of nutritive value in sorghum and pearl millet residues in South Asia through genetic enhancement. Field Crop Res 84:3–15

Zhao HJ, Liu QL, Fu HW, Xu XH, Wu DX, Shu QY (2008) Effect of non-lethal phytic acid mutations on grain yield and seed viability in rice. Field Crop Res 108:206–211

Zhu C, Naqvi S, Gomez-Galera S, Pelacho AM, Capell T, Christou P (2007) Transgenic strategies for the nutritional enhancement of plants. Trends Plant Sci 12(12):548–555

Zimmermann MB, Wegmueller R, Zeder C, Chaouki N, Biebinger R, Hurrell RF, Windhab E (2004) Triple fortification of salt with microcapsules of iodine, iron, and vitamin A. Am J Clin Nutr 80 (5):1283–1290

Global Poverty, Hunger, and Malnutrition: A Situational Analysis

2

Purushottam Sharma, Sudhakar Dwivedi, and Deepak Singh

Abstract

The world continued to face problems of poverty, hunger, and malnutrition, although good progress has been made in this direction by national governments and international development institutions. About 52 % of the population in the developing world thrives on less than $1.25 per day during 1981 which has declined significantly to 17 % during 2011. Despite over 59 % increase in population in the developing world, the people living in extreme poverty have significantly declined from 1.96 million in 1981 to 1.01 million in 2011. Sub-Saharan African and South Asian regions are home to most number of poor people. Poverty is both a cause and a consequence of hunger. Still about 805 million people are suffering from chronic hunger and nearly two billion people worldwide are affected by micronutrient deficiencies. Among children, about 162 million (one fourth) under five years of age are chronically malnourished (stunted), about 99 million are underweight, and nearly 55 million are acutely malnourished (wasted). In this context, a situational analysis is carried out involving an integrated approach including agricultural development as it is the key for eradicating poverty and reducing the prevalence of hunger and malnutrition. In addition, biofortification of food crops is a feasible alternative to reduce malnutrition.

Keywords

Hunger • Malnutrition • Poverty • Role of Agriculture

P. Sharma (✉)
ICAR-Directorate of Soybean Research, Indore, Madhya Pradesh 452 001, India
e-mail: purushottamji@gmail.com

S. Dwivedi
Division of Economics, Faculty of Agriculture, Sher-e-Kashmir University of Agricultural Sciences and Technology of Kashmir, Chatha, Jammu and Kashmir 208 024, India
e-mail: dwivedi.sudhakar@gmail.com

D. Singh
Division of Social Sciences, ICAR-Indian Institute of Pulses Research, Kanpur, Uttar Pradesh 208 024, India
e-mail: deepaksingh2112@gmail.com

U. Singh et al. (eds.), *Biofortification of Food Crops*, DOI 10.1007/978-81-322-2716-8_2

2.1 Introduction

Poverty and food and nutrition insecurity have been some of the most pressing global challenges of the twentieth century the world has been facing and will continue to be facing in the twenty-first century. Many development programs in several countries have been targeted toward alleviating poverty, hunger, and malnutrition and are being continued. Though the efforts paved the way toward reducing the intensity of these ills, it could not improve on regional imbalance and inequity. The World Bank Position Paper on Poverty and Hunger defined food security as "food security must assure access by all people at all times to enough food for an active and healthy life" (World Bank 1986). The 1996 World Food Summit added nutrition aspects and defined food security as "when all people, at all times, have physical and economic access to sufficient, safe and nutritious food to meet their dietary needs and food preferences for an active and healthy life" (FAO 1996; DFID 2003). Thus, food insecurity was defined as the lack of access to enough, safe, and nutritious food for a healthy and active life. The complex interrelationship between food security, hunger, and poverty gained recognition from developmental organizations and governments globally and started making serious efforts to alleviate poverty and ensure food security.

During the 1996 World Food Summit, it was targeted to halve the number of chronically undernourished people by 2015. This goal was at the heart of the Rome Declaration on World Food Security and formed the basis of the first Millennium Development Goal (MDG). To eradicate extreme poverty and hunger, the aim was set to halve, between 1990 and 2015, the proportion of people whose income is less than $1.25 a day. At that time, nearly 788 million people were chronically undernourished and about two billion people were suffering from different types of micronutrient deficiencies, including vitamin A, iron, and iodine deficiencies (FAO 2001). Though good progress has been made across the globe toward alleviating poverty, providing food and nutritional security, unfortunately, many countries are not on track to achieve the goal of reducing the number of people suffering from hunger, malnutrition, and poverty (Gómez et al. 2013). The UN Secretary-General recently announced meeting the challenge of "zero hunger" to be achieved by 2025.

About one third of preschool children in developing countries suffer from malnutrition—causing the death of 5–10 million of these children every year (FAO 2014; Andersen 2015). Inequality is also posing a serious problem as the richest one percent of the world's population earns as much as 57 % than the rest (UNDP 2002), and relative global income distribution is getting worse. It is evident that poverty and inequality contributes to national instability and armed conflict (Messer Ellen et al. 2001). The linkage between poverty and inequality, on the one hand, and instability and crime, on the other, particularly in urban areas is well known (Caldeira Teresa 2000). In 1960, average per-capita incomes in industrialized countries were nine times the average per capita in sub-Saharan Africa, and the gap has widened to more than 20 times. The condition is deteriorating mainly in low-income and sub-Saharan countries. The governments and their development partners are now refocusing on the plight of the poor and the hungry and continue to put efforts toward reducing these ills from the society.

To ensure adequate and nutritious food for everyone, the capacity and resources in the world are sufficient enough. Nevertheless, the large difference in availability of food across the regions of the world still persists. The efforts of governments and development agencies during the last two to three decades resulted in a decline in poverty and hunger. But about 805 million people still suffer from chronic hunger and nearly two billion people worldwide are affected by micronutrient deficiencies. Among children, the estimates indicated that about 162 million (one in four globally) under five years of age are chronically malnourished (stunted, inadequate length or height for age), almost 99 million are underweight (about 15 % of total children), and about 55 million are acutely malnourished (wasted) (UN 2014; IFPRI 2014a). Apart from the ethical dimensions of this complex problem, there are huge human, social, and economic costs to society at large in terms of lost productivity, health, well-being, decreased learning ability, and reduced fulfillment of human potential.

Poverty, hunger, and malnutrition are interrelated phenomena. The circumstances of poor people reinforce them to hunger and malnutrition. Poor people do not have enough money to buy or produce enough and nutritious food for themselves and their families and, in turn, tend to be weaker and cannot produce/earn enough to buy more food. The persons chronically undernourished are gripped in a vicious cycle—not getting adequate and nutritious food on a regular basis and therefore not being able to lead a healthy and active life and earn for their livelihood, not having access to health care, and thus not being able to either produce or procure required nutritious food. Poverty is both a cause and a consequence of hunger. Thus, there is close and complex interaction between hunger, malnutrition, and poverty, and this phenomenon portrays poverty trap—the poor are hungry and their hunger traps them in poverty (Ingutia et al. 2009).

2.2 Poverty Reduction: A Peep into Progress

Poverty is a social evil affecting the quality of human life, food intake, and other surroundings the people are suffering from. The poor household, unable to access quality and nutritious food and health services, is not able to produce more and support the family. There are different measures of poverty; the most prevalent is income or expenditure measure. The world is now growingly accepting poverty as a multidimensional concept encapsulating deprivation in different dimensions limiting opportunities for a happy, healthy, and productive life. The key deprivations include income poverty, hunger, malnutrition, gender bias, social exclusion, and lack of access to education, health services, and housing (Grewal et al. 2012; IFPRI 2014a).

The Herculean task of reducing poverty by the international developmental organizations and national governments has made dramatic effects. Extreme poverty, less than US$1.25 of income per day PPP (2005), in the world has decreased considerably in the last three decades (Table 2.1 and Fig. 2.1). About 52 % of the population in the developing world lived on less than $1.25 per day in 1981. This has declined significantly to 17 % in 2011. Furthermore, despite over 59 % increase in population in the developing world, the people living in extreme poverty have significantly declined from 1.96 million in 1981 to 1.01 million in 2011. The global poverty reduction has been mainly due to growth progress in the fast-growing economies of East Asia and to a lesser extent of South Asia (DFID 2005). However, the persons living in extreme poverty (1.01 billion) are still very high, requiring incessant efforts in this direction.

The regional pattern of decline in persons in extreme poverty line shows a disturbing picture. Though the percentage of the poor has declined in all the regions, the number of poor people in extreme poverty is steadily increasing in sub-Saharan Africa (Table 2.1) between 1981 and 2011. The people living below $1.25 a day in the sub-Saharan African region has almost doubled in 2011 (415.4 million) compared to 1981 (210.11 million). As a result, the share of extremely poor people in sub-Saharan Africa has increased from merely 11 % of the world in 1981 to about 41 % presently.

The dramatic decline in the number of extremely poor people (from 1107.4 million in 1981 to 160.76 million in 2011) as well as poverty rate (from 78 % to 7.9 %) was observed in East Asia. This was mainly due to the high growth of economy particularly in China where the rate of poverty has declined from more than 84.27 % in 1981 to merely 6.26 % presently (Fig. 2.1). India accounts for about 29.8 % of the world's extreme poor (higher than 21.8 % in 1981); Nigeria comes next, contributing 9.8 %

Table 2.1 Population below extreme poverty line

	At $1.25 a day (PPP)			At $2 a day (PPP)		
Regions	1981	1990	2011	1981	1990	2011
Number of poor people (millions)						
East Asia and the Pacific	1,107.44	939.11	160.76	1,312.87	1,333.79	459.58
China	837.55	689.51	84.14	972.12	960.82	394.62
Europe and Central Asia	12.55	7.15	2.35	35.75	31.48	10.35
Latin America and the Caribbean	42.49	55.17	27.63	86.63	97.61	55.38
Middle East and northern Africa	15.26	13.01	5.64	51.82	52.88	38.69
South Asia	570.33	620.47	398.95	810.65	958.75	979.47
India	427.42	446.26	301.26	620.70	721.93	862.42
Sub-Saharan Africa	210.11	287.07	415.4	287.58	389.20	616.58
Developing world	1,958.27	1,921.74	1,011.37	2,585.29	2,863.78	2,160.05
Poverty head count ratio (% of population)						
East Asia and the Pacific	77.96	57.01	7.93	92.41	80.79	22.67
China	84.27	60.74	6.26	97.81	84.64	29.79
Europe and Central Asia	2.88	1.54	0.49	8.32	6.78	2.16
Latin America and the Caribbean	12.58	12.63	4.63	23.77	22.36	9.28
Middle East and northern Africa	8.79	5.77	1.69	30.06	23.46	11.59
South Asia	61.35	54.09	24.50	87.20	83.58	60.15
India	59.77	51.36	24.67	86.63	82.62	72.42
Sub-Saharan Africa	52.76	56.02	46.81	72.20	75.95	69.48
Developing world	52.71	43.35	16.99	69.59	64.60	36.18
Share (in the world)						
East Asia and the Pacific	56.55	48.87	15.90	50.78	46.57	21.28
China	42.77	35.88	8.32	37.60	33.55	18.27
Europe and Central Asia	0.64	0.37	0.23	1.38	1.10	0.48
Latin America and the Caribbean	2.17	2.87	2.73	3.35	3.41	2.56
Middle East and northern Africa	0.78	0.68	0.56	2.00	1.85	1.79
South Asia	29.12	32.29	39.45	31.36	33.48	45.34
India	21.83	23.22	29.79	24.01	25.21	39.93
Sub-Saharan Africa	10.73	14.94	41.07	11.12	13.59	28.54
Developing world	100.00	100.00	100.00	100.00	100.00	100.00

Source: http://iresearch.worldbank.org/PovcalNet/index.htm

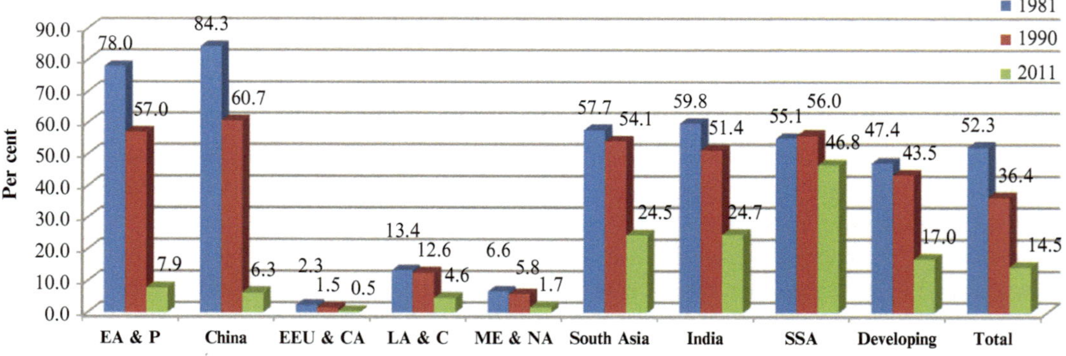

Fig. 2.1 Level of extreme poverty in different regions

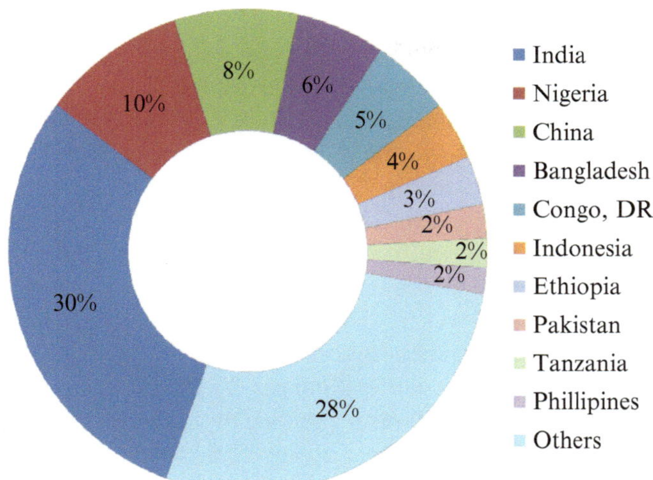

Fig. 2.2 Top ten countries with largest share of global extreme poor, 2011 (%)

(higher than 1.8 % in 1981) followed by China, accounting for about 8.3 % (down from 42.77 % in 1981).

During 1981 and 1990, the magnitude of extreme poverty was highest in East Asia, whereas presently about 80 % of the world's poor lives in South Asia and sub-Saharan Africa, and the share has doubled from 40 % in 1981. In South Asia, however, the number of people living below extreme level has decreased from 570.33 million in 1981 to 398.95 million in 2011, presently housing nearly 40 % of the world's poor (higher than 29 % in 1981). Growth is the major driver of poverty reduction and was instrumental in halving extreme poverty between 1990 and 2010 (World Bank 2015).

The extreme poverty head count in Latin America and the Caribbean region, after remaining stable at approximately 12 % for the last two decades of the twentieth century, has declined to merely 4.6 % in 2011. About three-fifths of the world's extreme poor are concentrated in just five countries: India, Nigeria, China, Bangladesh, and the Democratic Republic of the Congo (Fig. 2.2). Adding five more countries (Indonesia, Ethiopia, Pakistan, Tanzania, and the Philippines) would comprise just over 72 % of the extreme poor.

China and India, the most populous countries in the world, have played a great role in the global reduction of extreme poverty. The two countries together lifted about 876.6 million people out of poverty from 1981 to 2011, about 92 % of the total people lifted from extreme poverty in the world during the period. The large reduction in poverty rates at the global level is mainly due to growth in China (Chen and Ravallion 2012). Even though there is a tremendous increase in global population during the period, the poverty rate has declined significantly from 52.7 % to merely 17 % over the period, a great achievement. There are countries, mainly in sub-Saharan Africa, where the number of people living in extreme poverty is equal to or more than 40 % of the population in 2011 and have a population of less than 30 million people. Nonetheless, reducing poverty in these countries is a moral imperative and as important as poverty reduction in any other country (World Bank 2015).

Although there is significant progress made in reducing the proportion of people below extreme poverty line, the progress has been slower at the higher poverty lines. In total about 2.16 billion people (a marginal decline from 2.59 billion in 1981) are living on less than US$2 per day in 2011 (Table 2.1), the average poverty line in developing countries and another common measurement of deep deprivation (World Bank 2015a). Sumner (2012) argues that the international poverty line of US$2 a day is conceptually stronger as it is the median average of poverty lines of all developing countries (Chen and

Ravallion 2008) and close to the poverty lines defined by poor people (Narayan et al. 2009).

The extreme inequality is prevalent in the distribution of income across the world population: the average income of the richest 5 % is estimated to be nearly 200 times that of the poorest 10 % (UNCTAD 2013). The gap between the rich and poor is widening in some of the developing countries (Milanovic 2011). The poverty level is not only extreme in low-income countries but also prevalent in medium-income countries (Sumner 2010) with higher inequality. Poverty profiling by Olinto et al. (2013) revealed that about 78 % of the poor resides in rural settings and most likely to earn from agriculture (63 %)—most of them are smallholders and agricultural laborers. The gender gap in education is concentrated among the poor. The poor have lower access to basic services like water, electricity, and sanitation. The national governments are mainly responsible for achieving the goal of reducing poverty (Besley and Burgess 2003).

2.3 Reduction in Hunger and Malnutrition: Progress to Date

Hunger and malnutrition are two interrelated challenges the world is facing continuously and need increased attention toward providing food and nutritional security to the vulnerable sections. There are regions where a section of the people doesn't even get enough food to eat on a regular basis, making them weaker and less productive and in turn unhealthy and poor. The reduction in hunger around the globe continues incessantly with the interventions from governments as well as the international development community. According to latest FAO estimates, about 805 million people are estimated to be chronically undernourished in 2012–2014, more than 100 million lower over the last decade, and 209 million lower than in 1990–1992 (FAO, IFAD, and WFP 2014). The incidence of undernourishment has declined from 18.7 % to 11.3 % globally and from 23.4 % to 13.5 % for the developing countries during the corresponding period (Table 2.2 and Fig. 2.3).

The target of halving proportion of people undernourished by 2015 from the level of 1990–92 set under UN Millennium Development Goal 1c seems attainable, even many countries have already achieved, although the World Food Summit target of halving the number of undernourished people by 2015 seems far from reach. Of the total number of people undernourished globally, about 98.2 % resides in developing countries. Nearly 63 developing countries not only achieved the target, but even in about 11 countries, the proportion of undernourished people is below 5 % since 1990–1992. In this direction, the growing concern is that 805 million people in the world still do not get enough food on a regular basis.

Notwithstanding the good progress made in this direction, differences are apparent across the regions. Nearly one fourth of people are chronically undernourished in sub-Saharan Africa, while Asia (the world's most populous region) is home to the majority of hungry people (526 million—about two thirds of the total hungry people in the world). The persons not getting enough food has increased in the sub-Saharan African region by 21 %. Although the prevalence of hunger is consistently low in northern Africa (less than 5 %), the number of people not getting enough food to eat is estimated to be doubled (from 6 million in 1990–1992 to 12.6 million in 2012–2014). Similarly, in western Asia the number of chronically undernourished people has increased by 131 % (Table 2.2). Fast reduction in the prevalence of undernourishment has been achieved in East and Southeast Asia, whereas progress was slow and insufficient in South Asia. The main concern here is that the share of South Asia in the total number of undernourished people has increased from 28.8 % in 1990–1992 to 34.3 % in 2012–2014. Similarly sub-Saharan Africa accounts for about 26.6 % of the hungry people in the world, increasing from 17.3 %.

The highest progress in hunger reduction was achieved in Latin America and the Caribbean region, where the prevalence of hunger falls nearly by two thirds since 1990–1992 and the region is close to five percent hunger prevalence.

Table 2.2 Prevalence of undernourishment around the world

Regions	Number of undernourished *(millions)* and prevalence *(%)* of undernourishment						
	1990–1992		2000–2002		2012–2014[a]		
	Number	%	Number	%	Number	%	% Change in number
Developed regions	20.4	<5	21.1	<5	14.6	<5	−28.4
Developing regions	994.1	23.4	908.7	18.2	790.7	13.5	−20.5
Africa	182.1	27.7	209.0	25.2	226.7	20.5	24.5
Northern Africa	6.0	<5	6.5	<5	12.6	6.0	110.0
Sub-Saharan Africa	176.0	33.3	202.5	29.8	214.1	23.8	21.6
Asia	742.6	23.7	637.5	17.4	525.6	12.7	−29.2
Caucasus and Central Asia	9.6	14.1	10.9	15.3	6.0	7.4	−37.5
East Asia	295.2	23.2	222.2	16.0	161.2	10.8	−45.4
East Asia excluding China	6.4	9.7	10.4	14.5	10.4	13.5	62.6
China	288.9	23.9	211.7	16.1	150.8	10.6	−47.8
Southeast Asia	138.0	30.7	117.7	22.3	63.5	10.3	−54.0
South Asia	291.7	24.0	272.9	18.5	276.4	15.8	−5.2
South Asia excluding India	81.0	24.5	86.7	21.0	85.8	17.3	5.9
India	210.8	23.8	186.2	17.6	190.7	15.2	−9.5
Western Asia	8.0	6.3	13.8	8.6	18.5	8.7	131.3
Latin America and the Caribbean	68.5	15.3	61.0	11.5	37.0	6.1	−46.0
Caribbean	8.1	27.0	8.2	24.4	7.5	20.1	−7.4
Latin America	60.3	14.4	52.7	10.7	29.5	5.1	−51.1
Oceania	1.0	15.7	1.3	16.5	1.4	14.0	40.0
World	1,014.5	18.7	929.9	14.9	805.3	11.3	−20.6

Source: FAO, IFAD, and WFP (2014)
[a]Estimates

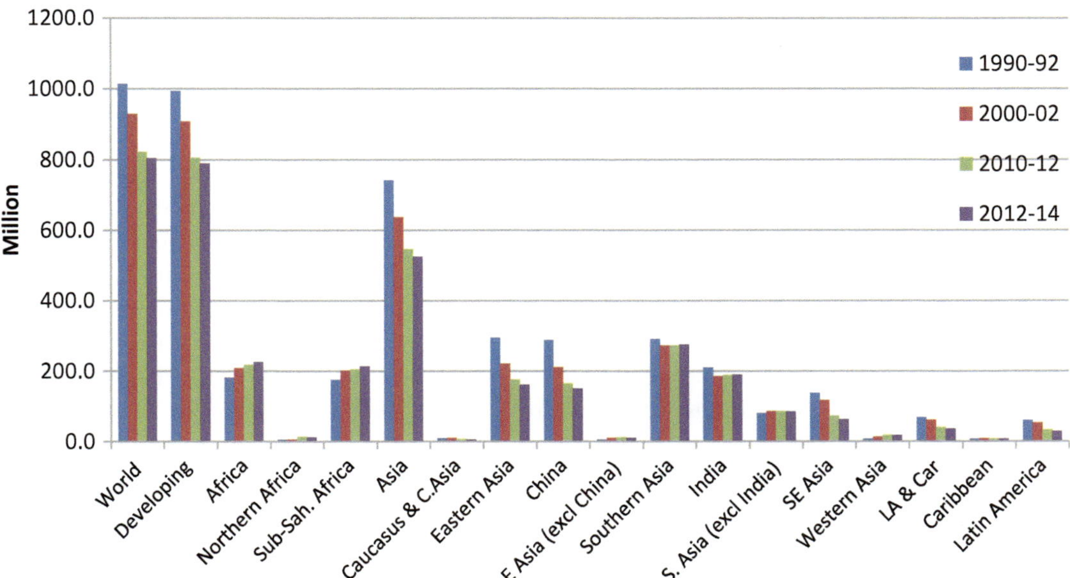

Fig. 2.3 Number of undernourished people across regions of the world

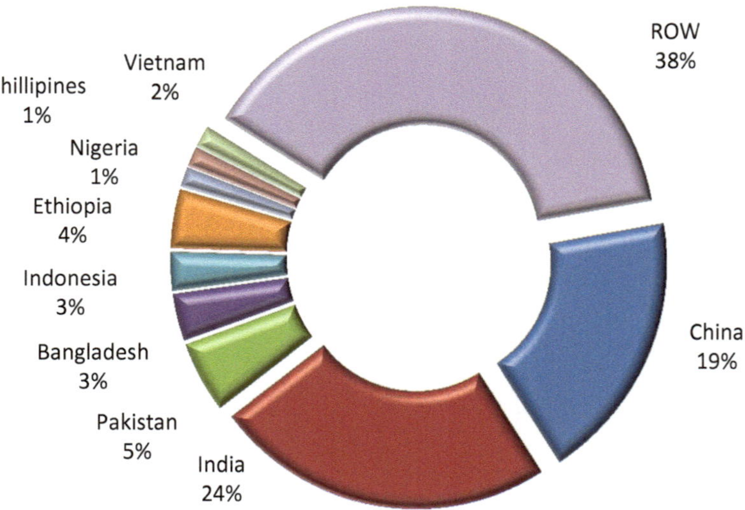

Fig. 2.4 Top countries with higher numbers of undernourished people

Concerted efforts of the national governments in the region and political commitment paved the way to this achievement. The Oceania region has the lowest number of undernourished people presently. But despite the low burden of hunger in the region, the number of hungry people has increased by 40 % over the period. Furthermore, the rising undernourishment is accompanied by an increasing burden of overweight and obesity, exposing the region to the double burden of malnutrition (FAO, IFAD, and WFP 2014).

Overnutrition and undernutrition coexist in most countries (Gillespie and Haddad 2001). The double burden of malnutrition is growing over the globe, as about 17 % of preschool children are underweight, one third suffer from iodine deficiency, 40 % of women of reproductive age are anemic (UNSCN 2010), and about one fourth of the global population is overweight (Finucane et al. 2011). India is the home of about 24 % of critically undernourished people globally (Fig. 2.4), followed by China (19 %), Pakistan (5 %), Ethiopia (4 %), and Bangladesh (3 %).

IFPRI in its global nutrition report revealed that of the 117 countries for which data for three indicators of malnutrition are available (stunting, wasting, and overweight), 17 countries have all three types of malnutrition problem in children under five years of age (IFPRI 2014b). Only 43 countries have only one type of child growth problem, while 64 countries have multiple types of under-five malnutrition. Only ten countries do not have these three indicators as a serious issue. The report also highlighted that there is lack of equity in nutrition (Black et al. 2013), and the malnourished children need extra support and finances to be out of this problem which their families are not able to support (Table 2.3).

2.4 Role of Agriculture

In spite of overwhelming progress in reducing poverty, hunger, and malnutrition during the recent decades, there are still about 1.01 billion people who are extremely poor, nearly 805 million people do not get enough food to eat regularly, and nearly two billion people suffer from micronutrient deficiency. More than two thirds of the poor live in rural areas; majority of them depends on agriculture for livelihood support. It is evident that agricultural growth has a high poverty-reduction payoff (Cleaver 2012). The rapid agricultural growth is significantly correlated with the reduction in poverty rates and incidence of undernourishment (World Bank 2007) in developing countries. Studies reported that every 1 % increase in per capita agricultural output led to a 1.61 % increase in the incomes of the poorest 20 % of the population (Gallup et al. 1997). A cross-country analysis by Thirtle et al. (2001) concluded that, on average,

Table 2.3 Countries with overlapping stunting, wasting and overweight in Children under five years of age

Overlap/indicator group	Number of countries	Total population (millions)	Countries
Stunting only	12	212	Democratic People's Republic of Korea, El Salvador, Guatemala, Honduras, Liberia, Nauru, Nicaragua, Solomon Islands, Togo, Uganda, Vietnam, Zimbabwe
Wasting only	6	68	Guyana, Oman, Saudi Arabia, Senegal, Sri Lanka, Suriname
Overweight only	25	603	Algeria, Argentina, Belarus, Belize, Bosnia and Herzegovina, Brazil, Chile, Costa Rica, Dominican Republic, Gabon, Georgia, Kazakhstan, Kuwait, Kyrgyzstan, Mexico, Mongolia, Montenegro, Morocco, Paraguay, Peru, Serbia, The former Yugoslav Republic of Macedonia, Tunisia, Uruguay, Uzbekistan
Stunting and wasting only	38	2,462	Bangladesh, Burkina Faso, Burundi, Cambodia, Cameroon, Central African Republic, Chad, Congo, Côte d'Ivoire, Democratic Republic of the Congo, Eritrea, Ethiopia, Gambia, Ghana, Guinea, Guinea-Bissau, Haiti, India, Kenya, Lao People's Democratic Republic, Maldives, Mali, Mauritania, Myanmar, Namibia, Nepal, Niger, Nigeria, Pakistan, Philippines, Somalia, South Sudan, Sudan, Tajikistan, Timor-Leste, United Republic of Tanzania, Vanuatu, Yemen
Stunting and overweight only	7	45	Armenia, Bolivia, Equatorial Guinea, Lesotho, Malawi, Rwanda, Swaziland
Wasting and overweight only	2	70	Republic of Moldova, Thailand
Stunting, wasting, and overweight	17	468	Albania, Azerbaijan, Benin, Bhutan, Botswana, Comoros, Djibouti, Egypt, Indonesia, Iraq, Libya, Mozambique, Papua New Guinea, São Tomé and Príncipe, Sierra Leone, Syrian Arab Republic, Zambia
Below cutoff for all three indicators	10	1,914	China, Colombia, Germany, Jamaica, Jordan, Republic of Korea, Saint Lucia, Tuvalu, United States of America, Venezuela

Source: IFPRI (2014)

every 1 % increase in agricultural yields reduced the number of people living on less than US$1 a day by 0.83 %.

Since the poor are concentrated in rural areas in many countries, the key to growth and poverty reduction lies in the rapid growth of agriculture and rural development (Wiggins 2005; Dorward et al. 2004). Notwithstanding the role of agricultural development on reduction in poverty and hunger in developing countries, investment in agriculture has declined since the 1980s, both by developing country governments and development institutions. The share of agriculture in total bilateral and multilateral aid decreased from 22.5 % in 1979–1981 to a low 5.4 % in 2003–2005 though improved marginally recently to 6 % (Cleaver 2012; Grewal et al. 2012).

2.5 Micronutrient Malnutrition and Biofortification

Although the issue of serious concern is that about one billion people are hungry in the world, the problem of micronutrient malnutrition, or hidden hunger (von Grebmer et al. 2014), has not been given much attention (Kristof 2009; The Economist 2011). Micronutrient deficiency in human diet can pose a serious concern for human health and can result in lack of stamina, impaired physical and cognitive development, morbidity, and blindness (Stein 2015). The estimates indicated that about seven billion people are suffering from either form of malnutrition and nearly five billion people from micronutrient deficiency; about two billion people worldwide are anemic, many due to iron deficiency (WHO 2015a), two billion people suffer from iodine deficiency (de Benoist et al. 2008), and about 17.3 % of the population is affected by zinc deficiency (Wessells and Brown 2012). It is estimated that 250 million preschool children are vitamin A deficient, and a substantial proportion of pregnant women in at-risk areas are suffering from vitamin A deficiency (WHO 2015b).

Linking agriculture with health and nutrition by appropriate agriculture, health, and nutrition policies in place seems to be a sustainable solution to malnutrition (Rouse and Davis 2004; Graham et al. 2007; Hawkes and Ruel 2006; World Bank 2007a). Growing trend worldwide now is the use of new agricultural technologies like "biofortification" to address micronutrient malnutrition. Biofortification is "the process of adding nutritional value to a crop" (Montagnac et al. 2009). Stein (2015) reported that by far the most frequently named crop is rice, followed by maize and wheat, and then pulses, vegetables and fruits, cassavas, sweet potatoes, sorghums, and model plants. The primary focus is on the conventional breeding to enhance staple food crops with sufficient levels of minerals and vitamins to meet the needs of at-risk populations (Hotz et al. 2007; White and Broadley 2009). Evidence suggests that biofortification is a cost-effective means to provide micronutrients to the most vulnerable people (Bouis 1999; Nestel et al. 2006; Pfeiffer and McClafferty 2007; Qaim et al. 2007).

2.6 Conclusions and the Road Ahead

Eradication of poverty, hunger, and malnutrition requires sustained political commitment at the highest level. It necessitates creating an enabling environment for improving food and nutrition security through policies, adequate investments, legal frameworks, and stakeholder participation. Institutional reforms are also inevitable for promotion and sustainable progress in this direction (FAO, IFAD, and WFP 2014). An integrated approach is required to reduce the prevalence of hunger, including public and private investments to raise agricultural productivity; better access to inputs, services, technologies, and markets; measures to promote rural development; social protection for the most vulnerable, including strengthening their resilience to conflicts and natural disasters; and specific nutrition programs, especially to address micronutrient deficiencies in mothers and children under five years of age (FAO 2014).

Sub-Saharan Africa poses the utmost food security challenges, where the progress in improving access to food is slow, along with sluggish income growth, high poverty rates, and poor infrastructure. All these hamper physical and distributional access. In South Asia, the home of the largest number of hungry people, food management remains the greatest challenge and calls for reducing the wastages along with efforts to improve productivity of crops and access to food through research support and institutional reforms. Structural economic transformation, social transformation to a low level of inequality, and political transformation are essential elements of long-term poverty reduction (Sumner 2010).

A development of a single sector can't solve the complex problems of poverty, hunger, and malnutrition; a broad-based approach has to be followed to deal with the issues. Renewed interest and actions are required in enhancing productivity of agriculture and allied sectors like fisheries, forestry, rural development, social protection, public works, trade and markets, resilience to shocks, education and health, and other areas. There are newer and high-impact technologies available now (like biotechnology, biofortification, and nanotechnology), which offer further opportunities for boosting agricultural productivity and enhancing food quality and nutritional value. The use of these technologies along with appropriate related policies and institutions can usher in the speed of progress in this direction (von Braun 2010).

References

Anderson PP (2015) Eradicating poverty and hunger as a national security issue for the United States. www.wilsoncenter.org/sites/default/files/ACF59B4.doc. Accessed 31 Jan 2015

Besley T, Burgess R (2003) Halving global poverty. J Econ Perspect 17(3):3–22

Black RE, Victora CG, Walker SP, Bhutta ZA, Christian P, de Onis M, Ezzati M et al (2013) Maternal and child undernutrition and overweight in low-income and middle-income countries. Lancet 382(9890):427–451

Bouis HE (1999) Economics of enhanced micronutrient density in food staples. Field Crop Res 60:165–173

Caldeira Teresa PR (2000) City of walls: crime, segregation, and citizenship in São Paulo. University of California Press, California/London

Chen S, Ravallion M (2008) The developing world is poorer than we thought, but no less successful in the fight against poverty. Policy research working paper 4703. World Bank, Washington, DC

Chen S, Ravallion M (2012) An update to the World Bank's estimates of consumption poverty in the developing world. World Bank, Washington, DC

Cleaver K (2012) Scaling up in agriculture, rural development and nutrition: investing in agriculture to reduce poverty and hunger, 2020 Vision for food, agriculture and the environment brief 2. International Food Policy Research Institute, Washington, DC

de Benoist B, McLean E, Andersson M, Rogers L (2008) Iodine deficiency in 2007: global progress since 2003. Food Nutr Bull 29:195–202. http://www.who.int/nutrition/publications/micronutrients/FNBvol29N3sep08.pdf?ua=1

DFID (2003) Eliminating hunger: strategy for achieving the Millennium Development Goal on hunger. Department for International Development, London

DFID (2005) Growth and poverty reduction: the role of agriculture. A DFID policy paper, Department for International Development, London

Dorward A, Kydd J, Morrison J, Urey I (2004) A policy agenda for pro-poor agricultural growth. World Dev 32(1):73–89

FAO (1996) Rome declaration on World Food Security and World Food Summit plan of action. World Food Summit 13–17 Nov 1996. FAO, Rome

FAO (2001) The World Food Summit Goal and the Millennium Development Goals. Food and Agriculture Organization of the United Nations. Rome, http://www.fao.org/DOCREP/MEETING/003/Y0688E.HTM. Accessed January 2015

FAO (2014) The state of the food insecurity in the world 2014. Food and Agricultural Organisation of the United Nations, Rome

FAO, IFAD, WFP (2014) The state of food insecurity in the world 2014: strengthening the enabling environment for food security and nutrition. Food and Agricultural Organisation of the United Nations, Rome

Finucane MM, Stevens GA, Cowan M, Danaei G, Lin JK, Paciorek CJ, Singh GM, Gutierrez H, Lu Y, Bahalim AN, Farzadfar F, Riley LM, Ezzati M (2011) National, regional and global trends in body-mass index since 1980: systematic analysis of health examination surveys and epidemiological studies with 960 country years and 9.1 million participants. Lancet 377(9765):557–567

Gallup J, Radelet S, Warner A (1997) Economic growth and the income of the poor. CAER II discussion paper no. 36. Harvard Institute for International Development, Boston

Gillespie S, Haddad L (2001) Attacking the double burden of malnutrition in Asia and the Pacific. Asian Development Bank/International Food Policy Research Institute, Manila/Washington, DC

Gómez MI, Barrett CB, Raney T, Pinstrup-Andersen P, Janice M, Croppenstedt A, Carisma B, Thompson B (2013) Post-green revolution food systems and the triple burden of malnutrition. Food Policy 42:129–138

Graham RD, Welch RM, Saunders DA, Monasterio I, Bouis HE, Bonierbale M, de Hann S, Burgos G, Thiele G, Liria R, Meisner CA, Beebe SE, Potts MJ, Kadiajn M, Hobbs PR, Gupta RK, Twomlow S (2007) Nutritious subsistence food systems. Adv Agron 92:1–74

Grewal B, Grunfeld H, Sheehan P (2012) The contribution of agricultural growth to poverty reduction. ACIAR impact assessment series report number 76. Australian Centre for International Agricultural Research, Canberra

Hawkes C, Ruel MT (2006) Understanding the links between agriculture and health. International Food Policy Research Institute, Washington, DC

Hotz C, McClafferty B, Hawkes C, Ruel M, Babu S (2007) From harvest to health: challenges for developing biofortified staple foods and determining their impact on micronutrient status. Food Nutr Bull 28: S271–S279

IFPRI (2014) 2014 Global nutrition report dataset. International Food Policy Research Institute [Distributor] V1 [Version] http://dx.doi.org/10.7910/DVN/27857

IFPRI (2014a) Global nutrition report 2014: actions and accountability to accelerate the world's progress on nutrition. International Food Policy Research Institute, Washington, DC. http://ebrary.ifpri.org/cdm/ref/collection/p15738coll2/id/128484

Ingutia RA, Zahidul Islam KA, Hossain MM (2009) Analysis of the relation between rural poverty, malnutrition and hunger, and the interlinks with agricultural production growth and productivity growth with special reference to Mozambique and Bangladesh. Discussion paper no. 32. Department of Economics and Management, University of Helsinki, Helsinki

Kristof ND (2009) The hidden hunger. New York Times, May 23. http://www.nytimes.com/2009/05/24/opinion/24kristof.html. Accessed January 2015

Messer E, Cohen MJ, Marchione T (2001) Conflict: a cause and effect of hunger. ECSP report 7, environmental change and security program. http://www.wilsoncenter.org/sites/default/files/ECSP7-featurearticles-1.pdf

Milanovic Branko (2011) Global inequality: from class to locations, from proletarians to migrants. Policy Research working paper 5820. Development Research Group, World Bank, Washington, DC

Montagnac JA, Davis CR, Tanumihardjo SA (2009) Nutritional value of cassava for use as a staple food and recent advances for improvement. Comp Rev Food Sci Food Safety 8:181–194. http://onlinelibrary.wiley.com/doi/10.1111/j.1541-4337.2009.00077.x/pdf

Narayan D, Pritchett L, Kapoor S (2009) Moving out of poverty: success from the bottom up. Palgrave Macmillan, New York

Nestel P, Bouis HE, Meenakshi JV, Pfeiffer W (2006) Biofortification of staple food crops. J Nutr 136:1064–1067

Olinto P, Beegle K, Sobrado C, Hiroki U (2013) The state of the poor: where are the poor, where is the extreme poverty harder to end, and what is the current profile of the world's poor? PREM economic premise note no. 125. World Bank, Washington, DC

Pfeiffer WH, McClafferty B (2007) HarvestPlus: breeding crops for better nutrition. Crop Sci 47:S88–S105

Qaim M, Stein AJ, Meenakshi JV (2007) Economics of biofortification. Agric Econ 37:119–133

Rouse TI, Davis DP (2004) Exploring a vision: integrating knowledge for food and health. a workshop summary. Board on Agric. and Nat. Resources, Division on Earth and Life Studies, National Research Council of the National Academies, The National Academies Press, Washington, DC. http://www.nap.edu/openbook.php?record_id =10936

Stein AJ (2015) The poor, malnutrition, biofortification, and biotechnology. In: Herring R (ed) The Oxford handbook of food, politics and society. Oxford University Press, New York. www.ajstein.de/cv/Stein2014_OUP_Malnutrition.pdf

Sumner A (2010) Global poverty and the new bottom billion: what if three-quarters of the world's poor live in middle-income countries?. IDS working paper 349. Institute of Development Studies, University of Sussex, Brighton

Sumner A (2012) Global poverty reduction: the last 20 years and the next 20 years. EADI policy paper series, European Association of Development Research and Training Institutes

The Economist (2011) Hidden hunger: how much can farming really improve people's health? The Economist, March 24. http://www.economist.com/node/18438289

Thirtle C, Irz X, Lin L, McKenzie-Hill V, Wiggins S (2001) Relationship between changes in agricultural productivity and the incidence of poverty in developing countries. Report commissioned by Department for International Development, London

UN (2014) The Millennium Development Goals report 2014. United Nations, New York

UNCTAD (2013) Growth and poverty eradication: why addressing inequality matters, post-2015. Policy brief No. 02, United Nations conference on trade and development

UNDP (2002) Human development report 2002: deepening democracy in a fragmented world. Oxford University Press for UNDP, New York

UNSCN (2010) Sixth report on the world nutrition situation: progress in nutrition. United Nations System, Standing Committee on Nutrition, Geneva

von Braun J (2010) Food insecurity, hunger and malnutrition: necessary policy and technology changes. N Biotechnol 27(5):449–452

von Grebmer K, Saltzman A, Birol E, Wiesman D, Prasai N, Yin S, Yohannes Y, Menon P, Thompson J, Sonntag A (2014) 2014 global hunger index: the challenge of hidden hunger. International Food Policy Research Institute/Concern Worldwide/Welthungerhilfe, Washington, DC/Dublin/Bonn

Wessells KR, Brown KH (2012) Estimating the global prevalence of zinc deficiency: results based on zinc availability in national food supplies and the prevalence of stunting. PLoS One 7(11), e50568. doi:10.1371/journal.pone.0050568

White PJ, Broadley MR (2009) Biofortification of crops with seven mineral elements often lacking in human diets- iron, zinc, copper, calcium, magnesium, selenium and iodine. New Phytol 182:49–84

WHO (2015a) Micronutrient deficiencies: iron deficiency anaemia. World Health Organization, Geneva http://www.who.int/nutrition/topics/ida/en/. Accessed 1 February 2015

WHO (2015b) Micronutrient deficiencies: vitamin A deficiency. World Health Organization, Geneva. http://www.who.int/nutrition/topics/vad/en/. Accessed 19 Jun 2015

Wiggins S (2005) Southern Africa's food and humanitarian crisis of 2001–04: causes and lessons, Nottingham. http://www.sarpn.org.za/documents/d0001240/P1388-Wiggins_May2005.pdf

World Bank (1986) Poverty and hunger: Issues and options for food security in developing countries. A World Bank Policy Study, World Bank, Washington, DC

World Bank (2007) Agriculture for development: world development report 2008. World Bank, Washington, DC

World Bank (2007a) From Agriculture to Nutrition: Pathways, Synergies, and Outcomes. Report No. 40196-GLB, Agriculture and Rural Development Department, World Bank, Washington, DC

World Bank (2015) Global monitoring report 2014/2015: ending poverty and sharing prosperity. World Bank, Washington, DC

World Bank (2015a) Poverty overview. http://www.worldbank.org/en/topic/poverty/overview. Accessed January 2015

Zinc in Crops and Human Health

3

Soumitra Das and Andrew Green

Abstract

Zinc has emerged as the most widespread micronutrient deficiency in soils and crops worldwide, resulting in severe yield losses and nutritional quality. Almost half of the soils in the world are deficient in zinc. Since cereal grains have inherently low concentrations, growing these on the potentially zinc-deficient soils further decreases grain zinc concentration. There is a high degree of correlation between zinc deficiency in soils and that in human beings. Zinc is an essential nutrient for human health. There is no life without zinc. Zinc deficiency is the fifth leading cause of death and disease in the developing world. According to the World Health Organization (WHO), about 800,000 people die annually due to zinc deficiency, of which 450,000 are children under the age of five. About one-third of the world's population suffers from zinc deficiency. The paper describes the role of zinc in crop production as well as human health. It highlights the initiatives of the International Zinc Association in addressing zinc deficiency in soils, crops and humans through increased use of zinc fertilisers.

Keywords

Crop yield • Human health • Nutritional quality • Zinc deficiency • Zinc fertilisers

S. Das (✉)
Zinc Nutrient Initiative – India, International Zinc Association, FAI House, 10 Shaheed Jit Singh Marg, New Delhi 110 067, India
e-mail: sdas@zinc.org

A. Green
Zinc Nutrient Initiative and Environment, Health & Sustainability, International Zinc Association, 1822 E NC Highway 54, Suite 120, NC 27713 Durham, USA
e-mail: agreen@zinc.org

3.1 Introduction

Zinc is one of the 17 essential elements necessary for the normal growth and developments of plants. It is among eight micronutrients essential for plants. Zinc plays a key role in plants as a structural constituent or regulatory cofactor of a wide range of different enzymes and proteins in

U. Singh et al. (eds.), *Biofortification of Food Crops*, DOI 10.1007/978-81-322-2716-8_3

many important biochemical pathways. These are mainly concerned with carbohydrate metabolism, both in photosynthesis and in the conversion of sugars to starch, protein metabolism, auxin (growth regulator) metabolism, pollen formation, maintenance of the integrity of biological membranes and resistance to infection by certain pathogens.

Zinc deficiency in plants retards photosynthesis and nitrogen metabolism; reduces flowering and fruit development; prolongs growth periods, resulting in delayed maturity; results in lower yield and poor produce quality; and results in suboptimal nutrient-use efficiency. Some of the common deficiency symptoms of zinc in plants are light green, yellow or bleached spots in the interveinal areas of older leaves, the emerging leaves are smaller in size and often termed as "little leaf", and the internodal distance in case of severe deficiency becomes so short that all the leaves appear to come out from the same point, termed as "rosetting".

3.2 Zinc Status in Soils

Zinc has emerged as the most widespread micronutrient deficiency in soils and crops worldwide (Fig. 3.1), resulting in severe yield losses and deterioration in nutritional quality. It is estimated that almost half of the soils in the world are deficient in zinc. Since cereal grains have inherently low concentrations, growing these on the potentially zinc-deficient soils further decreases grain zinc concentration.

India is not an exception. About 50 % soil samples analysed for available zinc were found deficient in India (Fig. 3.2). There is a significant response to applied zinc in the soils deficient in zinc. In India, zinc is considered the fifth most important yield-limiting nutrient after N, P, K and S in upland crops, whereas in lowland crops like rice, it is next to N.

The reasons responsible for the increase of incidences of zinc deficiency include large zinc removals due to high crop yields and intensive

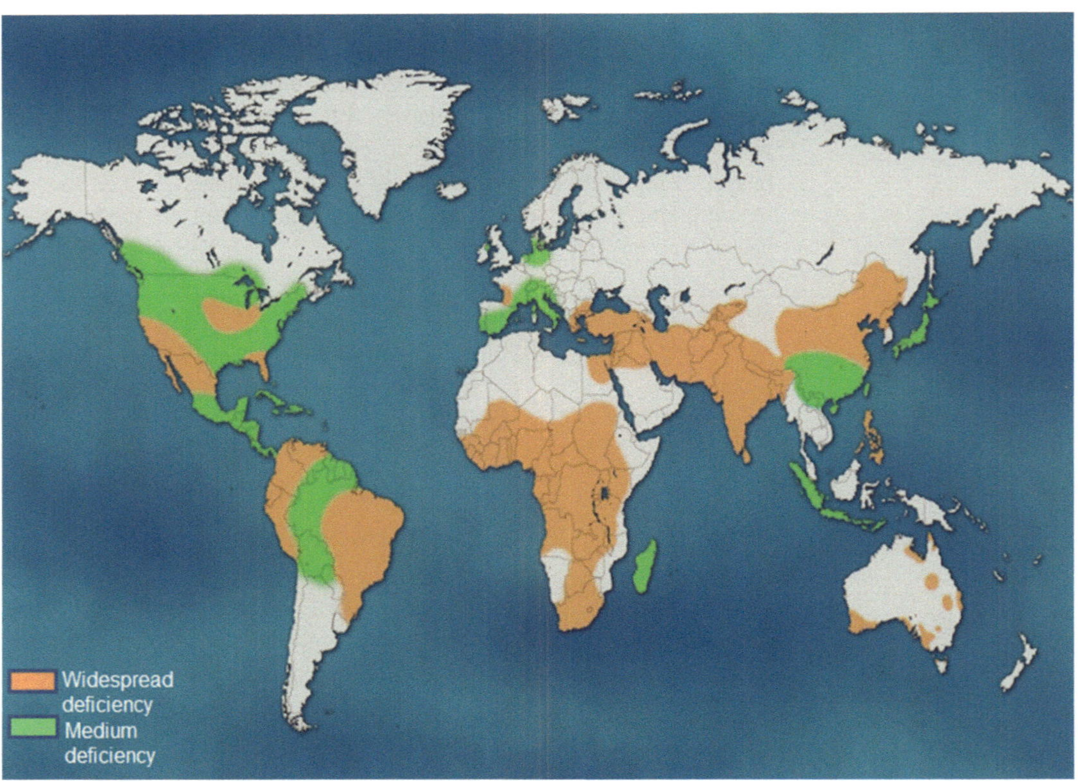

Fig. 3.1 Worldwide zinc deficiency in soils (Alloway 2008)

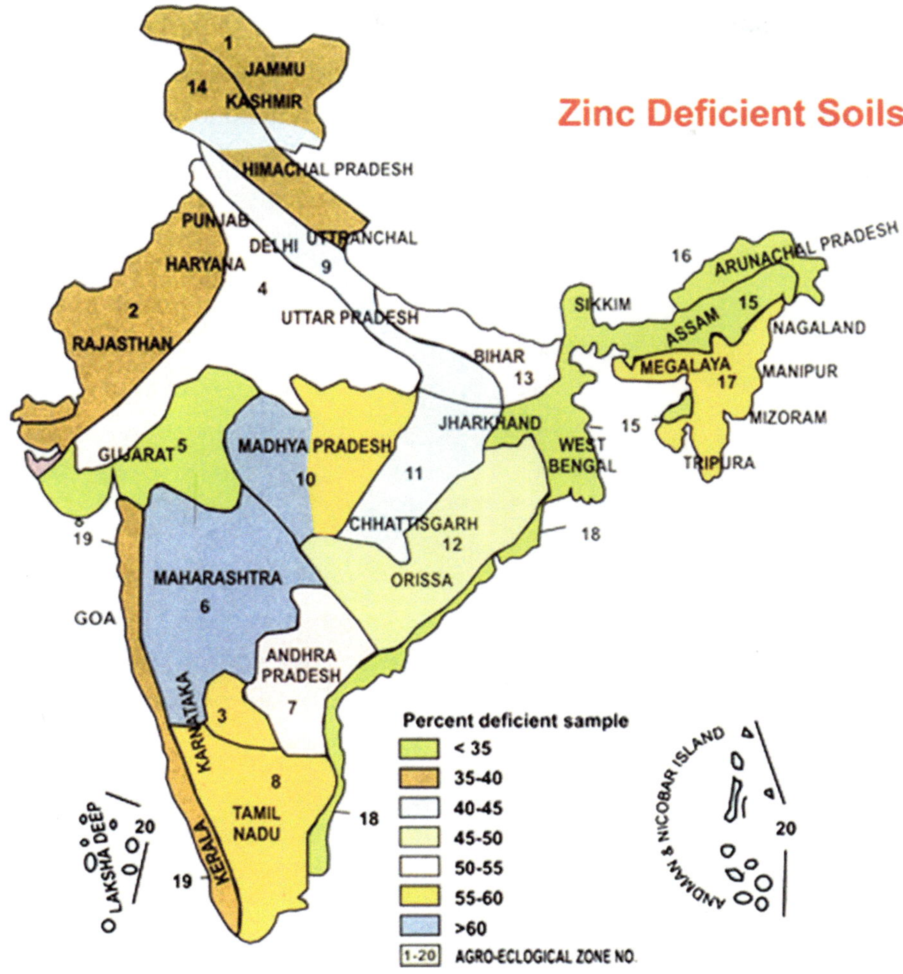

Fig. 3.2 Extent of zinc deficiency in India (Singh 2009)

cropping systems, lesser application of organic manures, use of high analysis fertilisers, increased use of phosphatic fertilisers resulting in P-induced zinc deficiency and the use of poor-quality irrigation water.

The zinc deficiency in India is expected to increase from the present level of around 50 to 63 % in 2025 if the trend continues. This is also because more and more areas of marginal lands are brought under intensive cultivation without adequate micronutrient supplementation. According to an estimate, a huge zinc deficit of about one million tonnes exists in Indian soils. Against the crop removal of about 1.2 million tonnes, the zinc fertiliser use is to the tune of only 0.25 million tonnes. Multi-nutrient deficiencies are fast emerging in many areas in India (Fig. 3.3), resulting in lower yield and quality.

3.3 Zinc: Essential for Life

Zinc is an essential nutrient for human health. There is no life without zinc. Recently, zinc deficiency – especially in infants and young children under 5 years of age – has received global attention. Zinc deficiency is the fifth leading cause of death and disease in the developing world. According to the World Health Organization (WHO), about 800,000 people die annually

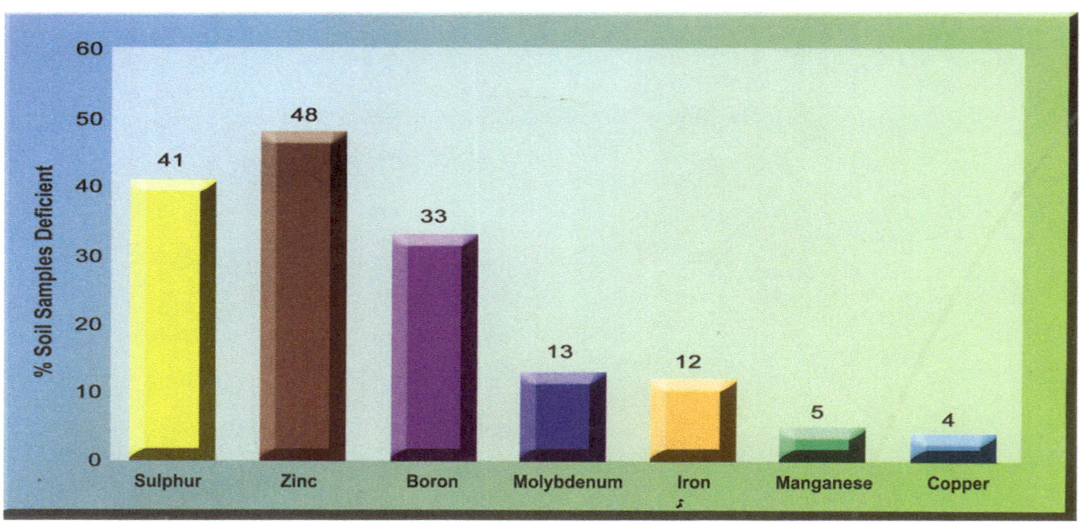

Fig. 3.3 Multi-nutrient deficiency status in India (Singh 2009)

due to zinc deficiency, of which 450,000 are children under the age of 5.

It is estimated that 60–70 % of the population in Asia and sub-Saharan Africa could be at risk of low zinc intake; in absolute numbers, this translates into about two billion people in Asia and 400 million people in sub-Saharan Africa (Prasad 2006). There is a high degree of correlation between zinc deficiency in soils and that in human beings (Fig. 3.4). It is estimated that about one-third of the world's population suffers from zinc deficiency.

Zinc is vital for many biological functions in the human body. The adult body contains 2–3 g of zinc. It is present in all parts of the body, including organs, tissues, bones, fluids and cells. It is vital for more than 300 enzymes in the human body, activating growth – height, weight and bone development, cell division, immune system, fertility, taste, smell and appetite, skin, hair and nails and vision.

Some of the reported symptoms due to zinc deficiency in humans, especially in infants and young children, are diarrhoea, pneumonia, stunted growth, weak immune system, retarded mental growth and dwarfism, impaired cognitive function, behavioural problems, memory impairment, problems with spatial learning and neuronal atrophy.

3.4 Zinc Malnutrition: Possible Solution is Biofortification

The possible solution to the zinc malnutrition in humans may be (i) food supplementation, (ii) food fortification or (iii) biofortification. The former two programmes require infrastructure, purchasing power, access to market and healthcare centres and uninterrupted funding, which have their own constraints. In addition, such programmes will most likely reach the urban population, which is easily accessible, especially in the developing countries. Alternatively, the latter programme, biofortification – fortification of crops especially food crops with zinc – is the best option for alleviating zinc deficiency. It will cater to both the rural and urban populations. It could be achieved through two approaches, genetic biofortification and agronomic biofortification.

There is a developing field of research on the biofortification of plant foods with zinc. This involves both the breeding of new varieties of crops with the genetic potential to accumulate a high density of zinc in cereal grains (genetic biofortification) and the use of zinc fertilisers to increase zinc density (agronomic biofortification). Although the plant breeding route is

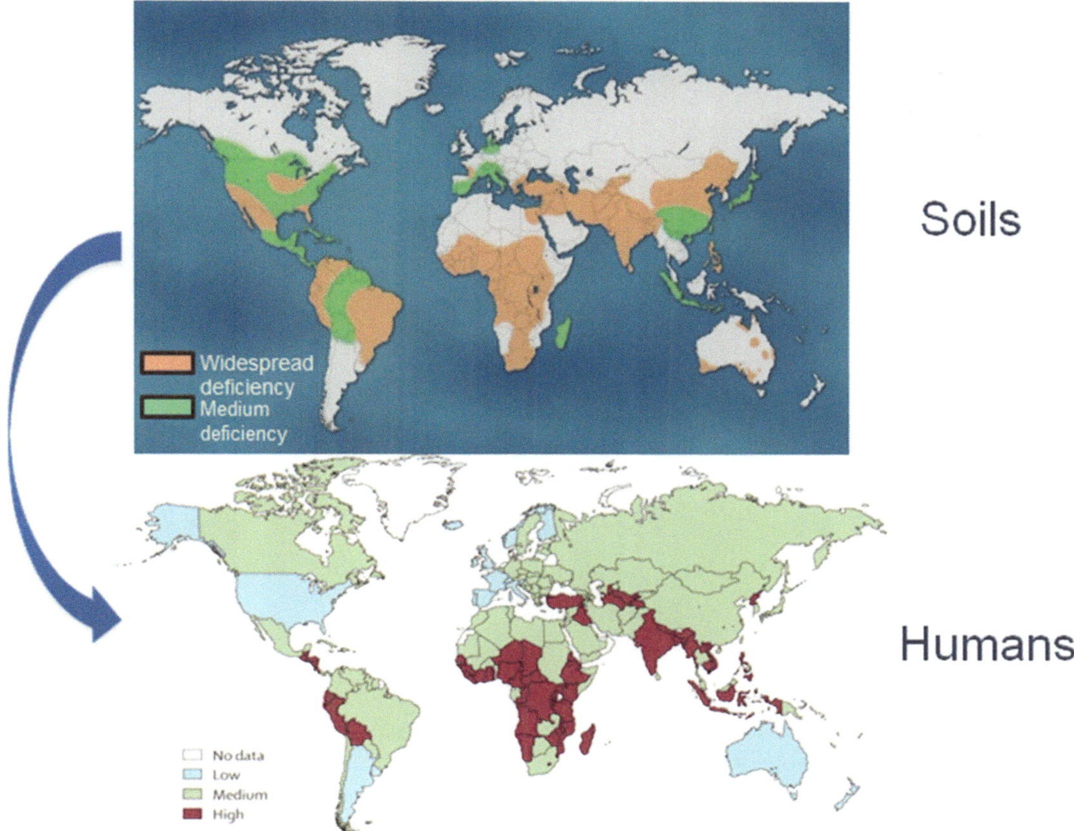

Fig. 3.4 Worldwide zinc deficiency in soils and humans (Alloway 2008)

likely to be the most cost-effective approach in the long run, the use of fertilisers is the fastest route to improve the zinc density in diets. In order to replenish the zinc taken up by the improved cultivars, higher and sustainable use of fertilisers is inevitable.

Ideally, the cereal grains should contain 40–60 mg zinc/kg, whereas currently, it is only 10–30 mg zinc/kg (Cakmak 2008). This needs to be rectified urgently.

3.4.1 Zinc Uptake and Removal by Crops

Zinc uptake by some pulses and oilseed crops are given in Table 3.1. The proportion of zinc absorbed by these crops which were present in the grains is also mentioned. In the pulse crops, the zinc uptake varied from 38 to 72 g ha^{-1} by pigeon pea and green gram for the corresponding grain yields of 1200 and 1000 kg ha^{-1}, whereas the zinc uptake in grain varied from 34 % to 62 % by pigeon pea and lentil for the grain yields of 1200 and 2000 kg ha^{-1}, respectively. These data show that a considerable proportion of absorbed zinc remains in the straw. Lentil and soybean seem to be able to transfer greater proportion of absorbed zinc to the seed than many other crops (Tandon 2013).

Table 3.2 shows zinc removal by pulses, cereal and oilseed-based cropping systems. The pigeon pea – wheat cropping system – removed 287 g Zn ha^{-1}, equivalent to 48 g Zn t^{-1}, for the grain yield of 6 t ha^{-1} (Prasad 2006).

3.4.2 Zinc Fertiliser Recommendations

Zinc fertiliser recommendations are based on crop–soil–climate-specific situations. However, wherever situation-specific recommendations

Table 3.1 Zinc uptake by some pulses and oilseeds and their proportions in grains

Crop	Grain yield (kg ha^{-1})	Uptake (g ha^{-1})	Uptake in grain (%)
Chickpea	1500	59	49
Pigeon pea	1200	38	34
Lentil	2000	52	62
Green gram	1000	72	58
Groundnut	1900	208	45
Mustard	1500	154	34
Sesame	1200	201	31
Soybean	2500	196	68

Table 3.2 Zinc removal by pulses, cereal and oilseed-based cropping systems

Cropping system	Grain yield (t ha^{-1})	Zinc removed (g ha^{-1})	Zinc removed (g t^{-1})
Pigeon pea–wheat	6.0	287	48
Rice–wheat	8.9	1153	129
Rice–rice	8.0	320	40
Maize–wheat	8.0	744	93
Soybean–wheat	6.5	416	64

are not readily available, blanket recommendation of 25 kg ha^{-1} $ZnSO_4$ heptahydrate, or 15 kg ha^{-1} $ZnSO_4$ monohydrate, equivalent to 5 kg ha^{-1} zinc are generally advocated for every year/alternate years for soil application. For foliar spray, 0.5 % $ZnSO_4$ is advocated. The state government recommends crop-specific zinc fertiliser doses.

3.4.3 Crop Response to Zinc

The response of pulses to soil applied Zn in 28 field trials conducted in different soils of Bihar showed that the response range varied from 33 to 860 kg ha^{-1} in lentil and chickpea, whereas the average response was from 50 to 450 kg ha^{-1} in green gram and peas, respectively (Singh et al. 2011) (Table 3.3).

Zinc biofortification trials conducted at Kanpur on pigeon pea during *kharif* 2009–2010 revealed that application of Zn enhanced the pigeon pea grain yield by 13 % over no Zn. In Bhopal, the increase in yield was 22 % with soil Zn application, whereas it was 64 % with soil + foliar Zn application (Shukla and Tiwari 2014) (Table 3.4).

Crop response to zinc has been observed in all crops under almost all types of soils and agroclimatic conditions. While the response was found to be higher in grain crops like rice, fruit and vegetable crops also respond well to applied zinc. Extent of crop response depends on the status of zinc in that soil. The higher the zinc deficiency in soils, the higher the crop response would be to applied zinc. Based on over 15,000 on-station field trials conducted all over India, the overall range of crop response to zinc was of the following scale (Singh 2008):

Cereals: 420–550 kg ha^{-1} (15.7–23.0 %)
Pulses: 170–460 kg ha^{-1} (7.3–28.2 %)
Oilseeds: 110–360 kg ha^{-1} (11.4–40.0 %)
Fodders: 90–4620 kg ha^{-1} (5.0–34.0 %)

3.4.4 Zinc Content in Crops

In a multilocation study in China, India, Lao PDR, Thailand and Turkey (averaged over nine site years), Zn concentration in unhusked rice grain was about 69 % higher with foliar application than with soil application; at some centres, it

Table 3.3 Response of pulses to Zn application on farmers' fields

Crop	No. of field trials	Response range (kg ha^{-1})	Average response (kg ha^{-1})
Chickpea	9	130–860	390
Peas	2	180–710	450
Green gram	1	50	50
Black gram	6	100–520	330
Lentil	9	33–440	240
Broad bean	1	250	250

Table 3.4 Effect of Zn biofortification strategies on grain and zinc concentration in different cultivars of pigeon pea at Kanpur and Bhopal

	Grain yield (t ha^{-1})		Zinc concentration (mg kg^{-1})	
Cultivar	−Zn	+Zn	−Zn	+Zn
Kanpur				
LRG 38	1.90	2.00	49.0	63.0
JKM 7	1.80	1.90	43.0	66.0
BSMR 736	1.80	1.90	22.0	34.0
Pusa 9	2.40	2.60	18.0	23.0
Bhopal				
ICPL 87119	1.84	1.99	27.8	41.8
T 15–15	1.46	1.55	32.3	44.6
Virsa Arhar 1	1.53	1.76	32.8	38.7

Table 3.5 Relative zinc concentration in unhusked, brown and white (polished) rice

Characteristic	Control (no Zn)	Soil Zn	Foliar Zn	Soil + foliar Zn
Grain yield (t ha^{-1})	6.7	7.0	6.9	7.0
Zn in unhusked rice (mg kg^{-1})	18.7	19.1	32.3	34.7
Zn in brown rice (mg kg^{-1})	19.1	20.8	24.4	25.5
Zinc in polished rice (mg kg^{-1})	16.1	16.2	17.7	18.4

Table 3.6 Zinc content in some pulses

Crop	Grain (in %)	Straw (in %)
Chickpea	19	10
Pigeon pea	11	4
Lentil	16	10
Green gram	42	5
Black gram	29	11

was almost twice that of with soil application (Prasad et al. 2014) (Table 3.5).

An analysis of the nutrient content of some pulse crops by Aulakh (1985) shows the zinc concentration in grain and straw (Table 3.6).

3.4.5 Effect of Zinc on Crop Quality

Application of zinc not only increases the crop yield but also improves its quality. In potato, it increased ascorbic acid in tubers, reduced phenol content and enhanced reducing sugars, sucrose and total sugar. Zinc was also found to increase the phenol tannin content of leaves, kernels and seed coat of cotton. An increase in the energy value, as well as total lipids, crude protein and carbohydrate content in rice, maize, wheat, mustard, chickpea and black gram, was accounted for zinc application. Improvement in amino acids in cereals was also observed. Sucrose recovery and juice quality were improved in sugarcane (Kalwe et al. 2001).

3.4.6 Interaction of Zinc with Other Nutrients

Zinc interacts positively with N and K and negatively with P. Antagonistic effect of Zn x P interaction has been the subject of intensive study in several countries. This implies that farmers should not overuse phosphatic fertilisers, or else it would lead to reduced uptake of zinc by the crops. Zinc interacts antagonistically with all three secondary nutrients S, Ca and Mg as well as other micronutrients, like, Fe, Mn, Cu and Mo.

3.4.7 Economics of Zinc Fertiliser Use

Trials carried on farmers' fields reveal that zinc application to crops is remunerative on zinc-deficient soils. Over 7000 trials carried out in cultivators' fields with cereal crops, rice, wheat, maize and barley showed an 8.3 % increase in mean grain response over NPK. The B/C ratio varied from 2.8:1 to 3.5:1, which showed that it was profitable, as B/C ratio over 2.5 is generally considered as remunerative to the farmers.

3.4.8 Zinc Fertilisers in FCO

The fertilisers containing zinc, mentioned in the Fertiliser Control Order (FCO), are enumerated below:

Schedule 1 Part A

1 (f) Micronutrients

- Zinc sulphate heptahydrate – 21 % Zn
- Zinc sulphate monohydrate – 33 % Zn
- Chelated zinc (Zn-EDTA) – 12 % Zn

1 (g) Fortified fertilisers

- Zincated urea – 2 % Zn
- Zincated phosphate (suspension) – 19 % Zn
- NPK Zn – 3.5 % Zn

Clause 20 A

- Micronutrient mixture – crop specific
- Zinc-fortified DAP/SSP/NPKs
- Zinc polyphosphate – 16 % Zn

Clause 20 B

- Customised fertilisers

3.4.9 Fertiliser Policy in India

The government is concerned with the problem of imbalanced use of plant nutrients, declining crop response ratio, stagnant crop productivity and rising subsidy bill. Switch over to the nutrient-based subsidy (NBS) scheme as proposed in the Union Budget for 2009–2010 provided the direction to the future fertiliser policies.

Keeping in view the agricultural situation and widespread soil degradation, the time was ripe for ushering into nutrient-based subsidy to promote balanced and efficient use of plant nutrients. Such policies encouraged development and use of customised fertilisers. Nutrient-based subsidy rather than product-based subsidy allows all fertilisers covered under the Fertiliser Control Order (FCO) to get the subsidy as per their nutrient content. This encourages development of new fertiliser products and use in the country depending upon the requirement of different soils and crops.

In addition, the large fertiliser players are coming into zinc sulphate manufacturing and marketing. It is generally felt that entering of the larger players into zinc fertiliser business will open up new vistas in the field of zinc fertiliser manufacturing, marketing and usage.

Under the NBS scheme, the role of zinc has been specially targeted through additional subsidy for the zinc-fortified products @ Rs. 500 per tonne. The Govt. of India is promoting the use of zinc under the National Food Security Mission also and providing an additional subsidy to the farmers @ Rs. 500 per hectare for use of micronutrient fertilisers. Of late, inclusion of urea in the NBS scheme is under consideration by the Govt. of India. Efforts are also being made to address the pricing issue of zincated urea, so that it can be considered for production and marketing by the fertiliser industry.

Fig. 3.5 Zinc Saves Kids Initiative of the International Zinc Association in support of UNICEF

Such steps will lead towards balanced fertilisation resulting in higher productivity and efficient use of fertilisers. Zinc fertiliser use has been significantly increased in the last couple of years and is bound to witness an upward trend further in the days ahead in India.

3.5 Role of International Zinc Association (IZA)

The International Zinc Association (IZA) is a non-profit organisation based in Brussels, Belgium, and was founded in 1991. IZA has regional offices in India, China, Europe, Latin America, North America, Middle East and Southern Africa.

IZA is the only global industry association dedicated exclusively to the interests of zinc and its users. Operating internationally and locally through its regional affiliates, IZA helps sustain the long-term global demand for zinc and its markets by promoting such key-end uses as corrosion protection for steel and the essentiality of zinc in human health and crop nutrition.

IZA helps grow and protect the global markets for zinc by promoting the essentiality of zinc in present and potential applications, human health and crop nutrition and by highlighting contribution of zinc to sustainable development. IZA's main programme areas are Technology and Market Development, Environment and Health, Zinc for Life and Communications.

3.5.1 Zinc Nutrient Initiative (ZNI)

The Zinc Nutrient Initiative (ZNI) is a programme of the International Zinc Association (IZA) which addresses zinc deficiency in soils, crops and humans through increased use of zinc fertilisers with the objectives to raise awareness on increased crop yield, improved nutritional quality of crops, improved human nutrition, increased farmers' income and improved food and nutritional security (Das and Green 2011).

The various approaches undertaken by the initiative are coordinating crop demonstration trials; organising international, national and regional conferences, workshops and seminars; conducting training programmes; publication of communication materials; and networking with the various stakeholders in zinc sector.

There are a number of global initiatives in which IZA is actively involved. Zinc Saves Kids is one such initiative in support of UNICEF of the United Nations (Fig. 3.5).

3.6 Challenges

The key challenges in zinc in balanced fertiliser use are:

- Availability of zinc fertilisers at the time of need of farmers
- Quality of zinc fertilisers available

- Information on soil–crop–animal–human continuum study (multidisciplinary approach)
- Awareness level of the extension and promotional workers
- Lack of awareness of the farmers – last mile delivery

3.7 Conclusions

Zinc deficiency in crops and humans is a critical issue and a global challenge. The sustainable solution is increased use of zinc in balanced fertiliser use, so that the soil health, food security as well as nutritional security are ensured.

References

Alloway BJ (2008) Zinc in soils and crop nutrition. IFA, Paris and IZA, Brussels, p 135

Aulakh MS (1985) Content and uptake of nutrients by pulse and oilseed crops. Indian J Ecol 12(2):238–242

Cakmak I (2008) Enrichment of fertilizers with zinc: an excellent investment for humanity and crop production in India. In: Proceedings of FAI annual seminar, New Delhi, SII-2, pp 1–8

Das S, Green A (2011) Zinc in crops and human health – IZA initiatives. Indian J Fert 7(10):152–170

Kalwe SP, Das S, Biswas BC (2001) Role of zinc in balanced fertilisation. In: Proceedings of national conference on zinc in health and agriculture, ILZDA, New Delhi, pp 1–10

Prasad R (2006) Zinc in soils and in plant, human and animal nutrition. Indian J Fert 2(9):103–119

Prasad R, Shivay YS, Kumar D (2014) Agronomic biofortification of cereal grains with iron and zinc. Adv Agron 125:55–91

Shukla AK, Tiwari PK (2014) Micro and secondary nutrients and pollutant elements research in India. Progress report 2011–2013. AICRP – MSPE (ICAR), IISS, Bhopal, pp 1–155

Singh MV (2008) Micronutrient deficiencies in soils and crops of India. In: Alloway BJ (ed) Micronutrient deficiencies in global crop production. Springer Sciences Business Media, Dordrecht, pp 93–125

Singh MV (2009) Micronutrient nutritional problems in soils of India and improvement for human and animal health. Indian J Fert 5(4):11–26

Singh KK, Praharaj CS, Choudhary AK, Kumar N, Venkatesh MS (2011) Zinc response to pulses. Indian J Fert 7(10):118–126

Tandon HLS (2013) Micronutrient handbook – from research to application. FDCO, New Delhi, p 234

Food Legumes for Nutritional Security and Health Benefits

4

N P Singh and Aditya Pratap

Abstract

In India, over a dozen of pulse crops, viz. chickpea, pigeon pea, cowpea, mung bean, urdbean, lentil, French bean, horse gram, field pea, moth bean, lathyrus, etc., are grown in one or on the other part of the country. Pulse seeds contain 16–50 % protein and provide one third of all dietary protein nitrogen. Therefore, pulses in combination with cereals provide one of the best solutions to protein calorie malnutrition, especially in the developing countries. These are also an important source of the 15 essential minerals needed by humans. Despite their importance in human health and nutrition, the global food legume production has witnessed only a marginal annual increase of 0.95 % with fluctuations only from 40.78 to 70 million tonnes. The slow growth in global pulses production along with rising population, diversified uses for end products and improved purchasing capacity of people has put tremendous pressure on per capita availability of pulses. In addition, decreasing cultivable land and factor productivity, biotic and abiotic stresses and unavailability of quality seeds have further strained the pulse supply chain. Nevertheless, this may be compensated to some extent by proper storage and pre- and postharvest management of pulses, value addition and manufacturing of value-added products. This chapter discusses the role of pulses for nutritional quality and health benefits.

Keywords

Biofortification • Cancer • Health • Nutrition • Pulses

N P Singh
Director, ICAR-Indian Institute of Pulses Research, Kanpur, Uttar Pradesh 208 024, India
e-mail: director@iipr.ernet.in

A. Pratap (✉)
Division of Crop Improvement, ICAR-Indian Institute of Pulses Research, Kanpur, Uttar Pradesh 208 024, India
e-mail: adityapratapgarg@gmail.com

4.1 Introduction

Pulses – with high protein content and 15 essential minerals – are the most important constituents of cereal-based vegetarian diets of

U. Singh et al. (eds.), *Biofortification of Food Crops*, DOI 10.1007/978-81-322-2716-8_4

the Indian subcontinent. Owing to their diverse uses as atmospheric nitrogen fixing agents, green manure and cover crops, catch crops in short season cropping windows, breakfast grains and ingredients of specialty diets, pulses are rightly a subject of agricultural, environmental and biotechnological research. In India, over a dozen of pulse crops including chickpea, pigeon pea, cowpea, mung bean, urdbean, lentil, French bean, horse gram, field pea, moth bean, lathyrus, etc. are grown in one or on the other part of the country. The most important pulse crops grown here are chickpea (41 %), pigeon pea (15 %), urdbean (10 %), mung bean (9 %), cowpea (7 %) lentil (5 %) and field pea (5 %).

Pulse seeds contain 16–50 % protein and provide one third of all dietary protein nitrogen. Therefore, pulses in combination with cereals provide one of the best solutions to protein calorie malnutrition, especially in the developing countries. Besides proteins, these are also an important source of the 15 essential minerals needed by humans (Wang et al. 2009). Consumption of pulses on a regular basis has been associated with lower risks for the development of type 2 diabetes, coronary heart disease and some forms of cancer (Chibbar et al. 2010). Those who consume pulses also tend to have lower rates of obesity and metabolic syndrome (Rizkalla et al. 2002). Therefore, a number of countries recommend that people should consume pulses as part of a healthy diet (USDA 2013).

Despite their so much importance in human health and nutrition, the global food legume production except groundnut and soybean has witnessed only a marginal annual increase of 0.95 % with fluctuations only from 40.78 to 70 million tonnes. The slow growth in global pulse production along with rising population, diversified uses for end products and improved purchasing capacity of people has put tremendous pressure on per capita availability of pulses. In addition, decreasing cultivable land and factor productivity, biotic and abiotic stresses and unavailability of quality seeds have further strained the pulse supply chain. Nevertheless, this may be compensated to some extent by proper storage and pre- and postharvest management of pulses, value addition and manufacturing of quality nutritional products.

4.2 Statistics of Pulses

Globally, the major pulses are grown over an area of 80.75 million ha with a production of 73.07 million tonnes. The major pulse-growing countries of the world include Myanmar, Brazil, China, Canada and Australia besides India which is the largest producer as well as consumer of pulses with 25 % share in the global production. Among different pulses, dry beans are the most widely cultivated pulses in the world, grown over 29.23 million ha with a production of 23.14 million tonnes. This is followed by chickpea (13.54 million ha), cowpeas (dry, 11.32 million ha), peas (6.38 million ha) and pigeon pea (6.22 million ha) (Table 4.1). However, as far as productivity is concerned, field pea (1721 kg/ha) records highest productivity globally followed by lentil (1140 kg/ha) and chickpea (968 kg/ha).

India has 23.47 million ha area under pulses during 2012–2013 (Table 4.2). Over a dozen

Table 4.1 Area, production and productivity of different pulses in the world

Crops	Area (m. ha)	Prod. (m.ton)	Yield (kg/ha)
Beans (dry)	29.23	23.14	792
Chickpea	13.54	13.10	968
Cowpeas (dry)	11.32	5.72	505
Peas (dry)	6.38	10.98	1721
Pigeon pea	6.22	4.74	762
Lentil	4.34	4.95	1140
Others	9.72	10.44	1074
Total	80.75	73.07	904

Table 4.2 Country-wise area, production and productivity of pulses in the world during 2012–2013

Countries	Area (m. ha)	Production (m.ton)	Yield (kg/ha)
India	28.17	19.27	781
Myanmar	3.88	5.44	1398
Brazil	2.85	2.95	1032
China	2.87	4.46	1550
Canada	2.42	6.10	2520
Australia	1.92	2.70	1410
USA	1.09	2.23	2038
Others	37.55	29.88	796
Total	80.75	73.07	904

Table 4.3 Crop wise area, production and productivity of pulses in India

Crops	Area (m. ha)	Production (m.ton)	Yield (kg/ha)
2013–2014			
Chickpea	9.19	9.53	1075
Pigeon pea	3.88	3.17	817
2012–2013			
Mung bean	2.75	1.19	432
Urdbean	3.19	1.90	596
Lentil	1.42	1.13	797
Peas	0.76	0.84	1105
Lathyrus	0.58	0.43	742
2011–2012			
Horse gram	0.24	0.10	485
Moth bean	1.37	0.47	346

pulse crops including chickpea (*Cicer arietinum*), pigeon pea (*Cajanus cajan*), mung bean (*Vigna radiata*), urdbean (*Vigna mungo*), cowpea (*V. unguiculata*), lentil (*Lens culinaris* ssp. *Culinaris*), lathyrus (*Lathyrus sativus* L.), French bean (*Phaseolus vulgaris*), horse gram (*Macrotyloma uniflorum*), field pea (*Pisum sativum*), moth bean (*V. aconitifolia*), etc. are grown in one or on the other parts of the country. However, the most important pulse crops grown here are chickpea (48 %), pigeon pea (15 %), mung bean (7 %), urdbean (7 %), lentil (5 %) and field pea (5 %). Crop wise, chickpea (9.19 million ha) occupies the largest area in the country with a production of 9.53 million ha and productivity of 1075 kg/ha (Table 4.3). This is followed by pigeon pea (3.88 million ha) and urdbean (3.19 million ha).

The production of pulses in India has tremendously improved during the last 2 years recording an all-time high record of 19.27 million tonnes during 2013–2014 (Table 4.4). Still, to meet the current demand, about two to three million tonnes of pulses need to be imported every year. Due to a production gap, increasing population and diversity of uses of pulses, the per capita availability of pulses has declined progressively from 41.6 g/day/capita in 1991 to 38 g in 2009 (against ICMR recommendation is 65 g/day/capita). Therefore, the production has to be increased to meet the domestic requirement. In order to ensure self-sufficiency, the pulse requirement in the country is projected to be about 39 million tonnes by 2050 which necessitates an annual growth rate of 4.0 %. This requires a paradigm shift in research, technology generation and dissemination and commercialization along with capacity building in frontier areas of research. To meet the projected requirement of pulses, the productivity needs to

Table 4.4 Trends in area production and productivity of pulses in India

Year	Area (m. ha)	Production (m. ton)	Yield (kg/ha)
2007–2008	23.63	14.76	625
2008–2009	22.09	14.57	659
2009–2010	23.28	14.66	630
2010–2011	26.28	18.24	694
2011–2012	24.78	17.09	690
2012–2013	23.47	18.34	781
2013–2014	NA	19.27	NA
2014–2015	NA	18.43[a]	NA

[a]Advance estimates

be raised to about 1200 kg per ha, and about three to six million ha additional area has to be brought under pulses.

4.3 Constraints in Increasing Pulse Production

The existing varieties of pulses have tremendous yield potential as demonstrated in experimental field and demonstrations. Nevertheless, the realized yield in most of the pulses is far from the potential yield ultimately leading to a shortfall in their total production. This shortfall has been attributed to a number of factors: geographical shift, abrupt climatic changes, complex disease-pest syndrome, socioeconomic conditions of the farmers and less market opportunities. The major constraints that limit the realization of potential yield of pulses include biotic and abiotic stresses prevalent in the pulse-growing areas besides socioeconomic factors. Among biotic stresses, *Fusarium* wilt coupled with root rot complex is probably the most widespread disease causing substantial losses to chickpea. While *Fusarium* wilt, sterility mosaic and phytophthora blight cause substantial losses in pigeon pea, yellow mosaic, *Cercospora* leaf spot and powdery mildew are considered as the most important diseases in both mung bean and urdbean. In lentil, the rust, powdery mildew and wilt cause considerable damage. Among key insect pests, gram pod borer (*Helicoverpa armigera*) in chickpea and pigeon pea, pod fly in pigeon pea and whitefly, jassids and thrips in dry beans cause severe damage to the respective crops. Weeds also cause substantial loss to pulses. Recently, nematodes have emerged as potential threat to the successful cultivation of pulses in many areas.

Among abiotic stresses, drought and high temperature at terminal stage, cold and sudden drop in temperature coupled with fog during the reproductive phase and salinity/alkalinity throughout the crop period inflict major yield losses and instability in production. All these stresses make pulse crops less productive with unstable performance in one or the other way.

4.4 Nutritional Composition of Pulses

Numerous studies have been conducted for estimating genetic variability for protein content as well as other micro- and macronutrients among a number of accessions of various pulses. A large amount of genetic variability was found to be available for these traits (Table 4.5). For example, protein content in different pulses varied from 26.5 to 57 % in soybean (Dwivedi et al. 1990), 20.9 to 29.2 % in common bean, 15.8 to 32.1 % in pea, 22 to 36 % in faba bean, 19 to 32 % in lentil, 16 to 28 % in chickpea, 16 to 31 % in cowpea, 21 to 31 % in mung bean and 16 to 24 % in pigeon pea (Burstin et al. 2011).

Pulses provide from 1040 to 1430 kJ per 100 g (similar to cereal grains), mostly by carbohydrates rather than fat. The carbohydrate content in pulses is about 60 %. The mono- and oligosaccharides represent only a small per cent of total carbohydrate in pulses, whereas starch is

Table 4.5 Range of variation (% of seed weight) in principal constituents of seeds of pulses

	Content (%)				
Protein	Oil	Starch	Fibre	Sucrose	Reference
Soybean					
35.1–42	17.7–21	1.5	20	6.2	Hedley (2001)
0–45	19–21.5	–	–	–	Hyten et al. (2004)
41.8–49.4	15.2–20.7	–	–	–	Chung et al. (2003)
40.4–50.6	13.4–21.2	–	–	–	Brummer et al. (1997)
26.5–47.6	–	–	–	–	Vollman et al. (2000)
Common bean					
20.9–27.8	0.9–2.4	41.5	10	5	Hedley (2001)
23–29.2	–	–	–	–	Coelho et al. (2009)
Pea					
18.3–31	0.6–5.5	45	12	2.1	Hedley (2001)
24–32.4	–	45.5–54.2	8.9–11.9	–	Gabriel et al. (2008)
20.6–27.3	–	–	–	–	Burstin et al. (2007)
Faba bean					
26.1–38	1.1–2.5	37–45.6	7.5–13.1	0.4–2.3	Duc et al. (1999)
22.4–36	1.2–4	41	12	3.3	Hedley (2001)
26–29.3	–	42.2–51.5	–	–	Avola et al. (2009)
Lentil					
23–32	0.8–2	46	12	2.9	Hedley (2001)
25.1–29.2	–	46–49.7	13.1–14.7	2.1–3.2	Wang et al. (2009)
Chickpea					
15.5–28.2	3.1–7	44.4	9	2	Hedley (2001)
18.7–21.1	–	42–45.1	–	–	Frimpong et al. (2009)
17.1–19.8	–	48–54.9	–	–	Frimpong et al. (2009)
Cowpea					
23.5	1.3	–	–	–	Hedley (2001)
24.8	1.9	–	6.3	–	Kabas et al. (2006)
20.9–36	2.6–4.2	–	–	–	Oluwatosin (1997)
16–31	2.4–4.3	–	–	–	Adekola and Oluleye (2007)
Mung bean					
21–31.3	1.2–1.6	–	8.9–12.9	–	Anwar et al. (2007)
23.7–31.4	–	–	–	–	Lawn and Rebetzke (2006)
Pigeon pea					
19.5–22.9	1.3–3.8	44.3	10	2.5	Hedley (2001)
15.9–24.1	–	–	–	–	Upadhyaya et al. (2007)

Modified from Burstin et al. 2011

the most abundant carbohydrate. Pulses generally have high amylose content. Among the sugars, raffinose, stachyose, verbascose, ajugose and pentosans predominate in most of the pulses. These sugars escape digestion in the gut and are fermented in the large bowel and can cause abdominal discomfort and flatulence if pulses are consumed in abundance. Soaking pulses overnight in plenty of water helps to reduce the content of these sugars. Pulses are low in fats (1–6 %), most of which is provided by polyunsaturated and monounsaturated fatty acids (Table 4.6). They are also fairly good sources of thiamine and niacin and provide calcium, phosphorus and iron. When consumed with the seed coat, pulses are much higher in dietary fibre than those that have been dehulled prior to consumption. There is a significant increase in nutrients in sprouted pulses when compared to their dried embryo. In the process of sprouting,

Table 4.6 Nutritional profile of pulse grains (per 100 g)

Pulses	Energy kcal	Protein (g)	Fat (g)	Carbohydrate (g)	Total dietary fibre (%)
Chickpea (*Cicer arietinum* L.)	368	21.0	5.7	61	22.7
Pigeon pea (*Cajanus cajan* L.)	342	21.7	1.49	62	15.5
Lentil (*Lens culinaris* Medik)	346	27.2	1.0	60	11.5
Mung bean (*Vigna radiata* L.)	345	25.0	1.1	62.6	16.3
Urdbean (*Vigna mungo* L.)	347	24.0	1.6	63.4	–
Field pea (*Pisum sativum* L.)	345	25.1	0.8	61.8	13.4
Rajmash (*Phaseolus vulgaris* L.)	345	23.0	1.3	62.7	17.7
Cowpea (*Vigna unguiculata*)	346	28.0	0.3	63.4	18.2
Horse gram (*Macrotyloma uniflorum*)	321	23.6	2.3	59.1	15.0
Moth bean (*Vigna aconitifolia*)	330	24.0	1.5	61.9	

the vitamins, minerals and protein increase substantially. Sprouting of the pulses not only improves nutritive value but also digestibility (Health Canada 2015). The nutritional profile of major pulses is described in Table 4.6.

4.4.1 Minerals and Vitamins

Pulses supply adequate vitamins and minerals. They are rich in iron, zinc, magnesium and phosphorus and contain more calcium than cereals. Pulses also contain more potassium which is important for the management of hypertension. They also contain some amount of manganese and copper. Beans and lentils have the highest iron content. Lentil contains 425–673 μg/kg of Se depending on location, soil characteristics and growing conditions. This potentially provides 80–120 % of the minimum recommended daily Se intake in only 100 g of dry lentil.

Pulse grains are a good source of B vitamins, especially thiamine, riboflavin, niacin, pyridoxal and pyridoxine (Table 4.7). The embryo of pulses contains vitamin E, a strong antioxidant. Dried pulses contain some amount of vitamin C, but pulse sprouts are good source of vitamin C. The vitamin C content rises from negligible levels to 12 mg/100 g after 18 h of germination. Pulses are also a good source of thiamine, riboflavin, niacin, pyridoxamine, pyridoxal and pyridoxine. Tocopherol content is higher in pulses than cereals. Peas contain greater amounts of α-than β + γ-tocopherols, and chickpea contains similar levels of α- and β + γ-tocopherols. The following table summarizes the content of vitamins in pulse grains.

4.4.2 Minerals

Beans and lentils have the highest iron and zinc contents, respectively (Table 4.8). Selenium (Se) is another essential micronutrient in human nutrition and is involved in important regulatory and protective mechanisms. It has been reported that lentil contains 425–763 ug/kg of Se depending on location, soil characteristics and growing conditions. This potentially provides 80–120 % of the minimum recommended daily Se intake in only 100 g of dry lentil. Chromium is involved in carbohydrate and lipid metabolism; the most frequent sign of Cr deficiency is altered

Table 4.7 Vitamin content in pulses (per 100 g)

Pulses	Thiamine (mg)	Riboflavin (mg)	Niacin (mg)	Pantothenic acid (mg)	Vitamin B6 (mg)	Folate (ug)	Vitamin C (mg)	Vitamin E (mg)
Chickpea	0.5	0.2	1.5	1.6	0.5	557	4	0.8
Pigeon pea	0.6	0.18	2.9	1.26	0.28	456	–	–
Lentil	0.8	0.2	2.6	2.21	0.54	479	4.4	0.3
Mung bean	0.6	0.2	2.2	1.9	0.4	625	4.8	0.5
Urdbean	0.6	0.2	2.3	–	0.2		4.8	
Field pea	0.7	0.2	2.9	1.8	0.2	274		0.3
Rajmash	0.53	0.22	2.08	0.79	0.4	399	1.8	–
Cowpea	0.94	0.22	2.36	1.39	0.44	545	4.6	–
Horse gram	0.4	0.2	1.5	–	–	–	–	–
Moth bean	0.4	0.09	1.5	–	–	–	–	–

Table 4.8 Mineral content in pulses (mg/100 g dried weight)

Pulses	Iron (mg)	Zinc (mg)	Calcium (mg)	Magnesium (mg)	Potassium (mg)	Sodium (mg)	Selenium (μg)
Chickpea	6.2	3.4	105	115	875	24	8.2
Pigeon pea	5.2	2.7	130	183	1392	17	–
Lentil	7.5	4.7	56	122	955	6	8.2
Mung bean	6.7	2.7	110	189	1246	15	8.2
Urdbean	8.4	3.5	55	–	–	–	–
Field pea	4.4	3.0	186	115	981	15	1.6
Rajmash	3.4	1.9	80.3	188	1316	18	12.9
Cowpea	7.54	3.77	287	250	1450	23	–
Horse gram	7.0	–	202	–	–	–	–
Moth bean	9.6	–	–	–	–	–	–

glucose tolerance. This nutrient has also been associated with diabetes and cardiovascular diseases. Provisionally, daily intake of 0.05–0.2 mg Cr has been recommended for adults. The level of Cr in pulses generally ranges between 0.05 and 0.60 Cr/g. Table 4.7 elaborates the content of different minerals in pulses.

4.4.3 Dietary Fibres

Pulse grains typically contain more dietary fibre (15–32 TDF) than cereals, and of this approximately one third to three quarters is insoluble and the remaining is soluble fibre. Insoluble fibre is associated with faecal bulking through its water-holding capacity, whereas soluble fibre ferments, positively affecting colon health through production of SCFA, lowered Ph and helps to slow gastric emptying rate.

4.4.4 Folic Acid

Beans have the highest folate content followed by mung bean, lentil and pea. Folate contents in raw chickpea and pea are 149.7 and 101.5 μg/100 g, respectively, and 78.8 and 45.7 μg/100 g in boiled chickpea and pea respectively, indicating that some folates might get leached in the water used for processing.

4.4.5 Nonnutritive Bioactive Components

A variety of phytochemicals are increasingly being recognized for their potential benefits for human health, which includes polyphenolic compounds, lectins, phytates and trypsin inhibitors, among others. Lignans and isoflavones have anticarcinogenic, weak oestrogenic and antioxidant properties. Phenolic compounds, including tannins found mainly in the seed coat, have antioxidant activity. Phytoestrogens in pulses may play a role in the prevention of hormone-related cancers, such as breast and prostate cancer. The lectins or haemagglutinins in some pulses are toxic when taken orally. They can cause vomiting, diarrhoea, nausea and bloating in humans. The enzyme inhibitors and lectins can even reduce protein digestibility and nutrient absorption, respectively, but both have little effect after cooking. Phytic acid can diminish mineral bioavailability. Some phenolic compounds reduce protein digestibility and mineral bioavailability, while galacto-oligosaccharides may induce flatulence. The lathyrus toxin in certain drought-resistant chickpeas can cause lathyrism, a neurological disorder, when consumed in large amounts. On the other hand, the same compounds may have protective effects against cancer. Phytic acid has antioxidant and DNA protective effects.

4.5 Health Benefits

4.5.1 Cardiovascular Health

Consumption of pulses has been shown to decrease serum LDL cholesterol and triglycerides (two major risk factors for CHD) as well as other risk factors, such as hypertension, diabetes and obesity. Among the food legumes, chickpea is the most hypocholesterolaemic agent; germinated chickpea was reported to be effective in controlling cholesterol level in rats. Further epidemiological data also suggest that pulse consumption could reduce the risk of cardiovascular disease.

4.5.2 Diabetes Management

Inclusion of pulses in a healthy diet can benefit those with diabetes and help prevent healthy people from becoming diabetic. Epidemiological studies strongly support the suggestion that high intakes of wholegrain foods protect against the development of type 2 diabetes mellitus. People who consume three or more servings of wholegrain foods per day are less likely to develop type 2 diabetes mellitus than low consumers with a 20–30 % risk reduction. Pulses share several qualities with whole grains of potential benefit to glycaemic control including slow release carbohydrate and a high fibre content. Pulses are low GI foods with GI values ranging from 28 to 52. Consumption of low GI foods (<55) results in moderate levels of glucose as opposed to high GI foods (>70), which causes rapid elevation in blood glucose.

4.5.3 Weight Management

Pulses are high in fibre and protein, low in fat and moderate in calories. About 100 g of cooked lentils or dry peas contains about half of the daily fibre recommendation for adults. Therefore, a combination of pulses with cereals helps people feel satisfied for a long time and therefore may help with weight management.

4.5.4 Celiac Disease

Celiac disease is the one in which the absorptive surface of the small intestine is damaged by glutens. Glutens are generally found in wheat, rye and barley. As compared with gluten-free alternatives of cereals, pulses are an excellent source of dietary proteins, minerals, vitamins and fibres. Pulses being rich in iron also supplement the celiac patients who otherwise suffer with iron and vitamin B deficiencies. Therefore, pulse flours are an excellent alternative to gluten-containing flour of cereals and provide valuable proteins and fibres in the form of baked foods prepared using the flours of pulses.

4.5.5 Pulses and Cancer Risk

Inverse correlations between pulse consumption and colon cancer mortality and risks of prostate cancer, gastric cancer and pancreatic cancer have been reported in several epidemiological studies. Many phytochemicals and bioactive components in pulses, viz. dietary fibre, oligosaccharides, folate, selenium, protease inhibitors, phytic avid, lignans, phenolics, saponins and isoflavones, are associated with anticarcinogenic activity.

4.6 Conclusions

Food legumes are the best option after cereals, not only because of their importance for humans and animals but also due to their soil ameliorative values and their ability to thrive under harsh and fragile environments. Keeping in view their importance for diversification and intensification of contemporary agriculture, systematic national and international efforts towards their genetic improvement have been taken which led to the development of a number of improved cultivars and development of their production technologies and also of crop management and protection. With the advent of modern techniques and availability of alien variations, precise and target-oriented research in genetic improvement of pulses has been underway globally towards the development of high-yielding, input-responsive, early maturing and high-nutrition varieties in pulses. This has led to production expansion of major pulses throughout the world. However, this is not sufficient to meet the ever-increasing demand. Intensified efforts must be initiated not only to increase the area and production of pulses but also to increase their nutritional quality better postharvest management and value addition besides suppressing their anti-nutritional factors. There is still a strong need of developing new references on nutritional aspects and health promoting values of pulses to promote their cultivation and industrialization throughout the world. Pulses in combination with cereals offer a miraculous solution to protein calorie malnutrition in the developing nations and, if given due attention, may definitely help in eradication of this prolonged predicament from the world.

References

Adekola OF, Oluleye F (2007) Influence of mutation induction on the chemical composition of cowpea *Vigna unguiculata* (L.) Walp. Afr J Biotechnol 6:2143–2146

Anwar F, Latif S, Przybylski R, Sultana B, Ashraf M (2007) Chemical composition and antioxidant activity of seeds of different cultivars of mungbean. J Food Sci 72:503–510

Avola G, Gresta F, Abbate V (2009) Diversity examination based on physical, technological and chemical traits in a locally grown landrace of faba bean (*Vicia faba* L. var. major). Int J Food Sci Technol 44:2568–2576

Brummer EC, Graef GL, Orf J, Wilcox JR, Shoemaker RC (1997) Mapping QTL for seed protein and oil content in eight soybean populations. Crop Sci 37:370–378

Burstin J, Marget P, Huart M, Moessner A, Mangin B, Duchene C, Desprez B, Munier-Jolain N, Duc G (2007) Developmental genes have pleiotropic effects on plant morphology and source capacity, eventually impacting on seed protein content and productivity in pea. Plant Physiol 144:768–781

Burstin J, Gallardo K, Mir RR, Varshney RK, Duc G (2011) Improving protein content and nutritional quality. In: Pratap A, Kumar J (eds) Biol and breed food legumes. CABI, Oxfordshire, pp 314–328

Chibbar RN, Ambigaipalan P, Hoover R (2010) Molecular diversity in pulse seed starch and complex carbohydrates and its role in human nutrition and health. Cereal Chem 87:342–352

Chung J, Babka HL, Graef GL, Staswisk PE, Lee DJ, Cregan PB, Shoemaker RC, Specht JE (2003) The seed protein, oil, yield QTL on linkage group I. Crop Sci 43:1053–1067

Coelho RC, Faria MA, Rocha J, Reis A, Oliveira MPPP, Nunes E (2009) Assessing genetic variability in germplasm of *Phaseolus vulgaris* L. collected in Northern Portugal. Sci Hortic 122:333–338

Duc G, Marget P, Esnault R, Le Guen J, Bastianelli D (1999) Genetic variability for feeding value of faba bean seeds (*Vicia faba* L.). Comparative chemical composition of isogenics involving zero-tannin and zero-vicine genes. J Agric Sci 133:185–196

Dwivedi SL, Jambunathan R, Nigam SN, Raghunath K, Shankar KR, Nagabhushanam GVS (1990) Relationship of seed mass to oil and protein contents in peanut (*Arachis hypogaea* L.). Peanut Sci 17:48–52

Frimpong A, Sinha A, Tar'an B, Warkentin TD, Gossen BD, Chibbar RN (2009) Genotype and growing environment influence chickpea (Cicer arietinum L.) seed composition. J Sci Food Agric 89:2052–2063

Gabriel I, Lessire M, Juin H, Burstin J, Duc G, Quillien L, Thibault JN, Leconte M, Hallouis JM, Ganier P, Meziere N, Seve B (2008) Variation in seed protein digestion of different pea (*Pisum sativum* L.) genotypes by cecectomized broiler chickens:

l. endogenous amino acid losses, true digestibility and in vitro hydrolysis of proteins. Livest Sci 113:251–261

Health Canada (2015) Eating well with Canada's food guide. Available online: http://www.hc-sc.gc.ca/fn-an/food-guide-aliment/index-eng.php. Accessed on 10 Apr 2015

Hedley CL (2001) Carbohydrate in grain legume seeds. CAB International, Wallingford, pp 1–11

Hyten DL, Pantalone VR, Sams CE, Saxton AM, Landau-Ellis D, Stefaniak TR, Schmidt ME (2004) Seed quality QTL in a prominent soybean population. Theor Appl Genet 109:552–561

Kabas O, Yilmaz E, Ozmerzi A, Akinci I (2006) Some physical and nutritional properties of cowpea seed (Vigna sinensis L.). J Food Eng 79:1405–1409

Lawn RJ, Rebetzke GJ (2006) Variation among Australian accessions of wild mungbean (*Vigna radiate* ssp. sublobata) for traits of agronomic, adaptive, or taxonomic interest. Aust J Agric Res 57:119–132

Oluwatosin OB (1997) Genetic and environmental variation for seed yield, protein, lipid and amino acid composition in cowpea (*Vigna unguiculata* (L.) Walp). J Sci Food Agric 74:107–116

Rizkalla SW, Bellisle F, Slama G (2002) Health benefits of low glycaemic index foods, such as pulses, in diabetic patients and healthy individuals. Br J Nutr 88S:255–262

Upadhyaya HD, Reddy KN, Gowda CLL, Silim SN (2007) Patterns of diversity in pigeonpea (Cajanus cajan (L.) Millsp.) germplasm collected from different elevations in Kenya. Genet Resour Crop Evol 54:1787–1795

USDA (2013) Beans and peas are unique foods, United States Department of Agriculture (USDA). Available online: http://www.choosemyplate.gov/food-groups/vegetables-beans-peas.html. Accessed on 14 Mar 2013

Vollman J, Fritz CN, Wagentristl H, Ruckenbauer P (2000) Environmental and genetic variation of soybean seed protein content under Central European growing conditions. J Sci Food Agric 80:1300–1306

Wang N, Hatcher DW, Toews R, Gawalko EJ (2009) Influence of cooking and dehulling on nutritional composition of several varieties of lentils (*Lens culinaris*). Food Sci Technol 42:842–848

Biofortification of Food Legumes and Bioavailability of Nutrients

5

Jagdish Singh

Abstract

Growing sufficient food will not in itself assure adequate nutrition and healthy productive lives. The diets of over two-thirds of the world's population lack one or more essential mineral elements; as a result, micronutrient malnutrition, the so-called hidden hunger, affects more than one-half of the world's population, especially women and preschool children. In developing countries, the rise in micronutrient deficiencies is linked to the shift in cultivation with dominance by cereals. High cereal productivity, the result of extensive research, has ensured that cereal production is relatively profitable with a relatively low risk of failure due to biotic and abiotic stresses. Food systems dominated by cereals are low in micronutrients. To address micronutrient deficiencies in the comprehensive way, several new approaches are needed simultaneously. Despite past progress in controlling micronutrient decencies through supplementation and food fortification, new approaches are needed to expand the reach of food-based interventions. Biofortification, a new approach that relies on conventional plant breeding and modern biotechnology to increase the micronutrient density of staple crops, holds great promise for improving the nutritional status and health of poor populations in both rural and urban areas of the developing world. Available genetic variation influences the level of micronutrient increments that can be achieved through breeding, but contribution to nutritional status largely depends on factors related to bioavailability that have to be considered when setting nutritional target levels for breeding. Critical information is needed on how much nutrient is retained after storage, processing, and cooking; on micronutrient bioconversion and bioavailability once the nutrient is ingested; and on micronutrient requirements of the target population. Many of these parameters are interrelated in a

J. Singh (✉)
Division of Basic Sciences, ICAR-Indian Institute of Pulses Research, Kanpur, Uttar Pradesh 208 024, India
e-mail: jagdish1959@gmail.com

U. Singh et al. (eds.), *Biofortification of Food Crops*, DOI 10.1007/978-81-322-2716-8_5

highly complex manner, since human micronutrient status, dietary composition, and health status affect bioavailability and its components' bioaccessibility, bioconversion, and bioefficacy. The bioavailability of Fe and Zn is associated with the presence of antinutrients and/or the lack of promoter substances for micronutrients. Since an increase in bioavailability translates into a proportional decrease in the nutritional target increment (increasing Fe bioavailability from 5 % to 10 % reduces the target increment by 50 %), breeding strategies for micronutrient density should consider indirect breeding for increased bioavailability, increased retention, or reduced postharvest micronutrient deterioration. Although not well understood, breeding for increased bioavailability offers tremendous potential.

Keywords
Food legumes • Micronutrient malnutrition • Phytochemicals • Pulses

5.1 Introduction

Micronutrient malnutrition ("hidden hunger") now afflicts over 2 billion people worldwide, resulting in poor health, low work productivity, high rates of mortality and morbidity, increased rates of chronic diseases (coronary heart disease, cancer, stroke, and diabetes), and permanent impairment of cognitive abilities of infants born to micronutrient-deficient mothers. Evidence is growing that our global food systems are failing to deliver adequate quantities of healthy, nutritionally balanced food especially to underprivileged people globally (Mason and Garcia 1993; McGuire 1993). The consequences are affecting human health, well-being, productivity, and livelihood and contributing to stagnating national development efforts in many developing nations. More than 2 billion people worldwide are iron deficient (ACC/SCN 2000). Iron deficiency anemia is by far the most common micronutrient deficiency in the world. Iron deficiency during childhood and adolescence impairs physical growth, mental development, and learning capacity. In adults, iron deficiency anemia reduces the capacity to do physical labor. Iron deficiency increases the risk of women dying during delivery or in the postpartum period. Prevalence among children exceeds 50 % in South and Southeast Asia, where 1.8 billion out of the approximately 4.5 billion people in developing countries live. Prevalence is equally high in Africa, although the number of persons affected is smaller. Prevalence is consistently highest for pregnant women and lowest for adult males. Even in India, about 79 % of the kids between 6 and 35 months of age and 56 % of women between 15 and 49 years of age are anemic. Zinc deficiencies have equally serious consequences for health. For example, meta-analyses of recent randomized controlled trials show that zinc supplementation can reduce morbidity from a number of common childhood infections, especially diarrhea, pneumonia, and possibly malaria, by one-third (Bhutta et al. 2000; Black 1998). In addition, zinc deficiency is an important cause of stunting (Umeta et al. 2000). Over 450 thousand infant deaths in the world during 2004 were ascribed to Zn deficiency. It is estimated that 60–70 % of the population in Asia and sub-Saharan Africa could be at the risk of low Zn intake; in absolute numbers this translates into 2 billion people in Asia and 400 million people in sub-Saharan Africa. Even in India Zn deficiency is a highly relevant health problem and is responsible for a loss of 2.8 million. Childhood stunting may be considered as an indicator of the human Zn nutritional deficiency and about 61 million children under the age of 15 are stunted. Pregnant women are also susceptible to Zn deficiency and a survey in Haryana on 285 pregnant women showed that 65 % of them had Zn deficiency. The primary underlying cause of micronutrient malnutrition is poor quality diets, characterized by high intake of food staples, but low consumption of animal and fish products, fruits, grain legumes, and vegetables, which are rich sources of bioavailable minerals and vitamins. As such, most of the malnourished are those who cannot afford to purchase high-quality, micronutrient-rich foods.

Modern agricultural systems are providing calories, but in the process, they have increased "hidden hunger" among the world's poor by displacing acreage allotted to traditional crops such as pulses, making many micronutrient-rich plant foods less available and more expensive to low-income families (Combs et al. 1996). The apparent consequences of "green revolution" cropping systems on micronutrient malnutrition are clearly demonstrated in several world regions. For example, in South Asia, the introduction of modern wheat and rice production practices (which resulted in about 200 % increase in rice production and 400 % increase in wheat production over the past 30 years) is associated with time trends in the growth of iron deficiency anemia among nonpregnant, premenopausal women and negatively related to time trends in the iron density (mg Fe per kcal of available food) of diets. Typically, populations and individuals at risk for iron and Zn deficiency are those that consume high levels of grains such as rice, wheat, and maize as their major source of calories. In regions where the cropping systems have changed and over the past 30 years, the cultivation and consumption of pulse crops have declined because high-yielding grain varieties have displaced them in the farming system. Essentially, the diversity of staple food crops was reduced, and pulse crops, which are relatively high in iron, declined in production and consumption. In addition, processing of grains such as rice and wheat removes much of the iron from these foods, and unfortunately, many populations prefer the processed products, lowering their consumption even more. Research efforts are now underway to improve the food supply in many countries through a process known as biofortification. Genetic biofortification is a sustainable method of naturally enriching food crops by conventional breeding and modern biotechnology. Biofortification is a focus in a wide range of food crops. In pulses, micronutrients such as iron, zinc, selenium, and folates are more important.

Compounding this problem is the utility or "bioavailability" of essential micronutrients. Many factors influence the degree to which ingested micronutrients are absorbed (bioavailability) and utilized (bioefficacy). Dietary components like phytic acid, polyphenols/tannins, dietary fiber, and ascorbic acid influence the bioavailability of minerals (Kies et al. 1983). Biofortification will have greater impact if the biofortified nutrients are highly bioavailable.

5.2 Biofortification in Pulses: A Sustainable Agricultural Strategy for Reducing Micronutrient Malnutrition

The permanent solution to micronutrient malnutrition in developing countries is a substantial improvement in dietary quality higher consumption of pulses, fruits, vegetables, fish, and animal products, which the poor already desire but cannot presently afford. Meanwhile, breeding staple foods that are dense in minerals and vitamins provides a low-cost, sustainable strategy for reducing levels of micronutrient malnutrition. The prospects of using conventional breeding techniques, biotechnology, and micronutrient-containing fertilizers to improve the micronutrient quality of staple crops and the micronutrient status of the poor have been put forth in recent years; this new strategy is referred to as "biofortification." Biofortification is a relatively recent addition to breeding goals in plants based on improving the nutritional quality of the edible portion of the plant through traditional or transgenic approaches. Biofortification proposes to use agricultural modifications as a public health intervention and, as a result, has the potential to reach more effectively to the rural poor—those who are often at highest risk for micronutrient deficiencies. This strategy is being developed and implemented through the international alliance of HarvestPlus to improve iron, zinc, and vitamin A status in low-income populations. Results from germplasm screening suggest that the iron and zinc content of staple foods can be doubled through conventional breeding. This result, in turn, implies that iron and zinc intakes in poor people's diets can be increased by 50 %. Under HarvestPlus, progress in breeding for biofortified crops has been steady, with promising levels of zinc and provitamin A carotenoids and, to a lesser extent, iron being achieved in several staple food crops, including maize, rice, wheat, pearl millet, potatoes, and bananas or plantains. However, limited pulse biofortification research were also conducted during the last decade which indicates that breeding for micronutrient-rich pulses with high bioavailability may be possible. The use of molecular marker-assisted selection in pulse crop biofortification and research efforts have been initiated in the USA, Canada, and CGIAR around the world. For development of molecular markers linked with high concentration of micronutrient loci, initial large-scale evaluation of available germplasm sets of different food legumes is essential as this in turn is a prerequisite to develop suitable mapping population. Few studies conducted so far to map and tag the gene(s)/QTL(s) controlling the micronutrients status in legumes were mostly found to be having quantitative mode of inheritance and resulting in identification of gene(s)/QTL(s) capable of explaining moderate amount of phenotypic variation for micronutrient concentration (Sompong et al. 2012) for phytic acid in mung bean and (Blair et al. 2010) for Fe and Zn in common bean.

5.3 Exploiting Genetic Variation in Nutrient Composition and Genetics of Nutritional Traits

Germplasm resources can be exploited to improve the nutrient output of food systems in several ways. The simplest way is by designing new cropping systems with better nutrient output, for example, by introducing micronutrient dense pulse species into cereal-dominated production systems. There is an urgent need to reduce the risk to farmers associated with pulse production and to enhance the yield and reliability of most pulses while maintaining or further enhancing their nutrient density. Within any food system, selecting micronutrient dense lines among existing varieties is the first and easiest approach. Not much work has been done in pulse crops as compared to wheat, rice, and maize; however, in beans considerable genetic variation in mineral concentration among wild beans as well as modern cultivars has been demonstrated. The concentration of zinc in beans is one of the highest among vegetable sources. Unlike, wheat and rice, the concentrations of iron are generally even higher than those of zinc, but like the cereals, iron and zinc are positively correlated. However, unlike cereals there is no correlation with calcium. The concentration of iron in most of the beans and in other pulse crops around 70–100 mg kg^{-1} is much higher than in the cereals, although relatively, the increase that is possible through selection is less. Again, like wheat and rice, the high density traits are fairly stable across environments, although, as in all crops, the environmental effect is also highly significant (Graham et al. 2001). The first genetic study of a micronutrient efficiency factor was conducted by Weiss (1943) on iron efficiency in soybeans, in which it was shown that efficiency was due to a single, major, dominant gene. Since then, several minor additive genes have been discovered to contribute to iron efficiency in this crop (Fehr 1982).

5.4 Bioavailability

The total amount of micronutrient in a plant food does not represent the actual micronutrient content of the food that is utilizable by the consumer. In human nutrition terms, the bioavailability is commonly defined as the amount of a nutrient that is absorbable and utilizable by the person concerned. There are multiple interactions occurring between micronutrients in plant foods and other plant substances once the food is consumed. There are several interacting nutrients and chemical substances that can either inhibit (i.e., antinutrients) or enhance (i.e., promoters—that can increase absorption and/or utilization) micronutrient bioavailability (Table 5.1). Thus, determining micronutrient bioavailability is a difficult and a complex issue. Various methods have been developed to determine micronutrient bioavailability in plant foods to humans that encompasses in vitro or in vivo models or combination of both. Unfortunately, none of these models are ideal for all foods, nutrients, and circumstances. Presently *Caco-2* cell culture developed from an adenocarcinoma isolated from a human large intestine is being used to screen large number of lines for bioavailable Fe and Zn as it is a human cell model and is relatively inexpensive, rapid, and free of the need of radioisotopes or stable isotope-labeled plants. *Caco-2* cell monolayers are morphologically and physiologically very similar to the layer of the mucosal epithelial cells that line the surface of the small intestine and are responsible for most micronutrient absorption from the gut.

Bioavailability issues must be addressed when employing plant breeding and/or transgenic approaches to reduce micronutrient malnutrition. Enhancing substances (e.g., ascorbic acid, S-containing amino acids, etc.) that promote micronutrient bioavailability and decreasing antinutrient substances (e.g., phytate, polyphenolics, etc) that inhibit micronutrient bioavailability (Table 5.2) are both options that could be pursued. Given this complexity, three breeding sub-strategies may be applied individually or in various combinations. These are (1) increasing the mineral and vitamin content, (2) reducing the level of antinutrients in food staples that inhibit the bioavailability of minerals and vitamins, and (3) increasing the levels of nutrients and compounds that promote the bioavailability of minerals and vitamins. For example, phytates and tannins inhibit iron and zinc absorption. With respect to compounds that promote bioavailability, certain amino acids (such as cysteine) enhance iron and/or zinc bioavailability (Hallberg et al. 1989). These amino acids occur in many staple foods, but their concentrations are lower in grain legumes than those found in animal products. A modest increase in the concentrations of promoter amino acids in grain

Table 5.1 Various factors affecting micronutrient bioavailability

Production practices	Meal components	Processing	Individual's characteristics
Nutritional efficiency	Antinutrients	Raw	Age
Fertilizer practices	Promoters	Cooking	Sex
Soil amendments	Protein quality	Fermentation	Ethnicity
Cropping system	Quantity of other interacting major and micronutrients	Malting	Economic status
Variety	Supplements	Extraction	Physiological status
Soil fertility and health		Soaking	Nutritional status
Other interacting elements		Drying	Disease status
		Polishing	Education and awareness
		Milling	Genetic propensity

Source: Graham et al. (2001)

Table 5.2 Examples of dietary substances that inhibit (antinutrients) or enhance (promoters) the bioavailability of nutrients to humans

Antinutrients	Promoters
Phytic acid	Some organic acids (ascorbate, fumarate, malate, citrate)
Fibers (cellulose, hemicellulose, lignin, cutin, suberin, etc.)	Phytoferritin (plant ferritin)—more bioavailable Fe
Tannins and other polyphenols	Some free amino acids (methionine, cysteine, histidine, lysine)
Oxalic acid	Long-chain fatty acids (palmitic acid) promotes Zn bioavailability
Hemagglutinins (lectins)	Inulin (oligofructose, β-2, 1-fructooligosaccharides) promotes calcium bioavailability
Goitrogens	Riboflavin (FAD) promotes bioavailability of iron and Zn
Toxic heavy metals (Cd, Hg, Pb, Ag)	Selenium promotes bioavailability of iodine

Source: Graham et al. (2001)

legumes may have a positive effect on iron and zinc bioavailability in humans. Iron and zinc occur only in micromolar amounts in plant foods, so only micromolar increases in the amounts of these amino acids may be required to compensate for the negative effects of antinutrients on iron and zinc bioavailability. These amino acids are essential nutrients for plants as well as for humans, so relatively even small increase in their concentrations in plant tissues should not have adverse consequences on plant growth (Graham et al. 2001). While most vitamins are very well absorbed, most essential minerals are not. Normally absorption of minerals ranges from less than 1 % to over 90 %.

5.4.1 Bioavailability of Iron from Grain Legumes

Iron bioavailability typically ranges from 5 % to 15 % of total iron intake (FAO/WHO 2002). Direct measurement of iron absorption from meals representing different types of largely plant-based diets suggests that the bioavailability of iron can be even less than 5 % (FAO/WHO 2002). This presents a particular challenge for the biofortification strategy, because a very large amount of additional iron would be needed to compensate for its very low bioavailability. Phytate is also a potent inhibitor of iron bioavailability. Whereas, for zinc the dose-response inhibitory effect of phytate is more gradual, for iron it is more acute, so that even relatively small amounts of phytate can have a substantial inhibitory effect on iron bioavailability (Hallberg et al. 1989). Polyphenolic compounds are also strong inhibitors of iron bioavailability. Studies using in vitro models of iron bioavailability have shown that even after reduction of the phytate content of staple foods such as sorghum and millet, iron bioavailability remains low in the presence of polyphenolic compounds (Rao and Deosthale 1988; Lestienne et al. 2005).

There are several plant food components that interact with iron and zinc during digestion to alter their bioavailability. For iron, staple food components that enhance the bioavailability of nonheme iron include ascorbic acid, and those that inhibit iron bioavailability include phytate and some polyphenolic compounds. The bioavailability of Fe is not only a question of the Fe concentration in a given food, but also whether the Fe is plant or animal derived and whether other biochemical factors are present within the food matrix (Hunt 2003; Bravo 1998). For example, an increase of 67 % Fe bioavailability in hybrid corn (*Zea mays* L.) over the control was associated with a mere 12 % increase in total Fe concentration, thus demonstrating the impact of other food matrix factors (Hoekenga et al. 2011). Promoters and inhibitors of Fe absorption within the food matrix must be considered with respect to the bioavailability of non-heme Fe in a food crop (Cook et al. 1972). The iron and zinc from vegetarian diets are generally less bioavailable than from non vegetarian diets because of more phytic acid and other plant-based inhibitors. Phytic acid (PA) myoinositol 1,2,3,4,5,6-hexakisphosphate, nearly ubiquitous in plants and used as the primary phosphorus (P) storage, inhibits absorption of Fe in the gut (Turnbull et al. 1962). Inhibition is achieved by chelation of Fe $^{2+/3+}$ by PA, but this action can be prevented by ascorbic acid (AA), depending on dosage (Siegenberg et al. 1991). Other notable inhibitors include fiber, heavy metals, and certain polyphenols and tannins (Welch and Graham 2004). Iron absorption from vegetarian diets can likely be somewhat improved by the consumption of iron-containing foods concurrently with sources of ascorbic acid-containing foods while limiting low-phytate foods or the use of preparation methods that reduce phytic acid (Gibson et al. 1997). Although a vegetarian diet is likely to contain iron in amounts equivalent to amounts in a non vegetarian diet, the iron from a vegetarian diet is likely to be substantially less available for absorption because of differences in the chemical form of iron and the accompanying constituents that enhance or inhibit iron absorption. The chemical form of iron is an important factor affecting the iron availability. Heme iron is better absorbed (≈15–40 %) than non-heme iron (≈1–15 %) (11–15). The Dietary Reference Intakes recently proposed for iron (Food & Nutrition Board 2001) suggest that vegetarians need to increase dietary iron by 80 % to compensate for an estimated lower iron bioavailability of 10 % from a vegetarian diet.

5.4.2 Bioavailability of Zinc

The main staple food component that inhibits the bioavailability of zinc is phytate. The proportion of zinc that is absorbed from typical diets appears to range from about 18 % to 34 %, where lower bioavailability is associated with a higher molar ratio of phytate: zinc in the diet. Plant foods rich in zinc, such as legumes, are also high in phytic acid, an inhibitor of zinc bioavailability (Harland and Oberleas 1987). Because the phytate content of whole grains and legumes tends to be very high, it is expected that breeding for lower phytate content would lead to improved zinc bioavailability and hence increase dietary zinc adequacy. A World Health Organization publication (WHO 1996) categorized phytate-zinc molar ratios of 5–15 as moderate in zinc bioavailability (30–35 % absorption). In comparison, high-zinc bioavailability (50–55 % absorption) diets were described as refined, low in cereal fiber, with a phytate-zinc molar ratio of <5, and with adequate protein. Low-zinc-bioavailability diets (15 % absorption) were listed as phytate-zinc ratio >15. Bioavailability of zinc is enhanced by dietary protein (Sandström et al. 1980), but plant sources of protein are also generally high in phytic acid.

The bioavailability of zinc from vegetarian diets is also likely to be less than that of nonvegetarian diets. High levels of calcium fortification may also reduce zinc bioavailability (WHO 1996). The description of the new Dietary Reference Intakes for zinc (Food and Nutrition Board 2001) suggested that, because of lower absorption of zinc, those consuming vegetarian diets, especially with phytate-zinc molar ratios >15, may require as much as 50 % more zinc than nonvegetarians.

5.4.3 Bioavailability of Other Trace Elements

Much less information is available about the bioavailability of other trace elements besides iron and zinc. Because plant foods are generally good source of elements such as copper and manganese, vegetarian diets are likely to provide greater amounts of these trace elements than non vegetarian diets. Plasma copper was not significantly different between vegetarians and nonvegetarians in a cross-sectional study (Krajcovicova-Kudlackova et al. 1995). As assessed by monitoring the fecal excretion of a stable copper isotope, copper was absorbed less efficiently from the vegetarian diet, but more total copper was absorbed because of the greater copper content of a vegetarian diet, compared with a nonvegetarian diet (Hunt and Vanderpool 2001). The selenium content of foods varies greatly with the selenium in the soil where the food is grown. Furthermore, the retention and use of selenium from the diet likely depends on the chemical form of selenium in foods. By analysis, the total selenium content of vegetarian diets was similar or lower than that of nonvegetarian diets (Drobner et al. 1997); in contrast, dietary selenium increased by about 40 % when subjects switched to a vegetarian diet (Srikumar et al. 1992). In those healthy subjects who switched to self-selected vegetarian diets, plasma selenium decreased by 11 %.

5.5 Bioavailability and Bioefficacy of Folate and Folic Acid

Folates represent an important group among the B vitamins, participating in one-carbon transfer reactions required within the cell, especially for purine and pyrimidine biosynthesis (DNA and RNA) and amino acid interconversions. Health benefits of folates regarding their prevention of neural tube defects in babies and occlusive vascular diseases caused by elevated plasma homocysteine, their link to mental fitness, and possibly certain forms of cancer have already been recognized. Low folate intake causes mild hyperhomocysteinemia, which is a potential risk factor for cardiovascular disease (Danesh and Lewington 1998). Supplementation with folic acid leads to a significant reduction in plasma homocysteine concentrations in healthy subjects, even at low doses (Ward et al. 1997). An increased intake of dietary folate lowers plasma homocysteine concentrations (Brouwer et al. 1999).

The analyses of food folates are tedious because of a lack of validated methods for characterization and quantitation of the great number of native folate forms but also due to a lack of adequate methods for sample pretreatment. Therefore, the assessment of folate losses through industrial and household food processing is still incomplete, as well as knowledge on folate bioavailability in humans.

The bioavailability and bioefficacy of dietary folate appears to be less than those of folic acid (Sauberlich et al. 1987). *Bioavailability* is defined as the fraction of folate that is absorbed and can be used for metabolic processes or storage as measured by changes in folate status, whereas *bioefficacy*, or more correctly *functional bioefficacy*, is defined as the fraction that has a positive effect on a functional parameter, for instance, the lowering of homocysteine concentrations (Brouwer et al. 2001). Folate functions as a coenzyme in single-carbon transfers in the metabolism of nucleotides and amino acids. It is essential for the formation of thymidylate (TMP) for DNA synthesis, so that without folate, living cells cannot divide. The need for folate is higher when cell turnover is increased, such as in fetal development. It is also involved in purine synthesis, in the generation of formate, and in amino acid interconversions. The bioavailability of folate has been a topic of active investigation for many years. Much of the interest in this area originated from reports by Tamura and Stokstad (1973) that showed a wide range of bioavailability of endogenous folate in a wide variety of common foods. It is widely recognized that for typical mixed diets, the bioavailability of naturally occurring folate is incomplete. The results of a long-term controlled dietary study with human subjects by Sauberlich et al. (1987) indicated that the bioavailability of folate in a typical mixed diet was not more than 50 %. Cuskelly et al. (1996) conducted a similar study with free-living subjects and observed that fortified foods and folic acid in supplements were substantially more effective than high-folate foods provided to the subjects.

5.6 Conclusion

Most of the agricultural systems in the developing world do not provide enough micronutrients to meet the human needs. Even though micronutrients are required in minute quantities, they have tremendous impact on human health. Insufficient dietary intake of these nutrients impairs the functions of the brain, the immune and reproductive systems, and energy metabolism. These deficiencies result in learning disabilities, reduced work capacity, serious illnesses, and even death. Micronutrient malnutrition is a serious global problem that limits the work capacity of people and seriously hinders economic development. Finding sustainable solution to this developing global nutritional crisis and to address micronutrient deficiencies in the comprehensive way needs several approaches. In the past, supplementation and fortification programs have shown some positive results in treating the symptoms of micronutrient malnutrition. Breeding staple foods that are dense in minerals and vitamins provides a low-cost, sustainable strategy for reducing levels of micronutrient malnutrition. Conventional breeding techniques, biotechnology and micronutrient-containing fertilizers to improve the micronutrient quality of staple crops and the micronutrient status of the poor have been put forth in recent years; this new strategy is referred to as "biofortification." Genetic biofortification is a sustainable method of naturally enriching food crops by conventional breeding and modern biotechnology. "Bioavailability" of essential micronutrients is another important aspect which needs attention. Iron bioavailability typically ranges from 5 % to 15 % of total iron intake, whereas the proportion of zinc that is absorbed from typical diets appears to range from about 18 % to 34 % and lower bioavailability is associated with a higher molar ratio of phytate: zinc in the diet. The bioavailability issues must be addressed when employing plant breeding and/or transgenic approaches to reduce micronutrient malnutrition.

References

ACC/SCN (UN Administrative Committee on Coordination, Subcommittee on Nutrition) (2000) Fourth report on the world nutrition situation. Geneva: ACC/SCN in collaboration with the International Food Policy Research Institute

Bhutta ZA, Bird SM, Black RE, Brown KH, Gardner JM, Hidayat A, Khatun F, Martorell R, Ninh NX, Penny ME, Rosado JL, Roy SK, Ruel M, Sazawal S, Shankar A (2000) Therapeutic effects of oral zinc in acute and persistent diarrhoea in children in developing countries: pooled analysis of randomized controlled trial. Zinc investigators' collaborative group. Am J Clin Nutr 72:1516–1522

Black RE (1998) Therapeutic and preventive effects of zinc on serious childhood infectious diseases in developing countries. Am J Clin Nutr 68:S476–S479

Blair MW, Knewton SJB, Astudillo C, Li C, Fernandez AC, Grusak MA (2010) Variation and inheritance of iron reductase activity in the roots of common bean (*Phaseolus vulgaris* L.) and association with seed iron accumulation QTL. BMC Plant Biol 10:215

Bravo L (1998) Polyphenols: chemistry, dietary sources, metabolism, and nutritional significance. Nutr Rev 56:317–333

Brouwer IA, van Dusseldorp M, West CE et al (1999) Dietary folate from vegetables and citrus fruit decreases plasma homocysteine concentrations in humans in a dietary controlled study. J Nutr 129:1135–1139

Brouwer IA, van Dusseldorp M, West CE, Steegers-Theunissen RPM (2001) Bioavailability and bioefficacy of folate and folic acid in humans. Nutr Res Rev 14:267–293

Combs GF, Welch RM, Duxbury JM, Uphoff NT, Nesheim MC (1996) Food-based approaches to preventing micronutrient malnutrition: an international research agenda. Cornell International Institute for Food, Agriculture, and Development, Cornell University, Ithaca

Cook JD, Layrisse M, Martinez-Torres C, Walker R, Monsen E, Finch CA (1972) Food iron absorption measured by an extrinsic tag. J Clin Investig 51:805–815

Cuskelly GJ, McNulty H, Scott JM (1996) Effect of increasing dietary folate on red-cell folate: implications for prevention of neural tube defects. Lancet 347:657–659

Danesh J, Lewington S (1998) Plasma homocysteine and coronary heart disease: systematic review of published epidemiologic studies. J Cardiovasc Risk 5:229–232

Drobner C, Röhrig B, Anke M, Thomas G (1997) Selenium intake of adults in Germany depending on sex, time, living area and type of diet. In: Fischer PWF, L'Abbé MRL, Cockell KA, Gibson RS (eds) Trace elements in man and animals: proceedings of the ninth international symposium on trace elements in man and animals. NRC Press, Ottawa, pp 158–159

FAO/WHO (2002) Food and Agriculture Organization/ World Health Organization-Human vitamin and mineral requirements. Report of a joint FAO/WHO expert consultation, Bangkok

Fehr WR (1982) Control of iron deficiency chlorosis in soybeans by plant breeding. J Plant Nutr 5:611–621

Food and Nutrition Board (2001) Dietary reference intakes for vitamin A, vitamin K, arsenic, boron, chromium, copper, iodine, iron, manganese, molybdenum, nickel, silicon, vanadium, and zinc. National Academy Press, Washington, DC

Gibson RS, Donovan UM, Heath AL (1997) Dietary strategies to improve the iron and zinc nutrition of young women following a vegetarian diet. Plant Foods Hum Nutr 51:1–16

Graham RD, Welch RM, Bouis HE (2001) Addressing micronutrient malnutrition through enhancing the nutritional quality of staple food: principles, perspectives and knowledge gaps. Adv Agron 70:77–142

Hallberg L, Brune M, Rossander L (1989) Iron absorption in man: ascorbic acid and dose-dependent inhibition by phytate. Am J Clin Nutr 49:140–144

Harland BF, Oberleas D (1987) Phytate in foods. World Rev Nutr Diet 52:235–259

Hoekenga OA, Lung'aho MG, Tako E, Kochian LV, Glahn RP (2011) Iron biofortification of maize grain. Plant Genet Resour 9:327–329

Hunt JR (2003) Bioavailability of iron, zinc, and other trace minerals from vegetarian diets. Am J Clin Nutr 78:633S–639S

Hunt JR, Vanderpool RA (2001) Apparent copper absorption from a vegetarian diet. Am J Clin Nutr 74:803–807

Kies C, Young E, McEndree L (1983) Zinc bioavailability from vegetarian diet: influence of dietary fiber, ascorbic acid, and past dietary practices. In: Inglett GE (ed) Nutritional bioavailability of zinc. American Chemical Society, Washington, DC, pp 115–126

Krajcovicova-Kudlackova M, Simoncic R, Babinska K et al (1995) Selected vitamins and trace elements in blood of vegetarians. Ann Nutr Metab 39:334–339

Lestienne I, Caporiccio B, Besancon P, Rochette I, Treche S (2005) Relative contribution of phytates, fibers, and tannins to low iron and zinc in vitro solubility in pearl millet (*Pennisetum glaucum*) flour and grain fractions. J Agric Food Chem 53:8342–8348

Mason JB, Garcia M (1993) Micronutrient deficiency – the global situation. SCN News 9:11–16

McGuire J (1993) Addressing micronutrient malnutrition. SCN News 9:1–10

Rao Udayasekhara P, Deosthale YG (1988) In vitro availability of iron and zinc in white and coloured ragi (*Eleusine coracana*): role of tannin and phytate. Plant Foods Hum Nutr 38:35–41

Sandström B, Arvidsson B, Cederblad A, Bjorn-Rasmussen E (1980) Zinc absorption from composite meals, I: the significance of wheat extraction rate, zinc, calcium, and protein content in meals based on bread. Am J Clin Nutr 33:739–745

Sauberlich HE, Kretsch MJ, Skala JH, Johnson HL, Taylor PC (1987) Folate requirement and metabolism in non pregnant women. Am J Clin Nutr 46:1016–1028

Siegenberg D, Baynes RD, Bothwell TH, Macfarlane BJ, Lamparelli RD, Car NG, MacPhail P, Schmidt U, Tal A, Mayet F (1991) Ascorbic acid prevents the dose-dependent inhibitory effects of polyphenols and phytates on nonheme-iron absorption. Am J Clin Nutr 53:537–541

Sompong U, Somta P, Raboy V, Srinivas P (2012) Mapping of quantitative trait loci for phytic acid and phosphorous contents in seed and seedling of mungbean (*Vigna radiata* (L.) Wilczek.). Breed Sci 62:87–92

Srikumar TS, Johansson GK, Öckerman P, Gustafsson J, Åkesson B (1992) Trace element status in healthy subjects switching from a mixed to a lactovegetarian diet for 12 months. Am J Clin Nutr 55:885–890

Tamura T, Stokstad ELR (1973) The availability of food folate in man. Br J Haematol 25:513–532

Turnbull A, Cleton F, Finch CA (1962) Iron absorption. IV. The absorption of hemoglobin iron. J Clin Investig 41:1897–1907

Umeta M, West CE, Haidar J, Deurenberg P, Hautvast JG (2000) Zinc supplementation and stunted infants in Ethiopia: a randomised controlled trial. Lancet 355:2021–2026

Ward M, McNulty H, McPartlin J, Strain JJ, Weir DG, Scott JM (1997) Plasma homocysteine, a risk factor for cardiovascular disease, is lowered by physiologic doses of folic acid. Q J Med 90:519–524

Weiss MG (1943) Inheritance and physiology of efficiency of iron utilization in soybeans. Genetics 28:253–268

Welch RM, Graham RD (2004) Breeding for micronutrients in staple food crops from a human nutrition perspective. J Exp Bot 55:353–364

World Health Organization (1996) Trace elements in human nutrition and health. WHO, Geneva

Nutritional Benefits of Winter Pulses with Special Emphasis on Peas and Rajmash

6

A.K. Parihar, Abhishek Bohra, and G.P. Dixit

Abstract

Currently, half of the global population is experiencing the severity of nutritional insecurity. In this context, food legumes lie central to the strategies established to combat against the problem of micronutrient malnutrition. Among *rabi* food legumes, peas and rajmash offer a wide range of health benefits which are attributed to their inherent qualities including higher protein content and enhanced concentration of essential nutrients. The health benefits of these two crops are also evident from the fact that a regular supplementation of diets with peas and rajmash helps in reducing the risks associated with coronary heart diseases and cardiovascular health problems. Similarly, higher fibre content in these pulses prevents unreasonable rising of blood sugar level in the human beings. This article outlines the growing importance of peas and rajmash especially in combating global malnutrition with their potential health benefits through consumption and biofortification and also explores the possibilities to shift the dietary pattern/habits to enhance the nutritional status of the population worldwide.

Keywords

Biofortification • Health benefits • Nutritional composition • Peas • Rajmash

A.K. Parihar (✉)
AICRP on MULLaRP, ICAR-Indian Institute of Pulses Research, Kanpur, Uttar Pradesh 208 024, India
e-mail: ashoka.parihar@gmail.com

A. Bohra
Division of Crop Improvement, ICAR-Indian Institute of Pulses Research, Kanpur, Uttar Pradesh 208 024, India
e-mail: abhi.omics@gmail.com

G.P. Dixit
AICRP on Chickpea, ICAR-Indian Institute of Pulses Research, Kanpur, Uttar Pradesh 208 024, India
e-mail: gpdixit1966@rediffmail.com

6.1 Introduction

Peas (*Pisum sativum* L.) are one of the important cool season food legume crops in the world with 6.59 million hectares area and 9.83 million tons annual production (FAOSTAT 2013). Peas, more specifically the yellow or green cotyledon varieties, are known as dry peas or field peas and

U. Singh et al. (eds.), *Biofortification of Food Crops*, DOI 10.1007/978-81-322-2716-8_6

are grown around the world for human and animal consumption (Dahl et al. 2012). The dry grains are consumed in various forms such as *soup*, *chat*, *chhola*, *dal*, *stew*, *snacks*, vegetables and flour. Dry pea flour is valued not only as an important dietary protein source but also for its industrial utilization like a thickening agent in certain food industries. Green seeds are used as fresh, frozen or canned vegetables. Pea is among the oldest pulse crops that underwent human-mediated domestication and is recognized as an inexpensive readily available source of protein, complex carbohydrates, vitamins and minerals. The higher nutrient density in dry peas makes them a valuable international food commodity, capable of fulfilling the dietary requirements of millions of undernourished individuals worldwide (FAOSTAT 2011).

Like peas, kidney bean (*Phaseolus vulgaris*) is another important food legume which is popularly known as rajmash, French bean, field bean, wax bean, etc. Rajmash is grown worldwide in different climatic regimes. It is a well-known food legume that is pervasively used globally in the preparation of varieties of dishes, mainly in rice, curries, salads and toppings. In particular, kidney beans are the prime constituents in the Mexican food, which is most popular worldwide. Indeed, like other food legume crops, it is also a potential source of high-quality protein, which offers an easily accessible and affordable alternative to meat or other animal proteins especially concerning the resource-poor people in the developing world. In addition to a source of lysine- and tryptophan-rich protein for human consumption, it also supplements adequate quantities of other important minerals and micronutrients including folate, potassium, iron, manganese, copper and zinc (Kutos et al. 2003; Costa et al. 2006; Singh et al. 2013).

From the human health perspectives, peas and rajmash offer a range of health benefits that are not restricted just to a part of the body but improve the overall health. The extended knowledge of health benefits of peas and rajmash encourages people to use these as integral constituents in daily diets. Taking the above facts into consideration, the present article aims to provide an overview of the potential health benefits associated with pea and rajmash consumption.

6.2 Peas

6.2.1 Nutritional Composition

6.2.1.1 Protein

Given the higher protein content coupled with the absence of major anti-nutritional factors, peas remain one of the most preferred sources of digestible protein for both human and livestock use. However, the protein content of seeds is reported to be influenced by various factors including genetic as well as environmental effects. In general, the protein concentration in peas ranges from 21.2 % to 32.9 % of seed DM (Table 6.1). The majority of proteins in round-seeded peas belong to storage or globulins type, and the rich amino acid profile (particularly high Trp and Lys) of these proteins determines their greater nutritional value (Bourgeois et al. 2011; Boye et al. 2011). Also, it is a good source of arginine, valine and methionine for humans (Tomoskozi et al. 2001).

6.2.1.2 Carbohydrates, Starch and Fibre

Carbohydrates are the major component found in peas and accounting up to 56–74 % of the dry matter. Starch content in peas ranges from 36.9 % to 49.0 % of the seed dry matter. Concerning the elemental composition, pea starch contains 20.7–33.7 % amylase, which imparts higher levels of enzyme resistance and is responsible for slow digestion of starch. The extent to which pea starch is slowly digestible is remarkable. However, to improve the digestibility, various treatments have been reported so far like annealing and heat-moisture treatment and gelatinization of starch that convert it to a rapidly digestible starch (Chung et al. 2009, 2010). Peas are also source of both soluble and insoluble fibre (Table 6.1). The seed coat and cotyledon are the dietary fibre-rich part of seed. The seed coat contains largely water-insoluble polysaccharides, primarily cellulose, whereas

Table 6.1 Nutritional composition of peas and rajmash

S.N.	Component	Concentration (%)	
		Peas (*Pisum sativum* L.)	Rajmash (*Phaseolus vulgaris*)
1	Protein (%N × 6.25)	21.2–32.9	17.5–28.7
2	Total carbohydrates	56–74	54.4–76
3	Starch	36.9–49.0	31.8–45.3
4	Amylose	20.7–33.7	19.9–29.6
5	Total dietary fibre	14–26	23–32
6	Insoluble fibre	10–15	20–28
7	Soluble fibre	2–9	3–6
8	Total lipid	1.2–2.4	2.0
9	Ash	2.3–3.4	3.6–4.28

the cotyledon fibre comprises of polysaccharides, viz. hemicelluloses, pectins and cellulose, that have various degrees of solubility (Reichert and MacKenzie 1982; Guillon and Champ 2002; Tosh and Yada 2010).

6.2.1.3 Minerals and Vitamins

Among the major minerals in peas, potassium (1.04 % of dry, dehulled weight) is the most prominent element, followed by phosphorus (0.39 %), magnesium (0.10 %) and calcium (0.08 %). The quantities of other seven trace minerals are reported as 97 (ppm) iron, 42 ppm selenium, 41 ppm zinc, 12 ppm molybdenum, 11 ppm manganese, 9 ppm copper and 4 ppm boron (Reichert and MacKenzie 1982). Importantly, the high selenium content of peas may be extremely beneficial in those areas where Se deficiency is prominent (Reichert and MacKenzie 1982). Additionally, peas supplement considerable quantity of folate with 101 μg per 100 g (Dang et al. 2000). Likewise, Hedges and Lister (2006) have reported the presence of substantial amount of vitamins in pea, viz. vitamin C, thiamine (vitamin B1) and vitamins B6, B3 and B2.

6.2.1.4 Other Phytochemicals

Like other pulse crops, peas contain a variety of phytochemicals, viz. carotenoids (including lutein and zeaxanthin) and β-carotene, chlorophyll, phenolic compounds, some flavonoids, saponins and oxalates (Campos-vega et al. 2010).

Carotenoids

In addition to carotenoids like xanthophylls, lutein and zeaxanthin that are well documented, peas also contain the carotenes, α- and β-carotene. The carotenoids are a group of yellow-orange-red pigments, and it cannot be synthesized in the human body and hence become available solely as a result of intake either from a plant source itself or product of animal that has consumed that plant source. The interesting point remains the boiled frozen peas have much more β-carotene than is usually found in raw peas, the reason being freezing and boiling processes break down cell structure, thereby releasing the compounds that were previously bound to other components. Interestingly, the concentrations of lutein and zeaxanthin in peas are several folds higher than that of observed in other food legumes (Hedges and Lister 2006).

Phenolics

Phenolics represent a group of over 4000 compounds occurring vastly in the plant kingdom. Among these, the two classes which are found in substantial amount and have immense dietary relevance are flavonoids and phenolic acids. The phenolic compounds are mostly present in the seed coat and cotyledon of peas. Several researchers have noticed enhanced phenolic compounds and their marked antioxidant activity in coloured seed coat (Duenas et al. 2004; Xu et al. 2007; Campos-vega et al. 2010). On the other hand, tannins having very high antioxidant activity property are detected only in dark seed coat (Hagerman et al. 1998). Isoflavones constitute a subgroup of the flavonoid class compounds which is also present in pea in appreciable amount (Hedges and Lister 2006).

Saponins

Saponin is a diverse group of biologically active glycosides distributed widely in the plant kingdom (Curl et al. 1985). A wide range of saponins have been isolated in peas with soyasaponin I being the most predominant (Murakami et al. 2001). Savage and Deo (1989) have noticed the saponin content of green peas as 2.5 g/kg. In a similar way, Daveby et al. (1997) evaluated several varieties of field peas and found that soyasaponin I content ranged from 0.82 to 2.5 g/kg in different varieties. Though saponins are reported to be heat-sensitive and water-soluble compounds (Shi et al. 2004), a short time cooking with minimal water would enable maximum retention of these compounds.

6.2.2 Health Benefits

6.2.2.1 Prevention and Management of Type 2 Diabetes

The high fibre content of peas may mediate the glycaemic response as compared with low-fibre foods having the equal carbohydrate proportions. The quality of both starch and dietary fibre makes peas a low glycaemic index (GI) food, and hence it is beneficial in the prevention and management of type 2 diabetes (Marinangeli et al. 2009). The use of whole peas and fractioned peas flour has been reported to reduce the fasting insulin level by 13.5 and 9.8 %, respectively, compared with baseline (Marinangeli and Jones 2011). Similarly, the bread containing 17 % pea hull fibre was reported to cause significant reduction in glycaemic response (Lunde et al. 2011).

6.2.2.2 Cholesterol Control

The soluble dietary fibre of peas may also be useful in reducing and stabilizing blood cholesterol by means of decreasing the reabsorption of bile acids (Ekvall et al. 2006). The niacin content in peas helps to reduce the production of triglycerides and very low-density lipoprotein (VLDL), which results in less bad cholesterol, increased HDL (good) cholesterol and lowered triglycerides. Saponins are phenolic compound believed to have an advantageous impact on human health particularly in terms of reducing cholesterol level (Daveby et al. 1998). It is considered that saponins usually cause adsorption of bile acids onto the dietary fibre in the intestine, which is subsequently eliminated from the body with the faecal matter. To redress this loss, serum cholesterol is converted by the liver into bile acids, thus lowering levels of cholesterol in the blood (Savage and Deo 1989).

6.2.2.3 Cardiovascular Health

It is well documented that fibre-rich diets are shown to lower blood pressure, improve serum lipid levels and reduce indicators of inflammation (Slavin 2008). Some flavonoids are also beneficial against heart disease because they inhibit blood platelet aggregation and provide antioxidant protection to LDL (Frankel et al. 1993). Peas are a reliable source of omega-3 fats in the form of alpha-linolenic acid (ALA), which is also useful in the prevention and management of cardiovascular health (Singh et al. 2013).

6.2.2.4 Cancer

Antioxidant flavonoids may act to prevent cancer through inducing detoxification enzymes, inhibiting cancer cell proliferation and promoting cell differentiation (Kalt 2001). The predominant flavonoid in peas is quercetin, which is found in the form of glycoside (with an attached sugar molecule). Like other flavonoids, it is also believed to have a number of cancer-combating properties. For instance, it inhibits growth of malignant cells, boosts up self-destruction of the cancerous cells and is also known to interfere with proliferative activities of certain cancer cells. Compared with other legumes, peas contain low levels of isoflavones (Kleijn et al. 2001; Boker et al. 2002). Isoflavones are capable of reducing the risk for hormone-related cancers, since the proliferation of these kinds of cancer cells is oestrogen dependent (Steer 2006).

Saponins are also believed to combat against cancer, by breaking down the cholesterol-rich membranes of cancer cells. Because saponins are not well absorbed into the blood stream, they are thought to be most useful in exerting a localized effect in the intestinal tract, such as fighting colon cancer (Joseph et al. 2002).

6.2.2.5 Weight Management

Lunde et al. (2011) investigated the impacts of a hypoenergetic diet rich with pea fibre and found that consumption of bread enriched with pea fibre resulted in an increased duration of satiety as compared to regular bread.

6.2.2.6 Gastrointestinal Function and Homeostasis

The water-attracting ability of insoluble fibres of peas may assist in reducing the transit time through the gut. This quality of peas fibre is important in prevention of constipation, diverticulitis and bowel cancer. For example, the addition of pea hull fibre in food resulted in a significant increase (7.5–24 %) in bowel movement frequency (Dahl et al. 2003; Flogan and Dahl 2010). Like other legumes, peas have significant amount of raffinose family and other galactose-containing oligosaccharides which may exhibit prebiotic effects in the large intestine (Tosh and Yada 2010; Fernando et al. 2010). Pea proteins in human gastrointestinal tract significantly increase the intestinal autochthonic bacterial population (*Bacteroides, Lactobacillus* and *Bifidobacterium*) leading to an increase in their metabolic activity and production of short-chain fatty acids (SCFA). SCFAs offer several beneficial effects, including generating energy for colonic mucosa. Besides SCFAs provide protection against various diseases of the colon like cancer and prevent the transformation of primary bile acids to cocarcinogenic secondary bile acids by lowering the colonic pH.

6.2.2.7 Antioxidant Activity

Antioxidants cause deactivation of free radicals and other oxidants, thus rendering them harmless. By nature, free radicals are highly unstable molecules that are usually generated in body either as induced from external stimuli (e.g. pollution, smoking, carcinogenic compound) or from internal sources as an outcome of physiological processes. It is important to note that free radicals developed in uncontrolled way can severally damage cell components, thereby impairing vital life processes. For example, antioxidant-led DNA damage often causes cancer, or they oxidize fats in the blood, which ultimately leads to severe health problems like atherosclerosis and heart disease. However, the human body produces antioxidants by its own and it has other defence mechanisms. Antioxidants supplemented to human diets also play a crucial role.

Vitamin C, the carotenoids and various phenolic compounds constitute the major antioxidants found in peas. Phenolic compounds are natural antioxidants that provide protection against many diseases including cancer and various inflammatory-related diseases. By virtue of their chemical properties, they are very efficient scavengers of free radicals and are also metal chelators (Shahidi and Naczk 1995). In addition to the antioxidant characteristics of flavonoids, other potential health-promoting bioactivities include anti-allergic, anti-inflammatory, antimicrobial and anticancer properties (Cody et al. 1986; Harbourne 1993). Antioxidant activity of isoflavones is also known to trigger anticarcinogenic response including inhibition of free radical reactions, cell mutation, cell proliferation and angiogenesis (Steer 2006).

6.3 Rajmash

6.3.1 Nutritional Composition

6.3.1.1 Carbohydrates, Starch and Dietary Fibre

Up to 50–60 % of the dry matter of kidney bean is primarily composed of carbohydrates found in good amounts (Vargas-Torres et al. 2004; Reynoso-Camacho et al. 2006; Ovando-Martinez et al. 2011). Structurally, starch and non-starch polysaccharides constitute the major proportion of carbohydrate together with considerable amounts of carbohydrate derivates such as oligosaccharides (Bravo et al. 1998; Reynoso-Camacho et al. 2006). Amylose and amylopectin are the two major forms of starch available in kidney beans. Dietary fibres are also found in sizeable proportions and comprise of edible parts or plant analogous carbohydrates such as cellulose, hemicellulose, pectins, oligosaccharides and lignins. Collectively these compounds resist digestion and absorption in the small intestine but are eventually fermented (partially or completely) in the large intestine (Hughes and Swanson 1989; Hughes 1991; Costa et al. 2006).

6.3.1.2 Source of Protein

Owing to the presence of high-quality dietary proteins, kidney bean plays an important role in alleviation of human malnutrition by complementing the traditional cereal-based food (Butt and Batool 2010). Nutritionally, kidney beans are equivalent to animal protein with the seed protein content ranging between 17.5 % and 28.7 % (Table 6.1). In contrast to other food legume proteins which generally show low quantities of glutelins (7–15 %), rajmash contains considerably higher amounts of glutelins (20–30 %) (Seena et al. 2005; Slupski 2010). Concerning the protein composition, globulins constitute the major fraction, i.e. up to 50–70 % of total protein. Like other legume proteins, the greater quality of kidney bean protein is attributable to the presence of essential amino acids particularly lysine that is otherwise deficient in cereal-based diets.

6.3.1.3 Lipids

The lipid content in kidney bean is estimated to be up to 2 % (Table 6.1) with valuable composition of unsaturated fatty acids (Mabaleha and Yeboah 2004). Kidney bean is a rich source of polyunsaturated fatty acids consisting of 71.1 g/ 100 g.

6.3.1.4 Minerals and Vitamins

Compared to cereals, kidney beans have greater levels of micronutrients, viz. minerals and vitamins (Welch et al. 2000). Noticeably, kidney beans have the highest level of mineral content than the other legumes. The levels of iron, zinc, copper, phosphorus and aluminium are appreciable, while other minerals are also present in reasonable amounts (Broughton et al. 2003; Shimelis and Rakshit 2005). The iron content, mostly present in nonheme form, is the highest in beans that ranges from 62.0 to 150 μg/g (Elhardallon and Walker 1992; Vadivel and Janardhanan 2000).

Kidney bean is also considered an important source of vitamins and variations in vitamin concentration are observed (Augustin et al. 2000). It also provides considerable quantities of folate, tocopherols, thiamine, riboflavin, niacin, biotin and pyridoxamine (Broughton et al. 2003; Campos-Vega et al. 2010). Kadam and Salunkhe (1989) have envisaged that the folate content (400–600 μg) of beans is sufficiently high to meet 95 % of the daily requirement.

6.3.1.5 Phenolics Compounds

Given the antioxidant and anticarcinogenic properties of kidney bean phytochemicals, these have a great potential as functional and nutraceutical ingredients (Cardador-Martianez et al. 2002; Dinelli et al. 2006). Phenolic compounds include variety of flavonoids such as anthocyanins, flavonol, proanthocyanidins, tannins, glycosides as well as a wide range of phenolic acids (Beninger and Hosfield 2003; Aparicio-Fernandez et al. 2005).

6.3.2 Health Benefits

6.3.2.1 Cardiovascular Disease

There are plenty of health benefits related to consumption of kidney beans as daily diet since it keeps people healthy in the long run. As reported by Anderson and Major (2002), the higher magnesium content and rich dietary fibres in kidney beans are responsible for lowering blood cholesterol levels and also important in combating against risks of stroke, heart attack and peripheral vascular disease. Magnesium is regarded as nature's own calcium channel blocker. In the vicinity of enough magnesium, the resistance is lessened (within veins and arteries) and it eventually improves the flow of blood, oxygen and nutrients throughout the body. Moundras et al. (1997) observed that dietary fibres facilitate lowering of the cholesterol by restraining the intestinal absorption of neutral steroids, bile acids and total steroid excretions. Resistant starch and dietary fibre content of kidney bean are mainly responsible in the management of metabolic syndrome by delaying the degree of glucose as fuels, changing fat utilization and controlling appetite through increased satiety, thus lowering the risk of cardiovascular diseases (Anderson et al. 2002; Park et al. 2004). It has been reported that intake of 100 % of the daily value (DV) of folate would, by itself, reduce the number of heart attacks. Several epidemiological and clinical studies have portrayed positive effects of kidney bean consumption in reducing the risk of coronary heart diseases and cardiovascular diseases (Finley et al. 2007; Winham and Hutchins 2007).

6.3.2.2 Diabetes Mellitus

The consumption of wholegrain food like kidney bean and other legumes has shown protectiveness, not only in the development of diabetes but also in the management of people suffering from type 2 diabetes mellitus. The ingestion of three or more servings of wholegrain foods per week reduces the risk of diabetes mellitus by 20–35 % (Campos-Vega et al. 2010). The rich fibre content and slow digestibility due to fibre prevent elevated glucose levels leading to reduced insulinemic and glycaemic responses (Zhou et al. 2004; Campos-Vega et al. 2010). Several studies indicate that consumption of low glycaemic index (GI) foods is beneficial in the reduction of diabetes mellitus and obesity (Rizkalla et al. 2002; Jenkins 2007). In addition to lowering cholesterol, kidney beans' high fibre content prevents blood sugar levels from rising too rapidly after a meal, making these beans an especially good choice for individuals with diabetes, insulin resistance or hypoglycaemia. As a high-potassium, low-sodium food, it may also helpful in reduction of blood pressure.

6.3.2.3 Cancer

The fact that the incidence of cancer could be reduced by changing the dietary pattern is well established. Compelling evidences are there suggesting a relation between intake of bean-rich diets and reduced incidence of numerous types of cancer, viz. colon, breast and prostate cancer (Cardador-Martinez et al. 2006; Patterson et al. 2009). The anticarcinogenic activity of beans is related to the presence of resistant starch, soluble and insoluble dietary fibre, phenolic compounds as well as other micro-constituents such as phytic acid, protease inhibitors and saponins (Hangen and Bennink 2003; Patterson et al. 2009). It also contains saponins, which are known to restrict the growth of abnormal crypt foci in the colon (Koratkar and Rao 1997).

6.3.2.4 Source of Energy

Among the food legumes, rajmash serves as an excellent source of energy, since it contains high amount of carbohydrate and iron which are essentially required for increasing body metabolism and energy. Iron helps in the circulation of oxygen throughout the body. Kidney bean can increase energy by helping to replenish iron stores. Mainly for menstruating women, who are more at risk of iron deficiency, boosting iron stores with kidney beans is a good idea – especially because, unlike red meat, another source of iron, kidney beans are low in calories and virtually fat-free. The manganese content in kidney beans also contributes to the production of energy in the human body.

6.3.2.5 Weight Management/Calorie Count

The amount of calorie found in rajmash is moderate and can be eaten by all the age group. Ingestion of kidney beans as a salad or low-calorie soup during lunch will be a good choice. At the same time, it can be used for weight management, since beans α-amylase inhibitor is known to have antiobesity effect as α-amylase inhibitory action to starch digestion causes energy restriction resulting in mobilization of body fat reserves (Obiro et al. 2008).

6.3.2.6 Good for the Brain

This pulse offers outstanding benefits for the brain. It contains appreciable amount of vitamin K which provides essential nutrition for both the brain and nervous system. Kidney beans are also a good source of thiamine or vitamin B1, which is essential for brain cells. It nourishes the brain nerves and cells which prevents age-related disease like Alzheimer's.

6.3.2.7 Improvement in Digestion

The soluble as well as insoluble fibre present in kidney beans help in maintaining healthy bowel movements. If it is eaten in the right quantity, it helps in cleansing the digestive tract. Regular bowel movements are associated with a lowered risk of colon cancer (Costa et al. 2006). Thiamine participates in enzymatic reactions central to energy production and is also critical for brain cell/cognitive function.

6.3.3 Other Benefits

The kidney bean is a very good source of folate as compared to cereals. It may be useful in significant reduction in the incidence of neural defect, some cancer and stroke. In adult a particular type of anaemia can result from long-term folate deficiency. Therefore, the high folate diets and supplemental folic acids may help to overcome the problem. Kidney beans are an excellent source of the trace mineral, molybdenum, an integral component of the enzyme sulphite oxidase, which is responsible for detoxifying sulphites.

6.4 Conclusion

Peas and rajmash are important source of various nutritional components as well as polyphenolic compounds with potential health benefits and antioxidant activity. Despite their high nutritive value, bioavailability may be poor due to high phytate content and some other anti-nutritional factors. However, special emphasis deserves to be placed towards crop-based biofortification and to extend the understanding of the effects of food processing techniques on bioavailability. If phytate is degraded or reduced through biofortification and food processing techniques, both the above said pulses could be considered a significant source of various minerals. Potentially mitigating risks of chronic diseases such as cardiovascular disease, obesity, diabetes and cancer, consuming pea and rajmash could be a cost-effective approach for improving human health. Therefore, strategic scientific research needs to be directed to experimentally verify the available findings and to further explore the potential health benefits of pea and rajmash.

References

Anderson JW, Major AW (2002) Pulses and lipaemia, short- and long-term effects: potential in the prevention of cardiovascular disease. Br J Nutr 88:263–271

Anderson JW, Gustafson NJ, Spencer DB, Tietyen J, Bryant CA (2002) Serum lipid response of hypercholesterolemic men to single and divided doses of canned beans. Am J Clin Nutr 51:1013–1019

Aparicio-Fernandez X, Yousef GG, Loarca-Pina G, Gonzalez de Mejia E, Lila MA (2005) Characterization of polyphenolics in the seed coat of black jamapa bean (*Phaseolus vulgaris* L.). J Agric Food Chem 53:4615–4622

Augustin J, Beck CB, Kalbfleish G, Kagel LC, Matthews RH (2000) Variation in the vitamin and mineral content of raw and cooked commercial *Phaseolus vulgaris* classes. J Food Sci 46:1701–1706

Beninger CW, Hosfield GL (2003) Antioxidant activity of extracts, condensed tannin fractions, and pure flavonoids from *Phaseolus vulgaris* L. seed coat color genotypes. J Agric Food Chem 51:7879–7883

Boker LK, Van der Schouw YT, De Kleijn MJJ, Jacques PF, Grobbee DE, Peeters PHM (2002) Intake of dietary phytoestrogens by Dutch women. J Nutr 132 (6):1319–1328

Bourgeois M, Jacquin F, Casseculle F et al (2011) A PQL (protein quantity loci) analysis of mature pea seed proteins identifies loci determining seed protein composition. Proteomics 11:1581–1594

Boye J, Zare F, Pletch A (2011) Pulse proteins: processing, characterization, functional properties and applications in food and feed. Food Res Int 43:414–431

Bravo L, Siddhurahu P, Saura-Calixto F (1998) Effect of various processing methods on the in vitro starch digestibility and resistant starch content of Indian pulses. J Agric Food Chem 46:4667–4674

Broughton WJ, Hernandez G, Blair M, Beebe S, Gepts P, Vanderleyden J (2003) Beans (Phaseolus spp.)-model food legumes. Plant and Soil 252:55–128

Butt MS, Batool R (2010) Nutritional and functional properties of some promising legume protein isolates. Pak J Nutr 9(4):373–379

Campos-Vega R, Loarca-Pina G, Oomah BD (2010) Minor components of pulses and their potential impact on human health. Food Res Int 43:461–482

Cardador-Martianez A, Loarca-Pina G, Oomah BD (2002) Antioxidant activity in common beans (*Phaseolus vulgaris* L.). J Agric Food Chem 50:6975–6980

Cardador-Martinez A, Albores A, Bah B, Calderon-Salinas V, Castano-Tostado E, Guevara-Gonzalez R, Shimada-Miyasaka A, Lorca-Pina G (2006) Relationship among antimutagenic, antioxidant and enzymatic activities of methonolic extract from common beans (*Phaseolus Vulgaris* L). Plant Food Hum Nutr 61:161–168

Chung HJ, Liu Q, Hoover R (2009) Impact of annealing and heat-moisture treatment on rapidly digestible, slowly digestible and resistant starch levels in native and gelatinized corn, pea and lentil starches. Carbohydr Polym 75:436–447

Chung HJ, Liu Q, Hoover R (2010) Effect of single and dual hydrothermal treatments on the crystalline structure, thermal properties, and nutritional fractions of pea, lentil, and navy bean starches. Food Res Int 43:501–508

Cody V, Middleton EJ, Harborne JB (1986) Plant flavonoids in biology and medicine: biochemical, pharmacological and structure activity relationship. Alan R Liss, New York

Costa GEA, Queiroz-Monici KS, Reis SMPM, Oliveira AC (2006) Chemical composition, dietary fibre and resistant starch contents of raw and cooked pea, common bean, chickpea and lentil legumes. Food Chem 94:327–330

Curl CL, Price KR, Fenwick GR (1985) The quantitative estimation of saponin in pea (*Pisum-sativum L.*) and soya (*glycine-max*). Food Chem 18:241–250

Dahl WJ, Whiting SJ, Healey AD et al (2003) Increased stool frequency and fecal output occurs when finely processes pea hull fiber is added to usual foods consumed by elderly long term care residents. J Am Diet Assoc 103:1199–1202

Dahl WJ, Foster LM, Tyler RT (2012) Review of the health benefits of peas (*Pisum sativum* L.). Br J Nutr 108:S3–S10. doi:10.1017/S0007114512000852

Dang J, Arcot J, Shrestha A (2000) Folate retention in selected processed legumes. Food Chem 68:295–298

Daveby YD, Aman P, Betz JM, Musser SM, Obermeyer WR (1997) The variation in content and changes during development of Soyasaponin I in dehulled Swedish peas (*Pisum sativum* L.). J Sci Food Agric 73:391–395

Daveby YD, Aman P, Betz JM, Musser SM (1998) Effect of storage and extraction on soyasaponin to 2,3-dihydro-2-5-dihydroxy-6-methyl-4-pyrone-conjugated soyasaponin I in dehulled peas (*Pisum sativum* L.). J Sci Food Agric 78:141–146

Dinelli G, Bonetti A, Minelli M, Marotti I, Catizone P, Mazzanti A (2006) Content of flavonols in Italian bean (*Phaseolus vulgaris* L.) ecotypes. Food Chem 99:105–114

Duenas M, Estrella I, Hernandez T (2004) Occurrence of phenolic compounds in the seed coat and the cotyledon of peas (*Pisum sativum* L.). Eur Food Res Technol 219:116–123

Ekvall J, Stegmark R, Nyman M (2006) Content of low molecular weight carbohydrates in vining peas (*Pisum sativum*) related to harvest time, size and brine grade. Food Chem 94:513–519

Elhardallon SB, Walker AF (1992) Binding of iron by three starchy legumes in the presence of iron alone, with calcium or with calcium, zinc, magnesium and copper. J Food Sci Nutr 43:61–68

FAOSTAT (2011) Food and Agriculture Organization Statistics. Food security data and definitions 2005–2007

FAOSTAT (2013) Food and Agriculture Organization Statistics. http://faostat.fao.org. Accessed July 2014

Fernando WMU, Hill JE, Zello GA et al (2010) Diets supplemented with chickpea or its main oligosaccharide component raffinose modify fecal microbial composition in healthy adults. Benefic Microbes 1:197–207

Finley JW, Burrell JB, Reeves PG (2007) Pinto bean consumption changes SCFA profiles in fecal fermentations, bacterial populations of the lower bowel, and lipid profiles in blood of humans. J Nutr 137:2391–2398

Flogan C, Dahl WJ (2010) Fiber fortification improves gastrointestinal function and decreases energy intake in children with a history of constipation. Infant Child Adolescent Nutr 2:312–317

Frankel EN, Waterhouse AL, Kinsella JE (1993) Inhibition of human LDL oxidation by resveratrol. Lancet 341:1103–1104

Guillon F, Champ MM (2002) Carbohydrate fractions of legumes: uses in human nutrition and potential for health. Br J Nutr 88:S293–S306

Hagerman AE, Riedl KM, Jones A et al (1998) High molecular weight plant polyphenolics (tannins) as antioxidants. J Agric Food Chem 46:1887–1892

Hangen LA, Bennink MR (2003) Consumption of black beans and navy beans (*Phaseolus vulgaris*) reduced azoxymethane-induced colon cancer in rats. Nutr Cancer 44:60–65

Harborne JB (1993) The Flavonoids – Advances in research since. Chapman & Hall, London

Hedges LJ, Lister C (2006) The nutritional attributes of legumes. Crop & food research confidential report

no. 1745. A report prepared for horticulture New Zealand
Hughes JS (1991) Potential contribution of dry bean dietary fiber to health. Food Technol 45:122–126
Hughes JS, Swanson BG (1989) Soluble and insoluble dietary fiber in cooked common bean *(Phaseolus vulgaris)* seeds. Food Microstruct 8:15–21
Jenkins AL (2007) The glycemic index: looking back 25 years. Cereal Foods World 52:50–53
Joseph JA, Nadeau DA, Underwood A (2002) The color code: a revolutionary eating plan for optimum health. Hypericon, New York, p 308
Kadam SS, Salunkhe DK (1989) Minerals and vitamins. In: Salunkhe DK (ed) Handbook of world food legumes. CRC Press, Boca Raton, pp 117–121
Kalt W (2001) Interspecific variation in anthocyanins, phenolics and antioxidant capacity among genotypes of high bush and low bush blueberries (*Vaccinium* section *cyanococcus* spp.). J Agric Food Chem 49:4761–4767
Kleijn MJJ, Schouw YTVD, Wilson PWF, Adlercreutz H, Mazur W, Grobbee DE, Jacques PF (2001) Intake of dietary phytoestrogens is low in postmenopausal women in the United States: the Framingham study. J Nutr 131:1826–1832
Koratkar R, Rao AV (1997) Effect of soy bean saponins on azoxymethane-induced preneoplastic lesions in the colon of mice. Nutr Cancer 27:206–209
Kutos T, Golob T, Kac M, Plestenjak A (2003) Dietary fibre content of dry and processed beans. Food Chem 80:231–235
Lunde MS, Hjellset VT, Holmboe-Ottesen G et al (2011) Variations in postprandial blood glucose responses and satiety after intake of three types of bread. J Nutr Metab. doi.org/10.1155/2011/437587
Mabaleha MB, Yeboah SO (2004) Characterization and compositional studies of the oils from some legume cultivars, *Phaseolus vulgaris*, grown in Southern Africa. J Am Oil Chem Soc 81:361–364
Marinangeli CP, Jones PJ (2011) Whole and fractionated yellow pea flours reduce fasting insulin and insulin resistance in hypercholesterolaemic and overweight human subjects. Br J Nutr 105:110–117
Marinangeli CP, Kassis AN, Jones PJ (2009) Glycemic responses and sensory characteristics of whole yellow pea flour added to novel functional foods. J Food Sci 74:S385–S389
Moundras C, Behr SR, Remesy C, Demigne C (1997) Fecal losses of sterols and bile acids induced by feeding rats guar gum are due to greater pool size and liver bile acid secretion. J Nutr 127:1068–1076
Murakami T, Kohno K, Matsuda H, Yoshikawa M (2001) Medicinal foodstuffs. XXII. Structures of oleanane-type triterpene oligoglycosides, pisumsaponins I and II, and kaurane-type diterpene oligoglycosides, pisumosides A and B, from green peas, the immature seeds of *Pisum sativum* L. Chem Pharm Bull 49(1):73–77
Obiro WC, Tao Z, Bo J (2008) The nutraceutical role of the *Phaseolus vulgaris* α-amylase inhibitor. Br J Nutr 100:1–12
Ovando-Martinez M, Luis A, Bello-Pereza LA, Whitney K, Osorio-Diaz P, Senay S (2011) Starch characteristics of bean (*Phaseolus vulgaris* L.) grown in different localities. Carbohydr Polym 85:54–64
Park OJ, Kang NE, Chang MJ, Kim WK (2004) Resistant starch supplementation influences blood lipid concentrations and glucose control in overweight subjects. J Nutr Sci Vitaminol 50:93–99
Patterson CA, Maskus H, Dupasquier C (2009) Pulse crops for health. Cereal Foods World 54:108–113
Reichert RD, MacKenzie SL (1982) Composition of peas (*Pisum sativum*) varying widely in protein content. J Agric Food Chem 30:312–317
Reynoso-Camacho R, Ramos-Gomez M, Loarca-Pina G (2006) Bioactive components in common beans (Phaseolus vulgaris L.). In: Guevara-González RG, Torres-Pacheco I (eds) Advances in agricultural and food biotechnology. Research Signpost, Trivandrum, pp 217–236
Rizkalla SW, Bellisle F, Slama G (2002) Health benefits of low glycaemic index foods, such as pulses, in diabetic patients and healthy individuals. Brit J Nutr 88:255–262
Savage GP, Deo S (1989) The nutritional value of peas (*Pisum sativum*). A literature review. Nutr Abstr Rev Ser A Human Exp Nutr 59:65–87
Seena S, Sridhar KR, Bajia B (2005) Biochemical and biological evaluation of ANF unconventional legume, Canavalia maritima of coastal sand dunes of India. Trop Subtrop Agroecosys 5:1–14
Shahidi F, Naczk M (1995) Food phenolics: sources, chemistry, effects, applications. Technomic Publishing Company Inc, Lancaster, pp 231–245
Shi J, Arunasalam K, Yeung D, Kakuda Y, Mittal G, Jiang YM (2004) Saponins from edible legumes: chemistry, processing, and health benefits. J Med Food 7:67–78
Shimelis EA, Rakshit SK (2005) Proximate composition and physicochemical properties of improved dry bean (*Phaseolus vulgaris* L.) varieties grown in Ethiopia. LWT 38:331–338
Singh J, Nadarajan N, Basu PS et al (2013) Pulses for human health and nutrition. Technical bulletin, IIPR, Kanpur-208024
Slavin JL (2008) Position of the American dietetic association: health implications of dietary fiber. J Am Diet Assoc 108:1716–1731
Slupski J (2010) Effect of cooking and sterilization on the composition of amino acids on immature seeds of flageolet beans (*Phaseolus vulgaris* L.) cultivars. Food Chem 121:1171–1176
Steer TE (2006) Phytochemicals – a future in functional foods? Food Sci & Technol Bull: Funct Foods 3:23–29
Tomoskozi S, Lasztity R, Haraszi R et al (2001) Isolation and study of the functional properties of pea proteins. Nahrung/Food 45:399–401

Tosh SM, Yada S (2010) Dietary fibres in pulse seeds and fractions: characterization, functional attributes, and applications. Food Res Int 43:450–460

Vadivel V, Janardhanan K (2000) Nutritional and anti-nutritional composition of velvet bean: an under-utilized food legume in South India. Int J Food Sci Nutr 51:279–287

Vargas-Torres A, Osorio-Dıaz P, Islas-Hernandez JJ, Tovar J, Paredes-Lopez O, Bello-Perez LA (2004) Starch digestibility of five cooked black bean (*Phaseolus vulgaris* L.) varieties. J Food Compos Anal 17:605–612

Welch RM, House WA, Beebe S, Cheng Z (2000) Genetic selection for enhanced bioavailable levels of iron in bean (*Phaseolus vulgaris* L.) seeds. J Agric Food Chem 48:3576–3580

Winham DM, Hutchins MH (2007) Baked beans consumption reduces serum cholesterol in hypercholesterolemic adults. Nutr Res 27:380–386

Xu BJ, Yuan SH, Chang SKC (2007) Comparative analyses of phenolic composition, antioxidant capacity, and color of cool season legumes and other selected food legumes. J Food Sci 72:S167–S175

Zhou Y, Hoover R, Liu Q (2004) Relationship between a-amylase degradation and the structure and physicochemical properties of legume starches. Carbohydr Polym 57:299–317

7 Nutritional Benefits of Pigeonpea

I.P. Singh

Abstract

Pigeonpea is a rich source of food proteins that is generally grown under risk-prone marginal lands. It occupies an important place among pulses and has been rated the best as far as its biological value is concerned. It has been recommended for a balanced diet with cereals, especially to fill in the nutritional gap for proteins among the poorer section in developing economies that cannot afford a nonvegetarian diet. At present, the protein availability in developing countries is about one third of normal requirements, and with ever-growing population, various nutritional development programs are facing a tough challenge to meet the protein demand. In general, pigeonpea can be grown both as annual crop or perennial plants in homestead and is consumed either as decorticated splits or green seeds as vegetables. It has been found that vegetable pigeonpea is considered superior to dry splits in crude fiber, fat, protein digestibility, as well as trace elements and minerals. Besides its nutritional value, pigeonpea also possesses various medicinal properties due to the presence of a number of polyphenols and flavonoids. It is an integral part of traditional folk medicine in India, China, and some other nations. Pigeonpea is known to prevent and cure human ailments like bronchitis, coughs, pneumonia, respiratory infections, pain, dysentery, menstrual disorders, curing sores, wounds, abdominal tumors, and diabetes in traditional folk medicine.

Keywords

Cajanus cajan • Pigeonpea • Nutritional value • Medicinal value

7.1 Introduction

Pigeonpea (*Cajanus cajan* L.) is the sixth most important grain legume in the world and second most important pulse crop after chickpea in India

I.P. Singh (✉)
All India Coordinated Research Project (Pigeonpea), ICAR-Indian Institute of Pulses Research, Kanpur, Uttar Pradesh 208 024, India
e-mail: ipsingh1963@yahoo.com

U. Singh et al. (eds.), *Biofortification of Food Crops*, DOI 10.1007/978-81-322-2716-8_7

which is grown in the Indian subcontinent accounting for the 90 % of the world's crop. However, this crop is also grown in Southeast Asia, Africa, and the Americas. In Kenya, Uganda, and Malawi in eastern Africa and in the Dominican Republic and Puerto Rico in Central America, there is a substantial area under pigeonpea. It is grown in almost all the states of India, but the major states in terms of area and production are Maharashtra, Uttar Pradesh, Madhya Pradesh, Karnataka, Gujarat, and Andhra Pradesh which together account for world's 87 % of the area and 83.8 % of the production.

As split *dal* it is an important constituent of Indian diet. The dry seed is dehulled and the split cotyledons that are called *dal* are cooked to make a thick soup primarily for mixing with rice. Immature fully developed pods form a source of protein-rich vegetable. Green seeds are cooked as a vegetable, in the Indian states of Gujarat, Karnataka, and Maharashtra, and tender pods are cooked whole in Brazil, Thailand, and the eastern islands of Indonesia.

The seed husk and pod walls are commonly fed to cattle. Its dry stems are an important household fuel wood in many countries. Dry stems are also used for making field fences, huts, and baskets. Being a legume it fixes nitrogen, and the leaf fall at maturity not only adds to the organic matter in the soil but also provides additional nitrogen for the following crop.

It has been estimated that approximately 40 kg/ha of the residual nitrogen is left in the field by pigeonpea crop which resulted from the leaf fall and nitrogen fixation. In Caribbean, the crop is grown for its immature green seeds for fresh use of canning, while in the USA and Australia it is used as forage.

7.2 Nutritional Benefits

Health and nutritional benefits from pigeonpea can broadly be categorized into two, viz., (i) nutritive values and (ii) medicinal values.

7.2.1 Nutritive Values of Pigeonpea

As human food pigeonpea is used as *dhal* (split seed without seed coat), whole seed, and green vegetable to supplement cereal-based diets. Pigeonpea seed is composed of cotyledons (85 %), embryo (1 %), and seed coat (14 %). In contrast to the mature seeds, the immature seeds are generally lower in all nutritional values; however, they contain a significant amount of vitamin C (39 mg per 100 g serving) and have a slightly higher fat content. Research has shown that the protein content of the immature seeds is of a high quality. The dietary nutrient values in the green seed, dry seed, and *dal* (processed seeds) of pigeonpea are summarized in Tables 7.1 and 7.2.

7.2.2 Nutritional Value of Immature (Vegetable) Seeds

In general, green pigeonpea seeds (vegetable pigeonpea) are considered superior to dry splits in nutrition. The observations recorded at ICRISAT showed that pigeonpea dal was better than green peas with respect to starch and protein (Table 7.1). On the contrary, the green pigeonpea seeds had higher crude fiber, fat, and protein digestibility. As far as trace and mineral elements were concerned, the green pea was better in phosphorus by 28.2 %, potassium by 17.2 %, zinc by 48.3 %, copper by 20.9 %, and iron by 14.7 %. The dal, however, had 19.2 % more calcium and 10.8 % more manganese (Faris and Singh 1990). Singh et al. 1977 reported that the vegetable-type pigeonpea had high polysaccharides and low crude fiber content than dal, irrespective of their seed sizes. They also reported that crude fiber contents in vegetable pigeonpea and garden pea (*Pisum sativum*) were comparable. There was a vast range in size and color in immature pods and mostly the consumer preference was for large green pods;

Table 7.1 The dietary nutrients of pigeonpea

Constituents	Green seed	Mature	*Dal*
Protein (%)	21.0	18.8	24
Protein digestibility (%)	66.8	58.5	60.5
Trypsin inhibitor (units mg-1)	2.8	9.9	13.5
Starch (%)	48.4	53.0	57.6
Starch digestibility (%)	53.0	36.2	-
Amylase inhibitor (units mg-1)	17.3	26.9	-
Soluble sugars (%)	5.1	3.1	5.2
Flatulence factors (g 100 g- 1 soluble sugar)	10.3	53.5	-
Crude fiber (%)	8.2	6.6	1.2
Fat (%)	2.3	1.9	1.6
Minerals and trace elements (mg 1001 g dry matter)			
Calcium	94.6	120.8	16.3
Magnesium	113.7	122.0	78.9
Copper	1.4	1.3	1.3
Iron	4.6	3.9	2.9
Zinc	2.5	2.3	3.0
Vitamins (mg 100- 1 g fresh weight of edible portion)			
Carotene (vit A 100- 1 g)	469.0		
Thiamin (vit B1)	0.3		
Riboflavin (vit B2)	0.3		
Niacin	3.0		
Ascorbic acid (vit C)	25.0		

Source: Faris et al. 1987

Table 7.2 Nutritional profile of various amino acids within mature seeds of pigeonpea

Essential amino acid	Available mg/g of protein	Min. required mg/g of protein
Tryptophan	9.76	7
Threonine	32.34	27
Isoleucine	36.17	25
Leucine	71.3	55
Lysine	70.09	51
Methionine + cystine	22.7	25
Phenylalanine + tyrosine	110.4	47
Valine	43.1	32
Histidine	35.66	18

however, these traits were not related to any organoleptic property of seeds. In pigeonpea, seed and pod size are positively correlated, and the varieties with large pod invariability have large immature and dry seeds.

7.2.3 Nutritional Changes in Developing Seeds

In a commercial vegetable pigeonpea crop, it is essential that a balance be established between seed yield and its nutritional quality. To achieve this, the green pods have to be harvested at a proper stage of seed growth. At Patancheru (17°N), it takes about 45–50 days from flowering to seed maturity, while vegetable pods are ready to harvest in about 25–35 days from flowering. In pigeonpea, the pods grow rapidly and attain their full size in about 20 days. During this period, the young seeds (ovules) inside the pod remain intact but do not attain significant weight. Soon after achieving its normal pod size, the seeds grow rapidly for the next 10–12 days to achieve their optimum size.

From nutritional and marketing points of view, it is essential that the green pods are picked at a proper stage to reap maximum seed yield with highest nutritional quality. In this context, Singh et al. (1991) observed that in growing seeds, the starch content was negatively associated with their protein and sugar contents. The amount of crude fiber content in the growing seeds increased slowly with maturation. The soluble sugars and proteins decreased in the developing seeds, but the starch content increased rapidly between 24 and 32 days after flowering. Meiners et al. (1976) showed that the minerals and trace elements such as calcium, iron, zinc, magnesium, and copper remained more or less constant and did not change markedly during seed development in pigeonpea.

7.2.4 Nutritional Value of Split Peas (Dal)

Carbohydrates (67 %) and protein (22 %) are main constituents of pigeonpea seeds (Faris and Singh 1990). Hulse (1977) reported that the seed protein among cultivars ranged around 22 %. The quality of protein is determined by its quantity and digestibility and amino acid contents. In pigeonpea the amino acids such as lysine and threonines are in good proportions, while methionine and cystine are deficient (Singh and Jambunathan 1982). Pigeonpea cotyledons are also rich in calcium and iron.

Wild relatives of pigeonpea are used as high-protein donor parents and it was demonstrated that seed protein content in cultivated types can be enhanced through conventional breeding. Singh et al. (1990) assessed these high-protein lines for their chemical composition. They reported large differences between the levels of protein in high-protein (28.7–31.1 %) lines and control cultivars (23.1–24.8 %). As expected, the starch component in high-protein lines was relatively less (54.3–55.6 %) than that of controls (58.7–59.3 %). Also the high-protein lines were marginally lower (2.5–2.6 %) in fat content when compared with control cultivars (2.9–3.1 %). The differences in the major protein fractions of the high- and normal-protein lines were also large. In comparison to controls (60.3–60.5 %), the globulin fraction was higher (63.5–66.2 %) in the high-protein lines and the reverse was true for glutelin.

Godbole et al. (1994) reported protease inhibitors in 7-day-old seeds, while Ambekar et al. (1996) found that such inhibitors are either not synthesized or inactive up to 28 days of the seed development. Except seed no any other plant part exhibited trypsin or chymotrypsin inhibitors (Mutimani and Paramjyothi 1995). The white-seeded pigeonpea cultivars contain relatively less amounts of polyphenols. Such cultivars are preferred in many countries where dehulling facilities are not available and whole seeds are consumed. In comparison to the white-seeded cultivars, the red-seeded types contain three times greater quantity of polyphenols (Singh 1984). Similarly, the enzyme inhibition activity was also greater in the colored seeds of pigeonpea. Since in India almost the entire pigeonpea production (3.2 m tones) is dehulled and converted into dal for consumption, the tannins present in the colored seed coat pose no nutritional problem.

7.2.5 Cooking Quality

Pigeonpea seeds in the form of either dry, green, or split peas are invariably consumed after cooking. Therefore, besides various nutritional aspects, the cooking time and other related parameters assume importance. Consumers always prefer a dal that cooks fast and produces more volume upon cooking with high consistency and flavor. Cooking time of dal is independent of taste and flavor. Manimekalai et al. (1979) and Jambunathan and Singh (1981) studied various physicochemical characteristics of pigeonpea and reported that quick cooking trait of dal was associated with large seed size, high solid dispersal, more water absorption, and high nitrogen solubility. Narasimha and Desikachar (1978) and Pal (1939) reported a positive association of cooking time of pigeonpea seeds with their calcium and magnesium contents. According to Salunkhe (1982), cooking of pigeonpea improved the bioavailability of nutrients and at the same time destroyed some antinutritional factors. Heat treatment of pigeonpea seeds is also known to enhance their starch digestibility. The lines, which take long time to cook, generally face the danger of losing important vitamins from food. Cooking pigeonpea seed after germination enhances their starch digestibility (Salunkhe 1982) but reduces the levels of oligosaccharides (Iyengar and Kulkarni 1977). The fermentation of seeds helps in reducing inhibitory activity of digestive enzymes (Rajalakshmi and Vanaja 1967). Geervani (1981) reported that thiamine and riboflavine were destroyed by heat but niacin content was unaltered during boiling, pressure cooking, and roasting of pigeonpea seeds. She further found that the availability of lysine and methionine decreased on roasting but the available methionine increased on boiling and pressure cooking.

7.2.6 Nutrient Losses During Storage and Processing

Pigeonpea is predominantly cultivated by smallholder farmers, and they generally store the whole seeds for over 6–12 months round the year for consumption. In rural areas the farmers process small quantities of grain as and when needed, and they do dehulling by hand-operated traditional grinding stones called chakki or quern. Since in pigeonpea the cotyledons are attached tightly with seed coat by various gums, the dehulling involves the process of dissolving the gum layers by soaking whole seeds in water, heat treatment, or adding oil after surface scarification. This is followed by drying, dehusking, and splitting of cotyledons. During these processes, the losses of certain proportions of cotyledons are inevitable, and it is estimated that by using advanced processing technology, about 15–17 % of grain mass is lost, while with chakki such losses may shoot up to 20–25 %.

In rural areas, the seeds are generally stored in gunny bags or bins made of mud and husk. According to Daniel et al. (1977), food grains containing more than 10 mg uric acid per 100 g of food are unfit for human consumption. Pigeonpea seeds when stored for 8 months turned unfit for consumption as their total uric acid content crossed the safe prescribed limit (Vimala and Pushpamma 1983). Cooking time of pigeonpea in general increased with storage time (Vimala and Pushpamma 1985). Daniel et al. (1977) reported that lysine, threonine, and protein efficiency ratios were adversely affected in pigeonpea when the seeds were stored in jute bags. The storage of pigeonpea seeds also resulted in the loss of vitamins. Such losses were less (10–26 %) in the protected seed and high (32–49 %) in the infested seeds (Uma and Pushpamma 1981). Thiamine and niacin contents also registered decline during storage. Factors such as moisture, temperature, relative humidity, and seed hardness determined the extent of quality losses during storage (Saxena et al. 2008).

Reddy et al. (1978) reported that in comparison to inner layers of cotyledons, the outer layers are rich in protein. From a nutritional point of view, this is a matter of concern since dehulling not only removes protein-rich germ but also some proportion of the outer layers of the cotyledons. Singh and Jambunathan (1990) further observed that dehulling also removes about 20 % calcium and 30 % iron. According to Kurien (1981), the dal yield under controlled conditions achieves an efficiency of 80–84 %, but at commercial level the recovery remains around 70 %. Therefore, with a combination of a superior variety and efficient processing technology, the nutrient availability can be maximized.

7.3 Medicinal Values of Pigeonpea

Plant kingdom had been considered from a long time a reservoir of folk medicines. The herbal medicines, also called as phytomedicines, refer to using a plant or its parts such as leaves, flowers, fruits, bark, or seeds for medicinal purposes. Since ancient periods and long before any recorded history, China, India, and Egypt were the leaders in folk medicines. The ancient Chinese and Egyptian papyrus writings described various medicinal uses of some plant species. The native Africans and Americans also used different herbs in a number of healing rituals. The herbal medicine system has a long tradition of usage outside the boundaries of synthetic medicine system. With the advent of improved chemical analytical methods along with quality control technologies and advances in clinical research, the value of herbal medicines increased in treating and preventing some human diseases. Subsequently, effective traditional medical systems such as "Ayurveda" in India and "Traditional Chinese Medicine" in China were developed with a fairly good recognition. Slowly the people in different parts of the world also started using the common herbal plants for medicinal purposes.

In the early nineteenth century, when the first chemical analysis became available, the scientists began to extract the active ingredients from selected plant species for pharmaceutical usage, and some chemists started synthesizing targeted plant compounds in their laboratories. Gradually, the usage of herbal medicines declined in favor of synthetic drugs. However, the World Health Organization estimated that 80 % of people worldwide, particularly from developing and underdeveloped countries, still rely on herbal medicines for some part of their primary health care. In Germany, about 600–700 plant-based medicines are still prescribed by 70 % of the physicians. Recently, some of the developed countries have also shown interest in the natural or organic remedies (Steven and Ehrlich 2009).

7.3.1 Herbal Properties

Pigeonpea is being used as an integral part of traditional folk medicine in India, China, and some other developing countries (Morton 1976). The importance of pigeonpea plant in ethnical folk medicine is well known in the prevention and cure of certain human ailments and its brief account is given below.

Pigeonpea floral decoctions are traditionally used for treating ailments such as bronchitis, coughs, and pneumonia. Pigeonpea flowers are also prepared into a "tea" for treating upper respiratory infections and pain. The flowers, when prepared in an infusion, are used for treating dysentery and menstrual disorders.

Scorched seeds, when added to coffee, alleviate headache and vertigo. Fresh seeds are believed to help incontinence of urine in males, while immature seeds are recommended for treatment of kidney ailments (Duke 1981). Pigeonpea seeds are infused to make a diuretic "tea" for inflammation and blood disorders. In South America pigeonpea seeds are used for febrifuge, stabilization of the menstrual period, and dysentery (Duke and Vasquez 1994); in Africa pigeonpea seeds are used for treating hepatitis and measles (Abbiw 1990). In Mexico, pigeonpea seeds are used as styptic drug and laxative, while in China these are used to arrest bleeding, relieve pain and kill worms and as an expectorant and sedative drug (Yuan et al. 1984; Tang et al. 1999). Some herbal researchers are of the opinion that pigeonpea diminishes swelling of internal organs such as the liver, intestines, etc. Clinical studies have also shown that the seed extract of pigeonpea helps in inhibiting sickling of red blood cells and, therefore, has potential to treat the person suffering from sickle cell anemia (Prema and Kurup 1973).

Dried roots of pigeonpea are sometimes used as an alexeritic, anthelminthic, expectorant, sedative, and vulnerary.

In India pigeonpea leaves are used for curing sores, wounds, abdominal tumors, and diabetes (Amalraj and Ignacimuthu 1998; Grover et al. 2002). The leaves of the pigeonpea are decocted in Argentina for treating acute bronchial problems and genital and skin irritations. The young leaves of pigeonpea can be chewed for treating cough, diarrhea (Edbordo 1978), traumatism, burnt infection, bedsore, toothache,

Table 7.3 Worldwide ethnomedical uses of pigeonpea

Sl. no.	Country	Uses
1.	Argentina	Bronchitis, coughs, genital irritation, pneumonia, skin problems
2.	Brazil	Blood disorders, coughs, fevers, inflammation, pain, respiratory infections, sores, ulcers
3.	China	Antidote, expectorant, sedative, vermifuge, vulnerary; for tumors
4.	Cuba	Bronchitis, colds
5.	Dominican Republic	Chest problems, sores, sore throat, wounds
6.	Haiti	Antidote (*Manihot*), gargle, and vulnerary; for jaundice, urticaria, wounds
7.	India	Colic, convulsions, leprosy, tumors (abdomen)
8.	Malaysia	Abdominal pain, coughs, dermatosis, diarrhea, earache, enteritis, sores
9.	Mexico	Astringent, diuretic, laxative, vulnerary; for dysentery
10.	Peru	Anemia, diabetes, dysentery, hepatitis, menstrual disorders, urinary infections, yellow fever; as a diuretic
11.	Trinidad	Flu, strokes

mouthwash, sore gums, child delivery, and dysentery (Chen et al. 1985; Li et al. 2001). It was also found that pigeonpea leaves have notable anti-inflammatory and antibiotic effects and also inhibit capillary permeability (Sun et al. 1995). In China, pigeonpea is considered as an excellent "Traditional Chinese Medicine" for therapy of ischemic necrosis of femoral head. The leaves are prepared in an infusion for overcoming anemia, hepatitis, urinary infections, yellow fever, and ulcers. In Brazil, pigeonpea leaves are infused for coughs, fevers, and ulcers.

7.3.2 Chemical Constituents of Leaves

In order to know the major chemical constituents of pigeonpea leaves, efforts were made to isolate and identify various active chemical compounds. The research efforts revealed that some polyphenols, especially flavonoids, play an important role in curing certain human ailments (Liu et al. 2008; Zu et al. 2006; Yuan et al. 1999, 2004). The four major flavonoids identified in the extracts of pigeonpea leaves are quercetin, luteolin, apigenin, and isorhamnetin. These compounds are known for their important pharmacological activities (Chen et al. 1985; Paul et al. 2003; Lin et al. 1999). Flavonoids are polyphenolic compounds, which are widely found in plant kingdom. As intrinsic components of fruits, vegetables, and beverages, many of the 4000 different flavonoids known to date are present in a common regular diet (Crozier et al. 1997). Pigeonpea leaves also contain other compounds such as hordenine, juliflorine, betulinic acid, stigmasterol, beta-sitosterol, etc. In recent years, extensive research is being carried on various antibacterial, antifungal, antiviral, anticancer, and anti-inflammatory properties of these flavonoids (Matsuda et al. 2003; Srinivas et al. 2003; Kim et al. 1999) (Table 7.3).

7.4 Conclusion

Pigeonpea is a rich source of food proteins. It has been recommended for a balanced diet with cereals, especially to fill in the nutritional gap for proteins among the poorer section in developing economies that cannot afford a nonvegetarian diet. Vegetable pigeonpea is considered superior to dry splits in crude fiber, fat, protein digestibility, as well as trace elements and minerals. Pigeonpea also possesses various medicinal properties due to the presence of a number of polyphenols and flavonoids. Pigeonpea is known to prevent and cure many human ailments.

References

Abbiw DK (1990) Useful plants of Ghana. Richmond intermediate technology publications and royal botanic gardens. Kew, London

Amalraj T, Ignacimuthu S (1998) Evaluation of the hypoglycemic effect of *Cajanus cajan* (seeds) in mice. Indian J Exp Biol 36:1032–1033

Ambekar SS, Patil SC, Giri AP, Kachole MS (1996) Trypsin and amylase inhibitors in pigeonpea seeds. Int Chickpea Pigeonpea Newsl 3:106–107

Chen DH, Li HY, Lin H (1985) Studies on chemical constituents in pigeonpea leaves. Chin Tradit Herb Drugs 16:134–136

Crozier A, Jensen E, Lean MEJ, McDonald MS (1997) Quantitative analysis of flavonoids by reversed phase high performance liquid chromatography. J Chromatogr 761:315–321

Daniel VA, Rajan P, Sanjeevarayappa KV, Srinivasan KS, Swaminathan M (1977) Effect of insect infestation on the chemical composition and the protein efficiency ratio of the proteins of bengal gram and red gram. Indian J Nutr Diet 14:70–74

Duke JA (1981) Hand book of legumes of world economic importance. Plenum Press, New York

Duke JA, Vasquez R (1994) Amazonian ethnobotanical dictionary. CRC Press, Boca Raton

Edbordo CR (1978) Medicinal plant of the Philippines. Katha Publishing Company Inc, Quezon City

Faris DG, Saxena KB, Mazumdar S, Singh U (1987) Vegetable pigeonpea: a promising crop in India. International Crops Research Institute for the Semi-arid Tropics (ICRISAT), Patancheru, India

Faris DG, Singh U (1990) Pigeonpea: nutrition and products. In: Nene YL, Hall SD, Sheila VK (eds) The pigeonpea. CAB International, Wallingford, pp 401–434

Geervani P (1981) Nutritional evaluation of pigeonpea (Variety Hyderabad 3A) processed by traditional methods. In: Proceedings of the international workshop on Pigeonpeas, Patancheru, 15–19 Dec 1980, 2:427–434

Godbole SA, Krishna TG, Bhatia CR (1994) Changes in protease inhibitory activity from pigeonpea (*Cajanus cajan* (L.) Millspaugh) during seed development and germination. J Sci Food Agric 66:497–501

Grover JK, Yadav S, Vats VJ (2002) Medicinal plants of India with antidiabetic potential. Ethnopharmacology 81:81–100

Hulse JH (1977) Problems of nutritional quality of pigeonpea and chickpea and prospects of research. In: Hulse JH, Rachie KO, Billingsley LW (eds) Nutritional standards and methods of evaluation for food legume breeders. International Development Research Center, Ottawa, pp 88–100

Iyengar AK, Kulkarni PR (1977) Oligosaccharide levels of processed legumes. J Food Sci Technol 14:222–223

Jambunathan R, Singh U (1981) Grain quality of pigeonpea. In: Proceedings of the international workshop on Pigeonpeas, Patancheru, 15–19 Dec 1980, 1:351–356

Kim HK, Cheon BS, Kim YH, Kim SY, Kim HP (1999) Effects of naturally occurring flavonoids on nitric oxide production in the macrophage cell line RAW 264.7 and their structure-activity relationships. Biochem Pharmacol 58:759–765

Kurien PP (1981) Advances in milling technology of pigeonpea. In: Proceedings of the international workshop on Pigeonpeas, Patancheru, 15–19 Dec 1980, 1:321–328

Li ZH, Zhou CH, Gu Y, Zhang JY (2001) The present status of study and utilization of pigeonpea in China and its prospects. For Res 14:674–681

Lin L, Xie N, Cheng ZH (1999) Flavonoids from *Cajanus cajan* L. J China Pharma Univ 30:21–23

Liu HY, Qiu NX, Ding HH, Yao RQ (2008) Polyphenols contents and antioxidant capacity of 68 Chinese herbals suitable for medical or food uses. Food Res Int 41:363–370

Manimekalai G, Neelakantan S, Annapan RS (1979) Chemical composition and cooking quality of some improved varieties of red gram *dal*. Madras Agric J 66:812–816

Matsuda H, Morikawa T, Ando S, Toguchida I, Yoshikawa M (2003) Structural requirements of flavonoids for nitric oxide production inhibitory activity and mechanism of action. Bioorg Med Chem 11:1995–2000. doi:10.1016/S0968-0896(03)00067-1

Meiners CR, Derise NL, Lau HC, Ritchey SJ, Murphy EW (1976) The content of nine mineral and elements in raw and cooked mature dry legume. J Food Chem 24:1126–1130

Morton JF (1976) The pigeonpea (*Cajanus cajan* (L) Millspaugh), a high protein tropical bush legume. Hortic Sci 11:11–19

Mutimani VH, Paramjyothi S (1995) Protease inhibitors in some pigeonpea lines. Int Chickpea Pigeonpea Newsl 2:79–81

Narasimha HV, Desikachar HSR (1978) Objective methods for studying cooking ability of turpulse (*Cajanus cajan*) and factors affecting varietal differences in cooking. J Food Sci Technol 15:47–50

Pal RK (1939) A review of the literature on the nutritive value of pulses. Indian J Agric Sci 9:133–137

Paul WC, Philip CS, Monique SJ (2003) Phenolic compounds on the pod-surface of pigeonpea, *Cajanus cajan*, mediate feeding behavior of Helicoverpa armigera larvae. J Chem Eco 29:811–821

Prema L, Kurup PA (1973) Hypolipidaemic activity of the protein isolated from *Cajanus cajan* in high fat cholesterol diet fed rats. Indian J Biochem Biophys 10:293–296

Rajalakshmi R, Vanaja K (1967) Chemical and biological evaluation of the effects of fermentation on the nutritive value of foods prepared from rice and grams. Br J Nutr 21(2):467–473

Reddy LJ, Green JM, Singh U, Bisen SS, Jambunathan R (1978) Seed protein studies on *Cajanus cajan*, Atylosia spp. and some hybrid derivatives. In: Seed protein improvement. Cereals and grain legumes, vol 2. International Atomic Energy Agency, Vienna, pp 105–117

Salunkhe DK (1982) Legumes in human nutrition: current status and future research needs. Curr Sci 51:387–394

Saxena KB, Kumar RV, Rao RP (2008) Pigeonpea nutrition and its improvement. J Crop Prod 5:227–260

Singh U (1984) The inhibition of digestive enzymes by polyphenols of chickpea (L.) and pigeonpea [*Cajanus cajan* (L). Millspaugh]. Nutr Rep Int 29:745–753

Singh U, Jambunathan R (1982) Distribution of seed protein fractions and amino acids in different anatomical parts of chickpea (*Cicer arietinum* L.) and pigeonpea (*Cajanus cajan* L.). Plant Foods Hum Nutr 31:347–354

Singh U, Jambunathan R (1990) Pigeonpea: post harvest technology. In: Nene YL, Hall SD, Sheila VK (eds) Pigeonpea. CAB International, Wallingford, pp 435–455

Singh L, Singh N, Srivastava MP, Gupta AK (1977) Characteristics and utilization of vegetable types of pigeonpea (*Cajanus cajan* (L.) Millspaugh). Indian J Nutr Diet 14:8–10

Singh U, Jambunathan R, Saxena KB, Subramaniam N (1990) Nutritional quality evaluation of newly developed high protein genotypes of pigeonpea (*Cajanus cajan* L.). J Sci Food Agric 50:201–209

Singh U, Jambunathan R, Saxena KB, Singh L (1991) Chemical changes at different stages of seed development in vegetable pigeonpea (*Cajanus cajan*). J Sci Food Agric 57:49–54

Srinivas KVN, Rao SK, Mahender YI, Das B, Krishna KVSR, Kishore KH, Murty USN (2003) Flavonoids from Caesalpinia pulcherrima. Phytochemistry 63:789–793

Steven D, Ehrlich NMD (2009) Solutions acupuncture, a private practice specializing in complementary and alternative medicine, Phoenix, AZ. Review provided by VeriMed Healthcare Network

Sun SM, Song YM, Liu J (1995) Studies on the pharmacology of Cajanin preparation. Chin Tradit Herb Drugs 26:147–148

Tang Y, Wang B, Zhou XJ (1999) Effect of external application of herbal cajani preparation on the fibronection content during healing process of open wound. J Guangzhou Univ Tradit Chin Med 16:302–304

Uma RM, Pushpamma P (1981) Effect of insect infestation and storage on the nutritional quality of different varieties of pigeonpea. In: Proceedings of the international workshop on Pigeonpeas, Patancheru, 15–19 Dec 1980, 2:443–451

Vimala V, Pushpamma P (1983) Storage quality of pulses stored in three agroclimatic regions of Andhra Pradesh-quantitative changes. Bull Grain Technol 21:54–62

Vimala V, Pushpamma P (1985) Effect of improved storage methods on cook ability of pulses stored for one year in different containers. J Food Sci Tech 22:327–329

Yuan H, Yao LL, Chen LK (1984) Chinese journal of integrated traditional and western medicine 4:352–354 cited from Zu YG, Fu YJ, Liu W, Hou CL and Kong Y (2006) Simultaneous determination of four flavonoids in pigeonpea [*Cajanus cajan* (L.) Millsp.] leaves using RP-LC-DAD, Chromatographia. doi:10. 1365/s10337-006-0784-z

Yuan H, Li X, He W (1999) Chinese journal of traditional medicine trauma and orthopedic, 17: 4–8 cited from Zu YG, Fu YJ, Liu W, Hou CL and Kong Y (2006) Simultaneous determination of four flavonoids in pigeonpea [*Cajanus cajan* (L.) Millsp.] leaves using RP-LC-DAD, Chromatographia. doi:10.1365/s10337-006-0784-z

Yuan J, Zhang J, Lin J et al (2004) Traditional Chinese drug research & clinical pharmacology 1: 429–431 Cited from Zu YG, Fu YJ, Liu W, Hou CL and Kong Y (2006) Simultaneous determination of four flavonoids in pigeonpea [*Cajanus cajan* (L.) Millsp.] leaves using RP-LC-DAD, Chromatographia. doi:10.1365/s10337-006-0784-z

Zu YG, Fu YJ, Liu W et al (2006) Simultaneous determination of four flavonoids in pigeonpea [*Cajanus cajan* (L.) Millsp.] leaves using RP-LC-DAD. Chromatographia 63:499–505. doi:10. 1365/s10337-006-0784-z

Part II

Biofortification Through Biotechnology and Breeding Tools

Enriching Nutrient Density in Staple Crops Using Modern "-Omics" Tools

8

Abhishek Bohra, Uday Chand Jha, and Sushil Kumar

Abstract

A sizeable proportion of the global population faces nutritional disorders. Notably, the poorest regions in the developing world share considerably large segment of malnourished people. Given the rising prevalence of nutritional disorders, sustainable solutions urgently need to be in place in order to tackle the menace of hidden hunger. An array of improvement strategies is suggested to meet the growing challenge. These strategies involve dietary diversification, food supplementation/fortification, and biofortification using nutritional breeding approaches, genetic engineering, and agronomic interventions. The mounting concerns about environmental safety and poor economic status of the target population further put a limit on the large-scale use of micronutrient-rich fertilizers. Hence, crop biofortification via conventional and molecular breeding stands to be the most economic, readily accessible, and globally accepted strategy. For some obvious reasons, staple crops that serve the daily dietary needs of the maximum population in the developing world are targeted for nutritional enhancement. As a prerequisite, survey of the germplasm pools is needed to quantify the exploitable genetic variation that exists in the crop gene pool. Further, modern omics approaches like genomics, proteomics, metabolomics, and ionomics will definitely advance our knowledge about the genetic makeup, molecular networks, and physiological alternations involved in the process of mineral accumulation and subsequent partitioning of minerals to edible plant parts. Similarly, engineering

A. Bohra (✉)
Division of Crop Improvement, ICAR-Indian Institute of Pulses Research, Kanpur, Uttar Pradesh 208 024, India
e-mail: abhi.omics@gmail.com

U.C. Jha
Crop Improvement, Horticulture and Agricultural Botany (CIHAB), Visva-Bharati, Santiniketan, West Bengal 731 236, India
e-mail: uday_gene@yahoo.co.in

S. Kumar
Centre of Excellence in Biotechnology, Anand Agricultural University, Anand, Gujarat 388 110, India
e-mail: sushil254386@yahoo.com

U. Singh et al. (eds.), *Biofortification of Food Crops*, DOI 10.1007/978-81-322-2716-8_8

metabolic pathways through genetic modification holds great relevance for expediting the development of nutrient-dense food crops. We expect that the "omics" assisted nutritional breeding, as the most potential biofortification strategy, will be greatly helpful in achieving the nutritional security of over two billion nutrient-deficient people worldwide.

Keywords
Biofortification • Diversity • Genome • DNA marker • Linkage • QTL • Molecular map

8.1 Introduction

The number of people that populate the earth is constantly increasing, and based on the current trajectory, over two billion is to be additionally incorporated to the present figure of seven billion by the first half of this century, i.e., 2050 (Evans 2009; Varshney et al. 2015). Obviously, a dramatic increase is also expected in the size of population that faces undernourishment and malnutrition (in particular micronutrient-malnutrition)-related issues (Bohra et al. 2014a, b). A comprehensive report on "The State of Food Insecurity in the World 2014" documented recently by FAO, IFAD, and WFP indicates that about 800 million people encountered the problem of undernourishment during 2012–2014, a substantial proportion of which inhabits the developing world. The main reason behind the prevalence of this micronutrient deficiency in developing nations is the dependence of inhabiting people on staple food dominated by crops like wheat and rice, which are intrinsically deficient in mineral micronutrient that is essentially required to maintain the metabolism (Graham et al. 2001; White and Broadley 2009; Borill et al. 2014).

In a similar manner, elemental dietary deficiencies characterized by insufficient supply of essential minerals and micronutrients or trace elements are known to affect almost three billion people worldwide (Carvalho and Vasconcelos 2013; Tako et al. 2013). For example, iron (Fe) and iodine (I) deficiencies are shown by over 30 % and 17 %, respectively, of the current global population. Scarcity of these essential elements in turn predisposes deficient individuals to a range of diseases (Bohra et al. 2014a). A major fraction (~2 billion) of today's global population suffers from anemia which principally appears as a direct outcome of prolonged feeding on iron-limiting diets. Severity of iron-deficiency anemia (IDA) is much more evident in the case of developing countries where this nutritional disorder is reported to afflict half of the pregnant women and 40 % of the preschool children (Murgia et al. 2012).

To improve the nutritional status of the global population, a variety of strategies is proposed including dietary diversification, food fortification and supplementation, and biofortification through different means (Graham et al. 2001; White and Broadley 2009; Carvalho and Vasconcelos 2013). Widespread implementation of the methods like dietary diversification and food fortification/supplementation is constrained by the poor access of the target population to the market coupled with marginal nature of their income. Therefore, biofortification of crop plants is an attractive and economic strategy to develop nutrient-rich crops. Biofortification through breeding offers sustainable solution in comparison to indiscriminate application of fertilizers containing micronutrients since regular use of such fertilizers poses a considerable threat to the environment. In addition, growing concerns about environmental safety coupled with the increasing rigidity of the legislative regulations also limit the use of such agronomic interventions (Borill et al. 2014).

Cereals and food legumes constitute the dominant portion of human diets (particularly, in the developing nations) complementing each other in a synergistic fashion (Graham et al. 2001;

Bohra et al. 2014b, 2015). Further, concerning nutritional enrichment of crops, extensive examination of the natural variation for minerals/ micronutrients that is present in the crop gene pool sets the initial step while executing crop-based biofortification (Bohra et al. 2015). This is accompanied by accelerated use of exploitable genetic variation in the nutritional breeding programs. In this regard, high-throughput modern omics tools and technologies have been available in recent years to augment the crop biofortification program. The unprecedented amount of the information that has been generated using various omics platforms like proteomics, metabolomics, and ionomics is expected to greatly advance our understanding of elemental composition and the underlying causative genes and their elaborate network. Taken this into consideration, here we review the recent progress made in the area of crop biofortification (especially the staple crops like cereals and food legumes) along with exploring the prospects and challenges that lie ahead.

8.2 Identifying Nutrient-Dense Crop Genotypes Through Analyzing Crop Gene Pool

As has been described in various published reports, cultivated *Triticum aestivum* L. and *T. turgidum* L. ssp. *durum* (Desf.) Husn. species contain lesser quantities of grain iron (Fe) and zinc (Zn) than the wild *Triticum* and *Aegilops* species (Chhuneja et al. 2006; Cakmak et al. 2000; Monasterio and Graham 2000; Rawat et al. 2009; Velu et al. 2014). Importantly wild emmer wheat (*T. turgidum* ssp. *dicoccoides*) serves as rich reservoir of micronutrients especially Fe and Zn (Cakmak et al. 2004; Peleg et al. 2008a, b; Chatzav et al. 2011). Moreover, *T. dicoccoides*, *Ae. tauschii*, *T. monococcum*, and *T. boeticum* are also worth mentioning while considering the potential sources of Fe and Zn (Cakmak et al. 2000; Tiwari et al. 2009; Rawat et al. 2008). Notably, various researchers (Chhuneja et al. 2006; Rawat et al. 2009; Tiwari et al. 2010) have reported wild *Triticum* and *Aegilops* species as the storehouse of high Fe and Zn content offering 75 % and 60 % higher Fe and Zn, respectively, than the cultivated bread wheat cultivar. Likewise, wild *Ae. kotschyi* acc. 3790 contained three times higher grain Fe in comparison to *T. aestivum* Cv. WH291 and *T. aestivum* Cv. WL711 (Singh et al. 2013). Synthetic hexaploids derived from *T. turgidum* spp. *durum* × *Aegilops tauschii* showed 30 % higher grain Fe and Zn (Calderini and Ortiz-Monasterio 2003). Attempts are under way at International Maize and Wheat Improvement Center (CIMMYT), Mexico (Velu et al. 2011a, b), for introgressing beneficial genes that contribute high Fe and Zn in synthetics derived from *T. spelta* × *T. dicoccon* and *Ae. tauschii* × *T. dicoccon* to high-yielding cultivars in wheat. In rice, after screening 46 rice genotypes (including cultivated and wild accessions), Banerjee et al. (2010) have reported that wild accessions of rice harbor greater amount of Fe than found in the cultivated genotypes. Genetic variation for Fe content varying from 15 to 109 mg/kg (based on dry weight) was reported in wild emmer wheat (Cakmak et al. 2004). Therefore, given the existence of tremendous germplasm variability for grain micronutrients (especially Fe and Zn) in the wild species, focused strategies are required to be undertaken that enable faster transfer of these micronutrient accumulating genes/QTLs into high-yielding popular cultivars.

Considering the importance of landraces vis-à-vis grain Fe content, a set of 52 accessions (commercial cultivars + landraces) was surveyed for variation in Fe content revealing a wide range between 1.32 ppm and 100.45 ppm. The landrace 'Lal Gotal' was observed to contain the highest amount of Fe (100.45 ppm) (Jahan et al. 2013). Likewise, analysis of 126 brown rice genotypes uncovered ample variability (varying from 6.2 to 71.6 ppm) for Fe content, and interestingly, the local accession had the highest value of Fe content (Anuradha et al. 2012a). The greater genetic variation in landraces for Fe content was also supported by Anandan et al. (2011). A higher Fe content was noted in the case of traditional cultivars, viz., 'Kalabath', 'Noothipattu', 'Koomvalazhi', and 'KDML

105', than the improved rice cultivars. Earlier, traditional rice cultivars (CMU122, CMU123, and CMU124) among the Thai brown rice cultivars had greater Fe content ranging from 7 to 22 mg Fe/kg (Prom-teru-thai and Rerkasem 2001). Up to threefold variation in Fe content (7.5–24 mg Fe/kg) was reported in cultivated brown rice (Senadhira et al. 1998; Graham et al. 1999; Prom-u-thai and Rerkasem 2001), whereas the range was found to be between 3 and 11 mg Fe/kg in white rice (Prom-u-thai et al. 2007). Likewise, by screening 15 Fe-dense and Fe-normal genotypes of unpolished rice through in vitro digestion/Caco-2 cells, variation in Fe content (14–39 μg/g) was recorded (Glahn et al. 2002). Applying the similar method, significant variation for kernel Fe content (ranging from 15.5 to 19.1 mg/kg) was obtained in maize (Oikeh et al. 2003), and particularly, the two genotypes 'ACR90POOL16-DT' and 'ACR86TZESR-W' were found to be promising relating to kernel Fe content in maize. Meng et al. (2005) have suggested that black rice is superior to the other forms of rice such as red rice, sticky rice, and fine rice concerning Fe content. And also, husk and chaff contain more Fe content. The grain Fe content of white rice varied between 0.05 and 0.2 μg/grain (Prom-u-thai et al. 2007). In vitro digestion/Caco-2 model in wheat aided in the identification of *Aegilops* derivatives with a 1.5-fold increase in the bioavailable Fe. Further, positive correlation of Fe content with protein and phytate contents was also obtained (Salunke et al. 2011). In a similar way, a threefold increase in Fe bioavailability was recorded in 11 Chinese rice genotypes tested using Caco-2 cell culture model. Moreover, impacts of ascorbic acid application on Fe bioavailability were also examined, and enhanced Fe bioavailability was noticed in the polished rice due to ascorbic acid (He et al. 2013).

Up to threefold difference in Fe content was reported after elemental evaluation of a worldwide wheat collection (Oury et al. 2006). Similarly, analyzing the Fe content in core collection of wheat showed the range of Fe content varied from 26.26 to 68.78 mg/kg, and 'Andalucia 344' accession was found to carry highest Fe content (Velu et al. 2011b). While investigating Fe accumulation in 20 wheat genotypes in response to Fe treatment, it was reported that selenate enhanced the Fe accumulation (de Souza et al. 2013). Besides, two Fe-rich genotypes, viz., 'EMB 38' and 'BRS 264' were recovered with higher grain Fe content.

Noteworthy genetic variation for Fe content was observed across 109 sub-Saharan African inbreds, in particular mid-altitude (15–159 ppm) and lowland (14–134 ppm) inbreds (Maziya-Dixon et al. 2000). In a study based on 3-year trial data, Prasanna et al. (2011) analyzed variation for kernel Fe concentration in 30 maize genotypes and observed a range from 11.28 to 60.11 mg/kg. They also identified two genotypes 'HP2' and 'BAJIM 06-17' which remained stable across multiple environments. Recent studies at ICRISAT showed Fe content varying from 29.8 to 44.2 mg/kg in sorghum (Kumar et al. 2010) and 30.1–75.7 mg/kg in pearl millet (Velu et al. 2007). Similarly, Rai et al. (2012) also reported adequate Fe variability (ranging from 18 to 97 ppm) in advanced breeding lines of pearl millet.

In wheat, wild emmer, einkorn, and landraces are known as the richest source of grain Zn content (Cakmak et al. 2000; Ortiz- Monasterio et al. 2007). Further, Cakmak et al. (2000) reported that other wheat species such as *T. dicoccoides*, *Ae. tauschii*, *T. monococcum*, and *T. boeticum* also contain higher Zn content. High variation for Zn content ranging from 30 to 118 mg/kg was recorded after screening 825 accessions of *T. dicoccoides* (Cakmak et al. 2004). In hexaploid wheat, the Zn concentration differed (15–35 ppm) in wheat germplasm collection that comprised of elite breeding lines and germplasm collected across the globe (Oury et al. 2006). Significant variation was also reported from wheat core collection with the Zn content varying between 16.85 and 60.77 mg/kg. Specifically, Chinese spring bread cultivar (Hong Duan Mang) showed the highest Zn content (Velu et al. 2011b). Similarly, genotypic variation for grain Zn was also noticed among 20 Brazilian wheat genotypes manifesting a twofold difference within the sample studied.

Further, addition of selenium led to an increase in the grain Zn content (Souza et al. 2014). To elucidate the role of genotype × environment (G × E) interaction in determining grain Zn and Fe contents, elite lines of wheat from CIMMYT were tested in Eastern Gangetic Plains (EGP) of India under multiple environments. The results revealed significant G × E effects operating on grain Fe and Zn concentrations, and more variation across locations was noted in the case of Zn (Joshi et al. 2010). Multilocation testing of advanced wheat lines (developed from *T. spelta*, landraces, and synthetic wheat) in South Asia and Mexico showed high heritabilities for Zn and Fe contents across multiple sites. The greater extent of variation generates new avenues for selection of Zn- and Fe-dense lines for future use (Velu et al. 2012). Based on phytosiderophore release, Daneshbakhsh et al. (2013) recorded two Zn efficient wheat genotypes, viz., Cross and Rushan.

Regarding Zn content in other staple crops, grain Zn variation of 13.5–58.4 mg/kg was recorded in rice germplasm (Gregorio et al. 2000; Welch and Graham 2002), while it ranged between 29 and 37 mg/kg in the case of aromatic rice (Graham et al. 1999), 24.5–64.8 mg/kg in pearl millet (Velu et al. 2007), and 22.2–32.9 mg/kg in sorghum (Kumar et al. 2010). Recently, Rai et al. (2012) have reported large variation for Zn content (from 22 to 69 ppm) in advanced breeding lines of pearl millet. Considering genotypic difference in the micronutrient lost due to milling process, variable degrees of loss were reported for Fe (24–60 %) and Zn (10–58 %) in 15 Thai rice genotypes (Saenchai et al. 2013). In maize, high degree of variation for Zn content was observed across 109 inbreds of sub-Saharan African maize ranging from 12 to 96 ppm for mid-altitude inbreds and 24 to 96 ppm for lowland inbred line (Maziya-Dixon et al. 2000). Prasanna et al. (2011) also reported variation (15.14–52.95 mg/kg kernel Zn) in maize and also identified a genotype (IML467) due to its kernel Zn concentration remaining stable over the environments.

Concerning the genetic variability for calcium (Ca) content, Graham et al. (1999) reported the Ca content varying from 0.25 to 0.73 g/kg (on dry weight basis) among 132 wheat genotypes. In durum wheat, the Ca content ranging from 388 to 640 mg/kg was obtained through analyzing 84 Italian wheat for two years under multiple environments (Ficco et al. 2009). In hexaploid wheat, cultivar 'PBW-396' showed the highest Ca content (76.67 mg/100 g dry weight) among the 10 genotypes tested (Mallick et al. 2013). Based on six field trials, Feil and Fossati (1995) recorded the range of Ca from 0.35 to 0.50 g/kg in triticale grains.

In relation to provitamin A, orange maize varieties are known to contain higher provitamin A than the yellow varieties (Li et al. 2007; Ortiz-Monasterio et al. 2007; Menkir et al. 2008). Based on the visual scoring of orange kernel color (as a marker for high provitamin A), Chandler et al. (2013) noticed moderate heritability in nested association mapping (NAM) maize panel. Similarly, yellow endosperm varieties were also reported to carry higher amount of provitamin A carotenoids in sorghum (Fernandez et al. 2009). Greater genotypic differences for provitamin A (0.24–8.80 μg/g) were manifested across 1000 tropical maize germplasm at CIMMYT (Ortiz-Monasterio et al. 2007). Likewise, Babu et al. (2012) estimated the range of provitamin A (15–20 μg/g) in the grains of improved lines. Genetic variability for kernel β-carotene varying from 0.02 to16.50 μg/g was discovered from 105 maize inbreds from India and CIMMYT (Vignesh et al. 2012). The authors also suggested the wide variation due primarily to the allelic variation in *crtRB1* 3'TE gene. More recently, Azmach et al. (2013) reported that the tropical maize lines harboring the functional markers, viz., crtRB1-5′TE and 3′TE, could act as rich source of provitamin A. To decipher the gene action, both the general combining ability (GCA) and specific combining ability (SCA) for the grain yield and provitamin A were examined in hybrids, and the results revealed existence of significant GCA effects for provitamin A, suggesting the presence of an additive gene action (Suwarno et al. 2014). Notably, significant G × E effects were observed for the grain carotenoid concentration, and associated carotenoid traits in maize population derived from the cross DEexp × CI7. Further, the tested population showed higher variation for the carotenoid under subtropical condition than temperate condition (Kandianis et al. 2013). Similarly in wheat, 'RSP-561' genotype was reported to contain highest carotenoid content (105.67 μg/100 g) among the 10 genotypes tested (Mallick et al. 2013).

Focus needs to be given not only towards evaluation of nutrients but also to a variety of so-called anti-nutritional factors like phytates. Phytate content ranged from 1.98 to 2.46 g/kg in maize while evaluating 90 S1 families derived from the BS31 population at two locations (Lorenz et al. 2008). Average value of phytate content was recorded to be 2.91 g/kg from 54 land races of maize (Drinic et al. 2009). Phytate content was measured varying from 800 to 1000 mg/ 100 g seed in inbred lines and F_1 hybrids. Based on the phytic acid (PA) content, Ki inbred lines were grouped into low PA group (<900 mg/100 g seed) and medium/high group (≥900 mg/100 g seed) (Chiangmai et al. 2011). Additionally, the population stemming from "derived flints" showed lower (1.95 g/kg) phytate content. The "Mediterranean flint"-based populations had the lowest (0.46 g/kg) phytate content in contrast to the 'flinty dents'-derived population which exhibited the highest value (3.43 g/kg) for phytate (Drinic et al. 2012).

In wheat, PA content varied from 200 mg/ 100 g to 400 mg/100 g in refined flour and 600–1000 mg/100 g in the whole flour (Febles et al. 2002). Two Iranian cultivars 'Pavarus' and 'Niknejad' were reported to contain low PA, while 'Estar', 'S-78-11', 'S-79-10', and 'Niknejad' had maximum phytase activity (Tavajjoh et al. 2011). Mallick et al. (2013) reported PA content ranging from 0.35 to 1.60 mg 100 g^{-1} across 10 genotypes investigated, and they identified 'HD2687' genotype exhibiting the lowest PA content. Notably, by performing a diallel analysis, Ahmad et al. (2013) recorded the PA up to 3.43 % among the F_1 hybrids of wheat; on the other hand F_1s of crosses Ps-2005 × Ghaznavi, AUP-4006 × Ps-2004, Janbaz × Ps-2004, and Janbaz × Ps-2005 exhibited lower PA. Likewise in barley, Dai et al. (2007) observed genetic variation in PA ranging from 3.85 mg/g to 9.85 mg/g across 100 genotypes. They suggested the significant variation was primarily accounted to effects exerted by both location and time.

In relation to selenium (Se), immense variability in Se content (5–720 μg/kg) was reported in wheat germplasm (Lyons et al. 2005). The authors also reported that majority of this variation could be credited to the soil factor. Based on an extensive survey involving 88 different sites in Malawi, Se concentration of maize grain was observed to range between 0.005 mg/kg and 0.533 mg/kg (Chilimba et al. 2011).

Considering nutrient uptake efficiency, wheat genotype "Maris Butler" was reported to be highly efficient of extracting manganese (Mn) from Mn-deficient soil (Jiang and Ireland 2005). Further, 'Maris Butler' and 'C8MM' genotypes of wheat were also reported to be highly efficient for Mn uptake (Jiang 2008). During reproductive phase, the wheat genotypes PBW550, BW9178, and HD2967 were identified based on the utilization efficiency of Mn. Therefore, these genotypes can be potentially exploited for the development of Mn-rich wheat cultivars in the near future.

More recently, Pinson et al. (2014) from USDA and other collaborating institutes comprehensively examined a worldwide collection comprising 1763 greatly diverse germplasm accessions (majorly from USDA core collection) in rice for 16 mineral nutrients in flooded and non-flooded conditions. They used inductively coupled plasma mass spectrometry (ICP-MS) technology to assess the level of elemental variation. Resultantly, in view of moderate to high heritability estimates obtained for these mineral [except for nickel (Ni)], the authors strongly advocated for the possibilities of rice nutritional enhancement using selection and breeding. Similarly, a set of 72 pearl millet genotypes including landraces and open-pollinated varieties was examined for mineral densities across multiple environments at Niger (Pucher et al. 2014). This compositional survey revealed a greater range of variation for Ca, Fe, copper (Cu), and Zn existing within the material studied. On the other hand, moderate level of variation was noticed for Mg, P, K, and Mn. Owing to the considerable extent of G × E interactions recorded particularly Fe and Zn densities, authors advocated multi-environmental testing of diverse accessions while measuring variations in mineral and micronutrient contents.

8.3 Elucidating the Genetic Architecture of Nutrient Accumulation via QTL Mapping

Detecting the causative loci (genes or QTLs) that determine the mineral/nutrient content is a crucial step while understanding the genetic makeup of the nutrient-related traits. Molecular mapping of the genome segments that govern nutrient content/concentration has been practiced in many crops (Table 8.1). In this section, we update the existing literature about mapping of QTL associated with nutrient traits. By applying inductively coupled plasma optical emission spectroscopy (ICP-OES) technique to F_4 lines derived from the cross (B84 × Os6-2) in maize, Simić et al. (2012) detected significant QTLs controlling concentrations of four prime micronutrients, i.e., phosphorus (P), Fe, Zn, and Mg. The QTL analysis revealed occurrence of a total of 32 QTLs associated with seven different traits and the phenotypic variances (PVs) of these QTLs ranged between 6.7–19.9 %. More importantly, co-localization of some of these QTLs in chromosome 3 offers a promising marker-delimited chromosomal region for immediate utilization in nutritional breeding. Similarly, the QTLs were discovered from $F_{2:3}$ lines stemming from the crosses Mu6 × SDM and Mo17 × SDM that determined Zn and Fe concentration in maize kernels and cobs (Qin et al. 2012). More importantly, a joint QTL analysis was performed using data from the two populations which eventually provided a set of 12 QTLs, majority of which were observed in both joint- and single-environment analyses. Lung'aho et al. (2011) reported three QTLs for grain Fe concentration in a maize recombinant inbred line (RIL) population B73 × Mo17 (popularly known as IBM). On the other hand ten QTLs were found, controlling Fe bioavailability accounting over 50 % of the total PV. Similarly, a 178 × P53-based $F_{2:3}$ population in maize was examined using atomic absorption spectrophotometer (AAS), and consequently, five QTLs (four for Zn and one for Fe content) were detected which accounted to 17.5 % (Zn) and 16.8 % (Fe) of observed PV (Jin et al. 2013). Furthermore, QTL data collected from five different studies were subjected to an integrated analysis or, more appropriately, the meta-QTL analysis, and as a result, 10 meta-QTLs (M-QTLs) were detected which governed up to 28 % of the observed PV. The investigation highlighted the underlying importance of *bins* 2.07 and 2.08 in deciphering the genetic architecture of mineral concentration in maize. Given the significance of "ionome" characterization in maize leaves particularly concerning the silage preparation, Zdunic et al. (2014) performed ionome profiling of over 300 intermated RILs belonging to IBM population. The QTL analysis using a high-resolution genetic map (comprising 2161 DNA markers) facilitated mapping of major QTLs ($\geq$10 % PV) for cadmium (Cd), potassium (K), and strontium (Sr) accumulation, whereas QTLs related to other metals like Cu, Fe, and Mg showed lesser R^2 values ($<$10 % PV). As a result of elemental examination conducted across two different environments, significant QTL × environment interactions were revealed in this study. Importantly, QTLs for Cd were noted to be consistent with previous QTL studies. The same population was earlier analyzed by Baxter et al. (2012) to profile grain ionome in maize using ICP-OES. Based on the experimental data collected across multiple locations during different years, i.e., 2003, 2005, and 2007, potential QTLs were

Table 8.1 List of some important QTLs that are associated with mineral/nutrient traits in different crops

Crop	Nutrient trait	Potential genomic region	Mapping population	Associated DNA marker	References
Rice	Zn	One major effect of QTL on chromosome 8	IL: *O. sativa* × *O. rufipogon*	SSR	Garcia-Oliveira et al. (2009)
	Zn, Fe	14 QTLs for Zn and Fe on chromosomes 1, 3, 5, 7, and 12	RIL: Madhukar × Swarna	SSR	Anuradha et al. (2012b)
	Zn, Fe	OsYSL6, OsYSL8, OsYSL14, OsNRAMP1, OsNRAMP7, OsNRAMP8, OsNAS1, OsFRO1, and OsNAC5 genes	–	–	Sperotto et al. (2010)
	Zn, Fe	5 QTLs for Fe and 3 QTLs for Zn present on 2, 3, 8, and 12	BC: Swarna × *O. nivara*	SSR	
		7 QTLs for Fe and 5 QTLs for Zn present on 7 and 12	RIL: Madhukar × Swarna	SSR	
	Zn, Fe	*OsYSL1*,*OsNAC*, *OsYSL16*, *OsZIP4*, *OsYSL17*, and *OsNAAT1*	BC: Swarna × *O. nivara*	SSR	
	Zn, Fe	*OsYSL1*, *OsMTP1*, *OsNAS1*, *OsNAS3*, *OsNRAMP1* candidate genes	RIL: Madhukar × Swarna	SSR	
	Zn	*qZnc4*, *qZnc6* on chromosomes 4 and 6	JX17 × ZYQ8	RFLP, SSR	Zhang et al. (2011)
	Fe	qFe1, qFe3, qFe4, qFe7 on 1, 3, 4, 7	Bala × Azucena	–	Norton et al. (2010)
	Zn	qZn6, qZn7, qZn10.1, qZn10.2 on 6, 7, 10			
	Cu	qCu1, qCu2, qCu12.1, qCu12.2 on 1, 2, 12			
	Mn	qMn10, qMn11 on 10 and 11			
	Se	qSe1.1, qSe1.2, qSe3, qSe6, qSe7, qSe9 on 1, 3, 6, 7, 9			
	Mn	Four QTLs: qGMn1, qGMn2, qGMn7, qGMn12 on chromosomes 1, 2, 7, 12	Sasanishiki × Habataki	–	
	Fe	QTL (qGFe4) on chromosome 4			
Maize	Fe	Three FeGC QTLs, 10 QTLs for FeGB	RIL: B736 × Mo17 (IBM)		Lungaho et al. (2011)
	Fe/P, Zn/P, and Mg/P	Three QTLs on chromosome 3	F4: B84 × Os6-2	SNP, SSR	Simić et al. (2012)
	Zn, Fe	QTL for ZnK, ZnC, FeK, and FeC on chromosome 2	F2:3: Mu6 × SDM and Mo17 × SDM		Qin et al. (2012)
		QTL for Znk, FeK, and FeC on chromosome 9			
		QTL for ZnK and ZnC on chromosome 7			
	Zn, Fe	Five QTLs and 10 Meta-QTLs	F2:3: 178 × P53	SSR	Jin et al. (2013)
	Zn and Fe	17 QTLs on chromosomes 1, 2, 3, 4, 6, 7, 10	DH: DH8 × DH40 and DH86 × S137	–	Zhou et al. (2010)
	Fe		B73 × Mo17	–	

(continued)

Table 8.1 (continued)

Crop	Nutrient trait	Potential genomic region	Mapping population	Associated DNA marker	References
		Three candidate regions on chromosomes 3, 6, and 9			Tako
Wheat	Fe	*QFe.pau-2A and QFe.pau-7A.*	RIL: *Triticum boeticum* × *T. monococcum*	SSR and RFLP	Tiwari et al. (2009)
	Zn	*QZn.pau-7A* (Zn QTL)			
	Zn	Four QTLS on chromosomes 3D, 4B, 6B, and 7A	RAC875-2 × Cascades	RFLP, AFLP, SSR, and DArT	Genc et al. (2009)
	Fe	One QTL	RAC875-2 × Cascades	RFLP, AFLP, SSR, and DArT	Genc et al. (2009)
	Zn	Five QTLs on seven different chromosomes	RIL: *T. spelta* × *T. aestivum*	DArT, SNP	Srinivasa et al. (2014)
	Fe	Five QTLs on seven different chromosomes	RIL: *T. spelta* × *T. aestivum*	DArT, SNP	
	Fe, Zn, and Protein	Nine additive and four epistatic QTLs on chromosomes 4B and 5A	RIL	–	Xu et al. (2012)
	Zn, Fe, Mn, and Cu	Major QTL on chromosome 5	RIL	–	Ozkan et al. (2007)
	Zn and Fe	QTL on chromosome 6B	*T. turgidum* ssp. *dicoccoides*	–	Distelfeld et al. (2007)
	Zn	Six QTLs on 2A, 5A, 6B, 7A, 7B	Durum wheat × wild emmer	–	Peleg et al. (2008b)
	Fe	11 QTLs			
	Cu	10 QTLs			
	Mn	Two QTLs			
	Zn	Four QTLs chromosomes 4A, 4D, 5A, and 7A	Hanxuan10 × Lumai 14	–	Shi et al. (2008)
	Zn	Seven QTLs for grain Zn on chromosomes 1A, 2D, 3A, 4A, 4D, 5A, and 7A			
	Ca	Nine QTLs on 1A, 4A, 5B, 6B, 2B, 4B, 6A, 6B, 7B	Langdon × G18-16	–	Peleg et al. (2009)
	Fe	11 QTLs on 2A, 3B, 5A, 6B, 7A, 2B, 3A, 4B, 6A, 7B, 5A, 6A			
	Zn	Six QTLs on chromosomes 2A, 5A, 6B, 7A, 7B			
	Mn	Two QTLs			
	Cu	10 QTLs on 2A, 4A, 4B, 5A, 6B, 7A, 1A, 3B, 6A, 7B			
Barley	Zn	Five QTLs on 1HS, 1HL#1, 2HS, 2HL#2, and 5HL	DH: Clipper × Sahara	–	Lonergan et al. (2009)

discovered for a variety of elements including Ca, Cu, Mg, Fe, K, Mn, P, Zn, and S, which controlled up to 46 % of the PV.

A recent study discovered a total of 134 QTLs that influenced grain element concentration in two rice mapping populations including RILs and introgression lines developed from the cross Lemont × TeQing. It is important to note that 34 of these QTL-containing regions were found consistent in more than one experimental population. Noticeably, the authors suggested the presence of an intricate network of Mg, P, and K (Zhang et al. 2014). In a similar fashion, candidate simple sequence repeat (SSR) markers associated with grain Zn content were identified from a RIL population (IRRI38 × Jeerigesanna) comprising 160 individuals. The phenotypic contributions of these QTLs varied between 4.5 % and 19 %. Additional three of these candidate markers were experimentally confirmed in panel of 96 accessions (Gande et al. 2014). Using 110 DNA markers, a total of 14 QTLs associated with Fe and Zn concentrations in rice grains were identified from 168 RILs (Madhukar × Swarna). The QTLs were mapped to different chromosomes, viz., 1, 3, 5, 7, and 12 (Anuradha et al. 2012b). Later, two inbred lines from this RIL population, viz., RP Bio5478-185 M (high Fe and Zn) and RP Bio5478-270 M (low Fe and Zn), and one of the parents (Swarna) were used for quantitative real-time PCR (qRT-PCR)-based expression profiling using 15 candidate genes. The QTL allele leading to an increase in Fe concentration was also found to be associated with enhanced upregulation (Agarwal et al. 2014).

In wheat, QTL analysis of more than 100 RI individuals (Tabassi × Taifun) led to the discovery of six and two QTLs respectively for Fe and Zn grain concentrations. The Fe-QTLs explained 29 % of the PV, while Zn-QTLs accounted to 51 % of the total PV. A major QTL for Cl^- influencing 32 % of the variation was detected from a DH population (Berkut × Krichauff). Mapped on chromosome 5A, the underlying genomic region was delimited by SSR markers *gwm304*and *barc141* (Genc et al. 2014). The Cl^- was analyzed with ICP-OES technique. More recently, Pu et al. (2014) created two RIL populations in wheat, i.e., SHW-L1 × Chuanmai 32 and Chuanmai 42 × Chuannong 16, which showed quantitative variation for Se, Fe, Zn, Cu, and Mn. The QTL analysis in SHW-L1 × Chuanmai 32 population facilitated identification of a series of QTLs contributing up to 28.5 % of the phenotypic variance. Similarly, 13 QTLs were obtained from Chuanmai 42/Chuannong 16 population with the PV varying between 7.5 % and 35.1 %. Grain protein content B1 (Gpc-B1) is a QTL detected in wheat and it was found to be associated with enhanced grain protein, Zn, and Fe contents (Uauy et al. 2006). An increment of 10–34 % in concentrations of grain Zn, Fe, Mn, and protein was observed in cultivated wheat after introgression of Gpc-B1 locus from the wild tetraploid wheat *T. turgidum* ssp. *dicoccoides* into different recombinant chromosome substitution lines (Distelfeld et al. 2007). Authors proposed that the Gpc-B1 locus promoted remobilization of protein, Zn, Fe, and Mn from the leaves to the grains.

In soybean, QTLs were discovered for seed Ca content from $F_{2:3}$ families obtained by crossing low- and high-calcium-containing genotypes (Zhang et al. 2009). In a similar fashion, Jegadeesan et al. (2010) analyzed a RIL population (AC Hime × Westag-97) and they reported a major QTL on LG K that accumulates Cd in soybean seeds and accounted for 57.3 % of the total PV. Further, in order to validate the QTLs detected from the population 'AC Hime × Westag-97', another RIL population (Leo × Westag-97) was employed for QTL mapping. Notably, the SSR markers found linked with the Cd locus or *Cda1* gene were also mapped in the Leo × Westag-97 population, and the mapping positions of these SSR markers manifested the same marker order. In addition, three of these linked markers, viz., SatK 147, SatK 149, and SattK 152, successfully discriminated between the low- and high-Cd-containing lines, thereby offering robust DNA markers for exercising marker assisted selection (MAS) for isolating low Cd genotypes or for the precise introgression of the candidate genomic region into diverse genetic backgrounds. Likewise, QTL analysis of 144 RILs [OAC Bayfield (high vitamin E content) × Hefeng 25 (low vitamin E content)] revealed candidate chromosomal regions within the soybean genome that determine vitamin E content in soybean seeds. Major QTLs were detected for alpha-, gamma-, and delta-tocopherols as well as for the total vitamin E. The percent PV (% R^2) of these QTLs ranged between 2.4 and 16.7 (Li et al. 2010). The QTLs and other genomic-based tools that are currently being used for legume biofortification are reviewed recently by Bohra et al. (2015).

8.4 Genome-Wide Association Study (GWAS)-Based Dissection of Elemental Composition and Nutrient-Related Traits

The transformational potential of GWAS for elucidating the architecture of important nutritional quality traits was experimentally demonstrated in rice using 44,100 single nucleotide polymorphism (SNP) markers and 413 diverse accessions collected from 82 countries (Zhao et al. 2011). Likewise, to finely resolve the genetic makeup of four mineral nutrients (As, Cu, Mo, and Zn), genomes of ~300 rice cultivars were scanned with genome-wide DNA markers to build significant marker trait associations (SMTAs). The genotyping of the entire diversity panel was performed using 36,901 SNPs, while elemental compositions were investigated through ICP-MS. The SNP markers displaying SMTAs with the grain nutrient contents were detected on different rice chromosomes [As (chromosome 5), Cu (chromosomes 1 and 5), Mo (chromosome 10), and Zn (chromosomes 7 and 9)] with a greater extent of QTL × environment interactions influencing the traits under consideration. Interestingly, the candidate genomic segments near these SNP markers were observed to be residing in close association with mineral transporters especially for Cu and Mo (Norton et al. 2014). More recently, a panel of accessions composed of 298 barley landraces belonging to African countries (Ethiopia and Eritrea) was subjected to genome-wide marker genotyping with almost 8 K SNPs of iSelect SNP chip (Mamo et al. 2014). GWAS of the micronutrient variation existing in the association panel allowed mapping of QTLs conferring enhanced Zn concentration in barley genome, more precisely to the chromosome 6HL (*Zn-qtl-6H_SCRI_RS_*10655). It is interesting to note that even a high-density genotyping using 8 K DNA markers could not be able to detect the QTLs for Fe concentration which remains in concordance with the earlier published reported that could not map QTL related to Fe concentration (Mamo et al. 2014). Notably, positive correlation between Zn and Fe concentrations was observed in both field and greenhouse experiments.

A joint linkage analysis and GWAS using nested association mapping (NAM) population provides valuable insights about the carotenoids in maize grains. NAM panel comprises of a total of ~5000 RILs which were produced by crossing 25 diverse parents to a common parent (B 73) (Yu et al. 2008). A total of 28 million SNPs derived from ultra high-resolution maize HapMaps were used to execute the genome-scale association analysis (Lipka et al. 2013a). Given the immense health implications of vitamin E and antioxidants (particularly in the developing world), a set of 281 inbreds was chosen to perform GWAS in order to illuminate the genetic architecture of composition as well as the content of tocopherols and tocotrienols in maize grains. A total of 591,822 SNPs were initially targeted for analysis; however, 294,092 were finally chosen for conducting the GWAS. Besides advancing understanding about the genetic association between *ZmVTE4* haplotypes with α-tocopherol content, novel insights were gained about relationship between the *ZmVTE1* and tocotrienol composition. In addition, candidate gene-based approach was also implemented to generate additional experimental evidences about molecular mechanism that underlies tocochromanol biosynthesis. A set of 60 candidate genes including *VTE* genes was targeted for pathway-level analysis

(Lipka et al. 2013b). A similar strategy was used for detecting significant association for carotenoid composition in maize kernels. The GWAS, as well as pathway-level analysis, was performed. The GWAS was applied with 284,180 SNP and seven insertion-deletion (indel) markers, whereas 7 K SNPs and seven indels were chosen for pathway-level analysis. The genome-scale and pathway-level analyses provided novel association within *zep1* and *lut1* and *dxs2* and *lut5*, respectively, which were not reported earlier (Owens et al. 2014).

Concerning association mapping, mapping populations derived from multiple founders such as NAM and multi-parent advanced generation intercross (MAGIC) offer advantages over diversity panel as the latter owing to the population structure is prone to generate spurious associations or false positives (Cavanagh et al. 2008). The MAGIC and NAM populations are increasingly generated in several crops such as rice, wheat, sorghum, etc. The most recent examples of MAGIC populations include an eight-parent wheat MAGIC population developed at National Institute of Agricultural Botany (NIAB), Cambridge, UK, which comprises a total of 1091 lines (Mackay et al. 2014). Coupling multi-founder populations with GWAS promises to be an attractive means to overcome the caveats routinely faced with the GWAS. Multi-parent populations have also been found promising in generating genome-wide predictions (GWPs) in crop plants.

8.5 Genomic Selection (GS) or GWPs to Develop Nutritionally Enriched Staple Crops

GS was proposed by Meuwissen et al. (2001) as a genome-level improvement strategy, which unlike MAS does not target a specific set of significant markers but exploits the genome-wide LD using high-density genetic variants spanning the entire genome. In other words, GS resembles a *black box* approach that does not intend to elucidate the detailed genetic architecture of complex traits; instead GS aims to provide the estimated breeding values (EBVs) of individuals using genome-wide marker data (Hamblin et al. 2011; Jonas and Koning 2013). In conventional plant breeding, EBVs derived from best linear unbiased predictions (BLUPs) have always been important criteria for selecting worthy individual which is based on the premise that performance of progenies acts as a better indicator of one's genetic merit than its own performance (Goddard and Hayes 2007; Jonas and Koning 2013; Meuwissen et al. 2001; Heffner et al. 2010).

In GS scheme, phenotyping which has always been a costly and time-consuming operation in plant breeding is exercised only to train the models, i.e., training population(Varshney et al. 2015). On the other hand, genotyping of both training and breeding populations was exercised using high-density DNA markers (Goddard and Hayes 2007). Several factors are known to affect the accuracy of genomic EBVs which include genetic composition and size of the training population, types and optimum number of DNA markers to be assayed, appropriate statistical methods used to generate GWPs, and, indeed, the heritability of the traits (Hamblin et al. 2011). Though extensively implemented in livestock breeding, GS is in developing stage in plants (Hamblin et al. 2011). Nevertheless, encouraging empirical results are being increasingly made available from a wide range of crops including maize (Lorenzana and Bernardo 2009; Crossa et al. 2013), barley (Zhong et al. 2009), wheat (Heffner et al. 2011; Poland et al. 2012; Crossa et al. 2010, 2013), soybean (Jarquín et al. 2014), sugar beet (Würschum et al. 2013), etc. While exploring the potential of GS in plant breeding, Jonas and Koning (2013) highlighted the immense potential of the GEBVs in increasing the gain per unit time; however, the authors have also opined that the implementation of GS in plant breeding may not be as simple as it has been in the case of livestock breeding. In addition, authors have also offered a set of valuable questions that need to be taken into consideration while executing GWP-based selection for crop improvement.

Desta and Ortiz (2014) recently reviewed the various prediction models used in GS including ridge regression best linear unbiased prediction (RR-BLUP), least absolute shrinkage and selection operator (LASSO), elastic net (EN), random forest (RF), reproducing kernel Hilbert spaces regression (RKHS), Bayesian LASSO, and other Bayesian methods like Bayes A, B, C, and so forth. Recently, Heslot et al. (2012) have compared the predictive abilities of various models that are currently being used in GS. Concerning provitamin A biofortification, Owens et al. (2014) used ~200 maize lines for prediction analysis using three statistical approaches, viz., RR-BLUP, LASSO, and EN. Apart from using various methods, three different marker sets were considered for assessing the prediction accuracies: first set included genome-wide DNA markers [284,180 SNP and seven insertion-deletion (indel) markers] which provided genome-scale predictions. The second set comprised of markers specific to carotenoid biosynthesis in maize (7 K SNPs and seven indels), thus generating pathway-level predictions. The remaining third set aimed to deliver QTL-targeted predictions as it focused on eight candidate genes (944 SNPs and seven indels) that were found associated with the genomic regions controlling carotenoid content in maize grain. Interestingly, the three statistical approaches offered more or less similar prediction accuracies. As a result of prediction analysis, an average GWP accuracy of 0.43 was recorded, with the highest (0.71) obtained for β-xanthophylls (Owens et al. 2014).

Ever-increasing capacities of next-generation sequencing (NGS) assays would likely to bolster these genome-wide techniques including GWAS and GS that rely on extensive exploration of linkage disequilibrium spanning the entire crop genome (Poland et al. 2012; Poland and Rife 2012). Further, the next-generation phenotypic screens driven largely by high-throughput automated platforms would radically expand the potential of such whole-genome strategies (Cobb et al. 2013).

8.6 Rising "-Omics" Technologies to Reinforce Plant Nutritional Genomics and Breeding

In conjunction with the rapid development in the field of plant genomics, the recent technological advances driving the "omics" science are also noteworthy (Bohra 2013). It is anticipated that the growing fields of proteomics, metabolomics, ionomics, and so forth would spectacularly supplement the crop biofortification. Interrogating "metabolic phenotypes" or "metabotype" in detail via genome-wide genomics approaches intends to illuminate the underlying genetic mechanism and intricate molecular network (Fiehn et al. 2000; Wen et al. 2014). According to Fernie and Schauer (2009), metabolomics involves metabolite profiling which is defined as the detailed characterization of the entire metabolites extracted from the cell. An array of strategies is being adopted to quantify the plant metabolites which primarily are based on either mass spectrometry (MS) or nuclear magnetic resonance (NMR). Fernie and Schauer (2009) have provided a brief review on these metabolite-profiling methods and have proposed metabolomics-assisted breeding as a compelling supplement to various breeding strategies owing to its relatively cost-effective nature compared to transcriptomics and other emerging "-omics" technologies.

Although majority of the metabolite QTLs identified to date have used model plant systems like *Arabidopsis*, this approach is rapidly expanding towards other plant species especially the staple crops. Recently, over 1000 SMTAs were established through applying metabolite-based GWAS across 702 maize lines, and the detected loci were further validated by resequencing, expression analysis (e-QTL), and family-based linkage analysis in two RILs (Wen et al. 2014). Collaborative attempts such as *META*bolomics for *P*lants, *H*ealth and *O*ut*R*each (META-PHOR) (http://www.meta-phor.eu/) were initiated in recent years to establish community-driven platforms that enable large-scale and rapid metabolite profiling and to develop public databases that serve as a global inventory for a wide variety of plant metabolites. In a similar fashion, ionome profiling has been suggested by Salt et al. (2008) as a powerful and high-throughput technique to assist functional analysis of "ionome" influencing genes and their complex network and to offer deeper insights about the overall physiological mechanism. Increasing attention is being placed towards development of high-throughput ionomics platforms such as available at IPK, Gatersleben, Germany, and The Purdue Ionomics Information Management System (PiiMS) (Baxter et al. 2007).

8.7 Conclusion

Taken the context of expanding volume of malnourished people worldwide, the crop biofortification offers the economic, easily accessible, and reasonable solution to develop the improved crop variety with nutritionally dense grains. Obviously, staple crops that satisfy the dietary needs of the maximum population in the developing world have been targeted for biofortification using various modern tools and technologies. The recent advances in omics science including genomics, proteomics, metabolomics, and ionomics greatly improve our understanding about causative genomic regions and the associated molecular network and further developments are likely to refine the existing knowledge. As already mentioned earlier, greater extent of exploitable genetic variation has been documented across gene pools in a wide range of crops. An accelerated incorporation of the gene(s)/QTLs that enhance the nutrient content in staple crops will largely determine the success of various nutritional breeding schemes. Consideration should also be taken about the bioavailability of the enhance nutrient. Therefore, collaborative research efforts involving breeders, biotechnologists, physiologists, biochemists, and, importantly, the nutritionists are needed to strengthen the biofortification programs severalfold in order to meet the challenge of attaining the nutritional security worldwide.

Acknowledgment The authors acknowledge the support from the Indian Council of Agricultural Research (ICAR), New Delhi, India.

References

Agarwal S, Venkata VGNT, Kotla A, Mangrauthia SK, Neelamraju S (2014) Expression patterns of QTL based and other candidate genes in Madhukar × Swarna RILs with contrasting levels of iron and zinc in unpolished rice grains. Gene 546:430–436

Ahmad I, Mohammad F, Zeb A, Noorka IR, Farhatullah, Jadoon SA (2013) Determination and inheritance of phytic acid as marker in diverse genetic group of bread wheat. Am J Mol Biol 3:158–164

Anandan A, Rajiv G, Prakash M (2011) Genotypic variation and relationships between quality traits and trace elements in traditional and improved rice (*Oryza sativa* L.) genotypes. J Food Sci 76:H122–H129

Anuradha K, Agarwal S, Batchu AK, Babu AP, Swamy BPM, Longvah T, Sarla N (2012a) Evaluating rice germplasm for iron and zinc concentration in brown rice and seed dimensions. J Phytol 4:19–25

Anuradha K, Agarwal S, Rao YV, Rao KV, Viraktamath BC, Sarla N (2012b) Mapping QTLs and candidate genes for iron and zinc concentrations in unpolished rice of Madhukar × Swarna RILs. Gene 508:233–240

Azmach G, Gedil M, Menkir A, Spillane C (2013) Marker-trait association analysis of functional gene markers for provitamin A levels across diverse tropical yellow maize inbred lines. BMC Plant Biol 13:227

Babu R, Rojas NP, Gao S, Yan J, Pixley K (2012) Validation of the effects of molecular marker polymorphisms in LcyE and CrtRB1 on provitamin A concentrations for 26 tropical maize populations. Theor Appl Genet 126:389–399

Banerjee S, Sharma DJ, Verulkar SB, Chandel G (2010) Use of *in silico* and semiquantitative RT-PCR approaches to develop nutrient rich rice (*Oryza sativa* L.). Indian J Biotechnol 9:203–212

Baxter I, Ouzzani M, Orcun S, Kennedy B, Jandhyala SS, Salt DE (2007) Purdue ionomics information management system. An integrated functional genomics platform. Plant Physiol 143:600–6100

Baxter IR, Gustin JL, Settles AM, Hoekenga OA (2012) Ionomic characterization of maize kernels in the intermated B73 × Mo17 population. Crop Sci 53:208–220

Bohra A (2013) Emerging paradigms in genomics-based crop improvement. Sci World J 2013, 585467

Bohra A, Pandey MK, Jha UC, Singh B, Singh IP, Datta D, Chaturvedi SK, Nadarajan N, Varshney RK (2014a) Genomics-assisted breeding in the four major pulse crops of developing countries: present status and prospects. Theor Appl Genet 127:1263–1291

Bohra A, Jha UC, Kavi Kishor PB, Pandey S, Singh NP (2014b) Genomics and molecular breeding in lesser explored pulse crops: current trends and future opportunities. Biotechnol Adv 32:1410–1428

Bohra A, Sahrawat KL, Kumar S, Joshi R, Parihar AK, Singh U, Singh D, Singh NP (2015) Genetics- and genomics-based interventions for nutritional enhancement of grain legume crops: status and outlook. J Appl Genet 56:151–61

Borill P, Connorton JM, Balk J, Miller AJ, Sanders D, Uauy C (2014) Biofortification of wheat grain with iron and zinc: integrating novel genomic resources and knowledge from model crops. Front Plant Sci 5:53

Cakmak I, Ozkan H, Braun HJ, Welch RM, Romheld V (2000) Zinc and iron concentrations in seeds of wild, primitive and modern wheats. Food Nutr Bull 21:401e403

Cakmak I, Torun A, Özkan H, Millet E, Feldman M, Fahima T, Korol AB, Nevo E, Braun HJ, Ozkan H (2004) *Triticum dicoccoides*: an important genetic resource for increasing zinc and iron concentration in modern cultivated wheat. Soil Sci Plant Nutr 50:1047–1054

Calderini DF, Monasterio IO (2003) Are synthetic hexaploids a means of increasing grain element concentrations in wheat? Euphytica 134:169–178

Carvalho S, Vasconcelos M (2013) Producing more with less: 'omics' at the service of plant-based food biofortification. Food Res Int 54:961–971

Cavanagh C, Morell M, Mackay I, Powell W (2008) From mutations to MAGIC: resources for gene discovery, validation and delivery in crop plants. Curr Opin Plant Biol 11:215–221

Chandler K, Lipka AE, Owens BF, Li H, Buckler ES, Rocheford T, Gore MA (2013) Genetic analysis of visually scored orange kernel color in maize. Crop Sci 53:189–200

Chatzav M, Peleg Z, Ozturk L, Yazici A, Fahima T, Cakmak I, Saranga Y (2011) Genetic diversity for grain nutrients in wild emmer wheat: potential for wheat improvement. Ann Bot 105:1211–1220

Chhuneja P, Dhaliwal HS, Bains NS, Singh K (2006) *Aegilops kotschyi* and *Ae. tauschii* are the sources for high grain iron and zinc. Plant Breed 125:1–3

Chiangmai PN, Yodmingkhwan P, Nilprapruck P, Aekatasanawan C, Kanjanamaneesathian M (2011) Screening of phytic acid and inorganic phosphorus contents in corn inbred lines and F_1 hybrids in tropical environment. Maydica 54:331–339

Chilimba ADC, Young SD, Black CR, Rogerson KB, Ander EL, Watts MJ, Lammel J, Broadley MR (2011) Maize grain and soil surveys reveal suboptimal dietary selenium intake is widespread in Malawi. Sci Rep 1:72

Cobb JN, DeClerck G, Greenberg A, Clark R, McCouch S (2013) Next-generation phenotyping: requirements and strategies for enhancing our understanding of genotype–phenotype relationships and its relevance to crop improvement. Theor Appl Genet 126:1–21

Crossa J, de L Campos G, Pérez P et al (2010) Prediction of genetic values of quantitative traits in plant breeding using pedigree and molecular markers. Genetics 186:713–724

Crossa J, Pérez-Rodríguez P, Hickey J, Burgueño J, Ornella L, Cerón-Rojas J et al. (2013). Genomic prediction in CIMMYT maize and wheat breeding programs. Heredity 112:48–60

Dai F, Wang J, Zhang S, Xu Z, Zhang G (2007) Genotypic and environmental variation in phytic acid content and its relation to protein content and malt quality in barley. Food Chem 105:606–611

Daneshbakhsh B, Khoshgoftarmanesh AH, Shariatmadari H, Cakmak I (2013) Phytosiderophore release by wheat genotypes differing in zinc deficiency tolerance grown with Zn-free nutrient solution as affected by salinity. J Plant Physiol 170:41–46

de Souza GA, de Carvalho JG, Rutzke M, Albrecht JC, Guilherme LR, Li L (2013) Evaluation of germplasm effect on Fe, Zn and Se content in wheat seedlings. Plant Sci 210:206–213

Desta ZA, Ortiz R (2014) Genomic selection: genome-wide prediction in plant improvement. Trends Plant Sci 19:592–601

Distelfeld A, Cakmak I, Peleg Z, Ozturk L, Yazici AM, Budak H, Saranga Y, Fahima T (2007) Multiple QTL-effects of wheat *Gpc-B1* locus on grain protein and micronutrient concentrations. Physiol Plant 129:635–643

Drinic SM, Ristic D, Sredojevic S, Dragicevic V, Ignjatovic Micic D, Delic N (2009) Genetic variation of phytate and inorganic phosphorus in maize population. Genetika 41:107–115

Drinic SM, Andjelkovic V, Micic DI (2012) Genetic diversity of maize landraces as sources of favorable traits. In: Caliskan M (ed) The molecular basis of plant genetic diversity. InTech, Rijeka, pp 89–112

Evans A (2009) The feeding of the nine billion: global food security. Chatham House, London

Febles CI, Arias A, Hardisson A, Rodriguez-Alvarez C, Sierra A (2002) Phytic acid level in wheat flours. J Cereal Sci 36:19–23

Feil B, Fossati D (1995) Mineral composition of triticale grains as related to grain yield and grain protein. Crop Sci 35:1426–1431

Fernandez MGS, Kapran I, Souley S, Abdou M, Maiga IH, Acharya CB, Hamblin MT, Kreso- vich S (2009) Collection and characterization of yellow endosperm sorghums from West Africa forbio- fortification. Genet Resour Crop Evol 56:991–1000

Fernie AR, Schauer N (2009) Metabolomics-assisted breeding: a viable option for crop improvement? Trends Genet 25:39–48

Ficco DBM, Riefolo C, Nicastro G, De Simone V, DiGesu AM, Beleggia R, Platini C, Cattivelli L, De Vita P (2009) Phytate and mineral elements concentration in a collection of Italian durum wheat cultivars. Field Crop Res 111:235–242

Fiehn O, Kopka J, Dörmann P, Altmann T, Trethewey RN, Willmitzer L (2000) Metabolite profiling for plant functional genomics. Nat Biotechnol 18:1157–1161

Gande NK, Kundur PJ, Soman R, Ambati R, Ashwathanarayana R, Bekele BD, Shashidhar HE (2014) Identification of putative candidate gene markers for grain zinc content using recombinant inbred lines (RIL) population of IRRI38 X Jeerigesanna. Afr J Biotechnol 13:657–663

Garcia-Oliveira AL, Tan LB, Fu YC, Sun CQ (2009) Genetic identification of quantitative trait loci for contents of mineral nutrients in rice grain. J Integr Plant Biol 51:84–92

Genc Y, Verbyla AP, Torun AA, Cakmak I, Willsmore K, Wallwork H, McDonald GK (2009) Quantitative trait loci analysis of zinc efficiency and grain zinc concentration in wheat using whole genome average interval mapping. Plant Soil 310:67–75

Genc Y, Taylor J, Rongala J, Oldach K (2014) A major locus for chloride accumulation on chromosome 5A in bread wheat. PLoS One 9, e98845

Glahn RP, Cheng Z, Welch RM, Gregorio GB (2002) Comparison of iron bioavailability from 15 rice genotypes: studies using an in vitro digestion/caco-2 cell culture model. J Agric Food Chem 50:3586–3591

Goddard ME, Hayes BJ (2007) Genomic selection. J Anim Breed Genet 124:323–330

Graham R, Senadhira D, Beebe S, Iglesias C, Monasterio I (1999) Breeding for micronutrient density in edible portions of staple food crops: conventional approaches. Field Crop Res 60:57–80

Graham RD, Welch RM, Bouis HE (2001) Addressing micronutrient malnutrition through enhancing the nutritional quality of staple foods: principles, perspectives and knowledge gaps. Adv Agron 70:77–142

Gregorio GB, Senadhira D, Htut H, Graham RD (2000) Breeding for trace mineral density in rice. Food Nutr Bull 21:382–386

Hamblin MT, Buckler ES, Jannink JL (2011) Population genetics of genomics-based crop improvement methods. Trends Genet 27:98–106

He WL, Feng Y, Wei YY, Yang XE, Shi CH, He ZL, Stoffella PJ (2013) Differential iron-bioavailability with relation to nutrient compositions in polished rice among selected Chinese genotypes using Caco-2 cell culture model. Int J Food Sci Nutr 64:822–829

Heffner EL, Lorenz AJ, Jannink J-L, Sorrells ME (2010) Plant breeding with genomic selection: gain per unit time and cost. Crop Sci 50:1681–1690

Heffner EL, Jannink J-L, Iwata H, Souza E, Sorrells ME (2011) Genomic selection accuracy for grain quality traits in biparental wheat populations. Crop Sci 51:2597–2606

Heslot N, Yang HP, Sorrells ME, Jannink JL (2012) Genomic selection in plant breeding: a comparison of models. Crop Sci 52:146

Jahan GS, Hassan L, Begum SN, Islam SN (2013) Identification of iron rich rice genotypes in Bangladesh using chemical analysis. J Bangladesh Agric Univ 11:73–78

Jarquín D, Kocak K, Posadas L, Hyma K, Jedlicka J, Graef G, Lorenz A (2014) Genotyping by sequencing for genomic prediction in a soybean breeding population. BMC Genomics 15:740

Jegadeesan S, Yu KF, Poysa V, Gawalko E, Morrison MJ, Shi C, Cober E (2010) Mapping and validation of simple sequence repeat markers linked to a major gene controlling seed cadmium accumulation in soybean [*Glycine max* (L.) *Merr*]. Theor Appl Genet 121:283–294

Jiang WZ (2008) Comparison of responses to Mn deficiency between the UK wheat genotypes Maris Butler, Paragon and the Australian wheat genotype C8MM. J Integr Plant Biol 50:457–465

Jiang WZ, Ireland CR (2005) Characterization of manganese use efficiency in UK wheat cultivars grown in a solution culture system and in the field. J Agric Sci 143:151–160

Jin T, Zhou J, Chen J, Zhu L, Zhao Y, Huang Y (2013) The genetic architecture of zinc and iron content in maize grains as revealed by QTL mapping and meta-analysis. Breed Sci 63(3):317–324

Jonas E, Koning DJ (2013) Does genomic selection have a future in plant breeding? Trends Biotechnol 31:497–504

Joshi AK, Crossa J, Arun B, Chand R, Trethown R, Vargas M, Ortiz-Monasterio I (2010) Genotype × environment interaction for zinc and iron concentration of wheat grain in eastern Gangetic plains of India. Field Crop Res 116:268–277

Kandianis CB, Stevens R, Liu W, Palacios N, Montgomery K, Pixley K, White WS, Rocheford T (2013) Genetic architecture controlling variation in grain carotenoid composition and concentrations in two maize populations. Theor Appl Genet 126:2879–2895

Kumar AA, Reddy BVS, Sahrawat KL, Ramaiah B (2010) Combating micronutrient malnutrition: identification of commercial sorghum cultivars with high grain iron and zinc. SAT eJournal 8:1–5

Li S, Tayie FAK, Young MF, Rocheford T, White WS (2007) Retention of provitamin A carotenoids in high β-carotene maize (*Zea mays*) during traditional African household processing. J Agric Food Chem 55:10744–10750

Li H, Liu H, Han Y, Wu X, Teng W, Liu G, Li W (2010) Identification of QTL underlying vitamin E contents in soybean seed among multiple environments. Theor Appl Genet 120:1405–1413

Lipka AE, Gore MA, Magallanes-Lundback M, Mesberg A, Lin H, Tiede T, Chen C, Buell CR, Buckler ES, Rocheford T, DellaPenna D (2013a) Genome-wide association study and pathway-level analysis of tocochromanol levels in maize grain. G3: Genes Genomes Genet 3(8):1287–1299

Lipka AE, Magallanes-Lundback M, Mesberg A, Bradbury P, Angelovici R et al (2013b) High-resolution mapping of tocochromanol and carotenoid grain traits *via* NAM-GWAS reveals distinct genetic architectures in maize. Maize Genet Conf Abstr 54: P282

Lonergan PF, Pallotta MA, Lorimer M, Paull JG, Barker SJ, Graham RD (2009) Multiple genetic loci for zinc uptake and distribution in barley (*Hordeum vulgare*). New Phytol 184:168–179

Lorenz AJ, Scott MP, Lamkey KR (2008) Genetic variation and breeding potential of phytate and inorganic phosphorus in a maize population. Crop Sci 48:79–84

Lorenzana RE, Bernardo R (2009) Accuracy of genotypic value predictions for marker-based selection in biparental plant populations. Theor Appl Genet 120:151–161

Lung'aho MG, Mwaniki AM, Szalma SJ, Hart JJ, Rutzke MA et al (2011) Genetic and physiological analysis of iron biofortification in maize kernels. PLoS One 6(6), e20429

Lyons G, Ortiz-Monasterio I, Stangoulis J, Graham R (2005) Selenium concentration in wheat grain: is there sufficient genotypic variation to use in breeding? Plant Soil 269:369–380

Mackay IJ, Bansept-Basler P, Barber T, Bentley AR, Cockram J, Gosman N, Greenland AJ, Horsnell R, Howells R, O'Sullivan D, Rose GA, Howell P (2014) An eight-parent multiparent advanced generation inter-cross population for winter-sown wheat: creation, properties and validation. G3: Genes Genomes Genet 4:1603–6010

Mallick SA, Azaz K, Gupta M, Sharma V, Sinha BK (2013) Characterization of grain nutritional quality in wheat. Indian J Plant Physiol 18:183–186

Mamo BE, Barber B, Steffenson BJ (2014) Genome–wide association mapping of zinc and iron concentration in barley landraces from Ethiopia and Eritrea. J Cereal Sci 60:497–506

Maziya-Dixon B, Kling JG, Menkir A, Dixon A (2000) Genetic variation in total carotene, iron, and zinc contents of maize and cassava genotypes. Food Nutr Bull 21:419–422

Meng F, Wei Y, Yang X (2005) Iron content and bioavailability in rice. J Trace Elem Med Biol 18:333–338

Menkir A, Liu W, White WS, Maziya-Dixon B, Rocheford T (2008) Carotenoid diversity in tropical-adapted yellow maize inbred lines. Food Chem 109:521–529

Meuwissen THE, Hayes BJ, Goddard ME (2001) Prediction of total genetic value using genome-wide dense marker maps. Genetics 157:1819–1829

Monasterio I, Graham RD (2000) Breeding for trace minerals in wheat. Food Nutr Bull 21:392–396

Murgia I, Arosio P, Tarantino D, Suoave C (2012) Biofortification for combating 'hidden hunger' for iron. Trends Plant Sci 17:47–55

Norton GJ, Deacon CM, Xiong L, Huang S, Meharg AA et al (2010) Genetic mapping of the rice ionome in leaves and grain: identification of QTLs for 17 elements including arsenic, cadmium, iron and selenium. Plant Soil 329:139–153

Norton GJ, Douglas A, Lahner B, Yakubova E, Guerinot ML et al (2014) Genome wide association mapping of grain arsenic, copper, molybdenum and zinc in rice (*Oryza sativa* L.) grown at four international field sites. PLoS One 9, e89685

Oikeh SO, Menkir A, Maziya-Dixon B, Welch R, Glahn RP (2003) Assessment of concentrations of iron and zinc and bioavailable iron in grains of early-maturing tropical maize varieties. J Agric Food Chem 51:3688–3694

Ortiz-Monasterio JI, Palacios-Rojas N, Meng E, Pixley K, Trethowan R, Pena RJ (2007) Enhancing the mineral and vitamin content of wheat and maize through plant breeding. J Cereal Sci 46:293–307

Oury FX, Leenhard F, Remesy C, Chanliaud E, Duperrier B, Balfourier F, Charmet G (2006) Genetic variability and stability of grain magnesium, zinc and iron concentrations in bread wheat. Eur J Agron 25:177–185

Owens BF, Lipka AE, Magallanes-Lundback M, Tiede T, Diepenbrock CH, Kandianis CB, Kim E, Cepela J, Mateos-Hernandez M, Buell CR, Buckler ES, DellaPenna D, Gore MA, Rocheford T (2014) A foundation for provitamin a biofortification of maize: genome-wide association and genomic prediction models of carotenoid levels. Genetics 198:1699–1716

Ozkan H, Brandolini A, Torun A, Altintas S, Eker S, Kilian B, Braun H, Salamini F, Cakmak I (2007) Natural variation and identification of microelements content in seeds of einkorn wheat (*Triticum monococcum*). In: Buck HT, Nisi JE, Salomon N (eds) Wheat production in stressed environments. Springer, Berlin, pp 455–462

Peleg Z, Cakmak I, Ozturk E, Yazici A, Jun Y, Budak H, Korol AB, Fahima T, Saranga Y (2008a) Quantitative trait loci conferring grain mineral nutrient concentrations in durum wheat 3 wild emmer wheat RIL population. Theor Appl Genet 119:353–369

Peleg Z, Saranga Y, Yazici A, Fahima T, Ozturk L, Cakmak I (2008b) Grain zinc, iron and protein concentrations and zinc-efficiency in wild emmer wheat under contrasting irrigation regimes. Plant Soil 306:57–67

Peleg Z, Cakmak I, Ozturk L, Yazici A, Jun Y, Budak H, Korol AB, Fahima T, Saranga Y (2009) Quantitative trait loci conferring grain mineral nutrient concentrations in durum wheat wild emmer wheat RIL population. Theor Appl Genet 119:353–369

Pinson SRM, Tarpley L, Yan W, Yeater K, Lahner B, Yakubova E, Huang XY, Zhang M, Guerinot ML, Salt DE (2014) Worldwide genetic diversity for mineral element concentrations in rice grain. Crop Sci 55:1–18

Poland JA, Rife TW (2012) Genotyping-by-sequencing for plant breeding and genetics. Plant Genome 5:92–102

Poland J, Endelman J, Dawson J, Rutkoski J, Wu S, Manes Y, Dreisigacker S, Crossa J, Sánchez-Villeda H, Sorrells M, Jannink J-L (2012) Genomic selection in wheat breeding using genotyping-by-sequencing. Plant Genome 5:103

Prasanna BM, Mazumdar S, Chakraborti M, Hossain F, Manjaiah KM, Agarwal PK, Guleria SK, Gupta HS (2011) Genetic variability and genotype × environment interactions for kernel iron and zinc concentrations in maize (*Zea mays*) genotypes. Indian J Agric Sci 81:704–711

Prom-u-thai C, Rerkasem B (2001) Iron concentration in Thai rice germplasm. Dev Plant Soil Sci 92:351–352

Prom-u-thai C, Fukai S, Godwin ID, Huang L (2007) Genotypic variation of iron partitioning in rice grain. J Sci Food Agric 87:2049–2054

Pu Z, Yu M, He Q, Chen G, Wang J, Liu Y, Jiang Q, Li W, Dai S, Wei Y, Zheng Y (2014) Quantitative trait loci associated with micronutrient concentrations in two recombinant inbred wheat lines. J Integr Agric 13:2322–2329

Pucher A, Høgh-Jensen H, Gondah J, Hash CT, Haussmann BIG (2014) Micronutrient density and stability in West African pearl millet – potential for biofortification. Crop Sci 54:1709–1720

Qin H, Cai Y, Liu Z, Wang G, Wang J, Guo Y, Wang H (2012) Identification of QTL for zinc and iron concentration in maize kernel and cob. Euphytica 187:345–358

Rai KN, Govindraj M, Rao AS (2012) Genetic enhancement of grain iron and zinc content in pearl millet. Qual Assur Saf Crops Foods 4:119–125

Rawat N, Tiwari VK, Singh N, Randhawa GS, Singh K, Chhuneja P, Dhaliwal HS (2008) Evaluation and utilization of *Aegilops* and wild *Triticum* species for enhancing iron and zinc content in wheat. Genet Resour Crop Evol 56:53–64

Rawat N, Tiwari VK, Singh N, Randhawa GS, Singh K, Chhuneja P, Dhaliwal HS (2009) Evaluation and utilization of *Aegilops* and wild *Triticum* species for enhancing iron and zinc content in wheat. Genet Resour Crop Evol 56:53–64

Saenchai C, Prom-u-thai C, Jamjod S, Dell B, Rerkasem B (2013) Genotypic variation in milling depression of iron and zinc concentration in rice grain. Plant Soil 361:271–278

Salt DE, Baxter I, Lahner B (2008) Ionomics and the study of the plant ionome. Annu Rev Plant Biol 59:709–733

Salunke R, Neelam K, Rawat N, Tiwari VK, Randhawa GS, Dhaliwal HS, Roy P (2011) Bioavailability of iron from wheat *Aegilops* derivatives selected for high grain iron and protein contents. J Agric Food Chem 59:7465–7473

Senadhira D, Gregorio G, Graham RD (1998) Breeding iron and zinc dense rice. Paper presented at the international workshop on micronutrient enhancement of rice for developing countries, Rice Research and Extension Center, Stuttgart, pp 1–29, 3 Sept 1998

Shi RL, Li HW, Tong YP, Jing RL, Zhang FS, Zou CQ (2008) Identification of quantitative trait locus of zinc and phosphorus density in wheat (*Triticum aestivum* L.) grain. Plant Soil 306:95–104

Simić D, Drinić SM, Zdunić Z, Jambrović A, Ledenčan T, Brkić J, Brkić A, Brkić I (2012) Quantitative trait loci for biofortification traits in maize grain. J Hered 103:47–54

Singh SP, Vogel-Mikus K, Arcon I, Vavpetic P, Jeromel L, Pelicon P, Kumar J, Tuli R (2013) Pattern of iron distribution in maternal and filial tissues in wheat grains with contrasting levels of iron. J Exp Bot 64:3249–3260

Souza GA, Hart JJ, Carvalho JG, Rutzke MA, Albrecht JC, Guilherme LR, Kochian LV, Li L (2014) Genotypic variation of zinc and selenium concentration in grains of Brazilian wheat lines. Plant Sci 224:27–35

Sperotto RA, Boff T, Duarte GL, Santos LS, Grusak MA, Fett JP (2010) Identification of putative target genes to manipulate Fe and Zn concentrations in rice grains. J Plant Physiol 167:1500–1506

Srinivasa J, Arun B, Mishra VK, Singh GP, Velu G, Babu R, Vasistha NK, Joshi AK (2014) Zinc and iron concentration QTL mapped in a Triticum spelta × T. aestivum cross. Theor Appl Genet 127:1643–1651.

Suwarno WB, Pixley KV, Palacios-Rojas N, Kaeppler SM, Babu R (2014) Formation of heterotic groups and understanding genetic effects in a provitamin A biofortified maize breeding program. Crop Sci 54:14–24

Tako E, Hoekenga OA, Kochian LV, Glahn RP (2013) High bioavailability iron maize (*Zea mays* L.) developed through molecular breeding provides more absorbable iron in vitro (Caco-2 model) and in vivo (Gallus gallus). Nutr J 12:3

Tavajjoh M, Yasrebi J, Karimian N et al (2011) Phytic acid concentration and phytic acid: zinc molar ratio in wheat cultivars and bread flours, Fars province, Iran. J Agric Sci Technol 13:743–755

Tiwari VK, Rawat N, Chhuneja P, Neelam K, Aggarwal R, Randhawa GS, Dhaliwal HS, Keller B, Singh K (2009) Mapping of quantitative trait loci for grain iron and zinc concentration in diploid A genome wheat. J Hered 100:771–776

Tiwari VK, Rawat N, Neelam K, Kumar S, Randhawa GS, Dhaliwal HS (2010) Substitutions of 2S and 7U chromosomes of *Aegilops kotschyi* in wheat enhance grain iron and zinc concentration. Theor Appl Genet 121:259–269

Uauy C, Distelfeld A, Fahima T, Blechl A, Dubcovsky J (2006) A NAC gene regulating senescence improves grain protein, zinc, and iron content in wheat. Science 314:1298–1301

Varshney RK, Kudapa H, Pazhamala L, Chitikineni A et al (2015) Translational genomics in agriculture: some examples in grain legumes. Crit Rev Plant Sci 34:169–194

Velu G, Rai KN, Muralidharan V, Kulkarni VN, Longvah T, Raveendran TS (2007) Prospects of breeding biofortified pearl millet with high grain iron and zinc content. Plant Breed. doi:10.1111/j.1439-0523.2007.01322.x

Velu G, Ortiz-Monastero I, Singh RP, Payne T (2011a) Variation for grain micronutrients concentration in wheat core-collection accessions of diverse origin. Asian J Crop Sci 3:43–48

Velu G, Singh R, Huerta-Espino J, Pena J, Ortiz-Monasterio I (2011b) Breeding for enhanced zinc and iron concentration in CIMMYT spring wheat germplasm. Czech J Genet Plant Breed 47:S174–S177

Velu G, Singh RP, Huerta-Espino J, Pena RJ, Arun B, Mahendru-Singh A, Mujahid MY, Sohu VS, Mavi GS, Crossa J, Alvarado G, Joshi AK, Pfeiffer WH (2012) Performance of biofortified spring wheat genotypes in target environments for grain zinc and iron concentrations. Field Crop Res 137:261–267

Velu G, Ortiz-Monasterio I, Cakmak I, Hao Y, Singh RP (2014) Biofortification strategies to increase grain zinc and iron concentrations in wheat. J Cereal Sci 59:365–372

Vignesh M, Hossain F, Nepolean T, Saha S, Agrawal PK, Guleria SK, Prasanna BM, Gupta HS (2012) Genetic variability for kernel β-carotene and utilization of *crtRB1* 3'TE gene for biofortification in maize (*Zea mays* L.). Indian J Genet 72:189–194

Welch RM, Graham RD (2002) Breeding crops for enhanced micronutrient content. Plant Soil 245:205–214

Wen W, Li D, Li X, Gao Y, Li W et al (2014) Metabolome-based genome-wide association study of maize kernel leads to novel biochemical insights. Nat Commun 5:3438

White PJ, Broadley MR (2009) Biofortification of crops with seven mineral elements often lacking in human diets – iron, zinc, copper, calcium, magnesium, selenium and iodine. New Phytol 182:49–84

Würschum T, Reif JC, Kraft T, Janssen G, Zhao YS (2013) Genomic selection in sugar beet breeding populations. BMC Genet 14:85

Xu YF, An DG, Liu DC, Zhang AM, Xu HX, Li B (2012) Molecular mapping of QTLs for grain zinc, iron and protein concentration of wheat across two environments. Field Crop Res 138:57–62

Yu JM, Holland JB, McMullen MD et al (2008) Genetic design and statistical power of nested association mapping in maize. Genetics 178:539–551

Zdunic Z, Grljušić S, Ledenćan T, Duvnjak T, Šimić D (2014) Quantitative trait loci mapping of metal concentrations in leaves of the maize IBM population. Hereditas 151:55–60

Zhang B, Chen P, Shi A, Hou A, Ishibashi T, Wang D (2009) Putative quantitative trait loci associated with calcium content in soybean seed. J Hered 100:263–269

Zhang X, Zhang G, Guo L, Wang H, Zeng D et al (2011) Identification of quantitative trait loci for Cd and Zn concentrations of brown rice grown in Cd-polluted soils. Euphytica 180:173–179

Zhang M, Pinson SR, Tarpley L, Huang XY, Lahner B, Yakubova E, Baxter I, Guerinot ML, Salt DE (2014) Mapping and validation of quantitative trait loci associated with concentrations of 16 elements in unmilled rice grain. Theor Appl Genet 127:137–165.

Zhao K, Tung CW, Eizenga GC, Wright MH, Ali ML, Price AH, Norton GJ, Islam MR, Reynolds A, Mezey J, McClung AM, Bustamante CD, McCouch SR (2011) Genome-wide association mapping reveals a rich genetic architecture of complex traits in *Oryza sativa*. Nat Commun 2:467

Zhong S, Dekkers JCM, Fernando RL, Jannink JL (2009) Factors affecting accuracy from genomic selection in populations derived from multiple inbred lines: a barley case study. Genetics 182:355–364

Zhou JF, Huang YQ, Liu ZZ, Chen JT, Zhu LY, Song ZQ, Zhao YF (2010) Genetic analysis and QTL mapping of zinc, iron, copper and manganese contents in maize seed. J Plant Genet Resour 11:593–595

Transgenic Strategies Towards Nutritional Enrichment of Crops 9

A. Das, S. Thakur, K.R. Soren, S. Datta, and N P Singh

Abstract

All the essential nutrients in the human diet are ultimately derived from plants. However, most of the major food crops lack certain essential vitamins and minerals. Although a diversified diet regime provides adequate nutrition, majority of the population in developing countries relies on staple crops, such as rice, wheat, maize or cassava, which lacks full complement of essential nutrients. Malnutrition, thus, is a significant humanitarian issue in most of the developing world. A pertinent way to address this challenge is through *biofortification* of crops to increase their essential nutrient content. Transgenic approaches offer the most rapid and precise way to develop high-nutrient crops, thus complementing mineral fertilization and conventional breeding towards ameliorating the scourge.

Keywords

Biofortification • Minerals • Transgenic strategies

9.1 Introduction

Half of the human population worldwide has limited access to a healthy and fresh food. According to estimates for the period, 2011–2013, 842 million people worldwide (nearly one in eight people) suffered from chronic hunger (FAO 2014). In developing countries, the situation is further aggravated by limited fruit, vegetables, meat and fish, leading to reliance on a staple diet of cereals, such as rice, wheat, maize, cassava, etc. Milled cereal grains are poor sources of vitamins and minerals which are essential for normal growth and metabolism. Even in the developed countries, lifestyle choices

A. Das (✉) • S. Thakur • K.R. Soren
Division of Plant Biotechnology, ICAR-Indian Institute of Pulses Research, Kanpur, Uttar Pradesh 208 024, India
e-mail: alokbio@gmail.com; shallu.thakur85@gmail.com; sorenars@gmail.com

S. Datta
Division of Biotechnology, ICAR-Central Research Institute for Jute and Allied Fibres, Barrackpore, West Bengal 700 120, India
e-mail: subhojit@email.com

N P Singh
Director, ICAR-Indian Institute of Pulses Research, Kanpur, Uttar Pradesh 208 024, India
e-mail: npsingh.iipr@gmail.com

U. Singh et al. (eds.), *Biofortification of Food Crops*, DOI 10.1007/978-81-322-2716-8_9

and lack of education can lead to an improper diet, although this can be addressed to a certain degree by dietary supplements and fortification programs. Therefore, undernourishment/malnutrition and the deficiency diseases (e.g. anaemia, night blindness, beriberi, pellagra, scurvy, rickets, etc.) remain a significant public health challenge. With food security a major focus, it is timely to look at the strategies available to scientists to improve the nutritional value of food.

Dietary diversity and supplementation with vitamins and minerals are possibly the best ways to tackle malnutrition, but these appear to be impractical. An alternative approach is to tackle the problem of nutritionally poor crops at its source by increasing their nutritional qualities through a strategy known as 'biofortification', which should translate into improved diets. Several different strategies for biofortification have been advocated. The simplest strategy relies on an increase of the mineral content of plants through the addition of the appropriate mineral as an inorganic fertilizer. Though this has been reported to be successful in few instances, but it depends on the crop species and cultivar, the mineral itself and the quality and properties of the soil, making the strategy difficult to apply generally. Another strategy to improve the nutrient content of plants is by conventional breeding in combination with mutagenesis. The identification of genetically enriched varieties and the use of marker-assisted selection (MAS) to introgress such traits into widely cultivated, adapted germplasm have also been reported (e.g. golden sweet potato). Mutagenesis has also been used to enhance nutrient levels (e.g. lysine-enriched maize opaque-2 mutant). However, the strategies are slowed down by the time taken to identify useful traits and breed them into elite cultivars.

Transgenic strategies to enhance the nutritional value of crops are now being demonstrated (e.g. Golden Rice). Transgenic strategies differ from other approaches in that novel genetic information is introduced directly into the plant's genome, i.e. plants can be tailored to *green factories* for the synthesis of desired compounds. The approach depends on whether the nutritional compound is synthesized de novo by the plant or obtained from the environment. Organic molecules (amino acids, fatty acids and vitamins) are synthesized by the plant and hence increasing the nutritional value requires some form of *metabolic engineering* with the aim of increasing the amount of these desirable compounds, decreasing the amount of a competitive compound or even extending an existing metabolic pathway to generate a novel product. By contrast, mineral nutrients are obtained by the plant from the environment; therefore, mineral enhancement involves strategies to increase uptake, transport and/or accumulation in edible tissues. The following sections highlight reports of transgenic approaches utilized to enhance the nutrient content of crops.

9.2 Transgenic Approaches

9.2.1 Enrichment of Vitamins

Vitamins are small group of organic compounds that are required in the human diet. While considerable advances have been made in understanding vitamin metabolic pathways in plants, efforts to provide adequate levels are a grand challenge. In this section, two prominent examples of metabolic engineering of vitamins (vitamin A and vitamin E) are discussed.

Vitamin A (retinol) is an essential nutrient needed in small amounts by humans for the normal functioning of the visual system: growth and development, maintenance of epithelial cellular integrity, immune function and reproduction. Humans can synthesize vitamin A, if the precursor molecule β-carotene (also known as pro vitamin A), a pigment found in many plants but not in cereal grains, is provided. Therefore, a strategy was devised to introduce the correct metabolic steps into rice endosperm to facilitate β-carotene biosynthesis (Ye et al. 2000). An initial breakthrough was the development of a rice line expressing a daffodil (*Narcissus pseudonarcissus*) phytoene synthase, enabling the accumulation of the vitamin A precursor, phytoene in the endosperm, followed shortly thereafter by the original 'Golden Rice' variety, expressing two daffodil enzymes and one from *Erwinia uredovora*, which reconstituted the entire pathway enabling the rice endosperm to accumulate β-carotene, resulting in its eponymous golden colour (Fig. 9.1). In the best lines, the grain contained >1.5 mg of β-carotene per gram of dry weight. Subsequently, additional lines have been generated that contain only two recombinant enzymes (daffodil *phytoene synthase* and Erwinia *phytoene desaturase*). Later, 'Golden Rice II' variety was developed in which the daffodil *phytoene synthase* gene is replaced with its more efficient maize homolog, resulting in a 23-fold improvement in β-carotene content (up to 37 mg/g). This has led to similar progress in other crops, like '*super banana*', '*potato*', '*orange cauliflower*', carrots and tomatoes. Strategy for higher expression levels with the β-carotene metabolic pathway transferred to the plastids was also reported (Wurbs et al. 2007).

A recently developed potato variety containing the phytoene synthase, phytoene desaturase and lycopene β-cyclase from *Erwinia herbicola* contained 114 mg carotenoids per gram of dry weight and 47 mg β-carotene per gram of dry weight (Diretto et al. 2007). These studies showed that investigations into alternative gene sources and expression strategies can have a profound effect on achievable β-carotene levels. Although the original Golden Rice line was criticized because of the large amounts of rice that would need to be consumed, the latest fortified potato contains enough β-carotene to provide 50 % of the recommended daily allowance.

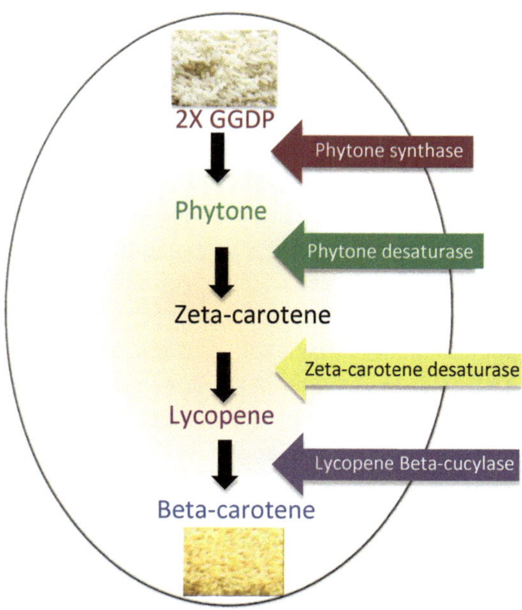

Fig. 9.1 Golden Rice for enrichment of vitamin A

Vitamin E is another group of extremely important compound. Vitamin E is a group of eight hydrophobic compounds, the most potent of which is α-tocopherol. Dietary vitamin E is obtained mainly from plant seeds and functions by preventing oxidation and polymerization of unsaturated fatty acids. Vitamin E deficiency leads to general wasting, kidney degeneration and infertility. In plants, tocopherol synthesis requires input from two metabolic pathways. Levels of vitamin E activity can be increased either by increasing the total amount of vitamin E or by shifting the metabolic flux towards α-tocopherol, as reported in *Arabidopsis*. Expression of *Synechocystis* PCC6803 and *Arabidopsis* genes encoding ϒ-tocopherol methyltransferase (ϒ-TMT) in *Arabidopsis* seeds, resulted in shift from ϒ/δ- to α/β-tocopherol, indicated that nutritional enhancement in plants was possible without altering total vitamin E levels (Shintani and Della-Penna 1998). However, the expression of *Arabidopsis* homogentisic acid prenyltransferase (HPT) produced twice the level of vitamin E found in normal seeds, whereas expression of the *Escherichia coli* tyrA gene, which encodes a dual-function enzyme (chorismate mutase and prephenate dehydrogenase), resulted in up to three times than the normal level of vitamin E (Savidge et al. 2002). Simultaneous expression of *Arabidopsis* genes encoding 2-methyl-6-phytylbenzoquinol (MPBQ) methyltransferase and ϒ-TMT in soybean showed a significant elevation in the total amount of vitamin E activity (fivefold greater than that of wild-type plants), which was attributable mainly to an eightfold increase in the levels of α-tocopherol, from its normal 10 % of total vitamin E to over 95 %.

Recently, plants have been engineered to accumulate several other vitamins like folate and ascorbate (vitamin C). Folate biofortification of lettuce by expression of chicken GTP cyclohydrolase I gene was reported (Nunes et al. 2009). Transgenic multivitamin corn through biofortification of endosperm with three vitamins (vitamin A, folate and ascorbate) representing three distinct metabolic pathways has also been reported (Naqvi et al. 2009).

9.2.2 Enrichment of Essential Amino Acids

Essential (indispensable) amino acids cannot be synthesized de novo by humans and therefore must be supplied by diet. The nine amino acids considered essential are phenylalanine, valine, threonine, tryptophan, methionine, leucine, isoleucine, lysine and histidine. Most crops are deficient in certain essential amino acids like cereal grains, rice and wheat are deficient in lysine and threonine, whereas legumes such as peas and beans lack methionine and cysteine. Since majority of the world's population relies on cereal and legume combination as a staple diet, there is a great interest in developing novel varieties with increased essential amino acid content.

A simple approach is the expression of recombinant storage proteins with desirable amino acid profiles. Expression of pea (*Pisum sativum*) legumin (high lysine content), in rice and wheat grains, and the expression of sunflower seed albumin (rich in methionine), in the laboratory model lupin (Molvig et al. 1997), are some of the examples. Similarly, AmA1 from the Prince's feather (*Amaranthus hypochondriacus*), which encodes seed albumin, was expressed in potato and was shown to double the protein content and increase the levels of several essential amino acids (Chakroborty et al. 2000). However, this approach has limited success as shown by the expression of sunflower seed albumin in rice (Hagan et al. 2003) and chickpea (Chiaiese et al. 2004). There was little impact on the content of sulphur-containing amino acids, perhaps because of the regulatory mechanism that enables seed storage-protein composition to be adjusted in response to sulphur and nitrogen levels (Tabe et al. 2002). Synthetic proteins (i.e. proteins designed from first principles) can also be expressed to boost the levels of particular amino acids. For example, a synthetic protein matched to human amino acid requirements was expressed in cassava (Zhang et al. 2003). The inability of heterologous proteins to change abruptly and predictably the essential amino acid content of target crops probably reflects the limited free amino acid pool, which provides the substrates for protein synthesis.

As far as metabolic engineering is concerned, an understanding of biosynthetic pathway is very essential. In all higher plants, lysine, threonine and methionine are synthesized from aspartic acid via a highly branched pathway under complex feedback control mechanism. Two key enzymes are aspartate kinase (AK), which functions early in the pathway and is inhibited by both lysine and threonine, and dihydrodipicolinate synthase (DHPS), which functions in the lysine-specific branch and is inhibited by lysine alone. Feedback-insensitive homologs from bacterial system have been expressed in plants resulting in increase in the free lysine content of *Arabidopsis* seeds or by knocking out the lysine catabolism pathway, resulting in 12-fold or fivefold gain in lysine content. However, where both the transgene and knockout were combined in the line, surprisingly, increase of 80-fold over wild-type levels were achieved (Zhu and Galili 2003). Similarly, the expression of DHPS in maize increased the levels of free lysine with concomitant increases in threonine. Analogous approaches have increased the lysine levels in canola and soybean (Falco et al. 1995). Expression of a feedback-insensitive subunit of rice anthranilate synthase in rice resulted in twice accumulation of tryptophan as compared to the wild-type level in the grain (Wakasa et al. 2006). Thus, combination of the heterologous protein and amino acid pool approaches in a single plant may be pivotal to boost the levels of essential amino acids.

9.2.3 Enrichment of Essential and Polyunsaturated Fatty Acids

Essential fatty acids (EFAs) are fatty acids [alpha-linolenic acid (omega-3 fatty acid) and linolenic acid (omega-6 fatty acid)] that humans and other animals require for good health, but cannot synthesize them. Fatty acids are also target for biofortification because some of them are essential nutrients with diverse roles in metabolism, cardiovascular health, inflammatory responses, blood pressure regulation, etc. The metabolic pathway of fatty acid biosynthesis has been demystified (Wu et al. 2005), and transgenic strategies for the modification of oil and fat content in plants are reported, either by enhancing the level of the essential fatty acid linoleic acid and α-linolenic acid or to synthesize polyunsaturated fatty acids (PUFAs) like arachidonic acid (ARA), eicosapentaenoic acid (EPA) and docosahexaenoic acid (DHA), which are usually sourced from fish.

Successful high level accumulation of fish oil omega-3 long-chain polyunsaturated fatty acids (EPA and DHA) was reported in oilseed crop, *Camelina sativa* (Ruiz-Lopez et al. 2014). The crop was transformed with five or seven genes for EPA and DHA biosynthesis: Δ6-desaturase gene from *O. tauri*, Δ6 fatty acid elongase gene from *P. patens*, Δ5-desaturase gene from *Thraustochytrium* spp., Δ12-desaturase gene from *P. sojae*, Δ3-desaturase from *P. infestans*, Δ5-fatty acid elongase gene from *O. tauri* and Δ5-desaturase gene from *E. huxleyi*. Genetically engineered seeds accumulate 12 % EPA and 14 % DHA, levels equivalent to those in fish oils, thus representing a sustainable source of fatty acids. However, the first breakthrough report of successful reconstitution of fatty acid synthesis pathway in *Arabidopsis* was obtained using the prokaryotic genes encoding D9-elongase, D8-desaturase and D5-desaturase (Abbadi et al. 2004). The elevated accumulation of VLC-PUFAs seemed to have no deleterious effect on plant growth. Another study exhibited lower enhancement by introducing genes from the conventional biosynthesis pathway under seed-specific promoter driven D6-desaturase, D6-elongase and D5-desaturase. This elevated the total level of D6-desaturated fatty acids, but EPA and ARA each accounted for <1 % of this fatty acid pool, which suggests a bottleneck during elongation; this bottleneck will be the target for future experiments.

Progress has also been made in the reconstitution of the DHA biosynthetic pathway in transgenic plants. *Arabidopsis* seeds with total fatty acids containing up to 0.5 % DHA were produced by expressing a bifunctional zebrafish D6/D5-desaturase and a dipteran D6-elongase in the seed, producing EPA for subsequent conversion into DHA by enzymes from the alga *Pavlova salina*. Utilizing similar strategy, DHA production in transgenic soybean seeds was obtained, but using a D6-elongase from the fungus *Mortierella alpina* and adding a v-3 microsomal desaturase from the fungus *Saprolegnia diclina* to maximize the accumulation of n-3 VLC-PUFAs. Another successful result has come from a study in which the VLC-PUFA biosynthetic pathway was reconstituted in *Brassica juncea* supplemented with a desaturase from *Phytophtora infestans*, a D12-desaturase from *Calendula officinalis* and an acyltransferase from *Thraustochytrium aureum*.

In each of these studies, there has been a striking increase in flux through the VLC-PUFA biosynthetic pathway, leading to the accumulation of EPA and DHA. Further investigation of pathway competition, feedback and branching will allow the development of plants with higher levels of DHA and other essential fatty acids in the future, presenting alternative to dwindling marine sources.

9.2.4 Enrichment of Minerals

Although *metabolic engineering* is the most suitable approach to enrich plants with organic nutrients, a different approach is required for minerals because they are not synthesized in the plant but obtained from the immediate environment. Transgenic strategies to increase the mineral content of crop plants have concentrated mainly on iron and zinc (which are most frequently deficient in human diets) and have used two distinct approaches, viz., increasing the efficiency of uptake and transport to harvestable tissues and increasing the amount of bioavailable mineral accumulating in the plant i.e. how much is accessible after digestion.

Iron, Fe (III), is the most abundant form of iron in the soil; unfortunately, plants cannot absorb iron in this state. Two different pathways are used to convert Fe (III) into Fe (II) for absorption: *Strategy I* involves the expression of Fe(III) reductases and the subsequent absorption of Fe (II) (all other plants), and *Strategy II* involves the secretion of chelating chemicals, called phytosiderophores, that bind to Fe(III) before absorption (graminaceous plants, i.e. grasses and cereals). Specific transport proteins are then used to absorb the minerals into the roots, and they are transported through the phloem to sink tissues, such as leaves in the form of complexes with nicotianamine, which specifically chelate Fe (II). The overexpression of these transport and chelating proteins promotes Fe accumulation. For example, efforts to increase iron uptake in roots by genetic modification have focused on the expression of iron transport proteins (Connolly et al. 2002). Alternatively, iron accumulation can be enhanced by the production of higher levels of phytosiderophores; for instance, the expression of the barley *naat*-A and *naat*-B genes, encoding nicotianamine aminotransferases (involved in phytosiderophore biosynthesis) in rice, resulted in increased iron uptake (Takahashi et al. 2001). Interestingly, there seems to be some crosstalk between the iron and zinc transport pathways because transgenic plants and mutants with overexpressed Fe (III) reductases and iron transporters also show increased zinc accumulation. This could reflect the enhanced synthesis of nicotianamine, which increases the mobilization of both metals in the vascular tissue. Thus, the overexpression of nicotianamine synthase also leads to iron and zinc accumulation; for example, the expression of barley *Hv*NAS1 in tobacco (*Nicotiana tabacum*) doubled the iron and zinc concentrations in leaves (Takahashi et al. 2003).

The second approach to mineral biofortification is to express recombinant proteins that enable minerals to be stored in a bioavailable form, for instance, overexpression of soybean ferritin in rice using an endosperm-specific promoter. Ferritin is an iron-storage protein. This produced rice grains with threefold levels of wild-type iron. Iron concentration was also measured in polished grains, because minerals are lost during polishing, but the levels of iron (and zinc) were still higher than in non-polished wild-type grains. By contrast, the use of a constitutive promoter to drive ferritin expression resulted in elevated iron levels in the leaves of transgenic rice plants, but not in grains, owing to higher expression levels of ferritin in vegetative tissues (Goto et al. 1999).

Another problem with mineral availability is its bioavailability, i.e. amount present in a form that can be utilized and absorbed by the human gut. Phytic acid (also known as phytate) is an anti-nutritional compound that chelates minerals and reduces their bioavailability in the gut. Therefore, a transgenic strategy has been developed that involves the expression of both ferritin and phytase (a fungal enzyme that breaks down phytate). Experiments with transgenic rice and maize indicated that the rice grains contained twice the wild-type amount of iron, and simulations of digestion and absorption using the maize kernels showed that the amount of bioavailable iron had also increased. The combined use of multiple strategies for iron fortification therefore provides the maximum levels of bioavailable iron (Drakakaki et al. 2000, 2005). Similar strategies can also be adopted for other minerals.

9.2.5 Improvement in Protein Quality in Pulses

Pulses are an important component of human nutrition because of high protein content (20–35 %). However, pulses are reported to have limited essential amino acids particularly methionine and cysteine. Strategies like metabolic engineering for increased methionine and cysteine content or/and expression of endogenous storage protein can provide potential alternative in this direction (Singh and Yadav 2010). Key regulatory proteins of essential amino acid biosynthesis pathway like cystathionine-γ-synthase (CGS), aspartate kinase (AK) and dihydrodipicolinate dehydrogenase (DHDPS) have already been engineered with insensitive homologs from other species, leading to increase in free Thr and Lys content (Fig. 9.2) (Zhu and Galili 2003). Alternatively, engineering of endogenous storage protein (like cereal *prolamins* and *glutamines*, legume *globulins*, etc.) by in vitro mutagenesis to mutate appropriate amino acid codons into essential amino acid codons (Met and Lys) or to insert stretches of additional codons of these amino acids appears promising. Inclusion of synthetic protein like ASP1 designed to have stable protein-like structure with desired amino acids profile can be another potential alternative. Single strategy or combination may be adopted for increased enrichment of pulses. However, potential biosafety (allergenicity) of the protein needs to be ascertained before such strategy is adopted.

9.3 Conclusion

Malnutrition is a big challenge particularly in the developing world, where measures like varied diet, fortification schemes and dietary supplements are difficult to realize. Transgenic biofortification strategies could help to alleviate malnutrition; however, regulatory and public perception issues need to be addressed. An extensive biosafety characterization of modified crops needs to be done addressing human health and allergenicity issues. Recent concepts of the use of *synthetic metabolons* could also greatly enhance the efficiency and increase the outcome of metabolic engineering in plants (Singleton et al. 2014). Engineering strategies have been implemented on a handful of plant species only and need to be transferred to highly consumed staple crops to maximally reach target population. Genetic engineering is not a silver bullet to eliminate malnutrition on its own; however, it can provide a significant component of integrated approaches towards alleviating malnutrition.

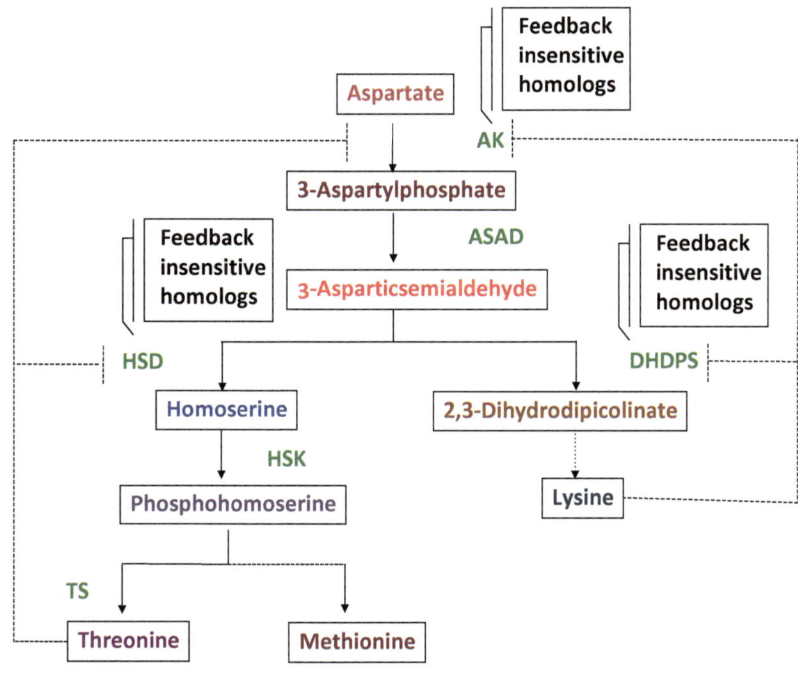

Fig. 9.2 Biosynthetic pathway for amino acid synthesis. Inclusions of feedback-insensitive homologs are the targets of metabolic engineering for enrichment of amino acids in plants

References

Abbadi A, Domergue F, Bauer J, Napier JA, Welti R, Zahringer U, Cirpus P, Heinz E (2004) Biosynthesis of very-long-chain polyunsaturated fatty acids in transgenic oilseeds: constraints on their accumulation. Plant Cell 16:2734–2748

Chakroborty S, Chakroborty N, Datta A (2000) Increased nutritive value of transgenic potato by expressing a nonallergenic seed albumin gene from *Amaranthus hypochondriacus*. Proc Natl Acad Sci U S A 97:3724–3729

Chiaiese P, Ohkama-Ohtsu N, Molvig L, Godfree R, Dove H, Hocart C, Fuliwara T, Higgins TJ, Tabe LM (2004) Sulphur and nitrogen nutrition influence the response of chickpea seeds to an added, transgenic sink for organic sulphur. J Exp Bot 55:1889–1901

Connolly EL, Fett JP, Guerinot ML (2002) Expression of the IRT1 metal transporter is controlled by metals at the levels of transcript and protein accumulation. Plant Cell 14:1347–1357

Diretto G, Al-Babili S, Tavazza R, Papacchioli V, Beyer P, Giovanni G (2007) Metabolic engineering of potato carotenoid content through tuber-specific overexpression of a bacterial mini pathway. PLoS One 2(4), e350

Drakakaki G, Christou P, Stoger E (2000) Constitutive expression of soybean ferritin cDNA in transgenic wheat and rice results in increased iron levels in vegetative tissues but not in seeds. Transgenic Res 9:445–452

Drakakaki G, Marcel S, Glahn RP, Lund EK, Pariagh S, Fischer R, Christou P, Stoger E (2005) Endosperm-specific co-expression of recombinant soybean ferritin and *Aspergillus phytase* in maize results in significant increases in the levels of bioavailable iron. Plant Mol Biol 59:869–880

Falco SC, Guida T, Locke M, Mauvais J, Sanders C, Ward RT, Webber P (1995) Transgenic canola and soybean seeds with increased lysine. Biotechnology (NY) 13 (6):577–582

FAO (2014) Food and Agriculture Organization of the United Nation. http://www.fao.org/

Goto F, Yoshihiro T, Shigemoto N, Toki S, Takaiwa F (1999) Iron fortification of rice seeds by the soybean ferritin gene. Nat Biotechnol 17:282–286

Hagan ND, Upadhyaya N, Tabe LM, Higgins TJ (2003) The redistribution of protein sulphur in transgenic rice expressing a gene for a foreign, sulphur-rich protein. Plant J 34:1–11

Molvig L, Tabe LM, Eggum BO, Moore AE, Craig S, Spencer D, Higgins TJV (1997) Enhanced methionine levels and increased nutritive value of seeds of transgenic lupins (*Lupinus angustifolius* L.) expressing a sunflower seed albumin gene. Proc Natl Acad Sci U S A 94:8393–8398

Naqvi S, Zhu C, Farre G, Ramessar K, Bassie L et al (2009) Transgenic multivitamin corn through biofortification of endosperm with three vitamins representing three distinct metabolic pathways. PNAS (USA) 106:7762–7767

Nunes AC, Kalkmann DC, Aragao FJ (2009) Folate biofortification of lettuce by expression of a codon optimized chicken GTP cyclohydrolase I gene. Transgenic Res 18:661–667

Ruiz-Lopez N, Haslam RP, Napier JA, Sayanova O (2014) Successful high level accumulation of fish oil omega-3 long-chain polyunsaturated fatty acids in a transgenic oilseed crop. Plant J 77:198–208

Savidge B et al (2002) Isolation and characterization of homogentisate phytyltransferase genes from *Synechocystis* sp. PCC 6803 and Arabidopsis. Plant Physiol 129:321–332

Shintani D, Della-Penna D (1998) Elevating the vitamin E content of plants through metabolic engineering. Science 282:2098–2100

Singh NP, Yadav IS (2010) Strategy for improving the nutritional quality of grain legumes through genetic engineering. Trends Biosci 3:1–4

Singleton C, Howard TP, Smirnoff N (2014) Synthetic metabolons for metabolic engineering. J Exp Bot 65:1947–1954

Tabe L, Hagan N, Higgins TJ (2002) Plasticity of seed protein composition in response to nitrogen and sulphur availability. Curr Opin Plant Biol 5:212–217

Takahashi M, Nakanishi H, Kawasaki S, Nishizawa NK, Mori S (2001) Enhanced tolerance of rice to low iron availability in alkaline soils using barley nicotianamine aminotransferase genes. Nat Biotechnol 19:466–469

Takahashi M, Terada Y, Nakai I, Nakanishi H, Yoshimura E, Mori S, Nishizawa NK (2003) Role of nicotianamine in the intracellular delivery of metals and plant reproductive development. Plant Cell 15:1263–1280

Wakasa K, Hasegawa H, Nemoto H, Matsuda F, Miyazawa H (2006) High-level tryptophan accumulation in seeds of transgenic rice and its limited effects on agronomic traits and seed metabolite profile. J Exp Bot 57:3069–3078

Wu G, Truska M, Datla N, Vriten P, Bauer J, Zank T, Cirpus P, Heinz E, Qiu X (2005) Stepwise engineering to produce high yields of very long-chain polyunsaturated fatty acids in plants. Nat Biotechnol 23:1013–1017

Wurbs D, Ruf S, Bock R (2007) Contained metabolic engineering in tomatoes by expression of carotenoid biosynthesis genes from the plastid genome. Plant J 49:276–288

Ye X, Al-Babili S, Klöti A, Zhang J, Lucca P, Beyer P, Potrykus I (2000) Engineering the provitamin A (beta-carotene) biosynthetic pathway into (carotenoid-free) rice endosperm. Science 287(5451):303–305

Zhang P, Jaynes JM, Potrykus I, Gruissem W, Puonti-Kaerlas J (2003) Transfer and expression of an artificial storage protein (asp1) gene in cassava (*Manihot esculenta* Crantz). Transgenic Res 12:243–250

Zhu X, Galili G (2003) Increased lysine synthesis coupled with a knockout of its catabolism synergistically boosts lysine content and also *trans* regulates the metabolism of other amino acids in Arabidopsis seeds. Plant Cell 15:845–853

Genomics-Enabled Breeding for Enhancing Micronutrients in Crops

10

K.R. Soren, P.S. Shanmugavadivel, Priyanka Gangwar, Pallavi Singh, A. Das, and N P Singh

Abstract

Improving nutritional health is one of the major socioeconomic challenges of the twenty-first century, especially with continuously growing population in developing country. Of the various nutrient intake choices, plant-based foods offer an array of nutrients which are essential for human nutrition and promote good health. Majority of staple food crops in the world are often deficient in the essential nutrients but with available of huge germplasm collections from diverse gene pool offer an opportunity to utilize in quality breeding program for improving micronutrients. However, pulses have got an added advantage over cereals and oilseeds of having enriched level of essential micronutrient in their diverse gene pool. Hence, utilizing these genetic resources using conventional breeding program and further adapting next-generation sequencing (NSG)-based novel genomic approaches to engineer designer crops with enhanced key micronutrient level in background of superior variety will be the future challenges before researcher to ensure the nutritional security in developing country. Here, we compiled the work done in cereals, oilseeds, and pulses for enriched nutrient status worldwide. These preliminary works and remarkable achievements in different crops established knowledge towards biofortified crops to revolutionize the entire community depending on plant-based food for essential nutrients to overcome the nutritional challenge in order to achieve malnutrition-free society.

K.R. Soren (✉) • P.S. Shanmugavadivel • P. Gangwar • P. Singh • A. Das
Division of Plant Biotechnology, ICAR-Indian Institute of Pulses Research, Kanpur, Uttar Pradesh 208 024, India
e-mail: sorenars@gmail.com; psshanmugavadivel@gmail.com; gangwar.priyanka8@gmail.com; pallavibiotech@yahoo.com; alokbio@gmail.com

N P Singh
Director, ICAR-Indian Institute of Pulses Research, Kanpur, Uttar Pradesh 208 024, India
e-mail: npsingh.iipr@gmail.com

U. Singh et al. (eds.), *Biofortification of Food Crops*, DOI 10.1007/978-81-322-2716-8_10

Keywords
Breeding • Genomic approaches • NGS • Biofortification • Nutri-genomics

10.1 Introduction

For proper metabolic activities, human requires at least 49 nutrients including protein, macroelement, lipids, and vitamins to meet their metabolic needs to develop healthy life. However, sometimes inadequate consumption of even one of these nutrients adversely affect the whole metabolic process leading to sickness, poor health, impaired growth and development, and irreversible damage to cognitive development in infants causing large economic cost to society (Ramakrishnan et al. 1999; Grantham-McGregor and Ani 1999; Branca and Ferrari 2002). As per the Food and Agriculture Organization (FAO) and the World Health Organization (FAO/WHO 2000), the daily required amounts for these nutrients for children and adults are well listed in their home site. In today's contest, agricultural products are prime source for human diet especially in developing countries to meet the daily requirements of all the nutrients including micronutrients. Today, more than three billion people are affected with micronutrient malnutrition, and again the numbers are increasing day by day (Mason and Garcia 1993; Welch et al. 1997; WHO 1999; World Bank 1994). It was estimated that nearly two-thirds of all deaths of children are associated with nutritional deficiencies, many of them suffering from micronutrient deficiencies (Caballero 2002). In the last decades, substantial progress has been made in controlling the severity of micronutrient deficiency after the intervention of the WHO. Micronutrient malnutrition is currently of alarming proportions in many developing nations (Mason and Garcia 1993; WHO 2002). In 2000, the World Health Report identified iron (Fe), vitamin A, zinc (Zn), and iodine (I) deficiencies as the most serious health constraints worldwide. Developing micronutrient-enriched (biofortified) staple foods, either through traditional plant breeding methods or via molecular breeding techniques, is a powerful intervention tool that targets the most vulnerable people, *viz.*, resource-poor women, infants, and children (Bouis 2000; Combs et al. 1996). The genetic potential for increasing the concentrations of bioavailable Fe, Zn, provitamin A, selenium, and iodine in edible portions of several major staple food crops has been reviewed recently. Mineral biofortification in cereals, pulses, and oilseed crops starts with the steps of breeding for the trait such as germplasm screening, inheritance, physiological, or bioavailability studies and finishes with product development in the form of new biofortified varieties. With this approach, staple foods enriched with micronutrients can be made available to millions of people through single-window system of public distribution system (PDS) reducing the blending cost of the processed food. Furthermore, it is a sustainable intervention, unlike traditional interventions that depend on supplementation and fortification programs that have not proved to be sustainable in many developing nations (Subbulakshmi and Naik 1999; Yip 1997). In addition, increasing the micronutrient metals stored in the seeds and grains increases crop productivity when these seeds are sown in micronutrient-poor soils (Welch 1999). Hence, there will be more benefits in developing world having significant areas of such soils (White and Zasoski 1999); also enhancing seeds with micronutrients will act as an incentive to farmers cultivating micronutrient-poor soils to adopt the micronutrient-enriched seeds for use on their farms (Graham et al. 2001).

10.2 Global and National Status of Malnutrition

Over more than 840 million people do not have enough food to meet their basic daily energy requirements, and even today, more than three billion people suffer from insidious effects of micronutrient deficiencies because of low purchasing power parity (PPP) to buy nutrient-rich food such as meat, fish, fruits, lentils, and vegetables. Under such economic conditions, mostly women and children from sub-Saharan Africa, South and Southeast Asia, Latin America, and the Caribbean are at high risk of disease, premature death, and impaired cognitive abilities because of diets poor in crucial nutrients, particularly Fe, vitamin A, I, and Zn. In India, it has the world's largest population of malnourished children, where each day, around 1500 children die of malnutrition. A government report titled "Children in India 2012-A Statistical Appraisal" notes: 48 % of children younger than 5 years of age are stunted. This social indicator indicates that half of the country's children are chronically malnourished and based on United Nations Children's Fund (UNICEF), one in 3 malnourished children in the world is Indian. It is estimated that reducing malnutrition could add some 3 % to India's GDP. India has already initiated different governmental interventions like, mid-day meals for children, public distribution system (PDS) and agricultural scientific intervention through biofortification.

10.3 Metabolic Importance of Major Micronutrients in Human

The macronutrients and micronutrients are essential to human growth and metabolic functions. The macronutrients provide chemical energy, while the micronutrients function as essential cofactors or coenzymes, required for obtaining the energy provided by macronutrients. Micronutrients are so named because they are required only in very small amounts compared to macronutrients. There are few examples of micronutrients and its biological function in human body cell and impact of its deficiency.

10.3.1 Iron (Fe)

Iron is an active constituent of the catalytic site of most of the heme and non-heme iron proteins. Due to iron deficiency, half of this world's population suffer from anemia (WHO/UNICEF/UNU 2001). Iron deficiency adversely affects normal body development, resistance to infection, productivity, work capacity, pregnancy, etc. Children of anemic mothers have low iron reserves, requiring more iron than is supplied by breast milk, and suffer from growth impairment. Iron-deficiency anemia (IDA) which affects more than 50 % of women and preschool children in developing countries is responsible for 20 % of deaths among women during childbirth and impairs physical and mental development in childhood and adolescence (Pfeiffer and McClafferty 2007). It is estimated that 8,00,000 deaths are attributable to iron-deficiency anemia annually. Endemic infectious diseases also intensify it in developing countries.

10.3.2 Iodine (I)

Iodine which regulates growth and development, and maintains basal metabolic rate is a component of thyroid hormones. The disorders of iodine deficiency are often quoted as the single greatest cause of preventable brain damage in the fetus and infants and retarded psychomotor development in young children. Goiter and cretinism are the most visible manifestations of iodine deficiency and almost more than half of one billion individuals suffer from goiter in Asia. It has been estimated that 16.5 million people worldwide suffer from cretinism, and it is likely that another 49.5 million suffer from less severe though still measurable forms of brain damage because of iodine deficiency (WHO 2004). India is one of the worst affected countries in the world; about more than 50 million and more than 2 million people are affected by goiter and cretinism, respectively.

10.3.3 Vitamin A

Vitamin A is a group of C_{20} carotenoid derivatives, which plays a vital role in vision, epithelial cell growth, immune response, reproduction, bone growth, and maintenance of the surface linings of the eyes, embryonic development, and regulation of adult genes. Vitamin A deficiency (VAD) is considered as a widespread public health problem and has adverse health problems like night blindness, corneal scaring/ Bitot's spots, and blindness. Preschool children and women of reproductive age are at high risk for VAD. An estimated figure of 127 million preschool children are affected by VAD, with 2,50,000–5,00,000 becoming blind every year, half of which die within 12 months of losing their sight. Ninety percent of the world's annual child deaths mostly occur in only developing countries (Jones et al. 2003).

10.3.4 Zinc (Zn)

Zinc is an essential functional component of many proteins and enzymes which are involved in DNA replication, gene expression, cellular growth, and differentiation processes (Hambridge 2000). This mineral is very important during periods of rapid growth, and insufficient intake during childhood and adolescence could adversely affect the growth, sexual development, and psychomotor development (Black 2003). Zinc deficiency is predominant in children and pregnant women, and its severe deficiency cause child and infant stunting, impairs immunity, vitamin A use, and vitamin D function, and leads to decreased health, higher mortality, and greater prevalence of some parasitic diseases (Bouis et al. 2003). The body of evidence on zinc deficiency has accumulated to the degree that zinc fortification has been jointly recommended by WHO and FAO (Allen et al. 2006).

10.3.5 Folate or Vitamin B_9

Folate is an essential coenzyme involved in carbon metabolism, together with vitamin B_{12}. Folate deficiency is associated with a higher risk of newborns with neural tube defects, spina bifida, and anencephaly and an increased risk of cardiovascular diseases, cancer, and impaired cognitive function in adults. Folate deficiency leads to widespread megaloblastic anemia during pregnancy and often exacerbates already existing iron-deficiency anemia (Rush 2000). Folate deficiency causes at least 2, 10,000 severe birth defects every year across the world.

10.4 Approaches to Develop Biofortified Crops

With the advancement of recent tools and technology, biofortification will implement a wide range of technologies including agronomic intervention, conventional breeding, molecular marker-assisted breeding, and genetic transformation. The use of this technology varies from crop to crop.

10.4.1 Genetic Biofortification

Micronutrient malnutrition can be overcome by any of these strategies like food fortification, supplementation, and biofortification. Concern on conventional food fortification and supplementation is that it does not reach many poor urban and rural populations especially in developing countries due to insufficient funding and poor distribution network (Nantel and Tontisirin 2002). Biofortification is considered as the most economically and environmentally sustainable solution because of it's cost effectiveness, and scientifically feasible for many micronutrients. Genetic biofortification is a strategy that uses plant breeding and advanced molecular breeding and transgenic techniques to produce food crops with enhanced micronutrient levels, reduced levels of antinutrient substances, and increased levels of promoters to increase the nutrient absorption/bioavailability (Bouis 2003). It includes both conventional plant breeding approach and modern genomics approaches that target the most vulnerable human populations. Success in genetic biofortification depends on the existence of genetic variation for the target traits in the gene pool. There are immense potentials to increase the micronutrient density in edible parts of staple food crops by breeding approach since the availability of vast genetic variation for considered traits, viz., β-carotene, iron, zinc, and other essential mineral nutrients, thus makes selection of genotypes with enhanced micronutrients possible (Graham and Welch 1996; Graham et al. 1999, 2001). A micronutrient trait is highly heritable and stable across the environmental locations and has a positive correlation with yield traits. Hence, breeding for enhanced micronutrient combining yield traits is possible (Graham et al. 2001). Micronutrient traits are controlled by many genes and definitely have high influence on trait expression by environment. Edible parts like grains/seeds and tubers widely used to prepare foods often contain either low level of micronutrients and promoters or high level of antinutrient compounds, thus reducing the bioavailability of micronutrients (Table 10.1). Grain concentration of Fe and Zn is quantitatively controlled in maize, rice, and wheat (Gregorio 2002; Long et al. 2004; Trethowan et al. 2005, 2007).

10.4.2 Molecular Marker-Assisted Breeding

Molecular breeding is now popularly been used for enhancing the efficiency of conventional breeding process by utilizing modern genomic tools and resources, including molecular markers. DNA markers have the potential to assist plant breeders in breeding program to select the genotypes precisely based on the information obtained from linked/flanked markers to the traits of interest even at the early vegetative stage, and also they have been utilized routinely for fingerprinting of genetic stocks and for assessing the available genetic diversity in a gene pool, thus helping breeder to choose genotypes wisely for further breeding works and genetic purity assessment. DNA markers are not influenced by environment; hence they have been widely utilized in plant system since 1988 to map the quantitative traits (i.e., controlled by many genes) by combining advanced statistical packages and mapping methods, viz., biparental population and association mapping panels. Using the successful deployment of novel marker system approaches, viz; SNP (single nucleotide polymorphism) genotyping, DArT (Diversity Arrays Technology) marker analysis, genotyping by sequencing (GBS), and association study analysis in next-generation populations (backcross population, multi-parent advanced generation intercross (MAGIC) population, natural population, Targeting Induced Local Lesions IN Genomes (TILLING) population, and Recombinant Inbred Line (RIL)) various Quantitative Trait Loci (QTL) have been established for important agronomic traits in different crop species. These well-established technologies should be utilized to harness its potential in biofortification program especially to map the genomic regions/QTLs for high mineral/micronutrient accumulating traits, genomic regions/QTLs responsible for controlling of enhancer, and antinutrient substances that determine the bioavailability of the available micronutrients in the edible parts. Once the QTLs/genes controlling nutrient levels have been identified, marker-assisted selection (MAS) can be effectively used to transfer targeted QTLs/genes with a help of flanking/linked or functional markers for high content of desired micronutrients into new superior/farmer adopted varieties.

10.5 Biofortification for Minerals and Amino Acids in Crops

10.5.1 Maize

The most important breakthrough achievements in maize breeding program is the development of

Table 10.1 List of the anti-nutrients/promoters responsible for bioavailability of minerals

S. no	Antinutrients/promoters	Dietary sources	Bioavailability
1	Phytic acid	Whole legume seeds and wheat grains	Reduce Fe and Zn
2	Fiber (cellulose, hemicellulose, lignin, cutin, suberin, etc.)	Whole grain products of wheat, rice, maize, oat, barley, and rye	Reduce Fe and Zn
3	Certain tannins and other polyphenolics	Beans, sorghum, tea, and coffee	Reduce Fe and Zn
4	Oxalic acids	Spinach leaves and rhubarb	Reduce Fe and Zn
5	Hemagglutinins (e.g., lectins)	Most legumes and wheat	Reduce Fe and Zn
6	Goitrogens	Brassica and alliums	Reduce Fe and Zn
7	Heavy metals (Cd, Hg, Pb, etc.)	Contaminated leafy vegetables and roots	Reduce Fe and Zn
8	Organic acids (ascorbic acid, fumarate, malate, and citrate)	Fresh fruits and vegetables	Increase Fe and/or Zn
9	Hemoglobin	Animal meats	Increase Fe
10	Amino acids (methionine, cysteine, histidine, lysine)	Animal meats	Increase Fe and/or Zn
11	Long chain fatty acids (palmitate)	Human breast milk	Increase Zn
12	Fats and lipids	Animal and vegetable fats	Increase vitamin A
13	Selenium	Sea foods and tropical nuts	Increase I
14	Iron and zinc	Animal meats	Increase vitamin A
15	Beta-carotene	Green and orange vegetables	Increase Fe
16	Inulin and other nondigestible carbohydrates	Chicory, garlic, onion, wheat, Jerusalem artichoke	Increase Ca, Fe, Zn
17	Vitamin E	Vegetable oil	Increase vitamin A

quality protein maize (QPM) after the discovery of maize mutant in the mid-1960s containing the opaque-2 gene (Mertz et al. 1964) which enhances levels of lysine and tryptophan in the endosperm protein. A natural mutant, *opaque 2* (o2), causes reduction of zeins, an increase of nonzein proteins, and, as a consequence, a doubling of lysine levels. Fortunately, there are quantitative trait loci, referred to as *o2 modifiers* (Mo2s), that convert soft endosperm into hard endosperm without losing the high lysine trait; this combination is called quality protein maize (QPM) (Vasal et al. 1980). The knockdown analysis of 27- and 16-kDa γ-zein through RNAi transgene found the important role of γ-zeins in endosperm modification by o2 modifiers (Wua et al 2010). For mineral enrichments, 11 significant QTLs were detected for six metal concentrations including Cu, Fe, and Mg. Of them, QTLs for Cu, Fe, and Mg were co-localized on chromosome 5 in the region of *ys1* (*yellow stripe 1*) gene which has significant dominant effect on trait expression (Soric et al. 2011), and also the gene *ys1* encodes the ferric-phytosiderophore transporter (Curie et al. 2001), but it is also involved in the intracellular transport of other metals such as Zn and Cu (Schaaf et al. 2004); hence it supports the involvement of *ys1* in accumulations of these metals. Some identified QTLs had no obvious candidate genes in QTL region and its vicinity thus offering the possibility of identifying unknown new genes that affect metal accumulation.

10.5.2 Common Bean

It is a valuable candidate for biofortification because it is rich in protein, iron, zinc, and certain other microelements that are generally found to be in low concentrations in most of starchy staples of cereals, other seed crops, and root/tubers. The baseline iron content of common bean is high at 55 ppm, and common bean has variable genetic diversity for this trait to the tune of up to 110 ppm (Blair et al. 2010a, b, c, Blair and Izquierdo 2012). The estimates for HarvestPlus challenge program in common bean mineral biofortification are additional 40 ppm iron to the baseline level to meet the recommended daily iron intake (Graham et al. 1999 and Welch et al. 2000). The amount of iron in common bean is naturally 4–10 times higher than what is present in milled rice, and also its zinc concentration is 2–3 times higher (Pfeiffer and McClafferty 2007). Beebe et al. (2000) evaluated a core collection of common bean, which has variation for Fe and Zn in the range of 34–89 μg/g and 21–54 μg/g, respectively (Graham et al. 1999; Beebe et al. 2000), and some of the lines from Peru were having even more than 100 μg/g Fe. They also found that traits responsible for improvement of Fe and Zn concentrations are stable across bean-growing environments and very high significant positive correlation between concentration of Fe and Zn across genotypes shows that genetic factors responsible for increasing Fe are co-segregating with factors responsible for increasing Zn concentration. But a quandary that exists in common bean biofortification is the distribution of variable concentration of iron between the principal parts of a seed, viz., cotyledonary tissue, embryo axis, and seed coat. The seed coat is rich of tannins and the cotyledon has starch and phytate which indirectly affect the bioavailability of Fe and Zn. With the help of advanced backcross population involving wild bean and cultivated bean, QTLs/genes for seed coat iron and cotyledonary iron and zinc were identified in two separate linkage groups (Blair et al. 2013).

There is a vast genetic variation for iron ranging from 30 to 110 ppm and 25 to 60 ppm for zinc in common bean core collection consisting of 1400 genotypes (Islam et al. 2002), and G14519 and G21242 are considered as high iron genotypes in common bean (Blair et al. 2010a, b, c). The inheritance study involving intergene, intragene pool populations and wild x cultivated species population found that iron and zinc concentration in seeds are controlled by many genes and have considerable G x E interactions (Blair et al. 2009; 2010a, b, c, 2011, 2012, Cichy et al. 2009; Gelin et al. 2007; Guzman-Maldonado et al. 2003). From intergene pool cross involving DOR364 and G19833, a total of 26 QTL_S were found mainly on linkage groups b03, b04, b06, b07, b08, b09, and b11 for both iron and zinc concentrations (Blair et al. 2009). The same way from intragene pool cross involving G14519 and G4825, the most important QTL for iron and zinc was identified and mapped on linkage group b06 which found to be overlapping for these traits (Blair et al. 2010a, b, c). Similarly Blair et al. (2011) mapped nine QTLs on five different linkage groups using a population derived from Andean gene pool genotypes, and some of the identified QTLs in this study overlap with QTLs identified in intergene pool population. This co-localization or overlapping implicates the possibility of pleiotropic effect and common physiology for iron and zinc mineral uptake and translocation process (Blair et al. 2013). Interestingly, one of the QTLs identified by Blair et al. (2009) for iron content on linkage group b07 is localized with phaseolin locus which encodes seed storage protein. On the other hand, ferritin is considered as candidate for iron accumulation which is a storage protein for iron and is co-localized with one of the QTLs on linkage group b08 (Blair et al. 2009, 2012). The QTLs for iron reductase activity in root were carried out by Blair et al. (2010a, b, c) who identified two QTLs on linkage groups b02 and b11, but both of these QTLs were not aligned in the chromosomal position of homologues iron reductase (FRO) found with in silico mapping based on common bean synteny with *Glycine max* and *Medicago truncatula* on b06 and b07.

10.6 Biofortification for Essential Fatty Acids

10.6.1 Palm Oil

The fatty acid composition of food crops has received considerable attention for its significance in human health. Intake of saturated fatty acid (SFA), such as C_{14}:0 and C_{16}:0, is considered to have an adverse effect on cardiovascular diseases (Keys et al. 1974). Therefore, improvement of fat quality traits will give rise to additional crop value in the consumer markets. Recently, high-density multiplex SNP genotyping arrays have been developed with availability of numerous SNPs through whole-genome sequencing. With the advance of SNP genotyping arrays, genome-wide association study (GWAS) has been practical for exploring QTLs, and QTLs associated with iodine value (IV), myristic acid (C_{14}:0), palmitic acid (C_{16}:0), palmitoleic acid (C_{16}:1), stearic acid (C_{18}:0), oleic acid (C_{18}:1), and linoleic acid (C_{18}:2) content were detected (Singh et al. 2009).

10.6.2 Maize

Apart from micronutrient, Genome Wide Association Study (GWAS) was done to dissect genetic architecture of oil biosynthesis in maize kernels and more than half of the identified loci localized in mapped QTL intervals, and one-third of the candidate genes were annotated as enzymes in the oil metabolic pathway. The 26 loci associated with oil concentration could explain up to 83 % of the phenotypic variation using a simple additive model (Li et al. 2013).

10.6.3 Rice

The oil trait is controlled by quantitative trait loci. The QTL analysis was performed using F_2 and $F_{2:3}$ progeny from a cross of an *indica* variety and a *japonica* variety. Gas chromatography–mass spectrometry (GC–MS) analysis revealed significant differences between parental lines in fatty acid composition of brown rice oil, and 29 associated QTLs in F_2 and/or $F_{2:3}$ populations were identified throughout the rice genome, except chromosomes 9 and 10. Eight QTLs were repeatedly identified in both populations across different environments. Five loci pleiotropically controlled different traits, contributing to complex interactions of oil with fatty acids and between fatty acids. Nine rice orthologs of genes encoding key enzymes in lipid metabolism are co-localized with 11 mapped QTLs. A strong QTL for oleic and linoleic acid was associated with a rice ortholog of a gene encoding acyl-CoA: diacylglycerol acyltransferase (DGAT) and another for palmitic acid mapped similarly to the acyl-ACP thioesterase (FatB) gene ortholog.

10.7 Recent Attempt Toward Biofortification in Pulses

10.7.1 Chickpea

The chickpea is a good source of iron, zinc, calcium, magnesium, copper, selenium, and phosphorus and it is having wide genetic variation for most of the mineral micronutrients (Thavarajah and Thavarajah 2012; Jukanti et al. 2012). The ranges for seed iron content vary from 3.0 to 14.3 mg and 2.2 to 20 mg of zinc in chickpea accessions (Wood and Grusak 2006; Ray et al. 2014). Diapari et al. (2014) identified a market-trait association for Fe, and Zn concentration using 94 diverse accessions of chickpea including Desi and Kabuli type. They identified eight SNP loci on chromosome no. 1, 4, 6, and 7 associated with seed iron and zinc concentration. The iron and zinc concentrations were found to be negatively correlated with grain yield and no correlation was observed between 100-seed weight. Recently Norton et al. (2014) reported that there is a significant effect of location on grain micronutrient content in rice, and Ray et al. 2014 also explained that the environment has a significant effect on the mineral micronutrient content of field pea, chickpea, common bean, and lentil cultivars. Zinc concentrations are highly dependent on soil properties but iron concentration is significantly affected by genotype and environment interactions (Chandel et al. 2010; Suwarto 2011). Diapari et al. 2014 identified three Kabuli-type chickpea accessions, namely., CDC Verano, ILC 2555, and FLIP85-1C which have high concentrations of Fe and Zn, and two Desi-type chickpea accessions, namely, FLIP97-677C and FLIP84-48C which have higher zinc concentration.

10.7.2 Field Pea

The US-grown field pea has a variation of iron and zinc concentration which varies from 46 to 54 mg/kg seeds and 39 to 63 mg/kg seeds, respectively (Amarakoon et al. 2012).

10.7.3 Lentil

Selenium is an essential micronutrient which is a component of 25 human enzymes, and 55 μg of selenium per day per adult is recommended for healthy living (Food and Nutrional Board 2000). Phytic acid, an anti-nutrient compound present mainly on seeds of legumes and cereals has the potential to bind mineral micronutrients and eventually reduce their bioavailability. Thavarajah et al. (2009) reported lentil genotypes which have lower phytic acid content (2.5–4.4 mg/g phytic acid) than present in mutants of corn, wheat, common bean, and soya bean.

10.7.4 Cowpea

Biofortified couple lines were generated for high protein and minerals, including potassium, calcium, iron, and zinc concentration (Singh 2007; Santos and Boiteux 2013).

10.7.5 Mung Bean

The mung bean genotypes of Indian origin have good variation for seed iron and zinc content which is varied from 1.79 to 6.58 mg/100 g seeds and 1.54 to 3.81 mg/ 100 g seeds, respectively (Singh et al. 2013).

10.7.6 Medicago and Lotus

Legume crops have the potential to serve essential micronutrient content to the human population. Klein and Grusak (2009) utilized the model legume crop *Lotus japonicus* for identification of QTLs responsible for mineral concentration and content, including iron and zinc, and identified 103 QTLs for various mineral concentrations and content. Similarly a total of 46 significant QTLs for mineral concentrations, including iron and zinc, and 24 QTLs for mineral content per seed were mapped following CIM approach in Medicago RIL population that could be utilized for improving other pulse crops (Klein and Grusak 2009). Among the QTLs identified, some of the QTLs were co-localized suggesting common transport process, loading, and translocation process (Klein and Grusak 2009). This information will be helpful to further identify the genomic regions in other pulse crops based on syntenic information and thus may help to breed better varieties for enhanced micronutrient contents.

10.8 New Prospects for Biofortification Approaches

10.8.1 Nutri-Genomics

Recently, large-scale genome sequencing data of plants, bacteria, fungi, and animals have been completed, and deep bioinformatics analysis of sequence for better understanding of basics in metabolic pathway is the reason for the genesis of nutri-genomics. As a component of the overall biofortification research program, nutrigenomics is a new approach to the study of complex biochemical pathways in plants. It seeks to elucidate the basic, underlying mechanisms involved in the synthesis and accumulation of essential vitamins and minerals in plant tissues. Because of the metabolic unity among organisms through evolution, the knowledge obtained in one organism by using this approach can be readily applied to many organisms. This comparative knowledge base will be used to increase specific micronutrient levels in crops that will be of most benefit to the needs of the developing world. Once the genes of interest are identified, they are moved into a crop species to demonstrate proof of concept to effect the desired change in nutritional content of the target tissues. This process can provide novel traits for breeding that are not available in the existing germplasm.

10.8.2 Utilization of Next-Generation Sequencing (NGS) Technology for Quality Improvement

Presently, NGS has occupied a central key position in a breeding pipeline to accelerate the precision of trait mapping and trait transfer. In last 5 years, in addition to classical Sanger sequencing technology, a range of different second-generation technologies (SGT), *viz.*, Roche/454 FLX Pyrosequencer, GS FLX Titanium/GS Junior, Genome Analyzer (Solexa/Illumina), and Solid Sequencer (Applied Biosystems), and third-generation sequencing (TGS) technologies, *viz.*, Ion Torrent PGM/Proton (Life Technologies), HiSeq/MiSeq from Illumina and Oxford Nanopore Technology has gained immense popularity in recent years because of its throughput, read lengths, and low sequencing cost.

NGS technologies are now widely being used for de-novo sequencing, whole-genome sequencing (WGS), whole-genome resequencing (WGRS), quantitative trait mapping, GWAS, TILLING study, mutational map (MutMap), genotyping by sequencing (GBS), genomic selection (GS), whole-genome bisulfite sequencing (WGBS), reverse and fast-forward genetics analysis, epiQTL analysis, transcriptomics/differential gene expression and epigenetic (MeDIP; ChIP)/ analysis, small RNA profiling, restriction-site-associated DNA sequencing (RAD-seq), SHORE map, Exome sequencing, *QTL-seq* technology (modified bulk segregant analysis) is being extensively used for developing genome-wide genetic markers and to develop fixed SNP genotyping arrays for marker-trait association study (Fig. 10.1).

Rapid identification approaches for SNPs/haplotypes, QTLs by whole-genome resequencing of DNAs from different mapping panels for successful identification of marker-trait association for agronomically important traits have been reported in different crop species.

10.9 Why We Need Biofortification in Pulse Crops? An Opportunity

Food security and malnutrition are the major challenge and concern in developing countries including India. In order to achieve the food production to feed the population, we need to increase the global food production and productivity by introducing high yielding variety. For that, it needs to adapt the technologies which support the sustainable production, ecological security, and nutritional security. Pulse crops have rich sources of genetic diversity for micronutrients, and hence they are highly amenable for biofortification through conventional and marker assisted breeding programmes of crops for combating malnutrition problem.

India is considered to be the hot spot for biodiversity in terms of animals, plants, insects, and many more. This variability is the future repository of gene/QTLs governing important trait for disease resistance, abiotic stresses, and nutritional quality. In India, the genetic stocks for plants are being maintained at the National Bureau of Plant Genetic Resources, New Delhi. As of today, this Bureau is maintaining about

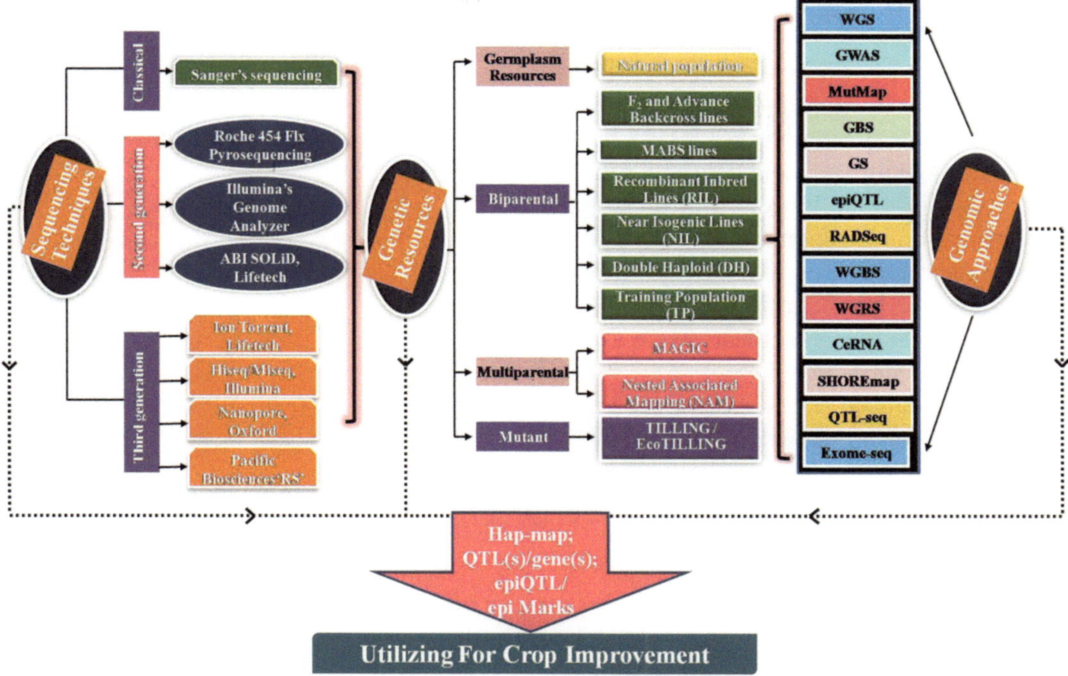

Fig. 10.1 Schematic representation of the multi-population multi-technology genomic approaches for crop improvement

Table 10.2 Mineral content present in major pulse crops

Pulses	Ca (mg/100 g)	P (mg/100 g)	Fe (mg/100 g)	Mg (mg/100 g)	Zn (μg/g)	Folic acid (μg/g)
Chickpea	114	387	6.2	168	8.6	5.5
Pigeon pea	124	304	5.8	133	6.4	-
Mung bean	124	326	7.3	1771	11.7	6.3
Urd bean	154	385	9.1	185	16.2	-
Pea	75	298	5.1	124	-	2.7
Lentil	69	293	4.8	94	11.8	4.3

Source: Gopalan et al. (1989)

4,04,945 cereal accessions conserved in 18 °C, 1,60,540 accessions of millets, 57,619 accessions of forage, and pseudocereals (6,876), grain legume (69,600), fiber crops (12,472), vegetables (25,439), fruit crops (530), medicinal and aromatics, and narcotics (7,002), spices and condiments (3,897), and agroforestry (2,443). In grain legume, NBPGR maintains chickpea (17,449), pigeon pea (11,692), lentil (7,712), and mung bean (3,714). This huge number of germplasm collection is the resources for various micronutrients and vitamins as given in Table 10.2, which can be judiciously utilized for biofortification.

Hence, biofortification of pulse crop with enhanced micronutrient elements (Fe, Zn, and Mn), vitamins, and essential fatty acid are necessary to mitigate widespread micronutrient deficiencies in humans.

10.10 Conclusion

The advantage of biofortifying pulse crops over cereals is the presence of different seed tissues/ compartments like seed coat, cotyledon, and embryo as a target for biofortification, and most of the pulses (not like cereals) are consumed as a whole after boiling, while the bran of cereal, which actually covers thin maternally derived aleurone layer surrounded with specialized endosperm tissue and less well-developed embryo is removed during milling process. Hence, there is vast scope to utilize pulses for biofortification of micronutrients with proper understanding of micronutrient distribution and its inheritance in each of the compartment/tissues. Our immediate goal is to achieve malnutrition free human population by feeding micronutrient enriched dietry food sources including pulsesderived through combining available genetic resources and recent genomic technologies.

References

Amarakoon D, Thavarajah D, McPhee D, Thavarajah P (2012) Iron, zinc and magnesium rich field peas (*Pisum sativum L.*) with naturally low phytic acid: a potential food based solution to global micronutrient malnutrition. J Food Compos Anal 27:8–13

Allen L, de Benoist B, Dary O, Hurrell R (2006) Guidelines on food with micronutrients (Internet). World Health Organisations and Food and Agricultural Organisation, Geneva. http://www.who.int/nutrition/publications/guide_food_fortification_micronutrients.pdf

Beebe S, Gonzalez A, Rengifo J (2000) Research on trace minerals in the common bean. Food Nutr Bull 21:387–391

Black MM (2003) The evidence linking zinc deficiency with children's cognitive and motor functioning. J Nutr 133:1473–1476

Blair MW, Astudillo C, Grusak M, Graham R, Beebe S (2009) Inheritance of seed iron and zinc content in common bean (*Phaseolus vulgaris* L.). Mol Breed 23:197–207

Blair MW, Knewtson SJB, Astudillo C, Li CM, Fernandez AC, Grusak MA (2010a) Variation and inheritance of iron reductase activity in the roots of common bean (*Phaseolus vulgaris* L.) and association with seed iron accumulation QTL. BMC Plant Bio 10:215

Blair MW, Medina JI, Astudillo C, Rengifo J, Beebe SE, Machado G, Graham R (2010b) QTL for seed iron and zinc concentrations in a recombinant inbred line population of Mesoamerican common beans (*Phaseolus vulgaris* L.). Theor Appl Genet 121:1059–1071

Blair MW, Monserrate F, Beebe SE, Restrepo J, Ortube J (2010c) Registration of high mineral common bean germplasm lines NUA35 and NUA56 from the red mottled seed class. J Plant Regul 4:1–5

Blair MW, Astudillo C, Rengifo J, Beebe SE, Graham R (2011) QTLs for seed iron and zinc concentrations in a recombinant inbred line population of Andean common beans (*Phaseolus vulgaris* L.). Theor Appl Genet 122:511–523

Blair MW, Izquierdo P, Astudillo C, Grusak M (2013) A legume biofortification quandary: variability and genetic control of seed coat micronutrient accumulation in common beans. Front Plant Sci 4:275

Bouis HE (2000) Special issue on improving human nutrition through agriculture. Food Nutr Bull 21:351–576

Bouis HE (2003) Micronutrient fortification of plants through plant breeding: can it improve nutrition in man at low cost? Proc Nutr Soc 62:403–411

Branca F, Ferrari M (2002) Impact of micronutrient deficiencies on growth: the stunting syndrome. Ann Nutr Metab 46:8–17

Caballero B (2002) Global patterns of child health: the role of nutrition. Ann Nutr Metab 46:3–7

Cakmak I, Pfeiffer WH, McClafferty B (2010) Biofortification of durum wheat with zinc and iron. Cereal Chem 87:10–20

Chandel G, Banerjee S, See S, Meena R, Sharma DJ, Verulkar SB (2010) Effects of different nitrogen fertilizer levels and native soil properties on rice grain Fe, Zn and protein contents. Rice Sci 17:213–227

Cichy KA, Caldas GV, Snapp SS, Blair MW (2009) QTL analysis of seed iron, zinc, and phosphorus levels in an Andean bean population. Crop Sci 49:1742–1750

Combs GF Jr, Welch RM, Duxbury JM, Uphoff NT, Nesheim MC (1996) Food-based approaches to preventing micronutrient malnutrition: an international research agenda. Cornell International Institute for Food, Agriculture, and Development, Cornell University, Ithaca, pp 1–68

Curie C, Panaviene Z, Loulergue C, Dellaporta SL, Briat JF, Walker EL (2001) Maize yellow stripe1 encodes a membrane protein directly involved in Fe(III) uptake. Nature 409:346–349

Diapari M, Sindhu A, Bett K, Deokar A, Warkentin TD, Tar'an B (2014) Genetic diversity and association mapping of iron and zinc concentrations in chickpea (*Cicer arietinum* L.). Genome 57:459–468

Food and Nutrition Board (2000) Dietary reference intakes for vitamin C, vitamin E, selenium and carotenoids. National Academy Press, Washington, DC, pp 311–316

Gelin JR, Forster S, Grafton KF, McClean P, Rojas-Cifuentes GA (2007) Analysis of seed-zinc and other nutrients in a recombinant inbred population of navy bean (*Phaseolus vulgaris* L.). Crop Sci 47:1361–1366

Gopalan C, Sastri BVR, Balasubramanian SC (1989) Nutritive values of Indian Foods. National Institute of Nutrition, Indian Council of Medical Research, Hyderabad, India. http://www.icmr.nic.in/pricepubl/content/1.htm Accessed 12 Jan 2015

Graham RD, Welch RM (1996) Micronutrient nutrition of plants. Breeding for staple-food crops with high micronutrient density: agricultural strategies for micronutrients working paper 3:1–72 International Food Policy Research Institute Washington, DC

Graham RD, Senadhira D, Beebe S, Iglesias C, Monasterio I (1999) Breeding for micronutrient density in edible portions of staple food crops: conventional approaches. Field Crop Res 60:57–80

Graham RD, Welch RM, Bouis HE (2001) Addressing micronutrient malnutrition through enhancing the nutritional quality of staple foods: principles, perspectives and knowledge gaps. Adv Agron 70:77–142

Grantham-McGregor SM, Ani CC (1999) The role of micronutrients in psychomotor sad cognitive development. Br Med Bull 55:511–527

Gregorio GB (2002) Plant breeding: a new tool for fighting micronutrient malnutrition, progress in breeding for trace minerals in staple crops. J Nutr 132:500–502

Guzman-Maldonado SH, Martínez O, Acosta-Gallegos J, Guevara-Lara FJ, Paredes-Lopez O (2003) Putative quantitative trait loci for physical and chemical components of common bean. Crop Sci 43:1029–1035

Hambridge KM (2000) Human zinc deficiency. J Nutr 130:1344–1349

Islam FMA, Basford KE, Jara C, Redden RJ, Beebe SE (2002) Seed compositional and disease resistance differences among gene pools in cultivated common bean. Genet Resour Crop Evol 49:285–293

Jones G, Steketee RW, Black RE, Bhutta ZA, Morris SS, Bellagio T (2003) Child survival group: how many child deaths can we prevent this year? Lancet 362:65–71

Jukanti AK, Gaur PM, Gowda CLL, Chibbar RN (2012) Nutritional quality and health benefits of chickpea (*Cicer arietinum L.*): a review. Br J Nutr 108:S11–S26

Keys A, Grand F, Anderson JT (1974) Bias and misrepresentation revised: perspective on saturated fat. Am J Clin Nutr 27:188–212

Klein MA, Grusak MA (2009) Identification of nutrient and physical seed trait QTL in the model legume *Lotus japonicus*. Genome 52:677–91. doi:10.1139/g09-039

Li H, Peng Z, Yang X, Wang W, Han Y, Chai Y, Guo T, Yang N, Liu J, Warburton ML, Cheng Y, Hao X, Zhang P, Zhao J, Liu Y, Wang G, Li J, Yan J (2013) Genome-wide association study dissects the genetic architecture of oil biosynthesis in maize kernels. Nat Genet 45:43–50

Long JK, Banziger M, Smith ME (2004) Diallel analysis of grain iron and zinc density in southern African-adapted maize inbreds. Crop Sci 44:2019–2026

Mason JB, Garcia M (1993) Micronutrient deficiency: the global situation. SCN News 9:11–16

Mertz ET, Bates LS, Nelson OE (1964) Mutant that changes protein composition and increase lysine content of maize endosperm. Science 145:279–280

Nantel G, Tontisirin K (2002) Policy and sustainability issues. J Nutr 132:839–844

Norton GJ, Douglas A, Lahner B, Yakubova E, Guerinot ML, Pinson SRM, Tarpley L, Eizenga GC, McGrath SP, Zhao FJ, Islam MR, Islam S, Duan G, Zhu Y, Salt DE, Meharg AA, Price AH (2014) Genome wide association mapping of grain arsenic, copper, molybdenum and zinc in rice (*Oryza sativa* L.) grown at four international field sites. PLOS ONE 9(2):1–12 (e89685). doi:10.1371/journal.pone.0089685

Ramakrishnan U, Manjrekar R, Rivera J, Gonzales-Cossio T, Martorell R (1999) Micronutrients and pregnancy outcome: a review of the literature. Nutr Res 19:103–159

Ray H, Bett K, Taran B, Vandenberg A, Thavarajah D, Warkentin TD (2014) Mineral micronutrient content of cultivars of field pea, chickpea, common bean, and lentil grown in Saskatchewan. Canada Crop Sci 54:1698–1708

Rush D (2000) Nutrition and maternal mortality in the developing world. Am J Clin Nutr 2:212S–240S

Santos CAF, Boiteux LS (2013) Breeding biofortified cowpea lines for semi-arid tropical areas by combining higher seed protein and mineral levels. Genet Mol Res 12:6782–6789

Schaaf G, Ludewig U, Erenoglu BE, Mori S, Kitahara T, von Wiren N (2004) ZmYS1 functions as a proton-coupled symporter for phytosiderophore- and nicotianamine-chelated metals. J Biol Chem 279:9091–9096

Singh BB (2007) Recent progress in cowpea genetics and breeding. Acta Horticult 752:69–75

Singh R, Tan SG, Panandam JM, Rahman RA, Ooi LCL, Low ETL, Sharma M, Jansen J, Cheah SC (2009) Mapping quantitative trait loci (QTLs) for fatty acid composition in an interspecific cross of oil palm. BMC Plant Biol 9:114

Singh R, Heusden AWN, Kumar R, Visser RGJ, Yadav RC (2013) Genetic diversity of mung bean (*Vigna radiata L.*), in iron and zinc content as impact by farmers varietal selection in northern India. Ecol Food Nutr 52:148–162

Soric R, Ledencan T, Zdunic Z, Jambrovic A, Brkic I, Loncaric Z, Kovacevic V, Maydica DS (2011) Quantitative trait loci for metal accumulation in maize leaf. Maydica 56:323–329

Subbulakshmi G, Naik M (1999) Food fortification in developing countries- current status and strategies. J Food Sci Technol 365:371–395

Suwarto N (2011) Genotype × environment interaction for iron concentration of rice in central Java of Indonesia. Rice Sci 18:75–78

Thavarajah P, Thavarajah D, Vandenberg A (2009) Low phytic acid lentils (*Lens culinaris L.*): a potential solution for increased micronutrient bioavailability. J Agric Food Chem 57:9044–9049

Thavarajah D, Thavarajah P (2012) Evaluation of chickpea (*Cicer arietinum L.*) micronutrient composition: Biofortification opportunities to combat global micronutrient malnutrition. Food Res Int 49:99–104

Trethowan RM, Reynolds MP, Sayre KD, Ortiz-Monasterio I (2005) Adapting wheat cultivars to resource conserving farming practices and human nutritional needs. Ann Appl Biol 146:404–413

Vasal SK, Villegas E, Bjarnason M, Gelaw B, Goertz P (1980) Genetic modifiers and breeding strategies in developing hard endosperm opaque 2 materials. In: Prollmer MG, Phillips RH (eds) Improvement of quality traits of maize for grain and silage use. Martinus Nijhoff, London, pp 37–73

Welch RM, Combs GF Jr, Duxbury JM (1997) Toward a Greener revolution. Issues Environ Sci Technol 14:50–58

Welch RM (1999) Importance of seed mineral nutrient reserves in crop growth and development. In: Rengel Z (ed) Mineral nutrition of crops. Fundamental mechanisms and implications. Food Products Press, New York, pp 205–226

Welch RM, House WA, Beebe S, Cheng Z (2000) Genetic selection for enhanced bioavailable levels of iron in bean (*Phaseolus vulgaris* L.). J Agric Food Chem 48:3576–3580

White JG, Zasoski RJ (1999) Mapping soil micronutrients. Field Crop Res 60:11–26

WHO (1999) Malnutrition worldwide. World Health Organization, Geneva. pp 1–13. http://www.who.int/nut/malnutrition_worldwide.htm. Accessed 16 Feb 2015

WHO (2000) The World health report 2000: health systems: improving performance. World Health Report, World Health Organization, 1211 Geneva 27, Switzerland. http://www.who.int/whr/2000/en/whr00_en.pdf?ua=1. Accessed 13 Apr 2015

WHO (2004) Iodine status worldwide. In: Benoist BD, Andersson M, Egli I, Takkouche B, Allen H (eds) WHO global database on iodine deficiency. World health organization, Geneva

WHO/UNICEF/UNU (2001) Iron deficiency anaemia: assessment, prevention and control: A guide for programme managers. Document WHO/NHD/01.03, Geneva

Wood JA, Grusak MA (2006) Nutritional value of chickpea. In: Yadav SS, Redden RJ, Chen W, Sharma B (eds) Chickpea breeding and management. Cromwell Press, Trowbridge

World Bank (1994) The challenge of dietary deficiencies of vitamins and minerals. In: Enriching lives: overcoming vitamin and mineral malnutrition in developing countries. World Bank, Washington, DC, pp 6–13

World Health Organization (2002) World health report. http://www.who.int/whr/2002/. Accessed 16 Feb 2012

Wua Y, Holdingb DR, Messinga J (2010) γ-Zeins are essential for endosperm modification in quality protein maize. PNAS 12810–12815

Yip R (1997) The challenge of improving iron nutrition: limitations and potentials of major intervention approaches. Eur J Clin Nutr 51:16–24

Zinc Transporter: Mechanism for Improving Zn Availability

11

Lalit Kumar, Nand Lal Meena, and Ummed Singh

Abstract

Zinc (Zn) is essentially required by plants for their growth and development. It plays very important role in various physiological procedures of plants such as photosynthesis, membrane integrity, protein synthesis, pollen formation, and immunity system. Although Zn is required by the plant in microconcentration, its bioavailable fraction in the soil is very low due to various soil factors. From soil solution it is absorbed by plants by root membrane transport mechanisms. After entering into plant system, it is neither oxidized nor reduced; but remains as divalent cation which has a great tendency to form tetrahedral complexes. From soil solution Zn reaches the plant root surface by three mechanisms, i.e., mass flow, diffusion, and root interception. Once it is absorbed, its transportation from roots to shoots occurs through the xylem and then easily retranslocated by phloem. This transport of ions and molecules from epidermal and cortical cell to xylem occurs through the symplastic or apoplastic route. The uptake of zinc into cells and its permeability into and out of intracellular organelles require some of the specific chemicals, generally known as transporter proteins. These proteins possess a quality to span the cell membranes which facilitate the movement of zinc. In recent years, a number of metal transporters have been identified in plants, including the P_{1B}-ATPase family, zinc-regulated transporter (ZRT), iron-regulated transporter (IRT)-like protein (ZIP), natural resistance-associated macrophage protein (NRAMP) family, and cation diffusion facilitator (CDF) family. The bioavailable content of Zn in the soil can be increased using both chemical and biological approaches. Mineral fertilizers are considered a good source of Zn, but it gets fixed quickly on soil matrix, resulting in poor availability to plants. It is crucial to

L. Kumar (✉) • N.L. Meena
Division of Basic Sciences, ICAR-Indian Institute of Pulses Research, Kanpur, Uttar Pradesh 208 024, India
e-mail: lalit1997@yahoomail.com; nd.iari09@gmail.com

U. Singh
Division of Crop Production, ICAR-Indian Institute of Pulses Research, Kanpur, Uttar Pradesh 208 024, India
e-mail: singhummed@yahoo.co.in

U. Singh et al. (eds.), *Biofortification of Food Crops*, DOI 10.1007/978-81-322-2716-8_11

increase bioavailability of Zn to plants by solubilizing fixed Zn and/or by reducing fixation of the applied Zn fertilizers. This can be achieved either by using organic amendments or potential Zn solubilizing bioinoculants. Organic amendments improve bioavailability of Zn by increasing microbial biomass, which not only enhance the rate of decomposition of organic matter (source of Zn) but also enhance the bioavailability of indigenous Zn by lowering the soil pH and by releasing chelating agents. Similarly, exogenous application of some potential Zn solubilizing microflora has shown huge capability to improve bioavailable Zn content in the soil and its uptake by plant roots. This manuscript critically reviews about the Zn transporters and the role of rhizosphere microflora as a potential tool in enhancing its bioavailability to higher plants.

Keywords
Zn • Zn fertilizers • Zn transporter

11.1 Introduction

Zinc (Zn), a transition metal, is essentially required by plants for their growth and development (McCall et al. 2000). It plays an important role in photosynthesis, membrane integrity, protein synthesis, pollen formation, and immunity system (Alloway 2008; Hajiboland and Amirazad 2010; Gurmani et al. 2012). It is also an important constituent of nucleic acids and Zn-binding proteins. It has been documented that higher plants possess about 3,000 proteins that contain Zn prosthetic groups (Tapiero and Tew 2003). Moreover, zinc is required as a cofactor for the activity of various enzymes (McCall et al. 2000) and enhances the level of antioxidants within plant tissues (Luo et al. 2010). Furthermore, it acts as an important hormonal regulating agent in plants (Marschner 1997). Apart from this, Zn finger proteins are also known for their involvement in the regulation of signal transduction events. These proteins affect transcription by binding to DNA/RNA or other proteins, leading to cell death (Englbrecht et al. 2004; Ciftci-Yilmaz and Mittler 2008).

Besides, Zn is also observed as a crucial substance for the synthesis of phytohormones such as auxin, abscisic acid, gibberellins, and cytokinins. Its deficiency reduces the level of these phytohormones in plant tissues, resulting in an impairment of cell growth. Thus, its deficiency in plant tissues adversely affects various vital processes occurring within the plant body. Though Zn is required by the plant in microconcentration, its bioavailable fraction in the soil is very low due to various soil factors (Alloway 2009). Some of the soils, in spite of having appreciable quantity of Zn, cannot support plant growth because of its poor bioavailability. It is further documented that around one third of the world's soils are Zn deficient (Kochian 2000). In this respect, scientifically, it has been estimated that about 90 % of the total available soil Zn has no relevance to bioavailability (Mandal et al. 1988). Even if Zn is supplemented though mineral fertilizers which are considered a good source of it, that to also gets fixed quickly on soil matrix, resulting in poor availability to plants (Zia et al. 1999). Therefore, it is crucial to increase bioavailability of Zn to plants either by solubilizing fixed Zn and/or by reducing fixation of the applied Zn fertilizers. The Zn bioavailable content in soil solution can be increased by using both chemical and biological approaches, and this can be achieved either by using organic amendments or potential Zn solubilizing bioinoculants. Organic amendments improve bioavailability of Zn by increasing microbial biomass, which does not only enhance the rate of decomposition of organic matter (source of Zn) but also enhance the bioavailability of indigenous Zn by lowering the soil pH and by releasing chelating agents after their decomposition. Similarly, exogenous application of some potential Zn solubilizing microflora has also shown huge capability to improve bioavailable Zn content in the soil and its uptake by plant roots (Tariq et al. 2007).

11.2 Functions of Zn in Plants

Zinc takes part in various vital biochemical reactions within the plants. Under Zn-deficient situations, some of the plants such as maize, sorghum, and sugarcane showed reduced photosynthetic carbon metabolism. Scientific studies conducted on these plants revealed that Zn modifies and/or regulates the activity of carbonic anhydrase enzyme which is known to regulate the conversion of carbon dioxide to reactive bicarbonate species required for fixation of carbohydrates. Zinc is also a part of several other enzymes such as superoxide dismutase and catalase, which is known to prevent oxidative stress in plant cells (Hacisalihoglu et al. 2003). Some of the important roles of zinc in plants are as follows:

(i) It helps in production of auxin which is known as an important essential growth hormone.
(ii) It regulates starch formation in plants and, hence, is responsible for proper root development.
(iii) It plays key role in formation of chlorophyll and carbohydrates
(iv) It provides strength to plants to withstand lower air temperatures.
(v) It also helps in the biosynthesis of cytochrome, a pigment, and maintains plasma membrane integrity and synthesis of leaf cuticle.

11.3 Zn Uptakes and Transport

11.3.1 Mechanism of Uptake and Factors Affecting Its Availability

Zinc is a transition metal essential for terrestrial life, as already described. However, it is only required at low concentrations, therefore making its function as a micronutrient. Some of the scientific reports indicate that concentrations between 30 and 100 μg zinc g^{-1} DW are sufficient to support adequate plant growth, whereas if concentrations exist above 300 μg g^{-1} DW, then zinc toxicity symptoms appear especially on those plant species that are not adapted to high-zinc exposure (Marschner 1995; Van de Mortel et al. 2006). Therefore, it is proven that for optimal growth, plants need to keep tight control over zinc homeostasis. Zinc homeostasis requires a complex network of cellular or tissue-specific functions to regulate metal uptake, accumulation, trafficking, and detoxification. The ability to take up Zn of higher plants depends much more on its bioavailability from the soil rather than the absolute soil concentrations. Zinc bioavailability is modulated by various physical and chemical soil factors as it is well known that zinc solubility in the soil decreases due to high levels of calcium carbonate, metal oxides, and pH and low levels of organic matter and soil moisture as well as high amounts of phosphate (Robson 1994; Cakmak 2011; Cakmak et al. 2010). When available in the soil solution, zinc is absorbed and transported in the divalent ion form (Zn^{+2}) from roots to shoots through the xylem, being easily retranslocated by phloem (Clemens 2001).

This transport of ions and molecules from epidermal and cortical cells to xylem can occur via symplastic or apoplastic route. Since zinc plays multiple roles in a number of biochemical and physiological processes of plants, hence, slight deficiencies can cause a decrease in growth, yield, by including the zinc content of edible parts. Therefore, it has become imperative to know about the molecular details of the element in relation to its uptake, translocation, and storage in plants. Some of the key factors which affect the Zn bioavailability are as follows:

(i) Though it is well known that Zn availability generally decreases as pH increases, this effect is not universal and can temporarily be overcome by using the band applications of acid-type fertilizers.
(ii) Zn availability is also affected by Zn:P balance. It is well documented that high levels of soil P adversely affect the Zn availability.
(iii) Presence of good amount of organic matter is reported to increase Zn availability because the organic compounds produced by decomposition of soil organic matter chelate the inorganic sources of Zn and enhance its availability.
(iv) Under N stress situations, generally, plants fail to take adequate amount of Zn because of the decreases in vigor of plants.
(v) Soil saturation situation is also reported to cause Zn deficiency because Zn does not undergo the valence changes as Mn does. Therefore, various research reports indicated that rice cannot take up Zn effectively under flooded conditions. The causes of this condition appear to be applicable to other crops too for some extent.
(vi) Zn:Cu balance is also reported to cause Zn deficiency because the absorption mechanism of both elements by plant roots is the same. Therefore, this causes interference in the uptake of one when the other is in excess in the root zone.
(vii) Contrary scientific reports are available for Zn:Mn balance Some results indicate an antagonistic relationship, while others indicate a sympathetic relationship.
(viii) Zn:Mg balance reported for their sympathetic relationship; various reports indicate that additions of Mg can increase the uptake of zinc.
(ix) As far as Zn:As balance is concern, it was observed that high levels of arsenic in the soil inhibit seriously both phosphorus and zinc uptake.

11.3.2 Zn Uptake from Soil

Generally, from soil solution and at neutral pH, zinc is mostly absorbed by most of the plants as a divalent cation (Zn^{++}), and at high pH, it can also be taken up as a monovalent ($ZnOH^{+}$) cation (Marschner 1995). Once taken up, it is neither oxidized nor reduced; therefore, the role of Zn in cells sticks to its behavior as a divalent cation, and in this valence form it is known for their strong tendency to form tetrahedral complexes. The ability to take up Zn of higher plants depends much more on its bioavailability from soil rather than the absolute soil concentrations. As we know the bioavailability of zinc is modulated by various physical and chemical properties of the soil. In most of the soil, zinc solubility is reported to decrease by high levels of calcium carbonate, metal oxides, and pH and low levels of organic matter and soil moisture as well as high amounts of phosphate (Robson 1994; Cakmak 2011).

From soil solution, uptake of Zn mainly depends on ion concentrations at the root surface, plant demand, and root absorption capacity. Mass flow, diffusion, and root interception mechanisms were reported by which Zn reaches to the plant root surface. Mass flow is considered as a passive nutrient transport from the soil to the roots and is driven by transpiration. Mass flow becomes the predominant mechanism when the soil solution contains a relatively large concentration of Zn, whereas diffusion operates when the Zn concentration is low, particularly in soils with low plant-available Zn. Contrary to mass flow, diffusion operates only in the immediate volume of soil surrounding by a root. Thus, root interception which includes root growth and root surface area is also an important factor in determining plant availability of Zn. It is seen if root interception is poor, it can limit Zn uptake even if granules of $ZnSO_4$ are banded in the soil; particularly, it has more relevance to the fact at a low rate of $ZnSO_4$ application.

11.3.2.1 Zn Movement in Plants

Transportation from Root Surface to Xylem and Phloem

When available in the soil solution, zinc is absorbed and transported in the divalent ion (Zn^{+2}) form from roots to shoots via the xylem where it is easily retranslocated by phloem (Clemens 2001). This kind of transportation of ions and molecules from epidermal and cortical cells to xylem can occur through the symplastic or apoplastic route. It is observed that accumulation of Zn in plant roots exhibits biphasic kinetics, with initial rapid entry and then binding within the root cell wall followed by a slower linear transport phase across the plasma membrane. Movement of Zn^{2+} from the external solution to the root cell wall-free space occurs via diffusion and then is subsequently transported across the plasma membrane by the influence of ion transport proteins. Since the cytoplasm of a plant cell is negatively charged, therefore, it favors entry of Zn^{2+} and other cations into the cell. Root Zn uptake is mediated by different transport systems which include a high-velocity, low-affinity membrane transporter system (K_m = 2–5 Î¼M) and a low-velocity, high-affinity system (K_m = 0.6–2 nM). Most likely the second transport system dominantly operates under low soil Zn conditions (Hacisalihoglu et al. 2001). It is generally observed that Zn-inefficient plants have lower rates of uptake as well as the comparable affinity values compared with Zn-efficient plants (Rengel and Wheal 1997; Hacisalihoglu et al. 2001). Most of the nutrients are known to transport into the root xylem via the epidermis, cortex, and endodermis (Marschner 1995). Zn may pass through the root to the xylem either through the extracellular spaces between root cells (apoplast) (White et al. 2002) or the cytoplasmic continuum of root cells linked by plasmodesmata (symplast). The apoplastic cation flux is largely dependent on the cation exchange capacity of the cell wall, water flows, and the existence of Casparian band; whereas the symplastic pathway involves specific transporters in the plasma membrane, particular cations can be selected for transport (Sattelmacher 2001). For example, in *Thlaspi caerulescens* and *T. arvense*, Zn reaches the xylem via the symplastic pathway, and the entry point for Zn accumulation is across the plasma membrane of root cells (Lasat et al. 1996; Lasat and Kochian 2000). With an increasing external Zn concentration, the apoplastic pathway plays a greater role in Zn uptake and influx to the xylem (White et al. 2002; Rengel and Wheal 1997).

Transportation from Xylem/Phloem to Shoot

Nutrients in the xylem move toward shoots in the transpiration stream. This driving force for transpiration arises from the gradient in water potential between stomatal cells and the atmosphere. An analysis of xylem fluid content showed that Zn moves as a complexed form (e.g., anionic) in the xylem. Zn in the xylem has been measured in both soluble and insoluble forms. In a soluble form it bound to small proteinous molecules, whereas Zn-phytate complexes exist as insoluble form of Zn. Metal complexes and ionic activities of micronutrients differ in the xylem and phloem saps (Welch 1995). The activity of cationic micronutrients, such as Mn, Zn, Fe, Cu, and Ni, is low in the phloem sap due to a high phosphate level and high pH. Hence, these micronutrients are likely to form metal complexes to move easily in the phloem stream (Welch 1995). Regardless of the chosen path, the solutes that reach the xylem parenchyma cells are transferred to the xylem elements in a tightly controlled process mediating membrane transport. Via the xylem sap, the minerals are released in the apoplast of the leaves and are subsequently distributed intracellularly (Clemens et al. 2002).

11.3.2.2 Zinc Transporters (Transportation to Cytoplasm)

Since Zn cannot diffuse across cell membrane, therefore for these purposes, some of the specific chemicals generally known as transporters are required to transport Zn into cytoplasm. Zn transporters are proteinous molecules and known to span cell membranes to facilitate the movement of zinc into cells and its transport into and out of intracellular organelles. Similarly, few P-type ATPases have also been demonstrated to play roles in eukaryotic zinc transport. In recent years, a number of metal transporter families such as P_{1B}-ATPase, zinc-regulated transporter (ZRT), iron-regulated transporter (IRT)-like protein (ZIP), natural resistance-associated macrophage protein (NRAMP), and cation diffusion facilitator (CDF) have not only been identified in plants, archaea, bacteria, fungi, and mammals, but also their involvement in relation to metal uptake and transport was demonstrated thoroughly. ZIP proteins are generally known to contribute toward metal ion homeostasis by means of transporting cations into the cytoplasm. Functional complementation in yeast indicated that ZIP proteins are able to transport various divalent cations, including Fe^{2+}, Zn^{2+}, Mn^{2+}, and Cd^{2+}. The ZIP proteins are reported to consist about 309–476 amino acid residues which consist eight potential transmembrane domains and a similar membrane topology. Known eukaryotic zinc transporters come largely from two families, the ZIP (SLC39) and CDF/ZnT (SLC30) proteins.

The ZIP Family

Plants, bacteria, fungi, and humans are reported to contain the ZIPs (ZRT, IRT-like proteins). These proteins are said to be involved in the transport of Fe, Zn, Mn, and Cd with a number of family members differing in their substrate range and specificity (Guerinot 2000). Structurally, the ZIP proteins are said to contain eight transmembrane domains with the amino- and carboxyl-terminal ends situated on the outer surface of the plasma membrane between transmembrane domains III and IV (Guerinot 2000; Colangelo and Guerinot 2004). In most of the cases, the expression of genes encoding the plant ZIP transporters appears to be induced by zinc or iron deprivation (Eide et al. 1996; Eide and Guarente 1992; Grotz et al. 1998). Some of the *ZIP* family members, such as *irt* genes, are also constitutively expressed, but in this case also expression levels increase with iron deprivation (Eckhardt et al. 2001; Vert et al. 2002). Presently about 85 ZIP family members have been identified from bacteria, archaea, and all types of eukaryotes, including 15 genes in *Arabidopsis*. The function of several members of the ZIP family in *Escherichia coli* (Grass et al. 2002), yeast (*Saccharomyces cerevisiae*; Zhao and Eide 1996a, b), plants (Eide et al. 1996; Grotz et al. 1998; Guerinot 2000; Pence et al. 2000; Eckhardt et al. 2001; Vert et al. 2002; Moreau et al. 2002), and humans (Gaither and Eide 2000, 2001) has been studied. In plant community, the dicots, i.e., pea (*Pisum sativum*), *Arabidopsis*, and *Thlaspi caerulescens*, were mainly reported to have ZIP transporters. Many members of ZIP family like the zinc-regulated transporter (ZRT) and iron-regulated transporter (IRT)-like protein were thought to be involved in transporting zinc into the cytosol across the plasma membrane, which is an important process for plant zinc uptake (Palmer and Guerinot 2009; Song et al. 2010).

As per the alignment of the predicted amino acid sequences, the ZIPs can be grouped into four subfamilies, although all of the higher plant genes appear to fall into a single group. They range widely in overall length, this being due to a variable region between TM-3 and TM-4. This region is predicted to be on the cytoplasmic side and is a potential metal-binding domain rich in histidine residues. ZIPs 1, 3, and 4 are expressed in the roots of Zn-deficient plants, while ZIP4 is also found in the shoots and is predicted to have a chloroplast targeting sequence (Grotz et al. 1998; Guerinot 2000). The proposed role of ZIP transporters in Zn nutrition is supported by the characterization of homologues from other species. ZRT3 is proposed to function in the mobilization of stored Zn from the vacuole (MacDiarmid et al. 2000).

A ZIP gene homologue, TcZNT1, from the Zn/Cd-hyperaccumulating plant, *Thlaspi caerulescens*, was shown to mediate high-affinity Zn^{2+} uptake and low-affinity Cd^{2+} uptake following expression in yeast (Pence et al. 2000; Zhao et al. 2002). Assuncao et al. (2010) reported that TcZNT1 and TcZNT2 were predominantly expressed in roots although their expression was seldomly Zn responsive. In contrary to that in case of the non-hyperaccumulator, i.e., *T. arvense*, these genes were exclusively expressed under conditions of Zn deficiency (Assuncao et al. 2010). Recently, a member of the ZIP family, GmZIP1, has been identified in soybean (Moreau et al. 2002). By functional complementation of the ZRT1 and ZRT2 yeast cells, GmZIP1 was observed to be highly selective for Zn, while yeast Zn uptake was inhibited by Cd. It was found that GmZIP1 was specifically expressed in the nodules and not in roots, stems, or leaves, and the protein was localized to the peribacteroid membrane, which indicates the possibility of its role in the symbiosis (Moreau et al. 2002) (Table 11.1).

Cation Diffusion Facilitator (CDF) Proteins

Members of the cation diffusion facilitator (CDF) family are known to control cation concentrations in cells through sequestration into internal compartments and efflux from cell and hence play an important role in living organisms (Gustin et al. 2011). These proteins belong to a family of heavy metal efflux transporters that presumably play an essential role in homeostasis and tolerance to metal ions. Some of the members of the cation diffusion facilitator (CDF) family such as MTP1 (metal tolerance protein, previously ZAT) are thought to transport heavy metals into the vacuoles of leaf epidermal cells in the metal ion-hyperaccumulating plants (hyperaccumulators) (Kupper et al. 1999; Persans et al. 2001; Delhaize et al. 2003). The Zn tolerance of *A. halleri* has been found to be due to an increased copy number of the MTP1 gene and an elevated level of transcription (Becher et al. 2004; Drager et al. 2004).

Zn transport activity of MTP1 in *Arabidopsis thaliana* (AtMTP1) was demonstrated by using reconstituted proteoliposomes of the protein expressed in *Escherichia coli* (Bloß et al. 2002) and by a yeast complementation assay with a Zn-hypersensitive double mutant of ZRC1 and COT1 (*zrc1/cot1*) (Drager et al. 2004; Kim et al. 2009). AtMTP1 has been found to retain about 92 % identity with *A. halleri* AhMTP1-3 (Becher et al. 2004; Drager et al. 2004). Ectopic expression of AtMTP1 in *A. thaliana* caused in enhancement of Zn resistance (van der Zaal et al. 1999). The increased tolerance to Zn is thought to be due to sequestration of Zn into vacuoles through the ectopically expressed AtMTP1. However, neither the subcellular location of AtMTP1 nor the Zn sequestration organelle has been reported.

P_{1B}-ATPases Family

Another family of transporters involved in zinc efflux is the P_{1B}-ATPase. *Arabidopsis* has been demonstrated with eight genes encoding P_{1B}-ATPases with difference in their structure, function, and regulation (Eren and Arguello 2004). Among these, HMA1, HMA2, HMA3, and HMA4 were observed to involve in zinc transport (Hussain et al. 2004). AtHMA1 is present in the chloroplast envelop and can contribute to Zn detoxification under excess zinc conditions (Kim et al. 2009). The AtHMA2 gene encodes a Zn^{+2}-ATPase located in the plasma membrane. The gene expression is induced by cadmium and zinc (Eren and Arguello 2004).

Since zinc hyperaccumulator species were observed to show higher expression of HMA3 than the non-hyperaccumulator ones, such as *Arabidopsis*, therefore, the AtHMA3 protein possibly mediates zinc hyperaccumulation (Becher et al. 2004; van de Mortel et al. 2006; Hassan and Aarts 2011). Recently, HMA3 has been cloned from *Thlaspi caerulescens* Alpine Penny-cress (Ueno et al. 2011) and rice (Ueno et al. 2010); therefore, it is a vacuolar influx transporter, important for cadmium tolerance in both species. This suggests that different homologues of HMA3 may have different metal-substrate specificity. AtHMA4, similar to AtHMA2 and acting alike, plays an important role in translocation of zinc, specifically in loading of zinc into the xylem (van de Mortel et al. 2006; Waters and Sankaran 2011). Hussain et al. (2004) reported that both HMA2 and HMA4 are essential to zinc homeostasis and they show a functional redundancy.

Table 11.1 Proposed specificity and location of the main transition metal transporters

Name	Family members in *A. thaliana*	Proposed specificity	Cellular location	Main tissue expression
Heavy metal ATPases(P_{1B})	8			
RANI		Cu	Post-Golgi	Whole plant
PAA1		Cu	Chloroplast	
Ca^{2+}-ATPases (P_{2A})	4			
ECA1		Mn^{2+}	ER	Roots
Nramp	6 (+EIN2)			
AtNramp 1		Fe, Mn		Roots
AtNramp 2		–		Root
AtNramp 3		Fe, Cd, Mn	Vacuole	Roots/shoot
AtNramp 4		Fe, Mn		Roots/shoots
OsNramp 1		Mn, Fe		Roots
OsNramp 2		Mn		Leaves
OsNramp 3		Mn		Roots/leaves
LeNramp1		Fe		Root vascular parenchyma
CDF	8			
ZAT (AtMTP1)		Zn	Vesicular/vacuolar	All tissues
TcZTP1		Zn	Intracellular membranes	Leaves, roots
TgMIP1		Cd, Co, Zn, Ni	Vacuole	Leaves
ShMTP1		Mn	Intracellular membranes	Roots, leaves
ZIP	15			
IRT1		Fe, Zn, Mn, Cd	PM	Roots
IRT2		Fe, Zn		Roots
OsIRT1		Fe		Roots
LeIRT1		Fe (broad)		Roots
LeIRT2		broad		Roots
TcZNT1		Zn, Cd		Roots, shoots
TcZNT2		±		Roots
ZIPs1-3		Zn		Roots
ZIP4		Zn	Plastids	Roots, shoots
GmZIP1		Zn	Peribacteroid membrane	Root nodules
Cation/H+ antiporters	12			
CAX 2		Ca, Mn, Cd	Vacuole	Roots
ABC transporters	128	PC-Cd, GS-Cd		
ID17		Fe		
COPT1-5	5	Cu		
GNGC channels	20	Ni, Pb, PM		

Source: Williams and Hall (2000)

11.4 Phytosiderophores and Metal Chelators

Roots are well known to change the rhizosphere chemistry by altering the rhizosphere pH (Wang et al. 2006) and/or releasing PS that can chelate soil Zn. Therefore, by both of the ways, it increases Zn availability (Cakmak et al. 1994). This root-mediated decrease in pH caused increases in Zn availability by solubilizing the Zn present in inorganic and organic soil complexes (Hacisalihoglu and Kochian 2003). The role of root exudates in relation to Zn efficiency has been demonstrated by a number of scientific evidences. Generally, PS are chemically identified as non-proteinogenic amino acids released from roots of various plants especially of graminaceous species under Fe- (Marschner 1995) and Zn-deficiency (Zhang et al. 1991; Kochian 1993) situations. These release compounds were found highly efficacious in mobilizing and complexing Zn in calcareous soils. Zhang et al. 1991 reported that PS release from the root apoplast of wheat plants caused mobilization of Zn and may also be involved in the translocation and solubility of Zn within plants (Welch 1995). It was also observed that under Zn-deficiency situations, release of PS in Zn-inefficient durum wheat is approximately 6–8 times lower than that in Zn-efficient bread wheat (Cakmak et al. 1996, 1998).

In the case of rice when grown in nutrient solution, the efficiency of Zn uptake was found highly to be correlated with the exudation rates of low molecular weight organic anions (Hoffland et al. 2006). Some of the root-exuded metal chelators were observed to play important role in metal homeostasis. One of the most important examples of this kind of compound is nicotianamine (NA) made by the action of NA synthase; this is known as a strong metal-chelating compound for binding of zinc, iron, copper, as well as nickel (Curie et al. 2009; Delhaize et al. 2003). NA is thought to be involved in long-distance transport, perhaps also playing a role in the entry of metals into the phloem or xylem through metal-NA chelate transporters of the yellow stripe-like (YSL) family (Gendre et al. 2007). NA synthase is encoded by four genes in *Arabidopsis*, AtNAS1–AtNAS4, which act functionally redundant, although they show different expression patterns, suggesting that each NAS gene may have a specialized function (Klatte et al. 2009). As an example, only AtNAS2 and AtNAS4 are highly expressed in roots under zinc deficiency (Van de Mortel et al. 2006).

11.5 Mechanism of Improving Zn Availability

Zn is an essential micronutrient for all living organisms. Approximately 40 % of the world's human population is zinc deficient; moreover this problem is worse in developing countries where people depend on cereal-rich-based diets for their sustenance. The improvement of plants for their ability to mitigate zinc deficiency and in turn to improve crop yield under zinc-limiting conditions (Reid et al. 1996) has thus far been hampered by lack of knowledge of the mechanisms and regulation of the zinc homeostasis network in plants. Therefore, by properly understanding the zinc homeostasis network in plants, bio-fortification of crops with zinc by using effective application of fertilizers, plant breeding, and other genetic approaches can offer a sustainable solution to this global problem. Various strategies such as application of chemical fertilizers, effectively chelated Zn, organic amendments, and bioinoculants are advocated to increase Zn concentration in the rhizosphere.

11.5.1 Effective Application of Zinc

Except some of the very insoluble materials, all zinc carriers are effective sources of zinc for crops, provided they are properly applied (Hodgson 1963). Zinc-containing materials may either be broadcast on the soil or thoroughly incorporated or used in a band near the seed at planting time. These materials are generally advocated to apply at a rate enough to grow the current crop; however, with this element it may be more feasible to increase the soil zinc content in order to ensure a multiyear supply. Granular zinc sulfate or finely divided zinc oxide or carbonates are cost-effective carriers of Zn and can easily be broadcast. These materials can also be applied as band application beside the seed at planting time. It was always found profitable to include a dose of zinc if we are going to apply approximately 10 lb of nitrogen per acre. Phosphorus is also an essential nutrient needed for good crop, and Zn can be applied effectively with liquid phosphatic fertilizers. For band application of Zn, carrier selection is more crucial; moreover, selection of suitable carriers plays more role under severe zinc-deficient situations, because the material is not well distributed in the soil. Under severe Zn-deficient conditions, if dry, granular fertilizer is chosen as a source of Zn as band application, then the organic zinc chelates should be preferred because chelated materials were found more effective than the other zinc carriers. To give satisfactory results with liquid polyphosphate fertilizers, finely divided zinc carriers should be used; however, under very extreme zinc-deficiency conditions, the role of zinc carrier does not show any difference when applied in a band with liquid polyphosphate fertilizers. Under these extreme deficiency conditions, it is difficult to supply enough zinc in a band; thus, broadcast treatment with thoroughly incorporation is likely to be more effective.

Zn deficiency can be mitigated by applying Zn-containing materials properly. Generally, three basic types of compounds are available for Zn fertilization; these include inorganic mineral compounds, synthetic chelates, and natural organic materials; however, their performance greatly depends on their water-solubility properties. Some of the common sources of Zn fertilizers are listed below in Table 11.2.

11.5.2 Organic Amendments

Soil organic matter is considered a very important factor in nutrient mobility in soil. Various organic amendments, such as compost, farmyard manure (FYM), poultry manure, olive husk, etc., are applied to soil for improving its health, fertility, as well as crop yields. The organic material can improve the availability of Zn by releasing Zn with time and by altering the physicochemical properties of soil. These properties may increase soluble/available fraction of Zn in soil for plant uptake. Moreover, application of organic amendments also improves biological properties of the soil (Tejada et al. 2006). For instance, microbial biomass and soil enzyme activities are substantially increased with application of organic amendments (Blagodatsky and Richter 1998; Liang et al. 2003). This increase in microbial population and activities is an important indicator of soil health and soil productivity. Soils having more microbial biomass and microbial activities are supposed to be productive soils as they have good nutrient mobility and availability to plants. Organic amendments improve soil microbial biomass carbon (Cmic) and the Zn content in soil and plant tissues.

11.5.3 Rhizosphere Microflora

Rhizosphere comprises the narrow zone of the soil around roots that is directly influenced by root secretions and is considered a hot spot of microflora, having manifold increase in microbial population than bulk soil. All the microbial communities residing in this region constitute rhizosphere microflora. The rhizosphere microflora may benefit plants through multifarious mechanisms including fixation of atmospheric nitrogen; mobilization of nutrients; production of phytohormones; altering indigenous level of phytohormones; improving plant stress tolerance to salinity, toxicity, drought, metal, and pesticide load; and also acting as a biocontrol agent (Glick and Bashan 1997; Lucy et al. 2004; Khalid et al. 2009).

Although, each and every mechanism has its own significance, but mobilization of nutrients by microflora has considered the most crucial function they perform in order to improve nutrient content in plant tissues. There is a good deal of research on mobilization of phosphorus in the rhizosphere through these tiny creatures, but

Table 11.2 Commonly used Zn supplements

Name of supplements and their chemical type	Chemical formula	Zinc content (%)
Inorganic compounds		
Zinc sulfate monohydrate	$ZnSO_4.H_2O$	36
Zinc sulfate heptahydrate	$ZnSO_4.7H_2O$	22
Zinc oxysulfate	$xZnSO_4.xZnO$	20–50
Basic zinc sulfate	$ZnSO_4.4Zn(OH)_2$	55
Zinc oxide	ZnO	50–80
Zinc carbonate	$ZnCO_3$	50–56
Zinc chloride	$ZnCl_2$	50
Zinc nitrate	$Zn(NO_3)_2.3H_2O$	23
Zinc phosphate	$Zn_3(PO_4)_2$	50
Zinc frits	Fritted glass	10–30
Ammoniated zinc sulfate solution	$Zn(NH_3)_4SO_4$	10
Organic compounds		
Disodium zinc EDTA	$Na_2ZnEDTA$	8–14
Sodium zinc HEDTA	NaZnHEDTA	6–10
Sodium zinc EDTA	NaZnEDTA	9–13
Zinc polyflavonoid		5–10
Zinc lignosulfonate		5–8

increase in Zn bioavailable fraction in the rhizosphere due to the activities of rhizosphere microbes has not been well explored yet. However, there are sufficient reports indicating substantial potential of these microbes in improving Zn bioavailable fraction in the rhizosphere of plants and Zn content in plant tissues. Among microbes, both bacteria and fungi have shown tremendous ability to improve plant-available Zn in the rhizosphere and also increase Zn in plant parts (Whiting et al. 2001; Fasim et al. 2002; Biari et al. 2008; Subramanian et al. 2009). The ways through which rhizosphere microflora may cause mobilization/solubilization of Zn include reduction in soil pH (Koide and Kabir 2000; Subramanian et al. 2009), chelation (Whiting et al. 2001), or through improving root growth and root absorptive area (Burkert and Robson 1994). These mechanisms vary from one microorganism to other acquisition/uptake in plant tissues. Importance and examples of these mechanisms have been discussed in detail here.

11.5.3.1 Reduction in pH

Availability of micronutrients in the soil is very much sensitive to soil. A little change in soil pH may have a great impact on micronutrient mobility/solubility in soil. It has been reported that availability of Zn decreases 100 times with one unit increase in pH (Havlin et al. 2014). Thus by decreasing the pH of alkaline soil, bioavailable fraction of Zn can be enhanced to an appreciable level. Rhizosphere microflora has been reported to lower the soil pH to a good extent (Wu et al. 2006), which may occur due to secretion of some organic acids and protons extrusion (Fasim et al. 2002). For instance, *Pseudomonas fluorescens* secreted gluconic acid and 2-ketogluconic acid in the culture during solubilization of Zn phosphate. In addition, concentration of protons was also found higher in the culture after incubation period (Simine et al. 1998). Likewise, Fasim et al. (2002) observed that solubilization of Zn oxide and phosphate was accompanied by proton extrusion and production of 2-ketogluconic acid. Martino et al. (2003) documented that ericoid mycorrhizal fungi secreted organic acid to solubilize Zn from insoluble ZnO and Zn_3 $(PO_4)_2$. A change in pH was observed when *Pseudomonas* and *Bacillus* spp. were used to solubilize ZnS, ZnO, and ZnCO3 in broth culture (Saravanan et al. 2004). Koide and Kabir (2000) proposed that mycorrhizal plants facilitate Zn availability by lowering the pH of the soil by the release of some organic acids.

11.5.3.2 Zn Chelation

Zinc ions have high interaction with the soil constituent due to which its persistency in the soil solution is very low (Alloway 2009). Due to low persistency/high reactivity of Zn in soil solution, plant-available fraction of Zn in the soil is poor. However, bioavailability of Zn could be increased by means of Zn-chelating compounds (Obrador et al. 2003). These compounds are either synthetic or synthesized and released by the plant roots and potential rhizosphere microflora into the rhizosphere to chelate the Zn and improve its bioavailability. The chelates of microflora are the metabolites, which form complexes with metal cations like Zn^{2+} (Tarkalson et al. 1998), which reduces their reaction with the soil. These Zn chelates subsequently move toward the roots and release chelating ligand (Zn^{2+}) at the root surface, making them free to chelate another Zn ion from the soil solution.

In some microorganisms, chelation has been observed as dominant phenomena to improve bioavailability and uptake by plant roots. For instance, Whiting et al. (2001) suggested that possible mechanism used by bacteria (*Microbacterium saperdae*, *Pseudomonas monteilii*, and *Enterobacter cancerogenes*) for increasing water soluble Zn (bioavailable) in soil was the production of Zn-chelating metallophores. In another report, Tariq et al. (2007) found that *Azospirillum lipoferum* (JCM-1270, ER-20), *Pseudomonas* sp. (96-51), and *Agrobacterium* sp. (Ca-18) mobilized Zn and made it bioavailable for longer period of time when they were applied as a biofertilizer to rice by producing chelating agent like ethylenediaminetetraacetate (EDTA).

11.5.3.3 Changes in Root Architecture

Zinc is immobile in soil and is taken up by plant mainly by diffusion (Imran et al. 2014). Due to poor native bioavailable Zn and low exogenous supply, depletion zones are formed around roots. Therefore, to improve Zn uptake, it should be in close proximity to roots. This can be achieved either by application of more Zn or improving root growth and surface area so that roots can take nutrients beyond the depletion zone. A rhizosphere microflora especially mycorrhizal fungus is widely known for its impact on root architecture. Mycorrhizal plants uptake Zn over more distances, crossing the depletion zone. According to Burkert and Robson (1994), arbuscular mycorrhizae can acquire Zn from a distance of 40 mm from the root surface. Jansa et al. (2003) noted that *Glomus intraradices* can take up Zn from a distance of 50 mm from the roots of maize. In the absence of Zn fertilization, Subramanian et al. (2009) observed that mycorrhizal fungus significantly increased root length, spread, and volume compared to the plants without fungal inoculation, and this increased the Zn concentration in the grain up to 4 %.

11.6 Bioinoculants

Several studies have revealed that bioinoculants help in mitigation of Zn deficiency in plants through improving mobilization of Zn in the soil. Many bacterial and fungal strains have been found capable of solubilizing fixed Zn and consequently increasing its uptake by plants. The role of fungal and bacterial inoculants in improving availability of Zn to plants is comprehensively discussed below.

11.6.1 Fungal Inoculants

Among the fungal inoculants, AM fungus is considered highly effective in improving the availability and absorption of immobile nutrients by higher plants (Ikram et al. 1992; Tarafdar and Marschner 1994; Liu et al. 2000). AM fungi are well known in improving the availability of phosphorus to plant roots. It has also been reported that mycorrhizal symbiosis is also very effective in improving availability of Zn to plants (Ortas et al. 2002; Gao et al. 2007; Ryan et al. 2007; Subramanian et al. 2009). There are good reports about increase in Zn uptake by the application of bioinoculants (Smith and Read 1997; Liu et al. 2000; Ryan and Angus 2003;

Subramanian et al. 2009), which might have occurred through the increase in bioavailable Zn in soil. For instance, Swaminathan and Verma (1979) observed a great improvement in bioavailable Zn fraction in the soil through fungal (*Glomus macrocarpum*) treatment which subsequently increased the Zn concentration in the leaves of wheat, maize, and potato grown on Zn-deficient soils. Likewise, Subramanian et al. (2009) found that inoculation of *Glomus intraradices* caused an overall increase of 43 % in bioavailable Zn in the soil after 75 days. Response of AM fungus inoculation has also been found very promising in terms of Zn accumulation in shoot and leaves. Giri et al. (2005) found about 15 times more Zn in the shoots of + AM compared to -AM plants. Chen et al. (2003) also recorded substantial increase in shoot Zn concentration through fungal inoculation. Similarly, mycorrhizal infection also improved concentration of Zn in the leaves of AM plants compared to uninoculated plants (Swaminathan and Verma 1979; Wu et al. 2011).

11.6.2 Bacterial Inoculants

Like fungi, bacterial bioinoculants are also helpful in increasing solubilization/availability of Zn in the soil and its further uptake by plants to improve plant Zn content. Several bacterial species have been reported that are able to solubilize insoluble Zn compounds in liquid medium (Simine et al. 1998; Fasim et al. 2002; Saravanan et al. 2007) and in the soil (Whiting et al. 2001; Tariq et al. 2007). For instance, in an in vitro study, Saravanan et al. (2004) found that *Pseudomonas* and *Bacillus* can solubilize various Zn compounds like ZnS, ZnO, and $ZnCO_3$ to a good extent in liquid medium. Likewise, in another study, Saravanan et al. (2007) demonstrated Zn solubilizing potential of *Glomus diazotrophicus* PA15. Inoculation with *G. diazotrophicus* PA15 resulted in 41, 15.7, and 60 times increase in soluble Zn content in the cases of ZnO, $ZnCO_3$, and $Zn_3(PO_4)_2$, respectively, after 48 h of incubation compared to uninoculated control. Similarly, Fasim et al. (2002) found a high potential of *Pseudomonas aeruginosa* to solubilize ZnO in liquid medium. Bacteria have also shown high mobilization of soil Zn as Tariq et al. (2007) observed almost 5.6 time higher bioavailable Zn in inoculated soil compared to uninoculated soil. Whiting et al. (2001) have also documented about 0.45-fold increase in bioavailable Zn in rhizosphere soil through bacterial inoculation. It has also been widely reported that bacterial inoculation improves plant Zn content (Whiting et al. 2001; Sadaghiani et al. 2008; Biari et al. 2008).

11.7 Improved Genotypes for Zn Use Efficiency

Plant genotypes vary widely in their tolerance to soils with low plant-available Zn with respect to both Zn uptake and utilization. Tolerance of plant genotypes to Zn deficiency, as a genetic trait, is usually referred to as Zn efficiency and defined as the ability of a cultivar to grow and yield well in soils that are too deficient in Zn to support a standard cultivar. The physiological and molecular mechanisms of Zn-deficiency tolerance are just beginning to be understood, and these mechanisms can be exploited in crop breeding programs (Hacisalihoglu and Kochian 2003) For example, Zn-efficient genotypes with better Zn utilization may contain higher amounts of chelators that bind Zn and increase its physiological availability at the cellular level. A better understanding of the physiological, morphological, and genetic bases of Zn efficiency is needed for the development of fast, simple, and reliable screening procedures for identifying and breeding genotypes for Zn efficiency. The first step in breeding for Zn efficiency is the assessment of a large number of segregating populations from crosses of Zn-efficient and Zn-inefficient parents.

11.8 Conclusion

Since Zn is involved in various plant physiological processes, therefore, it is a very important micronutrient for plant health and yield, and to get maximum yield, it should be available to plants in appropriate quantity. Although most of the soils have a fair quantity of Zn, unfortunately most of the cases posses extremely low plant-available fraction due to various soil factors. Therefore, recommendations were always made to apply Zn from exogenous sources in order to maintain the appropriate concentration of it in soil solution. Various chemical fertilizers, chelated Zn, and organic fertilizers are available for this purpose. Potential of microbial biotechnology has also been tested to improve the indigenous Zn availability and reduce the fixation of applied Zn. Application of various fungal and bacterial bioinoculants to the soil had shown very promising results in terms of improving Zn content in soil and plant tissues and improving yield. Apart from all these factors, high Zn efficiency in crops also appears to be related to various morphological and physiological traits, such as root surface area, Zn-mobilizing root exudates, and better utilization of Zn at the cellular level. In terms of understanding the roles of genes involved in uptake and translocation of zinc in plants, a lot has been achieved, but information on where in the plant each transporter functions and how each one is controlled in response to nutrient availability remains still unclear. The identification of transcription factors involved in the control of zinc-deficiency response offers interesting opportunities to modulate zinc-deficiency responsive gene expression to make plants less sensitive to zinc deficiency or to confer a constitutive zinc-deficiency response, which can induce plants to overaccumulate metals. Therefore, understanding the regulator identity of the zinc homeostasis network in plants should provide new insights for the development of crops in areas suffering from low zinc bioavailability and for biofortification strategies. Furthermore, understanding how zinc interacts with other metals, such as cadmium and lead, during its absorption is important to be known, avoiding undesirable accumulation of such heavy metals in plants.

References

Alloway BJ (2008) Zinc in soils and crop nutrition. International zinc Association and International Fertilizer Industry Association, Brussels/Paris, p 130

Alloway BJ (2009) Soil factors associated with zinc deficiency in crops and humans. Environ Geochem Health 31:537–548

Assuncao AGL, Herrero E, Lin YF, Huettel B, Talukdar S, Smaczniak C, Immink RGH, Eldik MV, Fiers M, Schat H, Aarts MGM (2010) *Arabidopsis thaliana* transcription factors bZIP19 and bZIP23 regulate the adaptation to zinc deficiency. Proc Natl Acad Sci U S A 107:10296–10301

Becher M, Talke IN, Krall L, Kramer U (2004) Cross-species microarray transcript profiling reveals high constitutive expression of metal homeostasis genes in shoots of the zinc hyperaccumulator *Arabidopsis halleri*. Plant J 37(2):251–268

Biari A, Gholami A, Rahmani HA (2008) Growth promotion and enhanced nutrient uptake of maize (*Zea mays* L.) by application of plant growth promoting rhizobacteria in arid region of Iran. J Biol Sci 8:1015–1020

Blagodatsky SA, Richter O (1998) Microbial growth in soil and nitrogen turnover: a theoretical model considering the activity state of microorganisms. Soil Biol Biochem 30:1743–1755

Bloß B, Clemens S, Nies DH (2002) Characterization of the ZAT1p zinc transporter from Arabidopsis thaliana in microbial model organisms and reconstituted proteoliposomes. Planta 214:783–791

Burkert B, Robson A (1994) 65Zn uptake in subterranean clover (*Trifolium subterraneum* L.) by three vesicular – arbuscular mycorrhizal fungi in a root – free sandy soil. Soil Biol Biochem 26:1117–1124

Cakmak I (2011) Breeding, transformation and physiological strategies for the development of wheat with high zinc and iron grain concentration. In: Bonjean A (ed) The world wheat book. Lavoisier, Paris, New York

Cakmak I, Gulut KY, Marschner H, Graham RD (1994) Effect of zinc and iron deficiency on phytosiderophore release in wheat genotypes differing in zinc efficiency. J Plant Nutr 17:1–17

Cakmak I, Sary N, Marschner H, Ekiz H, Kalaycy M, Yilmaz A, Braun HJ (1996) Phytosiderophore release in bread and durum wheat genotypes differing in zinc efficiency. Plant Soil 180:183–189

Cakmak I, Torun B, Erenoglu B, Ozturk L, Marschner H, Kalayci M, Ekiz H, Yilmaz A (1998) Morphological and physiological differences in the response of cereals to zinc deficiency. Euphytica 100:349–357

Cakmak I, Pfeiffer W, Mcclafferty B (2010) Biofortification of durum wheat with zinc and iron. Cereal Chem 87:10–20

Chen BD, Li XL, Tao HQ, Christie P, Wong MH (2003) The role of arbuscular mycorrhiza in zinc uptake by red clover growing in a calcareous soil spiked with various quantities of zinc. Chemosphere 50:839–846

Ciftci-Yilmaz S, Mittler R (2008) The zinc finger network of plants. Cell Mol Life Sci 65:1150–1160

Clemens S (2001) Molecular mechanisms of plant metal tolerance and homeostasis. Planta 212:475–486

Clemens S, Palmgren M, Kramer U (2002) A long way ahead: understanding and engineering plant metal accumulation. Trends Plant Sci 7:309–315

Colangelo E, Guerinot M (2004) Put the metal to the petal: metal uptake and transport throughout plants. Curr Opin Plant Biol 9:322–330

Curie C, Cassin G, Couch D, Divol F, Higuchi K, Le Jean M (2009) Metal movement within the plant: contribution of nicotianamine and yellow stripe 1-like transporters. Ann Bot 103:1–11

Delhaize E, Kataoka T, Hebb DM, White RG, Ryan PR (2003) Genes encoding proteins of the cation diffusion facilitator family that confer manganese tolerance. Plant Cell 15:1131–1142

Drager DB, Desbrosses-Fonrouge AG, Krach C, Chardonnens AN, Meyer RC, Saumitou-Laprade P, Kramer U (2004) Two genes encoding *Arabidopsis halleri* MTP1 metal transport proteins co-segregate with zinc tolerance and account for high *MTP1* transcript levels. Plant J 39:425–439

Eckhardt U, Mas Marques A, Buckhout TJ (2001) Two iron-regulated cation transporters from tomato complement metal uptake-deficient yeast mutants. Plant Mol Biol 45:437–448

Eide D, Guarente L (1992) Increased dosage of a transcriptional activator gene enhances iron-limited growth of *Saccharomyces cerevisiae*. J Gen Microbiol 138:347–354

Eide D, Broderius M, Fett J, Guerinot M (1996) A novel iron-regulated metal transporter from plants identified by functional expression in yeast. Proc Natl Acad Sci U S A 93:5624–5628

Englbrecht CC, Schoof H, Bohm S (2004) Conservation, diversification and expansion of C2H2 zinc finger proteins in the *Arabidopsis thaliana* genome. BMC Genomics 5:39–55

Eren E, Arguello J (2004) *Arabidopsis* HMA2, a divalent heavy metal-transporting P_{IB}-Type ATPase, is involved in cytoplasmic Zn^{2+} homeostasis. Plant Physiol 136:3712–3723

Fasim FN, Ahmed R, Parsons GM (2002) Solubilization of zinc salts by bacterium isolated by the air environment of tannery. FEMS Microbiol Lett 213:1–6

Gaither LA, Eide DJ (2000) Functional expression of the human hZIP2 zinc transporter. J Biol Chem 275 (8):5560–5564

Gaither LA, Eide DJ (2001) Eukaryotic zinc transporters and their regulation. Biometals 14:251–270

Gao X, Kuyper TW, Zou CF, Zhang HE (2007) Mycorrhizal responsiveness of aerobic rice genotypes is negatively correlated with their zinc uptake when nonmycorrhizal. Plant Soil 290:283–291

Gendre D, Czernic P, Conejero G, Pianelli K, Briat JF, Lebrun M, Mari S (2007) TcYSL3, a member of the YSL gene family from the hyperaccumulator *Thlaspi caerulescens*, encodes a nicotianamine-Ni/Fe transporter. Plant J 49:1–15

Giri B, Giang PH, Kumari R, Prasad R, Varma A (2005) Microbial diversity in soils. In: Buscot F, Varma S (eds) Micro-organisms in soils: roles in genesis and functions. Springer, Heidelberg, pp 195–212

Glick BR, Bashan Y (1997) Genetic manipulation of plant growth-promoting bacteria to enhance biocontrol of phytopathogens. Biotechnol Adv 15:353–378

Grass G, Wong MD, Rosen BP, Smith RL, Rensing C (2002) ZupT is a Zn(II) uptake system in *Escherichia coli*. J Bacteriol 184:864–866

Grotz N, Fox T, Connolly E, Park W, Guerinot ML, Eide D (1998) Identification of a family of zinc transporter genes from *Arabidopsis* that respond to zinc deficiency. Proc Natl Acad Sci U S A 95:7220–7224

Guerinot ML (2000) The ZIP family of metal transporters. Biochim Biophys Acta 1465:190–198

Gurmani AR, Khan SU, Andaleep R, Waseem K, Khan A (2012) Soil application of zinc improves growth and yield of tomato. Int J Agric Biol 14:91–96

Gustin JL, Zanis MJ, Salt DE (2011) Structure and evolution of the plant cation diffusion facilitator family of ion transporters. BMC Evol Biol 11:76

Hacisalihoglu G, Kochian LV (2003) How do some plants tolerate low levels of soil zinc? Mechanisms of zinc efficiency in crop plants. New Phytol 159(2):341–350

Hacisalihoglu G, Hart JJ, Kochian LV (2001) High-and low-affinity zinc transport systems and their possible role in zinc efficiency in bread wheat. Plant Physiol 125:456–463

Hacisalihoglu G, Hart JJ, Wang Y, Cakmak I, Kochian LV (2003) Zinc efficiency is correlated with enhanced expression and activity of Cu/Zn superoxide dismutase and carbonic anhydrase in wheat. Plant Physiol 131:595–602

Hajiboland R, Amirazad F (2010) Growth, photosynthesis and antioxidant defense system in Zn-deficient red cabbage plants. Plant Soil Environ 56:209–217

Hassan Z, Aarts M (2011) Opportunities and feasibilities for biotechnological improvement of Zn, Cd or Ni tolerance and accumulation in plants. Environ Exp Bot 72:53–63

Havlin JL, Tisdale SL, Nelson WL, Beaton JD (2014) Soil fertility and nutrient management: an introduction to nutrient management, 8th edn. Pearson, Upper Saddle River, p 505

Hodgson JF (1963) Chemistry of the micronutrient elements in soils. Adv Agron 15:119–159

Hoffland E, Wei C, Wissuwa M (2006) Organic anion exudation by lowland rice (*Oryza sativa* L.) at zinc and phosphorus deficiency. Plant Soil 283(1):155–162

Hussain D, Haydon MJ, Wang Y, Wong E, Sherson SM, Young J, Camakaris J, Harper JF, Cobbett CS (2004) P-type ATPase heavy metal transporters with roles in essential zinc homeostasis in *Arabidopsis*. Plant Cell 16:1327–1339

Ikram A, Mahmud AW, Ghani MN, Ibrahim MT, Zainal AB (1992) Field nursery inoculation of *Hevea brasiliensis* Muell. Arg. seedling rootstock with vesicular–arbuscular mycorrhizal (VAM) fungi. Plant Soil 145:231–236

Imran M, Arshad M, Khalid A, Kanwal S, Crowley DE (2014) Perspectives of rhizosphere microflora for improving Zn bioavailability and acquisition by higher plants. Int J Agric Biol 16:653–662

Jansa J, Mozafar A, Frossard E (2003) Long–distance transport of P and Zn through the hyphae of an arbuscular mycorrhizal fungus in symbiosis with maize. Agron 23:481–488

Khalid A, Arshad M, Shaharoona B, Mahmood T (2009) Plant growth promoting rhizobacteria and sustainable agriculture. In: Khan MS, Zaidi A, Musarrat J (eds) Microbial strategies for crop improvement. Springer, Berlin/Heidelberg, pp 133–160

Kim YY, Choi H, Segami S, Cho HT, Martinoia E, Maeshima M, Lee Y (2009) AtHMA1 contributes to detoxification cation of excess Zn(II) in *Arabidopsis*. Plant J 58:737–753

Klatte M, Schuler M, Wirtz M, Fink-Straube C, Hell R, Bauer P (2009) The analysis of *Arabidopsis nicotianamine* synthase mutants reveals functions for nicotianamine in seed iron loading and iron deficiency responses. Plant Physiol 150:257–271

Kochian LV (1993) Zinc absorption from hydroponic solutions by plant roots. In: Robson AD (ed) Zinc in soils and plants. Kluwer, Dordrecht, pp 45–57

Kochian LV (2000) Molecular physiology of mineral nutrient acquisition, transport, and utilization. Biochem Mol Biol Plants 20:1204–1249

Koide RT, Kabir Z (2000) Extra radical hyphae of the mycorrhizal fungus *Glomus intraradices* can hydrolyse organic phosphate. New Phytol 148:511–517

Kupper H, Zhao FJ, McGrath SP (1999) Cellular compartmentation of zinc in leaves of the hyperaccumulator *Thlaspi caerulescens*. Plant Physiol 119:305–311

Lasat MM, Kochian LV (2000) Physiology of Zn hyperaccumulation in *Thlaspi caerulescens*. In: Terry N, Banuelos G (eds) Phytoremediation of contaminated soil and water. CRC Press/LLC, Boca Raton, pp 159–169

Lasat MM, Baker AJ, Kochian LV (1996) Physiological characterization of root Zn^2+ absorption and translocation to shoots in Zn hyperaccumulator and nonaccumulator species of *Thlaspi*. Plant Physiol 112:1715–1722

Liang YC, Chen Q, Liu Q, Zhang WH, Ding RX (2003) Exogenous silicon (Si) increases antioxidant enzyme activity and reduces lipid peroxidation in roots of salt – stressed barley (*Hordeum vulgare* L.). J Plant Physiol 160:1157–1164

Liu A, Hamel C, Hamilton RI, Ma BL (2000) Acquisition of Cu, Zn, Mn and Fe by mycorrhizal maize (*Zea mays* L.) grown in soil at different P and micronutrient levels. Mycorrhiza 9:331–336

Lucy M, Reed E, Glick BR (2004) Applications of free living plant growth – promoting rhizobacteria. Anton Leeuw 86:1–25

Luo ZB, He XJ, Chen L, Tang L, Gao S, Chen F (2010) Effects of zinc on growth and antioxidant responses in *Jatropha curcas* seedlings. Int J Agric Biol 12:119–124

MacDiarmid CW, Gaither LA, Eide D (2000) Zinc transporter that regulate vacuolar Zinc storage in *Saccharomyces cerevisiae*. EMBO J 19(12):2845–2855

Mandal B, Hazra GC, Pal AK (1988) Transformation of zinc in soils under submerged condition and its relation with zinc nutrition of rice. Plant Soil 106:121–126

Marschner H (1995) Mineral nutrition of higher plants, 2nd edn. Academic, London

Marschner H (1997) Mineral nutrition of higher plants. Academic, San Diego, p 889

Martino E, Perotto S, Parsons R, Gadd GM (2003) Solubilization of insoluble inorganic zinc compounds by ericoid mycorrhizal fungi derived from heavy metal polluted sites. Soil Biol Biochem 35:133–141

McCall KA, Huang C, Carol AF (2000) Function and mechanism of zinc metalloenzymes. J Nutr 130:1437–1446

Moreau S, Thomson RM, Kaiser BN, Trevaskis B, Guerinot ML, Udvardi MK, Puppo A, Day DA (2002) GmZIP1 encodes a symbiosis-specific zinc transporter in soybean. J Biol Chem 277:4738–4746

Obrador A, Novillo J, Alvarez JM (2003) Mobility and availability to plants of two zinc sources applied to a calcareous soil. Soil Sci Soc Am J 67:564–572

Ortas I, Ortakçi D, Kaya Z, Cinar A, Onelge N (2002) Mycorrhizal dependency of sour orange in relation to phosphorus and zinc nutrition. J Plant Nutr 25:1263–1279

Palmer C, Guerinot M (2009) Facing the challenges of Cu, Fe and Zn homeostasis in plants. Nat Chem Biol 5:333–340

Pence NS, Larsen PB, Ebbs SD, Letham DLD, Lasat MM, Garvin DF, Eide D, Kochian LV (2000) The molecular physiology of heavy metal transport in the Zn/Cd hyperaccumulator *Thlaspi caerulescens*. Proc Natl Acad Sci U S A 97:4956–4960

Persans MW, Nieman K, Salt DE (2001) Functional activity and role of cation-efflux family members in Ni hyperaccumulation in *Thlaspi goesingense*. Proc Natl Acad Sci U S A 98:9995–10000

Reid RJ, Brookes JD, Tester MA, Smith FA (1996) The mechanism of zinc uptake in plants. Planta 198:39–45

Rengel Z, Wheal MS (1997) Kinetic parameters of zinc uptake by wheat are affected by the herbicide chlorsulfuron. J Exp Bot 48:935–941

Robson AD (1994) Zinc in soils and plants. Springer, New York

Ryan MH, Angus JF (2003) Arbuscular mycorrhizae in wheat and field pea crops on a low P soil: increased

Zn-uptake but no increase in P uptake or yield. Plant Soil 250:225–239

Ryan MH, McCully ME, Huang CX (2007) Relative amounts of soluble and insoluble forms of phosphorus and other elements in intraradical hyphae and arbuscules of arbuscular mycorrhizas. Funct Plant Biol 34:457–464

Sadaghiani MR, Barin M, Jalili F (2008) The effect of PGPR inoculation on the growth of wheat. International meeting on soil fertility land management and agroclimatology, Turkey, pp 891–898

Saravanan VS, Subramoniam SR, Raj SA (2004) Assessing in vitro solubilization potential of different zinc solubilizing bacterial (zsb) isolates. Brazil J Microbiol 35:121–125

Saravanan VS, Madhaiyan M, Thangaraju M (2007) Solubilization of zinc compounds by the diazotrophic, plant growth promoting bacterium *Gluconacetobacter diazotrophicus*. Chemosphere 66:1794–1798

Sattelmacher B (2001) The apoplast and its significance for plant mineral nutrition. New Phytol 149:167–192

Simine CDD, Sayer JA, Gadd GM (1998) Solubilization of zinc phosphate by a strain of *Pseudomonas fluorescens* isolated from a forest soil. Biol Fertil Soils 28(1):87–94

Smith S, Read DJ (1997) Mycorrhizal symbiosis. Academic, London, pp 11–33

Song WY, Choi KS, Kim Y, Geisler M, Park J, Vincenzetti V, Schellenberg M, Kim SH, Lim YP, Noh EW, Lee Y, Martinoia E (2010) *Arabidopsis* PCR2 is a zinc exporter involved in both zinc extrusion and long-distance zinc transport. Plant Cell 22:2237–2252

Subramanian KS, Tenshia V, Jayalakshmi K, Ramachandran V (2009) Role of arbuscular mycorrhizal fungus (*Glomus intraradices*)–(fungus aided) in zinc nutrition of maize. Agric Biotechnol Sustain Dev 1:29–38

Swaminathan K, Verma BC (1979) Response of three crop species to vesicular arbuscularmycorrhizal infection on Zinc deficient Indian soils. New Phytol 114:1–38

Tapiero H, Tew KD (2003) Trace elements in human physiology and pathology: zinc and metallothioneins. Biomed Pharmacother 57:399–411

Tarafdar JC, Marschner H (1994) Efficiency of VAM hyphae in utilization of organic phosphorus by wheat plants. Soil Sci Plant Nutr 40:593–600

Tariq M, Hameed S, Malik KA, Hafeez FY (2007) Plant root associated bacteria for zinc mobilization in rice. Pak J Bot 39:245–253

Tarkalson DD, Jolley VD, Robbins CW, Terry RE (1998) Mycorrhizal colonization and nutrition of wheat and sweet corn grown in manure-treated and untreated topsoil and subsoil. J Plant Nutr 21(9):1985–1999. doi:10.1080/01904169809365538

Tejada M, Hernandez MT, Garcia C (2006) Application of two organic amendments on soil restoration: effects on the soil biological properties. J Environ Qual 35:1010–1017

Ueno D, Yamaji N, Kono I, Huang CF, Ando T, Yano M, Feng J (2010) Gene limiting cadmium accumulation in rice. Proc Natl Acad Sci U S A 107:16500–16505

Ueno D, Milner MJ, Yamaji N, Yokosho K, Koyama E, Clemencia Zambrano M, Kaskie M, Ebbs S, Kochian LV, Feng J (2011) Elevated expression of TcHMA3 plays a key role in the extreme Cd tolerance in a Cd-hyperaccumulating ecotype of *Thlaspi caerulescens*. Plant J 66:852–862

van de Mortel JE, Almar Villanueva L, Schat H, Kwekkeboom J, Coughlan S, Moerland PD, Themaat EVL, Koornneef M, Arts MGM (2006) Large expression differences in genes for iron and zinc homeostasis, stress response, and lignin biosynthesis distinguish roots of *Arabidopsis thaliana* and the related metal hyperaccumulator *Thlaspi caerulescens*. Plant Physiol 42:1127–1147

van der Zaal BJ, Neuteboom LW, Pinas JE, Chardonnens AN, Schat H, Verkleij JAC, Hooykaas PJJ (1999) Overexpression of a novel *Arabidopsis* gene related to putative zinc transporter genes from animals can lead to enhanced zinc resistance and accumulation. Plant Physiol 119:1047–1055

Vert G, Grotz N, Dedaldechamp F, Gaymard F, Guerinot ML, Briat JF, Curie C (2002) IRT1, an *Arabidopsis* transporter essential for iron uptake from the soil and for plant growth. Plant Cell 14:1223–1233

Wang P, Bi S, Ma L, Han W (2006) Aluminum tolerance of two wheat cultivars (Brevor and Atlas66) in relation to their rhizosphere pH and organic acids exuded from roots. J Agric Food Chem 54:10033–10039

Waters B, Sankaran R (2011) Moving micronutrients from the soil to the seeds: genes and physiological processes from a biofortification perspective. Plant Sci 180:562–574

Welch RM (1995) Micronutrient nutrition of plants. Crit Rev Plant Sci 14:49–82

White PJ, Whiting SN, Baker AJM, Broadley MR (2002) Does zinc move apoplastically to the xylem in roots of *Thlaspi caerulescens*? New Phytol 153:201–207

Whiting SN, Desouza M, Terry N (2001) Rhizosphere bacteria mobilize Zn for hyperaccumulator by *Thlaspi caerulescens*. Environ Sci Technol 35:3144–3150

Willams DJ, Hall KB (2000) Experimental and computational studies of the G (UUCG)C RNA tetraloop. J Mol Biol 297(5):1045–1061

Wu SC, Cheung KC, Luo YM (2006) Wong effects of inoculation of plant growth promoting rhizobacteria on metal uptake by *Brassica juncea*. Environ Pollut 140:124–135

Wu QS, Li GH, Zou YN (2011) Roles of arbuscular mycorrhizal fungi on growth and nutrient acquisition of peach (*Prunus persica* L. Batsch) seedlings. J Anim Plant Sci 21:746–750

Zhang FS, Romheld V, Marschner H (1991) Diurnal rhythm of release of phytosiderophores and uptake rate of zinc in iron-deficient wheat. J Plant Nutr Soil Sci 37:671–678

Zhao H, Eide D (1996a) The *ZRT2* gene encodes the low-affinity zinc transporter in *Saccharomyces cerevisiae*. J Biol Chem 271:23203–23210

Zhao H, Eide DJ (1996b) The yeast *ZRT1* gene encodes the zinc transporter protein of a high-affinity uptake system induced by zinc limitation. Proc Natl Acad Sci U S A 93:2454–2458

Zhao FJ, Hamon RE, Lombi E, McLaughlin MJ, McGrath SP (2002) Characteristics of cadmium uptake in two contrasting ecotypes of the hyper-accumulator *Thlaspi caerulescens*. J Exp Bot 53:535–543

Zia MS, Aslam M, Baig MB, Ashraf A (1999) Fertility issues and fertilizer management in rice–wheat system: a review. Quart Sci Vis 5:59–73

12 Biofortification Through Breeding Interventions in Lentil

Jitendra Kumar, Rohit Kant, Syed Mohd Quatadah, Shiv Kumar, and Ashutosh Sarker

Abstract

Micronutrient deficiency affects more than two billion population worldwide especially in sub-Saharan Africa and South Asia. The deficiency in human body is commonly described as "hidden hunger" leading to a range of health hazards including low birth weight, anemia, learning disabilities, increased morbidity and mortality rates, and high health-care costs. Biofortification of food crop varieties with essential micronutrients is one of the potential means to address micronutrient deficiencies through crop breeding techniques. Grain legumes constitute the prime source of vegetarian diet in the developing world. Among grain legumes, lentil is a rich source of protein and other minerals including iron, zinc, selenium, folates, carotenoids, and vitamins. A wide range of genetic variability has been reported among the lentil germplasm with regard to nutrient-related traits. Therefore, lentil has been identified an ideal crop for micronutrient biofortification and a possible whole food solution to the global micronutrient malnutrition. Here, we discuss the current status of breeding interventions in nutritional enhancement of lentil varieties.

Keywords

Biofortification • Lentil • Nutrient • Genetic variability • Variety

J. Kumar (✉) • R. Kant • S.M. Quatadah
Division of Crop Improvement, ICAR-Indian Institute of Pulses Research, Kanpur, Uttar Pradesh 208 024, India
e-mail: jitendra73@gmail.com; rohitkant.gpb@gmail.com; squatadah@gmail.com

S. Kumar
International Center for Agricultural Research in the Dry Areas, Beirut 1108-2010, Lebanon
e-mail: sk.agrawal@cgiar.org

A. Sarker
International Center for Agricultural Research in the Dry Areas, CGIAR Block, NASC Complex, New Delhi 110 012, India
e-mail: A.Sarker@cgiar.org

U. Singh et al. (eds.), *Biofortification of Food Crops*, DOI 10.1007/978-81-322-2716-8_12

12.1 Introduction

More than 22 mineral nutrients are required by the human body for its normal function (White and Broadley 2005). These minerals include calcium (Ca), phosphorus (P), sodium (Na), chlorine (Cl), potassium (K), magnesium (Mg), iron (Fe), copper (Cu), cobalt (Co), iodine (I), zinc (Zn), manganese (Mn), molybdenum (Mo), fluoride (Fl), chromium (Cr), selenium (Se), sulfur (S), boron (B), silicon (Si), arsenic (As), and nickel (Ni) (Murray et al. 2000; Eruvbetine 2003). Based on the quantity required by our body, these minerals can be classified as macro- and micronutrients or ultra micro elements. Micronutrient deficiency affects more than two billion population worldwide (Welch and Graham 1999), which is rampant in sub-Saharan Africa and South Asia. Micronutrient malnutrition is commonly known as "hidden hunger" and causes many health hazards including low birth weight, anemia, learning disabilities, increased morbidity and mortality rates, low work productivity, and high health-care costs especially in developing nations (Welch and Graham 1999; Batra and Seth 2002). Although deficiency of any of these essential elements can be seen in world population, Fe, Zn, Se, and I deficiencies are predominant in rural areas of Southeast Asia. Recent estimates indicate that about 60 % of world population are Fe deficient and 33 % are Zn deficient (Hotz et al. 2004) and 15 % are Se deficient (FAOSTAT 2011). Zn deficiency is generally prevalent in areas where soil is generally poor in available form. These areas include India, Pakistan, China, Iran, and Turkey (Hotz et al. 2004; Khan et al. 2008). Preschool children and pregnant women are mostly affected by Fe and Zn deficiencies due to excessive reliance on cereal-based diets, which are low in these essential micronutrients. In India, every year, 330,000 children die due to vitamin A deficiency and 22,000 people, mainly pregnant women, from severe anemia. Besides, 6.6 million children are born mentally impaired every year due to I deficiency, intellectual capacity is reduced in 15 % of people due to I deficiency, and 200,000 babies are born every year with neural tube defects due to folic acid deficiency (Kotecha 2008). Therefore, it is essentially required to overcome these deficiencies by supplying the enriched diet with these minerals and vitamins. Development of staple food cultivars with enriched micronutrients is one of the viable options for combating global micronutrient malnutrition (Thavarajah et al. 2011). Biofortification of food crop varieties with essential micronutrients is one of the means to combat micronutrient deficiencies through classical plant genetic improvement.

12.2 Crop Biofortification Using Breeding Techniques

In genetic biofortification, plant breeding techniques are used to produce staple food crops with higher micronutrient levels, reducing levels of antinutrients and increasing the levels of substances that promote nutrient absorption in addition to higher yield (Bouis 2003). Plant breeders screen existing accessions in global germplasm banks to determine whether sufficient genetic variation exists to breed for a particular trait. They then selectively breed nutritious cultivars of major staples rich in Zn and Fe concentrations and with substances that promote the bioavailability of Zn and Fe. Biofortification of plant foods through genetic improvement has been introduced as a long-term and sustainable solution for increasing the bioavailability of minerals especially for lentils. This strategy has been widely accepted as a cost-effective way to minimize the extent of mineral deficiencies.

12.3 Lentil: A Nutritious Pulse Crop and Solution of Malnutrition

Lentil (*Lens culinaris* subsp. *culinaris* Medikus) is a cool season food legume. It is mostly cultivated in warm temperate, subtropical, and tropical regions of 52 countries on 3.6 million

Table 12.1 Nutritional value of lentil seeds

Protein	20–25 %
Carbohydrate	50–60 %
Fat	0.7–0.8 %
Ca	60–70 mg/100 g
Fe	7–9 mg/100 g
Zn	4–5 mg/100 g
Se	42–67 μg/100 g
Folate	261–290 μg/100 g

ha area with annual production of 3.6 million tons (FAOSTAT 2011). India is the major lentil producer in the world with 0.9 million tons of production harvested from 1.6 million ha area. Research done in Canada and the USA has revealed suitability of lentil crop for micronutrient biofortification as it holds the potential to provide solution to menacing problem of global micronutrient malnutrition (Thavarajah et al. 2011). It is a rich source of protein and other minerals including iron, zinc, selenium, folates, carotenoids, and vitamins (Thavarajah et al. 2011; Johnson et al. 2013; Gupta et al. 2013). Previous studies provided the proportions of micronutrients in lentil crop which is presented in Table 12.1. It has been shown that these values of micronutrients in 100 g lentil seeds could provide a minimum of 41–113 % of the recommended daily allowance (RDA) of Fe, 40–68 % of the RDA of Zn, and 77–122 % of the RDA of Se. Besides, lentils have been shown to be rich in beta-carotene and found 2–12 μg/g in lentils grown in the USA. Phytic acid (PA) is an antinutrient and high PA levels reduce mineral bioavailability. Lentils are naturally low in PA (2.5–4.4 mg/g), and these levels are lower than the level of PA found in other crops such as rice (1.22–2.23 mg/g), soybean (1.77–4.86 mg/g), wheat (1.24–2.51 mg/g), maize (3.3–3.7 mg/g), and common bean (0.52–1.38 mg/g). Low PA is a favorable factor to increase lentil mineral bioavailability. The genetic variability has also been observed among lentil genotypes, and certain lentil genotypes (e.g., CDC Milestone and CDC Redberry) are observed rich in highly bioavailable Fe and Se (Thavarajah et al. 2011). Lentils cook faster (less than 10 min) and are therefore a better solution to reduce energy and cooking time demands. These observations clearly show that lentils could be an ideal crop for micronutrient biofortification and a possible whole food solution to the global micronutrient malnutrition (Thavarajah et al. 2011).

12.4 Geographical Variation for Micronutrients in Lentil

Mineral levels of lentils are known to vary with growing location and soil conditions. It has been shown among international lentil samples that concentration of micronutrients varies from one geographical region to other. For example, high concentrations of Fe has been observed in seed sample of Syria (63 mg/kg), Turkey (60 mg/kg), the USA (56 mg/kg), and Nepal (50 mg/kg), while it was observed low in seed sample of Australia (46 mg/kg) and Morocco (42 mg/kg). Similarly, higher Zn was found in lentils grown in Syria (36 mg/kg), Turkey (32 mg/kg), and the USA (28 mg/kg) and lowest in Australia (18 mg/kg) and Morocco (17 mg/kg). Examination of Se concentrations in lentil genotypes grown in eight major world regions indicated considerable variation. For example, Nepal and Australia showed greater Se concentrations of 180 and 148 μg/kg, respectively. Syria, Morocco, and Turkey had the lowest values of 22, 28, and 47 μg/kg, respectively. Turkey landraces showed higher amount of Ca in their seeds (0.48–1.28 g/kg), and analysis of Indian samples showed high to low Fe (>60 mg/kg) and Zn (20–40 mg/kg) in their seeds. Concerning genotypes, CDC Greenland had higher selenium content with low iron and zinc content compared to Merritt. However, Red Chief was low in selenium and high in iron, zinc, calcium, and potassium compared to the other cultivars. Spanish brown cultivar Pardina was rich in iron and zinc but low in selenium as it was mostly grown in low-selenium soils. Generally, green lentil cultivars were low in iron and zinc compared to selenium.

12.5 Presence of Substantial Genotype × Environment (G × E) Interactions

Soils of growing location, weather, and other environmental conditions affect mineral levels in lentil. A study conducted in the USA for green and red lentil genotypes for over 2 years for a number of micronutrients revealed significant differences in concentration of micronutrients over the years. The evaluation of seven lentil genotypes over four locations along with a farmer's filed survey conducted in Bangladesh revealed significant genotype and location differences for seed selenium concentration but genotype × location interaction was nonsignificant. Similar results were also observed when 12 genotypes of lentil were evaluated over seven locations in Australia (Rahman et al. 2014). In other crops such as wheat, breeding for high Zn concentration is complicated by environmental conditions, particularly soil composition (Trethowan 2007). Therefore, despite advances in breeding for uptake efficiency or mobilization to the grain, grain Zn concentration is limited by Zn availability in the soil (Ortiz-Monasterio et al. 2007, 2011). Significant genotype × location interactions were observed for Zn and Fe in wild and improved wheat cultivars (Oury et al. 2006; Ortiz-Monasterio et al. 2007; Trethowan 2007; Gomez-Becerra et al. 2010). However, a recent study tested biofortified wheat lines at multiple locations in South Asia and revealed high heritability and high genetic correlation between locations for grain Zn suggesting that G × E interactions may not be a serious issue in breeding high Zn wheat genotypes (Velu et al. 2012). However, mineral characteristics of the crop plants are largely influenced by the genetic and environmental factors such as soil fertility, soil type, seed characteristics, seed composition, and climatic factors; hence further studies are required to conduct experiments under different environmental conditions to validate the results.

12.6 Breeding for Improved Nutrient Content

Naturally, lentil is a storehouse of most of the essential nutrients for human health. However, there is still scope to enhance the nutritive value of its edible part because a range of variability has been observed among the lentil germplasm including cultivars, breeding lines, landraces, and wild species. This gives an opportunity to make selection of nutritionally appropriate breeding materials, which can be used directly for cultivation at farmer's fields or can be used as donors in breeding programs. Development of improved variety with enhanced nutritional value is a onetime investment and cost-effective strategy to overcome the malnutrition. Its cost is estimated on average of about $400,000 per year per crop over a 10-year period, globally. Once biofortified varieties have been developed, in country trials and local adaptation research costs are incurred, after which routine maintenance breeding to ensure the trait remains stable is put in place. Development of lentil cultivars enriched with minerals and vitamins makes this as a biofortified crop. These cultivars can be developed by using the different approaches of breeding involving both conventional and modern techniques for increasing concentration and bioavailability of these essential elements in lentil (Welch 2002). Therefore, focus has been made on lentil biofortification, and different breeding strategies are being utilized for developing biofortified varieties as given below.

12.6.1 Screening of Existing Variability of Cultivated Species for Identification of Nutritionally Enriched Genotypes

Identification of natural variants having favorable alleles for enhanced micronutrient content in the germplasm is one way to develop the improved cultivars for cultivation. Earlier significant variability has been observed for micronutrients in the germplasm from Turkey, Syria, Canada, and Pakistan (Thavarajah et al. 2011). These studies have identified lentil germplasm with favorable nutritional traits including folates, macro-, and micronutrients. Many landraces of lentil are grown in particular regions due to their some specific characters including taste preferences. These landraces are a potential genetic resource for biofortification of lentils with increased micronutrients. Initially, under the HarvestPlus Challenge Program, more than 1600 germplasm, breeding lines, released cultivars, and wild relatives of lentil were analyzed for iron and zinc contents. A wide variation of 43–132 ppm of iron and 22–95 ppm of zinc was found in these materials. Earlier studies mainly exploited the variability for micronutrients in the breeding materials generated from ICARDA, and no source of variability was identified in Indian germplasm (Thavarajah et al. 2011; Gupta et al. 2013). However recently we studied the variability among breeding lines, varieties and parental lines, and exotic lines. We observed a range of variability for ten macro- and micronutrients. In Indian germplasm, the amount of Ca and Mg was recorded much lower than Turkish germplasm, while the availability of P (3907.0–7261.58 mg) and K (4771.43–9633.90 mg) was observed higher compared to Turkish landraces (2860–5330 mg for P and 6380–9500 mg for K). Higher amount of Fe, Zn, and Se was observed in Indian varieties and breeding lines compared to exotic lines. These findings encouraged breeders to improve iron and zinc contents in lentil through recombination breeding having significant amount of all macronutrients. Also, ICARDA research with partners has identified several released varieties of lentil in Bangladesh, Ethiopia, Nepal, Morocco, Turkey, Syria, Lesotho, and Portugal that had high iron and zinc contents. These varieties are being disseminated to farmers on a fast-track mode through national program. In Bangladesh, the government has taken a massive dissemination program to promote promising lentil varieties such as Barimasur 4 (86 ppm Fe and 63 ppm Zn) and Barimasur 5 (86 ppm Fe, 59 ppm Zn) and Barimasur 6 (86 ppm Fe and 63 ppm Zn) in traditional and nontraditional areas. Similarly, in Nepal lentil varieties such as Sishir (98 ppm Fe, 64 ppm Zn), Khajurah-2 (94 ppm Fe, 59 ppm Zn), Khajurah-1 (zn 58 ppm), and shital (59 ppm Zn) are spreading fast in the Terai region. In India, variety Pusa Vaibhav contains 102 ppm of Fe, and it is being grown by farmers in the northwest plain zone (ICARDA 2012).

12.6.2 Lentil Biofortification Through Pre-breeding

Wild species are a rich reservoir of useful alien genes, which are no longer available within the cultivated gene pool (Hawkes 1977; Doyle 1988; Tanksley and McCouch 1997). Continuous efforts have been underway to collect and conserve wild relatives of various food legume crops in national and international gene banks (Plucknett et al. 1987; FAO 1996). Over the years, ICARDA has collected and conserved 587 accessions representing six wild *Lens* species from 26 countries. In the past years, efforts have been made to search for genes imparting resistance to various stresses and other traits within the cultivated species and their wild relative. However, success of introgression of alien genes from wild relatives has been limited to a few diseases and insect pests, which are controlled by major gene(s), and could be available in crossable primary gene pool (Knott and Dvorak 1976; Stalker 1980; Ladizinsky et al. 1988; Prescott-Allen and Prescott-Allen 1986, 1988; Hajjar and Hodgkin 2007). To diversify and broaden the genetic base of cultivated germplasm, introgression of alien genes from wild species needs to be persuaded vigorously not only to minimize the risk of stress epidemics but also to make discernible yield advances in lentil. The advances in tissue culture techniques can now help to make feasible alien gene introgression easy. Moreover, evolutionary forces can make changes in accessions of wild species toward the possible cross-compatibility with cultivated species. Therefore earlier cross-incompatible wild species may now cross with cultivated species. Therefore, pre-breeding efforts are urgently required involving particularly those wild species, which carry useful alien genes for improving yield, quality, and stress resistance.

12.6.2.1 Crossability Group and Gene Pool

Knowledge of crossability between cultivated and wild species is essential for successful pre-breeding activities. Previously on the basis of crossability between the lentil species, crop species of lentil have been grouped into primary (*Lens culinaris* ssp. *culinaris*, *L. culinaris* ssp. *orientalis*, *L. odemensis*), secondary (*L. ervoides*, *L. nigricans*), and tertiary gene pools (*L. lamottei*, *L. tomentosus*) (Ladizinsky et al. 1984; Ladizinsky 1999; Muehlbauer and McPhee 2005). Crossing between species of primary gene pool and *L. nigricans* and *L. ervoides* produces nonviable seed in hybrids due to irregular meiosis (Ladizinsky et al. 1984, 1985). However, use of embryo rescue can produce viable seed in hybrids derived from crossing between *L. culinaris* and *L. ervoides* (Ladizinsky et al. 1985). *L. tomentosus* was put in tertiary gene pool, as a single species group, and this species does not produce viable seed in hybrids derived by crossing with other group of species. Although useful genes identified in the primary gene pool have readily been used for making genetic improvement by generating new variability, frequency of occurrence of the useful genes has been observed much more in the species of the secondary and tertiary gene pools (Collard et al. 2001; Mallikarjuna et al. 2006; Tullu et al. 2006). To use these gene pools in the improvement of lentil, it requires much effort on deployment of novel techniques. During the past years thus Fratini and Ruiz (2006) suggested that hybrid between *L. culinaris* ssp. *culinaris* and *L. nigricans*, *L. ervoides*, and *L. odemensis* can be viable through embryo rescue with a rate of 3–9 %. Based on these observations, *L. odemensis* has been considered secondary gene pool and *L. nigricans* and *L. ervoides* were classified in the tertiary gene pool (Ladizinsky 1993). However, Cubero et al. 2009 suggested placing *L. odemensis* in secondary gene pool, and *L. nigricans* and *L. ervoides* can be part of secondary gene pool by means of embryo rescue, while there is need to study hybridization in order to establish place of *L. tomentosus* and *L. lamottei* in secondary or tertiary gene pool.

Releasing of new hidden variability present in the background of wild species for nutritional traits depends upon the cross-compatibility of these species with cultivated species. Therefore, efforts have been made over the past years in lentil to study the cross-compatibility of wild species with cultivated species. The wild species *L. culinaris* ssp. *orientalis* and *L. odemensis* have been shown crossable with cultivated lentil (Ladizinsky et al. 1984; Abbo and Ladizinsky 1991, 1994; Fratini et al. 2004; Fratini and Ruiz 2006; Muehlbauer et al. 2006), although the fertility of the hybrids depends on the chromosome arrangement of the wild parent (Ladizinsky 1979; Ladizinsky et al. 1984). Most accessions of *L. culinaris* ssp. *orientalis* cross readily with *L. culinaris*, and both are genetically isolated from the other species. *Lens nigricans* and *L. ervoides* are not readily crossable with the cultivated lentil using conventional crossing methods due to hybrid embryo breakdown (Abbo and Ladizinsky 1991, 1994; Gupta and Sharma 2005). Crosses are possible between *L. culinaris* and the remaining species, but they are characterized by a high frequency of hybrid embryo abortion, albino seedlings, and chromosomal rearrangements that result in hybrid sterility, if these seedlings reach maturity (Abbo and Ladizinsky 1991, 1994; Ladizinsky 1993; Gupta and Sharma 2005).

12.6.2.2 Introgression of Alien Genes for Nutritional Traits from Wild Relatives

The wild relatives are important source for generation of new variability through recombination breeding approaches. The evidences from other crops such as wheat indicated that wild relatives have substantial variation for nutritional traits. For example, >3000 germplasm accessions, including hexaploid, tetraploid, and diploid sources from the International Maize and Wheat Improvement Center (CIMMYT) gene bank, have been screened for Zn and Fe variation (Monasterio and Graham 2000). Highest Zn and Fe concentrations are observed among accessions belonging to progenitors of wheat and landraces (Cakmak et al. 2000; Ortiz-Monasterio et al. 2007) and little variation in improved adapted wheat varieties. Therefore more focus has been given on a more in-depth evaluation of wheat landraces (Monasterio and Graham 2000) and secondary gene pool in wheat for identification of the most promising sources of high Fe and Zn grain concentration (Cakmak et al. 2000). In lentil, however, limited efforts were invested to identify the wild relatives enriched with minerals and vitamins. The study conducted at IIPR, Kanpur, with ten accessions of lentil wild species revealed protein content of 29.7 %, phenol content 8.9 mg/100 g, and higher total antioxidant activity (16.17 μmole TE/g), which was higher than that found among the cultivated accessions. Therefore, concerted efforts are required to screen more wild resources of lentil for minerals and other nutritional traits. These identified resources can be utilized in wide hybridization for transferring useful genes from alien species into cultivated species. It is well recognized that gene transfer through wide crosses is a long and tedious process due to lack of homology between chromosomes of participating species in the cross and pre- and post-zygotic crossability barriers between wild and cultivated species. Utilizing wild gene pool in breeding program is also constrained by collection gaps in wild species with poor or no information about genome relationship, poor/limited screening of wild species, linkage drag, and genetic complexity of the traits. Therefore, improvement through distant hybridization often takes longer time in order to recover genotypes with acceptable agronomic background, and thus, it requires a long-term approach to be in place.

12.6.2.3 Limitations in Pre-breeding

In lentil, strong crossability barriers limit utilization of wild gene pool for lentil improvement. Among these, problem of chromosome pairing was observed between the participating genomes in the cross between *L. culinaris* and *L. tomentosus* (Ladizinsky 1979). Only retardation of the growth of hybrid embryo was noticed, while no sign was detected concerning endosperm disintegration in some *L. culinaris* × *L. culinaris* ssp. *orientalis* crosses (Ladizinsky 1993). Later on, Abbo and Ladizinsky (1991) observed either absence or abnormal endosperm in *L. culinaris* × *L. culinaris* ssp. *orientalis* crosses. Hybrids showed varying degrees of fertility usually due to chromosome translocations and subsequent problems with chromosome pairing at meiosis in *Lens culinaris* × *L. nigricans* (Goshen et al. 1982; Ladizinsky et al. 1984). Fertility is often very low with less viable pollen produced in anthers and the extent varied depending on the accession in *L. culinaris* × *L. culinaris* ssp. *orientalis* crosses from 2 % to 69 % (Ladizinsky et al. 1984). These problems can occur in the F_1 and also persist in later generations causing partial or complete sterility. Additionally, albino seedlings can also occur in the F_1 generation and thus prevent hybridization success (Ladizinsky and Abbo 1993). Another common problem is that hybrid embryos cease to grow about 7–14 days after pollination due to endosperm degeneration and thus need rescuing in order to obtain viable hybrids (Ahmad et al. 1995; Ladizinsky et al. 1985). Hence, *L. culinaris* × *L. ervoides* or *L. culinaris* × *L. nigricans* crosses need embryo rescue techniques in order to develop mature hybrid plants (Abbo and Ladizinsky 1991; Cohen et al. 1984).

12.7 Markers Assisted Breeding for Biofortification in Lentil

Marker-assisted selection (MAS) is an effective alternative to genetic engineering strategies for biofortification of crops. Successful examples include increased levels of provitamin A (beta-carotene) in maize; a team of plant geneticists and crop scientists recently identified genetic markers that are associated with higher levels of provitamin A in maize. PCR-based DNA markers distinguishing alleles of three key genes of maize endosperm carotenoid biosynthesis (PSY1, lcyE, and crtRB1) have been developed, which can help to assess provitamin A concentration in maize kernels easily. Thus maize varieties with increased provitamin A contents to combat vitamin A deficiencies can easily be bred using MAS.

Significant progress can be made through conventional breeding; 85 % of HarvestPlus resources are currently devoted to conventional breeding. However, transgenic approaches are in some cases necessary and, in some cases, potentially advantageous compared with conventional breeding. The best-known example is Golden Rice; b-carotene has not been identified in the endosperm of any rice variety, and an advanced transgenic line containing 37 mg/g carotenoid, of which 31 mg/g is b-carotene, is now available (Paine et al. 2005). Ongoing transgenic research is exploring the use of an endosperm-specific promoter to deposit iron within the endosperm of rice so that it is not milled away (Goto et al. 1999). This approach is necessary because most of the iron in rice grains is deposited in the aleurone layer, which is removed when rice is milled to produce polished rice, a practice widely used in many countries (Gregorio et al. 2000). The QTL has also been identified in rice for biofortification of Zn. The functional markers associated with Zn content in RIL populations showed significant variation among RIL population with a phenotypic variation of 4.5 %, 19.0 %, 5.1 %, and 10.2 %, respectively. However no significant efforts were made in lentil. Initially we screened the parental lines of a mapping population that identified significant differences for Fe and Zn in the mapping populations.

12.8 Future Directions

As discussed above, more than three billion people worldwide suffer from mineral deficiencies. Lentil is an indispensable supplementary food in many countries, particularly in South Asia, West Asia, North and East, and North Africa. In South Asia, particularly in Pakistan, Nepal, Bhutan, India, and Bangladesh, red lentil is an integral part of daily diet, most importantly among rural population. Owing to its abundant nutritional and functional components, red lentil is very popular in Turkey and other Mediterranean countries. Thus, even a small increase in the nutritive value of lentil seed can have remarkable impacts on human nutrition. In summary, considerable variation in the macro- and micronutrient contents of lentil landraces and cultivars was observed. Identification of genetic variation is essential for achieving improvements in the mineral content of crops. Our results provide a useful foundation for the development of new cultivars of lentil with high mineral content. In particular, some of the landraces that were studied could be used to develop lentil varieties with more nutrient contents to address mineral deficiencies in developing countries. The observed genetic variation can also be used to identify quantitative trait loci associated with mineral uptake and transport. The data provided in this study gives us an important basis for improvement of micronutrient contents in lentil.

12.9 Conclusion

Lentil being a nutritious grain legume crop provides an appropriate candidate crop for nutritional enrichment. To serve the purpose, a range of scientific interventions were proposed of which crop breeding techniques remain cost effective and accessible, providing solutions to the growing problem of food insecurity worldwide. In the context, reasonable amount of genetic variation for nutrient traits has been reported so far in lentil, which constitutes the robust base for the downstream breeding methods. Among various conventional methods, utilization of wild relatives offers the most potential researchable area, and significant progress has been made toward this end. More importantly, unlike transgenic technologies, enriching nutrients in edible parts of crops using breeding protocols does not face any legislative or socio-political challenge. Once investment has been done in developing a nutrient-dense cultivar, the nutritional benefits can be reaped in a sustainable and economically viable manner.

References

Abbo S, Ladizinsky G (1991) Anatomical aspects of hybrid embryo abortion in the genus *Lens* L. Bot Gaz 152:316–320

Abbo S, Ladizinsky G (1994) Genetical aspects of hybrid embryo abortion in the genus *Lens* L. Heredity 72:193–200

Ahmad M, Fautrier AG, McNeil DL, Burritt DJ, Hill GD (1995) Attempts to overcome post-fertilization barrier in interspecific crosses of the genus *Lens*. Plant Breed 114:558–560

Batra J, Seth PK (2002) Effect of iron deficiency on developing rat brain. Indian J Clin Biochem 17:108–114

Bouis HE (2003) Micronutrient fortification of plants through plant breeding: can it improve nutrition in man at low cost? Proc Nutr Soc 62:403e411

Cakmak I, Ozkan H, Braun H-J, Welch RM, Romheld V (2000) Zinc and iron concentrations in seeds of wild primitive and modern wheats. Food Nutr Bull 21: e401–e403

Cohen D, Ladizinsky G, Ziv M, Muehlbauer FJ (1984) Rescue of interspecific Lens hybrids by means of embryo culture. Plant Cell Tissue Org 3:343–347

Collard BCY, Ades PK, Pang ECK, Brouwer JB, Taylor PWJ (2001) Prospecting for sources of resistance to ascochyta blight in wild *Cicer* species. Australas Plant Path 30:271–276

Cubero JI, Perez de la Vega M, Fratini R (2009) Origin phylogeny domestication and spread. In: Erskine W, Muehlbauer FJ, Sarker BS (eds) The lentil: botany production and uses. CABI Oxfordshire, UK, pp 13–33

Doyle JJ (1988) 5S ribosomal gene variation in the soybean and its progenitor. Theor Appl Genet 75:621–624

Eruvbetine D (2003) Canine nutrition and health. A paper presented at the seminar organized by Kensington Pharmaceuticals Nig Ltd Lagos. Accessed 21 Aug 2003

FAO (1996) State of the world's plant genetic resources for food and agriculture. Food and Agriculture Organization, Rome, p 131

FAOSTAT (2011) http://faostat.fao.org//. Accessed 14 Mar 2015

Fratini R, Ruiz ML (2006) Interspecific hybridization in the genus *Lens* applying in vitro embryo rescue. Euphytica 150:271–280

Fratini R, Ruiz ML, Perez de la Vega M (2004) Intra-specific and inter-sub-specific crossing in lentil (*Lens culinaris* Medik). Can J Plant Pathol 84:981–986

Gomez-Becerra HF, Yazici A, Ozturk L, Budak H, Peleg Z, Morgounov A, Fahima T, Saranga Y, Cakmak I (2010) Genetic variation and environmental stability of grain mineral nutrient concentrations in *Triticum* dicoccoides under five environments. Euphytica 171:39–52

Goshen D, Ladizinsky G, Muehlbauer FJ (1982) Restoration of meiotic regularity and fertility among derivatives of *Lens culinaris* × *L nigricans* hybrids. Euphytica 31:795–799

Goto F, Yoshihara T, Shigemoto N, Toki S, Takaiwa F (1999) Iron fortification of rice seed by the soybean ferritin gene. Nat Biotechnol 17:282–286

Gregorio GB, Senadhira D, Htut H, Graham RD (2000) Breeding for trace mineral density in rice. Food Nutr Bull 21:382–386

Gupta D, Sharma SK (2005) Embryo-ovule rescue technique for overcoming post-fertilization barriers in interspecific crosses of Lens. J Lentil Res 2:27–30

Gupta DS, Thavarajah D, Knutson P, Thavarajah P, McGee RJ, Coyne CJ, Kumar S (2013) Lentils (*Lens culinaris* L) a rich source of folates. J Agric Food Chem 61:7794–7799

Hajjar R, Hodgkin T (2007) The use of wild relatives in crop improvement a survey of developments over the last 20 years. Euphytica 156:1–13

Hawkes JG (1977) The importance of wild germplasm in plant breeding. Euphytica 26:615–621

Hotz C, Brown KH, Rivera JA et al (2004) Assessment of the risk of zinc deficiency in populations and options for its control. Food Nutr Bull 25:S94–S204

ICARDA (2012) Biofortified lentils to enhance nutritional security in South Asia. ICARDA-South Asia and China regional program, New Delhi, p 2

Johnson C, Thavarajah D, Thavarajah P (2013) The influence of phenolic and phytic acid food matrix factors on iron bioavailability potential in 10 commercial lentil genotypes (*Lens culinaris* L). J Food Compos Anal. 31:82–86

Khan MA, Fuller MP, Baloch FS (2008) Effect of soil applied zinc sulphate on wheat (*Triticum aestivum* L) grown on a calcareous soil in Pakistan. Cereal Res Commun 36:571–582

Knott DR, Dvorak J (1976) Alien germplasm as a source of resistance to diseases. Annu Rev Phytopathol 14:211–235

Kotecha PV (2008) Micronutrient malnutrition in India: let us say "no" to it now. Indian J Community Med 33:9–10

Ladizinsky G (1979) The origin of lentil and its wild gene pool. Euphytica 28:179–187

Ladizinsky G (1993) Wild lentils. Criti Rev Plant Sci 12:169–184

Ladizinsky G (1999) Identification of the lentil's wild genetic stock. Genet Resour Crop Evol 46:115–118

Ladizinsky G, Abbo S (1993) Cryptic speciation in *Lens culinaris*. Genet Resour Crop Evol 40:1–5

Ladizinsky G, Braun D, Goshen D, Muehlbauer FJ (1984) The biological species of the genus Lens L. Bot Gaz 145:253–261

Ladizinsky G, Cohen D, Muehlbauer FJ (1985) Hybridization in the genus Lens by means of embryo culture. Theor Appl Genet 70:97–101

Ladizinsky G, Pickersgill B, Yamamoto K (1988) Exploitation of wild relatives of the food legumes. In: Summerfield RJ (ed) World crops cool season food legumes. Kluwer, Dordrecht, pp 967–987

Mallikarjuna N, Jadhav D, Reddy P (2006) Introgression of Cajanus platycarpus genome into cultivated pigeon pea *C. cajan*. Euphytica 149:161–167

Monasterio I, Graham RD (2000) Breeding for trace minerals in wheat. Food Nutr Bull 21:393e396

Muehlbauer FJ, McPhee KE (2005) Lentil (*Lens culinaris* Medik). In: Singh RJ, Jauhar PP (eds) Genetic resources chromosome engineering and crop improvement grain legumes. Taylor & Francis, Boca Raton, pp 219–230

Muehlbauer FJ, Cho S, Sarker A, McPhee KE, Coyne CJ, Rajesh PN, Ford R (2006) Application of biotechnology in breeding lentil for resistance to biotic and abiotic stress. Euphytica 147:149–165

Murray RK, Granner DK, Mayes PA, Rodwell VW (2000) Harper's biochemistry, 25th edn. McGraw-Hill Health Profession Division, USA pp. 780–786

Ortiz-Monasterio I, Palacios-Rojas N, Meng E, Pixley K, Trethowan R, Pena RJ (2007) Enhancing the mineral and vitamin content of wheat and maize through plant breeding. J Cereal Sci 46:293e307

Ortiz-Monasterio I, Trethowan R, Holm PB, Cakmak I, Borg S, Tauris BEB, Brinch-Pedersen H (2011) Breeding transformation and physiological strategies for the development of wheat with high zinc and iron grain concentration. In: Bonjean AP, Angus WJ, Van Ginkel M (eds) The world wheat book: a history of wheat breeding, vol 2., pp 951–977

Oury FX, Leenhardt F, Rémésy C, Chanliaud E, Duperrier B, Balfouriera F, Charmet G (2006) Genetic variability and stability of grain magnesium zinc and iron concentration in bread wheat. Eur J Agron 25:177–185

Paine JA, Shipton CA, Chaggar S, Howells RM, Kennedy MJ, Vernon G et al (2005) A new version of golden rice with increased pro-vitamin A content. Nat Biotechnol 23:482–487

Plucknett DL, Smith NJH, Williams JT, Anishetty NM (1987) Gene banks and the world's food. Princeton University Press, New Jersey

Prescott-Allen C, Prescott-Allen R (1986) The first resource: wild species in the north American economy. Yale University, New Haven

Prescott-Allen C, Prescott-Allen R (1988) Genes from the wild: using wild genetic resources for food and raw materials. International Institute for Environment and Development, London

Rahman MM, Erskine W, Siddique KHM, Thavarajah P, Thavarajah D, Zaman MS, Materne MA, Mcmurray LM (2014) Selenium biofortification of lentil in Australia and Bangladesh. In: 6th international food legume research conference, TCU Place, Sakatoon, 7–11 July 2014

Stalker HT (1980) Utilization of wild species for crop improvement. Adv Agron 33:111–147

Tanksley SD, McCouch SR (1997) Seed banks and molecular maps unlocking genetic potential from the wild. Science 277:1063–1066

Thavarajah D, Thavarajah P, Wejesuriya A, Rutzke M, Glahn RP, Combs GF Jr, Vandenberg A (2011) The potential of lentil (*Lens culinaris* L) as a whole food for increased selenium iron and zinc intake: preliminary results from a 3 year study. Euphytica 180:123–128

Trethowan RM (2007) Breeding wheat for high iron and zinc at CIMMYT: state of the art challenges and future prospects. In: Proceedings of the 7th international wheat conference, Mar del Plata

Tullu A, Buchwaldt L, Lulsdorf M, Banniza S, Barlow B, Slinkard AE, Sarker A, Tar'an TD, Warkentin TD, Vandenberg A (2006) Sources of resistance to anthracnose (*Colletotrichum truncatum*) in wild *Lens* species. Genet Resour Crop Evol 53:111–119

Velu G, Singh RP, Huerta-Espino J, Peña-Bautista RJ, Arun B, Mahendru-Singh A et al (2012) Performance of biofortified spring wheat genotypes in target environments for grain zinc and iron concentrations. Field Crop Res 137:261–267

Welch MR, Graham DR (1999) A new paradigm for world agriculture: meeting human needs productive sustainable nutritious. Field Crop Res 60:1–10

Welch RM (2002) Breeding strategies for biofortified staple plant foods to reduce micronutrient malnutrition globally. J Nutr 132:495S–499S

White PJ, Broadley MR (2005) Biofortifying crops with essential mineral elements. Trends Plant Sci 10:588–593

Part III

Biofortification with Improved Agronomy

Soil Factors Associated with Micronutrient Acquisition in Crops- Biofortification Perspective

13

Anil Kumar, Anil K. Choudhary, Vijay Pooniya, Vinod K. Suri, and Ummed Singh

Abstract

The introduction of high-yielding varieties, intensive cultivation systems, micronutrient-free fertiliser application, non-addition of organic manures and imbalanced plant nutrition has led to multi-micronutrient deficiencies in soils in many parts of the globe. Importance of micronutrients can be realised from their incredible functions in plants that result in quality produce, as each essential micronutrients play some specific functions in plants. Availability of micronutrients to plants is regulated by various soil factors such as texture, soil reaction, organic matter, clay content, soil moisture, nutrient interactions in soil, microbial activity, redox potential and aeration, etc. Research has, however, clearly indicated a sharp increase in micronutrient uptake and yield of crops by alleviating soil conditions through proper management practices, liming and applying micronutrients directly in soil or as foliar application. Moreover, exploitation of soil microbes such as micronutrient solubilisers and AM fungi has proven as boon in micronutrient uptake and improving soil quality. Maintenance of optimum soil organic matter status and balanced fertilisation or soil test-based fertiliser application also lead to biofortified farm produce, eliminate micronutrient deficiency and improve soil and plant health.

A. Kumar
Farm Science Centre, GAD Veterinary and Animal Sciences University, Booh, Tarn Taran 143 412, Punjab, India
e-mail: anilkumarhpkv@gmail.coml

A.K. Choudhary (✉) • V. Pooniya
Division of Agronomy, ICAR-Indian Agricultural Research Institute, New Delhi 110 012, India
e-mail: anilhpau2010@gmail.com; vpooniya@gmail.com

V.K. Suri
Former Vice Chancellor, Chandra Shekhar Azad University of Agriculture and Technology, Kanpur 208 002, Uttar Pradesh, India
e-mail: surivk13jan@gmail.com

U. Singh
Division of Crop Production, ICAR-Indian Institute of Pulses Research, Kanpur, Uttar Pradesh 208 024, India
e-mail: singhummed@yahoo.co.in

U. Singh et al. (eds.), *Biofortification of Food Crops*, DOI 10.1007/978-81-322-2716-8_13

Keywords
Biofortification • Food quality • Micronutrients • Soil microbes • Soil organic matter

13.1 Introduction

There are seventeen essential plant nutrients which are required for proper plant growth and development, each one of which is equally important to the plants, and they are required in vastly different amounts. These differences have led to the grouping of these essential elements into three broader classes, viz., primary (macro) nutrients, secondary nutrients and micronutrients. *Primary nutrients* are nitrogen (N), phosphorus (P) and potassium (K). They are the most frequently required in a crop fertilisation programme. Also, they are needed in greatest quantity by the plants through chemical fertilisers. *Secondary nutrients* are calcium (Ca), magnesium (Mg) and sulphur (S). For most crops, these three are needed in comparatively lesser amounts than that of primary nutrients. *Micronutrients* consist of eight essential elements required by the plants, viz., zinc (Zn), iron (Fe), copper (Cu), manganese (Mn), molybdenum (Mo), boron (B), nickel (Ni) and chlorine (Cl). The term micronutrient or trace element is quite often used interchangeably, though most of the scientists prefer to use the term micronutrients to denote the elements which are essential for the plant but are required in small amounts. These elements occur in very small amounts both in soils and plants, but their role is equally vital as that of primary or secondary plant nutrients. They are constituents of enzymes and coenzymes. A deficiency of one or more of the micronutrients can lead to severe depression in growth, yield and crop quality. In such cases, supplemental micronutrient application in form of commercial fertilisers and foliar sprays or application of appropriate microbial inoculants in soil becomes highly essential (Suri et al. 2011a).

During initial years of introduction of high-yielding varieties (HYVs), micronutrient deficiencies were subsequently noticed in many pockets of the globe owing to high yield potential and high nutrient requirements of these germplasm (Singh 2008; Choudhary et al. 2008; Choudhary and Suri 2009; Rattan et al. 2009). Micronutrient deficiencies were discovered as an obstacle to higher crop yields and quality produce. These micronutrient deficiencies were attributed to introduction of HYVs, micronutrient-free high analysing fertiliser, non-addition of organic manures, imbalanced fertilisation, etc. (Paul et al. 2011, 2014; Yadav et al. 2013, 2015). In fact, high-yielding crop varieties have greatly increased the crop production and thereby heavy micronutrient removal (Brady 2002; Singh 2008; Deb et al. 2009; Rattan et al. 2009). The micronutrients are not only important for better crop productivity but also essential for sustaining human and animal health. Inadequate consumption or lack of any one of essential micronutrients limits plant growth due to adverse metabolic disturbances, leading to reduced yield and quality, even when all other nutrients are present in adequate amounts (Rattan et al. 2009; Kumar et al. 2014; Yadav et al. 2015). Thus, *eight* micronutrient elements, viz., Fe, Zn, Cu, Mn, B, Mo, Ni and Cl, are categorised as essential for plants. Above nutrients are directly involved in the metabolism of plants and cannot complete their life cycle without their absence. All above essential micronutrients are metals except B and Cl. Micronutrients are needed by plants in significantly lesser amounts in comparison to macronutrient elements. Besides essential elements, there are certain elements that help optimise the growth and development of plants, but are not absolutely necessary for the completion of the plant life cycle. These elements are known as *beneficial elements*. Aluminium (Al), cobalt (Co), sodium (Na), selenium (Se) and silicon (Si) are considered as beneficial elements for plants. They may be essential for particular plant species. When they are absent in the soil, plants can still live a normal life (Pilon–Smits et al. 2009).

The soil is a good reservoir for all plant nutrients including micronutrients. However, some soils may contain good amounts of certain micronutrients while small ones of others depending upon the parent material (*rocks or minerals*) from which they are formed. Total micronutrients content in soil does not indicate that whole amounts are available for plant uptake as many factors such as soil (Stevenson 1986; Lake et al. 1984; Lindsay 1991; Suri et al. 2006), plant (Barber 1995; Marschner 1995; Suri et al. 2011a; Suri and Choudhary 2013a, b), microbes and environment (Romheld and Marschner 1986; Clark and Zeto 2000; Suri et al. 2011b; Suri and Choudhary 2013c, 2014), management practices (Choudhary et al. 2010, 2013; Suri and Choudhary 2014; Dass et al. 2015), etc. may affect their availability in crop plants.

The demand for increasing crop production requires a thorough knowledge of 'soil factors' that regulate the supply and availability of micronutrients in soils as soil–plant interrelations are dynamic and are affected by both inputs (*fertilisers, pollutants, soil chemistry*) and losses (*erosion, leaching, harvesting*). Nutrients are released into soil solution via weathering and solubilisation of soil minerals and through decomposition of organic matter (Choudhary and Suri 2009, 2013, 2014a). Due to the importance of micronutrients in crop productivity and quality, it is imperative that scientists, farming community and extension functionaries must recognise the factors that lead to micronutrient deficiencies and affect overall crop growth and development.

13.2 Micronutrients in Plant Health

Research studies have shown that critical plant functions are severely affected if any of essential micronutrient is unavailable in soil or is not adequately balanced with other nutrients resulting in poor plant growth and various abnormalities and directly affecting both crop yield and quality. Micronutrients are essential elements that are absorbed by plants in lesser quantities, but they serve many important and critical functions in plant metabolism, growth and overall development of the plant. One of the most critical functions of micronutrients in plant is to serve as catalysts or coenzymes in various metabolic biochemical reactions. With the exception of Zn, these are capable of acting as 'electron carriers' in the enzyme systems, which are responsible for oxidation-reduction reactions in plants (Deb et al. 2009). The key functions of essential trace elements are presented in Table 13.1. From Table 13.1, it can be realised that micronutrients play a significant physiological role in growth and development of plants and microorganisms. Further, micronutrients play a key role in disease control. The plants become susceptible to various diseases, if there is lack in the concentrations of Zn, B, Mn, Mo, Ni, Cu and Fe in plant tissues (Huber 1980; Jones et al. 1990; Engelhard 1990; Graham and Webb 1991; Fageria et al. 1997a; Baligar et al. 1998).

Optimum contents of micronutrients in plants create resistance towards various stresses including infectious pathogens (Huber and Wilhetm 1988). Application of Cu as fungicide suppresses many soil-borne diseases; likewise, B sufficiency in plants reduces the incidence and severity of diseases, while B deficiency enhances them (Gupta 1993; Graham and Webb 1991). Further, optimum Mn concentrations in plant also decrease incidence of disease occurrence. Similarly, beneficial elements benefit the plant in terms of growth and resistance to diseases and environmental stresses (*drought, salinity and nutrient toxicity or deficiency*). The beneficial elements prevent the toxic effect of certain

Table 13.1 Functions of essential micronutrients in higher plants

Micronutrients	Functions in higher plants
Zinc	Present in several dehydrogenase, proteinase and peptidase enzymes Promotes growth hormones and starch formation Promotes seed maturation and production Necessary for chlorophyll production Necessary for carbohydrate and starch formation
Iron	Present in several peroxidise, catalase and cytochrome oxidase enzymes Found in ferredoxin, which participates in oxidation–reduction reactions (e.g. NO_3^- and SO_4^- reduction, N fixation) Important in chlorophyll formation Acts as an oxygen carrier
Copper	Present in laccase and several other oxidase enzymes Important in photosynthesis, protein and carbohydrate metabolism and probably N fixation Improves flavour of fruits and vegetable Indirect role in chlorophyll production
Manganese	Activates decarboxylase, dehydrogenase and oxidase enzymes Important in photosynthesis, nitrogen metabolism and nitrogen assimilation Aids in chlorophyll synthesis
Boron	Activates certain dehydrogenase enzymes Facilitates sugar translocation and synthesis of nucleic acids and plant hormones Essential for germination of pollen grains and growth of pollen tubes Essential for seed and cell wall formation Promotes maturity
Molybdenum	Present in nitrogenise (nitrogen fixation) and nitrate reductase enzymes Essential for N fixation and assimilation Play an important role in plant nodulation Needed to convert inorganic phosphates to organic forms in the plant
Nickel	Benefited the growth of nitrogen-fixing species (nodule weight and seed yield (Bertrand and Wolf 1967)) Component of nitrogen-fixing hydrogenase bacteria (Cammack 1995) Impart plant disease resistance Component of the plant urease (Dixo et al. 1995)
Chlorine	Turgor regulation (Stomatal regulation of water loss) Electrical charge balance Resisting diseases and photosynthesis reactions

(Source: Brady 2002)

elements in plants and act as substitute for essential nutrients under certain circumstances. The beneficial elements like silicon (Si) have a significant effect on soil texture, water holding capacity, adsorption capacity and soil erosion stability. The key functions of beneficial elements are presented in Table 13.2.

Beside above facts, certain soil conditions induce micronutrient deficiencies or toxicity to growing plants. Generally, micronutrient deficiencies occur on soils with very high pH, light textured soil with low pH, parent material originally low in micronutrient content, intensive cultivation practices receiving excessive levels of high analysing NPK fertilisers, high organic matter content, etc. (Singh 1996; Deb et al. 2009). The deficiency of Mo appeared at low pH soil conditions, whereas high pH led to Zn, Fe, Mn and Cu deficiencies. Further, excess addition of $CaCO_3$ (lime) to acid soils induces deficiencies of micronutrient cations. The general micronutrient deficiency symptoms that appear in crop plants are given in Table 13.3.

Excess supply of any nutrients may cause a reduction in the growth and induces toxicity in plants. Excess supply of Mn causes chlorosis of young leaves (*symptoms are different from Fe deficiency*). Similarly, excess supplies of Cu, Zn, Mo, B and Cl produce specific toxicity symptoms in plants. Besides excess application of micronutrients, there are certain soil factors that induce toxicity of said nutrients. Acid

Table 13.2 Functions of beneficial elements in higher plants

Micronutrient	Functions in higher plants
Silicon	Provides resistance against pathogen and pests Drought and heavy metal tolerance Quality and yield enhancement
Cobalt	Essential for the growth of *Rhizobium* bacteria (enhances the nitrogen-fixing ability of legumes) Key constituent of vitamin B_{12} and propionate Substituting for molybdenum, selenium and sodium
Sodium	Involved in the regeneration of phosphoenolpyruvate in CAM and C_4 plants Substitute for potassium under certain circumstances (*replacing K functions*) Increases leaf area and stomata and improves the water balance, thus promoting growth
Aluminium	Prevents toxic effects of Cu, Mn or Fe and promotes P uptake Suppresses root rot disease Increases antioxidant enzyme activity in tea
Selenium	Provides resistance against harmful pathogens May prevent P toxicity and enhance plant growth May act as an antagonist against phosphate toxicity in hyper-accumulators

Table 13.3 General micronutrient deficiency symptoms

Element	Deficiency symptoms
B	Light general chlorosis, death of growing point, deformed leaves with areas of discoloration
Cl	Chlorosis and wilting of young leaves, deficiency rarely seen on crop plants in field
Cu	Light overall chlorosis, leaf tips die back and tips are twisted, loss of turgor in young leaves
Fe	Chlorosis or yellowing between the veins of new leaves
Mo	Similar to those of ordinary nitrogen deficiency – general chlorosis (yellowing) of young plants, chlorosis of oldest leaves
Mn	Chlorosis or yellowing between the veins of new leaves (much like Fe deficiency)
Zn	Stunted growth, reduced internode length, young leaves are smaller than normal
Ni	In legumes, suppression in plant growth with development of leaf tip necrosis on typically pale green leaves. In graminaceous species, interveinal chlorosis and patchy necrosis in the youngest leaves (like Fe deficiency) (Eskew et al. 1983; Walker et al. 1985)
Cu	Young leaves become dark green, twisted and deformed Appearance of necrotic spots on leaves Seed heads become white and empty (seed production reduces)
Co	Root nodules in legumes remain small Appearance of pale green to yellow leaves on older leaves; however, some crops may develop red leaves, stems or petioles Plant remains stunted and tops may be less leafy
Si	Reduces the number of panicles per square metre and percentage of filled grains (Source: IRRI 1993) Leaflets of the new leaves showed malformations such as curving to the outside, warping and hardening (Miyake and Takahashi 1978)

(Source: Bennett 1993)

sulphate soils generally led to micronutrient toxicity that limits crop growth under upland soil conditions. In rice, Fe toxicity appears in soil rich in Fe (Deb et al. 2009; Dass et al. 2015). Effluents containing Cu and Zn, when applied to soil, produce toxicity of these micronutrients in growing crops. The general micronutrient toxicity symptoms that appear in crop plants are presented in Table 13.4.

13.3 Sources of Micronutrients in Soil

Inorganic micronutrients occur naturally (*inherited from rocks and minerals*) in soil minerals as their oxides, sulphides and silicates. However, natural sources of micronutrients in soil vary from soil to soil. All micronutrients

Table 13.4 General micronutrient toxicity symptoms

Element	Toxicity symptoms
B	Young leaves develop chlorosis between veins which starts from near base and spreads to apex that later become reddish brown in colour Similar symptoms are also found on leaf sheath, petioles and stems
Cl	Reduced size and number of leaves having bronze or yellow colouration with brown or scorched leaf margins
Cu	Interveinal chlorosis in young leaves Older leaves develop orange or pink colouration followed by severe rolling of leaf margins due to loss in turgidity
Fe	Bronzing and stippling of leaves Iron toxicity is not common, but some plants secrete acids from the roots
Mo	Tints of golden yellow or blue colour o leaves
Mn	Chlorosis of young leaves Leaf sheaths and lower part of stems in cereal normally consist of minute brown spots Legumes develop brown or purplish spots over leaf margins
Zn	Interveinal mild chlorosis in young leaves starting from base and spread towards apex Followed by interveinal reddish-brown colouration Rolling of leaves margins
Ni	In early stages, no clearly visible symptoms develop, though shoot and root growth suppresses (Brown et al. 1987) Interveinal chlorosis and restricted leaf expansion Severe toxicity results in complete foliar chlorosis with necrosis advancing in from the leaf margins, followed by plant death (Wood et al. 2004)
Al	Development of thick, stubby and distorted root systems Restricted cell division and cell expansion in the roots Yellow spots or pale flecking on the leaves of grasses or cereals Reduces the availability of P and S, through the formation of Al–P and Al–S, respectively, compounds

(Source: Singh 1996)

have been found in varying quantities in igneous rocks (Brady 2002). Fe and Mn have prominent structural positions in certain of original silicate minerals, whereas Zn also may occupy structural positions as minor replacements for the major constituents of silicates minerals (Brady 2002). During soil development process, a lot of transformations take place in soil minerals, and micronutrients are released after decomposition in the reaction products (Kumar et al. 2014). The micronutrient cations like Fe, Zn, Cu and Mn released during above transformations are either adsorbed on soil colloids or become part of secondary silicate minerals by isomorphous substitutions, while anions like borate and molybdate may undergo adsorption or reaction in soils similar to that of phosphate ions and form new reaction products (Brady 2002; Deb et al. 2009). Chlorine is added in soils in considerable quantities each year through rainwater. However, incidental addition of chlorine in soil through fertilisers and in other ways helps prevent its deficiency under field conditions.

Organic matter is an important secondary source of trace elements in soil. Most micronutrients are held tightly in complex organic compounds and may not be readily available to plants. However, they can be an important source of micronutrients when they are slowly released into a plant available form as organic matter decomposes (Choudhary et al. 2008; Paul et al. 2014). Incorporation of soil amendments and fertilisers to soil also contribute to micronutrient pools. Selenium (Se) is effectively bound to soil organic matter and thus less available to plants (Johnsson 1991). Presence of oxalate and citrate inhibited adsorption of selenate by competition for the anion binding sites and thus increased Se availability to plants (Wijnja and Schulthess 2000). The dominant forms of micronutrients that occur in soil solution are given in Table 13.5.

Table 13.5 Forms of micronutrients dominant in soil solution

Micronutrient	Source	Dominant soil solution forms
Boron	Soil and fertiliser	H_3BO_3, $H_2BO_3^-$
Chlorine	Soil and fertiliser	Cl^-
Copper	Soil and fertiliser	Cu^{2+}, $Cu(OH)^+$
Iron	Soil and fertiliser	Fe^{2+}, Fe^{3+}, $Fe\ (OH)^{2+}$, $Fe\ (OH)^+$
Manganese	Soil and fertiliser	Mn^{2+}
Molybdenum	Soil and fertiliser	MoO_4^{2-}, $HMoO_4^-$
Zinc	Soil and fertiliser	Zn^{2+}, $Zn(OH)^+$

(Source: Lindsay 1972)

13.4 Soil Factors Influencing Micronutrient Acquisition

The availability of micronutrients to plants depends on several factors such as soil texture, soil reaction, organic matter, clay content, soil moisture, nutrient interactions in soil, microbial activity, redox potential and aeration, etc. It is also not possible to select the most important soil factors influencing micronutrient availability to plants, because all factors are equally important and their importance can vary between elements.

13.4.1 Soil Texture

Soil texture exerts a significant effect on micronutrients availability in soil. Coarse-textured (sandy) soils are most often deficient in micronutrients, whereas fine-textured (clay) soils are less likely to be low in plant usable amounts, because clays and organic soils hold nutrients and water much better than sandy soils (Choudhary and Suri 2014b). As water drains from sandy soils, it often carries micronutrients along with it, which is generally known as leaching, and leached nutrients are not available for plants to use. Further, mineral soils with low organic matter tend to have lower micronutrient availability. Strongly leached acid soils are low in most micronutrients because their parent materials are originally deficient in elements, and acid leaching has removed large amount of micronutrients originally present (Choudhary and Suri 2014b; Kumar et al. 2014).

13.4.2 Temperature and Soil Moisture

Micronutrients availability decreases at low temperature and moisture content because of reduced root activity, low rates of dissolution and diffusion of nutrients. In north India, when the soil temperature goes down during winter months, Zn deficiency appears in wheat due to freezing. This Zn deficiency is attributable to decrease solubility of native Zn at low soil temperature (Deb et al. 2009). Likewise wheat, in early spring, Zn deficiencies in maize may be more severe because of cold, wet soil conditions. However, Zn deficiency symptoms often disappear when the weather gets warmer and drier soil conditions prevail.

Wet soil conditions at low temperature further reduce the availability of micronutrient cations. Further, soil moisture content influences mineralisation of micronutrients from soil organic matter (Almendros et al. 2013; Choudhary and Suri 2014c). Prolonged submergence of soils increases pH and reduces redox potential (Eh) which in turn decreases the availability of S, B, Cu and Zn in these acid soils (Karan et al. 2014). Zinc content tends to increase with soil temperature. Cu also increases with soil temperature, but manganese tended to decrease at the high temperature. Boron varied for species only and Al decreased with increasing temperature (Wallace et al. 1969). The general soil conditions that induce micronutrient deficiencies in different crop are presented in Table 13.6.

Table 13.6 Soil conditions lead to micronutrient deficiencies

Micronutrient	Soil characteristics	Crop
B	Sandy soils or highly weathered soils low in organic matter	Alfalfa, clover
Cl	Occasionally on sandy soils in areas, high rainfall very rare	–
Cu	Acid peats or mucks with pH < 5.3 and black sands	Wheat, corn
Fe	Soils with high soil pH, soluble salts and/or calcium carbonate levels	Maize, soybean
Mn	Peats and mucks with pH > 5.8, black sands and lakebed/low-lying soils with pH > 6.2	Soybean, wheat, sugar beets, corn
Mo	Acid prairie soils	Soybean
Zn	Peats, mucks and mineral soils with pH > 6.5	Maize, soybean

[Source: Tri-State Fertilizer Recommendations for Corn, Soybeans, Wheat and Alfalfa (1995)]

13.4.3 Soil Reaction (pH)

Soil reaction (pH) is the most important factor influencing availability of micronutrients to plants. Soil reaction is a measure of soils' acidity or alkalinity. The pH represents a measure of H^+ activity in a soil solution which is in a dynamic equilibrium with a negatively charged solid phase. H^+ ions are strongly attracted to these negative sites and have sufficient power to replace other cations from them. A diffuse layer in the vicinity of a negatively charged surface has higher H^+ activity than the bulk soil solution.

Generally, soil pH regulates the solubility, mobility, concentration of ions in solution and acquisition of elements by plants (Fageria et al. 1997b). In low pH (acid) soils, most of the micronutrients are at their peak availability (Fig. 13.1). Further, low pH favours free metal cations and protonated anions and higher pH favours carbonate or hydroxyl complexes. Thus, availability of micronutrient and toxic ions, which are present as cations in soil solution, increases with increasing soil acidity. Whereas, availability of those present as anions (MoO_4^{2-}, CrO_4^{2-}, SeO_4^-, SeO_3^- and $B(OH)_4^-$) increases with increasing alkalinity. The availability of B, Cu, Fe, Mn and Zn usually decreases and Mo increases as soil pH increases.

The micronutrient cations are most soluble and available under acid conditions. In very acid soils, there is a relative abundance of the ions of iron, manganese, zinc and copper. In fact under acid conditions, the concentration of one or more of these elements often is sufficiently high to be toxic to common plants (Brady 2002). Cations are held more strongly (less reversibly) when pH increases from 5 to 7. Cu, Zn and Ni become significantly less soluble and less exchangeable when pH increases from 5 to 7. At low pH, most of the B compounds are soluble and remain available to plants as boric acid (Deb et al. 2009). Moreover, availability of B is less in acid soils as compared to alkaline soils (higher pH). In acid soils, Mo availability is a serious problem as it gets fixed with Fe and Al compounds as well as on silicates, thus unavailable for plant use (Choudhary and Suri 2014c, d). An increase in soil pH from pH 5.9 to 6.4 involving loam soil with relatively low organic matter content leads to an increase in Se availability (Eich-Greatorex et al. 2007).

13.4.4 Nutrient Interactions

Nutrient interactions may take place in soil as well as within the plant. The Ability of soil to supply micronutrients to plants is not only a factor that influenced micronutrients acquisition, but it also depends on the number of other factors (Das 2011); rather their availability is influenced due to balance among micronutrients in the soil as well with other nutrients (Deb et al. 2009). Interactions occur between micronutrients as well as with primary nutrient elements (Das

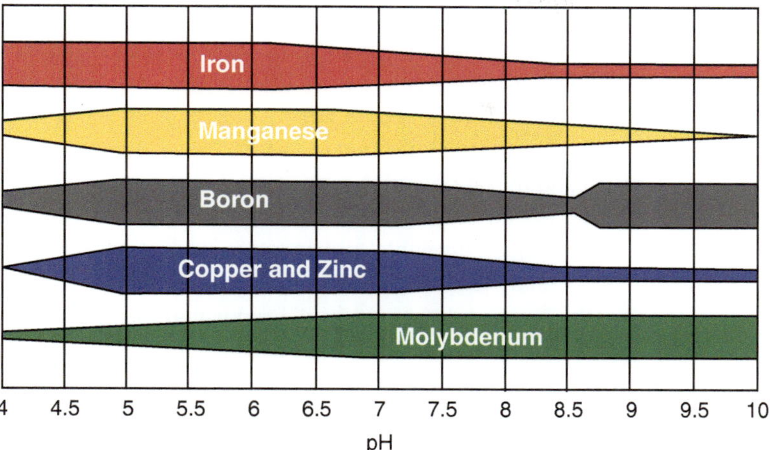

Fig. 13.1 Relative availability of micronutrients at different soil pH (Source: Truog 1946)

2011). The soils high in phosphorus levels or excessive application of P fertiliser may reduce zinc acquisition and reduce yields (Das 2011). Similarly, application of K consistently decreases the Mn and Fe content in rice plant. Amelioration of acid soils with dolomite ($CaCO_3$ and $MgCO_3$) reduces the availability of Zn to crop plants as pH rises. Further, increase in soil Ca^{2+} content following liming decreases heavy metal uptake by roots because of physiological antagonism of Ca^{2+}/heavy metal ions.

Olsen (1972) stated that micronutrient interactions among the various nutrients also influence and change nutrient concentrations in plants as well as growth responses. Similar opinion on the above aspect was also given (Moraghan and Mascagni 1991 and Fageria et al. 2002). The concentration of one micronutrient in soil–plant system may affect the level of other micronutrients through a process called antagonism. Too much Fe may produce Mn and Zn deficiencies, while high levels of Mn may induce Fe and Zn deficiencies. Excessive Zn addition in soils induces Cu deficiency in cereal crops like wheat and barley. Hence, Cu and Zn also showed antagonistic interactions. Application of Zn has been found to depress the content of extractable Fe and increased Mn content under submerged conditions. High Cu in soil causes Fe chlorosis in citrus plants. Soil application of Fe fertiliser eliminates toxic effect of Cu element. Under waterlogged conditions, ferrous (Fe^{2+}) can compete with zinc (Zn^{2+}) in the uptake process for the formation of chelates or other reactions. The Mn interferes with the translocation of Fe from roots to shoots. Absorption of Fe by plant roots increases with increasing concentration of Mn in soils (Das 2011).

The Cu and Mo have antagonistic effects in many crops. Fe and Mo interactions have been found frequently in plants. Further, Mo accentuates Fe deficiency due to formation of a Fe molybdate precipitate in the roots. Again, application of Mo combined with higher doses of Fe increases the yield of crops. Thus, two effects of Mo and Fe interaction are observed – one beneficial and other detrimental (Das 2011). High Fe concentration in soils suppresses copper absorption by rice. Mn inhibited Zn absorption by rice roots, but favoured the translocation of Zn within the plants, suggesting both antagonistic and synergistic effects (Das 2011).

13.4.5 Organic Matter

Organic matter is a reservoir for essential plant nutrients which continuously supply nutrients to the crop upon decomposition over the time. This reservoir is especially important for anions like boron, which do not bind to soil particles and are therefore subject to losses. Soil organic matter

(SOM) has been related to increased, decreased and no effects on micronutrients availability to crop plants (Stevenson 1982; Tisdale et al. 1985; Mortvedt 2000). Reduced availability of micronutrients in the presence of organic matter is attributable to complexation with humic acids, lignin and other organic compounds of high molecular weight; thus, insoluble precipitates are formed (Choudhary and Suri 2009). Likewise, increased availability results from solubilisation and mobilisation of micronutrients by the action of low molecular weight organic ligands such as amino acids, short-chain organic acids and other organic compounds (Stevenson 1982; Tisdale et al. 1985; Mortvedt 2000).

Soils that receive regular additions of organic residues or manures rarely show micronutrient deficiencies. However, imbalanced application of macronutrients may cause deficiency of certain micronutrients, such as excessive application of P fertiliser in overly manured soils, leading to Zn deficiency (Das 2011; Kumar et al. 2014). Soils rich in organic matter content accumulate higher amount of heavy metals in sorption complex and are released slowly in soil solution as compared to mineral soils, which is attributable to high affinity of soil organic compounds to heavy metals. Further, extremely high organic matter content in soil (*muck or peat*) also induces micronutrient deficiencies due to strong natural chelation, i.e. combination of a micronutrient with an organic molecule, which makes some of the micronutrients unavailable, especially copper, manganese and zinc. Organic soils retain large quantities of micronutrient cations in insoluble forms by complexation (Deb et al. 2009). In an organic matter-rich soil, lower pH resulted in higher plant Se availability (Eich-Greatorex et al. 2007).

13.4.6 Oxidation State

The degree of oxidation or reduction in soil is indicated by the redox potential measurement. In soils, reducing conditions are brought by anaerobic conditions as a result of waterlogging, while oxidised conditions are normally found in well-drained aerobic soils (Dass et al. 2015). The reduced states of Fe, Mn and Cu are more soluble than higher oxidation states at normal soil pH range. Generally, high pH favours oxidation, whereas acid conditions are more conducive to reduction (Brady 2002; Deb et al. 2009). At neutral soil conditions, oxidised states of micronutrient cations are generally less soluble than at reduced states. At low pH and poor aeration, micronutrient cations are somewhat more available at restricted drainage; thus, flooded soils generally show higher availabilities than well-aerated ones (Dass et al. 2015). At high pH range, well-drained, aerated, calcareous soils are sometimes deficient in available Fe, Zn or Mn that exists in oxidised state, even though adequate quantities of these elements are present in soil; therefore, plant suffers from micronutrient deficiency (Brady 2002; Deb et al. 2009).

13.4.7 Clay Content

Clay particles are negatively charged and therefore surrounded by a swarm of positively charged cations. Soil colloids that possess protonated functional groups (−OH, −COOH) contribute to cation exchange capacity of soils. Generally, cation exchange capacity of soil increases with increase in pH as dissociation of −OH and −COOH groups (*especially those on organic matter*) is pH dependent. With increased cation exchange capacity, metal cations are attracted towards negative sites on soil colloids, and thus, soil solution is depleted resulting in reduced metal availability to plants.

13.4.8 Rhizosphere

Rhizospheric and surrounding conditions play an important role towards micronutrient availability to plants. Nutrient availability in the rhizosphere is controlled by the combined effects of soil properties, interactions between plant roots and various soil microbes in the surrounding soil. Chemical conditions in the rhizosphere are usually very much different from those in the bulk soil. Microorganisms in the rhizosphere continuously produce chelating agents during the decay of plant and animal residues that have ability to transform solid phase micronutrient cations into soluble metal complexes and thereby increase the availability of insoluble micronutrients to plants (Deb et al. 2009). Arbuscular mycorrhizal fungi (AMF) is known to induce many favourable changes in the crop rhizosphere by way of exudation/secretion of organic acids/chelating agents (Rovira 1969; Suri and Choudhary 2012; Kumar et al. 2014), which lead to efficient solubilisation and mobilisation of plant nutrients from organic and inorganic complexes (Pare et al. 1999; Suri and Choudhary 2013a). Further, AMF greatly influence the absorption of micronutrients through extended rhizospheric exploratory area due to hyphal network (George et al. 1992; Harrier and Watson 2003). Thus, AMF has the ability to increase micronutrient availability to crop plants. Like AMF, there are a number of microbes in rhizosphere that play an important role in micronutrient regulation.

13.4.9 Other Factors

Application of irrigation water contains varying levels of micronutrients as well as other nutrients. When land is levelled to accommodate irrigation, micronutrient deficiencies, particularly Zn, often occur because the available Zn is removed with the surface soil (*organic matter*). Fixation of micronutrients by soil clays may cause serious problems for Zn and Cu and is less significant for Fe and Mn because of their high total content in soils (Deb et al. 2009). The general soil conditions that influenced micronutrient availability and acquisition by plants are given in Table 13.7.

Nutrient acquisition by plant is a dynamic process in which nutrients are continuously replenished in soil solution from soil solid phase followed by transportation to roots for absorption by plants (Suri and Choudhary 2013a). In soil system, nutrients move to plant roots by mass flow, diffusion and root interception (Barber 1995). More than 90–95 % of B, Cu and Zn and considerable quantity of Fe (65 %) are supplied to plants by mass flow. Considerable quantities of Mn, B and Fe (>20 %) move by diffusion, whereas root interception plays a significant role in B, Zn and Mn absorption. In recent years, increased anthropogenic activities such as burning fossil fuels, application of sewage and industrial sludge, use of amendments (fertilisers, manures, lime), application of pesticides and deposition of atmospheric particles induce addition of toxic trace elements like Cd, Cr, Ni, Pb, Cu, Zn, As, Co and Mn (*including some essential micronutrients*) in soil–plant system (Alloway 1995). Excessive levels of trace elements may pose phytotoxicities to crop plants. Liming of acid soils reduces N availability due to lime-induced B fixation by freshly precipitated Al and Fe hydrous oxides; overliming further induces B deficiency in crops. Liming of acid soils increases the availability of native Mo (Deb et al. 2009; Kumar 2012).

13.5 Agronomic Interventions for Micronutrients' Supply vis-à-vis Biofortification

Balanced and integrated use of micro- and macronutrients is essential for enhancing their use efficiency, maximising crop productivity and sustaining fertility of the soils (Choudhary et al. 2013). Thus, it is important to find out best packages or techniques for ameliorating and enhancing micronutrient use efficiencies. The choice of corrective measure for combating micronutrients deficiencies is largely determined by the nature of disorder, growth stage, condition and nutritional status of soil and plant (Singh 2008).

Organic matter helps in maintaining good soil structure, increases water holding capacity and exchangeable ions and reduces leaching of nutrients and toxicities (Choudhary et al. 2008). Adding crop residue, green manures, composts

Table 13.7 Chemical properties of micronutrients

Cations	*Positively charged – bind to soil particles*
Copper	Solubility is greatest under acid conditions
Iron	Most likely deficie006Et on calcareous soils or soils extremely high in organic matter where strong chelation decreases availability
Manganese	
Zinc	
Anions	*Negatively charged – subject to leaching*
Boron	In short supply in areas where they are readily leached and not being replenished by organic matter decomposition
Chlorine	
Molybdenum	

(Source: Truog 1946)

and animal manures, growing cover crops, reducing tillage and avoiding burning of crop residues can significantly lead to improved plant growth and acquisition of micronutrients (Fageria et al. 2002; Paul et al. 2014; Pooniya and Shivay 2015). The accumulation of micronutrients varies among plant species and cultivars within species (Marschner 1995), which have been attributed by genetic, environmental, physiological or biochemical mechanism, in response to agronomic management practices and tolerance to pest and diseases.

Ferti-fortification offers a rapid solution for increasing Zn concentration in grain and straw. Green manuring and Zn-coated fertilisers increased nutrient concentration and their uptake in grain and straw of basmati rice. Foliar fertilisation of 0.2 % zinc sulphate recorded higher Zn concentration in rice, whereas Zn-coated urea (ZCU) as $ZnSO_4.H_2O$ registered highest total Zn uptake (Pooniya and Shivay 2013, 2015). Residue recycling of summer green-manuring crops and Zn-coated urea significantly enhanced soil microbial activities, which are vital for the nutrient turnover and long-term productivity of soil, leading to enhanced productivity of basmati rice (Pooniya et al. 2012).

The AM fungi symbiosis is generally beneficial to the growth and development of host plants and healthy and productive agricultural ecosystem (Harrier and Watson 2003). The use of AMF in agricultural crops is highly desirable to meet out micronutrient needs economically, rationalising water use and maintaining soil health (Suri and Choudhary 2013b). In a study, integrated application of AMF at varying P and irrigation regimes significantly enhanced the concentrations and uptake of various micronutrients (Fe, Zn, Cu, Mn, B and Mo) in okra and pea crops (Kumar 2012). In okra, integrated application of AMF, P and irrigation regimes on an average enhanced total N, P, K, Ca, B and Mo uptake by 8, 24, 5, 14, 8 and 40 %, respectively, over their non-AMF counterpart treatments (Kumar 2012). Likewise, in pea, a higher amount of total N (8 %), P (19 %), K (12 %), Ca (22 %), Mg (12 %), Fe (10 %), Cu (28 %), Zn (22 %), Mn (10 %), B (11 %) and Mo (38 %) uptake was also registered in AMF imbedded treatments over non-AMF counterparts (Kumar 2012).

Enhanced acquisition has been reported for AMF inoculated crop plants in comparison with non-AMF ones of a number of micronutrients. A number of researchers have registered enhanced acquisition of micronutrients in plants such as Zn and Cu (Marschner and Dell 1994), Cl (Ellis et al. 1995), B (Kumar 2012), Mo (Raju et al. 1987), nickel (Rogers and Williams 1986), etc. following AMF inoculation in comparison with non-AMF ones. Further, AMF-colonised plants tend to have lower Mn acquisition compared to non-colonised plants (Kothari et al. 1990; Azaizeh et al. 1995; Suri et al. 2013; Kumar 2012). Thus, AMF use in Himalayan production systems is of tremendous significance to harvest nutritionally rich farm produce for Himalayan communities suffering from malnutrition like anaemia and Zn deficiency and equally to resource-poor Himalayan farmers who ill-afford expensive chemical fertilisers.

Azcon et al. (2003) reported that high availability of N and P in the soil reduced the content of micronutrients in AMF inoculate plants. High application of N and P to the soil reduced the uptake of Mn and Zn in AMF inoculation involving plants than non-AMF lettuce plants. Actually, higher levels of inorganic N and P fertilisers inhibited extra-radical colonisation, which in turn reduces nutrient uptake per unit of root mass. So, results demonstrated negative effects of high N and P application in soil on the acquisition of Fe, Mn and Zn for mycorrhizal plants.

Singh (2008) and Kumar et al. (2011), while working with maize, groundnut, sunflower, wheat and cauliflower, reported that *granubor II* having 15 % boron content in many cases was superior or statistically at par to that of borax as boron source. Thus either source of B can be applied to soils to correct boron deficiency in crops. Further, efficiency of applied boron in cauliflower was significantly higher with *granubor II* in comparison to borax, and it increased by 10.3 % with *granubor II* over borax (Kumar et al. 2011). He further reported that B use efficiency decreased with increasing boron levels from 0.5 to 1.5 kg ha^{-1}, and the highest value (9.2 %) was obtained at lower level of applied B (0.5 kg ha^{-1}) and lowest (4.2 %) with B applied at 1.5 kg ha^{-1}. Moreover, *granubor II* increased B uptake in cauliflower leaf and curd by 7.5 % and 22 %, respectively, over borax. Similarly, B uptake in cauliflower increased with increasing B levels (Kumar et al. 2011).

Another alternative to enhance micronutrient availability in plants is build-up in soil organic matter status. Organic matter is a vital soil component in rebuilding depleted soil as it ensures a continuous energy source for soil biomass (microbes, fungi, algae, protozoa, etc.), which transforms organic molecules into mineral elements that are readily available to plants. Moreover, above organisms maintain good soil structure by transforming organic matter into humus and producing compounds that cement small soil particles together, promoting both increased drainage and moisture retention. Beneficial microbes in soil convert organic and insoluble forms of micronutrients into available ones, thus increasing their availability in plants. Microbes in soil further reduce nutrient losses and help degrade toxic compounds.

Application of Zn and Fe fertilisers (agronomic biofortification) represents useful complementary approach to genetic biofortification. There are increasing evidences showing that 'foliar application' or combined 'soil + foliar application' of micronutrients under field conditions is highly effective and practical to maximise uptake and accumulation of micronutrients in grains of cereal crops (Deb et al. 2009).

Kumar and Babel (2013) reported that micronutrient availability (Fe, Cu, Zn, Mn and B) is positively and significantly correlated with silt, clay and organic carbon content of soils but negatively and significantly correlated with sand, calcium carbonate and pH of the soils. Application of organic manure led to significant increment of solubility of micronutrient elements for plant absorption (Mahmoodabadi and Ronaghi 2014).

Availability of Cu, Zn, Fe and Mn reduced significantly following liming of acid laterite and alluvial soils, whereas liming increases availability of B irrespective of soil moisture regimes (Karan et al. 2014). However, continuous flooding during rice-growing season decreases plant availability of Cu and Zn but increases that of B, Fe and Mn in both limed and unlimed acid laterite and alluvial soils. Alternate flooding and drying further were found beneficial to rice in comparison with continuous flooding under above soil conditions as it significantly increases the availability of B, Cu and Zn nutrients to plants; Fe and Mn availability decreases. In a study under acid soil, Kaur (2012) reported significant increase in micronutrient (Zn, Fe, Mn, Cu and B) acquisition in wheat following application of 100 % P, 10 kg Zn and 1 kg B ha^{-1}. Similarly in acid soil, Suri et al. (2011a) reported significant increase in micronutrients (Zn, Fe, Mn, Cu) acquisition in wheat following integrated application of P and AM fungi.

In a study, micronutrient combinations resulted in significant increase in chickpea grain yield. The percent increase in yield due to micronutrient application was in the range of 14.3–39.5 % with maximum response of 39.5 % following combined application of 5 kg Zn ha^{-1} + 1.0 kg B + 40 kg S ha^{-1} over NPK control (Singh 2008). Soil application of Zn to annual crops is a preferred method over less efficient foliar sprays. Biweekly foliar sprays with 0.5 % $ZnSO_4$ + 0.25 % lime suspension are recommended using 500 l of water ha^{-1} on crops exhibiting Zn deficiency symptoms; spraying continues until the disappearance of deficiency symptoms (Rattan et al. 2009). Results of 'All India Coordinated Research Project' experiments conducted at several location of India involving chickpea, mustard, soybean and sunflower indicated that combined application of Zn + S + B gave maximum yield response, whereas green gram, groundnut, lentil, maize, black gram and wheat gave maximum yield response following Zn + S + B + Mo application (Singh 2008).

Eich-Greatorex et al. (2007) reported that after the first growing season, Se concentration in wheat grain increased following Se application in soil. The Se content in wheat grain ranged between 0.03 and 0.09 mg kg^{-1} with 2.25 g Se ha^{-1} (lowest rate) to 0.17–0.49 mg kg^{-1} following 9 g Se ha^{-1}. However, incorporation of 18 g Se ha^{-1} led to Se concentrations well above the recommended values for human consumption. The increasing Se application rates did not influenced wheat yield significantly. Se concentrations in the grain of barley and oats were in a similar range as obtained for wheat.

Silicon nutrition has significant effect on crop growth, physiological attributes and yield parameters. Actually, Si boosted up accumulation of more photoassimilates from source to sink and consequently led to higher grain yield (Mukhtar et al. 2012). Beneficial effects of Si on sorghum were also registered by Ahmed et al. (2011), who found significant increases of 30 %, 40 %, 30 %, 30 % and 27 % in leaf area index, chlorophyll content, leaf dry weight, root dry weight and total dry weight, respectively, following Si use with water over water-deficient treatment. However, a decrease in leaf water potential and shoot/root ratio in sorghum was reported over control. Raid et al. (1992) reported an average increase of 20 % in sugar cane yield with 'Ca silicate' applied at 7.5 t ha^{-1}. Application of 120 kg Si ha^{-1} increased rice yield to the tune of 17.1 %, 7.1 % and 2.0 %, respectively, over 0, 40 and 80 kg Si ha^{-1} (Jawahar and Vaiyapuri 2013). The above results are attributed to the fact that adequate silicon supply improved photosynthetic activity, which enables rice plant to accumulate sufficient photosynthates, and their efficient translocation resulted in more numbers of filled grains with increased test weight and ultimately higher grain yield (Rani and Narayanan 1994). Further, Si at 120 kg ha^{-1} significantly increased Si uptake of rice. Datta and Shinde (1985) also registered a positive increase in rice grain yield with Si application under both upland and waterlogged conditions. Rogers and Williams (1986) found enhanced acquisition of Cobalt (Co) following AMF inoculation. Grover and Purves (1975) reported a promotive effect of Co^{2+} application on cucumber hypocotyl elongation and plant development, which is attributable to inhibition of ethylene biosynthesis by Co^{2+}.

13.6 Conclusion

The micronutrient deficiencies in agriculture are limiting the crop production worldwide, and these deficiencies will likely to increase in the future. Micronutrients are not only important for better crop productivity but also essential for sustaining human and animal health through agricultural produce biofortification. The demand of increasing crop production to feed ever-increasing population from limited land resource requires thorough knowledge of soil–plant interactions, so that a quality food may be supplied to world population through appropriate agronomic interventions. The balanced fertilisation, recycling of crop residues, agronomic biofortification, application of micronutrients in soil, biofertilisers, etc. offer a good solution to overcome the micronutrient deficiencies.

References

Ahmed M, Hassen FU, Qadeer U, Aslam MA (2011) Silicon application and drought tolerance mechanism of sorghum. Afr J Agric Res 6(3):594–607

Alloway BJ (1995) Heavy metals in soils, 2nd edn. Wiley, New York

Almendros P, Gonzalez D, Alvarez JM (2013) The influence of moisture on the residual effects of natural zinc chelates applied to two different soils. J Soil Sci Plant Nutr 13(4):797–807

Azaizeh HA, Marschner H, Romheld V, Wittenmayer L (1995) Effects of a vesicular–arbuscular mycorrhizal fungus and other soil microorganisms on growth, mineral nutrient acquisition and root exudation of soil grown maize plants. Mycorrhiza 5:321–327

Azcon R, Ambrosano E, Charest C (2003) Nutrient acquisition in mycorrhizal lettuce plants under different phosphorus and nitrogen concentration. Plant Sci 165:1137–1145

Baligar VC, Fageria NK, Elrashidi MA (1998) Toxicity and nutrient constraints on root growth. Hortic Sci 33:960–965

Barber SA (1995) Soil nutrient bioavailability: a mechanistic approach, 2nd edn. Wiley, New York

Bennett WF (1993) Plant nutrient utilization and diagnostic plant symptoms. In: Bennett WF (ed) Nutrient deficiencies and toxicities in crop plant. American Phytopathological Society Press, St. Paul, pp 1–7

Bertrand D, Wolf A (1967) Nickel, a dynamic trace element for higher plants. C R Acad Sci 265:1053–1055

Brady NC (2002) The nature and properties of soils. Prentice Hall of India Private Limited, Inc, Upper Saddle River, p 621

Brown PH, Welch RM, Cary EE (1987) Nickel: a micronutrient essential for higher plants. Plant Physiol 85:801–803

Cammack R (1995) Splitting molecular hydrogen. Nature 373:556–557

Choudhary AK, Suri VK (2009) Effect of organic manures and inorganic fertilizers on productivity, nutrient uptake and soil fertility in wheat (*Triticum aestivum*)–paddy (*Oryza sativa*) crop sequence in western Himalayas. Curr Adv Agric Sci 1(2):65–69

Choudhary AK, Suri VK (2013) 'On-Farm' participatory technology development on resource conservation technologies in rainfed upland paddy in Himachal Pradesh, India. Commun Soil Sci Plant Anal 44 (17):2605–2617

Choudhary AK, Suri VK (2014a) Integrated nutrient management technology for direct-seeded upland rice (*Oryza sativa*) in north–western Himalayas. Commun Soil Sci Plant Anal 45(6):777–784

Choudhary AK, Suri VK (2014b) 'On farm' participatory technology development on forage cutting and nitrogen management in dual-purpose wheat (*Triticum aestivum*) in north–western Himalayas. Commun Soil Sci Plant Anal 45(6):741–750

Choudhary AK, Suri VK (2014c) Frontline demonstration programme-an effective technology transfer tool for adoption of oilseeds production technology in Himachal Pradesh, India. Commun Soil Sci Plant Anal 45(11):1480–1498

Choudhary AK, Suri VK (2014d) Scaling-up of pulses production under frontline demonstrations technology transfer programme in Himachal Himalayas, India. Commun Soil Sci Plant Anal 45(14):1934–1948

Choudhary AK, Thakur RC, Kumar N (2008) Effect of integrated nutrient management on soil physical and hydraulic properties in rice–wheat crop sequence in NW Himalayas. Indian J Soil Conser 36(2): 97–104

Choudhary AK, Rahi S, Singh A, Yadav DS (2010) Effect of vermi-compost and biofertilizers on productivity and profitability in potato in north–western Himalayas. Curr Adv Agric Sci 2(1):18–21

Choudhary AK, Thakur SK, Suri VK (2013) Technology transfer model on integrated nutrient management technology for sustainable crop production in high value cash crops and vegetables in north–western Himalayas. Commun Soil Sci Plant Anal 44 (11):1684–1699

Clark RB, Zeto SK (2000) Mineral acquisition by arbuscular mycorrhizal plants. J Plant Nutr 23:867–902

Das DK (2011) Introductory soil science. Kalyani Publishers, New Delhi, India, p 645

Dass A, Chandra S, Choudhary AK, Singh G, Sudhishri S (2015) Influence of field re-ponding pattern and plant spacing on rice root-shoot characteristics, yield, and water productivity of two modern cultivars under SRI management in Indian Mollisols. Paddy Water Environ. doi:10.1007/s10333-015-0477-z

Datta NP, Shinde JE (1985) Yield and nutrition of rice under upland and waterlogged conditions. Effect of nitrogen, phosphorus and silica. J Indian Soc Soil Sci 33:53–60

Deb DL, Sakal R, Datta SP (2009) Fundamentals of soil science. Indian society of soil science. Cambridge Printing Works, New Delhi, p 728. ISBN 8190379747

Dixo NE, Gazzola C, Blakeley RL, Zerner B (1995) Jack bean urease (EC 3.5.1.5). A metalloenzyme. A simple biological role for nickel. J Am Chem Soc 97:4131–4133

Eich-Greatorex S, Sogn TA, Ogaard AF, Aasen I (2007) Plant availability of inorganic and organic selenium fertiliser as influenced by soil organic matter content and pH. Nutr Cycl Agroecosyst 79:221–231

Ellis JR, Watson DMH, Varvel GE, Jawson MD (1995) Methyl bromide soil fumigation alters plant element concentrations. Soil Sci Soc Am J 59:848–852

Engelhard AW (1990) Soil borne plant pathogen management of disease with macro and micro elements. American Phytopathological Society Press, Saint Paul

Eskew DL, Welch RM, Norvell WA (1983) Nickel: an essential micronutrient for legumes and possibly all higher plants. Science 222:621–623

Fageria NK, Baligar VC, Jones CA (1997a) Growth and mineral nutrition of field crops, 2nd edn. Dekker, New York

Fageria NK, Valigar VC, Wright RJ (1997b) Soil environment and root growth dynamics of field crops. Recent Res Dev Agron 1:15–58

Fageria NK, Baligar VC, Clark RB (2002) Micronutrients in crop production. Adv Agron 77:185–268

George E, Haussfer KV, Vetterlien D, Gorgus E, Marschner H (1992) Water and nutrient translocations by hyphae of *Glomus mosseae*. Can J Bot 70:2130–2137

Graham RB, Webb MJ (1991) Micronutrients and disease resistance and tolerance in plants. In: Mortvedt JJ, Cox FR, Shuman LM, Welch RM (eds) Micronutrients in agriculture, 2nd edn. SSSA, Madison, pp 329–370

Grover S, Purves WK (1975) Cobalt and plant development: interaction with ethylene in hypocotyls growth. Plant Physiol 57:886–889

Gupta UC (1993) Factors affecting boron uptake by plants. In: Gupta UC (ed) Boron and its role in crop production. CRC Press, Boca Raton, pp 87–123

Harrier LA, Watson CA (2003) The role of arbuscular mycorrhizal fungi in sustainable cropping systems. Adv Agron 42:185–225

Huber DM (1980) The role of mineral nutrition in defense. In: Hosfall JG, Cowling EB (eds) Plant pathology: an advanced treatise. Academic Press, New York, pp 381–406

Huber DM, Wilhelm NS (1988) The role of manganese in resistance to plant diseases. In: Graham RD, Hannam R, Uren NC (eds) Manganese in soils and plant. Kluwer, Dordrecht, pp 155–173

IRRI (1993) Rice research in a time change: IRRI's Medium – Term Plan, 1994–1998. International Rice Research Institute, Los Banos

Jawahar S, Vaiyapuri V (2013) Effect of sulphur and silicon fertilization on yield, nutrient uptake and economics of rice. Int Res J Chem 3(1):35–43

Johnsson L (1991) Selenium uptake by plants as a function of soil type, organic matter content and pH. Plant Soil 133:57–64

Jones JP, Engelhard AW, Wolty SS (1990) Management of *Fusarium* of vegetables and ornamentals by micro and macro element nutrition. In: Engelhard AW (ed) Soil borne plant pathogens: management of diseases with macro and microelements. American Phytopathological Society Press, Saint Paul, pp 18–32

Karan AK, Kar S, Singh VK, Singh CV (2014) Effects of liming and soil moisture regimes on time changes of soil pH, redox potential, availability of native sulphur and micronutrients to rice (*Oryza sativa* L.) in acid soils. Int J Soil Sci 9(1):1–15

Kaur R (2012) Phosphorus and rain-harvested water economy through vesicular arbuscular mycorrhizae (VAM) in okra–pea sequence. MSc thesis, CSK HPKV, Palampur

Kothari SK, Marschner H, George E (1990) Effect of VA mycorrhizal fungi and rhizosphere microorganisms on root and shoot morphology, growth and water relations in maize. New Phytol 116:303–311

Kumar A (2012) Phosphorus and rain-harvested water economy through vesicular arbuscular mycorrhizae (VAM) in okra–pea sequence. PhD thesis, CSK HPKV, Palampur

Kumar IA, Babel A (2013) Extractable micronutrients status in relation to other soil properties in Jangargari, Yamaltu–Deba local government area, Gombe State. Asian J Agric Food Sci 1(5):217–221

Kumar A, Parmar DK, Suri VK (2011) Effect of boron fertilizers and organic manure on autumn cauliflower in western Himalayas. Ann Hortic 5(1):17–24

Kumar A, Suri VK, Choudhary AK (2014) Influence of inorganic phosphorus, VAM fungi and irrigation regimes on crop productivity and phosphorus transformations in Okra (*Abelmoschus esculentus* L.)–Pea (*Pisum sativum* L.) cropping system in an acid Alfisol. Commun Soil Sci Plant Anal 45:953–967

Lake DL, Kisk PWW, Lester JN (1984) Fractionation, characterization and speciation of heavy metals in sewage sludge and sludge–amended soils: a reviews. J Environ Qual 13:175–183

Lindsay WL (1972) Inorganic phase equilibria of micronutrients in soils. In: Mortvedt JJ, Giordano PM, Lindsay WL (eds) Micronutrients in agriculture. Soil Science Society of America, Madison

Lindsay WL (1991) Inorganic equilibria affecting micronutrients in soil. In: Mortvedt JJ, Cox R, Shuman LM, Welch RM (eds) Micronutrients in agriculture, 2nd edn. Soil Science Society of America, Madison, pp 89–112

Mahmoodabadi MR, Ronaghi AM (2014) Effects of manure and soil texture on soybean plants. Acta Adv Agric Sci 2(1):1–6

Marschner H (1995) Mineral nutrition of higher plants. Academic Press, San Diego

Marschner H, Dell B (1994) Nutrient uptake in mycorrhizal symbiosis. Plant Soil 159:89–102

Miyake Y, Takahashi E (1978) Silicon deficiency of tomato plant. Soil Sci Plant Nutr 24(2):175–189

Moraghan JT, Mascagni HJ (1991) Environment and soil factors affecting micronutrient deficiencies and toxicities. In: Mortvedt JJ, Cox FR, Shuman LM, Welch RM (eds) Micronutrients in agriculture, 2nd edn. SSSA, Madison, pp 371–425

Mortvedt JJ (2000) Bioavailability of micronutrients. In: Sumner ME (ed) Hank book of soil science, CRC Press, Boca Raton, FL, pp D71–D88

Mukhtar A, Asif M, Goyal A (2012) Silicon the non-essential beneficial plant nutrient to enhanced drought tolerance in wheat crop plant. In: Goyal A (ed) Crop plant. CBS Publishers & Distributors Pvt. Ltd, New Delhi, India

Olsen SR (1972) Micronutrient application. In: Mortvedt JJ, Cox FR, Shuman LM, Welch RM (eds) Micronutrients in agriculture, 2nd edn. SSSA, Madison, pp 243–264

Pare T, Gregorich EG, Nelson SD (1999) Mineralization of nitrogen from crop residues and N recovery by maize inoculated with vesicular arbuscular mycorrhizal fungi. Plant Sci 218:11–20

Paul J, Suri VK, Sandal SK, Choudhary AK (2011) Evaluation of targeted yield precision model for soybean and toria crops on farmers' fields under sub-humid sub-tropical north–western Himalayas. Commun Soil Sci Plant Anal 42(20):2452–2460

Paul J, Choudhary AK, Suri VK, Sharma AK, Kumar V, Shobhna V (2014) Bioresource nutrient recycling and its relationship with biofertility indicators of soil health and nutrient dynamics in rice–wheat cropping system. Commun Soil Sci Plant Anal 45(7):912–924

Pilon-Smits EAH, Quinn CF, Tapken W, Malagoli M, Schiavon M (2009) Physiological functions of beneficial elements. Curr Opin Plant Biol 12:267–274

Pooniya V, Shivay YS (2013) Enrichment of *Basmati* rice grain and straw with zinc and nitrogen through ferti-fortification and summer green manuring crops under Indo-Gangetic Plains of India. J Plant Nutr 36 (1):91–117

Pooniya V, Shivay YS (2015) Influence of green manuring and zinc fertilization on quality parameters of *Basmati* rice. Commun Soil Sci Plant Anal 46:382–392

Pooniya V, Shivay YS, Rana A, Nain L, Prasanna R (2012) Enhancing soil nutrient dynamics and productivity of *Basmati* rice through residue incorporation and zinc fertilization. Eur J Agron 41:28–37

Raid RN, Anderson DL, Ulloa MF (1992) Influence of cultivar and amendment of soil with calcium silicate slag on foliar disease development and yield of sugarcane. Crop Prot 11:84–88

Raju PS, Clark RB, Ellis JR, Maranville JW (1987) VAM infection effects on sorghum growth, P efficiency and mineral nutrient uptake. J Plant Nutr 10:1331–1339

Rani AY, Narayanan A (1994) Role of silicon in plant growth. Annu Rev Plant Physiol (B&A) 1:243–262

Rattan RK, Patel KP, Manjaiah KM, Datta SP (2009) Micronutrients in soil, plant, animal and human health. J Indian Soc Soil Sci 57(4):5446–558

Rogers RD, Williams SE (1986) Vesicular–arbuscular mycorrhizae influence on plant uptake of cesium and cobalt. Soil Biol Biochem 18:371–376

Romheld V, Marschner H (1986) Mobilization of iron in the rhizosphere of different plant species. Adv Plant Nutr 2:155–204

Rovira AD (1969) Plant root exudates. Bot Rev 35:35–37

Singh SS (1996) Soil fertility and nutrient management. Kalyani Publishers, Ludhiana, India, p 215

Singh MV (2008) Micro and secondary nutrient and pollutant elements research in India. Coordinator report – AICRP micro and secondary nutrient and pollutant elements in soils and plants, IISS, Bhopal, vol 31, pp 1–77

Stevenson FJ (1982) Humus chemistry. Wiley, New York

Stevenson FJ (1986) Cycles of soil carbon, nitrogen, phosphorus, sulfur and micronutrients. Wiley, New York

Suri VK, Choudhary AK (2012) Fertilizer economy through VAM fungi under STCR targeted yield model in maize–wheat–maize crop sequence in Himalayan acid Alfisol. Commun Soil Sci Plant Anal 43:2735–2743

Suri VK, Choudhary AK (2013a) Effect of vesicular arbuscular mycorrhiza and applied phosphorus through targeted yield precision model on root morphology, productivity and nutrient dynamics in soybean in an acid Alfisol. Commun Soil Sci Plant Anal 44:2587–2604

Suri VK, Choudhary AK (2013b) *Glycine–Glomus–Phosphate solubilizing bacteria* interactions lead to fertilizer phosphorus economy in soybean (*Glycine max* L.) in a Himalayan acid Alfisol. Commun Soil Sci Plant Anal 44:3020–3029

Suri VK, Choudhary AK (2013c) Effect of VA-mycorrhizal fungi and phosphorus application through STCR precision model on crop productivity, nutrient dynamics and soil fertility in soybean (*Glycine max*)–wheat (*Triticum aestivum*)–soybean crop sequence in an acid Alfisol. Commun Soil Sci Plant Anal 44(13):2032–2041

Suri VK, Choudhary AK (2014) Comparative performance of geographical isolates of Glomus mosseae in field crops under low input intensive P-deficient acid Alfisol. Commun Soil Sci Plant Anal 45 (1):101–110

Suri VK, Chander G, Choudhary AK, Verma TS (2006) Co-inoculation of VA-mycorrhizae and phosphate solubilizing bacteria in enhancing phosphorus supply to wheat in typic Hapludalf. Crop Res 31(3):357–361

Suri VK, Choudhary AK, Chander G, Gupta MK, Dutt N (2011a) Improving phosphorus use through co-inoculation of vesicular arbuscular mycorrhizal fungi and phosphate solubilizing bacteria in maize in an acid Alfisol. Commun Soil Sci Plant Anal 42 (18):2265–2273

Suri VK, Choudhary AK, Chander G, Verma TS (2011b) Influence of vesicular arbuscular mycorrhizal fungi and applied phosphorus on root colonization in wheat and plant nutrient dynamics in a phosphorus-deficient acid Alfisol of western Himalayas. Commun Soil Sci Plant Anal 42(10):1177–1186

Suri VK, Choudhary AK, Kumar A (2013) VAM fungi spore populations in different farming situations and their effect on productivity and nutrient dynamics in maize and soybean in Himalayan acid Alfisol. Commun Soil Sci Plant Anal 44(22):3327–3339

Tisdale SL, Nelson WL, Beaton JD (1985) Soil fertility and fertilizers, 4th edn. Mc Millan, New York

Tri-State Fertilizer Recommendations for Corn, Soybeans, Wheat and Alfalfa (1995), Ohio State University. http://www.ohioline.osu.edu/e2567/. Accessed 14 Feb 2015

Truog E (1946) Soil reaction influence on availability of plant nutrients. Soil Sci Soc Am Proc 11:305–308

Walker CD, Graham RD, Madison JT, Cary EE, Welch RM (1985) Effects of nickel deficiency on some nitrogen metabolites in cowpeas, *Vigna unguiculata*. Plant Physiol 79:474–479

Wallace A, Romney EM, Hale VQ, Hoover RM (1969) Effects of soil temperature and zinc application on yields and micronutrient content of four crop species grown together in a glasshouse. J Agron 61(4):567–568

Wijnja H, Schulthess CP (2000) Interaction of carbonate and organic anions with sulfate and selenate adsorption on an aluminum oxide. Soil Sci Soc Am J 64:898–908

Wood BW, Reilly CC, Nyczepir AP (2004) Mouse-ear of pecan: a nickel deficiency. Hort Sci 39 (6):1238–1242

Yadav DS, Thakur SK, Sood P, Choudhary AK (2013) Assessing training needs of extension functionaries about organic farming in north–western Himalayas. J Organic Syst 8(1):17–27

Yadav A, Suri VK, Kumar A, Choudhary AK, Meena AL (2015) Enhancing plant water relations, quality and productivity of pea (*Pisum sativum* L.) through AM fungi, inorganic phosphorus and irrigation regimes in an Himalayan acid Alfisol. Commun Soil Sci Plant Anal 46(1):80–93

Nutrient Mining: Addressing the Challenges to Soil Resources and Food Security

14

Kaushik Majumdar, Saroj Kumar Sanyal, Sudarshan Kumar Dutta, T. Satyanarayana, and V.K. Singh

Abstract

Increasing population demands higher production and productivity of crops. Consequently, maintenance of soil fertility is a must for sustainable agriculture and future food security. Agriculturally productive soils have a pool of indigenous nutrients at any given point of time that are stored within the soil and may be available for supporting plant growth. This pool of nutrients, along with nutrient inputs from other sources such as irrigation water, crop residues, etc., constitutes the inherent soil nutrient-supplying capacity. There are several sinks of the native nutrients in the soil, most notably plant removal. Sustenance of inherent fertility of soils depends largely on replenishment of plant nutrients to the soil that are removed through intensive cultivation. *Nutrient mining* or negative balance between nutrient input and output results when the crop nutrient removal and nutrient losses to other sinks become higher than the soil-inherent nutrient supply. Current nutrient management strategies adopted by most farmers promote nutrient mining, as nutrient applications are inadequate and imbalanced. The application of 4R Nutrient Stewardship Principles, i.e., application of right source of fertilizer at right rate and time through a right method, has the potential to reduce nutrient mining

K. Majumdar
International Plant Nutrition Institute-South Asia, Gurgaon 122 016, India
e-mail: kmajumdar@ipni.net

S.K. Sanyal
Former Vice Chancellor, Bidhan Chandra Krishi Viswavidyalaya, Nadia 741 252, West Bengal, India
e-mail: sarojsanyal@yahoo.co.in

S.K. Dutta (✉)
International Plant Nutrition Institute-South Asia, Kolkata 700 028, India
e-mail: SDutta@ipni.net

T. Satyanarayana
International Plant Nutrition Institute- South Asia, Hyderabad 500 085, India
e-mail: tsatya@ipni.net

V.K. Singh
ICAR – Indian Institute of Farming System Research, Modipuram, Meerut 250 110, Uttar Pradesh, India
e-mail: vkumarsingh_01@yahoo.com

U. Singh et al. (eds.), *Biofortification of Food Crops*, DOI 10.1007/978-81-322-2716-8_14

from soils. These core principles help manage nutrients in a manner that crop productivity is sustained or improved without soil fertility depletion, and farm production economics is improved while environmental impact of agricultural nutrients is minimized.

Keywords
Food security • Nutrient mining • Soil resources

14.1 Introduction

Sustainability is the overarching goal of intensive agriculture. It depends on how well we use our primary resources, such as land and water, to produce required crop yields while maintaining them for posterity. The capacity of the soil to function within natural or managed ecosystem boundaries, to sustain plant and animal productivity, maintain or enhance water quality, and support human health and habitation is in the core of sustainability (SSSA 1995). The capacity of the soil to function adequately, in terms of producing enough food, depends on its ability to support plant growth. Soils, through a diverse chemical processes involving its organic and inorganic constituents, adsorb and release water and essential plant nutrients to support plant life. The limitation imposed by inadequate nutrient status strips the soil off its "capacity to function or perform," and adequate availability of essential nutrients in the soil is critical for sustained productivity of the soil.

The basic principle of maintaining the fertility status of a soil under high-intensity crop production systems is to annually replenish those nutrients that are removed from the field. A *negative* input–output balance of nutrients in the soil will eventually limit crop yield, facilitate *nutrient mining*, and result in the depletion of soil fertility. Experimental evidences suggest that restoring the fertility status of a soil denuded of its native fertility is a much costlier process that maintains the inherent fertility status through external application of nutrients. A soil severely depleted of its native potassium fertility status is difficult to restore even with external application due to structural collapse of its parent material (Sarkar et al. 2013, 2014). Rattan Lal (2014) in his recent paper quoted Sir Albert Howard who strongly believed in the relationship between the rise and fall of civilizations and their agricultural practices and argued that "the real arsenal of democracy is a fertile soil, the fresh produce of which is the birthright of nations." Indeed a fertile soil is a unique resource that ensures continued food and nutritional security of any nation, particularly so when population increases across the world are continuously challenging the boundary between food security and famine. The global population increase challenge will be intense in Asia, and particularly so in South Asia, due to the large base population of the region. Maintaining the soil resource base and its quality in a fertile state would be critical for sustained food security in the region.

14.2 Present and Future Nutrient Requirement for Food Production

With a projected increase of global population to 9.6 billion by 2050 from the current 7.2 billion, there is a tremendous need for increased production of food, feed, fuel, and raw materials from limited land area available for cultivation. India has around 18 % of the world population, 15 % of the world livestock with only 2.3 % of the total geographical area, and 0.5 % of pasture and grazing lands. The per capita availability of land has fallen drastically from 0.91 ha in 1951 to about 0.32 ha in 2001, and it is projected to decline further to 0.09 ha by 2050. The pressure on the land is increasing rapidly and land degradation is on the rise (Venkateswarlu and Prasad

2012). The current Indian population of 1.2 billion is expected to rise to 1.4 billion by 2025 and the country will need to produce at least 300 million tonnes (mt) of food grains for ensuring food security of the growing population. Ganesh-Kumar et al. (2012) summarized several studies that showed food grain demand in India reaching 293 mt by 2020 and increasing to 335 mt by 2026. Limited scope for horizontal expansion of cultivated area further accentuates the need for an increasing intensity of agricultural production. Greater crop yields per unit land area will require advances in genetics and crop improvement, increased fertilizer nutrient use, and improved soil and crop management technologies. Indeed Amarasinghe et al. (2008) estimated that the total calorie requirement (kcal/person/day) was 2495 in the year 2000 that is expected to increase to 2775 in 2025 and 3000 in 2050, respectively. Another study projected the food consumption levels in India to increase from the current level of 2400 kcal/per capita/day to about 3000 kcal/per capita/day in 2050 and estimated that the demand for cereals will rise to 243 mt in 2050 (Singh 2009). The major determinants for achieving future food security goals will be driven by the use of quality seeds, fertilizers, and water and its timely supply and application of modern technologies supporting precision management of resources.

Among the resources, fertilizer is one of the important inputs, whose judicious use would trigger the process of accelerated growth in production. Fertilizer use has played a significant role in the development of the agriculture sector in India, and the contribution of fertilizers to total grain production in India has increased from 1 % in 1950 to 58 % in 1995 (Subba Rao and Srivastava 1998). The total fertilizer nutrient consumption ($N + P_2O_5 + K_2O$) increased from 70 thousand tonnes in the early 1950s to 5.5 mt in 1980–1981, reaching an all-time high of 28.1 mt in 2010–2011 (FAI 2013). It is also estimated that there will be an additional demand for 40–45 mt of nutrients (NPK) along with the secondary and micronutrients by the year 2025 in order to meet the growing food production demand (Katyal 2001). Estimates based on the sufficiency approach showed that the requirement for zinc, iron, copper, boron, and manganese will be 324, 130, 11, 3.9, and 22 thousand tonnes by 2025 (Venkateswarlu and Prasad 2012).

14.3 Native Fertility and Inherent Soil Nutrient-Supplying Capacity

Janssen et al. (1990) defined inherent nutrient-supplying capacity or potential indigenous supply of nutrients from a soil as the cumulative amount of a nutrient originating from all indigenous sources that circulates through the soil solution surrounding the entire root system during one complete crop cycle. The native fertility or indigenous nutrient-supplying (INS) capacity of a soil at any given point of time is the nutrient pool that is stored within the soil and may be available for supporting plant growth. The INS by definition precludes inclusion of any nutrient contribution from external sources. However, it must be understood that the INS is not strictly confined to nutrients that are geogenic in nature, such as potassium, which has a potential source in micaceous minerals. Rather, the INS is a combination of geogenic and externally supplied nutrients that have been historically stored by the soil through various mechanisms such as adsorption, absorption, fixation, etc. Nutrient contributions from external sources may include irrigation water, atmospheric deposition, biological nitrogen fixation, and crop residues. In the absence of externally applied nutrient, INS provides support to plant growth by supplying the required nutrients. A soil with high indigenous supply of nutrients would be able to support adequate plant growth without external application, at least for a short period of time, while soils with low indigenous fertility will not be able to support plant growth without application of nutrients from external sources such as manures and fertilizers.

The INS is a critical component for estimating the amount of externally supplied nutrients required for supporting crops and cropping systems. Scientists and other stakeholders are in continued pursuit to critically estimate indigenous nutrient-supplying capacity of agricultural soils so that manure or fertilizer application could be optimized. The primary objective of the task is to supply adequate nutrition to the plants for effectively completing their life cycle. However, the co-benefits such as higher nutrient use efficiency, better economics of crop production, and less environmental footprint of agricultural nutrient use are substantial. The final outcome is a sustainable production system.

Soil testing is by far the classical method of estimating INS in a soil. Soil testing entails careful sampling of soil from the target location, extraction and analysis of available nutrients, and using the data through appropriate correlation and calibration to come up with a number that describes the quantity of a particular nutrient that will be available to the plants during a crop season. Several authors recently questioned the chemical methods adopted for analyzing available nutrients in the soil (Prasad 2013), the lack of correlation between soil test-based data for available nutrients and crop yield (Dobermann et al. 2002), or the approaches for developing integrated nutrient recommendations from soil test data (Tandon 2012). In fact, there are infrastructural issues that often limit the accuracy of INS estimates, particularly in Asia. For example, in the case of potassium, the NH_4OAc extract is the standard procedure followed in most soil testing laboratories. This method precludes the estimation of non-exchangeable potassium that contributes slowly to the soil available pool of potassium for plant uptake during the growing season. From a different perspective, the above method, employing the NH_4OAc extract of soil, does not provide any information about potassium depletion from the non-exchangeable pool in soil.

A plant-based approach for estimating indigenous nutrient-supplying capacity of soils was later developed by the International Rice Research Institute (Dobermaan et al. 2004). Instead of direct quantification of the nutrient status in the soil, this method relied on plant responses as an indicator of soil nutrient-supplying capacity. In this method, the yield of a crop in a field plot supplied with ample amounts of all limiting nutrients was compared with the yield of the crop in a plot that received all limiting nutrients except one that was omitted from the fertilization schedule. The difference in crop yield between the ample nutrient-supplied plot and the nutrient-omitted plot (yield loss) was ascribed to the direct effect of the omitted nutrient. This provided an indirect estimate of the indigenous nutrient-supplying capacity of the soil for the omitted nutrient. In other words, the crop is giving an indication of how much yield the soil can sustain in the absence of the external application of the omitted nutrient. For example, indigenous potassium supply (IKS) can be measured as total K accumulation in plant raised in a 0-K plot that received N, P, and all other limiting nutrients. The advantage of using plant indicators as nutrient supply is that they integrate nutrient supply from all the indigenous sources under field conditions (Dobermaan et al. 2004). A disadvantage of this approach is that the nutrient uptake is often affected by genotype and environmental variation in harvest index, and rooting pattern of the crop, which is often influenced by local growing conditions.

14.4 Nutrient Mining

The nutrient balance, the difference between nutrient input and output in a single crop growth period or in a cropping system cycle, is a critical determinant in the estimation of external nutrient application requirement. Several authors (Janssen et al. 1990; Witt et al. 1999; Witt and Dobermann 2004; Buresh et al. 2010) have used this approach as a major step toward development of site-specific nutrient management (SSNM) guidelines. Singh et al. (2014), in a recent paper, estimated the nutrient balance in a rice–wheat cropping system using the following equation:

$$B_{n(rw)} = \{(IW_n \times Eff) + (CR_n \times Eff) + (RF_n \times Eff) + (S_n \times Eff)\}\{(GY_r \times RIE_{nr}) + (GY_w \times RIE_{nw})\}$$

where B_n is the nutrient balance (N or P or K; kg ha^{-1}), and the IW_n, CR_n, RF_n, and S_n are the nutrient (N or P or K) contribution from irrigation water, crop residue, rainfall, and soil during the entire rice–wheat cropping cycle, respectively. The term "Eff" is the efficiency (%) of different nutrients from various pools of INS in terms of their availability to the crops. GY_r and GY_w are attainable grain yields (t ha^{-1}) of rice and wheat, respectively. RIE_{nr} and RIE_{nw} are the reciprocal internal efficiencies (kg plant nutrient per 1000 kg grain yield) for rice and wheat for N or P or K, respectively. It may be noted that the authors used the soil test values to evaluate the indigenous nutrient supply (INS) from soil during the cropping cycle. These indigenous nutrient contributions were obtained from the corresponding omission plot results in the approach used by Witt and Dobermann (2004) and Buresh et al. (2010). Singh et al. (2014) included the respective nutrient efficiency (Eff) values of the nutrients concerned as part of the equation to recognize that soil available nutrients have other sinks besides crop uptake.

In the above equation, nutrient (N, P, and K) contributions from soil available pool, irrigation water (IW), and rainfall (RF), crop residues (CR), and their availability (%, efficiency) to the crop constitute the INS or input part of calculation (the first part of the right-hand side of the equation). The nutrient contributions from IW and RF (kg ha^{-1}) were estimated using total amount of irrigation water applied/rainfall received (ha-cm) during the rice–wheat cycle and their N, P, and K content. Average available soil N, P, and K content (kg ha^{-1}) at the start of the study across several locations was used as contribution from soil. The nutrient input from residues of a crop (CR_n) was determined from the amount and nutrient content of the aboveground crop biomass retained in the field after harvest and expressed in kg ha^{-1}. The second part of the right-hand side of the equation, crop yield multiplied by the corresponding reciprocal internal efficiency, constituted the crop removal (output) in the cropping system.

It is obvious from the above equation that a negative nutrient balance will result when the second part (crop removal) becomes higher than the first part (INS). This will lead to the depletion of soil native fertility, in other words "nutrient mining."

The reciprocal internal efficiency (RIE) is an important component in the nutrient balance equation as shown earlier. It has its origin in the modified QUEFTS (Quantitative Evaluation of the Fertility of Tropical Soils) model (Janssen et al. 1990; Witt et al. 1999) and is used to assess the nutrient removal by crops at different yield levels. For balanced nutrition, the QUEFTS model assumes a constant internal efficiency (of the major plant nutrients N, P, K) up to yield targets of nearly 70–80 % of the yield potential. In the QUEFTS model, two boundary lines described the minimum and maximum internal efficiencies (IEs, kg grain per kg nutrient in the aboveground plant biomass) of N, P, and K. For example, the minimum and maximum internal efficiencies for rice (*Oryza sativa* L.) were estimated at 42 and 96 kg grain kg^{-1} N, 206 and 622 kg grain kg^{-1} P, and 36 and 115 kg grain kg^{-1} K, respectively (Witt et al. 1999; Dobermann and Witt 2004) using on-farm data from over 2000 locations across a wide range of yields and nutrient status. The balanced N, P, and K uptake requirements for 1000 kg of rice grain yield were estimated from the above as being equivalent to 68.0 kg grain kg N^{-1}, 385 kg grain kg P^{-1}, and 69.0 kg grain kg K^{-1}, respectively, for the aforesaid linear phase (Dobermann and Witt 2004). Similarly, Chuan et al. (2013) also estimated minimum and maximum IEs for wheat (*Triticum aestivum* L.) as 28.8 and 62.6 kg grain kg N^{-1}, 98.9 and 487.4 kg grain kg P^{-1}, and 23.0 and 112.9 kg grain kg K^{-1}. In this case, the above-stated QUEFTS model predicted a linear–parabolic–plateau curve for the balanced nutrient uptake at several target yields. The linear phase in this case was noted to continue up to 60–70 % of the potential yield, and 22.8 kg N,

4.4 kg P, and 19.0 kg K were required to produce 1000 kg grain. These authors also estimated the relationship between the grain yield and the nutrient uptake for suggesting the fertilizer application, avoiding excess or deficient nutrient supply.

The above section highlights the fact that the RIE used for calculating fertilizer requirement is significantly lower than the minimum IEs of the major plant nutrients for both the crops. Thus, the minimum internal nutrient efficiency for K for the cited case of rice cultivation was noted to be about 50 % lower than that for K in the above-stated linear phase, while for N and P, the respective lowering of these values was by about 40 % and 46 %.

Indeed, under such scenario, the point that comes up is if the linear phase of the internal nutrient efficiency of N, P, and K persists up to as much as 70–80 % of the potential yield of the rice cultivars under cultivation in Asia (Dobermann and Witt 2004), one of the envelopes of the internal nutrient efficiency curves, bound by the two boundary lines (describing the minimum and maximum IEs), would be used for making further progress in estimating the fertilizer requirement to support a targeted yield of a crop in a given soil under a specific scenario of crop residue retention in the field, irrigation water source, and so on. Is there any possibility of under- or overuse of fertilizers in such cases? In other words, does this render the soil *poorer or else richer* than what one would think in terms of the available methodologies to estimate the withdrawals from the soil?

14.5 Mechanism of Nutrient Losses

The biggest contributor to nutrient mining is plant removal. Whenever the plant nutrient removal exceeds the combined nutrient inputs from indigenous sources and external application, the soil nutrient status gets depleted. This is a common occurrence in farmers' fields where suboptimal application of nutrients promotes nutrient mining in intensive cropping systems. The following table (Table 14.1) shows on-farm results across several locations and cropping systems that highlight the extent of nutrient mining in cultivators' fields.

The data in the above table highlight two very critical issues. The first is that potassium mining of different degrees is evident in all the locations and cropping systems, and potassium mining in most cases exceeds that of other nutrients. Generally no application of potassium by farmers or suboptimal K recommendation in the state recommendation promoted K mining. This is a common occurrence across India, mainly due to lack of awareness among the farmers and the common belief at the decision-making level that the Indian soils are rich in potassium and may not need external application to support adequate crop growth. The second issue highlighted by the above table is the difference in crop removal in the SR and SR + M treatment. The N, P, and K application rates are similar in both the treatments, but the SR + M treatment received application of deficient secondary and micronutrients. The results showed that addition of secondary and micronutrients to SR triggered higher removal of major nutrients in the on-farm trials. Application of limiting secondary and micronutrients, even under suboptimal application of major nutrients, may increase crop yield, leading to higher yields and hence *more mining* of major nutrients (Table 14.1).

Nutrient mining may also arise from other processes such as volatilization or leaching losses as well as soil erosion. For example, nitrogen (N) is highly mobile in the soil and has high probability to be lost from the soil system through volatilization and leaching. On a global scale, at least 50 % of the fertilizer N applied is lost from the agricultural systems (Ladha et al. 2005). Nitrogen can be lost from agricultural systems as ammonia, nitrous oxide (N_20), or nitrogen oxides (NOx). Sharma et al. (2008) summarized the results of N_2O emission from agricultural soils based on actual field measurements. These experiments reveal average N_2O–N emission of 0.0025 and 0.0055 kg kg^{-1} N applied from rice and wheat fields, respectively. Nitrogen oxide (NOx)

Table 14.1 Nutrient use and removal at cultivators' field in India

		Nutrient addition ($kg\ ha^{-1}$)			Nutrient removal ($kg\ ha^{-1}$)			Apparent balance ($kg\ ha^{-1}$)		
Cropping system/location	Treatment	N	P	K	N	P	K	N	P	K
Rice–wheat/Kaushambi, UP (24)	FFP	206	28.4	0.0	133	22	150	73	6.4	−150
	SR	220	48.0	74.7	186	35	160	34	13	−85.3
	SR + M	220	48.0	74.7	204	40	174	16	8	−99.3
Rice–rice/Warangal, AP (24)	FFP	298	60.3	70.6	168	47	172	130	13.3	−101.5
	SR	240	52.4	66.4	185	52	189	55	0.4	−135.6
	SR + M	240	52.4	66.4	199	56	202	41	−3.6	−135.6
Pearl millet–mustard/ Deesa, Gujarat (18)	FFP	114	37.1	0.0	171	45	104	−57	−7.9	−104
	SR	130	39.3	54.0	193	51	116	−63	−11.7	−62.1
	SR + M	130	39.3	54.0	200	54	122	−70	−14.7	−65.1
Pearl millet–wheat/ Thasra, Gujarat (18)	FFP	130	34.9	0.0	129	28	65	1	6.9	−65
	SR	200	43.7	83.0	194	43	95	6	0.7	−12
	SR + M	200	43.7	83.0	200	44	101	0	−0.3	−18
Maize–Bengal gram/ Gadag, Karnataka (24)	FFP	80	38.0	0.0	138	26	133	−58	12	−133
	SR	110	32.8	20.8	142	28	169	−32	4.8	−148.3
	SR + M	110	32.8	20.8	156	32	181	−46	0.8	−160.3
Rice–green gram/ Kakdwip, WB (18)	FFP	52.5	26.6	34.0	130	22	129	−77.5	4.6	−95
	SR	100	34.9	66.4	138	26	161	−38	8.9	−94.6
	SR + M	100	34.9	66.4	148	29	176	−48	5.9	−109.6
Maize–wheat/Kangra, HP (18)	FFP	50	14.0	21.6	678	17	53	−18	−30	−31.4
	SR	170	37.6	58.1	130	30	89	40	7.6	−30.9
	SR + M	170	37.6	58.1	135	33	97	35	4.6	−38.9
Cotton–pearl millet/ Deesa, Gujarat (18)	FFP	202	37.8	0.0	287	46	85	−85	−8.2	−85
	SR	320	43.7	83.0	324	52	91	−4	−8.3	−8
	SR + M	320	43.7	83.0	378	53	102	−58	−9.3	−19

Source: AICRP-IFS Reports (2011–12); *FFP* farmers' fertilizer practice, *SR* state recommendation, *SR + M* state recommendation + micro- and secondary nutrients

emission from soils is primarily a result of NO production by the microbial oxidation of ammonium, the process being known as nitrification. The NO production in the soils also occurs through microbial reduction of nitrate (denitrification). Estimates (Sharma et al. 2008) suggest that about 0.5 % of fertilizer N applied to agricultural fields was emitted to the atmosphere as NO. Application of fertilizers in the agriculture fields and the livestock population is mainly responsible for NH_3 emission. Although most emissions of ammonia are from manure or natural sources, experiments demonstrate that nitrogen losses to the atmosphere in the form of ammonia following the application of urea can amount to 20 % or more, under temperate conditions. Losses occur when the urea is not incorporated into the soil immediately after spreading, and they are particularly high in calcareous soils. Losses are even higher, up to 40 % or more, under tropical conditions, for flooded rice and perennial crops to which the urea is applied on the soil surface, such as banana, sugar cane, oil palm, and rubber. Some studies have shown leaching loss of N from soils in the Indo-Gangetic Plains (IGP) as 10-–15 kg N ha^{-1}, while ammonia volatilization loss is 20–30 kg N ha^{-1} with application of 120 kg N ha^{-1} in rice and wheat (Majumdar et al. 2014a). The Indian soils, being in the subtropical region coupled with preponderance of tillage practices, are rarely sufficient in nitrogen. The volatilization and leaching losses of N aggravate the situation. These losses constitute removal of nutrients from the soil available pools, in other words nutrient mining, and crop production relies more on adequate external application of N

Table 14.2 Effect of slope gradient on loss of available phosphorus from different tilled soil management treatments

	Loss of P in runoff			Loss of P in erosion		
Slope gradient	Bare fallow	Maize–maize	Maize–maize (mulch)	Bare fallow	Maize–maize	Maize–maize (mulch)
%	Kg ha^{-1} yr^{-1}					
1	3.61	0.58	0.0	1.97	0.26	<0.1
5	3.41	1.49	0.03	9.30	1.03	<0.1
10	3.72	0.62	1.0	20.25	0.87	<0.1
15	4.68	1.70	0.93	13.08	2.29	<0.1

through fertilizer/manure sources rather than on the native soil reserve of nitrogen.

Phosphorus (P) and potassium (K), on the other hand, have lower potential for loss from the soil than N. Except erosion (for P) and leaching (for K), the extent of loss by other means is negligible for these two nutrients. Under favorable growth conditions, most agricultural crops recover 20–30 % of the applied phosphorus depending upon the growth stage of P applications. A large portion of the unused P accumulates in the soil as the soil–fertilizer reaction products, and eventually a part of it is recovered by subsequent crops over time; a much smaller fraction of P is lost as runoff (both particulate and dissolved P) or through leaching that can cause secondary off-site impacts. All forms of P within the soil system are subjected to a variety of pathways of transport at the soil profile, hill slope, or catchment scale. Particulate and colloid P transport is most commonly associated with soil erosion, which arises from raindrop impact and overland flow. Additionally, when fertilizer or manure application is coincident with fast or energetic water flows, this will contribute to particularly high losses (Gburek et al. 2005). Singh and Lal (2005) stated that principal mechanism of P loss from soil is by erosion as P-enriched sediments. The authors showed (Table 14.2) that the loss of P in erosion is higher in plowed bare or continuous maize treatments compared to maize crop with mulch. Further, the loss of P increased with increase in slope gradient from 1 % to 15 % in plowed and bare treatments.

Potassium, being a component of soil minerals, mica, and feldspar, is unique among the three major nutrients. In contrast to N and P, presence of mica and feldspar in soils provides an abundant in situ source of this nutrient. Besides, potassium gets absorbed in the negatively charged interlayer space of soil clays, a process that prevents its loss from the soil and maintains the nutrient in available form for plant uptake. However, on the issue of nutrient mining, potassium is probably the most pertinent nutrient that has the potential of significantly affecting agricultural production in the country. Potassium input–output budgets in agriculture are nearly always highly negative. Estimates for India and Indonesia suggest annual K losses of about 20–40 kg K/ha that have been increasing steadily during the past 40 years (Majumdar et al. 2014a). Historically low application rates of potassium in crops have led to overdependence on the native soil reserve of potassium. Besides crop removal, substantial amount of soil K (and other basic cations) could be lost via leaching in fields with adequate drainage. Leaching losses of K can be substantial in highly permeable soils with low cation exchange capacities. Yadvinder-Singh et al. (2005a) found that leaching losses of K were 22 % and 16 % of the applied K, respectively, in sandy loam and loamy soils maintained at submerged moisture regimes. For Bangladesh, such losses can be as high as 0.1–0.2 kg K ha^{-1} d^{-1} (Timsina and Connor 2001).

14.6 Soil Organic Carbon Dynamics and Nutrient Mining

Carbon sequestration, the storage of carbon in soil organic matter, is a crucial process for mitigating global warming and climate change. Balanced and adequate nutrient management has been identified as crucial to soil organic C (SOC) sequestration in tropical soils (Bhattacharyya et al. 2007; Mandal et al. 2007). Several studies cited by Pathak et al. (2010) stated that adequate supply of nutrients in soil could enhance biomass production and SOC content in the soil and enhance crop productivity. Although SOC sequestration is a major challenge in soils of the tropics and subtropics, the authors (Pathak et al 2010), while analyzing data from 26 long-term fertilizer experiments in India, showed that the application of balanced fertilizer rates along with farmyard manure has good potential in C sequestration in Indian soils and mitigating GHG emission without any additional cost. Rather, it increased yield and net return in majority of the experiments. Increase in SOC in soil makes it more productive leading to increase crop yield. Majumder et al. (2008) also observed that balanced fertilization caused a net enrichment of both the total carbon and organic carbon content of the soils in rice–wheat–jute rotation because of a large amount of crop residues and root biomass C left over in the soil owing to the significantly higher yield of the crops grown under those treatments compared to the control.

The current authors believe that there is a distinct connection between carbon sequestration and nutrient mining. During the C sequestration process, plant nutrients present in the organic substrate are immobilized in the complex organic structure of the sequestered carbon. Once immobilized within the SOC, nutrients are not available to plants or are not prone to be lost from the soil by other mechanisms as stated in the previous section. However, as the organic carbon in the soil gets decomposed, the nutrients within the organic matter become available for plant uptake or loss. Mineralization and immobilization of nutrients in the soil are critical processes influencing the dynamics of nutrients within the soil system. Both the processes are biochemical in nature and are bound to the activities of the heterotrophic biomass. These two processes significantly influence the dynamics of several nutrients, namely, N, P, S, and the micronutrients in soil–plant system. Indeed, the process of mineralization is responsible for the fundamental transformation of organic nutrients in plant residues back into the simple inorganic forms originally used by plants in their metabolic activities (Sanyal et al. 2009a).

In the current context, importance of carbon (chiefly organic) sequestration in soil and co-sequestration of plant nutrients are of importance in the soil nutrient mining. It is apparent that the breakdown of soil organic matter (SoM) under the different land use pattern would adversely affect the soil's capacity to store the inorganic nutrients, thereby protecting them from several avenues of loss. In other words, such depletion of SoM will encourage the nutrient mining. The subtropical climate in India aggravates loss of carbon from soil. Besides, heavy tillage before growing crops also exposes soil organic matter and mineralization rates are high. Saha et al. (2011) suggested that the most significant factors affecting SOC pools in a watershed include land use, land use changes, and soil erosion. The authors quoted several studies that showed cultivation can reduce SOC storage and may promote erosion loss of SOC.

Hence, attempts to sequester the organic matter in soil will help reduce the deleterious effect (on soil fertility) associated with nutrient mining in Indian agriculture. The information on such correlations is worth looking for as to how the relevant soil management options (including the incorporation of the inorganics and reduced tillage) can best be related to such nutrient dynamics in soil under a crop. Indeed, the beneficial effects of carbon sequestration in soil, including the carbon credit and budgeting, will be greatly reinforced in case the former is noted to cause significant retention of the inorganic nutrients in soil matrix, thereby preventing them from being lost.

14.7 Current Situation of Nutrient Mining

Assessment of nutrient mining through estimating nutrient balances using information on nutrient additions and removals generates useful, practical information on whether the nutrient status of a soil (or area) is being maintained, built up, or depleted. Estimates of nutrient input and output, as discussed in the previous section, allow the calculation of nutrient balance sheets for both individual fields and geographical regions. Nutrient balance sheets calculated in most soils of India have shown the signs of nutrient mining as crop nutrient removal far exceeded the nutrient additions through manures and fertilizers. During 1999–2000, the crop removal of nutrients is estimated to be about 28 mt, while the fertilizer consumption was only 18 mt with an annual nutrient gap of 10 mt. Although a part of this nutrient gap is expected to be bridged from sources like organic manures and through biological processes, still there remains a distinct gap in nutrient removal and supply leading to nutrient mining from the native soil which, in turn, poses a serious threat to long-term sustainability of crop production (Hegde and Babu 2001). Tandon (2004) also estimated the deficit between removals and additions at 8–10 M t N + P_2O_5 + K_2O per year and reported this trend to continue in the future (see later).

In India, the state-wise approaches to crop nutrient balances have been developed in 2001 considering the nutrient additions and removal data either from 1998 to 1999 or 1999 to 2000. Since then, the information on nutrient balances has not been updated. Satyanarayana and Tewatia (2009) made an attempt to generate fresh information on nutrient balances in major agriculturally important states of India considering the information available from FAI (2008). Nutrient-balance calculations in most of the cases do not give real picture as they consider nutrient removal by crops and addition through fertilizers neglecting contribution from sources other than fertilizers such as organic manures, crop residues and stubbles, irrigation water, etc. However, in this paper, the authors have tried to overcome that limitation by considering nutrient additions through organic sources wherever possible from the information available in the published literature.

While calculating nutrient removal by crops at the present production levels, attempt was made to consider removal of nutrients by fruits and vegetables in all the states; tea, coffee, jute, rubber, and other plantation crops in states wherever applicable and the total production values have been multiplied with the nutrient uptake per tonne of produce and arrived at removal figures. While generating data for nutrient additions, apart from additions through fertilizers, nutrient contribution from other sources like organic manures, farmyard manure, crop residues, irrigation water, biological nitrogen fixation, etc. has also been considered for some of the states depending upon the availability of information. Where information is not available, nutrient additions through only fertilizer are considered. Similarly, wherever nutrient balance studies were well established by previous authors, efficiency factors were involved for calculating net nutrient balances.

The above study (Satyanarayana and Tewatia 2009) revealed that the nutrient use pattern in majority of the agriculturally important states of India is inadequate and mostly dominated by NP fertilization. The overall N balance seemed to be positive in India (Table 14.3), with highest positive N balance observed in the northern region, which could be attributed to significant addition of nutrients through both inorganic and organic sources. The overall P balance seemed to be negative due to an excessive mining of P to an extent of 4.2 mt observed in the western region. The mining of K is evident in almost all the states, which implied neglect in the use of K fertilizers. Potassium mining in India was estimated at 9.7 mt, and the highest K mining was noticed in the western region (−3.8 mt), followed by northern, southern, and eastern region, respectively. Dutta et al. (2013) estimated K input–output balances in different states of India using the IPNI NuGIS approach and reported negative K balances in most of the states suggesting deficit K application as compared to crop K uptake, contributing to mining of native K. The authors included the manure application, along with the potassic fertilizers, across the different states of India to capture the most recent K balance scenario in India (Fig. 14.1).

It is evident from the figure that K_2O depletion was more in 2011 than in 2007 in most of the

Table 14.3 Region-wise nutrient additions, removal by crops, and apparent balances in India. All units in '000 t

Region	Nutrient	Addition	Removal	Balance
East	N	2079	1733	346
	P_2O_5	794.6	782.4	12.2
	K_2O	517.6	2428.2	−1910.6
	NPK total	3391.1	4943.7	−1552.6
West	N	3955.6	3838.5	117.1
	P_2O_5	1797.5	2212.3	−414.8
	K_2O	756.7	4579	−3822.3
	NPK total	6510	10,630	−4120
Northern region	N	5016.8	2728.3	2288.5
	P_2O_5	1432.6	1258.4	174.2
	K_2O	918.1	3534	−2615.9
	NPK total	7367.5	7520.7	−153.2
Southern region	N	2212.9	1755.1	457.8
	P_2O_5	844.7	836.4	8.3
	K_2O	1118.1	2447.7	−1329.6
	NPK total	4175.8	5039.1	−863.3
India	N	13264.3	10054.9	3209.4
	P_2O_5	4869.4	5089.5	−220.1
	K_2O	3310.5	12988.9	−9678.4
	NPK total	21444.4	28133.5	−6689.1

northern (such as Punjab, Haryana, Uttar Pradesh), eastern (Assam, Bihar, Orissa, Jharkhand, and Chhattisgarh), and western (such as Gujarat, Rajasthan) states of India. This indicated that soils in these states typically received less than the required amount of K_2O. The states of West Bengal and Tamil Nadu showed positive K_2O balances in both 2007 and 2011, suggesting that no K mining has occurred in these two states. Indeed the negative K_2O balance for several states in the country, such as Uttar Pradesh, Punjab, Haryana, Rajasthan, and Madhya Pradesh, increased significantly during the period from 2007 to 2011, probably due to lesser fertilizer application and/or higher crop production per unit land holding. However, such changes in native soil K fertility under intensive cropping can also be quite abrupt, particularly when the vegetables are included in the rice-based cropping systems (Sanyal et al. 2009a). Sen et al. (2008), while assessing the changes in nutrient availability through the GIS-based fertility mapping, observed the K fertility in an intensively cultivated village in the alluvial zone of West Bengal to decrease perceptibly within 2 years. The range of available K_2O changed from 87–448 kg ha^{-1} in 2006 to 56–375 kg ha^{-1} in 2008 with the mean value declining from 166 kg ha^{-1} in 2006 to 88 kg ha^{-1} in 2008. Potassium fertility of the village was generally low to medium in 2006, but the frequency distribution shifted more toward the low fertility category with a substantial increase in sample number in the lowest category (Fig. 14.2). These authors noted that the lower application of K during this period due to unavailability of K fertilizers and high K uptake by the vegetable crops apparently contributed to this sharp decline in K fertility of soils over such a short period of time.

Potassium additions through the prevailing practices of manuring and residue recycling, as well as the meager inputs through K fertilizers, are not sufficient to match the K removal by different crops, and therefore, tremendous efforts are needed to promote K consumption through use of K-rich fertilizers.

The current trends of nutrient balances reveals that the gap between nutrient use and supply in

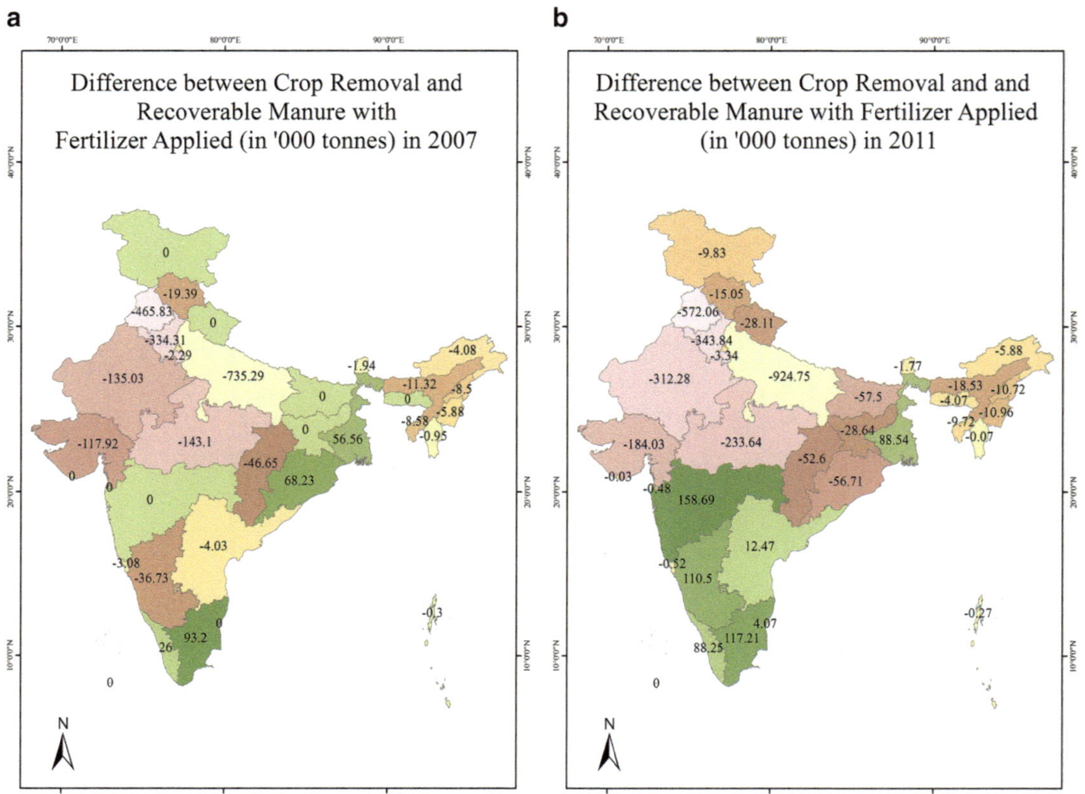

Fig. 14.1 The K_2O (applied fertilizer + manure – crop removal) balance for (**a**) 2007 and (**b**) 2011 across different states of India (Source: Dutta et al. 2013)

farming areas will continue to grow wide on account of intensive cropping, and therefore, there is a need to ensure proper and timely supply of major as well as secondary and micronutrients. Other than additions through fertilizer nutrients, practices like recycling of crop residues, rather than removing the residues from the field, and the use of animal manures through appropriate composting processes should be encouraged in place of diverting these resources for fuel and other secondary purposes.

Table 14.4 gives an example of the ground reality in the context of nutrient mining in the Murshidabad district of West Bengal as a case study (Sanyal et al. 2009b). Thus, taking only rice, wheat, mustard, and jute as the growing crops in the said district, and utilizing the state-recommended nutrient application rates and the ratio, this study demonstrated the deficit of N, P_2O_5, and K_2O running into 19,423 t, 10,669 t, and 11,438 t, respectively (Table 14.4). However, in Murshidabad, agriculture is much diversified. If one includes all other crops in such calculation, for example, 53,000 ha of pulses, more than 20,000 ha under fruits, and nearly 86,000 ha under different vegetables, a scenario of huge deficit of nutrients will show up. This deficit, however, is not confined to NPK alone: the corresponding levels of the secondary and the micronutrient deficits do also cause significant production loss. The calculation in the table was based on 100 % acceptance of the state recommendation, which is rarely the case at the ground level. This suggests that the total consumption of the district is being distributed among all the crops, fostering suboptimal nutrient application to most of them encouraging *nutrient mining* from soil to an unaccounted and alarming proportion.

Further, if the average use efficiency of fertilizers (N 50–60 %; P 15–25 %, K 60–70 %)

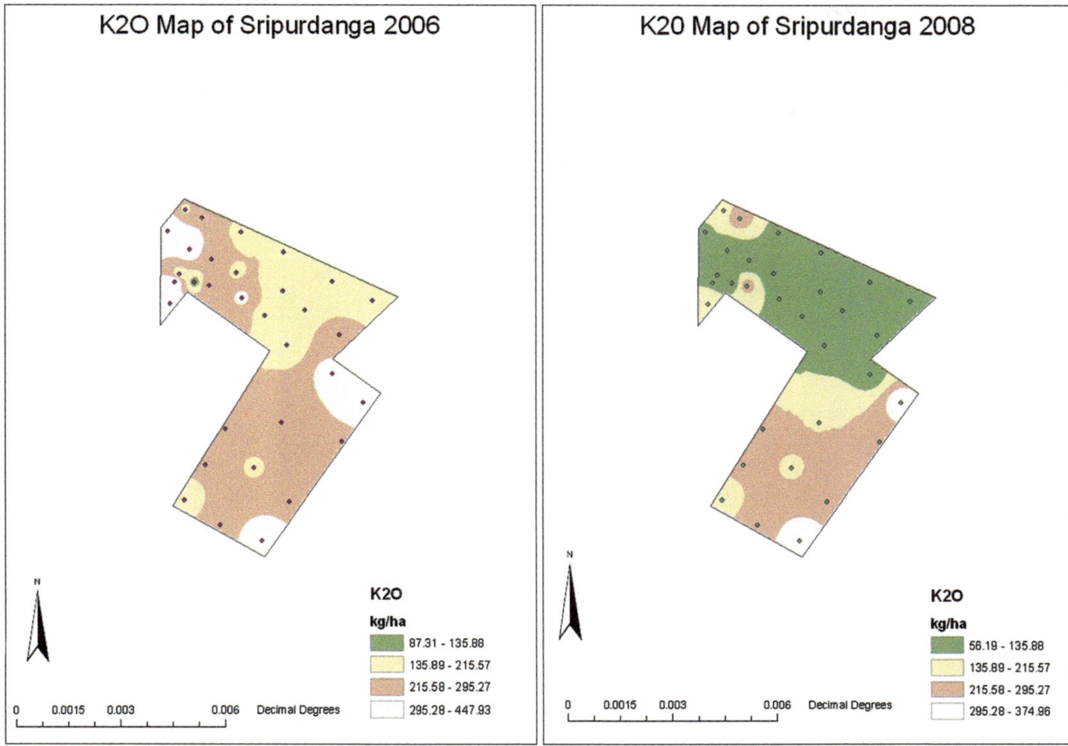

Fig. 14.2 Comparative maps of available potassium before and after four cropping seasons (Source: Sen et al. 2008)

Table 14.4 Theoretical consumption of nutrients at the state-recommended level in Murshidabad, West Bengal

		State recommendation (kg/ha)			Consumption at 100 % acceptance of state recommendation (t)		
Crop	Area (ha)	N	P_2O_5	K_2O	N	P_2O_5	K_2O
Rice	406,724	60	30	30	24,403	12,202	12,202
Mustard	62,031	80	40	40	4962	2481	2481
Wheat	133,961	120	60	60	16,075	8038	8038
Jute	143,852	50	25	25	7193	3596	3596
Calculated nutrient requirement for four crops at state-recommended fertilizer rates					52,633	26,317	26,317
Actual fertilizer consumption in the district					33,210	15,648	14,879
Difference					−19,423	−10,669	−11,438

is taken into account, the nutrient addition through fertilizers is further reduced, and therefore, the removal exceeds the consumption and the nutrient gap is widened. Nevertheless, the situation is balanced by addition of the nutrients through biofertilizers, FYM, compost, green manuring, or addition of crop residues in the field. The consumption data on secondary and micronutrient fertilizers are not available. There is a need to compute nutrient balances with respect to secondary and micronutrients, giving emphasis primarily to the most limiting nutrients like S, Zn, and B.

With the intensively grown production systems of India, heavy removal and inadequate replenishment of nutrients resulted in multiple nutrient deficiencies and depletion of soil nutrient reserves. For sustaining the crop productivity

Table 14.5 Initial soil K status, K addition, K removal, and subsequent K balance

Treatments	Initial soil K (kg/ha)	K added through fertilizer (kg/ha)	Total K removal (kg/ha)	K balance (kg/ha)	Postharvest soil K (kg/ha)
Kharif groundnut, Bhubaneswar site 1 (replicated trial)					
BFT (NPKSZnBCa)	113	60	28.3	144.7	98.5
RDF	113	40	27.4	125.6	74.6
FFP	113	57	26.2	143.8	70.5
Kharif groundnut, Deoda, Dharmasala site 2 (non-replicated trial, $n = 1$)					
BFT (NPKSZnBCa)	147	60	19.0	188	103.3
RDF	147	40	18.4	168.6	80.2
FFP	147	57	15.4	188.6	119
Rabi groundnut, Bhubaneswar site 3 (replicated trial)					
BFT (NPKSZnBCa)	127	60	46.1	140.9	110.3
RDF	127	40	34.4	132.6	110.7
FFP	127	57	38.6	145.4	80.6
Rabi groundnut, Sankaradiha and Bhuban, Dharmasala site 4 (non-replicated trial, $n = 7$)					
BFT (NPKSZnBCa)	104.4	60	49.1	115.3	158.7
RDF	104.4	40	41.6	102.8	143.1
FFP	104.4	57	30.0	131.4	105.4

and to restore the soil fertility, there is a need to arrest depletion of soil nutrient reserves. Indeed, tremendous efforts are needed to promote the right use of plant nutrients through promoting the concept of 4R Nutrient Stewardship (IPNI 2012).

As stated above, Tandon (2004) studied the nationwide scenario of nutrient balance and reported an annual mining of 9.7 M t of NPK (19 % N, 12 % P, and 69 % K) from the soil. He attributed higher K mining to higher crop removal, which is, on an average, 1.5 times greater than N, while K application through fertilizer is much lower than that of N or P.

14.8 Challenges for Nutrient Mining

If any plant nutrient, whether a major, secondary, or a micronutrient, is deficient in the soil, then crop growth is likely to be affected and nutrient mining would be promoted. Majumdar et al. (2014a) categorized the nutrient management approaches prevalent in the country and adopted by farmers into the following three groups based on their order of acceptance:

- Farmer applying fertilizer according to their own perception
- Ad hoc fertilizer recommendation
- Soil test-based fertilizer recommendation

According to the authors, 70–80 % of farmers, involved mainly in field crop production, apply fertilizer based on their own perception or as advised by their progressive peers. This has promoted over- or underuse of fertilizer, large-scale imbalance in nutrient application, and improper timing of fertilizer application. Scientists and policy makers have pointed out the declining nutrient use efficiency/fertilizer response, farm profitability, as well as sharp increase in areas with multiple nutrient deficiencies as clear indicators of inappropriate fertilization approaches adopted by farmers. The application of nitrogen fertilizers tends to be preferred by farmers, because of their relatively low cost per unit of nutrient, their widespread availability, and the quick and evident response of the plant.

P and K use is low as compared to N, and secondary and micronutrients are generally omitted from fertilization schedule, leading to possibility of nutrient mining. Such practices promote large-scale mining of nutrients that are under-applied.

Traditionally balanced fertilization in India means use of N, P_2O_5, and K_2O in a certain ratio, ideally 4:2:1, on a gross basis both in respect of areas and crops. Whatever may be the origin of 4:2:1 ratio, farmers in India do not follow it under most circumstances (Majumdar et al. 2014a). The ad hoc recommendations, developed by different state governments, are based on crop responses over large areas and provide recommendations for medium fertility soils. The recommendations are prescribed for large areas and do not take into account the spatial and temporal variability in soil nutrient-supplying capacity. This often results in over- or under-fertilization leading to yield and economic losses. Besides, the recommendations are generally for medium yield targets and often fall short for higher yield targets achievable through the use of better seeds and good management. In the case of soil test-based recommendations, only available nutrient levels are measured and used as basis for the recommendation. This may not affect the N recommendations as Indian soils are traditionally poor in available N and recommendations are not based on residual N in the soils. However, in the case of potassium, using only available or exchangeable K as the driver of fertilizer recommendation may lead to the depletion of non-exchangeable K. Presence of mica and feldspar in soils provides an in situ source of this nutrient that maintains the exchangeable fractions at the cost of depletion of non-changeable form.

With the realization of the lacunae in the current nutrient management approaches, the site-specific nutrient management (SSNM) strategies (Majumdar et al. 2014a) are now being advocated. SSNM strategies ensure that all the required nutrients are applied at proper rates and in proper ratios commensurate with the crop's nutrient needs. The universality of the principles of the SSNM approach has led to its application to different crops and agroecologies (Singh et al. 2014). The in-built algorithms of SSNM cut down over- and underuse of fertilizers and significantly reduce the probability of nutrient mining. Therefore, conceptually moving from a generalized nutrient management approach, based on some arbitrary ratio, to a rational site-specific approach would be the starting point of addressing the nutrient mining issue.

14.9 Fertilizer Best Management Practices for Nutrient Mining

Fertilizer best management practices (FBMPs) are agricultural production techniques and practices developed through scientific researches and verified in farmers' fields to maximize economic, social, and environmental benefits (Majumdar et al. 2014b). FBMP is aimed at managing the flow of nutrients in the course of producing affordable and healthy food in a sustainable manner that protect the environment and conserve natural resources at the same time profitable to producers.

The 4R Nutrient Stewardship approach, evolving from the conceptual framework of FBMPs, is an essential tool in the development of sustainable agricultural systems because its application can have multiple positive impacts. The basic principle behind fertilizer best management practices is simple, that is, the "4R" – using the right fertilizer source, at the right rate, right time, and right place which conveys how fertilizer applications can be managed to achieve economic, social, and environmental goals.

Application of 4R strategies to nutrient management in crops has the potential to reduce nutrient mining from soil. The following section provides examples of application of the 4R strategies in crops and cropping systems to reduce nutrient losses from the soil.

14.9.1 Right Source and Nutrient Mining

The form of added N plays a role in regulating N losses and influencing NUE. Nitrogen fertilizers predominantly contain N in the form of ammonia, NO_3^-, or urea. Among these forms, NO_3^- is the most susceptible to leaching, ammonia the least, and urea moderately susceptible. Ammonia and urea are more susceptible to volatilization loss of N than fertilizers containing NO_3^- (Ladha et al. 2005). Controlled release compounds have the potential to reduce losses of N from the system. A recent study (Halvorson and Del-Grosso 2012) commonly used granular urea (46 % N), liquid UAN (32 % N), a controlled-release polymer-coated urea (ESN®), stabilized urea and UAN products containing nitrification and urease inhibitors (SuperU and UAN + AgrotainPlus®), and a subsurface band ESN treatment (ESNssb) in maize and showed that the right choice of nitrogenous fertilizer significantly reduced N loss in irrigated maize (Fig. 14.3).

Singh et al. (2002) studied the effect of different sources of N on distribution and depletion of K in a rice–wheat cropping system grown on a vertisol for 8 years. The study revealed that continuous growing of rice–wheat for 8 years without FYM or green manure (GM) as a source of N caused a decline in the exchangeable K from 19.5 mg K 100 g^{-1} soil (initial) to 16.0 and 13.8 mg K 100 g^{-1} soil in the control and 90 kg N ha^{-1} treatments, respectively. However, the use of either FYM or GM together with 90 kg N ha^{-1} favored a buildup of solution and exchangeable K (Table 14.6). Application of FYM or GM alone as a source of N did not have a significant effect on exchangeable K. Depletion of K, which was measured as the difference of initial and final HNO_3 + $HClO_4$-K (after 8 years), led to a significant depletion of HNO_3 + $HClO_4$-extractable K (non-exchangeable) in the presence or absence of FYM or GM indicating a greater release of K to available pool. The magnitude of depletion was larger at 180 kg N ha^{-1}. The study indicated that choosing a right source of N resulted in better available K and the increase in the exchangeable K in the presence of FYM or GM as a source of N could be due to an increase in the exchange sites (CEC) as a result of addition of manure.

14.9.2 Right Rate and Nutrient Mining

Application of right fertilizer rate ensures balanced supply of plant nutrients with better crop uptake and minimal losses. Mohapatra et al. (2013) studied the response of groundnut to balanced fertilizer application and compared different rates of fertilizer application, namely, balanced fertilizer rate based on soil testing (BFT), state fertilizer recommendation (RDF), and farmer fertilizer practice (FFP). BFT varied across different locations involving variable rates of NPKSZnBCa, while RDF (only fixed rates of NPKSZnBCa) and FFP (fixed rates of NPK only) treatments were the same at all the experimental sites. The results revealed that the BFT recorded significantly higher pod yield of groundnut than any of the other treatments used across all locations and seasons and the average yield of groundnut with BFT was 2226 kg/ha followed by RDF (1885 kg/ha) and FFP (1611 kg/ha), respectively. The study indicated that choosing right fertilizer rate resulted in better groundnut yield while reducing the extent of K mining from soil (Table 14.5). With an average initial soil K status of 123 kg/ha, the average K balances after harvest of groundnut in BFT, RDF, and FFP treatments were 147.2, 132.4, and 152.3 kg/ha, respectively. However, the average postharvest soil K status in BFT, RDF, and FFP treatments was 117.7, 102.1, and 93.9 kg/ha, respectively. This indicated a corresponding K mining of 5.3, 20.9, and 29.1 kg/ha, respectively.

Singh et al. (2013) have also shown that not applying the right rate of potassium could cause nutrient mining in soils under intensive cropping system. Soil properties before the rice crop and after the wheat crop were used to determine change during one rice–wheat cropping cycle

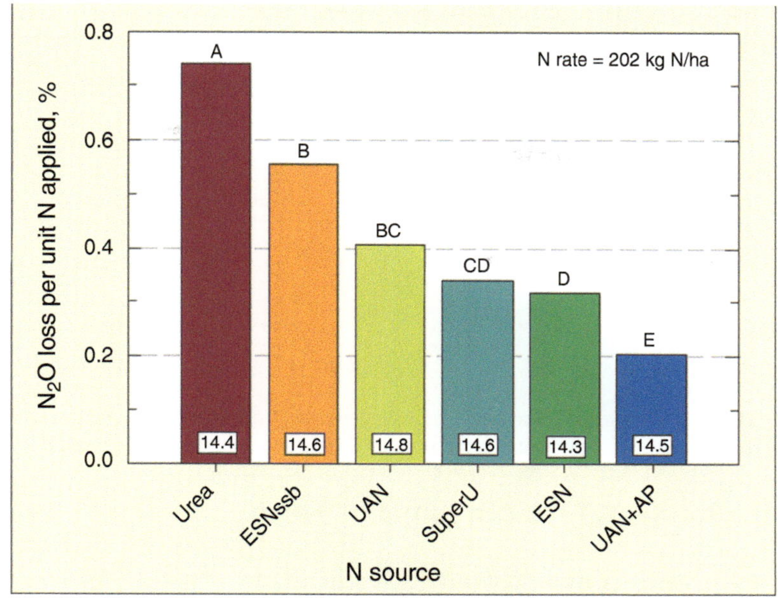

Fig. 14.3 Growing season soil nitrous oxide (NO) loss per unit of N applied as a function of N fertilizer source averaged over strip-till and no-till irrigated corn production systems in 2009 and 2010 near Fort Collins, Colorado. Average grain yields (t/ha) are shown in a *white box* within each bar

Table 14.6 Distribution of different forms of K as influenced by the use of different sources of N after 8 years of a rice–wheat system grown on a vertisol

Treatment	Water-soluble K (mg/100 g soil)	NH_4OAc–K (mg/100 g soil)	$HNO_3 + HClO_4$–K (mg/100 g soil)	Depletion of K (mg/100 g soil)
Initial	3.1 ± 0.2	19.5 ± 3.1	303 ± 8.2	–
Control	2.5	16	293	10
90 kg N ha^{-1}	3.2	13.8	289	13.5
180 kg N ha^{-1}	3.2	12.5	285	17.5
90 kg N + FYM 5 t ha^{-1}	4	23	288	15
90 kg N + GM 6 t ha^{-1}	4.1	24	288	14.1
5 t FYM ha^{-1}	3	21	296	6.5
6 t GM ha^{-1}	2.8	20.5	298	4.5
CD (P = 0.05)	0.4	1.1	9.7	–

(Table 14.7). In the absence of added K, exchangeable K decreased by 6–9 mg kg^{-1} and non-exchangeable K decreased by 18–30 mg kg^{-1} during one rice–wheat cropping cycle. With application of K, exchangeable K increased by 6–9 mg kg^{-1} and non-exchangeable K increased by 7–14 mg kg^{-1}. The difference between application of K and the farmer's practice without the added K after one rice–wheat cropping cycle ranged from 13–18 mg kg^{-1} for exchangeable K and 26–41 mg kg^{-1} for non-exchangeable K across the locations (Table 14.7). The decline in soil K without added K (Table 14.7) highlights the risk of rapid short-term mining of soil K with the farmer's current practice of using relatively high rates of N and P with little or no use of K. Long-term cropping with negative K balances has been associated with yield declines in the rice–wheat system in South Asia (Bijay-Singh et al. 2003). Although the K-supplying capacity of illite-dominated alluvial soils of the Indo-Gangetic Plains in India is relatively high, long-term intensive cropping with inadequate application of K can result in K mining leading to large negative balances and depletion of native K reserves (Yadvinder-Singh et al. 2005b).

Table 14.7 Change in soil potassium (K) during one cycle of rice–wheat cropping at five locations in northern India. The K × M × location interaction was not significant at $P \leq 0.05$ for all the listed parameters

Parameter	Fatehgarh Sahib	Meerut	Banda	Barabanki	Bhagalpur
Change in exchangeable K (mg kg^{-1})					
No K	−6	−6	−7	−8	−9
+K	6	7	7	8	9
Difference[a]	13***	13***	14***	17***	18***
Change in non-exchangeable K (mg kg^{-1})					
No K	−30	−22	−19	−26	−18
+K	11	9	7	14	14
Difference	41***	31***	26***	40***	33**

ns = not significant at $P \leq 0.05$, * = significant at $P \leq 0.05$, ** = significant at $P \leq 0.01$, and *** = significant at $P \leq 0.001$

[a]Difference between non-rounded means for two no-K treatments (no application of K, S, or Zn and application of S + Zn) and two + K treatments (application of K only and application of K with S + Zn)

14.9.3 Right Time and Nutrient Mining

Assessing crop uptake dynamics and patterns can be an important component in determining appropriate timing of nutrient application. Applications timed and targeted at specific growth stages may be beneficial to crop yield and/or quality in some production systems for some nutrients, most notably N. Timed and targeted applications may also be beneficial to reduce environmental impacts of nutrient loss from soil. The basic objective is to optimize the congruence of supply and demand of N (Giller et al. 2004). Inappropriate time of fertilizer application does not allow the plants to take up the nutrients from the soil, and the unutilized nutrients in the soil increases the probability of loss. Besides matching the physiological demand stages of crops, a rate X time combination becomes a major factor in reducing nutrient loss from the soil in on-farm situations. For example, fertilizer nutrient rates are generally decided before the cropping season, based on anticipated yield and soil nutrient-supplying capacity. In the case of nitrogen application in major cereals, the quantity of total N to be applied is generally split into 2–3 applications matching the physiological demand stages of the crop concerned. However, any in-season changes in climate, pest, disease attacks, etc. may change the yield targets and crop nutrient requirement. Application of previously decided rates in such scenarios could turn out to be lower or higher than the requirement. This is usually manifested as increase or decrease in nutrient use efficiencies (NUE), where decreased NUE suggests that the applied nutrient was less utilized for yield formation and may have been lost from the soil system. An unusually high NUE may mean that the crop may have experienced nutrient stress. Chlorophyll meters, leaf color charts (LCC), and GreenSeeker optical sensors are some of the promising tools developed in recent years that can help in corrective in-season N management. These tools have been extensively used by researchers to increase nitrogen use efficiency of crops (Ladha et al. 2005). Recently, Pasuquin et al. (2012) compared fixed rate of N application with standard splits and LCC-based N application and showed improved yield and nitrogen use efficiency in maize (Pasuquin et al. 2012) in Southeast Asia (Table 14.8). Research on the use of the optical sensor technology in South Asia has been carried out to evaluate its potential with the crops grown and management conditions used. The GreenSeeker sensor-based technology provides for a saving in N application of 10–20 % in comparison to blanket state recommendations, while maintaining crop yields (see http://nue.okstate.edu/GreenSeeker/India.htm and Bijay-Singh et al. 2011). Both higher N use efficiency and higher wheat grain protein were measured when the GreenSeeker was used to make N decisions. Similarly for rice, less N was used while yields were either maintained or increased. Both of these reports concluded that the optical sensor had potential to improve N use efficiency and profits for farmers through either increased productivity or profitability.

Table 14.8 Effect of N application time on yield and agronomic efficiency of irrigated maize

	Maize grain yield (Mg ha^{-1}) and agronomic efficiency (kg kg^{-1})			
	Maros (Indonesia)			O Mon (Vietnam)
Treatment	2008	2009	2008	2009
2-split fixed rate	11.2 (58.7)	10.6 (46.8)	5.4 (30.1)	6.6 (36.9)
3-split fixed rate	11.4 (62.8)	10.5 (45.8)	5.6 (31.4)	6.7 (37.6)
3-split LCC1	12.3 (64.8)	11.1 (47.0)	6.0 (30.3)	7.0 (34.7)
3-split LCC2	12.6 (65.7)	12.1 (46.4)	6.1 (30.4)	7.3 (32.4)
Mean of fixed rate	11.3 (60.7)	10.6 (46.3)	5.5 (30.7)	6.6 (37.3)
Mean of LCC	12.4 (65.4)	11.6 (46.7)	6.1 (30.4)	7.1 (33.5)
SE	1.17 (10.6)	0.93 (6.3)	0.55 (5.86)	0.36 (3.1)
Comparison of fixed rate with LCC	** (ns)	*** (ns)	** (ns)	*** (***)

*=$P < 0.05$, **=$P < 0.01$, ***=$P < 0.001$

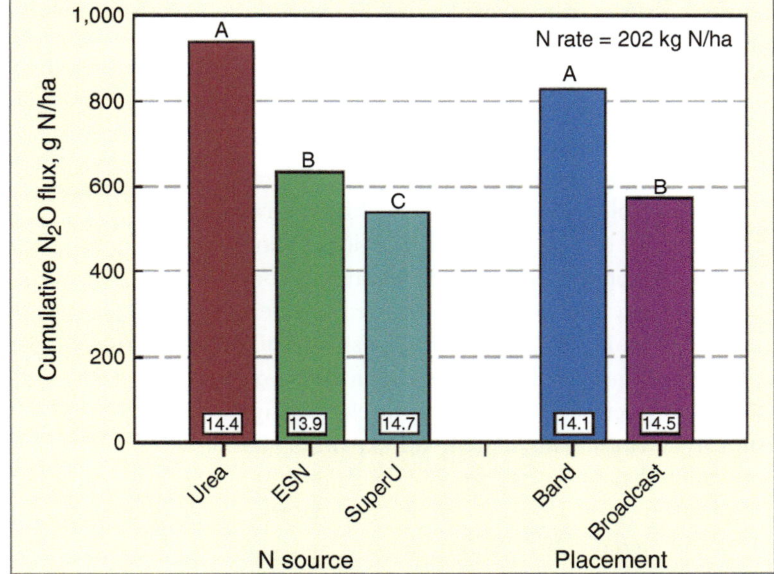

Fig. 14.4 Nitrogen source and placement effects on soil nitrous oxide (N_2O) emissions when averaged over strip-till and no-till systems (3 site years). Average grain yields (t/ha) are shown in a *white box* within each bar

14.9.4 Right Method and Nutrient Mining

Right place of fertilizer application suggests placing nutrients strategically in the soil so that a plant has access to them. Proper placement of fertilizer allows a plant to develop properly and realize its potential yield, given the environmental conditions in which it grows. Right place is, in practice, continually evolving. Plant genetics, placement technologies, tillage practices, plant spacing, crop rotation or intercropping, weather variability, and a host of other factors can all affect which placement strategy is appropriate (Majumdar et al. 2014b). In the case of phosphorus, surface applications are more prone to soluble P losses than incorporated or injected treatments (Bundy et al. 2001). Phosphorus is generally recommended for band placement in soils with high P-fixing capacity to reduce the contact with soil particles. Ladha et al. (2005) suggested that the common practice of surface broadcasting N fertilizers could entail large N losses, particularly through ammonia volatilization from the system and reduce NUE. In contrast, deep placement of urea or urea supergranules (USG) has been proven to improve NUE. Humphreys et al. (1992) noted that recovery efficiency of nitrogen was 37 % for broadcasting, 46 % for banding, and 49 % for deep point placement of USG in direct-seeded rice in Australia. In another study, Halvorson and Del-Grosso (2012) surface banded and broadcasted three N sources, urea, SuperU, and ESN, in maize to evaluate the effects of N placement on N_2O emissions under irrigated maize production. Band-applied N had a higher (45 %) N_2O emission than broadcast N averaged over 3 site years (Fig. 14.4).

14.10 Conclusion

Nutrient mining in soils have the potential to create food security challenges in South Asia. The increasing multi-nutrient deficiencies in soils across the region provide strong evidence of nutrient mining. Crop removal is the largest contributor to nutrient mining. However, there are several other avenues, such as leaching, erosion, and volatilization, through which native nutrients can be lost from soils. The loss of soil organic carbon due to high temperature and continuous tillage also makes plant nutrients vulnerable to loss in the region. Replenishing soils through adequate nutrient application based on the 4R Nutrient Stewardship approach, and rigorously factoring in nutrient offtake from the agricultural fields, is necessary to maintain soil fertility levels in intensive production systems. This will require overhauling existing nutrient management strategies prevalent in the region from a perception-based approach to a more science-based approach.

References

Amarasinghe UA, Shah T, Anand BK (2008). India's water supply and demand from 2025–2050: business-as-usual scenario and issues. In: Amarasinghe UA, Sharma BR (eds) Strategic analyses of the National River Linking Project (NRLP) of India, series 2. Proceedings of the workshop on analyses of hydrological, social and ecological issues of the NRLP, New Delhi, 9–10 Oct 2007. International Water Management Institute (IWMI), Colombo, pp 23–61

Annual Report of All India Coordinated Research Project on Integrated Farming Systems. 2006–2007. Project Directorate for Farming (earlier Cropping) Systems Research, Modipuram, Meerut

Annual Report of All India Coordinated Research Project on Integrated Farming Systems. 2010–2011. Project Directorate for Farming (earlier Cropping) Systems Research, Modipuram, Meerut

Bhattacharyya T, Pal DK, Easter M, Batjes NH, Milne E, Gajbhiye KS, Chandran P, Ray SK, Mandal C, Paustian K, Williams S, Killian K, Coleman K, Falloon P, Powlson DS (2007) Modeled soil organic carbon stocks and changes in the Indo-Gangetic Plains, India from 1980 to 2030. Agric Ecosyst Environ 122:84–94

Bijay-Singh, Yadvinder-Singh, Imas P, Xie JC (2003) Potassium nutrition of the rice-wheat cropping system. Adv Agron 81:203–259. doi:10.1016/S0065-2113(03)81005-2

Bijay-Singh, Sharma RK, Kaur J, Jat ML, Martin KL, Yadvinder-Singh, Varinderpal-Singh, Chandna P, Choudhary OP, Gupta RK, Thind HS, Singh J, Uppal HS, Khurana HS, Kumar A, Uppal RK, Vashistha M, Raun WR, Gupta R (2011) Assessment of the nitrogen management strategy using an optical sensor for irrigated wheat. Agron Sustain Dev 31:589–603

Bundy LG, Andraski TW, Powell JM (2001) Management practice effects on phosphorus losses in run-off in corn production systems. J Environ Qual 30:1822–1828

Buresh RJ, Pampolino MF, Witt C (2010) Field-specific potassium and phosphorus balances and fertilizer requirement for irrigated rice-based cropping systems. Plant Soil 335:35–64

Chuan L, He P, Pampolino MF, Johnston AM, Jin J, Xu X, Zhao S, Qiu S, Zhou W (2013) Establishing a scientific basis for fertilizer recommendations for wheat in China: yield response and agronomic efficiency. Field Crop Res 140:1–8

Dobermann A, Witt C (2004) The evolution of site-specific nutrient management in irrigated rice systems of Asia. In: Dobermann A, Witt C, Dawe D (eds) Increasing productivity of intensive rice systems through site-specific nutrient management. Science Publishers, Inc./International Rice Research Institute (IRRI), Enfield/Los Baños, p 75

Dobermann A, Witt C, Dawe D, Abdulrachman S, Gines HC, Nagarajan R, Satawathananont S, Son TT, Tan PS, Wang GH, Chien NV, Thoa VTK, Phung CV, Stalin P, Muthukrishnan P, Ravi V, Babu M, Chatuporn S, Sookthongsa J, Sun Q, Fu R, Simbahan GC, Adviento MAA (2002) Site-specific nutrient management for intensive rice cropping systems in Asia. Field Crop Res 74:37–66

Dutta S, Majumdar K, Khurana HS, Sulewski G, Govil V, Satyanarayana T, Johnston A (2013) Mapping potassium budgets across different states of India. Better Crops-South Asia 7:28–31

FAI (2008) 53rd fertilizer statistics. The Fertilizer Association of India, New Delhi

FAI (2013) 58th fertiliser statistics. The Fertiliser Association of India, New Delhi

Ganesh-Kumar A, Mehta R, Pullabhotla H, Prasad SK, Ganguly K, Gulati A (2012) Demand and supply of cereals in India: 2010–2025. IFPRI discussion paper 01158, Environment and Production Technology, IFPRI, New Delhi

Gburek WJ, Barberis E, Haygarth PM, Kronvang B, Stamm C (2005) Phosphorus mobility in the landscape. In: Sims JT, Sharpley AH (eds) Phosphorus agriculture and the environment, vol 46, Agronomy monograph. American Society of Agronomy, Crop Science Society of America, and Soil Science Society of America, Madison, pp 941–979

Giller KE, Chalk PM, Dobermann A, Hammond L, HeVer P, Ladha JK, Maene L, Nyamudeza P, Ssali H, Freney JR (2004) Emerging technologies to increase the efficiency of use of fertilizer nitrogen. In: Mosier AR, Syers JK, Freney JR (eds) Agriculture and the nitrogen cycle: assessing the impacts of fertilizer use on food production and the environment. Scientific Committee on Problems of the Environment (SCOPE), Paris, pp 35–51

Halvorson AD, Del-Grosso SJ (2012) Nitrogen source and placement affect soil nitrous oxide emissions from irrigated corn in Colorado. Better Crops 96(4):7–9

Hegde DM, Babu SNS (2001) Nutrient mining in agroclimatic zones of Karnataka. Fertil News 46 (7):55–72

Humphreys E, Chalk PM, Muirhead WA, White RJG (1992) Nitrogen fertilization of dry-seeded rice in southeast Australia. Fertil Res 31:221–234

IPNI (2012) The 4R nutrient strewardship concept. In: Bruulsema TW, Fixen PE, Sulewski GD (eds) 4R plant nutrition: a manual for improving the management of plant nutrition. International Plant Nutrition Institute, Norcross, pp 2.1–2.7

Janssen BH, Guiking FCT, Van der Eijk D, Smaling EMA, Wolf J, van Reuler H (1990) A system for quantitative evaluation of the fertility of tropical soils (QUEFTS). Geoderma 46:299–318

Katyal JC (2001) Fertilizer use situation in India. J Indian Soc Soil Sci 49:570–592

Ladha JK, Pathak H, Krupnik TJ, Six J, van Kessel C (2005) Efficiency of fertilizer nitrogen in cereal production: retrospects and prospects. Adv Agron 87:85–156. doi:10.1016/S0065-2113(05)87003-8

Lal R (2014) Societal value of soil carbon. J Soil Water Conserv 69(6):186A–192A

Majumdar K, Dey P, Tewatia RK (2014a) Current nutrient management approaches-issues and strategies. Indian J Fertil 10(5):14–27

Majumdar K, Dutta SK, Satyanarayana T (2014b) Fertilizer best management practices. In: Prasad R, Kumar D, Rana DS, Shivay YS, Tewatia RK (eds) Textbook of plant nutrient management. Indian Society of Agronomy, New Delhi, pp 361–385

Majumder B, Mandal B, Bandyopadhyay P (2008) Soil organic carbon pools and productivity in relation to nutrient management in a 20-year-old rice–berseem agroecosystem. Biol Fertil Soils 44:451–461

Mandal B, Majumder B, Bandyopadhyay PK, Hazra GC, Gangopadhyay A, Samantaray RN, Mishra AK, Chaudhury J, Saha MN, Kundu S (2007) The potential of cropping systems and soil amendments for carbon sequestration in soils under long-term experiments in subtropical India. Glob Chang Biol 13(2):357–369

Mohapatra AK, Parida P, Pattanayak SK, Satyanarayana T (2013) Response of groundnut to balanced fertilisation and omission of potassium. Better Crops-South Asia 7(1):8–11

Pasuquin JM, Saenong S, Tan PS, Witt C, Fisher MJ (2012) Evaluating N management strategies for hybrid maize in Southeast Asia. Field Crop Res 134:153–157

Pathak H, Byjesh K, Chakrabarti B, Agarwal PK (2010) Potential and cost of carbon sequestration in Indian agriculture: estimates from long-term field experiments. Field Crop Res. doi:10.1016/j.fcr.2010.09.006

Prasad R (2013) Soil test based fertilizer recommendation in India. How realistic? How reliable? Indian J Fertil 9 (3):16–19

Saha D, Kukal SS, Sharma S (2011) Landuse impacts on SOC fractions and aggregate stability in typic ustochrepts of Northwest India. Plant Soil 339:457–470

Sanyal SK, Gill MS, Majumdar K (2009a) Site-specific potassium management in rice-based cropping systems in India. In: Proceedings of the IPI-OUAT-IPNI international symposium on potassium pole and benefits in improving nutrient management for food production, quality and reduced environmental damages, Bhubaneswar, 5–7 Nov 2009, pp 389–405

Sanyal SK, Majumdar K, Chatterjee S (2009b) Influence of mineralogy on potassium availability in Indian soils. Indian J Fertil 5(11):49–52, 55–56, 57–58, 61–64

Sarkar GK, Chattopadhyay AP, Sanyal SK (2013) Release pattern of non-exchangeable potassium reserves in Alfisols, Inceptisols and Entisols of West Bengal, India. Geoderma 207–208:8–14

Sarkar GK, Debnath A, Chattopadhyay AP, Sanyal SK (2014) Depletion of soil potassium under exhaustive cropping in Inceptisol and Alfisols. Commun Soil Sci Plant Anal 45:61–72

Satyanarayana T, Tewatia RK (2009) State wise approaches to crop nutrient balances in India. In: The proceedings IPI-OUAT-IPNI international symposium on "potassium role and benefits in improving nutrient management for food production, quality and reduced environmental damages", OUAT, Bhubaneshwar, pp 467–485

Sen P, Majumdar K, Sulewski G (2008) Importance of spatial nutrient variability mapping to facilitate SSNM in small land-holding systems. Indian J Fertil 4:43–50

Sharma C, Tiwari MK, Pathak H (2008) Estimates of emission and deposition of reactive nitrogenous species for India. Curr Sci 94(11):1439–1446

Singh RB (2009) Towards a food secure India and South Asia making hunger a history. URL: http://www.apaari.org/wp-content/uploads/2009/08/towards-a-food-secure-india-makinghunger-history.pdf

Singh BR, Lal R (2005) Phosphorus management in low-input agricultural systems. In: Sims JT, Sharpley AH (eds) Phosphorus: agriculture and the environment. Agronomy monograph, vol 46. American Society of Agronomy, Crop Science Society of America, and Soil Science Society of America, pp 729–759

Singh M, Singh VP, Reddy DD (2002) Potassium balance and release kinetics under continuous rice-wheat cropping system in Vertisol. Field Crop Res 77:81–91

Singh VK, Dwivedi BS, Buresh RJ, Jat ML, Majumdar K, Gangwar B, Govil V, Singh SK (2013) Potassium fertilization in rice-wheat system on farmer's fields in India crop performance and soil nutrients. Agron J 105:471–481

Singh VK, Dwivedi BS, Tiwari KN, Majumdar K, Rani M, Singh SK, Timsina J (2014) Optimizing nutrient management strategies for rice–wheat system in the Indo-Gangetic Plains of India and adjacent region for higher productivity, nutrient use efficiency and profits. Field Crop Res 164:30–44

Soil Science Society of America (SSSA) (1995) SSSA statement on soil quality. Agronomy News, June 7. Soil Science Society of America, Madison

Subba Rao A, Srivastava S (1998) Role of plant nutrients in increasing crop productivity agro-chemicals news in brief, April-June 1998. FADINAP, Bangkok

Tandon HLS (2004) Fertilizers in Indian agriculture-from 20th to 21st century. FDCO, New Delhi, p 240

Tandon HLS (2012) Fertilizer research: some unanswered questions. Indian J Fertil 8(7):26–31

Timsina J, Connor DJ (2001) Productivity and management of rice-wheat cropping systems issues and challenges. Field Crop Res 69:93–132

Venkateswarlu B, Prasad JVNS (2012) Carrying capacity of Indian agriculture issues related to rainfed agriculture. Curr Sci 102:882–888

Witt C, Dobermann A (2004) Toward a decision support system for site-specific nutrient management. In: Dobermann A, Witt C, Dawe D (eds) Increasing productivity of intensive rice systems through site-specific nutrient management. Science Publishers Inc./International Rice Research Institute (IRRI), Enfield/Metro Manila, pp 359–395

Witt C, Dobermann A, Abdulrachman S, Gines HC, Wang G, Nagarajan R, Satawatananont S, Son TT, Tan PS, Tiem LV, Simbahan GC, Olk DC (1999) Internal nutrient efficiencies of irrigated lowland rice in tropical and subtropical Asia. Field Crop Res 63:113–138

Yadvinder-Singh, Pannu RPS, Bijay-Singh (2005a) Leaching of K from organic manures, crop residues and inorganic fertilizer in two soils under flooded and upland moisture regimes. J Indian Soc Soil Sci 53:207–213

Yadvinder-Singh, Bijay-Singh, Timsina J (2005b) Crop residue management for nutrient cycling and improving soil productivity in rice-based cropping systems in the tropics. Adv Agron 85:269–407. doi:10.1016/S0065-2113(04)85006-5

Improving Protein Density in Food Legumes Through Agronomic Interventions

15

C S Praharaj, Ummed Singh, S S Singh, and Narendra Kumar

Abstract

Agronomic biofortification is the easiest and fastest way for biofortification of cereals and pulses with Fe, Zn or other micromineral nutrients especially in developing Asian and African countries, where these are considered as the staple food. Agronomic biofortification is the only way to reach the poorest of the poor rural masses, those who will never have money to buy mineral supplements nor can afford to improve the components of their diet by incorporating animal products. It is suggested by scientific evidences that biofortification is feasible technically and even it can be achievable without negotiating for agronomic productivity. Cost–benefit analyses that are predicted also do support biofortification as being imperative in the armamentarium for monitoring micronutrient deficiencies and its control. However, the pertinent challenges facing all of us are about the accessibility of biofortified crop in large scale with proven purity and quality standards and to get producers/consumers acceptance for such biofortified crops and increase their intake. With the inception of good seed production and utilisation systems, the development of markets and market-oriented products and creation of demand for these, the commercial exploitation of crops with nutrient fortification can be possible and achievable. Amongst different types of malnutrition, protein malnutrition is the foremost one as it plays the pivotal role in health and nutrition of both human and animals. Besides protein malnutrition, deficiencies due to vitamin A, iron and zinc do affect over half of the population of the world. Although ample progress has been achieved to control these deficiencies through food fortification and supplementation, yet renewed and vigorous approaches are needed, especially to reach a sizeable mass of rural poor. It is true especially for reaching the

C S Praharaj (✉) • U. Singh • S S Singh • N. Kumar
Division of Crop Production, ICAR-Indian Institute of Pulses Research, Kanpur, Uttar Pradesh 208 024, India
e-mail: cspraharaj@hotmail.com; singhummed@yahoo.co.in; sssinghpatna@yahoo.co.in; nkumar.icar@gmail.com

U. Singh et al. (eds.), *Biofortification of Food Crops*, DOI 10.1007/978-81-322-2716-8_15

Millennium Development Goals (now, Sustainable Development Goals) that aims at reducing by at least two-thirds the under-five child mortality ratio and three-quarters the maternal mortality ratio by 2020. Therefore, biofortification of staple food crops including food legumes is a new public health approach to contain protein (and of course deficiencies of vitamin A, iron and zinc) in poor countries. This chapter gives a brief insight of the technology for improving protein density in food or seed legumes through agronomic interventions.

Keywords
Agronomic interventions • Food legumes • Grain legumes • Protein density • Sustainability

15.1 Introduction

India is recognised as the principal producer, consumer, processor and importer of pulses in the world. Paradoxically, the country's pulse production has reached at 19.5 million tonnes (Mt) from the acreage of 24.5 million hectares (M ha) during 2013–2014. Yet, the country is importing pulses to the extent of around 2–3 Mt every year for meeting its domestic demand. This is our major concern as an increase in the demand from India has shown to have surging effect on international prices, thereby draining the most sought-after precious foreign exchange. Expectedly, as the country's domestic requirements would be around 26.50 Mt by 2030, it further necessitates stepping up of production by almost 7.0 Mt of additional produce. This arduous task is to be accomplished even under more severe constraints in production environment including those related to biotic and abiotic stresses, climatic changes and accumulative secondary and micronutrient deficiencies in the soil. This necessitates a two-pronged proactive strategy that involves reducing cost of production and improving per unit productivity (Praharaj and Dhingra 2001).

Iron and zinc deficiencies in human nutrition are widespread and rampant in developing Asian and African countries where cereal grains are the staple food. Effects are therefore underway to develop cereal genotypes with grains denser in Fe and Zn by traditional plant breeding or using genetic engineering techniques (Grusak 2002) which require a long gestation period and adequate funds. However, the products of genetic engineering are not well accepted in many countries. Also, there is a trade-off between yield and grain biofortification. Therefore, agronomic biofortification offers to achieve this without sacrificing on yield and with no problem of product acceptance. From the viewpoint of biofortification, for example, foliar application has been reported to be better than the soil application of Fe and Zn, and for this purpose, chelated Fe and Zn fertilisers are better. When soil applied, water-soluble sources of Zn are better while soil application of Fe is not recommended. Therefore, agronomic biofortification depends upon management practices (tillage, water management, nutrient interactions and so on), soil factors (amounts of Zn present, *p*H, mechanisms of Zn fixation other than *p*H and others) and plant factors (root characteristics, excretion of phytosiderophores and organic acids by roots, utilisation of zinc at the cellular level, translocation within plant and mechanisms of Zn accumulation in grain). As genetic and agronomic biofortification are complementary to each other, once the genotypes having denser grains are developed, they will have to be adequately fertilised with Fe and Zn (Prasad et al. 2014).

15.2 Good Agronomic Practices

Good agronomic practices, popularly known as GAP, alone can help increase in productivity to the extent of 25–40 % in many field crops. Besides this, efficient production technologies are required with special importance on adequacy in crop nutritional aspects. Therefore, a need is arisen for developing appropriate production technologies for nontraditional areas involving crops and cropping systems involving pulses, for example, relay cropping and pulse-based crop rotations in rice fallows, off-season cultivation in summer/spring season and introduction of compatible crops in cropping systems. Since the availability of labour for various and diverse farm operations is reducing and cost of labour is increasing, appropriate crop management technologies need to be developed for reducing the cost of production. It also suits the two-pronged strategy of crop production, viz. increase in production/productivity of crops and/or decrease in the cost of cultivation/production (or increase in the farm income). Therefore, to harness the energy sources, efforts are already in place to assess and refine various production technologies with innovative agronomic practices (more emphasis on optimum plant population and adequate crop nutrition in pulses with desired control of both biotic and abiotic stresses). Depending upon the amount and distribution of rainfall, pulses – an important component of staple food crops – could be grown under double cropping (with annual rainfall of more than 75 cm), intercropping (60–75 cm) and monocropping (less than 60 cm rainfall) systems. In addition, efficient intercrops for pulses need to be screened/identified and popularised amongst the farmers so as to compensate against a possible crop loss through pure or monoculture (Praharaj et al. 2011).

In the recent past, food consumption pattern has also undergone considerable change owing to various factors like increase in income, urbanisation, change in consumer taste and preferences, awareness about safe and healthy food etc. As a result, the composition of diet and nutrition intake has changed considerably. It is evident from the fact that the dietary plan has shifted away from cereals and pulses towards fruits, vegetables, processed food and food items of animal origin. As a result, the consumption of pulses has come down due to various possible reasons like poor availability, high prices and availability of cheaper alternatives of animal origin. Although the shift in consumption towards horticultural crops and food items of animal origin has no doubt contributed towards higher intake of calories, the intake of protein at the same time has come down mainly due to decline in the consumption of pulses which is a major source of quality protein compared to other food items. The concern is reduction in consumption of pulses by predominantly vegetarian society and poor like that of India due to high price and fluctuation in supply of pulses. Moreover, pulses could act as a low-cost substitute during high prices of vegetables and food items of animal origin. Although the production of pulses has registered an impressive growth in the recent decade yet, it is not in pace with the increase in the population. Therefore, the need of the hour is to raise both production and availability of pulses by adopting various innovative measures. This will ensure food and nutritional security by bringing sustainability in agricultural production in the country.

In order to continue with the accelerated growth in production of pulses, institutional and policy supports are also required for enhancing area under pulses, development of high-yielding varieties, supply of quality inputs (Praharaj et al. 2011, 2014; Praharaj 2013), intercropping (Sankaranarayanan et al. 2011), proper extension of production technologies (Singh and Singh 2014), development of value chain etc. The supply of pulses can also be increased by having orderly marketing of pulses. The availability of information being a vital component will make farmers respond more effectively to the various initiatives of the government. With the advent of technology, the information flow could reach to the lowest level of farming community. Popularising low-cost technology of production, promotion of high-yielding varieties and marketing-related issues will be more effective using improved crop production techniques. The elasticity of the demand for high-value commodities is highly price sensitive, and hence, in the event in the rise in price of such commodities, pulses will act a substitute for cheaper protein. Also, considering the fact that widespread malnutrition is prevailing amongst children and women in India, there is a need to promote consumption of pulses by linking to programmes like midday meal and rural health mission by incorporating either free distribution of pulses or by subsidising the food (Shalendra et al. 2013).

15.3 Food Legumes in Human Nutrition

Legumes have not been emphasised properly during the last five decades of research in food science and human nutrition. It is still continued as the legume species are encountered with the deficiency of sulphur amino acids and the heat lability/stability of proteinase inhibitors and phytohaemagglutinins. A survey of literatures indicates that more research papers on the negativity of legumes have been published in different food journals such as *Journal of Agriculture and Food Chemistry*, *Journal of Food Science* and *Journal of the Science of Food and Agriculture*. This trend also indicates the repeatability of certain aspects of research on the importance of food legumes so far as human nutrition is concerned. Now the time has arrived to project and depict the image of legumes in proper perspective (Deshpande 1992). The validity of our continued patronage in defining the role of legumes in human nutrition especially at a time when the research dollars worldwide are becoming increasingly harder to come by is more pertinent now than before. One such aspect is supplementation of food legumes per se in human or animal protein nutrition and biofortification of such food/seed crops for raising its protein density.

15.3.1 Protein Malnutrition

Protein malnutrition is a worldwide phenomenon occurring in every nation around the globe as deficiencies of vitamin A, iron and zinc affect over one-half of the world's population. Biofortification is the development of protein or micronutrient-dense staple crops using the best traditional breeding practices, modern biotechnology and to some extent by appropriate agrotechnologies. This approach has multiple advantages. It takes advantage of a regular daily intake of a consistent and large amount of food staples by the family members. This strategy targets low-income households as staple foods predominate in the diets of the poor. After the one-time investment for developing seeds that fortify themselves, here the recurrent costs are low, and germplasm can be shared internationally that makes it cost-effective across time and distance. Moreover, once in place, the biofortified plant system is highly sustainable. Even if government attention and international funding for such (micronutrient) issues fade, nutritionally improved varieties will continue to be grown and consumed year after year.

In addition, biofortification provides a feasible and viable means of reaching undernourished populations in relatively remote rural areas, delivering naturally fortified foods to people with limited access to these. It is also critical that improved agronomy or breeding approach for biofortification in seeds should not incur a yield penalty due to the fact that biofortification and commercial fortification are highly complementary (Graham and Welch 1996; Graham et al. 2001). They may have either important spin-off effects for enhancing farm productivity or renders yield stability in food crops in an environmentally sound way. Trace minerals embedded in seed through biofortification should be essential in enabling plants to resist pests, diseases and other environmental stresses as these plants should survive under stress with rapid initial growth and vigour (and consequently produce similar or higher yields). The details of role of pulses in nutrition especially in relation to protein nutrition are given in subsequent sections.

15.3.2 Pulses and Protein Management

The word 'pulse' is derived from the Latin word 'puls', meaning pottage. The term 'pulse' is used for those leguminous plants containing small amount of fat. India is the leading country for high consumption of pulses with an average consumption of 35 g/person/day, but there is wide variation amongst the states. Intake per day ranged from 16 g in Tamil Nadu to 55 g in Madhya Pradesh. The highest intakes tended to be in northern states. The daily per capita availability of pulses in India has decreased, however, to a meagre of 35 g against FAO-recommended dose of 80 g/day.

Pulses are known for their cost-effectiveness and are in fact relatively a cheaper source of protein than milk, cheese, cashew, almonds, meat, fish etc. and thus valuable for developing countries. The seeds of pulses are the most commonly eaten/edible part, and most of them can be economically stored well for future use. The food value of seed of pulses is high as they have about the same calorific value per unit weight as cereals and are good sources of some vitamins and minerals. Pulses do contain about 18.0–32.0 % protein (Table 15.1 and 15.2) and about 1–5 % fat (Table 15.1). The protein density of pulses is generally higher than that of most cereals. In India, consumption of pulses is highest as compared to other pulse-growing countries due to religious restrictions on non-vegetarian diet and low purchasing power. The pulse protein in isolation has somewhat lower nutritive value than most other classes of protein like meat, fish and milk, but they contribute substantially in fulfilling the protein requirement when combined with other proteins in a mixed diet. Pulses are also a rich source of calcium than most cereals and contain about 100–200 mg of calcium/100 g of grain. They are also substantially rich in iron, thiamine, riboflavin and nicotinic acid as compared to cereals. Germinating sprouts of pulses like, mung bean, moth bean and chickpea are popular foods in many regions.

The protein density in pulses is mostly confined to the cotyledons that constitute the major proportion in the seed. Therefore, it contributes the major amount of protein to the whole seed. The seed coat of pigeon pea contains 5.6 % protein, while the cotyledon and embryonic axis have 24.3 and 48.1 %, respectively. Other pulses are also having more or less similar protein content in their seed coat, cotyledon and embryonic axis. Moreover, the pulses are subjected to various processing techniques like milling, dehulling, soaking, germination, fermentation and cooking to make it more value-added eatable products. Besides saving time, energy and fuel, these processing techniques have several nutritional advantages as these produce edible products with higher nutritional value and lower toxic compounds. In addition, the degree of elimination of toxic compound depends on the type of pulse and the processing technique. Proteins, carbohydrates, fats, minerals and vitamins are also important nutritional parameters of pulses.

Pulse proteins usually contain more than adequate levels of some of the nutritionally important amino acids such as lysine that are deficient in most cereals and other edible plant foods. Therefore, the combination of cereals and pulses provides a good balance of amino acids since cereals usually supply adequate amounts of methionine. Pulses are good source of dietary fibre also. Pulses do contain several antinutritional factors including trypsin and chymotrypsin inhibitors, lectins, polyphenols, flatulence factors, antivitamins, allergens, lathyrogens and saponins. The protease inhibitors, lectins and other antinutrients could also cause toxicity in human beings. Heat treatment is the well-established practice that destroys proteinaceous antinutrients, such as protease inhibitors and lectins, but heat treatment destroys some of the amino acids and vitamins as well. Therefore, for maintaining the nutritional value of food, it is prerequisite that heating temperature and length of processing do not exceed the optimum temperature and time required to eliminate the effect of inhibitors. Moreover, proteins in pulses are known to interact

Table 15.1 Comparative account of energy, protein and others in pulses (per 100 g edible part)[a]

Pulses (dal)	Energy (KCal)	Protein (g)	Carbohydrate (g)	Fat (g)
Chickpea	360	17.1	60.9	5.3
Black gram	347	24	59.6	1.4
Cowpea	323	24.1	54.5	1.0
Beans	347	24.9	60.1	0.8
Green gram	334	24	56.7	1.3
Horse gram	330	23.6	56.5	1.1
Peas (green)	93	7.2	15.9	0.1
Peas (dry)	315	19.7	56.5	1.1
Lentil	343	25.1	59	0.7
Red gram (arhar)	335	22.3	57.6	1.7
Rajmash	346	22.9	60.6	1.3
Soybean	432	43.2	20.9	19.5

[a]Source: IIPR (2009)

Table 15.2 Protein content (%) in seed of major pulses[a]

Pulses	Ranges
Chickpea	18.0–30.6
Pigeonpea	18.8–28.5
Mung bean	20.8–31.8
Urd bean	21.2–31.3
Lentil	20.4–30.5
Pea	21.2–32.0
Lathyrus	22.7–29.6

[a]Average content is listed based on random sampling and analysis

with other compounds such as lipids, tannins, phytates, flavanone compounds and pigments. These interactions do occur when pulses are processed and converted into edible products which decrease the bioavailability of proteins in pulses. In the similar fashion, tannins and phytates do interact with minerals and vitamins resulting in a decrease in bioavailability of minerals and vitamins.

Wide variability in seed protein in pulses also exists as chickpea has protein density which varies from 18.0 % to 30.6 % with an average of 21.5 %. Protein quality of a crop depends on its amino acid composition, and the most limiting amino acids determine the nutritive value of pulses. Pulses are deficient in sulphur-containing amino acids and tryptophan, but are rich in lysine in which cereals are relatively deficient. Lentil has the lowest amount of methionine, whereas pigeon pea has the lowest amount of tryptophan (Table 15.3).

The embryo is rich in methionine and tryptophan, but it contributes only about 2.5 % of their total quantity in seed. The cotyledon, being the major component of the seed, accounted for 93 % of methionine and tryptophan of the whole seed, while the seed coat was the poorest in these amino acids. Therefore, the application of P, Mo and N has been shown to increase the level of methionine. The applications of sulphur-containing fertilisers also increase the cystine content of chickpea protein. The pulses have low biological value. The biological value of pulse protein ranges from 32 % to 78 % and a large variation is observed within varieties of the same species. Low methionine in pulse protein is one of the reasons for lower biological value of pulses. Moreover, addition of 0.3 % methionine increases the protein efficiency ratio (PER) in most of the pulses such as chickpea, lentil and pea. As tryptophan is also a limiting amino acid in pulses, addition of tryptophan also improves PER of pulses.

The protein in pulses has low digestibility. The protein digestibility and nutritive value can be enhanced by subjecting pulses to cooking or some other form of heat treatment. Low protein value and digestibility is due to the presence of protease inhibitors and other antinutritional factors. Protein digestibility and availability of amino acids can be further improved by certain processing techniques like soaking, cooking, roasting, germination and fermentation. The

Table 15.3 Major essential amino acid content (g/100 g seed) of some important pulses[a]

Pulses	Lysine	Methionine	Tryptophan	Threonine
Chickpea	1.47	0.28	0.17	0.76
Pigeon pea	1.54	0.24	0.10	0.82
Mung bean	1.69	0.28	0.20	0.78
Lentil	1.57	0.20	0.20	0.91
Pea	1.75	0.29	0.26	0.91

[a]Source: IIPR (2009)

Table 15.4 Effect of processing on protein and starch digestibility of mung bean

Processing method	Protein digestibility (%)	Starch digestibility (g maltose released per 100 g meal)
Soaking (12 h)	64.2	2.8
Cooking (unsoaked seed)	63.0	12.5
Cooking (soaked seed)	70.2	13.8
Pressure cooking (unsoaked seed)	71.2	17.4
Pressure cooking (soaked seed)	83.0	20.8
Germination	78.8	8.7

Source: IIPR (2009)

effect of various processing techniques on protein and starch digestibility of pulses (mung bean) is also quite evident (Table 15.4).

15.3.3 Legumes Contain Antinutritional Proteins

Legume seed protein contains a number of components that can be classified as antinutritional or undesirable. Many of these belong to the albumin fraction which is often considered with a more favourable amino acid composition than the globulins although it is little understood to what extent these proteins are replaceable either to the seed or to the germinated plant. It is in fact due to some known cases in which appropriate mutants have been identified. So far as mutants are considered, for example, lipoxygenases in pea (*Pisum sativum*) and soybean (Forster et al. 1999), a concomitant loss of seedling vigour or seed has not been reported (Hedemann et al. 1999). Much attention is received by enzyme inhibitors as legume albumin proteins with negative attributes. It could have been inferred from many studies of near-isogenic pea lines with different quantity of trypsin inhibitor (TI) proteins that these benefit animal performance as a derivative of null mutants (Hedemann et al. 1999). Mutants and/or transgenic 'knockouts' also could provide answers to the pertinent question of whether or not these TI proteins are essential for viability of plant or seed.

Many compounds present in pulses have been found to have antinutritional effect. These include protease inhibitors, lectins, polyphenols, saponins, phytates and lathyrogen toxins. These antinutrients either reduce the digestibility of pulses or cause toxic effect on their consumption. It is therefore necessary to eliminate these substances by processing or remove them by genetic manipulation (Domoney et al. 2002). Processing plays an important role in the nutritive value of pulses. Cooking is commonly employed for heat treatment of pulses. Moist heat is effective in destroying trypsin inhibitor activity of pulses as application of dry heat to the meal or seed is not effective in inactivating the chymotrypsin inhibitory activity (CIA) and trypsin inhibitory activity (TIA). However, water soaking for 24 h followed by cooking for 20 min is effective in destroying both CIA and TIA. That is why germination (soaking in water and consequent physiological activities) also reduces the trypsin inhibitor activity in most of the pulses. In related processes where autoclaving is done for 15–20 min at 120 °C or extrusion cooking is undertaken at 150 °C or even microwave radiation is given at 107 °C for 30 min, these help in destroying most of the trypsin inhibitor activity of pulses and are equally effective as that of soaking cum cooking.

15.3.4 Some Other Nutritional Drawbacks

Some of the important drawbacks of dal – the finished product – include many important constituents that contribute to the human/animal nutrition. Due to the presence of oxalic acid in high proportion, black gram is not usually consumed by persons with rheumatic diseases. Therefore, those persons with uric acid in urine or rheumatism should not consume urad dal to avoid gout and arthritis. Moreover, pregnant women suffering from constipation should strictly avoid food dishes made from black gram so as to avoid stomach problems. While preparing it for medicinal purposes, since black gram may cause flatulence, other compatible products such as asafetida, pepper and ginger may be added to it so as to make its digestion better. Similarly, eating of mung beans can be avoided if one has diarrhoea or cold hands and feet. Green peas are an extremely low-fat food (with one-third gram of total fat per cup), but it supplies fat-soluble nutrients from this legume, including vitamin E and beta-carotene. In addition, lentils do have some antinutritional factors characterised by trypsin inhibitors and relatively high phytate content. Trypsin is an enzyme responsible for digestion, while phytates help reduce the bioavailability of many dietary minerals. Therefore, the phytates can also be reduced by soaking the lentils overnight in warm water (Crevieu et al. 1997).

15.3.5 Agronomic Feasibility in Increasing Protein Density

The potential to increase the micronutrient density and protein of staple foods by conventional breeding does exist (Graham and Welch 1996; Graham et al. 1999, 2001) as there exists an adequate genetic variation amongst cultivars in respect of concentrations of β-carotene, other functional carotenoids, iron, zinc and other minerals. Therefore, selection of nutritionally appropriate breeding materials is feasible. Moreover, micronutrient-density traits are stable across environments. Therefore, it is possible to combine the high-micronutrient-density trait with high yield in almost all the field crops including pulses. However, increasing protein density agronomically and any other means with increase in seed yield are rather difficult to realise as these are mostly negatively correlated. Therefore, although it is possible to improve the content of several limiting micronutrients together thereby pushing populations towards nutritional balance, yet enriching protein density with increase in seed yield in food crops is difficult.

Micronutrient supply to plants as per its need is more often limited and controlled. A lab incubation study undertaken for 140 days so as to measure potential release of Cu, Mn and Zn from city waste compost (CWC) and rock mineral flour (RMF) as compared to inorganic micronutrient fertilisers revealed that release of the micronutrients from CWC and RMF had different trends (Shivay et al. 2010a). It is because of the fact that restricted supply of micronutrients for plant growth is a common constraint worldwide. It is more so in organic farming systems where nutrient supply to crops is mostly dependent upon decomposition of applied manures and crop residues and mineralisation of native soil organic matter. The above study suggested that irrespective of the quantity of the RMF applied, about 4.6 % of Cu added as RMF was released. However, as the amount of compost added was increased, Cu release from CWC increased from 0.7 % to

3.5 %. The maximum recovery of copper (98 %) was from copper sulphate and that of zinc from zinc sulphate was 98 %. As the RMF level was increased, manganese release from RMF decreased from 114 % to 103 %, whereas the corresponding decrease in Mn release from CWC was from 14 % to −3 %. In the case of manganese sulphate, manganese recovery was maximum (100 %). Similarly, with an increase in the amount of RMF applied, zinc release from RMF increased from 5.8 % to 15.5 % with no Zn release from CWC. These results concluded that both CWC and RMF could be used for organically grown cereals to meet their Cu, Mn and Zn requirements. The results of this study have, however, general applicability in organic farming (Shivay et al. 2010a) with some limitations.

15.3.6 Mobilisation of Minerals to Seeds

Studies made on pulses indicated that there existed a differential, dynamic and selective mobilisation of minerals to seeds. A typical study was carried out for examination of the mineral nutrition of fruiting plants of *Pisum sativum* L., *Lupinus albus* L. and *Lupinus angustifolius* L. in sand cultures supplied with adequate and balanced quantity of essential nutrients. Changes in content of specific minerals in leaves, pods, seed coat and embryo were described while studying mobilisation of minerals to developing seeds of legumes (Hocking and Pate 1977). It was observed that about 60–90 % of the N, P and K were lost from the leaf, pod and seed coat during senescence, whereas 20–60 % of the Mg, Zn, Mn, Fe and Cu and less than 20% of the Na and Ca were lost during the stage. Relative to dry matter accumulation, P, N and Zn tend to increase in an organ. Other elements showed the trend more or less parallel with K, Mn, Cu, Mg and Fe or significantly behind Ca and Na as dry weight increased. Endosperm minerals were of only minor significance in embryo nutrition. Compared with 4–27 % for testa transfer to the embryo, mobilisation returns from pods were estimated to supply 4–39 % of the seeds' accumulations of specific minerals. The study was also made for comparisons of the mineral balance of plant parts of *Lupinus* spp. with that of stem xylem sap and fruit tip phloem sap. It was supporting the view that leaves and pods were principal recipients of xylem-borne minerals and that exported from these organs via phloem was the major source of minerals to the seeds. Since embryo and endosperm differ substantially in mineral composition from phloem sap, the study suggested that during seed development selective uptake occurred from the translocation stream. Substantial differences were also noticed between species in the effectiveness of transfer of specific minerals to the seeds and in mineral composition of plant organs as the said differences between species mostly related to Ca, Na and certain trace elements (Hocking and Pate 1977).

15.4 Agronomic Approaches

The approaches involved with improved agronomy are primarily concerned with enhancing both seed yield and protein density in pulses.

15.4.1 Agronomic Approaches for Biofortification

Most of the agronomic approaches for biofortification revolve around exogenous application of nutrient at critical stage of crop growth for absorption and subsequent mobilisation to seed from aerial parts. For example, application of sulphur and other micronutrients, viz. Zn, Mn, Cu and Mo, has been found to improve crop protein production and its quality significantly in respect of micronutrients. However, low solubility of micronutrients in soils (not the low total amount) is the major reason for the widespread occurrence of their deficiency in crop plants. Therefore, agronomic fortification through application of fertiliser nutrient alone may not be effective in many situations constrained with alkalinity or calcareousness, reduced levels of soil organic matter and soil moisture. On the other hand, large variation has been found in many crops with respect to underground structure/root characteristics and nutrient acquisition from relatively non-available pool. Therefore, genetic and agronomic fortification approaches when combined together may provide benefits both under long- and short-term strategy. Moreover, superiority of new technologies over the local practices is also shown by large-scale on-farm demonstrations (OFD) in India. Therefore, increase in pulse production to the tune of at least 13–42 % in the country (Table 15.5) is possible with adoption of these technologies. Alternatively, increase in yield also results in increase in protein substantially with relation to both quality and quantity.

In a large number of frontline demonstrations conducted across the country, it revealed that package technology (comprising improved varieties and integrated management of nutrients and pests) holds promise to increase productivity at least by 25–42 % in comparison to farmers' practice. However, improved varieties of different pulse crops caused at least 20–25 % yield advantage (Table 15.5) over the farmers' practices (Masood and Gupta 2012). Farmer-participatory testing or evaluation will also help in refining technologies, pinpoint and eliminate adoption constraints. Therefore, large-scale demonstrations on farmer's fields are considered important and crucial for bridging these yield gaps (Praharaj et al. 2013) and protein density with the involvement of extension agencies.

Similarly, improved production practices or agrotechnologies like ridge furrow planting, furrow irrigated raised bed (FIRB) planting or simply raised bed planting (50–120 cm raised beds), pre-emergence application of pendimethalin at 1–1.25 kg/ha followed by early post-emergence application of recommended herbicides, foliar spray of 2 % urea or DAP, seed treatment with *Rhizobium* and P solubiliser, application of sulphur at 20 kg/ha (for legumes and oilseeds) and bio-intensive IPM modules have been recommended or advocated for pulses. These were carefully made or recommended for farmers so as to increase in yield of both seed and protein after a lot of experimentation and large-scale frontline demonstrations. Similarly, the farmer participatory research (FPR) under complex rainfed areas needs to be developed and involve farmers more closely in on-farm research for better in situ understanding and execution of crop- and region-specific recommendation (Praharaj et al. 2013; Singh et al. 2011b). This is otherwise called site-specific recommendations. These efforts may further push average productivity from around 650–750 kg/ha, which means expected production of 17.7 Mt by 2020 (Masood and Gupta 2012) with ample increases in protein density in seed (percent basis and total). Fortunately, the country has produced 19.5 Mt pulses in 2013–2014 with collective application of these farmers' friendly technologies.

Other targets for legume seed protein improvement include removal of potential allergens, removal of antinutritional factors and activities that generate undesirable flavours, improved digestibility and improved functional behaviour for processing although the molecular basis of many of these has not been sufficiently well defined to enable directed improvement, either by genetic manipulation or by breeding using genetic variation. There have also been several attempts to alter the amino acid composition of the globulins through genetic manipulation and through the use of natural variation either to directly modify globulin amino acid sequence or to express exogenous sulphur-rich proteins (Krishnan 2000). There exists a great potential of these approaches as these are still largely unrealised and underutilised but should yield seed protein with enhanced quality in the future. Therefore, studies of model legumes can play an important role in biofortification per se through improved understanding of the effects of environment and genetic background on protein quality and of the regulation of the amounts of the individual seed proteins (Casey et al. 1986, 1993).

15.4.2 Agronomic Interventions

Biofortification is a novel approach that can lead to the development of micronutrient-dense staple crops. A study involving one bacterial

Table 15.5 Seed yield advantages (%) from improved techs of pulse crops under FLDs (2006–09)

Technology	Chickpea	Pigeon pea	Lentil	Mung bean	Urd bean	Field pea
Number of demonstrations	3480	3773	1454	1640	1098	686
Full package	24.9	34.6	41.9	33.9	27.8	40.1
Improved varieties	22.4	24.7	23.6	3.3	21.9	20.0
S application	15.4	17.4	20.3	19.0	19.9	24.1
Rhizobium inoculation	13.4	13.5	21.0	11.1	14.2	13.2
Weed management	40.0	30.0	24.7	29.6	18.3	26.4
Integrated pests management	19.9	28.1	13.1	20.4	17.6	20.2

Source: Masood and Gupta 2012

(*Providencia* sp. PW5) and three cyanobacterial strains CW1, CW2 and CW3 (*Anabaena* sp., *Calothrix* sp. and *Anabaena* sp., respectively) for enhancing the nutritional quality of wheat grains (protein, Fe, Cu, Zn and Mn) revealed that an enhancement to the tune of 18.6 % in protein density was measured with PW5 þ $N_{60}P_{60}K_{60}$ as compared to fertiliser control ($N_{60}P_{60}K_{60}$). Inoculation of *Providencia* sp. PW5 þ $N_{60}P_{60}K_{60}$ also resulted in an increase of 105.3 %, 36.7 % and 150.0 % in Fe, Mn and Cu, respectively. For reducing malnutrition in developing countries, the study clearly underlines the need for the inclusion of PGPR to complement the existing biofortification strategies (Rana et al. 2012).

As cereals are staple foods in most developing countries in Asia including India, Zn fortification of cereal grains benefits all including the poorest of the poor. Moreover, zinc deficiency has emerged as the fourth important micronutrient deficiency in humans (Singh et al. 2011a; Cakmak et al. 1990). Moreover, zinc fertilisation has great relevance to India, because about half of its soils are deficient in available Zn. Therefore, fertilisation of cereals with Zn is a faster, quicker and certain way to increase zinc content in cereals. This practice needs to be put in place without further delay. In a study involving an aromatic rice production, 1.0 % Zn-enriched urea ($ZnSO_4$) was most effective in realising higher grain yield and economic return. This study also suggests that $ZnSO_4$ is the better source so as to enrich prilled urea than ZnO (zinc oxide) (Yadav et al. 2010).

There was a paucity of information on biofortification in pulses. However, both cereal and pulses come under the gamut of food grains, and the agronomic approaches applicable for cereals are equally applicable to pulses. Moreover, many of these trials were made on rice in a rice–wheat cropping system as it covered about 24 million hectares in SE Asian countries including China, India, Pakistan, Nepal and Bangladesh, and zinc deficiency is widespread in rice–wheat belts of all these five countries. Keeping in view of low quality of nutrients in $ZnSO_4$ $7H_2O$ (zinc sulphate heptahydrate), the study suggested higher efficiency in Zn availability following coating with 2 % urea. Zinc sulphate was also a better coating material than ZnO. It was also revealed that various efficiency factors such as agronomic efficiency, apparent recovery, partial factor productivity and physiological efficiency of applied Zn decreased as the level of Zn coating was increased (Shivay et al. 2008).

Similarly, a confirmatory 2-year study showed that an increase in grain/straw yields along with zinc fortification of oat grains is feasible by Zn fertilisation. For higher grain and straw yields as well as Zn fortification of oat grains, coating oat seeds with Zn sulphate or Zn oxide before sowing was found to be the best method practised. This was followed by the next best method where deep placement of Zn fertilisers at sowing was made to garner higher nutrient use efficiency (Shivay et al. 2013). While studying protein, hardness score index, sedimentation value and Zn and Fe concentration in organically and conventionally grown wheat (both *T. aestivum* and *T. durum* cultivars), it revealed that protein density was more in conventionally grown cultivar than in organically grown one although grain hardness was not influenced by nutrient management, while sedimentation value was higher when wheat was conventionally grown (Shivay et al. 2010b). On the other hand, Zn concentration in both the cultivars was more in organically grown (with farmyard manure) than in conventionally grown wheat. Bread wheat cultivar 'HD 2733' had significantly more Fe in grain when organically grown compared to conventionally grown variety, while the reverse was true in the case of *Triticum durum* cultivar 'PDW 215'(Shivay et al. 2010a).

Another study on zinc levels (0, 2.5, 5.0 and 7.5 kg/ha) and chickpea variety (desi '*Pusa* 372' and '*Pusa* 5028' and Kabuli '*Pusa* 2024') revealed that '*Pusa* 372' had the highest grain yield (2.13 t/ha), Zn concentrations in both grain and straw (42.9 mg/kg grain and 36.8 mg/kg straw resp.), total Zn uptake (229.0 g/ha), protein yield (480.6 kg/ha) and protein density (22.7 %) in chickpea grains. Kabuli-type '*Pusa* 2024' chickpea had the lowest protein density (19.9 %). In addition, application of Zn significantly increased grain and straw yields, Zn concentrations in grain and straw and its total uptake and protein yield, protein density in grain, N uptake and net returns (Shivay et al. 2014c). It also increased the efficiency of applied fertiliser (26.6 kg grain/kg applied NPK with 7.5 kg Zn/ha and 22.3 kg grain/kg applied NPK with no Zn) (Shivay et al. 2014c).

On Fe nutrition, study also suggests that growing of aerobic rice with *Sesbania* mulch and Fe fertilisation produced higher grain and straw yield of aerobic rice. Moreover, application of three foliar sprays of 2.0 % $FeSO_4$ (at maximum tillering, pre-flowering and flowering) was found to increase higher Fe concentration in plant parts. This caused significant improvement in uptake in grain as well as in straw of aerobic rice (Yadav et al. 2013) . Moreover, Fe application increased Fe concentration to the extent of 5.4–19.9 % and 5.8–13.3 % in grain and straw of aerobic rice, respectively. Besides that, methods of rice production also increased Fe concentration from 5 % to 11 % and from 4.1 % to 6.3 % over bare soil (no mulch) in grain and straw, respectively (Yadav et al. 2013).

From the viewpoint of biofortification, foliar application is better and requires lesser amount of Fe and Zn fertilisers than their soil application. When cultivars or GM crops with grains denser in Fe and Zn are developed, adequate Fe and Zn fertilisation will be necessary. *The genetic and agronomic approaches are therefore complementary to each other and should progress in tandem.* However, a better understanding of the various reactions that micromineral nutrients undergo in soil and the mechanisms involved in their absorption and translocation in plants, especially to grains is required. This calls for adequate funding of agronomy, soil science and plant physiology departments in agricultural institutes in developing as well as in developed countries (Prasad et al. 2014).

Ferti-fortification offers a rapid solution for increasing Zn concentration in grain and straw. For example, green manuring and Zn-coated fertilisers increased nutrient concentration and their uptake in grain and straw. Foliar fertilisation of 0.2 % zinc sulphate led to higher Zn concentration in rice, while Zn-coated urea (ZCU) as monohydrated $ZnSO_4$ measured highest total Zn uptake (Pooniya and Shivay 2015). Similarly, residue recycling of summer green manuring crops and Zn-coated urea significantly enhanced soil microbial activities resulting in better nutrient turnover in addition to improved long-term productivity of soil and plant system (Pooniya et al. 2012; Singh et al. 2011a).

Studies were also carried out on nutrient accumulation through residues and its effect on nutrient fortification. While studying the effect of summer legumes grown as a sandwich crop in rice–wheat cropping system, it showed that cowpea residues accumulated significantly higher amount of N (97.7 and 4.56 kg/ha in shoot and root, respectively) over that in mung bean. Nutrients like, P, K, Mn and Zn accumulation in cowpea residues were significantly higher than those in mung bean although system productivity of aromatic hybrid rice–wheat–summer legume was highest with mung bean compared to cowpea and summer fallow (Jat et al. 2012). A field experiment to study the effect of genotypic variation and K nutrition on K accumulation and utilisation efficiency in barley (*Hordeum vulgare* L.) under rainfed conditions with low available soil potassium also revealed that successive increases in K nutrition had a significant effect on the dry matter and K accumulation either in the total or in various plant parts of barley at the tillering, stem elongation, heading and maturity growth stages (Shivay et al. 2003). K nutrition also led to significantly higher grain yield with each unit K application than without K application. Amongst 11 genotypes of barley tested, genotype 98–6 had the highest grain yield, and the K use efficiency of this genotype was 10.4 kg grain per kg K applied (Shivay et al. 2003). This shows the potential of existing genotypes for further use in breeding programmes.

Besides this, type of fertiliser materials also affect the nutrient uptake and its use efficiency. In an experiment carried out to study the effect of coating prilled urea with eco-friendly neem (*Azadirachta indica* A. Juss.) formulations for improving the efficiency of nitrogen use in hybrid rice (NDHR-3), it revealed that increasing levels of nitrogen significantly increased grain and straw yields and N uptake with a significant decline in agronomic nitrogen use efficiency (NUE). However, *Pusa* neem golden urea proved to be significantly superior to other nitrogen sources with regard to panicle length, grain yield, N uptake, NUE, apparent N recovery and (lesser) environmental hazards with reduced N loss (Singh and Shivay 2003).

In another study on the residual effect of N sources, S and B levels applied to sunflower on productivity, nutrient concentrations and their uptake by the succeeding crop of mung bean in sunflower–mung bean cropping system, it revealed that remarkable higher mung bean seed yield (961 kg/ha) was realised with 50 kg/ha S applied to preceding sunflower crop. Due to the residual effects of nutrients applied to the preceding sunflower crop, the concentrations and uptake of N, S and B were also greater in the succeeding mung bean crop (Shekhawat and Shivay 2012).

Similarly, a study on sulphur nutrition indicated that 40 kg/ha bentonite sulphur with 90 % pellets (or 20 kg/ha bentonite sulphur with 90 % pellets applied at 15 DAT + 20 kg/ha bentonite sulphur with 90 % pellets at 45 days after transplanting or DAT) significantly increased growth, yield attributes, productivity, sulphur and nitrogen concentrations in grain and straw and also their uptake in an aromatic rice in north-western plain zone of India (Shivay et al. 2014a). In a separate study, it was also observed that Zn application not only increased grain yield, net profit and B: C ratio but also protein concentration/density in grain and nitrogen uptake by chickpea. Three sprays of zinc sulphate heptahydrate (ZSHH) or Zn-EDTA (one each at maximum vegetative growth, flowering and grain filling stage) also yielded the best results (Shivay et al. 2014b).

Addition of organic matter maintains optimum soil structure, enhances water holding capacity and exchangeable ions and reduces leaching of nutrients and toxicities (Choudhary et al. 2008). Crop residue, green manures, composts, animal manures, growing cover crops, reducing tillage and avoiding burning of crop residues can significantly lead to improved plant growth and acquisition of micronutrients in final output (Fageria et al. 2002; Paul et al. 2014; Pooniya and Shivay 2015). Although accumulation of micronutrients varies amongst and within plant species yet, it is more or less attributed to genetics, environment, physiological or biochemical mechanism, agronomic management practices and pest and diseases.

Therefore, the application of nutrients such as N, P, K, S, Zn, Mo and Fe is known to enhance protein density in seeds of legumes. However, the efficiency of such enhancement depends upon many underlying conditions that have a bearing on increasing the level of different amino acid component of protein. For example, the application of P, Mo and N has been shown to increase the level of methionine and that of S-containing fertilisers increase the cystine content of chickpea protein. It is in fact low methionine in pulse protein that is one of the reasons for lower biological value of pulses. It is in this context, studies shows that addition of 0.3 % methionine increases the protein efficiency ratio (PER) in most of the pulses including chickpea, lentil and pea. Similarly, addition of tryptophan – another limiting amino acid in pulses – also improves PER of pulses.

15.4.3 Cost-Effectiveness of the Biofortification Strategy

The major cost components for biofortification relate to research and development needed to develop biofortified genotypes and its implementation. Since an internationally recognised agricultural research system is in place to develop modern varieties of staple food crops, the costs towards research are essentially the incremental costs for increasing or enhancing micronutrient density in food crops. These research costs are likely to be the single largest cost component of biofortification and are of course a one-time investment which is incurred at the start of this. It is also estimated that costs associated with plant breeding fall around an average of about $ 0.4 million/year/crop over a period of a decade globally. Therefore, once biofortified genotypes or varieties are developed with recurring costs on in-house trials and local adaptation research, thereafter the costs are minimal as it involves routine maintenance breeding costs to keep the trait stable. Moreover, where adequate and proper systems for dissemination of modern genotypes are in place (such as in South Asia), implementation costs are nil or negligible. Where such systems are underdeveloped or developing (such as in parts of sub-Saharan Africa), additional costs are incurred in establishing seed multiplication and delivery systems and creating both markets and consumer demand. In addition, the success of biofortification products available to consumers should be viable and continue in a sustainable manner. Therefore, biofortified crops should necessarily be incorporated into existing marketing chains and new market opportunities that are developed in due course of time.

Therefore, centres should facilitate in the dissemination of biofortified varieties to achieve this and create the demand for these varieties by linking producers and consumers with the strategy for product and market development. The strategy focuses on engaging and developing the capacity of users (processors/retailers, consumers and producers) and diffusers (institutions and people in organisations that interact directly with enablers to move a technology to implementation) to adopt the new technology. At the same time, they should transfer knowledge and create awareness of the new technology amongst enablers. Moreover, the people working in organisations and institutions who can create a favourable environment for the adoption, dissemination and increased consumption of biofortified crop varieties are important and crucial in developing certain leadership to spread the requisite message to people all around. Therefore, these considerations play vital for rapid spread of biofortification products in the masses which involve all the stakeholders in food chain.

15.4.4 HarvestPlus Biofortification Program: A Case study

Biofortification approach requires that agricultural research make direct linkages with the human health and nutrition sectors (Bouis 2003) that require an inter- or multidisciplinary research approach, an innovative funding strategy to support the research, adequate willingness amongst scientists and, of course, ultimate dissemination of the biofortified products or seeds. For example, in the *HarvestPlus Biofortification Program*, the major functional activity includes plant breeding at the NARES (National Agricultural Research and Extension Services) and CGIAR centres for maize, rice, sweet potato, common bean, cassava and wheat to develop varieties with the best nutritional and agronomic traits in these crops. The other activities include application of novel advances in biotechnology, genomics, genetics and molecular biology to identify and understand plant biosynthetic genes and pathways of nutritional importance, including those for nutrient absorption enhancers and inhibitors, screening of promising lines for micronutrient bioavailability, efficacy studies involving human subjects to evaluate nutritional impact of the most promising lines intended for release and, above all, food science and human nutrition research to measure the retention of nutrients in processing and cooking. In addition, although breeding may be a viable option, applying the above knowledge in marker-assisted selection for conventional breeding of crops and in the initial development (but not release) of transgenic lines is also very important. Moreover, the programme impact and policy research intends to target the following areas of the subject (biofortification):

1. Understand social and economic factors that determine the dietary quality of the poor and their micronutrient status, as well as policy advocacy based on that research
2. Target regions or states where biofortification will have the greatest social benefit and proven impact
3. Improving seed dissemination systems, demand creation, market and product development by reaching and, of course, fully engaging the end users
4. To provide support to internal project collaborators and external audiences (including donors, the academic and development communities, public officials and the general media) through coordinated communication activities

15.4.5 Biofortification for the Future

In addition to the traditional objectives of disease resistance, yield, drought tolerance etc. and based on micronutrient deficiency rates, there is compelling evidence that biofortification can be a key objective for plant breeders and agronomists in the future. Scientific evidences also suggest that biofortification is inevitable as it is technically feasible. Predictive cost–benefit analyses have also shown that biofortification is useful for controlling micronutrient deficiencies. The challenge is always there to get the desired consumer acceptance for biofortified crops, thereby increasing the intake of the target nutrients. Improved agronomy for seed protein enrichment could have biological impact. However, it is without compromising agronomic traits/yield as it has been amply demonstrated for food crops including pulses. On a positive note, this can become a reality with the advent of good seed supply systems, the development of markets and products and future demand creation. Therefore, from a given food crop(s), every effort is to be targeted at to yield more with proven quality and nutritional considerations.

15.5 Conclusion

The micronutrient deficiency is probably the main cause for hidden hunger. Biofortification or enriching the nutrition contribution of staple food crops through agronomic or plant breeding or any other appropriate approach is in fact the option for solving food-related malnutrition. Agronomic interventions for biofortification of food crops, food fortification/ supplementation through micronutrients and dietary diversification are very much required on priority basis. Improved agronomy has a significant role in productivity changes. It has a minor role in enhancing protein density as both are negatively correlated and protein is genetically controlled. However, total seed protein density is increased following productivity increases. Improved agronomy has a major role in micronutrient (Zn, S and Fe) biofortification which could possibly reduce protein and micronutrient malnutrition.

References

Bouis HE (2003) Micronutrient fortification of plants through plant breeding: can it improve nutrition in man at low cost? Proc Nutr Soc 62:403–11

Cakmak I, Kalayc M, Ekiz H, Braun HJ, Kilinc Y, Yilmaz A (1990) Zinc deficiency as a practical problem in plant and human nutrition in Turkey: a NATO-science for stability project. Field Crop Res 60:175–88

Casey R, Domoney C, Ellis N (1986) Legume storage proteins and their genes. Oxf Surv Plant Mol Cell Biol 3:1–95

Casey R, Davies DR, Casey R, Domoney C, Smith AM (1993) Biochemistry and molecular biology of seed products. In: Casey R, Davies DR (eds) Peas: genetics, molecular biology and biotechnology. CAB International, Wallingford, pp 141–163

Choudhary AK, Thakur RC, Kumar N (2008) Effect of integrated nutrient management on soil physical and hydraulic properties in rice–wheat crop sequence in NW Himalayas. Indian J Soil Conser 36:97–104

Crevieu I, Carre B, Chagneau A-M, Quillien L, Gueguen J, Berot S (1997) Identification of resistant pea (*Pisum sativum* L.) proteins in the digestive tract of chickens. J Agric Food Chem 45:1295–1300

Deshpande SS (1992) Food legumes in human nutrition. Crit Rev Food Sci Nutr 32:333–363

Domoney C, Welham T, Ellis N, Mozzanega P, Turner L (2002) Three classes of proteinase inhibitor gene have distinct but overlapping patterns of expression in *Pisum sativum* plants. Plant Mol Biol 48:319–329

Fageria NK, Baligar VC, Clark RB (2002) Micronutrients in crop production. Adv Agron 77:185–268

Forster C, North H, Afzal N, Domoney C, Hornostaj A, Robinson DS, Casey R (1999) Molecular analysis of a null mutant for pea (*Pisum sativum* L.) seed lipoxygenase-2. Plant Mol Biol 39:1209–1220

Graham RD, Welch RM (1996) Breeding for staple-food crops with high micronutrient density: agricultural strategies for micronutrients. Working paper no 3, International Food Policy Research Institute, Washington, DC

Graham RD, Senadhira D, Beebe SE, Iglesias C, Ortiz-Monasterio I (1999) Breeding for micronutrient density in edible portions of staple food crops: conventional approaches. Field Crop Res 60:57–80

Graham R, Welch R, Bouis H (2001) Addressing micronutrient malnutrition through the nutritional quality of staple foods: principles, perspectives, and knowledge gaps. Adv Agron 70:77–142

Grusak MA (2002) Enhancing mineral content in plant food products. J Am Coll Nutr 21:178S–183S

Hedemann MS, Welham T, Boisen S, Canibe N, Bilham L, Domoney C (1999) Studies on the biological responses of rats to seed trypsin inhibitors using near-isogenic lines of *Pisum sativum* L (pea). J Sci Food Agric 79:1647–1653

Hocking PJ, Pate JS (1977) Mobilization of minerals to developing seeds of legumes. Ann Bot 41:1259–1278

IIPR (2009) 25 years of pulses research at IIPR. Indian Institute of Pulses Research, Kanpur, pp 1–203

Jat SL, Shivay YS, Parihar CM, Meena HN (2012) Evaluation of summer legumes for their economic feasibility, nutrient accumulation and soil fertility. J Food Legum 25:239–242

Krishnan HB (2000) Biochemistry and molecular biology of soybean seed storage proteins. J New Seed 2:1–25

Masood A, Gupta S (2012) Carrying capacity of Indian agriculture: pulse crops. Curr Sci 102:874–881

Paul J, Choudhary AK, Suri VK, Sharma AK, Kumar V, Shobhna (2014) Bioresource nutrient recycling and its relationship with biofertility indicators of soil health and nutrient dynamics in rice–wheat cropping system. Commun Soil Sci Plant Anal 45:912–924

Pooniya V, Shivay YS (2015) Influence of green manuring and zinc fertilization on quality parameters of *Basmati* rice. Commun Soil Sci Plant Anal 46:382–392

Pooniya V, Shivay YS, Rana A, Nain L, Prasanna R (2012) Enhancing soil nutrient dynamics and productivity of *Basmati* rice through residue incorporation and zinc fertilization. Eur J Agron 41:28–37

Praharaj CS (2013) Managing precious water through need based micro-irrigation in a long duration pigeon pea under Indian Plains. In: International conference on policies for water and food security, Cairo, Egypt, 24–26 June 2013

Praharaj CS, Dhingra KK (2001) Growth, productivity and BNF in soybean (*Glycine max* L. Merr.) under *Bradyrhizobium* inoculation, pendimethalin and nitrogen schedules and their residual effect on succeeding wheat (*Triticum aestivum* L.) crop. Indian J Agron 46:635–642

Praharaj CS, Sankaranarayanan K, Kumar N, Singh KK, Tripathi AK (2011) Low-input technologies for increasing crop productivity and sustainability. Curr Adv Agric Sci 3:1–12

Praharaj CS, Sankaranarayanan K, Singh KK, Tripathi AK (2013) Village group approach for enhanced prosperity and livelihood security through conservation of resources – an overview. Curr Adv Agric Sci 5:167–174

Praharaj CS, Singh U, Hazra K (2014) Technological interventions for strategic management of water for conserving natural resources. In: 6th world congress on conservation agriculture, Winnipeg, Canada, 22–26 June 2014

Prasad R, Shivay YS, Kumar D (2014) Agronomic biofortification of cereal grains with iron and zinc. Adv Agron 125:55–91

Rana A, Joshi M, Prasanna R, Shivay YS, Lata N (2012) Biofortification of wheat through inoculation of plant growth promoting rhizobacteria and cyanobacteria. Eur J Soil Biol 50:118–126

Sankaranarayanan K, Praharaj CS, Nalayini P, Gopalakrishnan N (2011) Legume as companion crop for cotton. J Cotton Res Dev 24:115–126

Shalendra KC, Gummagolmath SP, Patil SM (2013) Role of pulses in the food and nutritional security in India. J Food Legum 26:124–129

Shekhawat K, Shivay YS (2012) Residual effects of nitrogen sources, sulfur and boron levels on mungbean (*Vigna radiata*) in a sunflower (*Helianthus annuus*)–mungbean system. Arch Agron Soil Sci 58:765–776

Shivay YS, Chen JH, Ding SR (2003) Genotypic variation for potassium accumulation and utilization efficiency in barley under rainfed potassium stress conditions. Acta Agron Hung 51:267–280

Shivay YS, Kumar D, Prasad R (2008) Relative efficiency of zinc sulfate and zinc oxide–coated urea in rice–wheat cropping system. Commun Soil Sci Plant Anal 39:1154–1167

Shivay YS, Krogstad T, Singh BR (2010a) Mineralization of copper, manganese and zinc from rock mineral flour and city waste compost for efficient use in organic farming. Plant Soil 326:425–435

Shivay YS, Prasad R, Rahal A (2010b) Studies on some nutritional quality parameters of organically or conventionally grown wheat. Cereal Res Commun 38:345–352

Shivay YS, Prasad R, Pal M (2013) Zinc fortification of oat grains through zinc fertilisation. Agric Res 2:375–381

Shivay YS, Kumar D, Prasad R, Pal M (2014a) Effect of cosavet fertis WG and bentonite sulphur pellets on growth, yield attributes, productivity and sulphur use efficiency in basmati rice. Indian J Fertil 10:32–38

Shivay YS, Prasad R, Pal M (2014b) Effect of zinc fertilisation on grain and protein yield, nitrogen uptake and fertiliser use efficiency in chickpea. Indian J Fertil 10:24–27

Shivay YS, Prasad R, Pal M (2014c) Effect of variety and zinc application on yield, profitability, protein content and zinc and nitrogen uptake by chickpea (*Cicer arietinum*). Indian J Agron 59:317–321

Singh S, Shivay YS (2003) Coating of prilled urea with ecofriendly neem (*Azadirachta Indica* a. juss.) formulations for efficient nitrogen use in hybrid rice. Acta Agron Hung 51:53–59

Singh AK, Singh L (2014) Technology adaptability for higher pulses productivity in varied conditions of Uttar Pradesh, India. Indian J App Res 4:4–7

Singh KK, Praharaj CS, Choudhary AK, Kumar N, Venkatesh MS (2011a) Zinc response in pulses. Indian J Fertil 7(10):118–126

Singh KK, Praharaj CS, Ghosh PK, Kumar N, Venkatesh MS (2011b) Nutrient management strategies for pulse based production systems in rainfed regions of India. Indian J Fertil 7:50–63

Yadav GS, Kumar D, Shivay YS, Singh H (2010) Zinc-enriched urea improves grain yield and quality of aromatic rice. Better Crops 94(2):6–7

Yadav GS, Shivay YS, Kumar D, Babu S (2013) Enhancing iron density and uptake in grain and straw of aerobic rice through mulching and rhizo-foliar fertilization of iron. Afr J Agric Res 8:5447–5454

16 Micronutrient Deficiencies in Humans and Animals: Strategies for Their Improvement

Anshu Rahal and Yashbir Singh Shivay

Abstract

Among the trace minerals, the most important for humans and animals comprised of Fe, Zn, Mn, Cu, B, and Mo which are physiologically and metabolically essential. In developing countries like India, amelioration of deficiency of iron and zinc in humans occupies prime priority as it results in malnutrition, while in animals Zn, Mn, and Cu need attention. Iron deficiency usually shows its effect in the form of anemia, fatigueness, reduction in intelligence, and work efficiency, while zinc deficiency is more prominent in the form of indigestion, respiratory infections, and growth retardation in both humans and animals. In order to combat against these deficiencies, nutritional intervention in the form of nutraceutical/mineral mixture and biofortification is required. Nutraceutical approach includes pharmaceutical or dietary Fe/Zn supplementation or diet diversification, while biofortification of cereal grains or other foods requires the role of nutritionist, biotechnologists, plant breeders, and agronomists. Global projects have focused on genetic biofortification of food and fodder crops like rice, wheat, maize, and pearl millet, etc., but still the gap is there as so far only two cultivars have become available for cultivation. Crops grown on above-listed micronutrient-deficient soil are treated with micronutrient fertilizer during agronomic biofortification. Success has been achieved in India for the biofortification of Zn in wheat, rice, oats, and chickpea which can help in eliminating malnutrition. Combined efforts are being needed involving agricultural scientists, physicians, planners, managers, and

A. Rahal
Department of Animal Nutrition, College of Veterinary and Animal Sciences, G. B. Pant University of Agriculture and Technology, Pantnagar, Uttarakhand 263 145, India
e-mail: anshurahal@rediffmail.com

Y.S. Shivay (✉)
Division of Agronomy, ICAR-Indian Agricultural Research Institute, New Delhi 110 012, India
e-mail: ysshivay@hotmail.com

U. Singh et al. (eds.), *Biofortification of Food Crops*, DOI 10.1007/978-81-322-2716-8_16

nutritionists (crops, animals, and human) for amelioration of micronutrient deficiencies in humans and animals.

Keywords
Animals • Biofortification • Deficiencies • Humans • Micronutrients • Symptoms

16.1 Introduction

Humans and animals require microelements such as Fe, I, Zn, Mn, Cu, Se, Mo, Cr, F, B, Ni, Si, V, As, Li, Sn, and Co and macroelement, that is, Cl (Welch 2005), while in plant microelements, namely, Fe, Mn, Zn, Cu, B, Mo, Cl, and Ni have been considered to be essential for normal growth and development (Prasad and Power 1997). Research indicates that elements like Ni (Dixon et al. 1975; Brown et al. 1987; Malavolta and Moraes 2007), Va (Lyalikova and Yurokova 1989; Guo 1987), and Co (Ahmed and Evans 1960) are also needed by plants. Mostly these elements are called micronutrients as required in very small quantity for nourishment. These are used in the form of chemical element, ion, compound, or molecule in plants, while in case of humans and animals, the plant and animal origin food products and their by-products serve as source of these micronutrients. Worldwide in human and animal nutrition, the term micronutrient refers to combined vitamins and minerals, and their deficiencies have been referred to as hidden hunger because the sufferer is unable to detect the deficiency of the micronutrients in primitive stage, which shows its effect only in the long run. This book chapter describes various essential micronutrients for humans and animals (Nielsen 1999) and the effect of biofortification in amelioration of their deficiency consequences.

16.2 Deficiency Symptoms and Recommended Daily Allowance

16.2.1 Iron

On an average human body contains 2.38 g iron in women and 3.8 g iron in men. Iron is present in all human cells. Hemoglobin (Hgb) makes up 96 % of the red blood cells on dry matter basis (Weed et al. 1963). Hgb is the iron containing oxygen transport metalloprotein in red blood cells of humans and most other vertebrates. This element has prime importance in humans and animals as it is involved in carrying oxygen throughout the body in the form of hemoglobin. Nearly about 85 % of the Fe in human body is present in two heme proteins, namely, hemoglobin and myoglobin, which are responsible for the transport and storage of oxygen in different body parts (Bell and Dell 2008). Some Fe in the form of nonheme proteins, such as ferritin and transferritin, is mainly responsible for storage and transport of Fe in the form of metalloflavoproteins, ferredoxins, or Fe–S proteins (Yip and Dallman 1996). Iron from animal food sources with exception of eggs is better absorbed than iron from plant sources. The poor availability of iron from egg yolk is due to the presence of phosphor-protein, phosvitin. Iron from hemoglobin is more available to human than nonheme iron. Iron from green vegetables is less available than the iron from bean, chick peas, and okra.

Survey data indicates that in developing countries, Fe-deficiency anemia is prevalent in around two billion people (Stolzfus and Dreyfuss 1998). Mostly sufferers are persons with fast growth like infants, toddlers, and pregnant women who have higher Fe requirement than normal. The deficiency appears in the form of pallor, fatigueness, reduced intelligence, and work efficiency (Lynch 2003). Once diagnosed iron supplementation has to be carefully monitored, because there is no mechanism of iron excretion from human body. Only 1–2 mg day^{-1} is lost as a result of sloughing of dead cells. In women about 0.006 mg iron kg^{-1} day^{-1} could be lost during menstruation (Schmeier and Petruzelli 2008). High amounts of iron could lead to iron poisoning especially in young children (Tenenbein 2005). Pain in the stomach, nausea, and vomiting including blood in the vomiting are the common symptoms of iron poisoning. Since green fodders are rich source of iron, farm animals rarely suffer from Fe deficiency (Underwood and Suttle 1999).

16.2.1.1 Recommended Dietary Allowance (RDA)

RDA (per day) for iron is as follows: infants 0–6 months 0.27 mg (average intake); infants 7–12 months 11 mg; children 1–3 years 7 mg, 4–8 years 10 mg, and 9–13 years 8 mg; adolescents 14–18 years 11 mg (males) and 15 mg (females); adults 19–50 years 8 mg (males) and 18 mg (females); pregnancy 27 mg; and lactating 14–18 years 10 mg and 19–50 years 9 mg (FNB 2001). The tolerable upper limit for iron is 45 mg day^{-1}. Heme iron content in some foods is as follows: mg 100 g^{-1}, clam (28.0), pork lever (18.0), lamb kidney and cooked oyster (12.0), beef lever (6.5), raw yellow beans (7.0), toasted sesame seeds (14.7), and spinach (3.5). Iron from meat being heme is more readily absorbed than from grains and vegetables, which is more non-heme. Phytates in cereal grains decrease the iron absorption, and low bioavailability of iron in the diet in developing countries is the primary cause of iron deficiency anemia in those countries (Berger and Dillon 2002; Yip and Ramakrishnan 2002). Efforts are underway to develop GM cereals with iron-rich grains, such as golden rice (Potrykus et al. 1996; Murray-Kolb et al. 2002).

16.2.2 Zinc

There is 2–4 g zinc in the human body (Rink and Gabriel 2000). Most zinc is present in the brain, muscles, bones, kidney, liver, prostate, and some parts of the eye (Wapnir 1990); the latter two body parts have the highest concentration. Semen is particularly rich in zinc, which is responsible for prostate function and reproductive organ growth (Berdnier et al. 2007). Zn is important for many metabolic processes in the body. As related with protein metabolism, Zn is responsible for repairment of wear and tear of body tissues, RNA synthesis, and DNA transcription (Shukla et al. 2009) and healing of wounds, blood formation, immunity, and overall growth (Walker and Black 2004). Zn is also involved in functioning of insulin hormone (Miccici 2000). There is some evidence that Zn helps in combating cancer (Ho 2004). Thymidine kinase enzyme is most sensitive to Zn depletion. Meat and other animal sources of protein are most reliable sources of zinc in the diet.

Zinc deficiency is widespread in the world and Prasad (1984) brought it to the focus, when he reported that zinc deficiency was the cause of dwarfism and hypogonadism among adolescents from the lowest social classes of Egypt and Iran. Dwarfism has also been reported in about 61 million children aged 0–5 years (Ringstad et al. 1990). The deficiency symptoms of Zn in children mainly include gastrointestinal disturbances and respiratory problems (Fischer Walker et al. 2009). According to USFNB, RDA for children (4–8 years) and adult males and females is 5, 11, and 8 mg day^{-1}, respectively. A survey in Haryana state in India showed that 65 % of the 258 pregnant women sampled had Zn deficiency (Pathak et al. 2008), which is a matter of concern in India. Stein et al. (2007) using disability adjusted life years (DALYs) observed that Zn deficiency in India is a highly relevant health problem and reported a loss of 2.8 million DALYs per year.

In animals the deficiency symptoms appear in the form of parakeratosis, dermatitis, hair loss, deformed limbs, stiff joints, reduced growth due to fall in appetite and reproductive problems, and swollen feet with open scaly lesions and impaired reproduction (Spears 1995; McDowell 2003; Miller and Miller 1962).

16.2.3 Manganese

Total Mn content of a 70 kg man is about 12–20 mg (Watts 1990). Manganese is distributed in tissue throughout the body with the highest concentrations in the liver, thyroid, pituitary, pancreas, kidney, and bones. In tissue Mn is largely located in mitochondria. Manganese is essential for the activity of several enzymes involved in mucopolysaccharide synthesis. In case of humans and animals, manganese is required in sufficient concentration for optimum growth, development, and reproduction. It is related with functioning of various enzymes like Mn-superoxide dismutase, pyruvate carboxylase, arginase, glycotransferases, glutamine synthetase, prolidase, etc. (Keen and Zidenberg-Cheor 1999). Usually Mn deficiency is uncommon in humans (Bell and Dell 2008) because requirement is easily met out by daily food items. However UNFNB (2001) has recommended 2.1–2.3 and 1.6–1.8 mg day^{-1} requirement for men and women, respectively. The largest content is in blueberries, wheat bran, nuts, and cocoa. Coffee and tea are also rich source of manganese. Calcium, phosphate, and iron interfere with absorption of manganese. Chelating agents like EDTA increase the excretion of manganese into the urine.

Manganese deficiency symptoms include hypocholesterolemia, impaired growth, weight loss, transient dermatitis, intermittent nausea (Watts 1990), bone demineralization (Norose et al. 1992), skin rash (Friedman et al. 1987), osteoporosis in women (Freeland-Graves and Llanes 1994), epilepsy (Carl and Gallagher 1994), and change of hair color (Doisy 1973). Manganese concentration in human hair has been reported to be related to certain diseases; lower levels have been found in Down's syndrome, epilepsy, and schizophrenia, while elevated levels were associated with multiple sclerosis, learning disabilities, and Parkinson's disease (Chatt and Katz 1988; Ashton 1980). For more information on the results of clinical trials on Mn, reference may be made to Drake (2015).

In livestock Mn deficiency appears in the form of reproductive disorders, skeletal deformation, and short tendons in newborn young ones (McDowell 2003).

16.2.3.1 Recommended Adequate Intake (AI) for Humans

There is no RDA for Mn in humans. The recommended adequate intake levels (mg day^{-1}) for Mn are as follows: infants 0–6 months 0.003 mg and 7–12 months 0.6 mg; children 1–3 years 1.2 mg, 4–8 years 1.5 mg, and 9–13 years 1.9 mg (males) and 1.6 mg (females); adolescents 14–18 years 2.2 mg (males) and 1.6 mg (females); adults 19+ years 2.3 mg (males) and 1.8 mg (females); pregnancy 2.0 mg; and breast-feeding 2.6 mg (FNB 2001).

16.2.4 Copper

Copper content in human adults varies from 50 to 120 mg (average 80 mg), with the highest concentrations being in the liver (3.6–5.1μgg^{-1}) and brain (3.3–6.3 μg g^{-1}) (WHO 1996; Uauy et al. 1998; Lech and Sadik 2007). Human hair is particularly rich in Cu and may contain 22.9–46.6 μg g^{-1} Cu (Angelova et al. 2011). Copper has a role to play in oxidation-reduction reactions and enzyme functions like Cu-superoxide dismutase, dopamine betahydroxylase and monoamine oxidase, Cu-metalloenzymes, cytochrome oxidase functioning of central nervous system, and genes regulated by Cu-dependent transcription factors (Uauy et al. 2008). Ferroxidase II is responsible for catalyzing the oxidation of Fe(II) to Fe(III), which facilitates its transport to sites of red blood cell formation (Bell and Dell 2008). Copper is necessary for iron absorption. Among the animal products, rich sources of copper include the liver, lobster, and oyster, while milk and eggs are poor.

Copper deficiency in humans is rarely found except in infants dependent on cow's milk (Turnlund 1999). USFNB (2001) has recommended RDA of 440 μg day^{-1} for children and 900 μg day^{-1} for adults. On the contrary Cu deficiency symptoms are widely noticeable in farm animals in the form of uncoordinated gait in lambs, hair around eyes of cattle, crimpled wool in sheep, and heart failure leading to death in cattle (McDowell 2003).

16.2.5 Boron

In the human body, the highest concentrations of boron are in the heart (28 mg kg^{-1}), followed by the ribs (10 mg kg^{-1}), spleen (2.6 mg kg^{-1}), and liver (2.3 mg kg^{-1}) (Devirian and Volpe 2003). A boron concentration in blood is 0.06 mg kg^{-1}, in plasma 0.02 mg kg^{-1}, in urine 0.75 mg kg^{-1}, and in bones, nails, and hair between 4.3 and 17.9 mg kg^{-1} (Hunt 2003, 2010). The boron content in the same tissue can differ depending on the health of the individual. For example, the concentration of boron was 3 mg kg^{-1} in arthritic bones compared to 56 mg kg^{-1} in the healthy bones (Scorei and Popa 2010). In human tissues and body fluids, most boron is present as boric acid (98.4 %), while the rest is as borate anion (1.6 %). Boron is mainly responsible for improving immune system (Hunt 2003) and in functioning of steroids, hormones, minerals, and vitamins in human and animal body system (Devirian and Volpe 2003). It is involved in binding process with the cell membrane (Verstaeten et al. 2005). Boron may have a role in immune function (Hunt 2003) and in the metabolism of steroids, hormones, some mineral nutrients, and vitamins (Devirian and Volpe 2003). Further research is needed for justifying the role of boron in metabolic reactions (Bell and Dell 2008). Some recent studies in Geochemistry point out the role of B in the synthesis of RNA (Periur 2001; Scorei and Cimpoiasu 2006; Saladino et al. 2011). It has been reported that ingested boron through diet helps in strengthening bone in case of animals like rats, chicken, and pigs (Armstrong and Spears 2001). No specific symptoms of B deficiency have been observed in animals and humans and no recommended dietary allowance (RDA) is recommended.

16.2.5.1 RDA for Boron

There is no RDA for B and the safe intake limit is 1–10 mg day^{-1}. Boron concentrations (mg kg^{-1}/mg L^{-1}) in some foods are as follows: raw avocado 14.3, peanut butter 5.87, salted dry roasted peanuts 5.83, dry roasted pecans 2.64 ppm, prune juice 5.64, grape juice 3.42, sweetened chocolate powder 4.29 ppm, and table wine (12.2 % alcohol) 3.64 (Hunt 2010). Boron concentrations are low in meat, poultry, or fish (Hunt 2010). Generally, the intake of boron is between 1 and 3 mg daily depending upon amounts and kind of foods (Dinca and Scorei 2013). Borax given the number E285 is also used as a food additive in some countries, such as, China, mainly for improving the texture of noodles.

16.2.6 Molybdenum

Molybdenum is now recognized as an essential micro-mineral in human nutrition (Rajagopalan 1988). Molybdenum is a cofactor present in the active site in four enzymes in humans (Coughlan 1983; Cohen et al. 1971), which are known as molybdoenzymes. These are sulfite oxidase, aldehyde oxidase, xanthine oxidase, and mitochondrial amidoxime reducing component (mARC). Sulfite oxidase catalyzes the transformation of sulfite to sulfate in the metabolism of sulfur-containing amino acids methionine and cysteine. This enzyme is known to be crucial for human health. Aldehyde oxidase and xanthine oxidase catalyze hydroxylation reactions that involve a number of different molecules with similar chemical structures. Xanthine oxidase catalyzes the breakdown of nucleotides (precursors to DNA and RNA) to form uric acid, which contributes to the plasma antioxidant capacity of the blood. The functions of mARC are not yet well understood.

No natural case of Mo deficiency in humans has been reported, although a case of diet-induced deficiency of Mo was reported by Abumrad et al. (1981). For more details on clinical aspects, reference may be made to Delage et al. (2015).

There are a number of enzymes other than those mentioned for humans with which Mo is associated in animals (Hille 1999). In ruminants, excess Mo in diet causes Cu deficiency due to the formation of insoluble complex in rumen which can be rectified by supplementation of sulfate; otherwise molybdenosis can occur in animals.

Excess dietary molybdenum has been found to result in copper deficiency in grazing animals. In the digestive tract of ruminants, the formation of compounds containing sulfur and molybdenum, known as thiomolybdates, prevents the absorption of copper and can cause fatal copper-dependent disorder (Suttle 2012).

16.3 Amelioration of Micronutrient Deficiencies in Humans and Animals

Amelioration of micronutrient deficiencies can be broadly grouped under two categories, namely, nutraceutical/mineral mixture and biofortification of cereals and other staple food crops.

16.3.1 Nutraceutical/Mineral Mixture

Nutraceutical/mineral mixture intervention of a micronutrient deficiency can be approached in three ways, namely, pharmaceutical supplementation, dietary supplementation, and dietary diversification.

16.3.2 Pharmaceutical Supplementation

India has made considerable success in pharmaceutical supplementation, and a variety of tablets, capsules, syrups, and tonics containing Fe are available off the shelf. Recently some Zn formulations have also entered the Indian market. However, more public awareness about these has to be created. Also people need to be advised that these supplements must be taken only under medical supervision, because dietary Fe overload can cause Fe poisoning, since the human body has no adjustable Fe excretory mechanism (Lynch 2003). There are also reports that Zn supplements over 100 mg day^{-1} taken over 10 years may increase the risk of prostate cancer (Litzman 2003).

16.3.3 Dietary Supplementation

Dietary supplementation can either be at industrial level or at home. Examples of industrial supplementation are iodized salt, vitamin A supplemented edible oils, Fe-enriched corn flakes, etc., which are already in the Indian market. Home supplementation with micronutrients is more difficult to achieve, because it involves training, specially housewives in making micronutrient supplemented foods. However, some success stories with Fe enrichment of foods such as khichri (rice cooked with a pulse) and laddoos (a popular Indian sweet) are available (Kanani and Poojara 2000; Sood and Sharda 2002; Varma et al. 2007).

16.3.4 Dietary Diversification

This involves adding or increasing the content of food articles rich in Fe, Zn, and other micronutrients. Animal products such as meat, fish, and poultry are rich sources of Fe and Zn, but per capita consumption of these in India is only 1.6, 1.9 and 1.28 kg year^{-1}, respectively, as against 5.9, 8.3 and 35.5 kg year^{-1} in China and 42.6, 45.4, and 29.7 kg year^{-1} in the USA. What to say of animal products, even the availability of pulses, which are richer in Fe and Zn, has declined from 68 g capita^{-1} day^{-1} in 1961 to 35 g capita^{-1} day^{-1} in 2006–2007 (Anonymous 2009). Thus, diversification of common man's diet in India is a difficult task.

Cereal grains such as rice and wheat, which are staple food in India, are low in Fe and Zn. For example, polished rice may contain only 5 mg Fe and 13 mg Zn per kg grain (Welch 2005). Wheat grains may contain 45–51 mg Fe and 30–43 mg Zn per kg grain (Dhaliwal et al. 2009). Further, cereal grains contain phytates, which reduce the bioavailability of Fe and Zn. A phytate to Zn ratio of 15 or less is considered good for zinc bioavailability (Gibson 2005). Some simple practices such as soaking the grains overnight or sprouting them can easily reduce phytate content and make Fe and Zn more bioavailable. Some green and fresh vegetables, such as cluster beans, French beans, ladies finger, green peas, carrots, okra, cabbage, etc., are rich in Fe (Jaggi 2007) and can help in meeting Fe needs. Beans are also rich in Zn.

All nutraceutical approaches require a well-organized infrastructure and trained staff to create the need awareness of micronutrients and continued funding, which is difficult to get. Further such programs, most of the time, would remain restricted to urban population, which is easy to reach and is more receptive due to a high level of women's education. People in rural areas, especially those difficult to reach by road or rail, thus remain deprived of benefits from the nutraceutical approach.

16.4 Biofortification of Cereal Grains/Straw and Other Foods with Micronutrients

16.4.1 Genetic Strategies

In an attempt to attain higher and higher yields, the concentration of micronutrients in grains has generally been overlooked. White et al. (2009) reported that increased yields in modern potato cultivars are often associated with reduced concentration of minerals. Similarly Cakmak et al. (2000) found higher Fe and Zn concentration in primitive wheat cultivars than in modern varieties. Efforts are therefore being made to breed food crop cultivars with higher micronutrient concentration (Graham et al. 2001; Bouis 2003).

16.4.2 Role of Biotechnology and Plant Genetic Diversity to Enhance Mineral Composition

Biotechnology refers to a set of medical, agricultural, and industrial techniques that use living organisms to create new, or to improve existing, products and processes. In particular, the use of the process of genetic modification is a very important aspect for increased density of the minerals in the economic produce of the crops. The capacity of genetic modification to produce plants with useful traits such as decreased pest resistance, reduced post-harvest losses, increased yield, reduced labor requirements, or enhanced content of particular desirable constituents is readily apparent (Tripp 2001). In the latter case, pure naturally occurring molecules can be produced economically and efficiently for use as pharmaceuticals, pesticides, or other purposes. Likewise efficient production of ingredients with specific qualities such as edible oils (Voelker 1997), pigments, and flavor ingredients supports the food industry in developed countries. Epidemiological, in-vitro, and clinical data affirm the positive relationship of a diversity of dietary components to good health. The biotechnology is already used to introduce, isolate, and enhance specific nutritional and functional properties in food, motivated by either consumer demand or entrepreneurial innovation (Bouis et al. 2003; Tucker 2003).

The use of biotechnology to create genetically modified organisms (GMOs) has the potential to design foods with specific attributes. Enhanced nutritional quality of crops may be achieved by enabling the capacity of the plant to synthesize vitamins, to take up minerals with greater efficiency, or by reducing anti-nutrient factors such as phytates or tannins that can make nutrients unavailable as well as lower food palatability (Raboy 2002; Bouis et al. 2003; Tucker 2003). Crops with enhanced qualities such as high β-carotene and lycopene tomatoes or oats with enhanced β-glucans that are available as whole foods can attract consumers to pay premium prices (Hasler 2002). Such products are demand driven in relation to their perceived benefit. However, futuristic scenarios of using genomics to match diets with individual genotypes (Kaput and Rodriguez 2004) bear little relevance to economic reality of developing country populations.

Forage crops show particular potential of nutritional enhancement with positive impact on livestock nutrition and productivity (Casler 2001; Cherney and Cherney 2002). Because animal-source foods generally provide more available iron, vitamin A, and protein than plants (Allen and Gillespie 2001; Allen 2003; Murphy and Allen 2003), even small increases in intake offer real benefits to the majority of the world's malnourished. In nutritional terms, considerable benefit will come from increasing animal productivity and consequently animal-source foods in the diet. However, animal-source foods are more expensive than plant foods and increased production can have negative environmental impact. Nonetheless, for poor households, keeping livestock is an important economic and dietary asset. Increasing consumption of animal-source foods in poor communities using improved fodders and other innovative means contributes to the economic benefits of animal ownership in poor communities and is a positive example of programs that have increased local production and consumption (Allen 2003).

While the technical achievements of nutritional enhancement of human foods and animal fodders may be analogous, the impacts on nutrition and health are not. The objectives of each exercise differ, as do the measures of health. In the case of animal fodder, the diet is consumed under controlled situations with no or little choice offered. Moreover, nutritional physiology and behavior of omnivorous humans are considerably more complex than herbivorous and provisioned animals.

For human populations, the benefits of consuming phytochemically enhanced foods (other than in terms of quantity of fruits and vegetables) for individual or public health have not been demonstrated. Neither will the long-term epidemiological studies required to show their impacts be soon forthcoming. Interventions with nutrient-enhanced foods in undernourished populations would have shorter-term impact but demonstrating the effectiveness and long-term sustainability of these interventions will also be challenging (Allen and Gillespie 2001; King 2002; Bouis et al. 2003). A less ambitious use of biotechnology is the marker-assisted selection for quality trait identification (Naylor et al. 2004). While accessions screened for markers of nutrient or phytochemical quality could be used in traditional breeding or biotechnology efforts, they can also be promoted directly to consumers as nutritionally valuable crop varieties.

A number of global projects on genetic biofortification of food crops are underway (Stein 2009). Some of these are listed below. In addition, there are several small projects on other crops (Stein 2009).

- *HarvestPlus, a Global Challenge Program* of the Consultative Group of the International Agricultural Research (CGIAR) focuses on breeding for higher levels of Fe, Zn, and beta-carotene in the major staple crops of the developing countries, namely, rice, wheat, maize, cassava, sweet potato, and beans.
- *The Golden Rice Project* focuses on genetic engineering approach to biofortify rice with beta-carotene, Fe, Zn, Vitamin E, and protein under the "Great Challenges in Global Health" scheme funded by Bill & Melinda Gate Foundation.
- African Biofortified Sorghum Project funded by Bill & Melinda Gates Foundation focuses to fortify sorghum with Fe, Zn, Vitamin A, and Vitamin E.
- Bio-Cassava Plus targets to increase Fe, Zn, vitamin A, vitamin E, and protein in cassava.
- Biofortification of bananas with Fe, pro-vitamin A, and vitamin E.

Genetic biofortification of food crops is beset with several problems. Biofortification is the process of generating genetically improved food crops that are rich in bioavailable micronutrients, either through conventional breeding or genetic modification. Some of these are:

- Initial genetic research requires heavy funding (Qaim et al. 2007), which at the moment has fortunately come from some philanthropic organizations, such as Bill and Melinda Gates Foundation. But such funds may not be available in the future.
- Genetic research needs time, may extend to several years.
- The cultivars fortified with micronutrients developed may not be as high yielding as the

present cultivars and may face problems in acceptance by the farmers.
- The newly developed cultivars fortified with micronutrients may differ in quality and the consumers may not accept these, at least initially (De Groote and Chege Kimenju 2008).
- Cultivars fortified with micronutrients using genetic engineering are still not accepted by a number of countries (Cohen and Pearlboy 2002; Conko and Miller 2002; Neilson et al. 2001).
- Even if a nation accepts genetically modified (GM) crop cultivars, it may not be possible to export the produce to other countries (Isaac 2002), which do not allow the consumption of GM foods.

Despite the enormous research activities, so far, only two biofortified crop cultivars have been successfully developed. So far no micronutrient-rich cultivar has been released in any food crop. It is generally opined that developing countries in Asia and Africa, where micronutrient problems are widespread, stand to gain from biofortified food crop cultivation.

16.4.3 Agronomic Strategies

Agronomic biofortification or ferti-fortification (Cakmak et al. 1999) of food crops with micronutrients involves application of micronutrient fertilizers to crops and is restricted to micronutrients that are essential for plant growth. Also it is applicable only to areas with soils deficient in micronutrients. It is worth mentioning that Zn-deficient soil areas are also the regions from where Zn malnutrition is most reported (Prasad 2009).

Shivay et al. (2008a) from New Delhi reported that an application of 5.2 kg Zn ha^{-1} as 2 % Zn (as zinc sulfate or zinc oxide) coated urea increased Zn content in rice grain from 30.2 to 47.7 mg kg^{-1}. Similarly an application of 2.6 kg Zn ha^{-1} as 2 % Zn (as zinc sulfate or zinc oxide) coated urea increased Zn content in wheat grain from 39.9 to 51.1 mg kg^{-1} (Shivay et al. 2008b). This increase in Zn content in rice or wheat grain was also associated with an increase in grain yield.

The major advantages of agronomic biofortification include the following:

1. It is done on crop cultivars already being cultivated by the farmers and the produce is acceptable to the consumers.
2. Farmer is saved from the investment on new seeds.
3. Gain in the micronutrient concentration in grain or other food products is obtained in the same year.
4. Application rates of mineral micro-nutrients (MMNs) are much smaller when these are applied to foliage.
5. Agronomic biofortification or ferti-fortification is thus a win–win approach for developing countries.

16.5 Conclusion

Iron and zinc deficiencies are widespread in India and the world. Since these are hidden hungers, creation of public awareness about the ill effects of Fe and Zn malnutrition is important. Research on micro-mineral nutrition in medical and other institutions needs to be encouraged. Agronomic biofortification of food crops is the fastest and safest way for the intervention of Fe and Zn deficiency. An integrated approach involving human and animal nutrition experts and agricultural scientists is suggested for ameliorating of micronutrients deficiencies in humans and animals.

References

Abumrad NN, Schneider AJ, Steel D, Rogers LS (1981) Amino acid intolerance during prolonged total parenteral nutrition reversed by molybdate therapy. Am J Clin Nutr 34:2551–2559

Ahmed S, Evans HJ (1960) Cobalt a micronutrient element for the growth of soybean plants under symbiotic conditions. Soil Sci 90:205–210

Allen LH (2003) Interventions for micronutrient deficiency control in developing countries: past, present and future. J Nutr 133:3875S–3878S

Allen LH, Gillespie SR (2001) What works? A review of the efficacy and effectiveness of nutrition interventions. ACC/SCN: Geneva in collaboration with the Asian Development Bank, Manila. ISBN 971-561-388-8

Angelova M, Asenova S, Nedkova V, Koleva-Kolarova R (2011) Copper in human organism. Trakia J Sci 9 (1):88–98

Anonymous (2009) Fertilizer statistics 2008–09. The Fertilizer Association of India, New Delhi

Armstrong TA, Spears JW (2001) Effect of dietary boron on growth performance, calcium and phosphorus metabolism and bone mechanical properties in growing barrows. J Anim Sci 79:3120–3127

Ashton BA (1980) Manganese and man. J Orthomol Psychiatry 9(4):237–249

Bell DW, Dell B (2008) Micronutrients for sustainable food, feed, fibre and bioenergy products. International Fertilizer Industry Association, Paris, p 175

Berdnier CD, Dwer JT, Feldman EB (eds) (2007) Handbook of nutrition and food. CRC Press, Boca Raton

Berger J, Dillon JC (2002) Control of iron deficiency in developing countries. Curr Opin Gastroenterol 25:129–135

Bouis HE (2003) Micronutrient fortification of plants through plant breeding: can it improve nutrition in man at low cost? Proc Nutr Soc 42:403–411

Bouis HE, Chassy BM, Ochanda JO (2003) Genetically modified food crops and their potential contribution to human nutrition and food quality. Trends Food Sci Technol 14:191–209

Brown PH, Welch RM, Cary EE (1987) Nickel: a micronutrient essential for higher plants. Plant Physiol 85:801–803

Cakmak I, Kalayci M, Ekiz H, Braun HJ, Yilmaz A (1999) Zinc deficiency an actual problem in plant and human nutrition in Turkey. A NATO – science for stability project. Field Crop Res 60:175–188

Cakmak I, Ozkan H, Graun HJ, Welch RM, Romheld V (2000) Zinc and iron concentration in seeds of wild, primitive and modern wheats. Food Nutr Bull 21:401–403

Carl GF, Gallagher BB (1994) Manganese and epilepsy. In: Klmis-Tavantzis DL (ed) Manganese in health and disease. CRC Press, Boca Raton, pp 133–157

Casler MD (2001) Breeding forage crops for increased nutritional value. Adv Agron 71:51–107

Chatt A, Katz SA (1988) Hair analysis: application in the biomedical and environmental sciences. VCH Publishers, New York

Cherney JH, Cherney DJR (2002) Legume forage quality. (Quality improvement in field crops). J Crop Prod 5:261–284

Cohen JI, Pearlboy R (2002) Explaining approval and avail-ability of GM crops in developing countries. New Biotechnol 4:1–6

Cohen HJ, Fridovich I, Rajgopalan VK (1971) Hepatic sulphite oxidase: a functional role for molybdenum. J Biol Chem 246:367–373

Conko G, Miller HI (2002) Food safety: the precautionary principles is the wrong approach. Eur Aff 2:122–128

Coughlan MP (1983) The role of molybdenum in human biology. J Inherit Metab Dis 6(Suppl 1):70–77

De Groote H, Chege Kimenju S (2008) Comparing consumer preference for colour and nutritional quality in maize: application of a semi-double bond logistic model on urban consumption in Kenya. Food Policy 33:362–370

Delage B, Mende RF et al (2015) Copper, molybdenum. Micronutrient Center, Linus Pauling Institute, Oregon State University. lpi.oregonstate.edu/mic/minerals/copper

Devirian TA, Volpe SL (2003) The physiological effects of dietary boron. Crit Rev Food Sci Nutr 43:219–231

Dhaliwal SS, Sadana US, Manchanda JS, Dhadli HS (2009) Biofortification of wheat grains with zinc and iron in *Typic Ustochrept* soils of Punjab. Indian J Fertil 5(11):13–20

Dinca L, Scorei R (2013) Boron in human nutrition and its regulations use. J Nutr Ther 2:22–29

Dixon NE, Gazzola C, Blackley RL, Zerner B (1975) Jackbean urease (EC3.5.1.5) a metalloenzyme: a simple biological role for nickel. J Am Chem Soc 97:4131–4133

Doisy CA Jr (1973) Micronutrient control on biosynthesis of clotting proteins and cholesterol. In: Hemphill DD et al (eds) Trace substances in environmental health VI. University of Missouri Press, Columbia

Drake VJ (2015) Micronutrient Information Center, Linus Pauling Institute, Oregon State University. http://lpi.oregonstate.edu/mic/micronutrients-health/cognitive-function

Fischer Walker CL, Ezzati M, Black RE (2009) Global and regional child mortality and burden of disease attributable to zinc deficiency. Eur J Clin Nutr 63:591–597

FNB (2001) Dietary reference intakes for vitamin A, vitamin K, boron, chromium, copper, iodine, manganese, molybdenum, nickel, silicon, vanadium and zinc. Food and Nutrition Board/Institute of Medicine/National Academy Press, Washington, DC

Freeland-Graves J, Llanes C (1994) Models to study manganese deficiency. In: Klimis-Tavantzis DL (ed) Manganese in health and disease. CRC Press, Boca Raton, pp 59–86

Friedman BJ, Freeland-Graves JH, Bales CW et al (1987) Manganese balance and clinical observations in young men fed a manganese-deficient diet. J Nutr 117 (1):133–143

Gibson RS (2005) Dietary strategies to enhance micronutrient adequacy: experiences in developing countries. In: Anderson P, Tuladhar JK, Kakri KB, Maskey SL (eds) Micronutrients in South and Southeast Asia. International Centre for Integrated Mountain Development, Kathmandu, pp 3–7

Graham RD, Welch RM, Bouis HE (2001) Addressing micro-nutrient malnutrition through enhancing the nutritional quality of staple foods: principles, perspectives and knowledge gaps. Adv Agron 70:77–142

Guo B (1987) A new application of rare earths in agriculture. Rare earths horizons. Australian Department of Industry and Commerce, Canberra, pp 237–246

Hasler CM (2002) Functional foods: benefits, concerns and challenges – a position paper from the American Council on Science and Health. J Nutr 132:3772–3781

Hille R (1999) Molybdenum enzymes. Essays Biochem 34:125–137

Ho E (2004) Zinc deficiency, DNA damage and cancer risk. J Nutr Biochem 15:572–578

Hunt CD (2003) Dietary boron: an overview for its role in immune function. J Trace Elem Exp Med 16:291–306

Hunt CD (ed) (2010) Boron: encyclopedia of dietary supplements, 2nd edn. Informa Healthcare, New York, pp 82–90

Isaac GE (2002) Agricultural biotechnology and transatlantic trade, regulatory barriers to GM crops. CABI Publishing, Wallingford

Jaggi RC (2007) Nutritional research in vegetables of Himachal Pradesh. Indian J Fertil 2(12):25–41

Kanani SJ, Poojara RH (2000) Supplementation with iron and folic acid enhances growth in adolescent Indian girls. J Nutr 130:452S–455S

Kaput J, Rodriguez RL (2004) Nutritional genomics: the next frontier in the postgenomic era. Physiol Genomics 16:166–177

Keen CL, Zidenberg-Cheor S (1999) Manganese in encyclopedia of food science and nutrition. Academic Press, Amsterdam, pp 3681–3691

King JC (2002) Evaluating the impact of plant biofortification on human nutrition. J Nutr 132:511S–513S

Lech T, Sadik JK (2007) Copper concentration in body tissue and fluids in normal subjects of Poland. Biol Trace Elem Res 118(1):10–15

Litzman MF (2003) Zinc supplement use and risk of prostate cancer. J Natl Cancer Inst 95:1004–1007

Lyalikova NN, Yurokova NA (1989) The influence of vanadium on microorganisms and their role in the transformation of the element. In: Anke M (eds) Molybdenum, vanadium, vol 1. Proceedings of 6th international trace elements symposium, Jena, pp 74–78

Lynch SR (2003) Iron deficiency anaemia. In: Encyclopedia of food sciences and nutrition. Academic Press, Amsterdam, pp 215–320

Malavolta E, Moraes M (2007) Nickel from toxic to essential nutrient. Better Crops 19(3):26–27

McDowell LR (ed) (2003) Minerals in animal and human nutrition, 2nd edn. Elsevier Science BV, Amsterdam

Miccici K (2000) Family common sense. Ohio State University, Columbus

Miller JK, Miller WJ (1962) Experimental zinc deficiency and recovery of calves. J Nutr 76:467–473

Murphy SP, Allen LH (2003) Nutritional importance of animal source foods. J Nutr 133:3932S–3935S

Murray-Kolb LE, Theil EC, Takaiwa F, Goto F, Yoshihara T, Beard JL (2002) Transgenic rice as a source of iron for iron-depleted rats. J Nutr 132:957–960

Naylor RL, Falcon WP, Goodman RM, Jahn MM, Sengooba T, Tefera H, Nelson RJ (2004) Biotechnology in the developing world: a case for increased investments in orphan crops. Food Policy 29:15–44

Neilson CP, Thierfelder K, Robinson S (2001) Genetically modified foods, trade and developing countries. International Food Policy Research Institute, Washington, DC (discussion paper 77)

Nielsen FH (1999) Ultratrace minerals in modern nutrition in health and disease. In: Shils ME, Olson JA, Shike M, Ross AC (eds) Modern nutrition in health and disease, 9th edn. Williams & Wilkins, Baltimore, pp 283–303

Norose N, Terai M, Norose K (1992) Manganese deficiency in a child with very short bowel syndrome receiving long-term parenteral nutrition. J Trace Elem Exp Med 5:100–101

Pathak P, Kapil U, Dwivedi SN, Singh R (2008) Serum Zn level amongst pregnant women in a rural block of Haryana state, India. Asia Pac J Clin Nutr 17:276–279

Periur BF (2001) Etude de l'activite' pre'biotiquepotentielle de l'activiteborique. Comptes Rendue Acad Sci Chimie 4:667–670

Potrykus I, Armstrong GA, Beyer P et al (1996) Transgenic *indica* rice or the benefit of less developed countries: toward fungal, insect and viral resistance and accumulation of beta carotene in the endosperm. Proceedings of the 3rd international rice genetics symp, International Rice Research Institute, Manila

Prasad AS (1984) Discovery and importance of zinc in human nutrition. Fed Proc 43:2829–2834

Prasad R (2009) Fertifortification of grains: an easy option to alleviate malnutrition of some micronutrients in human beings. Indian J Fertil 5(12):129–133

Prasad R, Power JF (1997) Soil fertility management for sustainable agriculture. CRC/Lewis, Boca Raton

Qaim M, Stein AJ, Meenakshi JV (2007) Economics of bio-fortification. Agric Econ 37:S119–S123

Raboy V (2002) Progress in breeding low phytate crops. J Nutr 132:503S–505S

Rajagopalan KV (1988) Molybdenum: an essential trace element in human nutrition. Annu Rev Nutr 8:401–427

Ringstad J, Aesth J, Alexander J (1990) Deficiency of mineral nutrients for mankind: Lag J (ed) Geomedicine. CRC Press, Boca Raton

Rink L, Gabriel P (2000) Zinc and the immune system. Proc Nutr Soc 59(4):541–552

Saladino R, Barontini M, Cosetti C, DiMano E, Crestini C (2011) The effects of borate minerals on the synthesis of nucleic acids, bases, amino acids and biogenic

carboxylic acids from formamide. Orig Life Evol Biosph 41:317–330
Schmeier AH, Petruzelli LM (2008) Hematology for medical students. Lippincott Williams & Wilkins, Philadelphia
Scorei R, Cimpoiasu VM (2006) Boron enhances the thermostability of carbohydrates. Orig Life Evol Biosph 36:1–11
Scorei R, Popa R (2010) Boron-containing compounds as preventive and chemotherapeutic agents for cancer. Anti Cancer Agents Med Chem 10:346–351
Shivay YS, Kumar D, Prasad R, Ahlawat IPS (2008a) Relative yield and zinc uptake by rice from zinc sulphate and zinc oxide coating onto urea. Nutr Cycl Agroecosyst 80:181–188
Shivay YS, Prasad R, Rahal A (2008b) Relative efficiency of zinc oxide and zinc sulphate enriched urea for spring wheat. Nutr Cycl Agroecosyst 82:259–264
Shukla VK, Doyon Y, Miller JC (2009) Precise genome modification in the crop species *Zea mays* using Zn-finger nucleases. Nature 459:437–441
Sood M, Sharda D (2002) Iron food supplement. Indian J Pediatr 69(11):943
Spears JW (1995) Improving cattle health through trace mineral supplements. Proceedings of the 14th range beef cow symposium, Gering, pp 1–7
Stein AJ (2009) Global impacts of human mineral nutrition. Plant Soil 321:117–152
Stein AJ, Nestel P, Meenakshi JV, Qaim M, Sachdev HPS, Bhutta ZA (2007) Plant breeding to control zinc deficiency in India: how cost effective in biofortification. Public Health Nutr 106:492–501
Stolzfus RJ, Dreyfuss ML (1998) Guidelines for the use of iron supplements to prevent and treat iron-deficiency anaemia. International Nutritional Anaemia Consultative Group (INACG)/ILSI Press, Washington, DC
Suttle NF (2012) Copper imbalances in ruminants and humans: unexpected common ground. Adv Nutr 3 (5):666–674
Tenenbein M (2005) Unit dose of packaging of iron supplements and reduction of iron poisoning in young children. Arch Pediatr Adolesc Med 159:557–560
Tripp R (2001) Can biotechnology reach the poor? The adequacy of information and seed delivery. Food Policy 26:249–264
Tucker G (2003) Nutritional enhancement of plants. Curr Opin Biotechnol 14:221–225
Turnlund JR (1999) Copper. In: Shils ME, Olson JA, Shike M, Ross AC (eds) Nutrition in health and disease, 9th edn. Williams and Wilkins, Baltimore, pp 241–252
Uauy R, Olivares M, Gonzles M (1998) Essentiality of copper in humans. Am J Clin Nutr 67 (5 Suppl):952S–959S
Uauy R, Olivares M, Gonzales M (2008) Essentiality of copper in humans. Am J Clin Nutr 67(Suppl 1):952S–959S
Underwood EJ, Suttle NF (1999) The mineral nutrition of livestock. CABI Publications, New York
USFNB (2001) Dietary reference intake for vitamin A, vitamin K, arsenic, boron, chromium, copper, iodine, iron, manganese, molybdenum, nickel, silicon, vanadium and zinc. US Food and Nutrition Board/National Academy Press, Washington, DC
Varma JL, Das S, Sankar R, Manna MG, Levinson FJ, Hamer DH (2007) Community level micronutrient fortification of a food supplement in India: a controlled trial in pre-school children aged 36–66 mo. Am J Clin Nutr 85:1127–1133
Verstaeten SV, Lanoue L, Kem CL, Oteiza PI (2005) Relevance of lipid polar headgroups on boron mediated changes in membrane physiological properties. Arch Biochem Biophys 438:103–110
Voelker T (1997) Transgenic manipulation of edible oilseeds. In: Johns T, Romeo JT (eds) Functionality of food phytochemicals. Plenum Press, New York
Walker CF, Black RE (2004) Zinc and the risk for infectious diseases. Annu Rev Nutr 5:233–249
Wapnir RA (1990) Protein nutrition and mineral absorption. CRC Press, Boca Raton
Watts DL (1990) The nutritional relationship of manganese. J Orthomol Med 5(4):219–222
Weed RI, Red CF, Berg G (1963) Is hemoglobin an essential structural component of human erythrocyte membranes? J Clin Invest 42(4):581–588
Welch RM (2005) Harvesting health: agricultural linkages for improving human nutrition. In: Anderson P, Tuladhada JK, Karki KB, Maskey SL (eds) Micronutrients in South and Southeast Asia. International Centre for Integrated Mountain Development, Kathmandu, pp 9–10
White PJ, Bradshaw JE, Finlay M, Ramsay G, Hammond JP, Broadley MR (2009) Relationship between yield and mineral concentrations in potato tubers. Hortic Sci 44:6–11
WHO (1996) Trace elements in human nutrition and health. World Health Organization, Geneva
Yip R, Dallman PR (1996) Iron in present knowledge in nutrition, 7th edn. ILSI Press, Washington, DC, pp 278–292
Yip R, Ramakrishnan U (2002) Experiences and challenges in developing countries. J Nutr 132:827S–830S

Micronutrient Fertilizers for Zinc and Iron Enrichment in Major Food Crops: A Practicable Strategy

17

Yashbir Singh Shivay, Rajendra Prasad, and Ummed Singh

Abstract

Zinc (Zn) and iron (Fe) deficiencies are well-documented public health issue and an important soil constraint to crop production. Generally, there is a close geographical overlap between soil deficiency and human deficiency of Zn and Fe, indicating a high requirement for increasing concentrations of micronutrients in food crops. Breeding new plant genotypes for high grain concentrations of Fe and Zn (genetic biofortification) is an effective strategy to address the problem, but this strategy is a long-term process. A rapid and complementary approach is therefore required for biofortification of food crops with Zn and Fe in the short term. In this regard, agronomic biofortification using micronutrient fertilizers represents a fast and effective strategy. Information generated by us on this aspect is presented in this chapter.

Keywords

Agronomic biofortification • Chickpea • Crops • Iron • Zinc

17.1 Introduction

Zinc and Fe deficiencies are a growing public health and socioeconomic issue, particularly in the developing world (Welch and Graham 2004). Dietary deficiency of essential micronutrients such as zinc (Zn) and iron (Fe) affects more than two billion people worldwide (White and Broadley 2009; WHO 2012), especially pregnant women and children below the age of five who suffer from severe acute malnutrition. Recent reports indicate that nearly 500,000 children under 5 years of age die annually because of Zn and Fe deficiencies (Black et al. 2008). Zinc and Fe deficiencies together with vitamin A deficiency have been identified as the top priority global issue to be addressed to achieve a rapid

Y.S. Shivay (✉) • R. Prasad
Division of Agronomy, ICAR-Indian Agricultural Research Institute, New Delhi 110 012, India
e-mail: ysshivay@hotmail.com; rajuma36@gmail.com

U. Singh
Division of Crop Production, ICAR-Indian Institute of Pulses Research, Kanpur, Uttar Pradesh 208 024, India
e-mail: singhummed@yahoo.co.in

U. Singh et al. (eds.), *Biofortification of Food Crops*, DOI 10.1007/978-81-322-2716-8_17

and significant return for humanity and global stability (www.copenhagenconsensus.com).

In many parts of the world, micronutrient deficiency is a more widespread problem than poor dietary quality and low energy intake (Stewart et al. 2010), and about 20 % of deaths in children under five can be attributed to vitamin A, Zn, Fe, and/or I deficiency (Prentice et al. 2008). In countries with a high incidence of micronutrient deficiencies, cereal-based foods represent the largest proportion of the daily diet (Cakmak et al. 2010a; Bouis et al. 2011). The Harvest Plus initiative of the CGIAR consortium (www.harvestplus.org) is working with national and international partners to alleviate deficiencies of these mineral nutrients by biofortifying staple food crops with essential minerals and vitamins, an approach considered to be the most economical solution to human micronutrient deficiency (Welch and Graham 2004; Bouis 2007; Cakmak 2008; Peleg et al. 2009). The biofortification program is focusing on three micronutrients that are widely recognized by the World Health Organization (WHO) as limiting: iron, zinc, and vitamin A. Full-time breeding programs are under way for six staple food crops, viz., rice, wheat, maize, cassava, sweet potatoes, and common beans. Pre-breeding feasibility studies are proposed for 11 additional staples: bananas, barley, cowpeas, groundnuts, lentils, millet, pigeon peas, plantains, potatoes, sorghum, and yams. Although plant breeding is the most sustainable solution to the problem, developing new micronutrient-rich plant genotypes is a protracted process, and its effectiveness can be limited by the low amount of readily available pools of micronutrients in soil solution (Cakmak 2008).

Cereal crops are inherently very low in grain Zn and Fe concentrations, and growing them on potentially Zn- and Fe-deficient soils further reduces Fe and Zn concentrations in grain (Cakmak et al. 2010a). Application of Zn- and Fe-containing fertilizers (i.e., agronomic biofortification) is a short-term solution and represents a complementary approach to breeding, which needs to be taken on priority at the global level to overcome these two essential micronutrient deficiencies in the food chain.

17.2 Agronomic Biofortification of Cereal Grains

Essentiality of Fe in plants was reported by Sachs in 1860, while that of Zn was established by Maze in 1916 (Bell and Dell 2008). Zn deficiency was later reported in citrus in the United States (Chapman et al. 1940). A number of reviews on Zn in crop nutrition are available (Hodgson 1963; White 1993; Alloway 2008). In India, Zn deficiency was first reported in rice by Nene (1966) and was followed by that in wheat in Punjab. Research on Zn in relation to crop production in India has been thoroughly reviewed (Katyal and Rattan 2003; Prasad 2006; Shukla et al. 2012). However, most work on Zn fertilization was done from the viewpoint of increasing crop yield. Work on Zn agronomic biofortification of wheat was started in Turkey by Cakmak (2004), while in India it was initiated on rice by the authors of this chapter (Shivay and Prasad 2012; Shivay et al. 2007, 2008a, b, c). Most information on biofortification of cereal grains with Zn is available on rice and wheat and is briefly discussed.

17.2.1 Rice

17.2.1.1 Method of Application

Zn can be applied to soil or foliage. The seed priming with Zn fertilizers and dipping of rice seedlings in Zn fertilizer solutions have been tested and recommended for increased yield, but no information are available on their effect on biofortification of rice grains. Shivay and Prasad (2012) from New Delhi showed that on Zn-deficient soils, application of Zn (as zinc sulfate heptahydrate or ZSHH) significantly increased grain yield of rice as well as Zn concentration in rice grain. Soil application of Zn also increased Zn harvest index by 2 %, although this was not statistically significant.

Shivay et al. (2010a, b) also reported that foliar application of only 1.2 kg Zn ha^{-1} as compared with 5.3 kg Zn ha^{-1} as soil application gave similar grain yield of rice but higher Zn concentration in grain. Agronomic efficiency of Zn with foliar application was about four times of that for soil application and rate of Zn application was much lower when applied on foliage. Dhaliwal et al. (2010) from Ludhiana, India, showed that averaged on five rice cultivars foliar-applied Zn (three sprays of 0.5 % ZSHH solution) recorded a Zn concentration of 47.0 mg kg^{-1} grain in brown rice when compared with 33.8 mg kg^{-1} grain in no Zn check. They also reported a Zn concentration of 29.1 mg kg^{-1} husk in Zn-sprayed crop as compared with 25.2 mg kg^{-1} husk in no Zn check.

In a multilocation study in China, India, Lao PDR, Thailand, and Turkey, Zn concentration in unhusked rice grain was about 69 % higher with foliar application than with soil application; at some centers, it was almost twice that of with soil application (Phattarakul et al. 2012). This study also provided data on relative Zn concentration in unhusked rice (whole grain with husk, known as paddy in India; most of the information on biofortification of rice are on unhusked rice), brown rice (whole caryopsis with husk removed by hand), and white rice (outer layer of pericarpsis including pericarp, testa, mucella, and part of aleurone layer along with embryo removed by polishing for 30 s in a standard laboratory mill). White rice is also known as polished rice (the form in which rice is mostly consumed). When Zn is foliar applied, only 53–54 % of that in unhusked rice is found in polished or white rice as compared with 84.8 %, when Zn is soil applied (Table 17.1). However, when Zn is soil applied, brown rice may contain a little more than in unhusked rice. Thus, a greater portion of foliar-applied Zn remained in husk. These data support the viewpoint of Jiang et al. (2007) that in rice Zn absorbed from the root plays a major role, while mobilization from the leaves plays a minor role. Using the data of Phattarakul et al. (2012) as the base, it worked out that although unhusked rice contained 52.6 % Zn, the polished rice from it is likely to contain only 28.8 % Zn, when Zn was foliar applied in the study of Shivay and Prasad (2012). As a contrast, when Zn was soil applied in sufficient quantity, Zn concentration in unhusked rice was 47.5 %, while it was 40.3 % in polished rice. Total Zn uptake by polished rice was also higher with soil-applied Zn; of course, much more Zn was applied to soil (25 kg ha^{-1}) as compared with that on foliage (1.2 kg ha^{-1}). Biofortification recovery (BRE_{Zn}), the term suggested by Impa and Johnson-Beebout (2012), with foliar application was about eight times of that obtained with soil application. Saenchai et al. (2012) from Thailand reported that the decrease in Zn concentration on milling of rice ranged from 16.2 % to 48.2 % in rice genotypes, being more in long and slender grain types. The range of Zn (mg kg^{-1}) in polished rice

Table 17.1 Grain yield and relative zinc concentration in unhusked, brown, and white (polished) rice (averaged over 9 site years in China, India, Lao PDR, Thailand, and Turkey)

Characteristic	Control (no Zn)	Soil Zn	Foliar Zn	Soil + foliar Zn	Significance
Grain yield (t ha^{-1})	6.7	7.0	6.9	7.0	NS
Zn in unhusked rice (mg kg^{-1})	18.7	19.1	32.3	34.7	$P < 0.01$
Zn in brown rice (mg kg^{-1})	19.1 (102.1)[a]	20.8 (108.9)	24.4 (75.5)	25.5 (73.5)	$P < 0.01$
Zinc in polished rice (mg kg^{-1})	16.1 (18.1)[b] (84.2)[c]	16.2 (84.8) (77.9)	17.7 (54.8) (72.5)	18.4 (53.0) (72.1)	$P < 0.01$

From Phattarakul et al. (2012)
[a]Zn in brown rice expressed as percentage of unhusked rice
[b]Zn in polished rice expressed as percentage of brown rice
[c]Zn in polished rice expressed as percentage of unhusked rice

was 9.6–40.2 (mean 20.6) when compared with 17.3–59.2 (mean 28.7) in brown rice.

It may be pointed out that in eastern India and in some other Asian countries, rice is parboiled before milling. Parboiling is a hydrothermal process to which unhusked rice is subjected before milling. It involves soaking in water, steaming, and drying: the degree of soaking and steaming differs considerably (Singh 1999). As parboiled rice gives much better head recovery, most millers practice parboiling to some degree. On soaking unhusked rice, nutrients from the outer layer of endosperm (pericarp, seed coat, nucellus, and aleurone layer) move into endosperm and the parboiled white rice is much richer in vitamins and minerals and has a better storage quality. Thus, the data on biofortification of rice grains depend very much on the milling process adopted. Recently, Prom-u-Thai et al. (2011) reported that when rice grains were soaked in Fe–EDTA + $ZnSO_4$ solutions, Fe and Zn penetrated across the husk and aleurone layer into endosperm, and when parboiled, the polished rice retained 70–80.5 % of Fe and Zn. Also, Fe and Zn polished rice was highly bio-accessible.

17.2.1.2 Sources of Zinc

Shivay et al. (2008a, c, d) and Shivay and Prasad (2012) from New Delhi reported that ZnSHH (zinc sulfate heptahydrate)-coated urea was significantly superior to ZnO-coated urea in increasing Zn concentration in unhusked rice (also in polished rice when calculated on the basis of Phattarkul's data). The superiority of ZnSHH was also recorded in succeeding wheat (Shivay et al. 2008a, c, d); Zn was applied to rice only. Water solubility of zinc sources is considered as an important criterion for Zn availability (Slaton et al. 2005). Westfall and Gangloff (2001) observed that the effectiveness of six granulated Zn fertilizers decreased as the percent of water-soluble Zn decreased in them and calculated that at least 50 % water-soluble Zn was considered desirable. In the United States, Zn fertilizer manufacturers are producing mixture of zinc sulfate and ZnO, which are known as Znoxysulfates. However, from the manufacturer's viewpoint, ZnO is easier to coat, because it forms a good emulsion with an oil. Kiekens (1995) suggested that ZnO, $Zn(OH)_2$, and $ZnCO_3$ are about 105 times more soluble than soil Zn and these materials could be used as fertilizers.

Naik and Das (2008) compared ZnSHH and Zn–EDTA for rice at Pakyong, Sikkim. ZnSHH was applied at 10 and 20 kg ha^{-1} as basal or in two equal splits (half basal and the rest half at grand tillering stage). Zn–EDTA was applied at 0.5 or 1.0 kg ha^{-1} in single application as basal; 1 kg ha^{-1} was also applied in two equal splits. Zn concentration in rice grain was significantly more (30.3 mg kg^{-1}) with 0.5 kg ha^{-1} Zn–EDTA than with 10 kg ha^{-1} ZnSHH (25.5 mg kg^{-1}). Split application was better than a single application in ZnSHH but not in Zn–EDTA. Zn–EDTA was better than ZnSHH, but more expensive.

17.2.2 Wheat

Soil Zn deficiency in major wheat growing areas leads to inherently low grain Zn concentration and is considered as a major factor in low human Zn intake (Alloway 2009). Compared to the breeding approach, agronomic biofortification (e.g., application of Zn fertilizers) represents a short-term solution to the problem (Cakmak 2008). Soil Zn applications are, however, less effective in increasing grain Zn, while foliar Zn applications result in remarkable increases in grain Zn concentration in wheat (Cakmak et al. 2010a, b). By optimizing the timing and the solute concentration of foliar Zn application, wheat grain Zn concentration could be further increased, not only in whole grains but also in the endosperm (Cakmak et al. 2010b; Kutman et al. 2010; Zhang et al. 2010).

Most Zn fertilization studies have focused on increasing grain yield, though grain Zn concentration is also starting to be addressed (Cakmak 2009). The various methods of Zn application may differentially influence yield and grain Zn concentration. Knowledge of the different forms of Zn fertilizer and timing of foliar Zn application is crucial for enhancing grain Zn. The most effective method for increasing grain Zn is the soil + foliar application method, which may result in an about threefold increase in grain Zn concentration (Cakmak et al. 2010a). When a high concentration of grain Zn is targeted, in addition to a high grain yield, combined soil and foliar application is recommended. Alternatively, using seeds with high Zn concentrations, together with foliar application of Zn, is also an effective way to improve both grain yield and grain Zn concentration. Applying Zn during the grain development stage contributes to increased grain Zn concentration (Zhang et al. 2010) as foliarly applied Zn can be absorbed by the leaf epidermis and then transported to other plant parts via the xylem and phloem (Haslett et al. 2001). McGarth et al. (2012) reported from Rothamsted, UK, that sewage sludge application to soil can increase Zn concentration in wheat grain in noncalcareous soils, but not on a calcareous soil for at least 2–8 years after application, and was similar in effectiveness to zinc carbonate.

The timing of foliar Zn application is an important factor determining its effectiveness in increasing grain Zn concentration; large grain Zn increases are most likely when foliar Zn fertilizers are applied to plants at a late growth stage. Ozturk et al. (2006) studied changes in grain Zn concentration in wheat during the reproductive stage and found that the highest concentration of grain Zn occurs during the milk stage of grain development. Foliar application of Zn during reproductive growth seems to be more effective in increasing grain Zn concentration than spraying of Zn at earlier growth stage. In addition to increasing the concentration of Zn in the whole grain, foliar application also increased the concentration in the starchy endosperm. Late season foliar application of Zn increased the concentration in the starchy endosperm by up to threefold. Since the concentration of phytate in the starchy endosperm (i.e., white flour) of wheat is very low, or even not measurable (Pomeranz 1988), such an increase in Zn implies a positive effect on the use of the grain for human nutrition. The increased Zn in the starchy endosperm resulting from foliar application should also be highly bioavailable due to the low phytate content.

Among the different forms of Zn fertilizer that were tested, the application of Zn as ZnSHH was most effective in increasing grain Zn, compared to other forms of Zn. The HarvestZinc (www.harvestzinc.org) initiative has been investigating different fertilizer strategies and the most efficient Zn application method for promoting Zn uptake and maximizing grain Zn accumulation. Increasing grain Zn by soil and/or foliar applications also provides additional positive impacts in terms of seed vitality and seedling vigor. Priming seeds in Zn-containing solutions is an alternative way to increase seed Zn prior to sowing. High-seed Zn concentrations ensure good root growth and contribute to better protection against soil-borne pathogens (Cakmak 2012). Preliminary studies showed that ZnSHH could be mixed with some wheat herbicides, insecticides, and fungicides without affecting the effectiveness of foliar application for increasing grain Zn concentration. This would increase the possibility that farmers may be willing to apply ZnSHH in their fields by reducing the cost and time of application. Cakmak et al. (2010b) and Cakmak (2012) using LA-ICP-MS (laser ablation inductively coupled plasma mass spectrometry) have reported increase in zinc concentration in endosperm by zinc sprays.

17.2.3 Maize

Much information on Zn and Fe agronomic biofortification in maize through fertilizers is not available in the literature. Small holder farmers in South Africa, Zimbabwe, and other African countries use very little amounts of chemical fertilizers, and the use of Zn fertilizers is a far cry. A study in Zimbabwe showed that the application of cattle manure (supplying 113 g Zn ha^{-1}) + NPK and leaf litter (supplying 430 g Zn ha^{-1}) + NPK significantly increased Zn concentration in corn grain over NPK (Manzeke et al. 2012). Recently, Shivay and Prasad (2014) from Indian Agricultural Research Institute, New Delhi, India, reported that Zn biofortification of corn grain and stover, foliar application of 1 kg Zn sulfate ha^{-1} (in two sprays at tasseling and initiation of flowering), or application of Zn-coated urea is better than soil application of Zn sulfate. However, the highest Zn concentration of 49.2 mg kg^{-1} of corn was recorded with 5 kg Zn to soil + 1 kg Zn as foliar, which was 22.4 % higher than control (no Zn). Welch and Graham (2004) observed that expected increase in Zn and iron from plant breeding is likely to be lesser in corn than in rice and wheat.

17.2.4 Oats

In a recent study, Shivay et al. (2013) reported that coating Zn as ZnO or zinc sulfate onto oat grains at 2 kg per 100 kg (required for sowing 1 ha) gave a Zn concentration of about 32 mg kg^{-1} as compared with about 25 mg kg^{-1} obtained with soil application at the same rate of application. For soil application, zinc sulfate was better than ZnO.

17.2.5 Chickpea

In a latest study, Shivay et al. (2014) reported that application of Zn as soil or foliar through ZnSHH or Zn–EDTA increased Zn concentration in grain and straw of chickpea. In the case of grain, three sprays of ZnSHH recorded significantly more Zn in grain than soil application or one or two sprays. As regards Zn–EDTA in both the years of study, application of three sprays recorded the highest Zn concentration (72.3 %), significantly more than two sprays, which in turn recorded significantly more than a single spray or soil application (Table 17.2). The two sources of Zn differed significantly, when two or three sprays were made; Zn–EDTA recorded significantly higher Zn concentration in grain than ZnSHH in both the years of study. With both the sources of Zn, different methods of application were in the following order: three foliar sprays > two foliar prays > one foliar spray or soil application; three foliar applications recorded the highest Zn concentration in straw. When soil applied or a single foliar application was made, Zn–EDTA recorded significantly more Zn in chickpea straw than ZnSHH straw.

17.3 Conclusions

Agronomic biofortification is the easiest and fastest way for biofortification of cereal grains with Fe, Zn, or other micro-mineral nutrients in developing Asian and African countries, where cereals are the staple food. Agronomic

Table 17.2 Effect of sources, time, and method of Zn application on Zn concentrations in grain and straw of chickpea

Treatment	Zn concentration (mg kg $grain^{-1}$) 2011–12	Zn concentration (mg kg $grain^{-1}$) 2012–13	Zn concentration (mg kg $straw^{-1}$) 2011–12	Zn concentration (mg kg $straw^{-1}$) 2012–13
Control	37.5	36.3	14.8	13.5
NPK	42.6	41.4	18.3	17.1
NPK + ZnSHH soil at 5 kg Zn ha^{-1}	51.9	50.7	22.6	21.3
NPK + ZnSHH, one spray	49.8	48.5	22.8	21.5
NPK + ZnSHH, two sprays	54.7	53.4	27.1	25.8
NPK + ZnSHH, three sprays	58.4	57.1	32.5	31.2
NPK + Zn–EDTA at 2.5 kg Zn ha^{-1}	52.6	51.3	24.6	23.4
NPK + Zn–EDTA, one spray	51.2	50.1	25.1	24.0
NPK + Zn–EDTA, two sprays	58.1	56.7	28.3	27.1
NPK + Zn–EDTA, three sprays	72.3	63.5	33.9	32.6
SE±	1.11	1.12	1.18	0.61
LSD (P = 0.05)	3.31	3.33	3.51	1.81

ZnSHH zinc sulfate heptahydrate

biofortification is the only way to reach the poorest of the poor rural masses, who will never have money to buy mineral supplements nor can afford to improve the components of their diet by incorporating animal products. From the biofortification viewpoint, foliar application is better and requires lesser amount of Fe and Zn fertilizers than their soil application. When cultivars or GM crops with grains denser in Fe and Zn are developed, adequate Fe and Zn fertilization will be necessary. The genetic and agronomic approaches are therefore complementary to each other and should progress in tandem.

References

Alloway BJ (2008) Zinc in soils and crop nutrition, 2nd edn. International Zinc Association, Brussels

Alloway BJ (2009) Soil factors associated with zinc deficiency in crops and humans. Environ Geochem Health 31:537–548

Bell DW, Dell B (2008) Micronutrients for sustainable food, feed, fibre and bio-energy products. International Fertilizer Industry Association, Paris, p 175

Black RE, Lindsay HA, Bhutta ZA, Caulfield LE, De Onnis M, Ezzati M, Mathers C, Rivera J (2008) Maternal and child under nutrition: global and regional exposures and health consequences. Lancet 371:243–260

Bouis HE (2007) The potential of genetically modified food crops to improve human nutrition in developing countries. J Dev Stud 43:79–96

Bouis HE, Hotz C, McClafferty B, Meenakshi JV, Pfeiffer WH (2011) Biofortification: a new tool to reduce micronutrient malnutrition. Food Nutr Bull 32:31S–40S

Cakmak I (2004) Identification and correction of widespread zinc deficiency in Turkey-a success story. The International Fertilizer Society, London

Cakmak I (2008) Enrichment of cereal grains with zinc: agronomic or genetic biofortification? Plant Soil 302:1–17

Cakmak I (2009) Enrichment of fertilizers with zinc: an excellent investment for humanity and crop production in India. J Trace Elem Med Biol 23:281–289

Cakmak I (2012) HarvestPlus zinc fertilizer project: HarvestZinc. Better Crops 96:17–19

Cakmak I, Kalayaci M, Kaya Y, Torun AA, Aydin N, Wang Y, Arisoy Z, Erdem H, Yazici A, Gokmen O, Ozturk L, Horst WJ (2010a) Biofortification and localization of zinc in wheat grain. J Agric Food Chem 58:9092–9102

Cakmak I, Pfeiffer WH, McClafferty B (2010b) Biofortification of durum wheat with zinc and iron. Cereal Chem 87:10–20

Chapman HD, Liebig GF, Vanselow AP (1940) Some nutrient relationships revealed by a study of micronutrient deficiency and excess symptoms in citrus. Soil Sci Soc Am Proc 4:196–200

Dhaliwal SS, Sadana US, Khurana MPS, Dhadli HS, Manchanda JS (2010) Enrichment of rice grains through ferti fortification. Indian J Fert 6(7):28–35

Haslett BS, Reid RJ, Rengel Z (2001) Zinc mobility in wheat: uptake and distribution of zinc applied to leaves or roots. Ann Bot 87:379–386

Hodgson JF (1963) Chemistry of the micronutrient elements in soils. Adv Agron 15:119–159

Impa SM, Johnson-Beebout SE (2012) Mitigating zinc deficiency and achieving high grain Zn in rice through integration of soil chemistry and plant physiology research. Plant Soil 361:3–41

Jiang W, Strik PC, Lingna J, van Keulen H, Ming Z, Stomph TJ (2007) Uptake and distribution of root-applied or foliar applied ^{65}Zn after flowering in aerobic rice. Ann Appl Biol 150:383–391

Katyal JC, Rattan RK (2003) Secondary and micronutrients—research gaps and future needs. Fert News 48(4):9–20

Kiekens L (1995) Zinc. In: Alloway BJ (ed) Heavy metals in soils. Blackie Academic and Professional, London, pp 284–305

Kutman UB, Yildiz B, Ozturk L, Cakmak I (2010) Biofortification of durum wheat with zinc through soil and foliar applications of nitrogen. Cereal Chem 87:1–9

Manzeke GM, Mapfumo P, Mtambanengwe F, Chikovo R, Tendayi TA, Cakmak I (2012) Soil fertility management effects on maize productivity and grain Zn content in small holder farming system of Zimbabwe. Plant Soil 361:57–69

McGarth SP, Chamber BJ, Taylor MJ, Carlton-Smith CH (2012) Biofortification in wheat grain by the application of sewage sludge. Plant Soil 361:97–108

Naik SK, Das DK (2008) Relative performance of chelated Zn and zinc sulphate for lowland rice (Oryza sativa L). Nutr Cycl Agroecosyst 81:219–227

Nene YL (1966) Symptoms, causes and control of khaira disease. Bull Indian Phyto Soc 3:99–101

Ozturk L, Yazici MA, Yucel C, Torun A, Cekic C, Bagci A, Ozkan H, Braun HJ, Sayers Z, Cakmak I (2006) Concentration and localization of zinc during seed development and germination in wheat. Physiol Plant 128:144–152

Peleg Z, Cakmak I, Ozturk L, Yazici A, Jun Y, Budak H, Korol AB, Fahima T, Saranga Y (2009) Quantitative trait loci conferring grain mineral nutrient concentrations in durum wheat – wild emmer wheat RIL population. Theor Appl Genet 119:353–369

Phattarakul N, Rerkasem B, Li LJ, Wu LH, Zou CQ, Ram H, Sohu VS, Kang BS, Surek H, Kalayci M, Yazici A, Zhang FS, Cakmak I (2012) Biofortification of rice grain with zinc through zinc fertilization in different countries. Plant Soil 361:131–141

Pomeranz Y (1988) Wheat chemistry and technology, vol 2. American Association of Cereal Chemists Inc, St. Paul

Prasad R (2006) Zinc in soils and plant, human and animal nutrition. Indian J Fert 2(9):103–119

Prentice AM, Gershwin ME, Schaible UE, Keusch GT, Victoria LG, Gordon JI (2008) New challenges in studying nutrition disease interactions in the developing world. J Clin Investig 118:1322–1329

Prom-u-Thai C, Humg L, Cakmak I, Rerkesam B (2011) Simultaneous fortification of iron and zinc in parboiled rice. Sci Asia 37:296–302

Saenchai C, Prom-u-tahi C, Jamjod S, Bernard D, Rerkasem B (2012) Genotypic variation in milling depression of iron and zinc concentration in rice grain. Plant Soil 361:271–278

Shivay YS, Prasad R (2012) Zinc coated urea improves productivity and quality of basmati rice (*Oryza sativa*) under zinc stress condition. J Plant Nutr 35:928–951

Shivay YS, Prasad R (2014) Effect of source and methods of zinc application on corn productivity, nitrogen and zinc concentrations and uptake by high quality protein corn (*Zea mays*). Egypt J Biol 16(1):72–78

Shivay YS, Kumar D, Ahlawat IPS, Prasad R (2007) Relative efficiency of zinc oxide and zinc sulphate coated urea for rice. Indian J Fert 3(2):51–56

Shivay YS, Kumar D, Prasad R (2008a) Effect of zinc enriched urea on productivity, zinc uptake and efficiency of an aromatic rice-wheat cropping system. Nutri Cycl Agroecosyst 81:229–243

Shivay YS, Kumar D, Prasad R (2008b) Relative efficiency of zinc sulphate and zinc oxide coated urea in rice-wheat cropping system. Commun Soil Sci Plant Anal 39:1154–1167

Shivay YS, Kumar D, Prasad R, Ahlawat IPS (2008c) Relative yield and zinc uptake by rice from zinc sulphate and zinc oxide coatings on to urea. Nutr Cycl Agroecosyst 81:181–188

Shivay YS, Prasad R, Rahal A (2008d) Relative efficiency of zinc oxide and zinc sulphate enriched urea for spring wheat. Nutr Cycl Agroecosyst 82:259–264

Shivay YS, Prasad R, Rahal A (2010a) Genotypic variation for productivity zinc utilization efficiencies, and kernel quality in aromatic rices under low available zinc conditions. J Plant Nutr 33:1835–1848

Shivay YS, Prasad R, Rahal A (2010b) Studies on some nutritional quality parameters of organically or conventionally grown wheat. Cereal Res Commun 38:342–352

Shivay YS, Prasad R, Pal M (2013) Zinc fortification of oats grain and straw through zinc fertilization. Agric Res 2:375–381

Shivay YS, Prasad R, Pal M (2014) Effect of source, level and method of zinc application on yield, zinc biofortification of grain, zinc uptake and zinc use efficiency in (*Cicer arietinum* L.). Commun Soil Sci Plant Anal (15 Sept 2014 published online). doi: 10.1080/00103624.2014.941472

Shukla AK, Behera SK, Shivay YS, Singh P, Singh AK (2012) Micronutrients and field crop production in India: a review. Indian J Agron 57:S123–S130

Singh A (1999) Storage and processing of rice. In: Prasad R (ed) A text book of rice agronomy. Jain Brothers, New Delhi, pp 203–210

Slaton NA, Norman RJ, Wilson CE Jr (2005) Effect of Zn source and application time on Zn uptake and grain yield of flood irrigated rice. Agron J 92:272–278

Stewart CP, Dewey KG, Ashoran P (2010) The under nutrition epidemic: an urgent health priority. Lancet 375:282

Welch RM, Graham RD (2004) Breeding for micronutrients in staple food crops from a human nutrition perspective. J Exp Bot 55:353–364

Westfall D, Gangloff B (2001) Zinc fertilizers-is there a difference. Agronomy News 21(6):8

White CL (1993) The chemistry of zinc. In: Robson AD (ed) Zinc in soils and plants. Kluwer, Dordrecht

White PJ, Broadley MR (2009) Biofortification of crops with seven mineral elements often lacking in human diets -iron, zinc, copper, calcium, magnesium, selenium and iodine. New Phytol 182:49–84

WHO (2012) The world health report. World Health Organization, Geneva

Zhang YQ, Shi RL, Karim MR, Zhang FS, Zou CQ (2010) Iron and zinc concentrations in grain and flour of winter wheat as affected by foliar application. J Agric Food Chem 58:12268–12274

18 Soil Test Crop Response: Concepts and Components for Nutrient Use Efficiency Enhancement

S.R. Singh

Abstract

The economy of the country is based on agriculture where 70 % of the total population is in rural areas dependent in agriculture. Fertilizers are one of the costly inputs but continue to exert significant contribution to produce additional food grains for the ever-increasing population. The promotion of balanced chemical fertilizers has become a necessity to meet the ever-increasing demands of food grains, pulses, oil crops, sugarcane production, fruits, vegetables, fibres, forage and grasses. It has been proved that imbalance use of fertilizer not only causes deterioration in soil quality but also affected nutrient use efficiency. Soil forms the basis for any crop production activity and is the most precious natural resources. Declining soil fertility is one of the important factors that directly affect the productivity. Blanket crop production technologies including fertilizer recommendations have accelerated the situation over three to four decades. Therefore, soil fertility management is crucial to ensure productivity and nutritional security while maintaining soil health and sustainability. To get a maximum benefit, enhanced nutrient use efficiency and reduced nutrient losses from fertilizers, they must be applied in the right quantity, sources and combination at the right time using the right methods. The package of practices recommended should cater to the need of soil variability. The response to fertilizers is greatly influenced by the soil type and spatial soil variability that has resulted from complex geological and paedological processes. Spatial variation of soil properties decreases the use efficiency of fertilizers applied uniformly at the field scale. Therefore, application of variable rates of fertilizers has been proposed.

S.R. Singh (✉)
IACR-Indian Institute of Sugarcane Research, Lucknow, Uttar Pradesh 226 002, India
e-mail: shivramsingh22@gmail.com

U. Singh et al. (eds.), *Biofortification of Food Crops*, DOI 10.1007/978-81-322-2716-8_18

Keywords
GRD • Integrated plant nutrient supply system (IPNSS) • Nutrient use efficiency • STCR

18.1 Introduction

The package of practices for crop production are uniformly adopted for all the crops in India. However, crop response to fertilizers is greatly influenced by soil type and spatial soil variability. Spatial variation in soil properties decreases the input use efficiency. Therefore, application of variables rather than uniform rates of fertilizers has been proposed to avoid application of excess fertilizers where it will not be properly utilized by crops.

18.2 Approaches of Fertilizer Recommendations

18.2.1 General Fertilizer Recommendation (GRD)

The general fertilizer recommendations are based on multi-locational trials conducted with different doses of N, P and K fertilizers and their economic evaluation obtained at an optimum dose for a particular crop. These recommendations are suitable for medium soil fertility condition irrespective of wide variation that occurs in soil fertility status which is not taken into consideration, and hence, under high or low soil fertility conditions, the applied nutrients prove often a wasteful expenditure and insufficient, respectively. In both the cases, optimum fertilizer use efficiency cannot be achieved. The package of general fertilizer recommendations (GRD) of different crops are given in Table 18.1.

18.2.2 Fertilizer Recommendations Based on Soil Fertility Rating

In this approach, medium soil fertility is equated with general recommended doses. However, under very low to low and high to very high fertility categories, the fertilizer doses are increased or decreased by 25–50 % of the GRD as per the fertility gradient. At present, most of fertilizer recommendations issued by soil testing laboratories are based on this approach. Similarly, soil fertility rating and soil rating were developed on the basis of actual soil test values in All India Coordinated Project on Soil Test Crop Response Correlation study and are passed on to the State Soil Testing Laboratory, which are under administrative control of the state for fertilizer recommendations of different crops (Table 18.2).

18.2.3 Fertilizer Recommendations Based on Critical Limit of Soil Available Nutrients

Table 18.1 Package of fertilizer recommendations for different crops

S. no.	Crops	Fertilizer dose (kg/ha)		
		N	P_2O_5	K_2O
i.	Lentil	20	40	40
ii.	Wheat	120	60	60
iii.	Rice	120	60	60
iv.	Pearl millet	80	60	40
v.	Chickpea	20	50	20
vi.	Pigeon pea	20	50	20
vii.	Maize	120	60	50

Table 18.2 Soil test rating based on soil fertility evaluation experiment

Parameters	Rating of soil		
	Low	Medium	High
Organic C (%)	<0.5	0.5–0.75	>0.75
Available N (kg/ha)	<280	280–560	>560
Available P (kg/ha)	<10	10.0–24.6	>24.6
Available K (kg/ha)	<108	108–280	>280
Recommendation	125 % of normal dose	100 % of normal dose	75 % of normal dose

This concept was developed by Cate and Nelson (1971) on the basis of graphical method for dividing the per cent yield versus soil test levels. Critical limit is the level of soil available nutrient above which that nutrient is not considered as a primary limiting factor. The probability of getting economic response to fertilizer application in soils having available nutrient above the critical limit is quite low, while in soils below the critical limit, the probability of getting economic response is quite high. Critical limit varies, depending on the soil types, crops and varieties, soil test methods used and seasonal variations. This concept separates the soils in responsive and non-responsive groups but it does not suggest quantification of fertilizer dose for individual situations in responsive groups. Hence, this concept may be more useful for fertilizer recommendation of micronutrients.

18.2.4 Fertilizer Recommendation Based on Mitscherlich-Bray's Concept for Maximum Yield

In this concept, an empirical relationship is developed between per cent yield, soil and fertilizer nutrient so that the fertilizer can be recommended for getting the percentage of maximum yield. Here it follows Mitscherlich-Bray's (Mitscherlich 1924; Bray 1958) equation which is:

$$\text{Log}\ (A - Y) = \text{Log}\ A - C_1 b - CX$$

where A is the theoretically calculated maximum yield, Y is the percentage of maximum yield, b is the soil test value, C_1 and C are the efficiency factors for soil and fertilizer nutrient, respectively, and X is the fertilizer dose. The limitation of this concept is that firstly the theoretical maximum yield is quite high than the actual obtained. Secondly, the calculated maximum yield is different for different nutrients and hence it becomes difficult to predict ultimate achievable yield.

18.2.5 Multiple Regression Based on Fertilizer Recommendation for Economic Yield

The fertilizer dose that gives higher profit as well as optimum economic return in rupees invested on fertilizers is called optimum dose for economic yield. Ramamoorthy et al. (1974) established a significant relationship between soil test, added fertilizers and crop yields by fitting a multiple regression of the quadratic form using linear, quadratic and interaction terms between soil and fertilizer form of nutrient as listed below:

$$\begin{aligned} Y = {} & A + b_1 SN + b_2 SP + b_3 SK \\ & + b_4 FN + b_5 FN^2 + b_6 FP \\ & + b_7 + FP^2 + b_8 F + b_9 FK^2 \\ & + b_{10} SNFN + b_{11} SPFP \\ & + b_{12} SKFK \end{aligned}$$

where Y is crop yield (kg/ha); 'A' is intercept (kg/ha); b_{1-12} are regression coefficients; SN, SP and SK are soil available N, P and K (kg/ha); and FN, FP and FK are fertilizer dose of nitrogen, phosphate and potash (kg/ha), respectively.

If this response to applied nutrient is curvilinear type following the law of diminishing return which is indicated by +, and – sign of linear, quadratic and interaction terms of the fertilizer nutrient, then fertilizer dose for economic yield can be calculated as:

$$\text{Fertilizer dose (kg/ha)} = \frac{\text{Coefficient of linear term}}{2 \times \text{Coefficient of quadratic term}} - \frac{\text{Coefficient of interaction term}}{2 \times \text{Coefficient of quadratic term}} \times \text{STV} - \frac{1}{2 \times \text{Coefficient of quadratic term}} \times \text{R}$$

For example: In case of nitrogen fertilizer,

$$\text{Fertilizer dose for N (kg/ha)} = \frac{b_4}{2 \times b_5} - \frac{b_{10}}{2 \times b_5} \times \text{STV} - \frac{1}{2 \times b_5} \times \text{R}$$

where STV is soil test value (kg/ha) and R represents ratio of the cost of nutrient (1 kg) to the cost of grain/economic produce (1 kg).

Fertilizer recommendation for economic yield includes the factor of modifying the fertilizer dose with changes in soil test value as well as changes in the cost ratio of fertilizers and produce. Allowance of this ratio makes soil test in recommendations more dynamic and responsive so as to ensure higher profitability from fertilizer investment. The main limitation of this approach is uncertainty of getting curvilinear response type for all the three nutrients N, P and K from the field experiment.

18.2.6 Fertilizer Recommendations Based on Targeted Yield Approach

Troug (1960) illustrated the possibility of prescription method of fertilizer used for attaining high yields of maize using empirical values of nutrient availability from soil and fertilizer. However, Ramamoorthy et al. (1974) established the concept on the basis of theoretical as well as field experimental proof after verification of the principles of fertilizer application for targeted yield of field crops during 1965–1967. Liebig's law of minimum (of plant nutrition) states that growth of plant is limited by the plant nutrient element present in the smallest amount, all others being in adequate quantities. From this, it follows that a given amount of a soil nutrient is sufficient for some yield with a given percentage nutrient composition. This forms the basis for fertilizer application for targeted yields.

This concept strikes a balance between fertilizing the crop and fertilizing the soil which provide the basis for balanced fertilization and balance between applied nutrients and soil available nutrients. It assume that there is a linear relationship established between grain yield and nutrient uptake by the crop for obtaining particular yield, a definite amount of nutrient is taken up by the crop. This is also borne out by the near constancy when the response is expressed in the form of unit grain production per unit of nutrient absorbed by the plant which is expressed as nutrient requirement in kg per quintal of grain production. Once nutrient requirement is established for a given yield, the fertilizer needed can be estimated taking into account the efficiency of contribution from the soil available nutrients. Farmers can choose a particular yield target depending upon the resources available with him and rates of fertilizer application on the basis of soil test to obtain economically profitable yields.

When resources of the farmers are limited or in case of unavailability of fertilizers, planning for low yield target is done so as to cover more area with available resources ensuring increased total production. Fertilizer nutrient efficiency and the total production are higher when fertilizer is applied for low yield targets. When fertilizers are applied for achieving lower targets, the excess of soil nutrients over the minimum yield possible by the limiting nutrient is advantageously exploited.

If soil fertility is maintained, heavier doses of fertilizers have to be used over the actual requirment that removed by the crops so that enough residues are left in the soil even after normal losses of nutrients. But this will not make the economic return from the investment of fertilizer. To get the maximum return from fertilizer investment, the turnover from it must be very quick as fertilizers are applied for immense contribution towards increased yield over seasons. However, exhausting the unutilized excess nutrients from the soil resulting in faster depletion of soil fertility. Although these two approaches seem to be pulling in different directions, fertilizer application and the fixed yield targets can be maintained so that both high profit from fertilizer investment and maintenance of soil fertility can be achieved. Therefore, it provides the fertilizer dose in balanced and quantitative terms in relation to soil test values and crop requirement which is necessary to optimize the response to added fertilizers, maximize the profit and achieve the desired yield target with ±10 % deviation.

18.2.6.1 Developing a Targeted Yield Equation

The essential basic data required for soil test crop response correlation from the field experiments under a given soil type, crop and agroclimatic conditions include the following:

1. Nutrient requirement in kg per 100 kg of economic produce (grain or other economic parts)
2. The per cent nutrient contribution from soil available nutrient to total uptake
3. The per cent nutrient contribution from the applied nutrient to the total uptake
4. The per cent contribution of nutrient from the organic sources to the total uptake under IPNSS

The linear relationship between yield and nutrient uptake implies obtaining a given yield. A definite quantity of nutrients (both from soil and fertilizers) must be taken up by the plant. It is borne out by the near constancy when the response is expressed in the form of units of grain production per unit of nutrient absorbed by the plant. Once the nutrient requirement is known for a given yield, quantity of fertilizer can be estimated taking into account the efficiency of contribution from the soil available nutrients and that from the fertilizer nutrients. Field experimental data can provide a range in soil test values, nutrient and yield levels which enable calculation of these essential basic parameters.

18.3 Calculation of Basic Parameters for Fertilizer Recommendations

So as to make efficient recommendation of fertilizers, arithmetic analysis is important and plays a crucial role in plant nutrition management strategies. Some of the important nutrient use efficiency indices have been given as follows.

18.3.1 Nutrient Requirement (NR)

(a) $\text{Kg N required/kg grain production} = \dfrac{\text{N uptake by grain and straw (kg/ha)}}{\text{Grain yield (kg/ha)}}$

(b) $\text{Kg P required/ kg grain production} = \dfrac{\text{P uptake by grain and straw (kg/ha)}}{\text{Grain yield (kg/ha)}}$

(c) $\text{Kg K required/kg grain production} = \dfrac{\text{K uptake by grain and straw (kg/ha)}}{\text{Grain yield (kg/ha)}}$

18.3.2 Per cent Nutrient Contribution from Soil to Total Nutrient Uptake (CS)

(a) Per cent contribution of N from soil $= \dfrac{\text{Uptake of N (kg/ha) by biomass from control plot}}{\text{Soil test value for available N (kg/ha) from control plot}}$

(b) Per cent contribution of Pfrom soil $= \dfrac{\text{Uptake of P (kg/ha) by biomass from control plot}}{\text{Soil test value for available P (kg/ha) from control plot}}$

(c) Per cent contribution of K from soil $= \dfrac{\text{Uptake of K (kg/ha) by biomass from control plot}}{\text{Soil test value for available K (kg/ha) from control plot}}$

18.3.3 Per cent Nutrient Contribution from Fertilizer to Total Uptake (CF)

(a) Per cent contribution of N from soil

$$= \frac{\text{[N uptake (kg/ha)]} - [\text{Soil test value for } \textit{available N} \times \text{Per cent contribution]} \atop \text{by biomass available N (kg/ha) of N from soil/100}}{\text{Fertilizer N applied in kg/ha}} \text{x100}$$

(b) Per cent contribution of P from soil∗

$$= \frac{\text{[Puptake (kg/ha)]} - [\text{Soil test value for } \textit{available P} \times \text{Per cent contribution]} \atop \text{by biomass available P (kg/ha) of P from soil/100}}{\text{Fertilizer P applied (kg/ha)}} \text{x100}$$

(c) Per cent contribution of K from soil∗

$$= \frac{\text{[Kuptake (kg/ha)]} - [\text{Soil test value for } \textit{available K} \times \text{Per cent contribution]} \atop \text{by biomass available K (kg/ha) of K from soil/100}}{\text{Fertilizer K applied in kg/ha}} \text{x100}$$

Note: Conversion factor may be used, if required. P × 2.29 = P_2O_5; K × 1.21 = K_2O

18.3.4 Formulae Used to Develop Targeted Yield Equations

18.3.4.1 Fertilizer Adjustment Equations/Targeted Yield Equations (Without IPNSS)

$$\text{FN (kg/ha)} = \frac{\text{NR}}{\text{CF}/100}\text{T (q/ha)} - \frac{\text{CS\%}}{\text{CF\%}} \times \text{SN (Alkaline KMnO}_4 - \text{N)}$$

$$\text{F K}_2\text{O (kg/ha)} = \frac{\text{NR}}{\text{CF}/100}\text{T (q/ha)} - \frac{\text{CS\%}}{\text{CF\%}} \times \text{SK (NH}_4\text{OAc.} - \text{K)} \times 1.21$$

18.3.4.2 Fertilizer Adjustment Equations/Targeted Yield Equations (with IPNSS)

(a) $$\text{FN(kg/ha)} = \frac{\text{NR}}{\text{CF}/100}\text{T (q/ha)} - \frac{\text{CS\%}}{\text{CF\%}} \times \text{SN(Alkaline KMnO}_4 - \text{N)} - \frac{\text{CO\%}}{\text{CF\%}} \times \text{ON}$$

(b) $$\text{FP}_2\text{O}_5\text{(kg/ha)} = \frac{\text{NR}}{\text{CF}/100}\text{T (q/ha)} - \frac{\text{CS\%}}{\text{CF\%}} \times \text{SP(Olsens} - \text{P)} \times 2.29 - \frac{\text{CO\%}}{\text{CF\%}} \times \text{OP}$$

(c) $$\text{FK}_2\text{O (kg/ha)} = \frac{\text{NR}}{\text{CF}/100}\text{T (q/ha)} - \frac{\text{CS\%}}{\text{CF\%}} \times \text{SK(NH}_4\text{OAc.} - \text{K)} \times 1.21 - \frac{\text{CO\%}}{\text{CF\%}} \times \text{OK}$$

where:

FN, FP_2O_5 and FK_2O are fertilizer N, P_2O_5 and K_2O dose (kg/ha).

SN, SP and SK are available soil test values (kg/ha).

NR = Nutrient requirement in kg/q of fibre/grain production.

CS% = Per cent contribution from soil available nutrient.

CF% = Per cent contribution from farmyard manure.

CO% = Per cent contribution of nutrients from FYM or any other organic resources.

T = Yield target (q/ha).

18.4 Fertilizer Recommendation Based on Soil Test and Targeted Yield Equation

The following procedure should be followed while calculating fertilizer doses for a specific yield target of a particular crop. An example is cited hereunder (Table 18.3):

- Soil order: Alluvial
- Crop: Lentil
- Yield target fixed: 15 q/ha

Soil test values:

- Alkaline $KMnO_4$-N = 250 kg/ha
- Olsen's P = 20 kg/ha
- Ammonium acetate –K = 160 kg/ha
- FYM (dry weight) contains N = 0.5 %, P = 0.30 % and K = 0.7 %

The All India Coordinated Research Project (AICRP) on Soil Test Crop Response (STCR), CRIJAF, Barrackpore unit has generated the basic data of nutrient requirements (NR) in kg/q of grain production, per cent contribution of nutrients from soil (CS%), fertilizer (CF%) and organic sources by conducting field experiments and developed targeted yield equations and ready reckoner for fertilizer recommendations of lentil (Table 18.4). Addition of 5 ton FYM/ha which contains 0.5 % N, 0.3 % P and 0.7 % K contributes about 25 kg N, 15 kg P and 35 kg K/ha. Hence, for achieving 15 q/ha fixed yield target of lentil, we have to apply 32, 44 and

Table 18.3 Fertilizer adjustment equations with and without IPNS

Fertilizer adjustment equations	(Without IPNSS) FN = 4.41 T – 0.13 SN FP = 3.46 T – 0.29 SP FK = 6.23 T– 0.23 SK	(With IPNSS) FN = 4.41 T – 0.13 SN – 0.07 ON FP = 3.46 T – 0.29 SP – 0.08 OP FK = 6.23 T– 0.33 SK – 0.18 OK
FN (kg/ha)	=4.41 × 15 – 0.13 × 250 =66.2– 32.5 =33.7 kg N/ha =33.7 × 2.17 = 73.2 kg urea/ha	=4.41 × 15 – 0.13 × 250 – 0. 07 × 30 =66.2– 32.5 – 2.10 =31.6 kg N/ha =31.6 × 2.17 = 69 kg urea/ha
F P_2O_5 (kg/ha)	=3.46 × 15– 0.27 × 20 =51.9 – 5.80 =46.1 kg P_2O_5/ha =46.1 × 6.25 = 288 kg SSP/ha	=3.46 × 15– 0.27 × 20 – 0. 08 × 15 =51.9 – 5.80 – 1.2 =43.7 kg P_2O_5/ha =43.7 × 6.25 = 273 kg SSP/ha
F K_2O (kg/ha)	=6.26 × 15– 0.33 × 150 =94.75–50 =45.0 kg K_2O/ha =45 × 1.67 = 75 kg MOP/ha	=6.26 × 15– 0.33 × 160 – 0.18 × 35 =94.75–52.8 – 6.30 =45.0 kg K_2O/ha =36 × 1.67 = 60 kg MOP/ha

Table 18.4 IPNSS targeted yield equations and ready reckoner of lentil (var. B 256)

Basic data					
Nutrients	NR (kg/q)	CS (%)	FS (%)	CFYM (%)	Targeted yield equations
N	6.61	19.47	149.7	10.51	FN (kg/ha) = 4.41 T – 0.13 SN – 0.07 ON
P	0.76	6.61	22.01	1.70	FP_2O_5 (kg/ha) = 3.46 T – 0.29 SP– 0.08 OP
K	3.89	20.61	62.47	11.22	FK_2O (kg/ha) = 6.23 T – 0.33 SK – 0.18 OK

Soil test values (kg/ha)			Fertilizer prescription (kg/ha)					
			15 q/ha target			20 q/ha target		
$KMnO_4$- N	Olsen's- P	NH_4OAc- K	N	P_2O_5	K_2O	N	P_2O_5	K_2O
200	10	100	43	47	48	65	64	79
225	15	120	40	45	41	62	63	73
250	20	140	37	44	35	59	61	66
275	25	160	33	42	28	56	60	59
300	30	180	30	41	22	52	58	53
325	35	200	27	40	15	49	57	46
350	40	220	24	38	8	46	55	40

36 kg N, P_2O_5 and K_2O, respectively. The requirment of N, P_2O_5 and K_2O as per soil test values and targeted yield equations, it should be converted to their fertilizer equivalents. For instance, the farmer has urea, single super phosphate (SSP) and muriate of potash (MOP) as fertilizer source for supplying N, P_2O_5 and K_2O, respectively. As urea contains 46 % N, single super phosphate contains 16 % P_2O_5, and muriate of potash contains 60 % K_2O, then 69, 273 and 60 kg urea, SSP and MOP, respectively, will be required to attain 15 q/ha targeted yield of lentil (Table 18.4).

18.5 Effect of ST-TY on Target Yield and Nutrient Use Efficiency

The relevance and value of soil testing increases by choosing the yield target at such a level so that the cost of fertilizer requirement becomes more or less the same as what was being practised by the farmer already. The results of such demonstration trials conducted at different locations in Nadia district of West Bengal (Tables 18.5 and 18.6) revealed that the response per unit fertilizer is higher than that

Table 18.5 Effect of ST-TY based on fertilizer application on yield and NUE of rice

Treatment	Fertilizer dose as per ST-TY (kg/ha)			Grain yield (t/ha)	Straw yield (t/ha)	Nutrient use efficiency (kg/kg)		
	N	P	K			N	P	K
Control	0	0	0	3.04	44.48	–	–	–
FYM	0	0.0	0.0	3.59	50.29	0.0	0.0	0.0
RDF	59	13.1	25.0	3.88	56.32	14.2	28.0	28.0
ST-TY-I	80	17.5	33.3	4.45	58.32	17.7	35.3	35.3
ST-TY-I + *Azot* + PSB	36	12.7	38.3	3.93	54.51	24.8	30.8	19.4
ST-TY-I + FYM	29	8.3	33.3	4.18	54.99	39.4	60.2	28.6
ST-TY-I + *Azot.* + PSB	36	10.0	40.0	4.49	56.80	40.3	63.1	30.2
ST-TY-II	87	14.4	49.2	4.59	58.35	17.8	46.9	26.2
CD ($P = 0.05$)				0.36	6.80			

Table 18.6 Effect of ST-TY based on fertilizer application on yield and NUE of lentil

Treatment	Fertilizer dose as per ST-TY (kg/ha)			Grain yield (t/ha)	Straw yield (t/ha)	Nutrient use efficiency (kg/kg)		
	N	P	K			N	P	K
Control	–	–	–	0.98	1.38	–	–	–
FYM	0	0.0	0.0	1.53	1.97	0.0	0.0	0.00
RDF	15	13.1	16.7	1.21	1.88	15.5	7.7	11.60
ST-TY-I	30	17.5	16.7	1.24	1.97	8.8	6.6	13.15
ST-TY-I + *Rhhiz.* + PSB	47	19.2	89.2	1.48	1.92	10.5	11.3	4.64
ST-TY-I + FYM	43	17.0	80.8	1.56	2.00	13.6	14.9	6.02
ST-TY-I + *Rhiz.* + PSB	47	20.5	86.7	1.81	2.51	17.5	17.7	7.96
ST-TY-II	73	29.7	127.5	1.65	2.20	9.1	9.8	4.37
CD ($P = 0.05$)				0.16	0.23			

with other practices when balanced fertilization is adopted for targeted yield. The targeted yields of rice and lentil were achieved within ±10 % yield deviations which validate the superiority of targeted yield equations. The lower targeted yields of rice and lentil were achieved with very narrow range of yield deviation as compared to higher targeted yield. Application of fertilizers based on soil test and targeted yield equations apparently enhanced nutrient use efficiency of N, P and K in most of the cases as compared to the application of fertilizers as per recommended dose of fertilizer (RDF) and farmers practice (FP). However, integrated application of fertilizer as per ST-TY with FYM and bio-inoculants has recorded highest improvement in terms of N, P and K use efficiency under most of the experimentations and field demonstrations.

18.6 Utility of ST-TY Approach and Conclusion

Sometimes unwarranted issues arise in practical implementation of this approach, viz.

- Lack of qualified manpower
- Inadequate linkage with SAUs/institutes
- Lack of improved and refined soil test methods and latest equipments
- Improper utilization of soil testing services by the farmers
- Soil sample collection and delayed test reports
- Cost of soil sample testing and prescription delivery

The above issues may be tackled through proper/effective implementation of policies by the government agencies. Keeping these facts aside, ST-TY approach is better over traditional/blanket recommendations so as to enhance nutrient use efficiency. Some of the advantages are listed below:

1. It ensures the achievement of desired yield target within ±10 % deviation under optimum management conditions.
2. Efficient use of fertilizers according to soil fertility and crop requirement ensures high profit and response to applied fertilizers.
3. It ensures maintenance of soil fertility at appropriate levels in cropping system for sustainable crop production.
4. It offers wide choice of fixing appropriate yield target according to the availability of resources and soil fertility.
5. Suitable crop rotations can be adopted from the point of view of relative ability of crops and crop varieties to utilize soil and fertilizer nutrient.

References

Bray RH (1958) The correlation of phosphorus soil test with response of wheat through a modified Mitscherlich equation. Soil Sci Soc Am Proc 22(4):314–317. doi:10.2136/sssaj1958.03615995002200040013x

Cate RB, Nelson LA (1971) A simple statistical procedure for partitioning soil test correlation data into two classes. Soil Sci Soc Am J 35(4):658–660. doi:10.2136/sssaj1971.03615995003500040048x

Mitscherlich EA (1924) The determination of the fertilizer requirement of the soil. Soil Sci 20:361–364

Ramamoorthy B, Velayutham M, Mahajan VK (1974) Recent trends in making fertilizer recommendations based on soil test under fertilizer response constraints in India. In: Proc FAI-FAO national seminar optimizing agricultural production under limited availability of fertilizers. New Delhi, pp 335–346

Troug E (1960) Fifty years of soil testing. Trans. 7th international congress of soil science. Vol. III. Commission IV, Paper no 7 pp 46–53

Part IV

Biofortification Through Soil-Plant Interactions

19 Biofortification with Microorganisms: Present Status and Future Challenges

Radha Prasanna, Lata Nain, Anuj Rana, and Yashbir Singh Shivay

Abstract

Biofortification is the process of adding essential micronutrients and other health-promoting compounds to crops or foods to improve their nutritional value. This is imperative as the diets of over two-thirds of the world's population lack one or more essential mineral elements and the three staple crops, rice, maize, and wheat, which provide nearly half of the calories consumed by humans, are deficient in micronutrients. A large body of information exists on augmentation through breeding approaches, both conventional and molecular, or through agronomic management practices. Other options include dietary diversification, mineral supplementation, food fortification, or increase in the concentrations and/or bioavailability of mineral elements in the produce. With the advent of metagenomic and next-generation sequencing tools and the development of the "holobiome" concept, the significance of microbiome in the productivity of soil and crops is becoming more evident. Plant growth-promoting rhizobacteria (PGPR) represent a wide variety of microorganisms, growing in association with plants. They lead to stimulation of growth of the host, due to the increased mobility, uptake, and enrichment of nutrients in the plant. Their significance in improving nutrient use efficiency of applied fertilizers and improving nutrient uptake in problematic soils or denuded lands is well established. However, they are less explored options in biofortification strategies and need to be included in agronomic and breeding approaches to develop effective biofortification strategies for the staple crops.

R. Prasanna (✉) • L. Nain • A. Rana
Division of Microbiology, ICAR-Indian Agricultural Research Institute, New Delhi 110 012, India
e-mail: radhapr@gmail.com; latarajat@yahoo.co.in; anuj23rana@gmail.com

Y.S. Shivay
Division of Agronomy, ICAR-Indian Agricultural Research Institute, New Delhi 110 012, India
e-mail: ysshivay@hotmail.com

U. Singh et al. (eds.), *Biofortification of Food Crops*, DOI 10.1007/978-81-322-2716-8_19

Keywords
Cyanobacteria • Nutrient mobilization • PGP bacteria • Phytosiderophores

19.1 Introduction

The inadequate dietary intake of essential micronutrients, especially the "big four," iron (Fe), zinc (Zn), vitamin A, and iodine, is a serious global problem, predominant in developing countries and affecting more than two million people (WHO 2011). Among these, deficiencies of Zn and Fe contribute to the most number of widespread nutritional disorders, as these two elements serve a multitude of biological functions (White and Bradley 2009). The application of specific fertilizers has neither been adequate nor effective in supplying the nutrient to the plant, because they are notoriously inefficient and tend to form complexes in the soil. Interventions in the past mainly focused on supplementation, food fortification, or dietary diversification, which had limited success. Biofortification is a strategy that aims to increase the content of bioavailable micronutrients in staple food crops such as rice, wheat, maize, pearl millet, and others, which can be attempted genetically or through agronomic or soil management practices. Traditional and molecular breeding are the major approaches being utilized for achieving this goal (Pfeiffer and McClafferty 2007; Ortiz-Monasterio et al. 2007, 2011; Velu et al. 2014). Maize hybrids and varieties that yield grain with 25–30 % more Fe and Zn than common cultivars have been developed as part of the Harvest Plus program (Banziger and Long 2000). These strategies consider achieving this goal by improving the uptake of minerals by plants from soil, followed by enhancing their movement and bioavailability in the edible parts of the plants (Rengel et al. 1999; Waters and Sankaran 2011).

Plants are known to be selective with their bacterial associations and recruit those that are beneficial for their growth. Most of these interactions influence growth and development of plants, by altering nutrient uptake dynamics and susceptibility to pathogens (Dobbelaere et al. 2001; Gopal et al. 2013). The use of microorganisms possessing multifarious nutrient sequestering and plant growth-promoting traits, which can load high levels of minerals in plant roots and translocate to the edible parts, can be another promising biofortification strategy (Rana et al. 2012a, b; Sharma et al. 2013; Adak et al. 2015). However, there is limited published literature available on this aspect.

19.2 Why Microbe Mediated Biofortification?

Cereals represent the staple food for a majority of the world's population, besides maize and wheat; however, they are all inherently poor in micronutrients. It is well recognized that humans require at least 22 mineral elements, which can be supplied through an appropriate diet (Graham et al. 2007), while plants require 14 mineral elements to complete their life cycle. These essential micronutrients act as cofactors or metabolic precursors and are required for specific biological processes, and their insufficient intake results in characteristic deficiency diseases. The major deficiency diseases in developing countries correspond to essential nutrients that tend to be present at low levels in milled cereal grains, for example, vitamin A, iron, iodine, zinc, vitamin C, and folic acid. Besides their requirement in specific metabolic processes, certain essential nutrients also act as antioxidants or promote the activity or availability of antioxidants. This will which help to prevent diseases that result from or that are exacerbated by the accumulation of oxidative damage to cells, including cancer, cardiovascular disease, and neurodegenerative disorders (Welch and Graham 2004). The biofortification of foods such as flour, bread, packaged cereals, dairy products, and salt is a public health policy in the industrialized world, aiming to reduce the number of people suffering from malnutrition and to increase general health in the population.

Large quantities of N fertilizers are commonly used to obtain high yields in many crops. But these chemical fertilizers compromise on human and animal health and pollute the environment. Microbes represent a promising input, which possess an array of mechanisms to sequester macro- and macronutrients from soil or water, which can be also made available to plants (Nain et al. 2010; Mader et al. 2011; Prasanna et al. 2012). Therefore, the application of microbial biofertilizers needs to be advocated as a possible strategy not only to increase nutrient concentrations in edible crops but also to improve yields on infertile or nutrient-poor soils.

In India, micronutrient deficiencies are more pronounced in the Indo-Gangetic Plains (IGP), due to the spread of intensive and chemical fertilizer-driven farming practices in this region (Joshi et al. 2010; Shukla et al. 2005; Singh 2009). Approximately 50 % of the global population is thought to be malnourished, but the vast majority of malnourished people are the rural poor in developing countries, subsisting on a diet of milled cereal grains lacking many essential nutrients and other health-promoting compounds (Farre et al. 2011). The biofortification strategy seeks to take advantage of the consistent daily consumption of large amounts of food staples by all family members, including women and children who are most at risk for micronutrient malnutrition. As a consequence of the predominance of food staples in the diets of the poor, this strategy implicitly targets low-income households. Therefore, biofortification is a more sustainable strategy, because this involves the fortification of crops at source to accumulate nutritionally important minerals, therefore avoiding the need to fortify processed food products. Apart from breeding programs worldwide (Morgounov et al. 2007; Bouis 2003; Kamran et al. 2014), including the *Harvest Plus*, biofortification of crops with micronutrients can also be achieved through microbial inoculants, organic practices, and proper soil management practices (Pooniya et al. 2012; Rana et al. 2012a, b).

19.3 Agronomic and Genetic Approaches of Mineral Enrichment

Biofortification provides a truly feasible means of reaching malnourished populations in relatively remote rural areas, delivering naturally fortified foods to people with limited access to commercially marketed fortified foods, by producing staple foods whose edible portions are denser in bioavailable minerals. Through this process, the farmers could develop the biofortified crops that naturally reduce anemia, cognitive impairment, and other nutritionally related health problems in hundreds of millions of people. Biofortification with minerals can be attempted genetically or through agronomic or soil management practices such as balanced or need-based fertilization or the use of slow release formulations. Agronomic strategies to increase the concentrations of mineral elements in edible tissues generally rely on the application of mineral fertilizers and/or improvement of the solubilization and mobilization of mineral elements in the soil. When crops are grown where mineral elements become immediately unavailable in the soil, targeted application of soluble inorganic fertilizers to roots or to leaves is practiced. Fertilization strategies targeted at improving the mineral content in cereals grains have been only partially successful in the case of zinc in rice and wheat (Cakmak 2008; Cakmak et al. 2010a, b; Shivay et al. 2008, 2010; Pooniya et al. 2012), while iron sequestration appears to be difficult in this regard. Therefore, even a small increase in the micronutrient content of rice can be highly significant for human nutrition. In situations where mineral elements are not readily translocated to edible tissues, foliar applications of soluble inorganic fertilizers are made. So it is successful for some minerals but not for all, and the success rate varies according to the environmental conditions. Also, iron (Fe) has a low mobility in soil, because Fe (II), which is given as $FeSO_4$, is bound by soil particles and converted into Fe (III). Pooniya et al. (2012) recorded higher grain yield (3.79 t ha^{-1}) as well as total content of N, P, K, and Zn in rice and also soil biological properties, through combined application of summer green manuring residues (*Sesbania aculeata*) and 2.0 % Zn-enriched urea (ZEU as $ZnSO_4 \cdot H_2O$).

Increasing the concentrations of essential nutrients in produce through the application of mineral fertilizers can be complemented by breeding crops with an increased ability to acquire and accumulate these minerals in their edible portions (Ortiz-Monasterio et al. 2007, 2011). Considerable genetic variation appears to exist in the concentrations of the mineral elements most frequently lacking in human diets in the edible portions of most crop species. The Food and Agriculture Organization (FAO) estimates that 22 % of India's population is undernourished. Every second child in India is malnourished; therefore, to halt the deteriorating "hidden hunger" situation in India, the government has proposed the "Mission Mode" program in the budget 2015. Molecular marker-assisted breeding programs are focusing on this aspect worldwide (Pfeiffer and McClafferty 2007; Velu et al. 2014).

Utilization of plant genetic resources to enhance the chemical composition of a few staple foods through biotechnology or conventional breeding appears cost-effective, given the large scale of crop production and the use of existing formal public and commercial seed supply systems. Agricultural research has already successfully produced crops with enhanced nutrient content. Maize varieties with enhanced levels of lysine and tryptophan developed in India and by the International Maize and Wheat Improvement Center (CIMMYT) are currently being adapted, grown, and promoted in several Latin American, Asian, and sub-Saharan African countries (Gupta et al. 2008; Prasanna et al. 2011; Shiferaw et al. 2011). Golden rice is being evaluated for its nutritional and dietary value in addressing vitamin A deficiencies globally. Bashir et al. (2013) reviewed the information available on iron biofortification of rice and concluded that despite understanding of the role of DMA (deoxymugenoic acid) in rice, lines overexpressing this gene have not been identified. It was proposed that regulating rice genes involved in the synthesis of subcellular Fe transporters may prove beneficial in developing plants with increased uptake efficiency.

As 40 % of the world's population relies on its own food production or limited outside resources for nutrition, fortified staples may be beyond the reach of the rural poor and malnourished. To meet the nutritional needs of fast-growing global population at the rate of 1.4 % per year, i.e., eight billion by 2030, there is a need for 50 % more food grains with higher and more stable yields (Yan and Kerr 2002). Micronutrient deficiencies are prevalent in most of the developing countries, and there is a decline in natural resources such as arable land and water. To meet these challenges, biofertilizers are a valuable tool to improve nutritional value in plants and crops, which can be accessible to poor farmers.

Genetic biofortification represents a sustainable strategy, with plant breeders exploiting the natural genetic variation existing in germplasm accessions for a particular trait and selectively breeding to enrich the specific trait. Screening of the germplasm of wheat and its wild relatives revealed a large amount of variation in terms of Zn and Fe concentrations (Monasterio and Graham 2000; Oury et al. 2006; Karami et al. 2009; Cakmak et al. 2004, 2010a, b; Cakmak 2012). The variation for Zn concentration ranged from 11.7 to 177 mg/kg among bread wheat, durum wheat, and their wild relatives, with a mean of 21–89 mg/kg. Agarwal et al. (2012) observed wide genetic variation for the concentration of Fe and Zn in 67 diverse genotypes and showed that kernel Fe was more affected by fluctuations in the environment over a period of 3 years. Iron is the fourth most abundant and a ubiquitous element in soils and sediments, usually found in the range of crustal abundance around 3.5 % of inorganic and organic complexes. Under Fe deficiency, graminaceous monocots release high-affinity Fe-chelating substances from the mugineic acid family called phytosiderophores. These substances solubilize Fe^{3+}, and the resulting Fe^{3+}- phytosiderophore complexes are taken up by the root cells via a specific plasma membrane transport system without reduction of the ferric ion. The use of MAS (marker-assisted selection), QTLs, and genome-wide association studies supplemented with the use of microarrays and metabolomics has generated a large body of information on stress-responsive/nutrient deficiency-induced pathways and expression cascades. The advent of high throughput screening methodologies, especially for elemental analyses such as EDXRF (energy-dispersive X-ray fluorescence spectrometry) or alternative colorimetric methods, besides ICP-based methods has revolutionized the analyses of samples from large multilocational trials in cost- and time-effective manner.

Precision phenotyping, involves a holistic approach in which the soil and environmental variables such as temperature, organic matter, soil texture, etc., are included; this is becoming an important strategy in breeding for stable and high Zn concentrations in wheat as it takes into consideration all factors which can influence the mobilization of the micronutrients in the root region (Joshi et al. 2010). It is well known that genotype x environment interactions, especially in relation to soil type and its composition, are significant for breeding high Zn genotypes. Rawat et al. (2009a, b) observed a high correlation of the Zn and Fe concentrations in flag leaf with those grains in *Aegilops* species, but further studies at different geographical locations are needed to confirm such relationships. The role of N nutritional status of plants and protein content of grains has a positive correlation with improved translocation of these micronutrients inside the plants. N may also influence the mobility and root uptake of Zn and Fe from soils (Kutman et al. 2010, 2011). In maize, increasing N application was effective in enhancing C partitioning into roots and promoting exudation of C-containing compounds from roots into rhizosphere (Liljeroth et al. 1994). Enhanced knowledge of the effects of N on the amount or expression levels of transporter proteins of Zn and Fe will contribute to better understanding the positive association between N, Zn, and Fe in seeds. Breeding crop cultivars with better ability to acquire nutrients from fertilizer or native sources is considered a promising approach for improving use efficiency in crop production.

Plants take up Zn from soil using transporters, particularly those belonging to the ZIP gene family. Phytosiderophores released by cereals are known to play a critical role in the uptake of Fe and Zn by the plants, which are found to be enhanced when nutritional status is optimal for growth (Cakmak et al. 1994; Murata et al. 2006; Cakmak et al. 2010a, b). Low molecular weight organic acids (LMWOAs) have been implicated in the acquisition of several nutrients, especially P and Zn in rice (Rose et al. 2013), and increased exudation was correlated with nutrient deficiency conditions. One ligand showing a promise as a breeding target for enhancing the rice plant's ability to acquire Zn is the phytosiderophore DMA (deoxymugenoic acid). Root traits, especially the architectural/configurational aspects such as degree of lateral/adventitious roots, their length, number, or root aerenchyma/root fineness, have been investigated by several researchers for enhancing acquisition of nutrients, and traits associated with crown root emergence and maintenance have been highlighted as potential traits for breeders (Wissuwa et al. 2009). Oxidative damage mediated by reactive oxygen species (ROS) is considered as one of the major mechanisms by which Zn deficiency damages crop plants, and breeding for related traits is considered as an important strategy (Cakmak et al. 2000; Rose et al. 2011, 2013).

Transgenic approaches to biofortification rely on improving the synthesis and accumulation of nutrients/vitamins in edible tissues. The use of biotechnology to create genetically modified organisms (GMOs) has the potential to design foods with specific attributes (Zhu et al. 2012). Enhanced nutritional quality of crops may be achieved by enabling the capacity of the plant to synthesize vitamins, to take up minerals with greater efficiency, or to reduce antinutrient factors (Bouis 2003). Gayen et al. (2013) generated transgenic high-iron rice lines by the insertion of ferritin gene and found that both the brown and milled rice grains from such lines had much higher nutritional quality.

19.4 Nutrient Sequestration and Mobilization by Microbes

Plants and microorganisms are known to have coevolved, and the linkages – both ecological and biological – between the aboveground and belowground biota are considered as the main drivers for maintaining soil health and productivity of crops (Barret et al. 2011; Hardoim et al. 2011). In agriculture, since decades, the practice of artificially inoculating specific microorganisms has been followed for improving yields. The "core microbiome" concept is gaining interest among researchers for developing suitable inoculants as nutrient management and disease management options (Gopal et al. 2013; Scheuring and Yu 2013).

The rhizosphere is a metabolically active site in which microorganisms sequester, mobilize, and make available both macro- and micronutrients to the plants. Conservative estimates suggest that more than 20, 000 plant species exhibit obligate symbiotic associations with microorganisms for their survival and proliferation (van Der Heijden et al. 2008). This is reflective of the significance of microorganisms in the nutrient cycling as drivers of plant productivity. Pfeiffer et al. (2013) analyzed the diversity and heritability of field-grown maize inbreds and illustrated the heritable variation in the rhizosphere microbial community composition. Biofortification of crops can therefore be achieved through the application of microbial inoculants/biofertilizers which mobilize/solubilize the essential mineral micronutrients in the soil and make it easily available to the plant. Among the various types of microorganisms, plant growth-promoting rhizobacteria (PGPR) are the most well-studied group. They are known to employ one or more direct and indirect mechanisms of action to improve plant growth and health, although the major mode of action is through increasing the availability of nutrients for the plant in the rhizosphere region (Antoun and Kloepper 2001; Glick 1995; Rana et al. 2012a, b).

Micronutrients are present in the soil as free ions or as ions adsorbed onto mineral or organic surfaces or in the form of precipitates within the soil biota. Uptake of these micronutrients from the rhizosphere is the first step in the process of accumulation in the plant, prior to translocation to seeds. PGPR play an important role in solubilization of nutrients from soil and enhancing their availability to plants. PGPR produce plant growth-promoting compounds and mineral-solubilizing enzymes and play a pivotal role in the cycling of macro- and micronutrients by modifying the root morphology, resulting in greater root surface area for the uptake of nutrients within the soil. The acquisition of mineral elements with restricted mobility in the soil, such as P, K, Fe, Zn, and Cu, can be improved by developing a more extensive root system, with the application of biofertilizers (Rana et al. 2012a, b) in wheat crop.

Another strategy employed by microorganisms for nutrient acquisition is the secretion of phytosiderophores by microorganisms and plants in drought- and salt-affected areas. This represents an efficient strategy for the uptake of micronutrients by plants from the rhizosphere. A complex iron cycle is found in the rhizosphere, involving multiple interactions between soils, plants, and microorganisms, which are promoted by the release of root exudates. Plants and microbes have evolved effective strategies of iron uptake in response to iron limitation, among which regulated exudation of organic ligands by plants and microorganisms is commonly observed. However, in cultivated soils at pH values compatible with plant growth, the solubility of iron is controlled at extremely low levels by stable hydroxides, oxyhydroxides, and oxides, and iron is known to enter the plant via the root from where it is distributed inside the plant. Generally, iron is present in concentrations of about 10–500 μg Fe/g dry weight in plant tissues. This mechanism is termed Strategy II, much resembling the microbial siderophore strategy.

Sharma et al. (2013) observed an enhancement in the iron content of rice grains using a set of PGPR exhibiting iron solubilization ability and found that the translocation efficiency of iron from roots to grains was twofold higher in the treatments involving the inoculation of PGP bacterial strains – *P. putida, P. fluorescens,* and *Azospirillum lipoferum.* Mishra et al. (2011) also reported that the iron concentration doubled in lentil seeds, when grown in association with *Pseudomonas* species.

Rana et al. (2012a, b) also observed that the Fe concentration in wheat grains in the different treatments ranged from 67.73 in the absolute control to 271.93 and 206.13 mg kg^{-1} in inoculated treatments. Analyses of micronutrients showed that the treatment involving inoculation of *Providencia* sp. + $N_{60}P_{60}K_{60}$ recorded the highest value for Zn accumulation (41.73 mg kg^{-1}), besides a threefold increase in Fe and Cu concentration, as compared to absolute control (Rana et al. 2012a). Inoculation with *Providencia* sp. alone as well as in combination with the other PGPR strains resulted in higher yield as well as growth, along with 40 kg ha^{-1} of N savings (Rana et al. 2012b). Significant increase in the concentration of Fe (143.6 %), Cu (193 %), Mn (63.7 %), and Zn (45.5 %) content as compared to treatment involving full-dose chemical fertilizer was also recorded. This *Providencia* strain had shown siderophore production and antifungal activity in an earlier study (Rana et al. 2011) and exhibited synergistic interactions with cyanobacterial and other rhizobacteria in wheat crop (Manjunath et al. 2011). The ecological fitness of this strain was evident from its significant influence on plant biomass, grain yield, and macro (NPK)- and micronutrient content. A significant positive correlation of root biomass with panicle ($r = 0.72$), plant biomass ($r = 0.70$), microbial biomass C, and organic C ($r = 0.75$) was observed. The production of siderophores by the bacterial strains has been reported in our earlier studies (Rana et al. 2011, 2012a, b), which can be responsible for the higher iron concentration in the root zone of the plants, leading to increased root activity and higher uptake of Fe in the plants, including translocation into the grains. Enhancement in the uptake of N, P, and micronutrients by bacterial inoculants and cyanobacteria was supported by the PGPR traits possessed by these strains which illustrated the positive involvement of such traits in improvement of yield, micronutrient concentration and their uptake in wheat grains. Tariq et al. (2007) demonstrated the efficiency of a commercial mixed PGPR consortium (containing *Pseudomonas* sp. and other strains of PGPR) acting as Zn solubilizer and increasing Zn concentration up to 157 % in rice. Inoculation with *Pseudomonas* and *Acinetobacter* strains led to enhanced uptake of Fe, Zn, Mg, Ca, K, and P by crop plants (Khan 2005).

Fungi, being heterotrophs (saprotrophs, necrotrophs, and biotrophs), are well known to play a major role in influencing soil fertility, decomposition, and cycling of minerals and organic matter, thereby improving plant health and nutrition. Finlay (2008) described seven different types of mycorrhizal symbiosis on the basis of their morphological characteristics and the fungal and plant species. Considered as among the most ancient and widespread below-ground associations of microorganisms with plants, mycorrhizal symbioses facilitate their plant hosts to the heterogeneously distributed nutrients required for their growth. This not only enables the continuous flow of energy-rich compounds required for nutrient mobilization but also provides a channel for the translocation of mobilized products back to their hosts involved. The most investigated associations are those of arbuscular mycorrhizal and ectomycorrhizal mycelia, which improve the acquisition of mineral nutrients, already available in solution by means of extraradical mycelia. Such mycelia represent physical extensions of the root system which increase the surface area across which nutrients can be taken up or bacteria to interact with, besides providing a direct pathway for translocation of photosynthetically derived carbon to microsites having better accessibility. In the last decade, speculation on their role in releasing nutrients from mineral particles and rock surfaces through weathering by themselves, or in association with bacteria or other fungi, is gaining a lot of interest among researchers (Landeweert et al. 2001). Actinomycetes have also been isolated from different surfaces of rocks and stones, and their resilience and ability to leach and reprecipitate metals are known to assist in interactions with bacterial/fungal communities. This can play a significant role to dissolve the primary rock-forming minerals to obtain essential nutrients and also act as nucleation sites for the precipitation of secondary minerals.

Another important function of mycorrhizal fungi is related to its role in the mobilization of N and P from various types of litter, including plant and micro-/mesofauna products, which are responsible for the diversity and richness of specific habitats. Under poor fertility conditions, plant growth-promoting rhizobacteria (PGPR), along with AMF, are known to improve plant growth and grain yield of rice, wheat, and other crops most efficiently. Eleiwa et al. (2012) reported a similar trend by inoculation with individual cultures of *Azospirillum brasilense*, *Azotobacter chroococcum,* or *Bacillus polymyxa.* Mader et al. (2011) found a substantial increase in Zn and Mn concentration through the use of natural AMF (arbuscular mycorrhizal fungi) consortium as well as with combined inoculation of two *Pseudomonas* strains. De Santiago et al. (2011) found a 1.5-fold enhancement in the Fe concentration in the aerial parts of wheat when treated with a siderophore producing strain of *Trichoderma asperellum.* Artursson et al. (2006) provided a comprehensive review of the potential of such interactions for stimulating plant growth.

Lehmann et al. (2014) evaluated the influence of 10 independent variables on AMF-mediated Zn concentration by performing a random effect meta-analysis, using a set of 104 articles, comprising 263 trials, and compared the data with those from non-mycorrhizal control plants in terms of their content in the above-, below-ground, fruit, and seed tissues. They concluded that AMF positively affected Zn concentration in various crop plant tissues, but exhibited a differential effect which was related to the environmental conditions. For example, although AMF had a positive overall impact on Zn concentration in all tissue types, it was modulated primarily by soil texture; soil pH and soil Zn concentration were found to affect AMF-mediated shoot Zn concentration, while soil P concentration influenced fruit Zn concentration. The micronutrients copper (Cu), manganese (Mn), and iron (Fe) are essential for crop plant development and productivity; however, no quantitative, data-based consensus has yet been reached on the role of arbuscular mycorrhizal fungi (AMF) in the nutrition in crops (Lehmann and Rillig 2015). In general, the effect was positive, but the duration of effect was influenced strongly by plant type, soil, and other environmental factors. Low molecular weight (LMW) organic acids produced by ectomycorrhizal fungi may play a role in weathering of minerals (Ahonen-Jonnarth et al. 2012). Among the several LMW organic acids released by plant roots, bacteria and fungi, oxalate, citrate, and malate have been reported as the strongest chelators of trivalent metals such as Al^{3+} and Fe^{3+}.

Rose et al. (2011, 2013) have speculated that there is a need for improving our understanding of the rhizosphere microbial dynamics and arbuscular mycorrhizal colonization as a prelude to enhancing the acquisition of nutrients by roots. Balakrishnan and Subramanian (2012) conducted field experiments using arbuscular mycorrhizal fungi (AM fungi) and observed the promise of its symbiotic association and its role in improving bioavailability of micronutrients in maize grains. Despite the availability of evidence for a significant role of AMF for Zn, Cu, Fe, and, to a limited extent, Mn in crop plant nutrition, the role of AMF in crop plant fortification is still in its infancy.

Cyanobacteria or blue-green algae represent one of the types of PGP agents which can also be key players in nutrient sequestration, improving nutrient use efficiency and crop yields (Watanabe and Yamamoto 1971; Prasanna et al. 2011, 2012, 2013a, b, 2014, 2015a, b; Gupta et al. 2013). Although cyanobacterial inoculation is known to increase rice crop yields (grain yields) to the extent of 10–24 % in diverse locations in the world, especially South Asia, they are also able to colonize the roots of plants, enter the root and shoot tissues, and enhance growth of various crops positively, employing direct or indirect mechanisms (Karthikeyan et al. 2007; Prasanna et al. 2009, 2013a, 2014; Manjunath et al. 2011; Bidyarani et al. 2015; Babu et al. 2015) including countering the deleterious effect of phytopathogenic microorganisms (Kloepper et al. 1991; Bashan 1998; Prasanna et al. 2013b, c). This has been attributed to a variety of other factors, besides nitrogen fixation, such as production of allelochemicals, IAA, and the secretion of extracellular polysaccharides which improve soil structure and increase the carbon and nitrogen status of soil (De Caire et al. 2000). Besides being diazotrophs, cyanobacteria can also improve the plant growth by sequestering nutrients and improving their mobilization into plants (Mandal et al. 1998; Irisarri et al. 2001; Karthikeyan et al. 2007; Maqubela et al. 2009; Prasanna et al. 2013b, c, 2014, 2015a, b). Their ability to synergistically interact with other microorganisms led to the development of consortia and the novel concept of cyanobacteria-based biofilms as biofertilizers (Nain et al. 2010; Manjunath et al. 2011; Prasanna et al. 2013c). This has shown promise and led to improved plant growth in wheat, cotton, vegetables, and legumes, besides rice (Prasanna et al. 2014, 2015a, b; Swarnalakshmi et al. 2013). The use of cyanobacterial inoculants (*Anabaena-Azotobacter* biofilm) and *Anabaena* sp.-*Providencia* sp. in a set of maize hybrids was found to enhance the activity of defense enzymes such as peroxidase, PAL, and PPO in roots, which also showed a positive correlation with Zn concentration in the flag leaf. Cyanobacterial inoculation enhanced the Zn mobilization to flag leaf in maize hybrids, without any negative effects on plant vigor and yields (Prasanna et al. 2015a).

The performance of *Anabaena*-based biofilm inoculants was also found to be superior under two methods of rice cultivation (conventional – flooded and SRI – System of Rice Intensification), leading to 13–46 % enhancement of iron and 15–41 % enhancement of zinc in rice grains over uninoculated controls. SRI cultivation methods appear to be contributing to better nutrient access and/or greater nutrient uptake by the rice plants, as evidenced by higher biometrical parameters and activity of enzymes. Issues which need in-depth investigation include a better understanding of plant-cyanobacteria interactions in aerobic soil and linking the mechanisms underlying nutrient translocation from root to grains. In general, the inoculants under SRI proved better in enhancing the concentration of Zn, Cu, Fe, and Mn, particularly in grains, showing significant correlation with the activity of defense- and pathogenesis-related enzymes and yield parameters (Adak et al. 2015). Manganese is known to be a cofactor for both phenylalanine ammonia lyase and peroxidase, along with other decarboxylase and dehydrogenases involved in Krebs cycle and photosynthetic oxygen evolution (Sukalovic et al. 2010). The positive correlation ($r = 0.62$) of manganese concentration with phenyl ammonia lyase activity is illustrative of this metabolic link. In both systems of rice cultivation, only in the *Anabaena-Pseudomonas* biofilm treatments, rice grains showed an increase in copper accumulation.

The role of copper in photosynthesis and its intracellular trafficking has been elucidated by Cavet et al. (2003), and its uptake has been found to be typically biphasic in the diazotrophic cyanobacterium *Nostoc calcicola* (Verma and Singh 1991). The uptake and mobilization of copper with cyanobacteria is well known, and the presence of metallothioneins in *Pseudomonas* has been reported (Turner and Robinson 1995). Therefore, it could be surmised that a consortium of both organisms, i.e., *Anabaena-Pseudomonas* biofilm, may therefore facilitate its translocation into roots and to leaves leading to significantly higher values in this treatment. It is established that Zn accumulation and translocation in plants increase through the activity of root-colonizing bacteria, as well as mycorrhiza, which play an important role in the mobilization of nutrients in soil (Cavet et al. 2003). With SRI practice, highest Zn accumulation of rice grains was recorded in *Anabaena-Trichoderma* biofilm formulation treatment (61.26 mg kg^{-1}) and *Trichoderma* formulation treatment

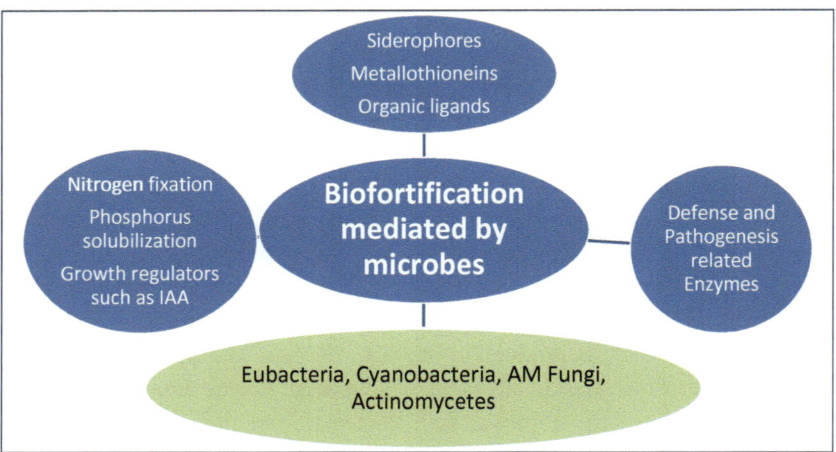

Fig. 19.1 Overview of mechanisms involved in microbe-mediated biofortification of crops

(55.13 mg kg^{-1}) with 32.6 % and 19.32 % enhancement as compared to control (50 % N + FD PK) (46.2 mg kg^{-1}). In conventional practice also, *Anabaena-Trichoderma* biofilm formulation (19.87 mg kg^{-1}) and *Trichoderma* formulation (16.27 mg kg^{-1}) treatment recorded the best Zn accumulation, which was 40.6 % and 15.14 % higher, as compared to control (14.13 mg kg^{-1}). However, a number of issues concerning SRI practices need in-depth investigation for a better understanding of plant-cyanobacteria interactions in aerobic soil and linking the mechanisms underlying nutrient translocation from root to grains. Figure 19.1 indicates the overview of mechanisms involved in microbe-mediated biofortification of crops.

19.5 Conclusion

Microbial formulations are a cost-effective and environment-friendly option which can provide savings in terms of chemical fertilizers. Harnessing the potential beneficial effects of these bacteria/fungi existing as free living or symbionts, for sustaining the yield and nutritional quality, seems essential, especially considering the increasing demands for food quality in the future. The complex interactions between soil, plants, and microbes in relation with micronutrient dynamics represent a unique opportunity for improving soil fertility. There is a need to deploy a synergistic set of practices involving targeted breeding and effective soil and agronomic management practices along with the inclusion of microbial inoculants for the development of micronutrient-dense staple crops and combating malnutrition.

Acknowledgments We are thankful for financial support from the Department of Biotechnology (DBT) and the Network Project on Microorganisms "Application of Microorganisms in Agriculture and Allied Sectors" by ICAR), New Delhi. Thanks are also due to the Division of Microbiology, IARI, New Delhi, for their infrastructural support.

References

Adak A, Prasanna R, Babu S, Bidyarani N, Verma S, Pal M, Shivay YS, Nain L (2015) Micronutrient enrichment mediated by plant-microbe interactions and rice cultivation practices. J Plant Nutr (In Press)

Agarwal PK, Jaiswal SK, Prasanna BM, Hossain F, Saha S, Guleria SK, Gupta HS (2012) Genetic variability and stability for kernel iron and zinc concentration in maize (*Zea mays* L.) genotypes. Indian J Genet 74:421–428

Ahonen-Jonnarth U, Van Hees PAW, Lundstrom US, Finlay RD (2012) Production of organic acids by mycorrhizal and non-mycorrhizal *Pinus sylvestris* L. seedlings exposed to elevated concentrations of aluminium and heavy metals. New Phytol 146:557–567

Antoun H, Kloepper JW (2001) Plant growth promoting rhizobacteria. In: Brenner S, Miller JH (eds)

Encyclopedia of genetics. Academic, New York, pp 1477–1480

Artursson V, Finlay RD, Jansson J (2006) Interactions between arbuscular mycorrhizal fungi and bacteria and their potential for stimulating plant growth. Environ Microbiol 8:1–10

Babu S, Prasanna R, Bidyarani N, Singh R (2015) Analysing the colonization of inoculated cyanobacteria in wheat plants using biochemical and molecular tools. J Appl Phycol 27(1):327–338. doi:10.1007/s10811-014-0322-6

Balakrishnan N, Subramanian KS (2012) Mycorrhizal symbiosis and bioavailability of micronutrients in maize grain. Maydica 57:129–38

Banziger M, Long J (2000) The potential for increasing the iron and zinc density of maize through plant-breeding. Food Nutr Bull 21:397–400

Barret M, Morrissey JP, O'Gara F (2011) Functional genomics analysis of plant growth-promoting rhizobacterial traits involved in rhizosphere competence. Biol Fertil Soils 47:729–743

Bashan Y (1998) Inoculants of plant growth promoting bacteria for use in agriculture. Biotechnol Adv 16:729–770

Bashir K, Nozoye T, Ishimaru Y, Nakanishi H, Nishizawa NK (2013) Exploiting new tools for iron biofortification of rice. Biotechnol Adv 31:1624–1633

Bidyarani N, Prasanna R, Chawla G, Babu S, Singh R (2015) Deciphering the factors associated with the colonisation of rice plants by cyanobacteria. J Basic Microbiol 55(4): 32#407–419. doi:10.1002/jobm.201400591

Bouis HE (2003) Micronutrient fortification of plants through plant breeding: can it improve nutrition in man at low cost? Proc Nutri Soc 62:403–411

Cakmak I (2008) Enrichment of cereal grains with zinc: agronomic or genetic biofortification? Plant Soil 302:1–17

Cakmak I (2012) Harvestplus zinc fertilizers project: Harvestzinc. Better Crops 96:17–19

Cakmak I, Gulut KY, Marschner H, Graham RD (1994) Effects of Zinc and iron deficiency on phytosiderophore release in wheat genotypes differing in zinc efficiency. J Plant Nutr 17:1–17

Cakmak I, Ozkan H, Braun HJ, Welch RM, Romheld V (2000) Zinc and iron concentrations in seeds of wild, primitive and modern wheats. Food Nutr Bull 21:401–403

Cakmak I, Torun A, Millet E, Feldman M, Fahima T, Korol A, Nevo E, Brau HJ, Ozkan H (2004) *Triticum dicoccoides*: an important genetic resource for increasing zinc and iron concentration in modern cultivated wheat. Soil Sci Plant Nutr 50:1047–1054

Cakmak I, Kalayci M, Kaya Y, Torun A, Aydin N, Wang Y, Arisoy Z, Erdem H, Yazici A, Gokman O, Ozturk L, Horst WJ (2010a) Biofortification and localization of zinc in wheat grain. J Agric Food Chem 58:9092–9102

Cakmak I, Pfeiffer WH, McClafferty B (2010b) Biofortification of durum wheat with zinc and iron. Cereal Chem 87:10–20

Cavet JS, Graham AI, Meng W, Robinson NJ (2003) A cadmium-lead-sensing ArsR-SmtB repressor with novel sensory sites. Complementary metal discrimination by NmtR and CmtR in a common cytosol. J Biol Chem 278:44560–44566

De Caire GZ, De Cano MS, Palma RM, De Mule CZ (2000) Changes in soil activities following additions of cyanobacterial biomass and exopolysaccharide. Soil Biol Biochem 32:985–987

De Santiago A, Quintero JM, Aviles M, Delgado A (2011) Effect of *Trichoderma asperellum* strain T34 on iron, copper, manganese, and zinc uptake by wheat grown on a calcareous medium. Plant Soil 342:97–104

Dobbelaere S, Croonenborghs A, Thys A, Ptacek D, Vanderleyden J, Dutto P, Labandera-Gonzalez C, Caballero-Mellado J, Aguirre JF, Kapulnik Y (2001) Responses of agronomically important crops to inoculation with *Azospirillum*. Funct Plant Biol 28:871–879

Eleiwa ME, Eman R, Hamed ER, Shehata HS (2012) The role of biofertilizers and/or some micronutrients on wheat plant (*Triticum aestivum* L.) growth in newly reclaimed soil. J Med Plant Res 6:3359–3369

Farre G, Twyman RM, Zhu C, Capell T, Christou P (2011) Nutritionally enhanced crops and food security: scientific achievements versus political expediency. Curr Opin Biotechnol 22:245–251

Finlay RD (2008) Ecological aspects of mycorrhizal symbiosis: with special emphasis on the functional diversity of interactions involving the extraradical mycelium. J Exp Bot 59:1115–1126

Gayen D, Sarkar SN, Datta SK, Datta K (2013) Comparative analysis of nutritional compositions of transgenic high iron rice with its non-transgenic counterpart. Food Chem 138:835–840

Glick BR (1995) The enhancement of plant growth by free-living bacteria. Can J Microbiol 41:109–117

Gopal M, Gupta A, Thomas GV (2013) Bespoke microbiome therapy to manage plant diseases. Front Microbiol 4:355. doi:10.3389/fmicb.2013.00355

Graham RD, Welch RM, Saunders DA, Ortiz-Monasterio I, Bouis HE, Bonierbale M, de Haan S, Burgos G, Thiele G, Liria R, Meisner CA, Beebe SE, Potts MJ, Kadian M, Hobbs PR, Gupta RK, Twomlow S (2007) Nutritious subsistence food systems. Adv Agron 92:1–74

Gupta HS, Agrawal PK, Mahajan V, Bisht GS, Kumar A, Verma P, Srivastava A, Saha S, Babu R, Pant MC, Mani VP (2008) Quality protein maize for nutritional security: rapid development of short duration hybrids through molecular marker assisted breeding. Curr Sci 96:230–237

Gupta V, Ratha SK, Sood A, Chaudhary V, Prasanna R (2013) New insights into the biodiversity and applications of cyanobacteria (blue-green algae) – prospects and challenges. Algal Res 2:69–97

Hardoim PR, Andreote FD, Reihold-Hurek R, Sessitch A, van Overbeek LS, van Elsas JD (2011) Rice root-associated bacteria: insights into community structures across 10 cultivars. FEMS Microbiol Ecol 77:154–164

Irisarri P, Gonnet S, Monza J (2001) Cyanobacteria in Uruguayan rice field diversity, nitrogen fixing ability and tolerance to herbicides and combined nitrogen. J Biotechnol 191:95–103

Joshi AK, Crossa I, Arun B, Chand R, Trethowan R, Varagas M, Ortiz-Monasterio I (2010) Genotype-environment interaction for zinc and iron concentration of wheat grain in eastern Gangetic plains of India. Field Crops Res 116:268–277

Kamran A, Kubota H, Yang RC, Randhawa HS, Spaner D (2014) Relative performance of Canadian spring wheat cultivars under organic and conventional field conditions. Euphytica 196(1):13–24. doi:10.1007/s10681-013-1010-3

Karami M, Afyuni M, Khoshgoftarmanesh AH, Papritz A, Schulin DR (2009) Grain zinc, iron, and copper concentrations of wheat grown in central Iran and their relationships with soil and climate variables. J Agric Food Chem 57:10876–10882

Karthikeyan N, Prasanna R, Lata, Kaushik BD (2007) Evaluating the potential of plant growth promoting cyanobacteria as inoculants for wheat. Eur J Soil Biol 4:23–30

Khan AG (2005) Role of soil microbes in the rhizospheres of plants growing on trace metal contaminated soils in phytoremediation. J Trace Elem Med Biol 18:355–364

Kloepper JW, Zablokovicz RM, Tipping EM, Lifshitz R (1991) Plant growth promotion mediated by bacterial rhizosphere colonizers. In: Keister DL, Cregan PB (eds) The rhizosphere and plant growth. Kluwer Academic Publishers, The Netherlands, pp 315–326

Kutman UB, Yildiz B, Ozturk L, Cakmak I (2010) Biofortification of durum wheat with zinc through soil and foliar applications of nitrogen. Cereal Chem 87:1–9

Kutman UB, Yildiz B, Cakmak I (2011) Effect of nitrogen on uptake, remobilization, and partitioning of zinc and iron throughout the development of durum wheat. Plant Soil 342:149–164

Landeweert R, Hoffland E, Finlay RD, Kuyper TW, Van Breemen N (2001) Linking plants to rocks: ectomycorrhizal fungi mobilize nutrients from minerals. Trends Ecol Evol 16:248–254

Lehmann A, Rillig M (2015) Arbuscular mycorrhizal contribution to copper, manganese and iron nutrient concentrations in crops – a meta-analysis. Soil Biol Biochem 81:147–158

Lehmann A, Veresoglou SD, Leifheit EF, Rillig MC (2014) Arbuscular mycorrhizal influence on Zinc nutrition in crop plants – a meta-analysis. Soil Biol Biochem 69:123–131

Liljeroth E, Kuikman P, Vanveen JA (1994) Carbon translocation to the rhizosphere of maize and wheat and influence on the turnover of native soil organic-matter at different soil-nitrogen levels. Plant Soil 161:233–240

Mader P, Kaiser F, Adholeya A, Singh R, Uppal HS, Sharma AK, Srivastava R, Sahai V, Aragno M, Wiemken A, Johri BN, Fried PM (2011) Inoculation of root microorganisms for sustainable wheat-rice and wheat black gram rotations in India. Soil Biol Biochem 43:609–619

Mandal B, Vlek PLG, Mandal LN (1998) Beneficial effect of blue green algae and Azolla excluding supplying nitrogen, on wetland rice fields a review. Biol Fertil Soils 27:329–342

Manjunath M, Prasanna R, Sharma P, Nain L, Singh R (2011) Developing PGPR consortia using novel genera- *Providencia* and *Alcaligenes* along with cyanobacteria for wheat. Arch Agron Soil Sci 57:873–887

Maqubela MP, Mnkeni PNS, Issa MO, Pardo MT, D'Acqui LP (2009) *Nostoc* cyanobacterial inoculation in South African agricultural soil enhances soil structure, fertility and maize growth. Plant Soil 315:79–92

Mishra PK, Bisht SC, Ruwari P, Joshi GK, Singh G, Bisht JK, Bhatt JC (2011) Bioassociative effect of cold tolerant *Pseudomonas spp.* and *Rhizobium leguminosarum*-PR1 on iron acquisition, nutrient uptake and growth of lentil (*Lens culinaris* L.). Eur J Soil Biol 47:35–43

Monasterio I, Graham RD (2000) Breeding for trace minerals in wheat. Food Nutr Bull 21:393–396

Morgounov A, Gomez-Becerra HF, Abugalieva A, Dzhunusova M, Yessimbekova M, Muminjanv H, Zelenskiy Y, Ozturk I, Cakmak I (2007) Iron and zinc grain density in common wheat grown in central Asia. Euphytica 155:193–203

Murata Y, Ma JF, Yamaji N, Ueno D, Nomoto K, Iwashita T (2006) A specific transporter for iron (III)-phytosiderophore in barley roots. Plant J 46:563–572

Nain L, Rana A, Joshi M, Jadhav SD, Kumar D, Shivay YS, Paul S, Prasanna R (2010) Evaluation of synergistic effects of bacterial and cyanobacterial strains as biofertilizers for wheat. Plant Soil 331:217–230

Ortiz-Monasterio I, Palacios-Rojas N, Meng E, Pixley K, Trethowan R, Pena RJ (2007) Enhancing the mineral and vitamin content of wheat and maize through plant breeding. J Cereal Sci 46:293–307

Ortiz-Monasterio I, Trethwan R, Holm PB, Cakman I, Borg S, Tauris BEB, Brinch-Pedersen H (2011) Breeding, transformation and physiological strategies for the development of wheat with high zinc and iron grain concentration. In: Bonjean AP, Angus WJ, Van Ginkel M (eds) The world wheat book-a history of wheat breeding, vol 2., pp 951–977

Oury FX, Leenhardt F, Remesy C, Chanliaud E, Dupperrier B, Balfouriera F, Charmet G (2006) Genetic variability and stability of grain magnesium, zinc and iron concentration in bread wheat. Eur J Agron 25:177–185

Pfeiffer WH, McClafferty B (2007) HarvestPlus: breeding crops for better nutrition. Crop Sci 47:88–105

Pfeiffer JA, Spor A, Koren O, Jin Z, Tringe SG, Dangi JL, Buckler ES, Ley RE (2013) Diversity and heritability of the maize rhizosphere microbiome under field conditions. Proc Natl Acad Sci 110:6548–6553

Pooniya V, Shivay YS, Rana A, Nain L, Prasanna R (2012) Enhancing soil nutrient dynamics and productivity of basmati rice through residue incorporation and zinc fertilization. Eur J Agron 41:28–37

Prasanna R, Jaiswal P, Nayak S, Sood A, Kaushik BD (2009) Cyanobacterial diversity in the rhizosphere of rice and its ecological significance. Indian J Microbiol 49:89–97

Prasanna BM, Mazumdar S, Chakraborti M, Hossain F, Manjaiah KM, Agarwal PK, Guleria SK, Gupta HS (2011) Genetic variability and genotype X environment interactions for kernel iron and zinc concentrations in maize (*Zea mays* L.) genotypes. Indian J Agric Sci 81:704–711

Prasanna R, Joshi M, Rana A, Shivay YS, Nain L (2012) Influence of co-inoculation of bacteria- cyanobacteria on crop yield and C- N sequestration in soil under rice crop. World J Microbiol Biotechnol 28:1223–1235

Prasanna R, Babu S, Rana A, Kabi SR, Chaudhary V, Gupta V, Kumar A, Shivay YS, Nain L, Pal RK (2013a) Evaluating the establishment and agronomic proficiency of cyanobacterial consortia as organic options in wheat-rice cropping sequence. Exp Agric 49:416–434

Prasanna R, Chaudhary V, Gupta V, Bab S, Kumar A, Shivay YS, Nain L (2013b) Cyanobacteria mediated plant growth promotion and bioprotection against Fusarium wilt in tomato. Eur J Plant Pathol 13:337–353

Prasanna R, Kumar A, Babu S, Chawla G, Chaudhary V, Singh S, Gupta V, Nain L, Saxena AK (2013c) Deciphering the biochemical spectrum of novel cyanobacterium based biofilms for use as inoculants. Biol Agric Hortic 29:145–158

Prasanna R, Triveni S, Bidyarani N, Babu S, Yadav K, Adak A, Khetarpal S, Pal M, Shivay YS, Saxena AK (2014) Evaluating the efficacy of cyanobacterial formulations and biofilmed inoculants for leguminous crops. Arch Agron Soil Sci 60:349–366

Prasanna R, Babu S, Bidyarani N, Kumar A, Triveni S, Monga D, Mukherjee AK, Kranthi S, Gokte-Narkhedhar N, Adak A, Yadav K, Nain L, Saxena AK (2015a) Prospecting cyanobacteria fortified composts as plant growth promoting and biocontrol agents in cotton. Exp Agric 51:42–65

Prasanna R, Bidyarani N, Babu S, Hossain F, Shivay YS, Nain L (2015b) Cyanobacterial inoculation elicits plant defense response and enhanced Zn mobilization in maize hybrids. Cogent Food Agric 1(1). doi:10.1080/23311932.2014.998507

Rana A, Saharan B, Joshi M, Prasanna R, Kumar K, Nain L (2011) Identification of multi trait PGPR isolates and evaluating their potential as inoculants for wheat. Ann Microbiol 61:893–900

Rana A, Saharan B, Nain L, Prasanna R, Shivay YS (2012a) Enhancing micronutrient uptake and yield of wheat through bacterial PGPR consortia. Soil Sci Plant Nutr 58:573–582

Rana A, Joshi M, Prasanna R, Shivay YS, Nain L (2012b) Biofortification of wheat through inoculation of plant growth promoting rhizobacteria and cyanobacteria. Eur J Soil Biol 50:118–126

Rawat N, Tiwari VK, Neelam K, Randhawa GS, Singh K, Chhuneja P, Dhaliwal HS (2009a) Development and characterization of wheat- *Aegilops kotschyi* amphiploids with high grain iron and zinc. Plant Genet Res 7:271–280

Rawat N, Tiwari VK, Singh N, Randhawa GS, Singh K, Chhuneja P, Dhaliwal HS (2009b) Evaluation and utilization of *Aegilops* and wild *Triticum* species for enhancing iron and zinc content in wheat. Genet Res Crop Evol 56:53–64

Rengel Z, Batten GD, Crowley DE (1999) Agronomic approaches for improving the micronutrient density in edible portions of field crops. Field Crops Res 60:27–40

Rose MT, Pariasca-Tanaka J, Rose TJ, Wissuwa M (2011) Bicarbonate tolerance of Zn-efficient rice genotypes is not related to organic acid exudation, but to reduced solute leakage from roots. Funct Plant Biol 38:493–504

Rose TJ, Impa SM, Rose MT, Pariasca-Tanaka J, Mori A, Heuer S, Johnson-Beebout SE, Wissuwa M (2013) Enhancing phosphorus and zinc acquisition efficiency in rice a critical review of root traits and their potential utility in rice breeding. Ann Bot 112:331–345

Scheuring I, Yu DW (2013) How to assemble a beneficial microbiome in three easy steps. Ecol Lett 15:1300–1307

Sharma A, Shankhdhar D, Shankhdhar SC (2013) Enhancing grain iron content of rice by the application of plant growth promoting rhizobacteria. Plant Soil Environ 59:89–94

Shiferaw B, Prasanna BM, Hellin J, Banziger M (2011) Crops that feed the world 6. Past successes and future challenges to the role played by maize in global food security. Food Security 3(3):307–327

Shivay YS, Kumar D, Prasad R, Ahlawat IPS (2008) Relative yield and zinc uptake by rice from zinc sulphate and zinc oxide coatings onto urea. Nutr Cycl Agroecosyst 80:181–188

Shivay YS, Prasad R, Rahal A (2010) Genotypic variation for productivity, zinc utilization efficiencies and kernel quality in aromatic rices under low available zinc conditions. J Plant Nutr 33:1835–1848

Shukla AK, Sharma SK, Tiwari R (2005) Nutrient depletion in the rice-wheat cropping system of the Indo-Gangetic plains. Better Crops 89:28–31

Singh MV (2009) Micronutrient nutrient problems in soils of India and improvement for human health and animal health. Indian J Fert 5:19–26

Sukalovic VHT, Vuletic MS, Veljovic J, Vucinic Z (2010) The effects of manganese and copper in vitro and in vivo on peroxidase catalytic cycles. J Plant Physiol 167:1550–1557

Swarnalakshmi K, Prasanna R, Kumar A, Pattnaik S, Chakravarty K, Shivay YS, Singh R, Saxena AK (2013) Evaluating the influence of novel cyanobacterial biofilmed biofertilizers on soil fertility and plant nutrition in wheat. Eur J Soil Biol 55:105–116

Tariq M, Hameed S, Malik KA, Hafeez FY (2007) Plant root associated bacteria for zinc mobilization in rice. Pak J Bot 39:245–253

Turner JS, Robinson NJ (1995) Cyanobacterial metallothioneins biochemistry and molecular genetics. J Indus Microbiol 14:119–125

van Der Heijden MGA, Bardgett RD, Van Straalen NM (2008) The unseen majority soil microbes as drivers of plant diversity and productivity in terrestrial ecosystems. Ecol Lett 11:296–310

Velu G, Ortiz-Monasterio I, Cakmak I, Hao Y, Singh RP (2014) Biofortification strategies to increase grain zinc and iron concentrations in wheat. J Cereal Sci 53:365–372

Verma SK, Singh HN (1991) Evidence for energy-dependent copper efflux as a mechanism of Cu^2+ resistance in the cyanobacterium *Nostoc calcicola*. FEMS Microbiol Lett 84:291–294

Watanabe A, Yamamoto Y (1971) Algal nitrogen fixation in the tropics. Plant Soil 35:403–413

Waters BM, Sankaran RP (2011) Moving micronutrients from the soil to the seeds genes and physiological processes from a biofortification perspective. Plant Sci 180:562–574

Welch RM, Graham RD (2004) Breeding for micronutrients in staple food crops from a human nutrition perspective. J Expl Bot 55:353–364

White PJ, Broadley MR (2009) Biofortification of crops with seven mineral elements often lacking in human diets – iron, zinc, copper, calcium, magnesium, selenium and iodine. New Phytol 182:49–84

WHO (World Health Organization) (2011) Micronutrient deficiencies: iron deficiency anaemia. URL: http://www.who.int/nutrition/topics/ida/en/index.html

Wissuwa M, Mazzola M, Picard C (2009) Novel approaches in plant breeding for rhizosphere-related traits. Plant Soil 321:409–430

Yan L, Kerr PS (2002) Genetically engineered crops: their potential use for improvement of human nutrition. Nutr Rev 60:135–141

Zhu C, Naqvi S, Gomez-Galera S, Pelacho M, Capell T, Christou P (2012) Transgenic strategies for the nutritional enhancement of plants. Trends Plant Sci 12:548–555

Plant Growth Promoting Rhizobacteria in Nutrient Enrichment: Current Perspectives

20

Poonam Sharma, K.C. Kumawat, and Sabhjeet Kaur

Abstract

The intensive application of chemical fertilizers to meet the growing demand of food for burgeoning population has led to an unprecedented perturbation of the soil environment. Environment degradation has had serious impacts on the nutrient use efficiency by the soil crop plants owing to irrational use of chemical fertilizers, thus warranting prioritization of the beneficial soil microorganisms for improving the soil health. Rhizosphere supports huge diversity of microbial community including beneficial plant growth promoting rhizobacteria (PGPR) as primary determinants of plant health and soil fertility. PGPR colonize roots of dicots and monocots and improve plant growth either by assisting nutrient acquisition (N, P, and essential nutrients) or modulating root system architecture by releasing phytohormones (direct effect) or reducing detrimental effects of various biotic and abiotic stress (indirect effect). The abilities of PGPR are of immense importance in sustainable agriculture in terms of improving crop production and soil health, therefore declining the negative effect of inorganic fertilizers on the environment. This chapter is an effort to illuminate the prevailing scenario on underlying mechanisms of nutrient management by PGPR. It is vital that soil microbiologists and agronomists pay due attention to strategies for nutrient management for enhancing crop production in a sustainable manner.

P. Sharma (✉) • S. Kaur
Department of Plant Breeding and Genetics, Punjab Agricultural University, Ludhiana, Punjab 141 004, India
e-mail: poonamsharma@pau.edu; poonam1963in@yahoo.co.in; sabhjeetk@pau.edu

K.C. Kumawat
Department of Microbiology, Punjab Agricultural University, Ludhiana, Punjab 141 004, India
e-mail: kailash-cobsmb@pau.edu

U. Singh et al. (eds.), *Biofortification of Food Crops*, DOI 10.1007/978-81-322-2716-8_20

Keywords
NUE • PGPR • Phytohormone • P-solubilization • Siderophore production • Zn uptake

20.1 Introduction

Green revolution has coaxed greater yield from food crops than ever before with the principle of fertility of soil and management of plant nutrients along with irrigation and mechanization of farm and genetic enhancement of plant crops. During the twentieth century, globally, chemical fertilizers are responsible for an increase of 50 % in crop yields (Borlaug and Dowswell 1994). Recently, agricultural production slowed down the world over; it is utmost urgent to maintain high productivity, with limited alteration in the environment. Further enhancement in agricultural productivity is the need of the hour due to increases in population the world over and is being expected to be approximately nine billion by 2050; threefold increase in nitrogen (N) and phosphorous (P) application is required if there is no increase in plant nutrient efficiency (NUE). Chemical fertilizers, no doubt, are an important source of nutrients, but their imbalanced and continuous uses have led to the nutrient pollution and deteriorate soil health. Overfertilization leads to poor uptake of nutrient by plant; only 30–50 % applied N and 45 % of P are taken up by crop plants. Improvement in NUE include the improved agronomic management techniques, development of more efficient crop plants through crop improvement, alternative management, and extensive research in the area of nutrient cycling by resident soil microflora. In this chapter, our main focus will be on the role of microorganisms present in soil and rhizosphere as a key player in improving NUE in crop plant.

This approach could head for environmentally sustainable agriculture while maintaining ecosystems and biodiversity. One potential way to reduce the negative environmental impact due to continued application of chemical fertilizers, viz., pesticides and herbicides are the use of plant growth promoting microorganisms (PGPM). This term was first coined by Kloepper and Schroth (1978) as soil microorganisms that colonize the rhizosphere of plants, growing in, on, or around plant tissues, and stimulate plant growth by several mechanisms. They can do so endophytically, in symbiosis or as free-living cells. PGPM may act directly by facilitating plant nutrient acquisition or influencing plant hormone level or indirectly by attenuating the inhibiting effects of pathogens (Glick 2012). Vesicular-arbuscular mycorrhiza (VAM) plant symbiosis is a well-established system present in over 90 % plant species with extensive colonization through external hyphae. VAM increased the fitness of host by scavenging minerals in turn of exchange for fixed carbon (Brundrett 2002). Root nodule-forming bacteria are the best understood prokaryote, which fix atmospheric nitrogen to ammonia in dedicated symbiotic organs in legume called nodules in exchange for plant-supplied sugars and microaerobic conditions. Rhizobia are bacterial symbionts of legumes that fix atmosphere nitrogen in a process known as biological nitrogen fixation (BNF). These are excellent examples of soil bacteria able to provide a macronutrient to plants. Large diversity of PGPR is still undiscovered in nature, and huge potential of PGPR can be exploited as novel mechanism for improvement in crop productivity. As N and P are the most limiting nutrients for plant growth, keeping in mind the focus on important micronutrients with the efficient mechanisms of microorganisms for their procurement will be highlighted in this chapter.

20.2 Rhizosphere

Overuse of chemical fertilizers has caused imbalance in supply of essential soil nutrient and degradation of soil health. Soil is the reservoir of microbes interacting immediately with emerging roots of plant for colonization. Rhizobacteria are the most abundant organisms in rhizosphere. Rhizospheric competing bacteria are rhizobacteria which progressively colonize root environment of plant in the prevalence of competing microflora. In addition to providing mechanical support and facilitating water and nutrient uptake, plant roots also synthesize, accumulate, and secrete a diverse array of compounds (Walker et al. 2003). Plant roots exude a huge diversity of organic nutrients (organic acids, phytosiderophore, sugars, vitamins, amino acids, nucleosides, mucilage, etc.) and signals to attract plant-beneficial microbial diversity which can metabolize compounds secreted by plant roots (Drogue et al. 2013). Microbial activity in the rhizospheres also affects the rooting patterns and supply of nutrients to plants, thereby altering the quality and quantity of root exudates. Further, a fraction of root exudates were also metabolized by microbes in the vicinity as C&N sources, and subsequently, plants reutilized microbe-oriented molecules for growth and development.

Root exudates are the largest source of C supply within the soil; the rhizosphere compartment houses a rich microbial community comprising up to 10^{10} bacteria per gram of soil (Roesch et al. 2008). Thus, rhizosphere can be defined as any volume of soil specifically influenced by plant roots and/or in association with root hairs and plant-produced materials. Three separate but interacting components are now recognized in the rhizosphere, and plant-associated bacteria can be classified as rhizospheric (in vicinity of root), rhizoplanic (on surface of root), and endophytic (inside the plant tissue) bacteria.

20.3 Plant Growth Promoting Microorganism (PGPM)

The aim of using soil microorganisms for enhancing nutrient availability for plant is an important practice for agriculture. PGPM association 400 millions year ago evolved in the form of mycorrhiza that assisted the first rootless land plants to absorb nutrients. Early studies were initiated in the 1950s on nitrogen-fixing bacteria. PGPM include both bacterial and fungal species. Application of plant growth promoting rhizobacteria for sustainable agriculture has increased tremendously the world over. Inoculation of PGPR has revealed positive response for growth and yield in various agronomically important crops. Studies according to plant growth promoting ability of PGPR have shown high specificity to species, cultivar, and genotypes of plant (Lucy et al. 2004). PGPM are characterized by three intrinsic characters: (i) they must colonize the root surface; (ii) multiply, survive, and compete with other microbiota at least for the time needed to express their plant promotion; and (iii) promote plant growth. Numerous investigations on PGPM have been focused on the effect of biotechnology along with environment and agriculture.

20.3.1 Plant Growth Promoting Rhizobacteria (PGPR)

According to Vessey (2003), soil bacterial species multiplying in plant rhizospheres with a plethora of mechanisms are known as PGPR (plant growth promoting rhizobacteria) (Table 20.1).

PGPR mostly occur in soil as 2–5 %, when reintroduced by plant infection in soil containing competing microflora, exert a positive effect on plant growth, and called as plant growth promoting rhizobacteria (PGPR) (Kloepper and Schroth 1978). PGPR have been classified by Somers et al. (2004) on the basis of plant growth promoting traits as (i) biofertilizers (improving the nutrient uptake to plant), (ii) phytostimulators (increasing the plant growth by phytohormone, viz., auxin (IAA), GAs, etc.), (iii) rhizoremediators which degrade organic pollutants, (iv) biopesticides (controlling diseases through antagonistic activity by the production of antibiotics, antifungal metabolites, and synthesis of extracellular enzymes) (Glick et al. 2007).

Further, in most studies, a PGPR has diversifying modes of action including plant growth promoting traits and antagonistic activities (Vessey 2003). Alternatively, Gray and Smith (2005) classified PGPR on the basis of plant tissue compartment, occupied as extracellular PGPR (ePGPR) and intracellular PGPR (iPGPR) that exist between the cell of root cortex in the rhizospheric and rhizoplanic space (Figueiredo et al. 2011). Few examples of ePGPR and iPGPR are *Agrobacterium*, *Flavobacterium*, *Klebsiella*, *Enterobacter*, *Enterococcus*, *Arthrobacter*, *Azotobacter*, *Bacillus*, *Pseudomonas*, *Burkholderia*, *Caulobacter*, *Chromobacterium*, *Erwinia*, *Micrococcus*, *Serratia*, etc. Similarly, some examples of the iPGPR mostly belong to the family *Rhizobiaceae* like *Azorhizobium*, *Allorhizobium*, *Bradyrhizobium*, *Mesorhizobium/Ensifer*, *Rhizobium*, etc. Mostly, all the rhizobacteria are Gram-negative, rods, cocci or pleomorphic creamy, white, translucent, and beneficial to plant growth and antagonistic activity for plants. In rhizosphere, major components of microbial diversity are actinomycetes with multiple PGP traits (IAA and P-solubilization) (Bhattacharyya and Jha 2012). Potential actinomycetes, viz., *Micromonospora* sp., *Streptomyces* sp., *Streptosporangium* sp., and *Thermobifida* sp., are potent biocontrol agents against root fungal pathogen.

20.3.2 Vesicular-Arbuscular Mycorrhiza (VAM)

The ancient obligate symbiotic fungal partner in majority of modern plants belongs to about 200 *Glomus* species. *Glomus* are propagated via soil spores and unable to be cultured without compatible host. Mycorrhizal symbioses can be characterized agronomically by increased growth and yield or ecologically by improved fitness or reproductive ability of host plant. VAM produce long filamentous hyphae to explore up to soil area available to root zone. The hyphae form bridge connecting plant roots with large soil area which serve as pipeline to funnel nutrients back to plant, getting in exchange 20 % of plant carbon for its growth and energy needs (Peterson et al. 2004). VAM have ability to colonize the roots and rhizosphere simultaneously. Arbuscular mycorrhizal fungi (AMF) have an extensive ramified hyphal network spread internally and externally in the roots of plants and have access to larger amount of water and minerals. AMs are capable of increasing supply of N, P, Cu, Zn, Fe, Ca, B, Mn, Ni, K, etc. (Clark and Zeto 2000).

AMF explore soil volume more efficiently than roots because hyphae grow faster, are thinner, and branch more extensively. Arbuscular mycorrhizal (AM) associations can increase the nutrient absorptive area up to 100 times relative to root length and external mycelia weighing as much as 3 % of total root. This is especially valuable for the plant in scavenging immobile minerals, viz., P, N, and other mineral nutrients efficiently. Distribution and function of extramatrical hypha play an important role in nutrient uptake. If mycorrhiza is to be effective in nutrient uptake, the hyphae must be distributed beyond the nutrient depletion zone that develops around the roots. For a poorly mobile ion, such as phosphate, a sharp and narrow depletion zone develops around the root. Mycorrhizal hyphae are known to readily bridge this depletion zone and grow in the soil with an adequate supply of P uptake including micronutrients, viz., Zn and Cu as these elements are also diffusion limited in many soils. Plant species vary in their dependence on AM fungi for nutrient acquisition as well for growth in AM-colonized soils inoculated with AM fungi. For instance, in two tropical soil uptake of P per root length was fourfold higher in comparison to non-AM plants. Whereas under similar condition, soybean had nearly threefold increase in P uptake compared to non-AM plants.

AMF are not very efficient in enhancing uptake of N by plants, as most of nitrogen is present in diffusible water soluble as nitrate and is easily absorbed by plant roots. Thus, it appears that AMs are unable to transfer N to plant. Contrary to earlier findings, Barea (2002) suggested that bioavailability of major nutrients, N and P

Table 20.1 [a]Response of PGPR inoculants on various agricultural crops

Crop	Bacteria/consortia	PGPR mechanism reported	Growth conditions	Results	Reference
Rice	*P. fluorescens* Aur6 *Chryseobacterium balustinum* Aur9 (individual and consortia experiments)	Biocontrol (*Magnaporthe grisea*) salinity	Field	Three-field experiments in different years. Each strain individually protected rice against rice blast, but the consortium was the most effective treatment (reaching 50 % of protection against disease)	Lucas et al. (2009)
Wheat and soybean	*Bacillus aryabhattai*	Biostimulation and biofortification of zinc Phytostimulation	Microcosm conditions Pot	MDSR7 and MDSR14 of *Bacillus aryabhattai* strains substantially influenced mobilization of zinc and its concentration in edible portion, enhanced yield of soybean and wheat	Ramesh et al. (2014)
	Arthrobacter sp. and *B. subtilis* (individual and consortia experiments)	Stress controller (salinity)	Greenhouse	Coinoculation increased dry weight up to 20 % and 40 % at 2 and 6 dS m^{-1} of salinity level	Upadhyay et al. (2012)
	A natural AMF *P. jessenii*, *P. synxantha* (individual and consortia experiments)	Biofertilization	Field	Dual inoculation of wheat with PGPR and AMF increased grain yield by 41 % as compared to uninoculated controls. Yield responses to the inoculants were highest at locations with previously low yields	Mader et al. (2011)
	Providencia sp. 2 strains of *Anabaena sp.* and *Calothrix* sp. (consortia experiment)	Biofertilization	Growth chamber, greenhouse and field	Enhancement 18.6 % protein content	Rana et al. (2012)
	Paenibacillus rhizosphaerae TGX5E, *Glomus intraradices* (consortia), *P. favisporus* TG1R2, *G. intraradices* (consortia), *P. rhizosphaerae* TGX5E, *P. favisporus* TG1R2, *G. intraradices* (consortia)	Biofertilization	Growth chamber, greenhouse and field	The frequency of mycorrhizal colonization and the extraradical mycelium increased significantly with time in plants inoculated with *G. intraradices* and both *Paenibacillus* strains. The highest dry biomass was found in soybean plants treated with *P. favisporus*, when this species was inoculated separately or in combination with *G. intraradices*	Fernandez-Bidondo et al. (2011)
Maize	*A. braziliense* Az39 *B. japonicum* E109 (consortium)	Phytostimulation	Pot	*A. braziliense* Az39 and *B. japonicum* E109, singly or in consortia, promoted easily seed germination and nodule formation	Cassan et al. (2009)

(continued)

Table 20.1 (continued)

Crop	Bacteria/consortia	PGPR mechanism reported	Growth conditions	Results	Reference
Bean	*R. tropici* CIAT899 *G. intraradices* (consortia experiments) *G. sinuosum*, *Gigaspora albida Ga*, *P. fluorescens* (consortia experiments)	Biofertilizer Biofertilizer	Greenhouse Field	Improvement in N and P content of soil with enhancement in nodulation and growth parameters Coinoculation *G. sinuosum* + *P. fluorescens* + mustard oil cake (MOC) and *G. albida* + *P. fluorescens* + mustard oil cake (MOC) reduced disease incidence along with improvement in growth and N and P content	Tajini et al. (2012) Neeraj (2011)
Sugarcane	*Azotobacter* sp. *Enterobacter* sp.	Biofertilizer Biofertilizer	Pot Micropropagation	Consortia (A17, A8, Azoto1, and Azoto2) improved root and shoot biomass over uninoculated plant 15 N isotope dilution technique revealed nitrogen fixation contribution (28 % of total plant N) in plantlets inoculated with isolate SC20	Ashraf et al. (2011) Mirza et al. (2001)
Cotton	*Pseudomonas aeruginosa* Z5 (AY548952) and *Bacillus fusiformis* S10 (AY548956)	Biofertilizer	Pot and field	Seed treatment of cotton plants with *Pseudomonas aeruginosa* Z5 (AY548952) and *Bacillus fusiformis* S10 (AY548956) improved growth and yield in cotton with reduced levels of chemical fertilizers	Yasmin et al. (2013)
Chickpea	*Mesorhizobium* sp. *Mesorhizobium* sp. and *Bacillus* sp. (single and dual inoculants)	Biofertilizer and biocontrol Biofertilizer	Glasshouse Pot experiment	Inoculation with *Mesorhizobium* sp. isolates significantly increased yield as well as antagonistic activity against *Fusarium* wilt Symbiotic effectiveness of *Mesorhizobium* sp. *Cicer* strain Ca181 was further improved on coinoculation application of six *Bacillus* strains as rhizobacteria, i.e., CBS9, CBS17, CBS20, CBS106, CBS127, and CBS155 with *Mesorhizobium* (Ca181) over *Mesorhizobium*-alone treatment improved plant growth	Jida and Assefa (2012), Sivaramaiah (2007)

[a]Adapted from Perez-Montano et al. (2014)

with N_2-fixing bacteria and PSM, also contribute to the AM role in nutrient acquisition, under low-input technology management. Such interaction plays a pivotal role in improving NUE in crop plants. Certain rhizobia are well known in improving spore germination, mycelial growth from the mycorrhizal propagules, and "crack formation" on developing root system on the host legume plant for mycorrhizal fungi (Jeffries et al. 2003). Enhancement in 15 N/14 N ratio in roots of *Medicago sativa* was recorded in *Rhizobium*-inoculated mycorrhizal plants, relative to N_2 fixation rates achieved by the same *Rhizobium* in non-mycorrhizal plants. Multi-inoculant (AMF, *Rhizobium* sp., and PGPR) combinations were also effective in improving plant development, nutrient uptake, N_2 fixation, or root system quality in cotton plant (Sultana and Pindi 2012). Interactive effect of phosphate-solubilizing bacteria (PSB) and AMF on plant use of soil P sources of low bioavailability (added as rock phosphate) has been evaluated in *Medicago sativa.* Dual inoculants significantly increased N and P accumulation in plant tissue in comparison to control plant, suggesting that the mycorrhizal and bacterized plants were using P sources otherwise unavailable to the plants. Whereas, ammonium is the most common form of synthetic fertilizer. Ammonium is fairly immobile in soil due to breakdown product of organic decomposition. There is an evidence that in maize, AM readily absorbed and translocated it with belowground hyphal network connection with neighboring root (He et al. 2003) which transfer N from plant to plant. But the exact mechanism is not well understood. AM are mostly considered with no saprophytic abilities; it has been reported that few species of AM are able to enhance decomposition due to their hydrolytic and oxidative enzymes, which depolymerizing organic nitrogen polymer (like protein, chitin) transferred liberated ammonium to the host plant (Hodge et al. 2001).

Effective absorption of nutrients by AMs is their narrow diameter relative to the roots. The steepness of diffusion gradient for nutrient is inversely related to the radius of absorbing unit. Therefore, soil solution is less depleted at the surface of a narrow absorbing unit such as hypha further, which can grow into small soil pores not accessible to roots not even root hairs. Various factors affect the NUE of AMs in agriculture production system. In well-fertilized soil with Zn and P, the ability of AM to improve NUE appears to be nonsignificant as the transfer of C to the fungus has been considered a drain on the host pant (Ryan and Graham 2002). Occasionally, plant growth suppression has been attributed to mycorrhizal colonization which usually occurs only under low-light (photosynthate limiting) or high-phosphorus condition. It can be easily predicted that a variety grown in nutrients rich soil will be less responsive to AM inoculation as compared to soil with poor quality of nutrients. An African variety grows better in P-deficient soil and was less sensitive to AM, whereas European variety has been greatly benefitted. There is a great variation among the modern varieties of crop plants for impact of AM (Sawers et al. 2008). It seems that root branching and P absorption ability are indirectly corrected with changing positive effects of AM fungi (Rengel 2002).

Another important benefit of the plant is the production of high molecular weight glycoprotein (glomalin) which is produced in abundance by AM fungi. This material accumulates in the soil and is positively correlated with aggregate stability. Glomalin is realized as carbon to soil and plays an important benefit of mycorrhizal colonization for the maintenance of healthy plant-soil system. In certain tropical soil, glomalin contributes up to 5 % of total soil carbon (Rilling et al. 2001). It further appears that mycorrhizosphere interactions played a key role in biogeochemical cycling of P and thus promoting plant fitness. These effects were further evaluated under field conditions in different crop plants like soybean, bell pepper, cotton, maize, etc. AM fungi are known for enhancing root branching, which might increase surface area of roots, due to secretion of a diffusible myc factor (Olah et al. 2005). The myc factors are capable of stimulating nodulation genes in plants along with formation of lateral roots without any inhibitory effect on elongation of primary root as it would be expected due to increase in levels of auxin production (Kosuta et al. 2003). Root morphology changes (Berta et al. 2002) triggered by AM colonization can be correlated with changing levels in plant auxin (Ludwig and Guther 2007), abscisic acid (Herrera et al. 2007), ethylene, and jasmonic acid (Tejeda et al. 2008), but the exact mechanism is still not clear.

20.3.2.1 *Piriformospora indica*

Sebacinales is the most basal *Basidiomycota* group with known mycorrhizal members which are ubiquitously distributed (Sharma et al. 2008). *P. indica* forms a novel type of mutualistic symbiosis with a broad spectrum of monocotyledonous and dicotyledonous plants (Weis et al. 2004) in contrast to AMF-Brassicaceae (Verma et al. 1999; Peskan-Berghoefer et al. 2004; Deshmukh et al. 2006). A novel plant promotional root-colonizing fungus, *Piriformospora indica* was isolated from Thar Desert soils in the northwestern part of India. *P. Indica* has potency to grow axenically on a variety of complex and semisynthetic media (Pham et al. 2004). According to Verma et al. (1998), *P. indica* enters the root cortex and forms inter- and intracellular hyphae. Within the cortical cells, the fungus often forms dense hyphal coils or branched structure intracellularly.

Interestingly, the host spectrum of *P. Indica* is very much alike AMF. It colonizes the roots of host plants as diverse as *Zea mays* L., *Nicotiana tabacum* L., *Petroselinum crispum* L., *Glycine max* L. Merr., *Populus tremula* L., *Setaria italica* L., *Oryza sativa* L., *Sorghum vulgare* L., *Triticum sativum* L., *Cicer arietinum* L., *Solanum melongena* L., *Artemisia annua* L., and *Bacopa monniera* L. Wett (Bagde et al. 2010). Also like AMF, *P. indica* does not colonize the members of Brassicaceae. It is a wide-host root, colonizing endophytic fungus which allows plants to grow under extreme physical and nutrient condition. It functions as a plant promoter and biofertilizer in nutrient-deficient soils, as a bioprotector against biotic and abiotic stresses. Studies *on P. indica* have shown fungal-mediated uptake of radiolabeled phosphorous from the medium and its translocation to the host in an energy-dependent process, evident by a sharp increase in its content in the shoot (Altomor et al. 1999). *P. indica* produces significant amounts of acid phosphatases for the mobilization of a broad range of insoluble, condensed, or complex forms of phosphate, enabling the accessibility of host plant for adequate phosphorous from immobilized reserves in the soil (Pham et al. 2004). *P. indica* contains substantial amount of an acid phosphatase with potential to solubilize phosphate in the soil and delivers it to the plant. It was also demonstrated that growth promotion of *Arabidopsis* seedling was associated with a massive uptake of phosphate from the growth medium (Shahollari et al. 2005).

20.4 Mechanism of Plant Growth Promotion

20.4.1 Phytohormones and Root Growth

20.4.1.1 Auxin (IAA)

A key mechanism of PGPR used to stimulate root growth is the release of auxin within the plant (Spaepen et al. 2007) as part of their colonization strategy. PGPR can modulate improvement in NUE by altering root architecture and growth due to release of phytostimulators: indole-3-acetic acid (IAA), gibberellins (GAs), cytokinin, ethylene, and certain volatiles (Perez-Montano et al. 2014) (Table 20.2). IAA is the most common plant auxin and can affect every aspect of plant development including cell enlargement, division, and tissue differentiation. A low concentration of IAA is known to stimulate primary roots whereas higher concentrations reduce primary root elongation. IAA concentrations simultaneously also induce initiation of lateral and adventitious roots with increased absorptive surface area. Many fungi and bacteria are capable of producing phytohormones (Tsavkelova et al. 2006). The ability to produce phytohormones is widely distributed among soil and plant-associated bacteria. Release of phytohormone is a bacterial strain-specific capacity

Table 20.2 [a]Plant growth promoting rhizobacteria with functional traits

PGPR	Plant growth promotional activity	References
Pseudomonas sp.; *Klebsiella* sp.; *Ralstonia metallidurans*	IAA, siderophores, HCN, ammonia, exopolysaccharides, phosphate and Zn solubilization, biocontrol potentials, heavy metal solubilization	Sharma et al. (2014), Ahemad and Khan (2011a, c, 2012a, c, e), Ma et al. (2011a), Naik and Dubey (2011), Braud et al. (2009), Tank and Saraf (2009), Poonguzhali et al. (2008), Shaharoona et al. (2008), Rajkumar and Freitas (2008), Ganesan (2008)
Bacillus sp., *Pseudomonas* sp., *Azotobacter* sp.	IAA production, ACC deaminase production, siderophore production	Kaur and Sharma (2013), Wani and Khan (2010)
Mesorhizobium sp.	IAA, siderophores, HCN, ammonia, exopolysaccharides	Bhagat et al. (2014), Ahemad and Khan (2009a, 2010d, e, h, 2012a), Wani et al. (2008)
Rhizobium sp.	IAA, siderophores, HCN, ammonia, exopolysaccharides	Ahemad and Khan (2009b, 2010c, f, g, 2011d, h, i, 2012b), Zahir et al. (2010)
Bradyrhizobium sp. *Pseudomonas* sp., *Ochrobactrum cytisi*	IAA, siderophores, HCN, ammonia, exopolysaccharides, heavy metal mobilization	Ahemad and Khan (2011g, j, 2012f, Dary et al. (2010)
Azotobacter sp., *Mesorhizobium* sp., *Pseudomonas* sp., *Bacillus* sp.	IAA, siderophore, antifungal activity, ammonia production, HCN	Ahmad et al. (2011e)
Enterobacter sp.	IAA, siderophores, HCN, ammonia, exopolysaccharides, phosphate solubilization, ACC deaminase	Ahemad and Khan (2010a, b), Kumar et al. (2008)
Klebsiella sp.	IAA, siderophores, HCN, ammonia, exopolysaccharides, phosphate solubilization	Ahemad and Khan (2011b, e, f)
Stenotrophomonas maltophilia	Nitrogenase activity, phosphate solubilization, IAA, ACC deaminase	Mehnaz et al. (2010)
Rahnella aquatilis	Phosphate solubilization, IAA, ACC deaminase	Mehnaz et al. (2010)
Proteus vulgaris	Siderophores	Rani et al. (2009)
Acinetobacter spp.	IAA, phosphate solubilization, siderophores	Rokhbakhsh-Zamin et al. (2011)
Azospirillum amazonense	IAA, nitrogenase activity	Rodrigues et al. (2008)
Psychrobacter sp. SRS8	Heavy metal mobilization	Ma et al. (2011b)
Serratia marcescens	IAA, siderophore, HCN	Selvakumar et al. (2008)
Paenibacillus polymyxa	IAA, siderophores	Pindi et al. (2014)
Acinetobacter sp., *Pseudomonas* sp.	ACC deaminase, IAA, antifungal activity, N_2 fixation, phosphate solubilization	Indiragandhi et al. (2008)
Burkholderia	ACC deaminase, IAA, siderophore, heavy metal solubilization, phosphate solubilization	Jiang et al. (2008)

[a]Adapted from Ahemad and Kibret (2014)

(Boiero et al. 2007) for enhancement of nutrient and water uptake from the soil by changing in root morphology and physiology. Root branching and root hair production are key traits of root for enhancement of NUE. Specifically, rhizobia, PGPR including *Pseudomonas*, *Bacillus*, *Gluconoacetobacter*, *Azotobacter*, *Azospirillum*, *Burkholderia*, and *Herbaspirillum* sp., have been reported to improve plant growth through manipulation of complex and balanced network of plant hormones through direct involvement by stimulating the root formation (Khalid et al. 2004).

Application of *Rhizobium* improved yield in rice due to increased root system with a significantly larger absorptive surface area and extensive root architecture (Biswas et al. 2000). This poses the question whether this benefit of rhizobia to rice may be due to their associative nitrogen-fixing activity or their ability to change the phytohormone balance. This phenomenon is well illustrated in auxin-resistant *Arabidopsis* mutant (axr 1 and axr 2) which produced lesser lateral roots than wild type, whereas overproducing *Arabidopsis* mutants dramatically increased the formation of lateral roots and root hairs. Auxin synthesis in *Agrobacterium* plays a role in gall formation and pathogenesis but many nonpathogenic fungi and bacteria are also known to synthesize auxin by six different genetic pathways. Tryptophan (amino acid), an important molecule, acts as the main precursor for IAA synthesis and thus helps in modulating the level of IAA biosynthesis (Zaidi et al. 2009). IAA productions by PGPR have been reported to promote plant growth in different crop plants. Microbial synthesis of the phytohormone auxin (IAA) has been known for a long time. Starting with tryptophan, at least five different pathways have been described in microorganisms for the synthesis of IAA, and most pathways show similarity to those described in plants, although some intermediates can differ.

Auxins (IAA) are also synthesized and released as secondary metabolites by 80 % of microbes isolated from rhizospheric and rhizoplanic spaces of different legume crops (Kaur and Sharma 2013; Kumawat et al. 2014). In several microorganisms, IAA also acts as reciprocal signaling molecules and affected interaction of the gene expression of microbes. IAA has an indispensable role in virtually all aspects of plant growth development and defense mechanism. Consequently, IAA plays an important role in plant-microbial interaction. Generally, IAA affects plant cell division, proliferation, and differentiation of root growth, enhancing seed and tuber germination, increases the rate of lateral and adventitious root development, and controls vegetative growth. IAA also affected the photosynthesis, pigment formation, biosynthesis of secondary metabolites, and resistance to stressful conditions due to plant auxin pool, which are mediated response to light, gravity, and flourescence. IAA produced by rhizobacteria likely interferes the above physiological processes of plants by altering the plant auxin pool. However, bacterial IAA interacts with plant and increases the surface area and length of root, thereby providing the greater amount of soil nutrient to the plants. IAA also provides additional nutrient for the growth of rhizospheric and rhizoplanic bacteria by enhancing the quantity of root exudation through loosening the cell wall of plants (Glick 2012). Thus, rhizobacterial IAA is identified as an effector molecule in plant-microbe interactions, both in pathogenesis and phytostimulation.

IAA is involved in multifarious process comprising, viz., multiplication, differentiation, and vascular bundle formation; all these processes are necessary for nodule formation. Therefore, IAA levels are essential for nodule formation (Glick 2012; Spaepen et al. 2007). It is also reported that the coinoculation of *Rhizobium leguminosarum* bv. *viciae* with introduced IAA biosynthetic pathway produced potential nitrogen-fixing root nodules containing up to 60-fold more IAA than nodules formed by the wild-type counterpart in *Vicia hirsuta* (Camerini et al. 2008). In different bacteria, IAA biosynthesis level is also altered by environmental stress conditions including acidic pH, osmotic potential, matrix stress, and carbon limitation and genetic factors like location of auxin biosynthesis genes in the bacterial genome (either plasmid or chromosomal) and the mode of expression (constitutive vs. induced) (Spaepen et al. 2007; Martinsez-Viveros et al. 2010). Mostly, plasmids are present in multiple copies so that the IAA level is affected by the location of the biosynthetic genes. This difference in the IAA level can be demonstrated between the rhizobacterial strains, *Pseudomonas savastanoi* pv. *savastanoi* and *P. syringae* pv. *syringae*. In the latter strain, the auxin biosynthesis genes were located on a plasmid, but later, it was found on the chromosomal DNA, and production of IAA was low. The application of low-copy plasmid-carrying IAA biosynthetic operon in *Pseudomonas syringae* pv. *syringae*; the production of IAA is increased by many times (Spaepen et al. 2007).

20.4.1.2 Gibberellic Acid and Cytokinins

Gibberellins (GAs) are secondary metabolites that can play a key role as signaling factors toward the host plant. Previous reports also documented GA production by *Azospirillum* or *Bacillus* sp. as growth promotional in plants (Bottini et al. 2004). PGPR also produced cytokinins that were involved in root initiation, cell division, cell enlargement, and increase in root surface area of crop plants through enhanced lateral and adventitious roots (Salamone et al. 2005). Few strains of *Azotobacter* spp., *Rhizobium* spp., *Pantoea agglomerans*, *Rhodospirillum rubrum*, *Pseudomonas fluorescens*, *Bacillus subtilis*, and *Paenibacillus polymyxa* are reported to produce cytokinins (Glick 2012). However, the exact role and regulation of PGPR-synthesized cytokinins is not clear. Root system optimization with the auxin- and GA-secreting endophyte inoculant, *Rhizobium leguminosarum* bv. *trifolii* E11, significantly increased grain yield, agronomic fertilizer N-use efficiency, and harvest index in certain varieties of rice (Yanni et al. 2001). Further results echoed root enhancement with GA-producing PGPR strains of *B. licheniformis* and *Bacillus pumilus*; isolates from alder rhizosphere resulted in a stimulated growth of roots and plant yield by production of GA (Gutierrez-Manero and Martinez-Romero 2001). Enhanced root biomass in maize seedlings inoculated with *Azospirillum brasiliensis* and *A. lipoferum* also well correlated with the increased GA3 levels (Lucangeli and Bottini 1997). Same results were recorded when maize and soybean seedlings were inoculated with *A. brasilense* (Az39) and *Bradyrhizobium japonicum* (E109) strains which were capable of secreting GA. However, the root biomass increased without increasing root branching of elongation (Cassan et al. 2009). Various nodule-forming *Rhizobium* have been known to produce IAA, gibberellins, and cytokinins along with abscisic acid (ABA) and ethylene, but the exact mechanism and purpose for production of these substances are yet unclear (Boiero et al. 2007).

20.4.1.3 Ethylene

Ethylene is another vital compound for PGPR-dependent root growth enhancement through the reduction of their volatile plant hormone. Ethylene is an essential metabolite for the normal development of plants (Khalid et al. 2006). Approximately, all plants endogenously produced ethylene as growth hormones, and it also induces multifarious physiological changes in plants. Depending upon the concentration of ethylene in root tissue, it has various effects on plant growth. For instance, high concentration of ethylene induces defoliation and may lead to reduced plant growth as well as senescence, chlorosis, and organ abscission, all of which lead to reduced crop performance (Bhattacharyya and Jha 2012).

Ethylene has been considered as a stress hormone, and in various biotic and abiotic stress conditions, viz., salinity, drought, water logging, heavy metals and pathogenicity, the endogenous level of ethylene increased which have shown detrimental effects on the plant growth. For instance, the high concentration of ethylene induces defoliation and other cellular processes that may lead to reduced crop performance (Bhattacharyya and Jha 2012). Accumulation of ethylene leads to a downward spiral effect and resulted in poor root growth with diminished ability to acquire water and nutrients. Under different stress conditions, plants synthesized (Glick et al. 2007) 1-aminocyclopropane-1-carboxylate (ACC), a precursor of ethylene (Glick et al. 2007). In rhizosphere, some amount of ACC is secreted and absorbed by the plant roots, where it is converted from one form of ACC to other form of ethylene. PGPR with ability to produce enzyme, 1-aminocyclopropane-1-carboxylate (ACC) deaminase, facilitate plant

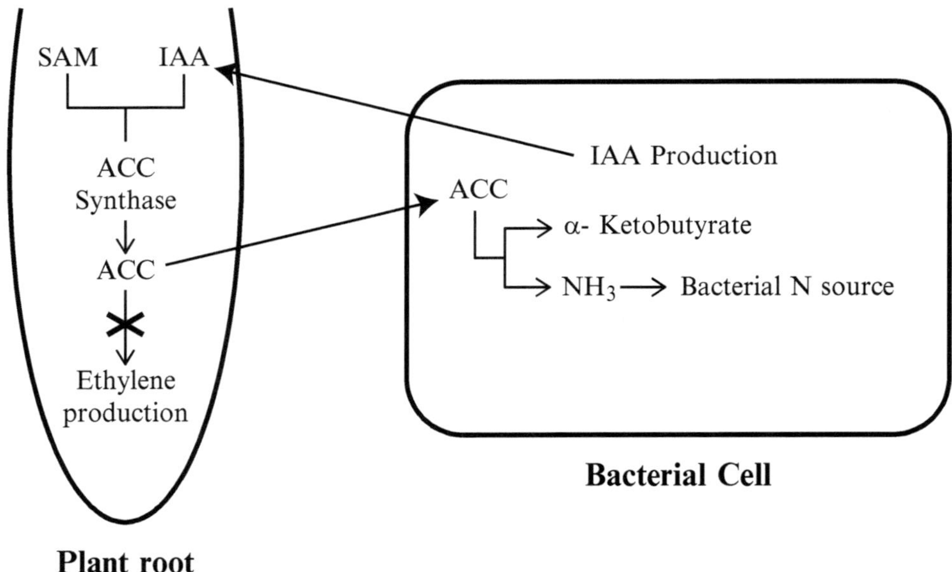

Fig. 20.1 Mechanism for stress control in plants by production of ACC deaminase by PGPR

growth and development by decreasing ethylene levels, inducing salt tolerance and reducing drought stress in plants (Zahir et al. 2010). Currently, a vast array of bacterial strains exhibiting ACC deaminase activity in a wide range of genera such as *Acinetobacter*, *Achromobacter*, *Agrobacterium*, *Alcaligenes*, *Azospirillum*, *Bacillus*, *Burkholderia*, *Enterobacter*, *Pseudomonas*, *Ralstonia*, *Serratia*, *Rhizobium*, etc. have been reported in the literature (Zahir et al. 2009; Kang et al. 2010).

Rhizosphere-inhabiting bacteria are known to affect plant ethylene by secreting ACC, the precursor of ethylene into two components, i.e., ammonia and α-ketobutyrate (Fig. 20.1); the former is used to reduce nitrogen source of PGPR (Bernard 2005) and improved plant growth in various crop plants (Glick 2012). *B. phytofirmans* PsJN isolated from onion infected with pathogen was an efficient plant growth promoting endophytic bacteria (Sessitsch et al. 2005). ACC deaminase gene in *B. phytofirmans* was mutated to decrease the ability of bacteria to enhance root elongation in canola seedlings (Sun and Cheng 2009).

PGPR with the ability to degrade ACC in the rhizosphere also reestablish an extensive root system required to survive under environmental stress improves NUE of plant. ACC deaminase helps in decreasing or preventing negative effects of elevated ethylene levels (Glick et al. 1998). Out of 233 strains of *R. leguminosarum* from Saskatchewan, Canada, 27 isolates possess this gene (Duan et al. 2009). Rhizobia expressing ACC deaminase are more effective in root nodule formation in legumes. *R. leguminosarum* bv. *viciae* contains one copy of ACC deaminase, which when mutated reduces its ability to nodulate *Pisum sativum* L. cv. Sparkle (pea) (Ma et al. 2003). ACC deaminase gene from *R. leguminosarum* bv. *viciae* was introduced into *Sinorhizobium meliloti* which does not have this enzyme; transgenic bacteria (RM BPC-2) enhanced 35–40 % for nodulation in alfalfa (Ma et al. 2004), and it being sold by an American company, Research Seed Inc., it is the first of its kind of genetically modified microorganisms that have been commercially released. Most of the measurable effects are root growth, enhancement of shoot growth, and improvement in rhizobial nodulation and NUE of N, P, and K acquisition as well as mycorrhizal colonization in different legume crops (Nadeem et al. 2009; Glick 2012). Various studies have documented that inoculation with ACC deaminase-producing PGPR capable of producing it helps in increasing resistance against environmental stress. Grichko and Glick (2001) reported that inoculation of tomato seeds with *Pseudomonas putida* and *Enterobacter cloacae* (ACC-producing bacteria) can lead to increased resistance for 9 consecutive days of flooding at 55 days of age. Root elongation in *Brassica campestris* was stimulated by inoculating three *Bacillus* sp., i.e., *B. firmus*, *B. circulans*, and *B. globispora* (Ghosh et al. 2003). Inoculation with *Achromobacter piechaudii* in tomato plants under saline and water stress conditions resulted in significant enhancement in fresh and dry weight of plants (Mayak et al. 2004). An increase in dry matter content of root was observed by Reed and Glick (2005) when the seeds were inoculated with *Pseudomonas asplenii*, an ACC deaminase-producing bacteria.

20.5 Role of Non-hormone Metabolites

Microbe's secret non-hormone metabolites which may alter the root architecture. Volatile glucose metabolites, 2,3-butanediol, and acetoin are released by rhizobacteria under oxygen-deficit conditions to stimulate *Arabidopsis* root growth (Ryu et al. 2003). N-acyl-homoserine lactones are another type of non-hormonal metabolites which hold great importance in cell to cell communication for many microorganisms including PGPR by affecting root development. 2,4-diacetylphloroglucinol (DAPG) is another important non-hormone metabolite produced by *P. fluorescens* isolates containing the phlD gene. It inhibits the growth of primary root, whereas it stimulates production of lateral roots in tomato through altered auxin signaling (Brazelton et al. 2008). Corn seed treatment with DAPG-producing pseudomonas in acidic soils results in increased vigor and enhancement in absorption of Mg and P while decreasing Al accumulation (Raudales et al. 2009).

20.6 Nutrient Acquisition

20.6.1 Nitrogen Uptake

Nitrogen (N) is the principal nutrient for plant growth and productivity. About 78 % N is present in the atmosphere but still limiting to the growing plants due to the high losses by emission or leaching in agricultural ecosystems. Diazotrophs include cyanobacteria, eubacteria, and actinomycetes, which use fixed C to break down the triple bond of dinitrogen (N_2) by energy consuming and oxygen-sensitive process to form ammonia through BNF. Microorganisms with ability to enhance plant growth by improving nutrient uptake of the plants are known as biofertilizers. Improvement in the nutrient status of host plants by diazotrophs were attributed due to multifunctional modes including nitrogen fixation, enhancing the root proliferation, increasing the availability of nutrients in the rhizosphere, and promoting beneficial symbiosis of the host. Plants mostly can uptake N from soil as nitrite (No_2^-), nitrate (No_3^-), or ammonia (NH_3). N is a scarce element in most soils, and the chemical N fertilizers are employed in agriculture to fulfill the requirement of synthetic N frequently lost during rainfall or by mineral leaching of the fertilizers.

Bacteria with ability to convert atmospheric N_2 to plant usable form by symbiotic and nonsymbiotic biological fixation process play a critical role. Symbiotic BNF is a well-defined mechanism of fixing most of atmospheric N_2 but restricted to leguminous plants and trees and shrubs forming actinorhizal roots in association with *Frankia*. The process of SNF is carried out in nodules. *Rhizobium*, *Mesorhizobium*, *Sinorhizobium*, and *Bradyrhizobium* are most extensively studied rhizobia. These bacteria are not yet considered as PGPR although their SNF quality is known except in case of association with other nonleguminous plants (Dobbelaere et al. 2003). N_2 fixation is a well-defined process in nodules of legume plants which are specialized structures on root or stem for microbial fermentation resulting in formation of infection thread and a signaling pathway between plant and the bacteria for enhanced availability of N to the plant. An oxygen-deficit high-sugar level environment is created to facilitate N_2 fixation. Depending on rhizosphere of soil and plants, both Gram-negative (rhizobial) or Gram-positive (actinorhizal) bacteria hosted it as symbiotic and nonsymbiotic N_2 fixers; however, nodule formation and function involve a similar set of plant-microbe signals coded by genes that appear to have been transferred horizontally between many different proteobacterial strains (Pedrosa et al. 2002). Legumes are the largest and most agronomically important group of plants, largely due to the ability of 88 % of the 19,000 species described to form nitrogen-fixing nodules making them largely nitrogen independent in agricultural settings (Graham and Vance 2003). Several studies have also demonstrated, free-living N_2-fixing bacteria and rhizobia can also stimulate the plant growth in nonlegumes and contributed in reducing dependence upon nitrogen-based chemical fertilizers (Bhattacharjee et al. 2008).

PGPR are present almost in every plant, but and their potential to fix biological nitrogen is not yet fully exploited. All crop plants except for legumes have high requirement for chemical nitrogen. Wheat, rice, and maize have nitrogen requirement of 20–40 kg ha^{-1} N for a tonne of grain produced. The main focus of research was to induce nodulation in cereals but the success is very limited in this area. PGPR have the ability to fix atmospheric nitrogen nonsymbiotically in nonlegume plants. Increased dry weight of some Mexican landraces of maize was possibly due to effective PGPR, *R. elti* present in nodules on *P. vulgaris* (Gutierrez-Zamora and Martinez-Romero 2001). Under low-nitrogen fertilization (50 kg N ha^{-1}), inoculation of wheat with *R. leguminosarum* and 15 N tracer techniques also suggested positive effect, and plants were capable of fixing 29 % of the nitrogen in shoots, whereas no nitrogen fixation was observed in uninoculated plants. In addition to rhizobia, nonsymbiotic bacteria have been found in several agricultural plant and also hosted several other diazotrophic endophytic bacteria as well: these include *Rahnella*, *Pantoea*, *Pseudomonas*, *Azospirillum*, *Brevundimonas*, *Klebsiella*, *Ideonella*, and *Herbaspirillum* in maize (Roesch et al. 2008) and *Herbaspirillum*, *Pantoea*, *Pseudomonas*, *Brevundimonas*, *Ideonella*, *Rheinheimera*, *Azospirillum*, and *Enterobacter* in rice (Mano and Morisaki 2008). A large diversity of diazotrophs (PGPR) were isolated from sugarcane, including *B. tropicalis*, *G. diazotrophicus*, *Herbaspirillum rubrisubalbicans*, and B. *brasilensis* in Brazil. These isolates were seemed to fix large amount of N in micropropagated sugarcane (Oliveira et al. 2002). *G. diazotrophicus* has been isolated from many crops such as tea, pineapple, mango, sweet potato, finger millet, and coffee and may be assumed to be playing an important role in biological nitrogen fixation in host plant (Muthukumarasamy et al. 2002). It grows at pH 5.5 and 10 % sucrose level conditions similar to inner tissue of sugarcane stem; however, it cannot survive well in soil. It is insensitive to nitrate and poorly sensitive to ammonium at high sugar (10 % sucrose) levels with ability to fix N_2, indicating that it may fix N_2 in tissue of cane even when the plant is taking N from soil. Most interestingly, in vitro growth of *G. diazotrophicus* with yeast shows that more than half the fixed N_2 is secreted and usable by the yeast, suggesting that this fixed nitrogen is bioavailable to its plant host as well (Cojho et al. 1993). Contrary to the above discussion, several researchers reported that nonsymbiotic bacteria have limitation to fix higher quantities of N due to more energy requirement for fixation of atmospheric N_2 and metabolic activity of free-living microorganisms is low and they have to compete for the root exudates released by plant in rhizosphere. The substantial contribution of diazotrophic bacteria of N_2 fixation and of N in plant growth is not considered. Although N in plant growth can be easily assessed in vitro, in vivo studies are more complex and variable. Dobbelaere et al. (2003) suggested that significant quantities of N can be provided by rhizobacteria in crop plants. However, inoculation of *Azospirillum* in wheat, sorghum, and maize contributed 5 kg N/ha/year. This quantity is very negligible in comparison to 150–200 kg N/ha/year as common practice in modern agriculture. Contribution of free-living rhizobacteria for crop plants in Australia is even less than 10 kg N/ha/year (Unkovich and Baldock 2008). Peoples et al. (2002) also pointed out that the range of N_2 fixation varied from 0 to 15 kg N/ha/year. Further, value of N_2 fixation ranged between 1 and 10 kgN/ha/year as suggested by Bottomley and Myrold (2007). The ability of plant growth promoting bacteria to fix atmospheric N_2 is no longer considered as an important criterion to classify them under as a biofertilizer.

20.6.2 Phosphorous Uptake

Phosphorus (P) is the second most vital nutrient after nitrogen for growth and productivity in crop plants. It is present in soil, plant, microorganisms, and a number of organic and inorganic compounds. However, the total P content in average soil is 0.05 %, and only 0.1 % of total P present in the soil is available to plants. This low availability of P to plants is because of its fixation and low solubility. Approximately 80 % of applied P fertilizers are immobilized due to the formation of complex with Al or Fe in acidic soils or Ca in calcareous soil. An alternative way to circumvent P deficiency in soil and to improve crop production is coupled with phosphate-solubilizing activity of microorganisms, referred as phosphate-solubilizing microorganism (PSM). PSM provides available form of P to plant by converting the fixed phosphates (organic and inorganic) into two soluble forms, the monobasic (H_2PO^{-4}) and the dibasic ($HPO_2^{-\ 4}$) ions (Bhattacharyya and Jha 2012). Recently, P content in soil is high due to long-term application of phosphatic fertilizers over several decades. Most of insoluble P is present as an inorganic mineral like as apatite and some organic forms involved, viz., inositol phosphate (soil phytate), phosphomonoesters, and phosphodiesters (Glick 2012).

Most of phosphate-solubilizing bacteria (PSB) are inhabiting in the rhizosphere, considered as potential biofertilizers, which liberate phosphorous; otherwise, they supply P either with mineralization or solubilization. Bacterial genera like *Bacillus*, *Enterobacter*, *Flavobacterium*, *Pseudomonas*, *Rhizobium*, *Enterococcus*, *Stenotrophomonas*, and *Serratia* are reported as the potential phosphate-solubilization rhizobacteria (Bhattacharyya and Jha 2012). Due to production of low molecular weight organic acids, viz., acetic, lactate, oxalic, succinic, citric, gluconic, and ketogluconic acids involved in improvement of inorganic P by PSB. PSB also release proton through the assimilation $NH^+{}_4$ (Zaidi et al. 2009). Mineralization of organic phosphorous takes place by breakage of phosphoester bond by phosphorous-dependent enzymes, i.e., phosphatase. Phosphate solubilization and mineralization simultaneously present in the same bacterial strain (Tao et al. 2008).

PGPR grow in Pikovaskaya's medium in vitro amended with tricalcium phosphate or similar insoluble sole phosphate source. These PGPR not only solubilize for their own nutritional requirement but also make it available to crop plants (Chen et al. 2006). Organic source of phosphate constitutes between 30 % and 50 % of total P present in soil (Turner et al. 2002). Phytate are mineralized by phytase-producing enzymes (Lim et al 2007). Higher population of PGPR dissolving insoluble P have been reported in rhizosphere and rhizoplane of different crops as compared to non-rhizosphere soil (Jorquera et al. 2008b). To date, very few reports are documented phytate mineralization by PGPR. Common culturable bacteria, i.e., *Enterobacter*, *Burkholderia*, *Bacillus*, *Pseudomonas*, *Staphylococcus*, and *Serratia* are reported to be phytase-producing rhizobacteria (Hussin et al. 2007; Shedova et al. 2008). P-solubilizing bacteria were isolated from rhizosphere of cultivated plants (*Triticum aestivum*, *Lolium perenne*, *Avena sativa*, *Trifolium repens*, and *Lupinus luteus*) (Jorquera et al. 2008a) and different mechanisms are reported for utilizing phosphorus. The application of PSB alone or in combination with other rhizospheric microbes as consortium (Ahemad and Khan 2012a) also had positive effects on growth and yield in mustard crop. PSB provide phosphorous to plant and also increases the efficiency of BNF and increases the availability of trace elements through the formation of plant growth promoting substances (Zaidi et al. 2009). As with diazotrophs, the production of growth hormones helping in improved root surface area may have indirect effects on P-solubilizing ability of the bacteria.

Presently, major drawback is the inconsistent effects on phosphate mobilization by PGPR in vivo are severely affected because of the environmental stresses and competition with native microflora (Ahmad and Khan 2010a, b). It appears that the evaluation of ability to solubilize P in vitro does not participate as efficiently under field conditions (Rengel 2008). Thus, it may be concluded that the biofertilizers with dual action might have been mediated by directly solubilizing inorganic source of P, mineralizing organic P, and stimulating root growth or formation of mycorrhiza.

20.6.3 Fe Uptake

Plant required Fe in minor quantity which is important for various biosynthesis pathways, viz., Kerb's cycle; photosynthesis and also acts as a cofactor of many intermediate biosynthetic processes (Kobayashi and Nishizawa 2012). Due to deficiency of iron, the vegetative part gets yellowish due to chlorosis, and ultimately, the plant dies and causes are loss of yield and quality of crop. Fe deficiency also leads to anemia in humans (Zheng 2010; Murgia et al. 2012). In plants, efficient utilization of Fe has induced two processes in angiospermic plants, i.e., non-graminaceous (dicot) and gramineae (monocot). In non-graminaceous plants, due to proton gradient channel (proton extrusion), the acidification of rhizosphere takes place across the plasma membrane due to more soluble forms of Fe, i.e., ferric to ferrous takes place via ferric chelate reductase (FRO_2), and this soluble forms of Fe transport into root cell by iron-regulated transporter 1(IRT1) (Ivanov et al. 2012; Kobayashi and Nishizawa 2012). In gramineae (monocot), Fe uptake is mediated due to release mugineic acid from the phytosiderophores (MAs) (Nozoye et al. 2011).

In most of aerobic microbial habitats, Fe^{2+} is present in oxidized state to Fe^{3+}, forming insoluble compounds, hence, unavailable to microorganisms. Under such situations, some bacteria and AMF produce low molecular compounds with very high affinity for Fe (III) as iron chelators known as siderophores (Machuca et al. 2007). Siderophore also mediate Fe solubilization from mineral organic compounds in iron-deficient condition under neutral to alkaline pH soils, due to low iron solubility at elevated pH (Sharma et al. 2003).

In aerobic calcareous soil, potential microorganisms liberate Fe-chelating compounds, i.e., siderophore; this Fe chelate overcomes low Fe availability, resulting in more solubility of Fe hydroxides (Schalk et al. 2011). Microorganisms form siderophores which increase the bioavailability of Fe by forming the siderophore-iron complexes (Lemanceau et al. 2009; Saha et al. 2012). There are almost 500 compounds identified as siderophores on basis of their ligand architecture and have been classified as hydroxamate, catecholate, and/or hydroxycarboxylic acid (Bonnefoy and Holmes 2012). Fe-chelate compound, i.e., siderophores produced in rhizospheric zone of plant, plays a pivotal role in efficient utilization of Fe through solubilization from Fe hydroxides (Hayat et al. 2012).

Fe acquisition by production of siderophores plays an important role in assessing for root colonization and outcompeting deleterious microorganisms by sequestering Fe^{3+} in plant rhizosphere (Siddiqui 2006). Reduction in Fe concentration around the plant roots does not significantly affect the plant growth as decreased Fe concentrations occur at the microsites having high microbial activity during the course of pathogen establishment. A number of plant species are reported to use bacterial siderophores as the source of Fe even when the concentration of Fe is quite low as compared to Fe uptake by plant. PGPR belonging to *Rhizobium* (Roy and

Chakrabartty 2000), *Bradyrhizobium* (Khandelwal et al. 2002), *Serratia* (Kuffner et al. 2008), and *Pseudomonas* sp. (Kaur and Sharma 2013) have been reported as siderophore producing by various researchers. Siderophore-mediated Fe uptake due to inoculation of siderophore-producing rhizobacteria vis-a-vis plant growth promotion has been reported earlier by Rajkumar et al. (2010). Inoculation of siderophore-producing *Pseudomonas*, *Rhizobium*, and *Azospirillum* showed positive effects on alfalfa plantlet growth and was recorded by Carrillo-Castaneda et al. (2002) and grown in iron-limited condition.

Siderophores also mediate Fe solubilization through chelation from organic compounds and minerals in Fe-deficient situation under neutral to alkaline pH soil due to low iron solubility at elevated pH (Crowley and Kraemer 2007). Rhizospheric microorganisms have the ability to utilize Fe-siderophore complexes. Similarly, *Pseudomonas fluorescens* C7 synthesize Fe pyoverdine complex which leads to an increase of iron inside plant tissues with improved plant growth in *Arabidopsis thaliana* (Vansuyt et al. 2007). Sharma et al. (2003) also assessed the role of the siderophore-producing *Pseudomonas* strain GRP3 on iron nutrition of *Vigna radiata*. After 45 days, plants with chlorotic symptoms with iron deficiency in *Vigna radiata* corrected by the application of *Pseudomonas* strain GRP3 with improvement Fe nutrition and chlorophyll a and chlorophyll b as compared to control treatment (Sharma et al. 2003). Nevertheless, the enhanced growth may be due to other PGP mechanism or combination of different mechanisms leading to enhanced availability of nutrients, pathogen suppression, or effect on root growth due to production of hormones.

Fe acts as a cofactor of auxin (phytohormone), which is an important chemical signaling compound enhancing Fe-deficiency inducible responses. Application of synthetic auxin exogenously like IAA, ACC, and α- NAA increase the Fe uptake in plants by increasing the surface area of root hairs as well as lateral roots with expression of FRO2 and IRT1 with Fe-deficiency induced reduction of Fe (Jin et al. 2008; Chen et al. 2010; Wu et al. 2012). In Fe-limited soils, the ethylene also participates to overcome iron deficiency by the microbial action. Symbiotic N_2 fixation in grain legume nodules by microorganisms having Fe-containing protons plays a significant role in iron uptake (Terpolilli et al. 2012). Requirement of Fe in nodulated legumes is higher as compared to non-nodulated legumes meet this increased damage of Fe; similar nodulated legumes have developed a mechanism to increase Fe-deficiency induced responses in root nodules. As an example, in peanut plants, the roots liberate proton and reductants of Fe-deficient which stimulate the population of *Rhizobium* nodulation, and results were documented by Terry et al. (1988), Soerensen et al. (1988) in soybean plants. In red clover plant root Fe deficient, the *Rhizobium* nodulation clearly stimulates the activity of Fe-chelate reductase (Jin et al. 2007).

The radioisotope studies reveal that the transportation of Fe takes place from roots to shoots in red clovers by *Rhizobium* nodulation. The Fe uptake and growth of red clover decreased in calcareous soil due to absence of *Rhizobium*, but the Fe uptake and growth of red clover increased in presence of *Rhizobium leguminosarum* bv. *trifolii* (Jin et al. 2006). Mishra et al. (2011, 2012) showed that the uptake of Fe content enhances in pea and lentil plants in the presence of root-nodulating *R. leguminosarum*-PR1. But the clear reason is still unknown how to improve the Fe uptake by *Rhizobium* by increasing Fe-deficiency induced responses. It showed that symbiotic nitrogen fixers and Fe enhancing are independent of each other. The nodules that devoid N_2 fixation still enhance the activity of ferric-chelate reductases in roots. Under iron-limitation condition, the leguminous plants secrete more phenolic compounds. The growth of nodulating rhizobia also performs as signal for rhizobial nod gene expression (Deryl and Skorupska 1992; Hassan and Mathesius 2012). Fe deficiency promotes the phenolic secretion which also provides favorable condition for nodulating rhizobia in leguminous plants. In Fe-limiting condition, the secretion of phenolic compounds is also known to promote the growth of *Rhizobium meliloti* in alfalfa (Masaoka et al. 1997). Thus, Fe-deficient legumes secrete the phenolic compounds which enhance the Fe uptake of legumes themselves by inducing *Rhizobium* nodulation by phenolics. Under Fe-limiting condition, the application of rhizobial strains form Fe-chelate siderophores for increasing the nodulation in legume roots (Arora et al. 2001). The mechanism by which siderophores involved in Fe uptake of nodules and finally enhance the formation of Fe-containing proteins. The inoculation of siderophore-producing rhizobia increases the Fe uptake in leguminous plants as comparison to nonproducing siderophore rhizobia (Terpolilli et al. 2012).

Mycorrhizae also secrete low molecular weight organic chelating compounds, viz., siderophores, citric acid, oxalic acid, and proton for mobilization of available Fe present in rhizospheric soil and might have involved in uptake of Fe in plants (Winkelmann 2007; Bharadwaj et al. 2012). Several studies show that the mycorrhizal symbiosis helps in Fe uptake in mycorrhiza-infected plants (Amanullah et al. 2012; Labidi et al. 2012). On the contrary, Clark et al. (1999) revealed decrease in Fe uptake to mycorrhizal application. These variable responses of mycorrhizal symbiosis in Fe uptake might be attributed due to fluctuation in soil properties and environmental conditions for plant growth. Treeby (1992) and Medeiros et al. (1993) revealed improved Fe uptake in mycorrhizal-infected plants under low pH than grown under high pH. In addition, Raju et al. (1990) also concluded *Glomus macrocarpum* (vesicular-arbuscular mycorrhizal fungi) at 25 °C or 30 °C enhanced 10 times Fe uptake in comparison to *Glomus* at 20 °C in sorghum. The symbiotic relationship between mycorrhiza and plant species contributes in Fe uptake. In a nutshell, the previously documented literature provided the information that mycorrhizal symbiosis might be beneficial for Fe uptake in agricultural crops (Clark and Zeto 2000).

20.6.4 Zinc Uptake

Zinc has an immense role in nutrition of both eukaryotic and prokaryotic organisms as cofactor or metal activator in various enzyme systems. According to Prasad (2010), in Indian soils under soybean-wheat cropping systems, harvest of 6.5 ton grain/ha/year removes 416 g Zn/ha/year and exhibits 50 % Zn deficiency, below the critical level of 1.5 ppm of available zinc. About 75 % exogenous application of zinc sources like $ZnSO_4$ get fixed in soil. Fixation of Zn in soils with pH >7.0 increases with increasing concentration of carbonates, thus becoming unavailable and can be reverted back to available form with Zn-solubilizing microorganism. Application of potential native Zinc solubilizers is known to enhance the availability of zinc for crop assimilation and can also reduce zinc malnutrition in human masses. Native zinc mineralizating and solubilizing bacteria can decrease the Zn deficiency by increasing the zinc availability to the crops (He et al. 2010). Bacteria are known to immobilize metal by precipitation and adsorption. The ability to dissolve immobilized zinc, viz., zinc phosphate, zinc oxide, and zinc carbonate in appreciable quantity is not a common feature among the cultivable bacteria on soil surface. Few Zn-solubilizing bacterial genera, viz., *Thiobacillus thiooxidans*, *Thiobacillus ferrooxidans*, *Acinetobacter*, *Bacillus*, *Gluconacetobacter*, and *Pseudomonas* and facultative thermophilic iron oxidizers have been reported as zinc solubilizers. Commonly used nitrogen-fixing bacteria like *Rhizobium*, *Azospirillum*, and *Azotobacter* are not reported as zinc solubilizers, but some endophytic bacteria, viz., *Acinetobacter*, *Bacillus*, and *Pseudomonas* (Saravanan et al. 2007) have been reported to solubilize zinc.

Various studies reveal the exchange of metal ions in rhizospheric soil. The concentration of metal ion increases which leads to reduction of pH in rhizospheric soil due to root exudations. Consistent enhancement in Fe uptake in soybean and wheat (panicle initiation) with three *B. aryabhattai* strains is well documented with earlier literature with bacteria of genera *Bacillus* and also increases the root and shoot dry weight (He et al. 2010; Zhao et al. 2011; Minaxi et al. 2012). The secretion of microbial metabolites due to intensive microbial activity which play a key role in soil zinc cycle changes into one form to another form of soil zinc. These metabolites increase microbial biomass and physiological and enzymatic processes (Neumann and Romheld 2000). Application of Zn solubilizer native bacteria increased zinc concentration in shoots and roots in soybean and wheat plants as reported by He et al. (2010) and Madhaiyan et al. (2010) in comparison to non-inoculated condition. Similarly, enhancement in dry matter accumulation and zinc acquisition through the inoculation of plant growth promoting rhizobacteria have been reported (Rana et al. 2012). Significant improvement in grain yield with *B. aryabhattai* strains MDSR7 and MDSR-14 in comparison to uninoculated control was increased due to microbial biomass C and enzymatic processes with significant reduction of rhizospheric pH which ultimately leads to improvement of crop production (Sharma et al. 2014).

Native potential strains can be applied as consortium inoculants to ameliorate zinc deficiency in soils after extensive assessment improved availability of zinc in soybean and wheat. Bacilli are gaining special interest early due to endospore formation that allows them to encounter with their adverse environmental conditions and can be mass produced, permit easy multiplied, and can be stored with longer shelf life ability to form nonspecific association with different host plants (Perez-Garcia et al. 2011). Recently, effectiveness of dual inoculants in mobilizing zinc and assimilation by crop plants with concomitant changes in rhizosphere properties has been conducted (Sharma et al. 2011).

20.7 Conclusion

Chemical fertilizer has largely increased global agricultural production to meet future demands in an unsustainable manner. This chapter has been focused on a heterogeneous group of PGPM naturally present in the rhizosphere. They live in close association with plant roots and have significant potential to improve nutrient use efficiency. Among PGPM bacteria and fungi discovered till now, the bacteria *Rhizobium*, *Azospirillum*, *Bacillus*, *Pseudomonas*, and fungus *Glomus* are the widely demonstrated in numerous published studies. PGPM improve plant nutrient use efficiency with a plethora of mechanisms with extensive root hyphae for nutrient absorption, release of phytohormone for stimulation of root growth (e.g., auxin, gibberellins, cytokinin, and ethylene), and alteration of plant metabolism for higher nutrient acquisition (N, P, Fe, and Zn). Currently, application of PGPR with multiple PGP traits is limited in agricultural crops due to their variation under different host and agroclimatic conditions. To achieve the desired goal with PGPR, we need matching technology for specific host plant with a given set of environmental conditions to improve growth and yield in agricultural crops.

PGPR may be advantageous in improving plant growth due to the existing reluctance worldwide to accept foods produced by transgenic plants. PGPM are the potential tools for agriculture and trend for the future. For this reason, research efforts should be intensified on the development of synergistic consortium PGPM inoculants compatible with agrochemicals as well as soil organic amendments under diverse agroecological conditions.

References

Ahemad M, Khan MS (2009a) Effect of insecticide-tolerant and plant growth promoting *Mesorhizobium* on the performance of chickpea grown in insecticide stressed alluvial soils. J Crop Sci Biotechnol 12:213–222

Ahemad M, Khan MS (2009b) Toxicity assessment of herbicides quizalafop-p-ethyl and clodinafop towards *Rhizobium* pea symbiosis. Bull Environ Contam Toxicol 82:761–766

Ahemad M, Khan MS (2010a) Influence of selective herbicides on plant growth promoting traits of phosphate solubilizing *Enterobacter asburiae* strain PS2. Res J Microbiol 5:849–857

Ahemad M, Khan MS (2010b) Plant growth promoting activities of phosphate-solubilizing *Enterobacter asburiae* as influenced by fungicides. Eur Asian J Biosci 4:88–95

Ahemad M, Khan MS (2010c) Comparative toxicity of selected insecticides to pea plants and growth promotion in response to insecticide-tolerant and plant growth promoting *Rhizobium leguminosarum*. Crop Prot 29:325–329

Ahemad M, Khan MS (2010d) Phosphate-solubilizing and plant-growth-promoting *Pseudomonas aeruginosa* PS1 improves greengram performance in quizalafop-p-ethyl and clodinafop amended soil. Arch Environ Contam Toxicol 58:361–372

Ahemad M, Khan MS (2010e) Ameliorative effects of *Mesorhizobium* sp. MRC4 on chickpea yield and yield components under different doses of herbicide stress. Pestic Biochem Physiol 98:183–190

Ahemad M, Khan MS (2010f) Insecticide-tolerant and plant growth promoting *Rhizobium* improves the growth of lentil (*Lens esculentus*) in insecticide-stressed soils. Pest Manag Sci 67:423–429

Ahemad M, Khan MS (2010g) Growth promotion and protection of lentil (*Lens esculenta*) against herbicide stress by *Rhizobium* species. Ann Microbiol 60:735–745

Ahemad M, Khan MS (2010h) Improvement in the growth and symbiotic attributes of fungicide-stressed chickpea plants following plant growth promoting fungicide-tolerant *Mesorhizobium* inoculation. Afr J Basic Appl Sci 2:111–116

Ahemad M, Khan MS (2011a) Toxicological assessment of selective pesticides towards plant growth promoting activities of phosphate solubilizing *Pseudomonas aeruginosa*. Acta Microbiol Immunol Hung 58:169–187

Ahemad M, Khan MS (2011b) Effects of insecticides on plant growth-promoting activities of phosphate solubilizing rhizobacterium *Klebsiella* sp. strain PS19. Pestic Biochem Physiol 100:51–56

Ahemad M, Khan MS (2011c) Assessment of plant growth promoting activities of rhizobacterium *Pseudomonas putida* under insecticide-stress. Microbiol J 1:54–64

Ahemad M, Khan MS (2011d) Ecotoxicological assessment of pesticides towards the plant growth promoting activities of Lentil (*Lens esculentus*) specific *Rhizobium* sp. strain MRL3. Ecotoxicology 20:661–669

Ahemad M, Khan MS (2011e) Biotoxic impact of fungicides on plant growth promoting activities of phosphate-solubilizing *Klebsiella* sp. isolated from mustard (*Brassica compestris*) rhizosphere. J Pestic Sci 85:29–36

Ahemad M, Khan MS (2011f) Toxicological effects of selective herbicides on plant growth promoting activities of phosphate solubilizing *Klebsiella* sp. strain PS19. Curr Microbiol 62:532–538

Ahemad M, Khan MS (2011g) Insecticide-tolerant and plant growth promoting *Bradyrhizobium* sp. (*vigna*) improves the growth and yield of greengram (*Vigna radiata* (L.) Wilczek) in insecticide stressed soils. Symbiosis 54:17–27

Ahemad M, Khan MS (2011h) Effect of tebuconazole-tolerant and plant growth promoting *Rhizobium* isolate MRP1 on pea-*Rhizobium* symbiosis. Sci Hortic 129:266–272

Ahemad M, Khan MS (2011i) Plant growth promoting fungicide tolerant *Rhizobium* improves growth and symbiotic characteristics of lentil (*Lens esculentus*) in fungicide-applied soil. J Plant Growth Regul 357:449–469

Ahemad M, Khan MS (2011j) Response of greengram (*Vigna radiata* (L.) Wilczek) grown in herbicide-amended soil to quizalafop- p-ethyl and clodinafop tolerant plant growth promoting *Bradyrhizobium* sp. (*vigna*) MRM6. J Agric Sci Technol 13:1209–1222

Ahemad M, Khan MS (2012a) Effect of fungicides on plant growth promoting activities of phosphate solubilizing *Pseudomonas putida* isolated from mustard (*Brassica compestris*) rhizosphere. Chemosphere 86:945–950

Ahemad M, Khan MS (2012b) Ecological assessment of biotoxicity of pesticides towards plant growth promoting activities of pea (*Pisum sativum*)-specific *Rhizobium* sp. strain MRP1. Emir J Food Agric 30:334–342

Ahemad M, Khan MS (2012c) Evaluation of plant growth promoting activities of rhizobacterium *Pseudomonas putida* under herbicide-stress. Ann Microbiol 62:1531–1540

Ahemad M, Khan MS (2012d) Effects of pesticides on plant growth promoting traits of *Mesorhizobium* strain MRC4. J Saudi Soc Agric Sci 11:63–71

Ahemad M, Khan MS (2012e) Alleviation of fungicide-induced phytotoxicity in greengram (*Vigna radiata* (L.) Wilczek) using fungicide-tolerant and plant growth promoting *Pseudomonas* strain. Saudi J Biol Sci 19:451–459

Ahemad M, Khan MS (2012f) Productivity of greengram in tebuconazole-stressed soil, by using a tolerant and plant growth promoting *Bradyrhizobium* sp. MRM6 strain. Acta Physiol Plant 34:245–254

Ahemad M, Kibret M (2014) Mechanisms and applications of plant growth promoting rhizobacteria: current perspective. J King Saud Uni Sci 26:1–20

Altomare C, Norvell WA, Bjorkman T, Harman GE (1999) Solubilization of phosphates and micronutrients by the plant growth promoting and biocontrol fungus *Trichoderma harzianum rifai*. Appl Environ Microbiol 144:1295–1322

Amanullah MM, Archana J, Manoharan S, Subramanian KS (2012) Influence of iron and AM inoculation on metabolically active iron, chlorophyll content and yield of hybrid maize in calcareous soil. J Agron 11:27–30

Arora NK, Kang SC, Maheshwari DK (2001) Isolation of siderophore producing strains of *Rhizobium meliloti* and their biocontrol potential against *Macrophomina phaseolina* that causes charcoal rot of groundnut. Curr Sci 81:673–677

Ashraf MA, Rasool M, Mirza MS (2011) Nitrogen fixation and indole acetic acid production potential of bacteria isolated from rhizosphere of sugarcane (*Saccharum officinarum* L.). Adv Biol Res 5:348–355

Bagde US, Prasad R, Varma A (2010) Interaction of Mycobiont: *Piriformospora Indica* with medicinal plants and plants of economic importance. Afr J Biotechnol 9:9214–9226

Barea JM (2002) Mycorrhizosphere interactions to improve plant fitness and soil quality. Anton Leeuw 81:343–351

Bernard RG (2005) Modulation of plant ethylene levels by the bacterial enzyme ACC deaminase. FEMS Microbiol Lett 251:1–7

Berta G, Fusconi A, Hooker JE (2002) Arbuscular mycorrhizal modifications to plant root systems: scale, mechanisms and consequences. In: Gianinazzi S, Schuepp H, Barea JM, Haselwandter K (eds) Mycorrhizal technology in agriculture. Birkhauser Press, Basel, pp 71–86

Bhagat D, Sharma P, Sirari A, Kumawat KC (2014) Screening of *Mesorhizobium* spp. for control of *Fusarium* wilt in chickpea *in vitro* conditions. Int J Curr Microbiol Appl Sci 3:923–930

Bharadwaj DP, Alstrom S, Lundquist PO (2012) Interactions among *Glomus irregulare*, arbuscular mycorrhizal spore-associated bacteria and plant pathogens under in vitro conditions. Mycorrhiza 22:437–447

Bhattacharjee RB, Singh A, Mukhopadhyay SN (2008) Use of nitrogen-fixing bacteria as biofertilizer for non-legumes: prospects and challenges. Appl Microbiol Biotechnol 80:199–209

Bhattacharyya PN, Jha DK (2012) Plant growth-promoting rhizobacteria (PGPR): emergence in agriculture. World J Microbiol Biotechnol 28:1327–1350

Biswas JC, Ladha JK, Dazzo FB (2000) Rhizobial inoculation influences seedling vigor and yield of rice. Agron J 92:880–886

Boiero L, Perriq D, Masciarelli O, Penna C, Cassan F, Luna V (2007) Phytohormone production by three strains of *Bradyrhizobium japonicum* and possible physiological and technological implications. Appl Microbiol Biotechnol 74:874–880

Bonnefoy V, Holmes DS (2012) Genomic insights into microbial iron oxidation and iron uptake strategies in extremely acidic environments. Environ Microbiol 14:1597–1611

Borlaug EN, Dowswell CR (1994) Feeding a human population that increasingly crowds a fragile planet. Paper presented at 15th world congress of soil science, World congress of soil science, Acapulco, 10–16 July 1994

Bottini R, Cassan F, Piccoli P (2004) Gibberellin production by bacteria and its involvement in plant growth promotion and yield increase. Appl Microbiol Biotechnol 65:497–503

Bottomley PJ, Myrold DD (2007) Biological N inputs. In: Paul E (ed) Soil microbiology, ecology and biochemistry. Academic, Oxford, pp 365–387

Braud A, Jezequel K, Bazot S, Lebeau T (2009) Enhanced phytoextraction of an agricultural Cr-, Hg- and Pb-contaminated soil by bioaugmentation with siderophore producing bacteria. Chemosphere 74:280–286

Brazelton JN, Pfeufer EE, Sweat TA, Gardener BB, Coenen C (2008) 2, 4-Diacetylphloroglucinol alters plant root development. Mol Plant-Microbe Interact 21:1349–1358

Brundrett MC (2002) Coevolution of roots and mycorrhizas of land plants. New Phytol 154:275–304

Camerini S, Senatore B, Lonardo E, Imperlini E, Bianco C, Moschetti G, Rotino GL, Compion B, Defez R (2008) Introduction of a novel pathway for IAA biosynthesis to rhizobia alters vetch root nodule development. Arch Microbiol 190:67–77

Carrillo-Castaneda G, Juárez Munos J, Peralta-Videa JR, Gomez E, Tiemannb KJ, Duarte-Gardea M, Gardea-Torresdey JL (2002) Alfalfa growth promotion by bacteria grown under iron limiting conditions. Adv Environ Res 6:391–399

Cassan F, Perrig D, Sgroy V, Masciarelli O, Penna C, Luna V (2009) *Azospirillum brasilense* Az39 and *Bradyrhizobium japonicum* E109, inoculated singly or in combination, promote seed germination and early seedling growth in corn (*Zea mays* L.) and soybean (*Glycine max* L.). Eur J Soil Biol 45:28–35

Chen YP, Rekha PD, Arun AB, Shen FT, Lai WA, Young CC (2006) Phosphate solubilizing bacteria from subtropical soil and their tricalcium phosphate solubilizing abilities. Appl Soil Ecol 34:33–41

Chen WW, Yang JL, Qin C, Jin CW, Mo JH, Ye T, Zheng SJ (2010) Nitric oxide acts downstream of auxin to trigger root ferric-chelate reductase activity in response to iron deficiency in *Arabidopsis*. Plant Physiol 154:810–819

Clark RB, Zeto SK (2000) Mineral acquisition by arbuscular mycorrhizal plants. J Plant Nutr 23:867–902

Clark RB, Zobel RW, Zeto SK (1999) Effects of mycorrhizal fungus isolate on mineral acquisition by *Panicum virgatum* in acidic soil. Mycorrhiza 9:167–176

Cojho EH, Reis VM, Schenberg ACG, Dobereiner J (1993) Interactions of *Acetobacter diazotrophicus* with an amylolytic yeast in nitrogen-free batch culture. FEMS Microbiol Lett 106:341–346

Crowley DE, Kraemer SM (2007) Function of siderophores in the plant rhizosphere. In: Pinton R et al (eds) The rhizosphere, biochemistry and organic substances at the soil-plant interface. CRC Press, Boca Raton, pp 73–109

Dary M, Chamber-Parez MA, Palomares AJ, Pajeuelo E (2010) In situ phytostabilisation of heavy metal polluted soils using *Lupinus luteus* inoculated with metal resistant plant-growth promoting rhizobacteria. J Hazard Mater 177:323–330

Deryl M, Skorupska A (1992) Rhizobial siderophore as an iron source for clover. Physiol Plant 85:549–553

Deshmukh S, Huckelhoven R, Schafer P, Imani J, Sharma M (2006) The root endophytic fungus *Piriformospora indica* requires host cell death for proliferation during mutualistic symbiosis with barley. Proc Natl Acad Sci 103:18450–18457

Dobbelaere S, Vanderleyden J, Okon Y (2003) Plant growth-promoting effects of diazotrophs in the rhizosphere. Crit Rev Plant Sci 22:107–149

Drogue B, Meynet EC, Loccoz MY, Dye FW, Cambaret PC (2013) Control of the cooperation between plant growth-promoting rhizobacteria and crops by rhizosphere signals. .In: deBruijn FJ (ed) Molecular Microbial Ecology of the Rhizosphere, 1&2, Wiley, Hoboken, pp 281–294

Duan J, Mullar KM, Charles TC, Vesely S, Glick BR (2009) 1-aminocyclopropane-1-carboxylate (ACC) deaminase genes in rhizobia from southern Saskatchewan. Microb Ecol 57:423–436

Fernandez-Bidondo L, Silvani V, Colombo R, Pergola M, Bompadre J, Godeas A (2011) Pre-symbiotic and symbiotic interactions between *Glomus intraradices* and two *Paenibacillus* species isolated from AM propagules. *In vitro* and *in vivo* assays with soybean (AG043RG) as plant host. Soil Biol Biochem 43:1866–1872

Figueiredo MVB, Selden L, de Araujo FF, Mariano RLM (2011) Plant growth promoting rhizobacteria: fundamentals and applications. In: Maheshwari DK (ed) Plant growth and health promoting bacteria. Springer, Berlin/Heidelberg, pp 21–42

Ganesan V (2008) Rhizoremediation of cadmium soil using a cadmium-resistant plant growth-promoting rhizopseudomonad. Curr Microbiol 56:403–407

Ghosh S, Penterman JN, Little RD, Chavez R, Glick BR (2003) Three newly isolated plant growth promoting bacilli facilitate the seedling growth of canola, *Brassica campestris*. Plant Physiol Biochem 41:277–281

Glick BR (2005) Modulation of plant ethylene levels by the bacterial enzyme ACC deaminase. FEMS Microbiol Lett 25:1–7

Glick BR (2012) Plant growth-promoting bacteria: mechanisms and applications. Scientifica 2012:1–15

Glick BR, Penrose DM, Li J (1998) A model for the lowering of plant ethylene concentrations by plant growth-promoting bacteria. J Theor Biol 190:63–68

Glick BR, Todorovic B, Czarny J, Cheng Z, Duan J, McConkey B (2007) Promotion of plant growth by bacterial ACC deaminase. Crit Rev Plant Sci 26:227–242

Graham PH, Vance CP (2003) Legumes: importance and constraints to greater use. Plant Physiol 131:872–877

Gray EJ, Smith DL (2005) Intracellular and extracellular PGPR: commonalities and distinctions in the plant-bacterium signaling process. Soil Biol Biochem 37:395–412

Grichko VP, Glick BR (2001) Amelioration of flooding stress by ACC deaminase-containing plant growth-promoting bacteria. Plant Physiol Biochem 39:11–17

Gutiérrez-Manero FJ, Romas-Solano B, Probanza A, Mehouachi J, Tadeo FR, Talon M (2001) The plant-growth-promoting rhizobacteria *Bacillus pumilus* and *Bacillus licheniformis* produce high amounts of physiologically active gibberellins. Physiol Plant 111:206–211

Gutierrez-Zamora ML, Martinez-Romero E (2001) Natural endophytic association between *Rhizobium etli* and maize (*Zea mays* L.). J Biotechnol 91:117–126

Hassan S, Mathesius U (2012) The role of flavonoids in root–rhizosphere signalling: opportunities and challenges for improving plant–microorganism interactions. J Exp Bot 63:3429–3444

Hayat R, Ahmed I, Sheirdil RA (2012) An overview of plant growth promoting rhizobacteria (PGPR) for sustainable agriculture. In: Ashraf M, Ozturk M, Ahmad MS, Aksoy A (eds) Crop production for agricultural improvement, part 3. Springer, Dordrecht, pp 557–579

He XH, Critchley C, Bledsoe C (2003) Nitrogen transfer within and between plants through common mycorrhizal networks (CMNs). Crit Rev Plant Sci 22:531–567

He CQ, Tan GE, Liang X, Du W, Chen YL, Zhi GY, Zhu Y (2010) Effect of Zn-tolerant bacterial strains on growth and Zn accumulation in *Orychophragmus violaceus*. Appl Soil Ecol 44:1–5

Herrera Medina MJ, Steinkellner S, Vierheilig H, Okampo-Bote JA, Garcia-Garrido JM (2007) Abscisic acid determines arbuscule development and functionality in the tomato arbuscular mycorrhiza. New Phytol 175:554–564

Hodge A, Compbell CD, Fitter AH (2001) An arbuscular mycorrhizal fungus accelerates decomposition and acquires nitrogen directly from organic material. Nature 413:297–299

Hussin ASM, Farouk AE, Greiner R, Salleh HM, Ismail AF (2007) Phytate degrading enzyme production by bacteria isolated from Malaysian soil. World J Microbiol Biotechnol 23:1653–1660

Indiragandhi P, Anandham R, Madhaiyan M, Sa TM (2008) Characterization of plant growth-promoting traits of bacteria isolated from larval guts of diamond back moth *Plutella xylostella* (Lepidoptera: Plutellidae). Curr Microbiol 56:327–333

Ivanov R, Brumbarova T, Bauer P (2012) Fitting into the harsh reality: regulation of iron-deficiency responses in dicotyledonous plants. Mol Plant 5:27–42

Jeffries P, Gianinazzi S, Perotto S (2003) The contribution of arbuscular mycorrhizal fungi in sustainable maintenance of plant health and soil fertility. Biol Fertil Soils 37:1–16

Jiang C, Sheng X, Quan M, Wang Q (2008) Isolation and characterization of a heavy metal-resistant *Burkholderia* sp. from heavy metal-contaminated paddy field soil and its potential in promoting plant growth and heavy metal accumulation in metal polluted soil. Chemosphere 72:157–164

Jida M, Assefa F (2012) Phenotypic diversity and plant growth promoting characteristics of *Mesorhizobium* species isolated from chickpea (*Cicer arietinum* L.)

growing areas of Ethiopia. Afr J Biotechnol 11:7483–7493

Jin CW, He YF, Tang CX, Wu P, Zheng SJ (2006) Mechanisms of microbially enhanced iron uptake in red clover. Plant Cell Environ 29:888–897

Jin CW, You GY, He YF, Tang C, Wu P, Zheng SJ (2007) Iron-deficiency-induced secretion of phenolics facilitates the reutilization of root apoplastic iron in red clover (*Trifolium pratense* L.). Plant Physiol 144:278–285

Jin CW, You GY, Zheng SJ (2008) The iron deficiency-induced phenolics secretion plays multiple important roles in plant iron acquisition underground. Plant Signal Behav 3:60–61

Jorquera M, Martınez O, Maruyama F, Marschner P, Mora ML (2008a) Current and future bio-technological applications of bacterial phytases and phytase-producing bacteria. Microbes Environ 23:182–191

Jorquera MA, Hernandez MT, Rengel Z, Marschner P, Mora ML (2008b) Isolation of culturable phosphorbacteria with both phytate mineralization and phosphate-solubilization activity from the rhizosphere of plants grown in a volcanic soil. Biol Fertil Soils 44:1025–1034

Kang BG, Kim WT, Yun HS, Chang SC (2010) Use of plant growth-promoting rhizobacteria to control stress responses of plant roots. Plant Biotechnol Rep 4:179–183

Kaur N, Sharma P (2013) Screening and characterization of native *Pseudomonas* sp. as plant growth promoting rhizobacteria in chickpea (*Cicer arietinum* L.) rhizosphere. Afr J Microbiol Res 7:1465–1474

Khalid A, Tahir S, Arshad M, Zahir ZA (2004) Relative efficiency of rhizobacteria for auxin biosynthesis in rhizosphere and non-rhizosphere soils. Aust J Soil Res 42:921–926

Khalid A, Akhtar MJ, Mahmood MH, Arshad M (2006) Effect of substrate-dependent microbial ethylene production on plant growth. Microbiology 75:231–236

Khandelwal SR, Manwar AV, Chaudhari BL, Chincholkar SB (2002) Siderophoregenic bradyrhizobia boost yield of soybean. Appl Biochem Biotechnol 102:155–168

Kloepper JW, Schroth MN (1978) Plant growth-promoting rhizobacteria on radishes. In: Proceedings of the IVth international conference on plant pathogenic bacteria, vol 2. Station de Pathologie Vegetale et Phyto-Bacteriologie, Angers, pp 879–882

Kobayashi T, Nishizawa NK (2012) Iron uptake, translocation and regulation in higher plants. Annu Rev Plant Physiol Plant Mol Biol 63:131–152

Kosuta S, Chabaud M, Lougnon G, Gough C, Narie JD, Barker DG, Becard G (2003) A diffusible factor from arbuscular mycorrhizal fungi induces symbiosis-specific MtENOD11 expression in roots of *Medicago truncatula*. Plant Physiol 131:952–962

Kuffner M, Puschenreiter M, Wieshammer G, Gorfer M, Sessitsch A (2008) Rhizosphere bacteria affect growth and metal uptake of heavy metal accumulating willows. Plant Soil 304:35–44

Kumar KV, Singh N, Behl H, Srivastava S (2008) Influence of plant growth promoting bacteria and its mutant on heavy metal toxicity in *Brassica juncea* grown in fly ash amended soil. Chemosphere 72:678–683

Kumawat KC, Sharma P, Gill BS, Kaur G (2014) Isolation and characterization of non-rhozobial endophytes bacteria in soybean (*Glycine max* (L.) Merill) In: Mahal GS, Chawla W (eds) Proceedings of international conference on crop productivity and sustainability – shaping the future held at Baba Farid Group of Institutes, Bathinda, pp 155–160

Labidi S, Jeddi FB, Tisserant B, Debiane D, Rezgui S, Ferjani AG, Sahraoui ALH (2012) Role of arbuscular mycorrhizal symbiosis in root mineral uptake under $CaCO_3$ stress. Mycorrhiza 22:337–345

Lemanceau P, Bauer P, Kraemer S, Briat JF (2009) Iron dynamics in the rhizosphere as a case study for analyzing interactions between soils, plants and microbes. Plant Soil 321:513–535

Lim BL, Yeung P, Cheng C, Hill JE (2007) Distribution and diversity of phytate mineralizing bacteria. ISME J 1:321–330

Lucangeli C, Bottini R (1997) Effects of *Azospirillum* on endogenous gibberellin content and growth of maize (*Zea mays* L.) treated with uniconazole. Symbiosis 23:63–72

Lucas JA, Solano BR, Montes F, Ojeda J, Megias M, Manero FJG (2009) Use of two PGPR strains in the integrated management of blast disease in rice (*Oryza sativa*) in Southern Spain. Field Crop Res 114:404–410

Lucy M, Reed E, Glick BR (2004) Application of free living plant growth-promoting rhizobacteria. Anton Leeuw 86:1–25

Ludwig-Muller J, Guther M (2007) Auxins as signals in arbuscular mycorrhiza formation. Plant Signal Behav 2:194–196

Ma W, Guinel FC, Glick BR (2003) *Rhizobium leguminosarum* biovar *viciae* 1-aminocyclopropane-1-carboxylate deaminase promotes nodulation of pea plants. Appl Environ Microbiol 69:4396–4402

Ma W, Charles TC, Glick BR (2004) Expression of an exogenous 1-aminocyclopropane-1-carboxylate deaminase gene in *Sinorhizobium meliloti* increases its ability to nodulate alfalfa. Appl Environ Microbiol 70:5891–5897

Ma Y, Rajkumar M, Luo Y, Freitas H (2011a) Inoculation of endophytic bacteria on host and non-host plants-effects on plant growth and Ni uptake. J Hazard Mater 195:230–237

Ma Y, Rajkumar M, Vicente JAF, Freitas H (2011b) Inoculation of Ni-resistant plant growth promoting bacterium *Psychrobacter* sp. strain SRS8 for the improvement of nickel phytoextraction by energy crops. Int J Phytoremediation 13:126–139

Machuca A, Pereira G, Aguiar A, Milagres AMF (2007) Metal-chelating compounds produced by ectomycorrhizal fungi collected from pine plantations. Lett Appl Microbiol 44:7–12

Mader P, Kaiser F, Adholeya A, Singh R, Uppal HS, Sharma AK, Srivastava R, Sahai V, Arangno M, Wiemken A, Johri BN, Fried PM (2011) Inoculation of root microorganisms for sustainable wheat-rice and wheat-black gram rotations in India. Soil Biol Biochem 43:609–619

Madhaiyan M, Poonguzhali S, Kang BG, Lee YJ, Chung JB (2010) Effect of co-inoculation of methylotrophic *Methylobacterium oryzae* with *Azospirillum brasilense* and *Burkholderia pyrrocinia* on the growth and nutrient uptake of tomato, red pepper and rice. Plant Soil 328:71–82

Mano H, Morisaki H (2008) Endophytic bacteria in the rice plant. Microbes Environ 23:109–117

Martinsez-Viveros O, Jorquera MA, Crowley DE, Gajardo G, Mora ML (2010) Mechanisms and practical considerations involved in plant growth promotion by rhizobacteria. J Soil Sci Plant Nutr 10:293–319

Masaoka Y, Koshino H, Arakawa Y, Asanuma S (1997) Growth promoting effect of root exudates of Fe-deficient alfalfa on *Rhizobium meliloti*. In: Proceedings of the XIII international plant nutrition colloquium, Plant nutrition for sustainable food production and environment, Tokyo, pp 505–506

Mayak S, Tirosh T, Glick BR (2004) Plant growth-promoting bacteria confer resistance in tomato plants to salt stress. Plant Physiol Biochem 42:565–572

Medeiros CAB, Clark RB, Ellis JR (1993) Effects of MES (2(N-morpholino) ethane sulfonic acid) and pH on mineral nutrient uptake by mycorrhizal and non mycorrhizal maize. J Plant Nutr 16:2255–2272

Mehnaz S, Baig DN, Lazarovits G (2010) Genetic and phenotypic diversity of plant growth promoting rhizobacteria isolated from sugarcane plants growing in Pakistan. J Microbiol Biotechnol 20:1614–1623

Minaxi LN, Yadav RC, Saxena J (2012) Characterization of multifaceted *Bacillus* sp. RM-2 for its use as plant growth promoting bioinoculant for crops grown in semi arid deserts. Appl Soil Ecol 59:124–135

Mirza MS, Ahmad W, Latif F, Haurat J, Bally R, Normand P, Malik KA (2001) Isolation, partial characterization, and the effect of plant growth-promoting bacteria (PGPB) on micro-propagated sugarcane in vitro. Plant Soil 237:47–54

Mishra PK, Bisht SC, Ruwari P, Joshi GK, Singh G, Bisht JK, Bhatt JC (2011) Bioassociative effect of cold tolerant *Pseudomonas* spp. and *Rhizobium leguminosarum*-PR1 on iron acquisition, nutrient uptake and growth of lentil (*Lens culinaris* L.). Eur J Soil Biol 47:35–43

Mishra PK, Bisht SC, Mishra S, Selvakumar G, Bisht JK, Gupta HS (2012) Coinoculation of *Rhizobium leguminosarum*- PR1 with a cold tolerant *Pseudomonas* sp. improves iron acquisition, nutrient uptake and growth of field pea (*Pisum sativum* L.). J Plant Nutr 35:243–256

Murgia I, Arosio P, Tarantino D, Soave C (2012) Crops biofortification for combating 'hidden hunger' for iron. Trends Plant Sci 17:47–55

Muthukumarasamy R, Revathi G, Seshadri S, Lakshminarasimhan C (2002) *Gluconacetobacter diazotrophicus* (syn. *Acetobacter diazotrophicus*), a promising diazotrophic endophyte in tropics. Curr Sci 83:137–145

Nadeem SM, Zahir ZA, Naveed M, Arshad M (2009) Rhizobacteria containing ACC-deaminase confers salt tolerance in maize grown on salt-affected fields. Can J Microbiol 55:1302–1309

Naik MM, Dubey SK (2011) Lead-enhanced siderophore production and alteration in cell morphology in a Pb-resistant *Pseudomonas aeruginosa* strain 4EA. Curr Microbiol 62:409–414

Neeraj KS (2011) Organic amendments to soil inoculated arbuscular mycorrhizal fungi and *Pseudomonas fluorescens* treatments reduce the development of root-rot disease and enhance the yield of *Phaseolus vulgaris* L. Eur J Soil Biol 47:288–295

Neumann G, Romheld V (2000) The release of root exudates as affected by the plant physiological status. In: Pinton R, Varanini Z, Nannipieri P (eds) The rhizosphere: biochemistry and organic substances at the soil–plant interface. Dekker, New York, pp 41–89

Nozoye T, Nakanishi H, Nishizawa NK (2011) Phytosiderophore efflux transporters are crucial for iron acquisition in graminaceous plants. J Biol Chem 286:5446–5454

Olah B, Briere C, Becard G, Denarie J, Gough C (2005) Nod factors and a diffusible factor from arbuscular mycorrhizal fungi stimulate lateral root formation in *Medicago truncatula via* the DMI1/DMI2 signalling pathway. Plant J 44:195–207

Oliveira ALM, Urquiaga S, Dobereiner J, Baldani JI (2002) The effect of inoculating endophytic N_2-fixing bacteria on micro-propagated sugarcane plants. Plant Soil 242:205–215

Pedrosa FO, Hungria M, Yates G (2002) Horizontal gene transfer in rhizobia: ecological implications. In: Nitrogen fixation: from molecules to crop productivity. Springer, Netherlands, pp 593–594

Peoples MB, Giller KE, Herridge DF, Vessey JK (2002) Limitations to biological nitrogen fixation as a renewable source of nitrogen for agriculture. In: Finan TM, O'Brian MR, Layzell DB, Vessey JK, Newton W (eds) Nitrogen fixation: global perspectives. CAB International, Wallingford, pp 356–360

Perez-Garcia A, Romero D, de Vicente A (2011) Plant protection and growth stimulation by microorganisms: biotechnological applications of bacilli in agriculture. Curr Opin Biotechnol 22:187–193

Perez-Montano F, Villegas CA, Bellogin RA, Cerro PD, Espuny MR, Guerrero IJ, Lopez-Baena FJ, Ollero FJ, Cubo T (2014) Plant growth promotion in cereal and leguminous agricultural important plants: from micro organism capacities to crop production. Microbiol Res 169:325–336

Peskan-Berghofer T, Shahollari B, Giong PH, Hehl S, Markert C, Blanke V (2004) Association of *Piriformospora indica* with *Arabidopsis thaliana* roots represents a novel system to study beneficial plant-microbe interactions and involves early plant protein modifications in the endoplasmic reticulum and at the plasma membrane. Plant Physiol 122:465–477

Peterson BM, Ferrarese L, Gilbert KM (2004) Central masses and broad-line region sizes of active galactic nuclei. II. A homogeneous analysis of a large reverberation-mapping database. Astrophy J 613:682–699

Pham GH, Kumari R, Singh A, Sachdev M, Prasad R, Kaldorf M (2004) Axenic cultures of Piriformospora indica. In: Varma A, Abbott L, Werner D, Hampp R (eds) Plant surface microbiology. Springer, Berlin, pp 593–616

Pindi PK, Sultana T, Vootla PK (2014) Plant growth regulation of Bt-cotton through Bacillus species. 3 Biotech 4:305–315

Poonguzhali S, Madhaiyan M, Sa TM (2008) Isolation and identification of phosphate solubilizing bacteria from Chinese cabbage and their effect on growth and phosphorus utilization of plants. J Microbiol Biotechnol 18:773–777

Prasad R (2010) Zinc biofortification of food grains in relation to food security and alleviation of zinc malnutrition. Curr Sci 98:1300–1304

Rajkumar M, Freitas H (2008) Effects of inoculation of plant growth promoting bacteria on Ni uptake by Indian mustard. Bioresour Technol 99:3491–3498

Rajkumar M, Ae N, Prasad MNV, Freitas H (2010) Potential of siderophore-producing bacteria for improving heavy metal phytoextraction. Trends Biotechnol 28:142–149

Raju PS, Clarck RB, Ellis JR, Maranville JW (1990) Effects of species of VA-mycorrhizal fungi on growth and mineral uptake of sorghum at different temperatures. Plant Soil 121:165–170

Ramesh A, Sharma SK, Sharma MP, Yadav N, Joshi OP (2014) Inoculation of zinc solubilizing *Bacillus aryabhattai* strains for improved growth, mobilization and biofortification of zinc in soybean and wheat cultivated in Vertisols of central India. Appl Soil Ecol 73:87–96

Rana A, Joshi M, Prasanna R, Shivay YS, Nain L (2012) Biofortification of wheat through inoculation of plant growth promoting rhizobacteria and cyanobacteria. Eur J Soil Biol 50:118–126

Rani A, Souche YS, Goel R (2009) Comparative assessment of *in situ* bioremediation potential of cadmium resistant acidophilic *Pseudomonas putida* 62BN and alkalophilic *Pseudomonas monteilli* 97AN strains on soybean. Int Biodeterior Biodegrad 63:62–66

Raudales RE, Stone E, Gardener BM (2009) Seed treatment with 2, 4-diacetylphloroglucinol-producing pseudomonads improves crop health in low-pH soils by altering patterns of nutrient uptake. Phytopathology 99:506–511

Reed MLE, Glick BR (2005) Growth of canola (*Brassica napus*) in the presence of plant growth-promoting bacteria and either copper or polycyclic aromatic hydrocarbons. Can J Microbiol 51:1961–1969

Rengel Z (2002) Breeding for better symbiosis. Plant Soil 245:147–162

Rengel Z (2008) Bioavailability of phosphorus and micronutrients in the soil-plant-microbe continuum. J Soil Sci Plant Nutr 8:84–91

Rillig MC, Wright SF, Kimball BA, Pinter PJ, Wall GW, Ottman MJ, Leavitt SW (2001) Elevated carbon dioxide and irrigation effects on water stable aggregates in a sorghum field: a possible role for arbuscular mycorrhizal fungi. Glob Chang Biol 7:333–337

Rodrigues EP, Rodrigues LS, De Oliveira ALM, Baldani VLD, Teixeira KRDS, Urquiaga S, Reis VM (2008) *Azospirillum amazonense* inoculation: effects on growth, yield and N fixation of rice (*Oryza sativa* L.). Plant Soil 302:249–261

Roesch LFW, Camargo FAO, Bento FM, Triplett EW (2008) Biodiversity of diazotrophic bacteria within the soil, root and stem of field-grown maize. Plant Soil 302:91–104

Rokhbakhsh-Zamin F, Sachdev D, Kazemi PN, Engineer A, Pardesi KR, Zinjarde S, Dhakephalkar PK, Chopade BA (2011) Characterization of plant-growth-promoting traits of *Acinetobacter* species isolated from rhizosphere of *Pennisetum glaucum*. J Microbiol Biotechnol 21:556–566

Roy N, Chakrabartty PK (2000) Effect of aluminum on the production of siderophore by *Rhizobium* sp. (*Cicer arietinum*). Curr Microbiol 41:5–10

Ryan MH, Graham JH (2002) Is there a role for arbuscular mycorrhizal fungi in production agriculture? Plant Soil 244:263–271

Ryu CM, Farag MA, Hu CH, Reddy MS, Wei HX, Pare PW, Kloepper JW (2003) Bacterial volatiles promote growth in *Arabidopsis*. Proc Natl Acad Sci USA 100:4927–4932

Saha R, Saha N, Donofrio RS, Bestervelt LL (2012) Microbial siderophores: a mini review. J Basic Microbiol 52:1–15

Salamone IEG, Hynes RK, Nelson LM (2005) Role of cytokinins in plant growth promotion by rhizosphere bacteria. In: Siddiqui ZA (ed) PGPR: biocontrol and biofertilization. Springer, Amsterdam, pp 173–195

Saravanan V, Madhaiyan M, Thangaraju M (2007) Solubilization of zinc compounds by the diazotrophic, plant growth promoting bacterium *Gluconacetobacter diazotrophicus*. Chemosphere 66:1794–1798

Sawers RJ, Gutjahr C, Paszkowski U (2008) Cereal mycorrhiza: an ancient symbiosis in modern agriculture. Trends Plant Sci 13:93–97

Schalk IJ, Hannauer M, Braud A (2011) New roles for bacterial siderophores in metal transport and tolerance. Environ Microbiol 13:2844–2854

Selvakumar G, Mohan M, Kundu S, Gupta AD, Joshi P, Nazim S, Gupta HS (2008) Cold tolerance and plant growth promotion potential of *Serratia marcescens* strain SRM (MTCC 8708) isolated from flowers of summer squash (*Cucurbita pepo*). Lett Appl Microbiol 46:171–175

Sessitsch A, Coenye T, Sturz AV, Vandamme P, Ait Barka E, Salles JF, Van Elsas JD, Faure D, Reiter B, Glick BR, Wang-Pruski G, Nowak J (2005) *Burkholderia phytofirmans* sp. nov. a novel plant-associated bacterium with plant-beneficial properties. Int J Syst Evol Microbiol 55:1187–1192

Shaharoona B, Naveed M, Arshad M, Zahir ZA (2008) Fertilizer-dependent efficiency of Pseudomonads for improving growth, yield and nutrient use efficiency of wheat (*Triticum aestivum* L.). Appl Microbiol Biotechnol 79:147–155

Shahollari B, Varma A, Oelmuller R (2005) Expression of a receptor kinase in *Arabidopsis* roots is stimulated by the basidiomycete *Piriformospora indica* and the protein accumulates in Triton X-100 insoluble plasma membrane microdomains. J Plant Physiol 162:945–958

Sharma A, Johri BN, Sharma AK, Glick BR (2003) Plant growth-promoting bacterium *Pseudomonas* sp. strain GRP3 influences iron acquisition in mung bean (*Vigna radiata* L. Wilzeck). Soil Biol Biochem 35:887–894

Sharma M, Schmid M, Rothballer M, Hause G, Zuccaro A (2008) Detection and identification of bacteria intimately associated with fungi of the order *Sebacinales*. Cell Microbiol 10:2235–2246

Sharma SK, Johri BN, Ramesh A, Joshi OP, Prasad SVS (2011) Selection of plant growth-promoting *Pseudomonas spp.* that enhanced productivity of soybean–wheat cropping system in central India. J Microbiol Biotechnol 21:1127–1142

Sharma P, Kumawat KC, Kaur S, Kaur N (2014) Assessment of zinc solubilisation by plant growth promoting rhizobacteria in legume rhizosphere. Ind J Appl Res 4:439–441

Shedova E, Lipasova V, Velikodvorskaya G, Ovadis M, Chernin L, Khmel I (2008) Phytase activity and its regulation in a rhizospheric strain of *Serratia plymuthica*. Folia Microbiol 53:110–114

Siddiqui ZA (2006) PGPR: prospective biocontrol agents of plant pathogens. In: Siddiqui ZA (eds) PGPR: biocontrol and biocontrol. Springer, Dordrecht, pp 112–142

Sivaramaiah N (2007) Improvement in symbiotic efficiency of chickpea (*Cicer arietinum*) by co-inoculation of *Bacillus* strains with *Mesorhizobium* sp. *Cicer*. Indian J Microbiol 47:51–56

Soerensen KU, Terry RE, Jolley VD, Brown JC, Vargas ME (1988) The interaction of iron-stress response and root nodules in iron efficient and inefficient soybeans. J Plant Nutr 11:853–865

Somers E, Vanderleyden J, Srinivasan M (2004) Rhizosphere bacterial signalling: a love parade beneath our feet. Crit Rev Microbiol 30:205–240

Spaepen S, Vanderleyden J, Remans R (2007) Indole-3-acetic acid in microbial and microorganism-plant signaling. FEMS Microbiol Rev 31:425–448

Sultana T, Pindi PK (2012) Role of arbuscular mycorrhizae (AM) fungi and multi bioinoculants in cotton plant growth. Br Microbiol Res J 2:123–130

Sun Y, Cheng Z (2009) The presence of a 1-aminocyclopropane-1-carboxylate (ACC) deaminase deletion mutation alters the physiology of the endophytic plant growth-promoting bacterium *Burkholderia phytofirmans* PsJN. FEMS Microbiol Lett 296:131–136

Tajini F, Trabelsi M, Drevon JJ (2012) Combined inoculation with *Glomus intraradices* and *Rhizobium tropici* CIAT899 increases phosphorus use efficiency for symbiotic nitrogen fixation in common bean (*Phaseolus vulgaris* L.). Saudi J Biol Sci 19:157–163

Tank N, Saraf M (2009) Enhancement of plant growth and decontamination of nickel-spiked soil using PGPR. J Basic Microbiol 49:195–204

Tao GC, Tian SJ, Cai MY, Xie GH (2008) Phosphate solubilizing and mineralizing abilities of bacteria isolated from. Pedosphere 18:515–523

Tejeda-Sartorius M, Martizez de la Vega O, Delano-Frier JP (2008) Jasmonic acid influences mycorrhizal colonization in tomato plants by modifying the expression of genes involved in carbohydrate partitioning. Physiol Plant 133:339–353

Terpolilli JJ, Hood GA, Poole PS (2012) What determines the efficiency of N_2-fixing *Rhizobium*–legume symbioses. Adv Microb Physiol 60:325–389

Terry RE, Hartzook A, Jolley VD, Brown JC (1988) Interactions of iron nutrition and symbiotic nitrogen fixation in peanuts. J Plant Nutr 11:811–820

Treeby MT (1992) The role of mycorrhizal fungi and non-mycorrhizal microorganisms in iron nutrition of citrus. Soil Biol Biochem 24:857–864

Tsavkelova EA, Klimova SY, Cherdyntseva TA (2006) Microbial producers of plant growth stimulators and their practical use. Appl Biochem Microbiol 42:117–126

Turner BL, Paphazy MJ, Haygarth PM, Mckelvie LD (2002) Inositol phosphates in the environment. Philos Trans R Soc Lond B Biol Sci 357:449–469

Unkovich M, Baldock J (2008) Measurement of asymbiotic N_2 fixation in Australian agriculture. Soil Biol Biochem 40:2915–2921

Upadhyay SK, Singh JS, Saxena AK, Singh DP (2012) Impact of PGPR inoculation on growth and antioxidant status of wheat under saline conditions. Plant Biol 14:605–611

Vansuyt G, Robin A, Briat JF, Curie C, Lemanceau P (2007) Iron acquisition from Fe-pyoverdine by *Arabidopsis thaliana*. Mol Plant Microbe Interact J 20:441–447

Varma A, Verma S, Sahay NS, Butehorn B, Franken P (1999) *Piriformospora indica* a cultivable plant growth promoting root endophyte with similarities to arbuscular mycorrhizal fungi. Appl Environ Microbiol 65:2741–2744

Verma S, Varma A, Rexer KH, Hassel A, Kost G, Sarbhoy A, Bisen P, Butehorn B, Franken P (1998) *Piriformospora indica*, a new root colonizing fungus. Mycologia 90:896–903

Vessey JK (2003) Plant growth promoting rhizobacteria as biofertilizers. Plant Soil 255:571–586

Walker TS, Bais HP, Grotewold E, Vivanco JM (2003) Root exudation and rhizosphere biology. Plant Physiol 132:44–51

Wani PA, Khan MS (2010) *Bacillus* species enhance growth parameters of chickpea (*Cicer arietinum* L.) in chromium stressed soils. Food Chem Toxicol 48:3262–3267

Wani PA, Khan MS, Zaidi A (2008) Chromium-reducing and plant growth-promoting *Mesorhizobium* improves chickpea growth in chromium-amended soil. Biotechnol Lett 30:159–163

Weiss M, Selosse MA, Rexer KH, Urban A, Ober-winkler F (2004) *Sebacinales*: a hitherto overlooked cosm of heterobasidiomycetes with a broad mycorrhizal potential. Mycol Res 108:1003–1010

Winkelmann G (2007) Ecology of siderophores with special reference to the fungi. Biometals 20:379–392

Wu T, Zhang HT, Wang Y, Jia WS, Xu XF, Zhang XZ, Han ZH (2012) Induction of root Fe (III) reductase activity and proton extrusion by iron deficiency is mediated by auxin-based systemic signalling in *Malus xiaojinensis*. J Exp Bot 63:859–870

Yanni YG, Rizk FK, El-Fattah A, Squartini A, Corich V, Giacomini A, Bruijn FD, Rademaker J, Flores JM, Ostrom P (2001) The beneficial plant growth-promoting association of *Rhizobium leguminosarum* bv. *trifolii* with rice roots. Aust J Plant Physiol 28:845–870

Yasmin S, Hafeez FY, Schmid M, Hartmann A (2013) Plant-beneficial rhizobacteria for sustainable increased yield of cotton with reduced level of chemical fertilizers. Pak J Bot 45:655–662

Zahir ZA, Ghani U, Naveed M, Nadeem SM, Asghar HN (2009) Comparative effectiveness of *Pseudomonas* and *Serratia* sp. containing ACC-deaminase for improving growth and yield of wheat (*Triticum aestivum* L.) under salt-stressed conditions. Arch Microbiol 191:415–424

Zahir ZA, Shah MK, Naveed M, Akhter MJ (2010) Substrate-dependent auxin production by *Rhizobium phaseoli* improves the growth and yield of *Vigna radiata* L. under salt stress conditions. J Microbiol Biotechnol 20:1288–1294

Zaidi A, Khan MS, Ahemad M, Oves M (2009) Plant growth promotion by phosphate solubilizing bacteria. Acta Microbiol Immunol Hung 56:263–284

Zhao L, Xu Y, Sun R, Deng Z, Yang W, Wei G (2011) Identification and characterization of the endophytic plant growth promoter *Bacillus cereus* strain MQ23 isolated from *Sophora alopecuroides* root nodules. Braz J Microbiol 42:567–575

Zheng SJ (2010) Iron homeostasis and iron acquisition in plants: maintenance, functions and consequences. Ann Bot 105:799–800

Role of Phytosiderophores in Acquisition of Iron and Other Micronutrients in Food Legumes

21

Lalit Kumar, Nand Lal Meena, and Ummed Singh

Abstract

Some of the micronutrients, such as iron (Fe), zinc (Zn), copper (Cu), calcium (Ca), magnesium (Mg), iodine (I), selenium (Se), etc., exist in most of the soils in appreciable amount, but unfortunately a little fraction of these elements are phyto-available. These mineral elements can be present in the soil in different forms such as free ions, or ions adsorbed onto mineral or organic surfaces, dissolved compounds or precipitates, part of lattice structures or contained within the soil biota. The availability of these minerals largely depends on the various soil physicochemical properties such as soil pH, redox conditions, cation exchange capacity, activity of microbes, soil structure, organic matter, and water content. Therefore, in spite of having high concentrations of Fe, Zn, and Cu in many soils, their phytoavailability is often restricted by some of the unsupportive properties of soil. Certain plant-mediated procedures such as the exudation of protons, phytosiderophores, and organic acids by roots are reported to increase Fe, Zn, and Cu phytoavailability in the rhizosphere. Therefore, root exudation and phytosiderophores play a key role in acquisition of mineral nutrients from the soil. Two strategies, viz., strategy I and strategy II, described by which the plant community fight to mitigate the deficiency of these elements. Under strategy I, plants follow "acidification/reduction" mechanism for enhancement of Fe solubility prior to uptake. Therefore, under this category, plants generally include some of the dicotyledons and non-graminaceous monocotyledons and respond to Fe deficiency by extruding both protons and the reducing substances (phenols) from the roots which enhance the ferric reduction activity at the root plasma membrane. The grasses, which include most of the

L. Kumar (✉) • N.L. Meena
Division of Basic Sciences, ICAR-Indian Institute of Pulses Research, Kanpur, Uttar Pradesh 208 024, India
e-mail: lalit1997@yahoomail.com; nd.iari09@gmail.com

U. Singh
Division of Crop Production, ICAR-Indian Institute of Pulses Research, Kanpur, Uttar Pradesh 208 024, India
e-mail: singhummed@yahoo.co.in

U. Singh et al. (eds.), *Biofortification of Food Crops*, DOI 10.1007/978-81-322-2716-8_21

world's staple grains, followed strategy II for sequestering Fe and other mineral ions. This strategy is best described as a "chelation" strategy, similar to that used by various bacteria and fungi. This manuscript briefly reviewed the role of phytosiderophores in acquisition of iron and other micronutrients in food legumes.

Keywords
Micronutrients • Phytosiderophores • Rhizosphere

21.1 Introduction

Generally, most of the soils contain large quantity of many micronutrients such as iron (Fe), zinc (Zn), copper (Cu), calcium (Ca), magnesium (Mg), iodine (I), selenium (Se), etc., but unfortunately little fraction of this quantity is available to the plants. These mineral elements can be present in the soil in different forms such as free ions, or ions adsorbed onto mineral or organic surfaces, as dissolved compounds or precipitates, and as part of lattice structures or contained within the soil biota. It is a well-known fact that plants can only absorb mineral elements in some of their specific chemical forms. For instance, some of the minerals such as Fe, Zn, Cu, Ca, and Mg can only be absorbed via roots of all plant species in their cationic forms, whereas graminaceous species can also take up Fe, Zn, and Cu as metal chelates (Marschner 1995; White 2003). Selenium can be absorbed by plant roots in form of selenate, selenite, or as organo-selenium compounds (Li et al. 2008), whereas iodine can be taken up either as iodide or iodate forms (Umaly and Poel 1971; Mackowiak and Grossl 1999). The occurrence of these chemical forms in soil solution largely depends on various soil physicochemical and biological properties, which in turn ultimately determine the phytoavailability of these elements to the plants. Some of the most important soil properties that determined the mineral availability are soil pH, redox conditions, cation exchange capacity, activity of microbes, soil structure, organic matter, and water content (Shuman 1998; Frossard et al. 2000).

This has been proved by various scientific reports that in spite of having high concentrations of Fe, Zn, and Cu in many soils, the phytoavailability of these mineral elements is often restricted by some of the unsupportive soil properties. Since Zn^{2+} and Cu^{2+} are known for their small diffusion coefficients and limited mobility in the soil which may be of their low concentrations in the soil solution, therefore, to acquire sufficient Zn and Cu for plant nutrition, plant roots must forage properly to cover the maximum volume of soil. Certain plant-mediated procedures such as the exudation of protons, phytosiderophores, and organic acids by plant roots are well described to increase Fe, Zn, and Cu phytoavailability in the rhizosphere, which ultimately increase the concentrations of these elements in crops. Therefore, root exudation and phytosiderophores play a key role in acquisition of mineral nutrients from the soil.

21.2 Phytosiderophores

Phytosiderophores generally described as "iron carrier" are small, high-affinity iron-chelating organic compounds such as nicotinamine, mugineic acids (MAs), avenic acid, etc. secreted by various microorganisms such as bacteria, fungi, and plants under Fe-deficient conditions. Siderophores are best known for their strongest soluble Fe^{3+}-binding properties. The PS not only solubilize but also known to mobilize some of the micronutrients such as Fe, Zn, Mn, and Cu from the soils to plant in deficient condition. Apart from iron, siderophores are also known to chelate some of the other metals such as aluminum, gallium, chromium, copper, zinc, lead, manganese, cadmium, vanadium, indium, plutonium, and uranium. As compared to other naturally occurring abundant metal ions, siderophores usually form a stable, hexadentate, octahedral complex most preferentially with Fe^{3+}. During formation of this kind of octahedral complex, if there are less than six donor atoms, then water can also take part for this coordination.

Some of the major characteristics of siderophores are as follows:

(i) These are molecules with high affinity for Fe^{3+} and remove the Fe^{3+} from minerals and contribute toward their dissolution.
(ii) These Fe chelates are highly soluble and stable over a wide pH range.
(iii) They are of crucial importance for the zinc and iron transport in soils and its supply to plants.
(iv) Zn-PS has also been reported for their similar structural confirmations as Fe-PS and an identical regulatory mechanism for the biosynthesis and/or release of PS under both Zn and Fe-deficient conditions.
(v) Most of the plants are reported to release PS at higher amounts within few hours of the onset of the light period, whereas if there is either continuous darkness or continuous light, the rate of release of PS slowed down.
(vi) The production of PS was found rising sharply within 3 h after onset of the light period and then goes down thereafter.

21.2.1 Type of Phytosiderophores and Their Efficiency

As per the chemical nature of the functional group or groups attached for Fe(III) coordination, the siderophores can be divided into three main classes, i.e., (1) catecholates (sensu stricto, catecholates and phenolates, better termed as "aryl caps"), (2) hydroxamates, and (3) (α-hydroxy)-carboxylates (Miethke and Marahiel 2007). Apart from these three, a fourth group also exists generally known as "mixed type" and is comprised of those siderophores that use a combination of any of the above types to chelate iron. Examples of each siderophore class with the iron-binding functional groups highlighted are shown in Fig. 21.1. The siderophores, those that possess three bidentate ligands per molecule and forming a hexadentate complex, are considered most effective ones because this fashion of arrangement of ligands cause a smaller entropic change as compared to the entropy changes occurred by chelating a single ferric ion with separate ligands.

21.2.1.1 Hydroxamates

The hydroxamate family of siderophores is the most important variety of siderophore, and under this, ferrichrome, which is an iron-transporting natural-occurring product from microbial sources, was recognized the first member of the series (Neilands 1984). The hydroxamate family members generally possess one or more hydroxamic acid groups derived from amines such as lysine or ornithine to bind ferric iron (Patel and Walsh 2001). Several other compounds of this group of siderophores such as exochelin, aerobactin, rhizobactin 1021, schizokinen, and alcaligin utilized two hydroxamate groups to bind iron. These are known as dihydroxamate siderophores and are noted to be produced by both gram-positive (e.g., exochelin) and gram-negative (e.g., alcaligin) bacteria.

21.2.1.2 Catecholate

The catecholate, or catechol type, siderophores are the second most common siderophore class aside from the fact that they have thus far only been found to be produced by bacteria (Dave et al. 2006). They have been found to contain either a mono- or dihydroxybenzoic acid residue that is used to chelate ferric iron and is derived from dihydroxybenzoic acid (Fischbach and

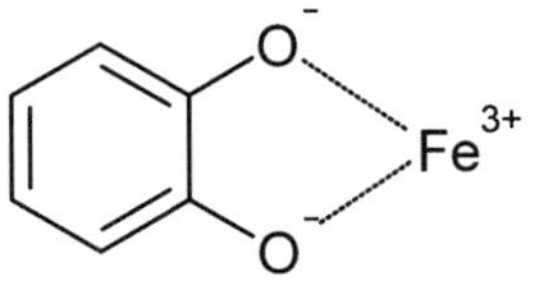

O⁻ ... Fe^{3+} ... O⁻

Fig. 21.1 Iron binding with siderophore

Walsh 2006). The best studied example of a catechol-type siderophore is enterobactin, sometimes called enterochelin, produced by *E. coli* (Brien and Gibson 1970).

21.2.1.3 Carboxylate

The carboxylate family of siderophores is comprised of those that rely solely on the oxygen donor atoms of hydroxyl and carboxyl functional groups as donor groups to bind ferric. Since Fe^{3+} is a hard Lewis acid, therefore it prefers hard Lewis bases such as anionic or neutral oxygen to coordinate with. From this kind of coordination, iron is being released by the procedure of reduction mediated through microbes since Fe^{2+} has little affinity to these ligands (Fig. 21.2).

21.3 Strategy of Mineral Nutrient Acquisition by Plants

Mechanisms of Fe and Zn acquisition in higher plants have been grouped into two categories generally known as strategy I and strategy II. Dicotyledons and non-graminaceous monocotyledons follow strategy I, whereas grasses which include most of the world's staple grains follow strategy II for acquisition of Fe and Zn. Under strategy I, plants generally utilized an "acidification/reduction" mechanism to enhance Fe solubility prior to uptake. These groups of plants are reported to respond to Fe deficiency by extruding both protons and the reducing substances such as phenols from their roots which enhance the ferric reduction activity at the root plasma membrane. The secretion of protons from plant roots brings down the pH of rhizosphere soil and as a result the solubility of Fe(III) is increased. Apart from this, the extruded protons also cause reduction of Fe(III) to Fe (II) by the action of root ferric chelate reductase procedure which further enhances Fe solubility, since Fe(II) is more soluble than Fe(III). In contrary to that, distinct mechanism was adopted by the plants that follow strategy II for acquiring Fe from the soil. This strategy is best described as a "chelation" strategy and is identical to that utilized by various bacteria and fungi (Miethke and Marahiel 2007) and may have arisen as an adaptation to alkaline soils where acidification of the rhizosphere is difficult to achieve. The siderophores which have the capability to chelate Fe ion strongly are generally called phytosiderophores (PS). The PS are synthesized by the plants and secreted via their roots into the rhizosphere for binding of Fe(III) ion. This Fe (III)–PS complex formed so is then taken up into root cells by the action of some of the specific transporters of the complex.

Chemically, phytosiderophores differ much from the siderophores obtained from bacterial and fungal inoculants and belong to a class of compounds called mugineic acids (Ma and Nomoto 1996). The biosynthetic pathway of mugineic acids is known to start from a condensation of three S-adenosylmethionine molecules to form the precursor nicotianamine (NA), (Kobayashi et al. 2001).

21.3.1 Mechanism of Acquisition of Iron and Other Micronutrients by Food Legume

Iron and Zn deficiency symptoms in leguminous plants appear in the form of chlorosis. These kind of nutritional disorder symptoms most commonly appear in plants grown on calcareous and/or alkaline soils because of an extremely low solubility of soil Fe. As stated above, mechanisms of Fe acquisition in higher plants grouped into two strategies, viz., strategies I and II. Food legumes are reported to follow strategy I for acquisition of Fe. This strategy is based on the solubilization of Fe by exudation of protons and reducing substances (phenols) via the roots of plants as exudates; as a result, these exuded substances enhance the ferric reduction activity at the root plasma membrane which increases the Fe solubility.

Leguminous plants are said to secrete both high and low molecular weight compounds from their roots as exudates. These secreted chemicals are not only reported to act as nutrients for soil microbes, but they may act as signal molecules in establishing plant–microbe

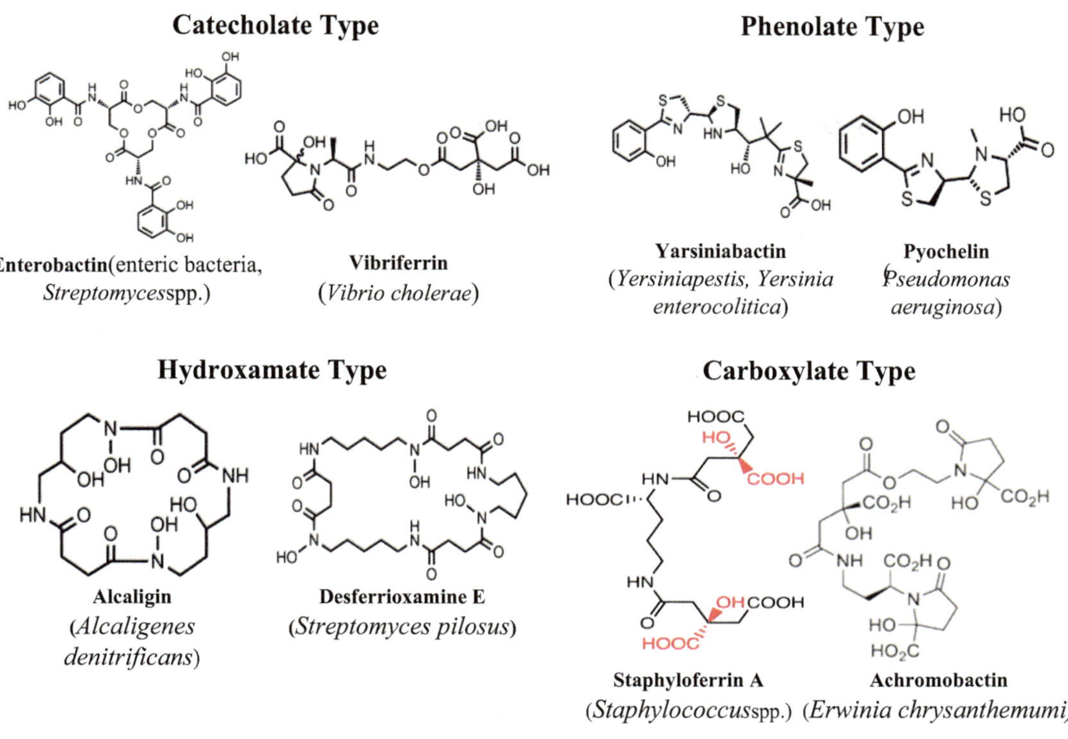

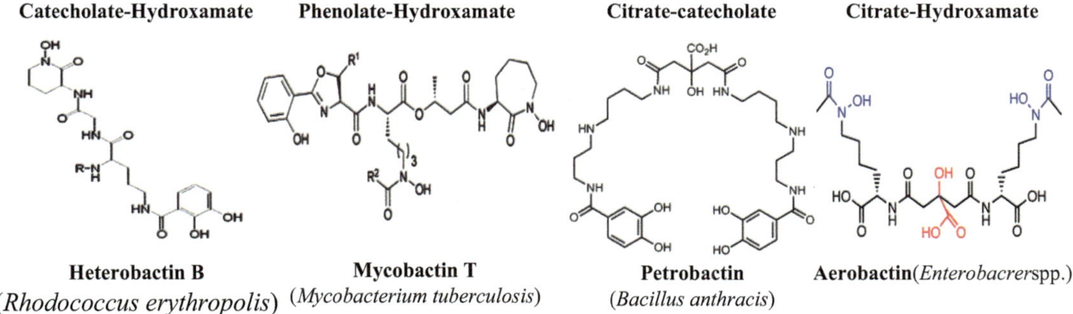

Fig. 21.2 Different types of siderophores

interactions. For example, flavonoid and strigolactone group of chemicals secreted by several legume plants as root exudates are said to act as signal molecules to establish the symbiotic interactions with rhizobia and arbuscular mycorrhizal fungi which are well known to help in acquisition of several nutrients such as nitrogen and phosphate. Some of the root-exuded chemicals of legume plants are acidic in nature; therefore, they acidify surrounding soils which help in acquiring phosphate. Some of the important functions of legume root exudates are given below in Table 21.1.

21.3.2 Root-Exuded Chemicals and Their Role in Mineral Acquisition

Root exudates, release via leguminous plant, are known to be a complex mixture of several organic compounds such as anionic molecules

Table 21.1 Root-exuded chemicals and their functional role in the rhizosphere

Class of chemical compounds	Functional role of rhizosphere
Phenolics	Nutrient source, chemoattractant signals to microbes, microbial growth promoters, *nod* gene inducers in rhizobia, *nod* gene inhibitors in rhizobia, resistance inducers against phytoalexins, chelators of poorly soluble mineral nutrients, detoxifiers of Al, phytoalexins against soil pathogens
Organic acids	Nutrient source, chemoattractant signals to microbes, chelators of poorly soluble mineral nutrients, acidifiers of soil, detoxifiers of Al, *nod* gene inducers
Amino acids and phytosiderophores	Nutrient source, chelators of poorly soluble mineral nutrients, chemoattractant signals to microbes
Vitamins	Promoters of plant and microbial growth, nutrient source
Purines	Nutrient source
Enzymes	Catalysts for P release from organic molecules, biocatalysts for organic matter transformation in soil
Root border cells	Produce signals that control mitosis, produce signals controlling gene expression, stimulate microbial growth, release chemoattractants, synthesize defense molecules for the rhizosphere, act as decoys that keep root cap infection-free, release mucilage and proteins

Compiled from Hawes et al. (1998) and Dakora and Phillips (1996)

of organic acids, phytosiderophores, sugars, vitamins, amino acids, purines, nucleosides, inorganic ions (e.g., HCO_3^-, OH^-, H^+), gaseous molecules (CO_2, H_2), enzymes, and root border cells (Tables 21.2 and 21.3). These release chemicals play their role, directly or indirectly in the acquisition of mineral nutrients essentially required for plant growth. N_2-fixing legumes are known to release phenolics and aldonic acids as exudates directly by their roots; these chemicals were demonstrated to act as major signals to Rhizobiaceae bacteria responsible for formation of root nodules where N_2 is reduced to ammonia. Some of the chemical compounds that belong to the same chemical group also affect development of mycorrhizal fungi which are known for their crucial role in phosphate uptake. Apart from these symbiotic signaling chemicals for microbes, leguminous plants also demonstrated to release some of the molecules which involved in nutrient procurement; this has been confirmed by growing the plants in low-nutrient environments. For example, the function of some of the extra cellular enzymes was demonstrated in relation to release of P from organic compounds; in addition to that, several types of molecules have also been detected which increase iron availability through chelation. Acidic root exudates can enhance the availability of Ca, Fe, and Al by solubilizing unavailable soil Ca, Fe, and Al phosphates.

Leguminous plants are also known to modify exudates as per soil environment/growing medium; for example, if they are growing on nitrate, then the requirement is to maintain electronic neutrality, and this is achieved by releasing an excess of anions such as hydroxyl ions. In contrary to that, the legumes which can grow well without nitrate, i.e., based on the advantage of N_2 reduction in the root nodules, must release a net excess of protons. These released protons can cause significantly decrease in rhizosphere pH; as a result, it does not only decrease the availability of some mineral nutrients but also associate with the effective functioning of some soil bacteria by including the rhizobial bacteria themselves. Thus, very acidic environments can not only pose a challenge to nutrient acquisition by plant roots but also not favorable for the survival of many beneficial microbes including the roots themselves. Some of the plants such as rooibos tea (*Aspalathus linearis* L.) are known to modify actively their rhizosphere pH by exudation of OH^- and HCO_3^- ions to facilitate growth in low soil pH (pH 3–5).

21.3.3 Role of Root Exudate Organic Compounds in Mobilization of Nutrients

Different root-exuded chemicals of leguminous plants possess different kind of mechanisms in acquisition of minerals which are as follows:

21.3.3.1 Phenolics

Some of the phenolic compounds released via root of plants as exudates play a great role in solubilizing of Fe, P, and the other nutrients from their unavailable sources for uptake by plants. Under Fe-deficient situation, some of the plants especially of dicots were demonstrated to release these kinds of molecules which influence Fe and P mobility in soil (Romheld 1987). Isoflavonoid phytoalexin [2-(3, 5-dihydroxyphenyl)-5, 6-dihydroxyben zofuran] is an example of such kind of compound which is being released by Fe-deficient alfalfa plants. This chemical not only enhances the availability of Fe and P by intensely dissolving ferric phosphate but also serves as a phytoalexin against pathogen attack (Masaoka et al. 1993). In contrary to that, Fe-sufficient plants were found to release root exudates having limited capability of ferric phosphate-dissolving capacity. Tomato plants were also demonstrated to release caffeic acid via their roots as exudates under Fe-deficient situation; this acid solubilizes Fe from insoluble Fe sources and makes it available to the plants (Olsen et al. 1981). Such kind of phenolic

Table 21.2 Identified organic compounds and enzymes in root exudates of different legumes

Amino acids	Organic acids	Sugars	Vitamins	Purines/ nucleosides	Enzymes	Inorganic ions and gaseous molecules
α-Alanine	Citric	Glucose	Biotin	Adenine	Acid/ alkaline	$HCO3^-$
β-Alanine	Oxalic	Fructose	Thiamin	Guanine	Phosphatase	OH^-
Asparagine	Malic	Galactose	Niacin	Cytidine	Invertase	H^+
Aspartate	Fumaric	Maltose	Pantothenate	Uridine	Amylase	CO_2
Cysteine	Succinic	Ribose	Riboflavin		Protease	H_2
Cystine	Acetic	Xylose				
Glutamate	Butyric	Rhamnose				
Glycine	Valeric	Arabinose				
Isoleucine	Glycolic	Raffinose				
Leucine	Piscidic	Desoxyribose				
Lysine	Formic	Deoligosaccharis				
Methionine	Aconitic					
Serine	Lactic					
Threonine	Pyruvic					
Proline	Glutaric					
Valine	Malonic					
Tryptophan	Aldonic					
Ornithine	Erythronic					
Histidine	Tetronic					
Arginine						
Homoserine						
Phenylalanine						
γ-Aminobutyric acid						
α-Aminoadipic acid						

Source: Dakora and Phillips (1996)

Table 21.3 Legume root exudates induce rhizobial nod genes and/or VA fungal development

Legume species	Inducer molecule	Reference
Rhizobium symbiosis		
Alfalfa	4,4′-Dihydroxy-2′-methoxychalcone, 4′-7-dihydroxyflavone, liquiritigenin	Maxwell et al. (1989)
Cowpea	Daidzein, genistein, coumestrol	Dakora (2000)
Common bean	Genistein genistein-3-O-glucoside, eriodictyol, naringenin, daidzein, coumestrol	Dakora et al. (1993), Hungria et al. (1991), Dakora et al. (1993)
Kersting's bean	Daidzein, genistein, coumestrol	Dakora (2000)
Soybean	Isoliquiritigenin, genistein, genistein-7-O-glucoside, genistein-7-O-(6″-O-malonylglucoside), daidzein, daidzein-7-O-(6″-O-malonylglucoside)	Kape et al. (1992), Smit et al. (1992)
Vetch	3,5,7,3′-Tetrahydroxy-4′-methoxyflavone; 7,3′-dihydroxy-4′-methoxyflavone; 2′,4′,4-dihydroxychalcone; 4′,4-dihydroxy-2′-methoxychalcone; naringenin; liquiritigenin; 7,4′-dihydroxy-3′-methoxyflavone; 5,7,4′-trihydroxy-3′-methoxyflavone; 5,7,3′-trihydroxy-4′-methoxyflavone	Recourt et al. (1991)
White clover	7,4′-Dihydroxyflavone, umbelliferone formononetin	Djordjevic et al. (1987)
Bambara groundnut	Daidzein, genistein, coumestrol	Dakora and Muofhe (1996)
Sesbania	Liquiritigenin	Messens et al. (1991)
Lupin	Erythronic acid tetronic acid	Gagnon and Ibrahim (1998)
Pea (Pisum sativum)	Apigenin, eriodictyol	Firmin et al. (1986)
Chickpea	Citric acid solubilize (Ca-P) vertisol soil	Ae et.al. (1990)
Pigeon pea	Piscidic acid, malonic acid, oxalic acid (Fe-P &A-P) alfisol soil	Otani et al. (1991)
VA fungal symbiosis		
Alfalfa	7-Dihydroxyflavone, 4′7-dihydroxyflavanone	Tsai and Phillips (1991)

Source: Dakora and Phillips (1996)

compounds are good in making relatively stable chelates with Fe, P, and Al present in insoluble Fe and Al phosphates, thereby increasing the solubility of Fe and P for plant uptake.

21.3.3.2 Organic Acids

Root exudates of a number of plant species of those suffering from nutrient starvation were found to accumulate markedly with different acids such as citric, malic, succinic, fumaric, and aconitic acids (Table 21.1) (Gardner et al. 1983; Hoffland et al. 1992; Jones 1998; Lipton et al. 1987; Ohwaki and Hirata 1992). Apart from these acids, root exudates of a number of plants were also demonstrated to have few more acids such as acetic, glycolic, malonic, oxalic, formic, and piscidic acid (Ae et al. 1990; Fox and Comerford 1990; Vancura and Hovadik 1965). These root-exuded acids are well known to play an important role in nutrient acquisition of certain nutrients such as P, Fe and Mn and this has been confirmed by growing plants in low nutrient soils (Table 21.2; Fig. 21.1). Under nutrient starvation situation, the release response in terms of both quantity and the spectrum of acid release was found differing between plant species. In the same way, legume species were also reported to respond differently to P deficiency, in terms of exudation rates as well as the spectrum of organic acids released. For example, chickpea was demonstrated to produce approximately 11 and

24 times more root exudates as compared to pigeon pea and soybean, respectively, whereas peanut produced only 8 and 17 times more than the two species (Ohwaki and Hirata 1992). Though the exudates of soybean and pigeon pea were reported to contain low concentration of fumaric, malic, and citric acids, but compared to soybean pigeon pea was found very efficient in P acquisition in acidic soils because of release of piscidic acid which is known as a strong chelator of iron, thereby helping in mobilization of P in acidic soil (Ae et al. 1990).

Organic Acid Anions

Various plants have also been reported to release certain organic acid anions via their roots as exudates. These root-exuded organic acid anions are also reported to chelate certain minerals such as Fe and Mn in their oxide forms (i.e., Fe_2O_3 and MnO_2) thereby making them available for plant uptake. Similarly, these acid anions are also demonstrated to liberate P for uptake by roots via forming the complexes with Ca, Al, and Fe present in the soil in the form of their insoluble phosphates which help in mobilization of P (Fig. 21.1; Marschner 1995). Apart from this, these acid anions can also desorb P from sesquioxide surfaces by the mechanism of anion exchange phenomenon (Bolan et al. 1994; Jones 1998; Jones and Darrah 1994; Parfitt 1979). Additionally, the acid anions are also known to maintain sulfate mobility in rhizosphere soil by the phenomenon of competitive displacement from adsorption sites (Evans and Anderson 1990).

Isoflavonoids

Isoflavonoids are very important class of chemicals secreted by root as exudates. These chemicals are not only known to act as phytoalexins against pathogen attack but also serve their function as signals to mutualistic soil microbes (Dakora and Phillips 1996). It is scientifically observed that at a particular concentration of N and some of the other nutrient elements such as P, S, Ca, Mg, Fe, and Cu, the isoflavonoid production in plants got elicited, for example, calcium which is not only known for their involvement in the closure of guard cells but also found to mediate the response of plants to ethylene. Ethylene is considered as a gaseous plant hormone and known to regulate and control several vital physiological processes such as seedling development, flowering, fruit ripening, and senescence in plants (Raz and Fluhr 1992).

Gaseous Root Exudates

Roots of several plants are also demonstrated to release some of the gaseous molecules such as CO_2 and H_2 as exudates. These gaseous exudates are also known to play an important role in acquisition of some of the mineral nutrients by the plants. For example, actively growing young roots were observed to release excess of CO_2 into the soil environment via carbohydrate respiration (Zwarun 1972), and this CO_2 release procedure is known to get stimulated by lumichrome which is a plant and bacterial exudate molecule (Phillips et al. 1999). By this procedure, CO_2 can accumulate up to the extent of 17.5 % in the root zone provided its escape is prevented by high soil water (Jackson 1979). Such a high amount elevated levels of CO_2 in the presence of carbonic acid which dissociates to form H+ at neutral to alkaline pH easily increase the dissolution of soil CaCO3; as a result, Ca^{2+} ion got produced which can easily be taken up by plants. In the same way, HCO_3^- excreted directly by plant roots into the soil can aid the dissolution of calcite to yield soluble supplies of $Ca^{2+}$for plant nutrition (Rendig and Taylor 1989). Moreover, as we know that an exogenous supply of CO_2 is required for better growth of rhizobia and VA fungi in the rhizosphere, therefore, CO_2 release by plant roots can also influence indirectly on the N and P nutrition of plants by making this kind of symbiotic relations with *Rhizobium* bacteria and VA fungi since the increase in micro symbiont population is known for providing adequate symbiotic N and P for the host plant (Becard et al. 1992; Lowe and Evans 1962).

Recently, the role of hydrogen gas (H_2) is revealed in net chemoautotrophic CO_2 fixation. H_2 gas is evolved as a major by-product of N_2 fixation especially in leguminous crops; for example, rhizobia with uptake hydrogenase

(HUP^+) system produced this gas during nitrogenase which on oxidation yields more energy. Though, in production process of this gas, about 5 % of net photosynthesis is said to be utilized; thereby, previously this procedure was thought to be a wasteful one for C, but recent studies have revealed that pretreating soil with H_2 can stimulate approximately 10–30 % growth in agricultural crops (Kettlewell et al. 2000). Same experiment results revealed that in the H_2-treated fields, about 60 % of the reducing power of this gas was flowing toward O_2, whereas 40 % to CO_2, hence resulting in net chemoautotrophic CO_2 fixation (Kettlewell et al. 2000). Though the specific elicitor which directly acts as plant growth promotion during this procedure still remains unknown, it is quite obvious that H_2 gas released by nodule exerts positive effect on the growth and nutrition of symbiotic legumes.

Apart from the active role of root exudates in mineral acquisition, they are also demonstrated to play their potential role in phytoremediation. For example, several studies (Banuelos et al. 1993; De Souza et al. 1999; Mench and Martin 1991; Nanda Kumar et al. 1995; Terry et al. 1992; Zayed et al. 1998) revealed the potentiality of some of the plant species to accumulate and release heavy metals as volatile root exudates; therefore, via this procedure, they reduce the concentrations of these heavy metals in the soil environment.

21.4 Conclusion

Nowadays, concrete efforts have been made in laboratory to establish the roles of some of the transport proteins in plant Fe homeostasis and the functions of root exudates in plant–microbe interactions for the acquisition of mineral nutrients from the soil by the plants including the leguminous plants. As we know that billions of microorganisms such as bacteria, fungi, algae, cyanobacteria, protozoa, etc. surrounded the root surfaces or the rhizosphere, and all soil-growing plants generally develop intimate relationships with these available organisms populating the rhizosphere. This symbiotic relationship reported to get established via the organic compounds exuded either directly from the roots themselves or the dead root epidermal cells which are good food for the soil microorganisms. The microorganisms feeding on these exuded carbon compounds along the root surfaces are also known to secrete certain chemicals such as enzymes, organic acids, antibiotics, growth regulators, hormones, and other substances which are absorbed by the roots and in turn benefit the plant in many ways; hence, a beautiful symbiotic relationship is established. Therefore, the complex mixture of several organic compounds secreted by plants as exudates and by the soil microorganism plays a great role either directly or indirectly in the acquisition of mineral nutrients essentially required for plant growth. As far as the leguminous species plants are concern, they may have a tri-interaction system mediated by root exudates including volatiles. But unfortunately, most of the studies in relation to plant–microbe interactions conducted so far have been focused only on two major soil microbes, i.e., *Rhizobium* and arbuscular mycorrhizal fungi; however, in nature, legume plants may have interactions with many microbes in the soil.

Therefore, for futuristic research, there is a great need to understand the interaction of legume plants with entire microbial communities of the soil. This kind of knowledge will definitely be of great use in expansion of our understanding about the plant–microbe interactions to the extent that could help for sustainable development of agriculture which can be achieved by using the various functions of microbes on the soil.

References

Ae N, Arihara J, Okada K, Yoshihara T, Johansen C (1990) Phosphorus uptake by pigeonpea and its role in cropping systems of the Indian subcontinent. Science 248(4954):477–480

Banuelos GS, Cardon G, Mackey B, Ben-Asher J, Wu L, Beuselinck P, Akohouse S, Zambrzuski S (1993) Boron and selenium removal in boron laden soils by four sprinkler irrigated plant species. J Environ Qual 22:786–792

Becard G, Douds DD, Pfeffer PE (1992) Extensive in vitro hyphal growth of vesicular-arbuscular mychorrhizal fungi in the presence of CO2 and flavonols. Appl Environ Microbiol 58:821–825

Bolan NS, Naidu R, Mahimairaja S, Baskaran S (1994) Influence of low-molecular-weight organic acids on the solubilization of phosphates. Biol Fertil Soils 18:311–319

Brien OIG, Gibson F (1970) The structure on enterochelin and related 2, 3-dihydroxy-N-benzoylserine conjugates from *Escherichia coli*. Biochim Biophys Acta 215:393–402

Dakora FD (2000) Commonality of root exudation signals and nitrogen assimilation in tropical grain legumes belonging to the tribe phaseoleae. Aust J Plant Physiol 27:885–892

Dakora FD, Muofhe ML (1996) Molecular signals involved in nodulation of the African Bambara groundnut (*Vigna subterranea* L. Verde). In: Proceedings of the international Bambara groundnut symposium, University of Nottingham, pp 171–179

Dakora FD, Phillips DA (1996) Diverse functions of iso-flavonoids in legumes transcend anti-microbial definitions of phytoalexins. Physiol Mol Plant Pathol 49:1–20

Dakora FD, Joseph CM, Phillips DA (1993) Common bean root exudates contain elevated levels of daidzein and coumestrol in response to Rhizobium inoculation. Mol Plant-Microbe Interact 6:665–668

Dave BP, Kena A, Hajela P (2006) Siderophores of halophilic archaea and their chemical characterization. Indian J Exp Biol 44:340–344

De Souza MP, Chu D, Zhao M, Zayed AM, Ruzin SE, Schichnes D, Terry N (1999) Rhizosphere bacteria enhance selenium accumulation and volatilization by Indian mustard. Plant Physiol 119:565–573

Djordjevic MA, Redmond JW, Batley M, Rolfe BG (1987) Clovers secrete specific phenolic compounds which either stimulate or express nod gene expression in *Rhizobium trifolii*. EMBO J 6:1173–1179

Evans A, Anderson TJ (1990) Aliphatic acids: influence on sulphate mobility in a forested cecil soil. Soil Sci Soc Am J 54:1136–1139

Firmin JL, Wilson KE, Rossen L, Johnston AWB (1986) Flavonoid activation of nodulation genes by Rhizobium reversed by other compounds present in plants. Nature 324:90–92

Fischbach MA, Walsh CT (2006) Assembly-line enzymology for polyketide and nonribosomal Peptide antibiotics: logic, machinery, and mechanisms. Chem Rev 106:3468–3496

Fox TR, Comerford NB (1990) Low-molecular weight organic acids in selected forest soils of the southeastern USA. Soil Sci Soc Am J 54:1139–1144

Frossard E, Bucher M, Machler F, Mozafar A, Hurrell R (2000) Potential for increasing the content and bioavailability of Fe, Zn and Ca in plants for human nutrition. J Sci Food Agric 80:861–879

Gagnon H, Ibrahim RK (1998) Aldonic acids: a novel family of nod gene inducers of *Mesorhizobium loti*, *Rhizobium lupini* and *Sinorhizobium meliloti*. Mol Plant-Microbe Interact 11:988–998

Gardner WK, Barber DA, Parbery DG (1983) The acquisition of phosphorus by *Lupinus albus* L. III. The probable mechanism by which phosphorus movement in the soil/root interface is enhanced. Plant Soil 70:107–124

Hawes MC, Brigham LA, Wen F, Woo HH, Zhu Y (1998) Function of root border cells in plant health: Pioneers in the rhizosphere. Annu Rev Phytopathol 36:311–327

Hoffland E, Van Den Boogaard R, Nelemans J, Findenegg G (1992) Biosynthesis and root exudation of citric and malic acids in phosphate-starved rape plants. New Phytol 122:675–680

Hungria M, Joseph CM, Phillips DA (1991) Rhizobium nod gene inducers exuded naturally from roots of common bean (*Phaseolus vulgaris* L.). Plant Physiol 97:759–764

Jackson MB (1979) Rapid injury to peas by soil waterlogging. J Sci Food Agric 30:143–152

Jones DL (1998) Organic acids in the rhizosphere – a critical review. Plant Soil 205:25–44

Jones DL, Darrah PR (1994) Role of root derived organic acids in the mobilization of nutrients from the rhizosphere. Plant Soil 166:247–257

Kape R, Parniske M, Brandt S, Werner D (1992) Isoliquiritigenin, a strong *nod* gene- and glyceollin-inducing flavonoid from soybean root exudate. Appl Environ Microbiol 57:316–319

Kettlewell B, Dong Z, Wu L, Layzell D (2000) Why do legume nodules evolve H_2 gas? Abstract C10, 17th North American conference on symbiotic nitrogen fixation. Quebec City, Canada

Kobayashi T, Nakanishi H, Takahashi M, Kawasaki S, Nishizawa NK, Mori S (2001) In vivo evidence that Ids3 from *Hordeum vulgare* encodes a dioxygenase that converts 2′-deoxymugineic acid to mugineic acid in transgenic rice. Planta 212:864–871

Li HF, McGrath SP, Zhao FJ (2008) Selenium uptake, translocation and speciation in wheat supplied with selenate or selenite. New Phytol 178:92–102

Lipton DS, Blanchar RW, Blevins DG (1987) Citrate, malate, and succinate concentration in exudates from P-sufficient and P-stressed *Medicago sativa* L. seedlings. Plant Physiol 85(2):315–317

Lowe RH, Evans HJ (1962) Carbon dioxide requirement for growth of legume nodule bacteria. Soil Sci 94:351–356

Ma JF, Nomoto K (1996) Effective regulation of iron acquisition in graminaceous plants: the role of mugineic acids as phytosiderophores. Physiol Plant 97:609–617

Mackowiak CL, Grossl PR (1999) Iodate and iodide effects on iodine uptake and partitioning in rice (*Oryza sativa* L.) grown in solution culture. Plant Soil 212:135–143

Marschner H (ed) (1995) Mineral nutrition of higher plants. Academic, London, p 889

Masaoka Y, Kojima M, Sugihara S, Yoshihara T, Koshino M, Ichihara A (1993) Dissolution of ferric phosphate by alfalfa (*Medicago sativa* L.) root exudates. Plant Soil 155(156):75–78

Maxwell CA, Hartwig UA, Joseph CM, Phillips DA (1989) A chalcone and two related flavonoids released from alfalfa roots induce nod genes in *Rhizobium meliloti*. Plant Physiol 91:832–842

Mench M, Martin E (1991) Mobilization of cadmium and other metals from two soils by root exudates of *Zea mays* L. *Nicotiana tabacum* L. and *Nicotiana rustica* L. Plant Soil 132:187–196

Messens E, Geelen D, Montagu MV, Holsters M (1991) 7, 4′-dihydroxyflavanone is the major *Azorhizobium* nod gene including factor in *Sesbania rostrata* seedling exudate. Mol Plant-Microbe Interact 4:262–267

Miethke M, Marahiel MA (2007) Siderophore-based iron acquisition and pathogen control. Microbiol Mol Biol Rev 71:413–451

Nanda Kumar PBA, Dushenkov V, Motto H, Raskin I (1995) Phytoextraction: the use of plants to remove heavy metals from soils. Environ Sci Technol 29:1232–1238

Neilands JB (ed) (1984) Methodology of Siderophores structure and bonding. Springer, Berlin Heidelberg, p 58

Ohwaki Y, Hirata H (1992) Differences in carboxylic acid exudation among P-starved leguminous crops in relation to carboxylic acid contents in plant tissues and phospholipid level in roots. Soil Sci Plant Nutr 38:235–243

Olsen RA, Bennett JH, Blume D, Brown JC (1981) Chemical aspects of the Fe stress response mechanism in tomatoes. J Plant Nutr 3:905–921

Otani K, Mihara K, Okada M, Kaneko S, Fukushima Y (1991) Crossover reaction between haloperidol and amoxapine for NMS. Br J Psychiatry 159:881–889

Parfitt RL (1979) The availability of P from phosphate-goethite bridging complexes. Description and uptake by rye grass. Plant Soil 53:55–65

Patel HM, Walsh CT (2001) In vitro reconstitution of the Pseudomonas aeruginosa nonribosomal peptide synthesis of pyochelin: characterization of backbone tailoring thiazoline reductase and N-methyltransferase activities. Biochemistry 40:9023–9031

Phillips DA, Joseph CM, Yang GP, Martinez-Romero E, Sonborn JR, Volpin H (1999) Identification of lumichrome as a Sinorhizobium enhancer of alfalfa root respiration and shoot growth. Proc Natl Acad Sci U S A 22:12275–12280

Raz V, Fluhr R (1992) Calcium requirement for ethylene dependent response. Plant Cell 4:1123–1130

Recourt K, Schripsema J, Kijne JW, Van Brussel AAN, Lugtenberg BJJ (1991) Inoculation of *Vicia sativa* subsp. Nigra roots with *Rhizobium leguminosarum* biovar viciae results in release of nod gene activating flavanones and chalcones. Plant Mol Biol 16:841–852

Rendig VV, Taylor HM (1989) Principals of soil plant interrelationship. McGraw-Hill Publishing Company, New York

Romheld V (1987) Different strategies for iron acquisition in higher plants. Plant Physiol 70:231–234

Shuman LM (1998) Micronutrient fertilizers. J Crop Prod 1:165–195

Smit G, Puvanesarajah V, Carlson RW, Barbour M, Stacey G (1992) *Bradyrhizobium japonicum* nod D1 can be specifically induced by soybean flavonoids that do not induce the nodYABCSUIJ operon. J Biol Chem 267:310–318

Terry N, Carlson C, Raab TK, Zayed AM (1992) Rate of selenium volatilization among crop species. J Environ Qual 21:341–344

Tsai SM, Phillips DA (1991) Flavonoids released naturally from alfalfa promote development of symbiotic Glomus spores in vitro. Appl Environ Microbiol 57:1485–1488

Umaly RC, Poel LW (1971) Effects of iodine in various formulations on the growth of barley and pea plants in nutrient solution culture. Ann Bot 35:127–131

Vancura V, Hovadik A (1965) Root exudates of plants. II. Composition of root exudates of vegetables. Plant Soil 22:21–32

White PJ (2003) Ion transport. In: Thomas B, Murphy DJ, Murray BG (eds) Encyclopaedia of applied plant sciences. Academic, London, pp 625–634

Zayed A, Lytle CM, Terry N (1998) Accumulation and volatilization of different chemical species of selenium by plants. Planta 206:284–292

Zwarun AA (1972) Microbial competition for glucose in excised root experiments. Soil Sci Soc Am Proc 36:968–969

Selenium Bioavailability Through Microbes

22

Santhappan Paulraj and M. Senthil Kumar

Abstract

Selenium is an important essential trace micronutrient for living systems. Selenium in trace concentrations are essential for the growth and reproduction of plants, animals and microorganisms; however, this essential trace micronutrient element easily become toxic at concentrations higher than the physiological level. Selenium deficiency is regarded as a major health problem for 0.5–1 billion people worldwide. Oxyanions of selenium, *viz.* selenite and selenate, are bioavailable; selenium in the form of selenate ion (SeO_4^{2-}) is more toxic to most organisms than selenite (SeO_3^{2-}). Contrarily, elemental selenium (Se^0) is insoluble, less toxic compared to other selenium forms. Nano-Se (Se^0) in the range of 100–500 nm has similar bioavailability to other selenium forms into plants, animals, humans and microorganisms. Microbial nano-selenium biosynthesis is an eco-friendly and potentially economically viable 'green synthesis' route towards synthesis of red elemental selenium and contributes to the application of selenium for human health. In the soil, applied selenium is rapidly reduced to insoluble forms, and usually the crop nutrient use efficiency was less than 10 % only. Selenium addition in commercial fertilizers may be a larger programme method that is too wasteful, as much of the Se used thereby will be lost for future utilization. Direct addition of selenium compounds to food (process fortification) can be undertaken by the food industry for judicial use of this essential trace micronutrient. Selenium is a non-renewable resource. So it should be a concern of the all stakeholders that the extracted selenium should be judiciously used and to be stockpiled for use as an essential nutrient over generations.

S. Paulraj (✉) • M.S. Kumar
Division of Basic Sciences, ICAR-Indian Institute of Pulses Research, Kanpur, Uttar Pradesh 208 024, India
e-mail: spaulraj.in@gmail.com; senthil_iari@yahoo.co.in

U. Singh et al. (eds.), *Biofortification of Food Crops*, DOI 10.1007/978-81-322-2716-8_22

Keywords
Biofortification • Bioavailability • Green biosynthesis • Nano-selenium

22.1 Introduction

Selenium is an important essential trace micronutrient for living systems. Selenium has special value to industrial element having the unique properties as a semiconductor, but it is also an essential nutrient for humans and animals and may promote plant growth and quality. About 0.5–1 billion people worldwide have major health problem related to selenium deficiency; even larger number may consume less selenium than required for optimal protection against cancer, cardiovascular diseases and severe infectious diseases like HIV disease. Selenium in trace concentrations are essential for the growth and reproduction of plants, animals and microorganisms; however, this element easily become toxic at concentrations higher than the physiological level.

22.2 Status of Selenium

In geosphere the natural elemental selenium that is worth to exploit with economic benefit is not enough; there are no deposits that can be mined for selenium alone. Usually, we can find relatively large amount of elemental selenium in ores, few carbon layers of the soil and the soils of volcanic areas. Selenium is extracted as a by-product of copper mining, and the world reserves of selenium at 93,000 tonnes only cover the estimated contents of economic copper deposits. Substantial resources also exist in association with other metals, coal deposits and uneconomic copper deposits (Table 22.1). The average world production of selenium is estimated at 3000–3500 tonnes per year. In 2010, the production of selenium metal in respect of chief producers was estimated to be 1924 tonnes (Table 22.2).

Selenium occurs in the environment at concentrations of 0.01–2.0 $\mu g \cdot g^{-1}$ of surface soil as selenate, Se(IV); selenite, Se(IV); selenide, Se(II); and elemental selenium, Se^0 (James et al. 1989; Wang and Gao 2001; Stolz et al. 2006). From a global perspective, a selenium cycle occurs as the oxidation states of selenium are changed in response to geochemical (Myneni et al. 1997) and biological processes (Frankenberger and Engberg 1998; Dhanjal and Cameotra 2010). The most common selenium forms in the soil environment are the inorganic selenium salts. These salts leach easily from the soil and become available for the plants (Cooke and Bruland 1987).

22.3 Selenium: An Essential Trace Element

Selenium is as an essential trace element for many biological systems from bacteria to mammals (Shamberger 1983). In plants selenate is much more easily transported than selenite or organic Se (Zayed et al. 1998). The principal form of Se in cereals and other plants are the Se proteins as Se-*Met* are common in Brazil nuts and onion, garlic, species of *Brassica* genus and

Table 22.1 World reserves of Selenium (by principal countries)

Country	Reserves[a] (Tonnes of metal content)
Canada	6000
Chile	20,000
Peru	13,000
Philippines	500
Russia	20,000
USA	10,000
Other countries	23,000
World total (rounded)	93,000

[a]Reserve based on identified copper deposits only (IBM 2012)

Table 22.2 World production of selenium metal (by principal countries)

Country	2008	2009	2010
Belgium	200	200	200
Canada	191	193	79
China	65	65	65
Finland	65	66	66
Germany	250	230	250
Japan	754	709	754
Kazakhstan	130	120	130
Poland	82	80	80
Russia	170	160	170
Sweden	139	129	130
Average world production (estimated)			3000

Source: IBM (2012)

Table 22.3 Recommended dietary allowances of selenium

Group		Intake (μg Se/day)
Infants	(0–0.5 years)	10–40
	(0.5–1 year)	20–60
Children	(1–3 years)	20–80
	(4–6 years)	30–120
Adolescents	(7–10 year)	50–200
	(+11 year)	50–200
Adults	(18 years and above)	50–200

Source: NAS (1980)

mushrooms. Plant species, developmental phase, physiological condition, form, concentration of the Se available and the presence of other substances, especially sulphates, determine the selenium distribution over several plant parts or compartments. Se-*Met* and its S-analogue *Met* cannot be distinguished by animal systems. So both amino acids are incorporated in proteins *via* the same enzymatic pathway (Dumont et al. 2006). The yeast, *Saccharomyces cerevisiae*, has the ability to accumulate and transform high concentration of selenium. Due to its low cost and its ability to synthesize selenoproteins, yeasts can be well utilized for nutritional supplementation.

The breakthrough discovery between supra-nutritional doses of Se and possible cancer protection came from Clark et al. (1996) and has resulted in considerable research to elucidate fully its mode of action and rapidly becoming recognized as one of the more promising cancer chemopreventative agents (Meuillet et al. 2004). The level at which Se is considered likely to be most effective in chemoprevention exceeds that in daily recommended allowances of 50–70 μg per day (approximately 70 μg in men and 50 μg in women) (Table 22.3). About 100–200 μg Se per day has been shown to inhibit cancer development in humans, with 400 μg Se per day being considered the upper limit (El-Bayoumy et al. 2002); at intake doses above 350 μg per day, it starts to exert toxic (Rayman 2000) or even mutagenic effects (Shamberger 1985). Nevertheless, the behaviour of selenium with particular respect to the health of humans and animals is dichotomous.

22.4 Global Selenium Budget

In selenium budget, the selenium production can be derived by the addition of the world's production of Se from mining and agricultural products and fisheries. A demand estimate of how much

Se is needed annually to cover the Se requirements for the world's human and livestock population is calculated by multiplying world population with the recommended Se intake using both the recommended amount of about 50 mg Se per day and the daily amount that may be recommended to prevent certain types of cancer, 250 mg Se per day (Table 22.4). This amount of Se if evenly distributed is enough to supply nutritional demand per day for each person and each head of livestock in the world, but it is not enough to supply the higher rate that may protect against cancer. The annual Se production via agriculture and fisheries may be too small to cover the human requirement and livestock needs, and supplementation of Se would be likely to be needed in many areas of the world.

22.5 Selenium Toxicity

Many elements in trace concentrations are essential for the life; however, these elements easily become toxic at concentrations higher than the physiological level. Nevertheless, the behaviour of selenium with particular respect to the health of humans and animals is dichotomous. The toxicity of selenium is related to its chemical form. Both oxyanions selenite and selenate are soluble and bioavailable; selenium in the form of Se IV, selenate ion (SeO_4^{2-}), is more toxic to most organisms than Se VI, selenite (SeO_3^{2-}). Selenates have better soluble but worse absorption properties than selenites. Contrarily, elemental selenium (Se^0) is insoluble and cannot be absorbed by the biological systems (Barceloux 1999), but elemental selenium is

Table 22.4 Estimated annual selenium budget – Se production, demand and scarcity

Particulars	Se production (tonnes)	Se consumption (tonnes)	Se scarcity (tonnes)
Se production			
Mining industries (IBM 2012)	3500		
Se produced from agriculture and fisheries (Haug et al. 2007)	400		
Total Se production	3900		
Se demand			
Human requirement, prevention of Se deficiency (0.05 mg/day/person for 7.0 billion people) (Combs 2001)		130	
If human requirement is higher for prevention of several diseases, 0.25 mg/day/person for 0.7 billion people (one tenth of population) (Combs 2001)		65	
#Domestic animal requirement (for 3.42×10^9 heads @0.1 mg/day) (FAO 2006)		125	
#Fertilizing one third of arable land of 1.40×10^8 ha @20 g Se/ha/year (FAO 2006)		935	
#Fertilizing one third of pasture land of 3.43×10^8 ha @20 g Se/ha/year (FAO 2006)		2290	
Industrial demand* (IBM 2012)			
Metallurgy (30 % of Se produced)		900	
Glass (30 % of Se produced)		900	
Chemical and pigments (10 % of Se produced)		300	
Electronics (10 % of Se produced)		300	
Other industries (10 % of Se produced)		300	
Total Se consumption		6245	
Net Se demand scarcity			−2345

less toxic compared to other selenium forms, and the inorganic forms of Se are accepted to be more acutely toxic than organic forms. However, a number of human studies show that up to 800 mg Se/day administered as Se yeast gave no symptoms of toxicity (Rayman 2004), and about 400 mg Se/day is considered a safe upper limit (Whanger 2004). Presently, researchers proved that elemental nano-Se has similar bioavailability to other selenium forms. Nano-size, in the range of 100–500 nm, also helps the better absorption of selenium into plants, animals, humans and microorganisms.

22.6 Soil Selenium Bioavailability for Crop Uptake

Soil Se deficiency is the most common cause of severe Se deficiency in poor countries, such as in parts of sub-Saharan Africa. Low Se concentration in the soil and poor availability of soil Se for uptake into the plant roots (or a combination of both factors) lead the agricultural crops as Se deficient. Poor uptake seems often to be the principal cause of Se deficiency in plants grown on cultivated lands in industrial countries (where there are large regions with medium to high soil Se concentrations). The low bioavailability of soil Se is likely to become even more prevalent in the future, as a consequence of population and economic growth.

22.7 Factors Influencing Se Bioavailability

22.7.1 Forms of Selenium

The most common selenium forms in the soil environment are the inorganic selenium salts as selenate, selenite, selenide and elemental selenium. Selenate is less strongly adsorbed to minerals in the soil and more readily taken up by plants than selenite. Selenate ion (SeO_4^{2-}) is thermodynamically stable form of selenium and is in the alkaline environment. Selenates have better soluble but worse absorption properties than selenites. Selenite (SeO_3^{2-}) occurs in neutral pH environment, and it is less soluble compared to selenate. Selenite can be reduced to elemental selenium (Se^0) by chemical or biological ways. Selenides (Se^{2-}) and the selenium-enriched sulphides occur in reductive or acidic environment, have weak soluble and oxidating properties and have limited bioavailability to plants and animals. Many factors such as dry climate, low organic matter concentration in the soil, high temperature, high pH and no water logging may give a high ratio between selenate and selenite in the soil. However, in the Nordic countries with high concentrations of organic matter, the selenite is the dominant form of inorganic Se in soils (because of low soil temperatures causing much slower degradation of soil organic matter than in tropical countries) and most likely also in waterlogged soils (during rice cultivation).

22.7.2 Selenium Fortification in Commercial Fertilizers

Fortification of commercial fertilizers used in the food production chain was chosen to ensure an increased Se intake for the whole population in Finland. The Finnish supplementation programme to add sodium selenate to multinutrient (NPK) fertilizers for all the field and horticultural crops at rates of 6 mg/kg fertilizer for grasses and 16 mg/kg fertilizer for cereal and horticultural crops (Aro et al. 1995; Aspila 2005) has provided experience with this approach to increase Se concentration in the diet, and it has been shown to be an effective and safe approach to raise Se levels in a human population.

In acid soil conditions, Se from sodium selenate was more readily taken up by the plants than Se from sodium selenite. Selenium supplementation to plants may also enhance the production and quality of edible plant products, by increasing antioxidant activity in tea leaves (Xu et al. 2003) and in rice (Xu and Hu 2004), alter glucosinolate and sulphoraphane content (Robbins et al. 2005) as well as increase plant growth and fertility (Graham et al. 2005). Therefore, Se fertilization may not only be beneficial for nutritive value in the food chain but also for crop quality and, under certain circumstances, yield.

22.7.3 Agronomic Selenium Use Efficiency by Fertilization

In the soil, Se is rapidly reduced to insoluble forms; usually less than 10 % of the applied Se was taken up by the crop. The Se not taken up by plants readily after application was apparently unavailable for crops growing later in the season or the next year (Yli-Halla 2005), and better recovery of Se applied in fertilizers to grain (18 %) was reported by Lyons et al. (2004). Se fertilization to spring season crops at when the crop is beginning to grow rapidly recorded higher efficiency (Curtin et al. 2006).

The foliar application of Se either as sodium selenate or sodium selenite has shown to be several times more efficient than application in fertilizers (Aspila 2005) but riskier as Se uptake by the crop depends on spraying conditions Curtin et al. (2006). The sodium selenate which is commonly used in foliar application is highly toxic. Health and safety precautions must therefore be taken during its on-farm use. However, Lyons et al. (2004) found foliar application to be less efficient than application to soil at planting (at application rates of 40 and 120 g Se/ha) in Australian trials.

Seed can be treated by applying sodium selenate directly on the seed surface (Gissel-Nielsen 1998). However, it has not been shown to give efficient recovery of Se (Curtin et al. 2006; Gissel-Nielsen 1998).

22.7.4 Selenium Interaction with Other Fertilizers

In soil environment that phosphorus renders much of the Se unavailable for uptake by plants (Hopper and Parker 1999; Dhillon and Dhillon 2000; Liu et al. 2004) may be due to co-precipitation of selenite ions with phosphate. When new phosphate fertilizers are added to the soil and new precipitation of phosphate minerals takes place, Se remains fixed in the precipitate and unavailable for uptake. Conversely, phosphate may also lead to desorption of selenite ions bound to minerals in the soil, as phosphate is bound more strongly to trivalent iron and aluminium than is selenite (Liu et al. 2004; Nakamaru et al. 2006).

In the prairie lands of USA and Canada where the soil has been so fertile, that application of large amounts of phosphate fertilizers has been unnecessary. Cereals grown on these lands have enriched with Se at high concentrations that can be explained by a combination of high natural Se concentration in the soil and excellent bioavailability of soil Se for uptake by the plant roots, due to lack of commercial P fertilizer application.

Sulphate fertilization further reduced the selenium concentrations in plants as a result of competition between sulphate and selenate for transporters in plant roots (Lyons et al. 2004). Selenium and sulphur (S) compete with each other in the biochemical pathways, leading to the synthesis of selenomethionine (Se-*Met*) and methionine in plant cells. So the concentration of Se-*Met* in plant seeds depends strongly on the ratio of Se uptake to S uptake in the roots.

The process that was used for the removal of arsenic from superphosphate fertilizers took away Se as well, and the colour of the fertilizer changed from pink (because of finely disseminated elemental Se) to white or grey that leads declined selenium status in plants and livestock in New Zealand.

22.7.5 Microbe-Selenium Plant Interactions

Plant-microbe interactions in Se hyper accumulators can range from pathogenic to symbiotic (Rogerson 1957; Wangeline and Reeves 2007), and such interactions appear to be strongly influenced by the distribution and accumulation of Se across the plant tissues (de Souza et al. 1999; Di Gregorio et al. 2006).

Microbes assimilate, accumulate and detoxify selenium through elemental reduction or volatilization (Hassoun et al. 1995; Chasteen and Bentley 2003; Vallini et al. 2005; Stolz et al. 2006) and that levels of tolerance varied among them and by above processes. Microbes assimilate selenium into seleno-enzymes using the similar pathways for sulphur (Bradfield et al. 1970; Cypionka 1987; Stolz et al. 2006; Lechenne et al. 2007). When exposed to 30 mg/kg $NaSeO_4$ or $NaSeO_3$, selenium-sensitive fungi can initially reduce most of the selenium and accumulate 90 % of it as organic Se and selenite. Selenium-tolerant fungi will accumulate between 60 and 70 % of the Se as organic selenium, 15–30 % as elemental selenium and 5–10 % as selenite. Selenium reduction and greater volatilization are among the many mechanisms that selenium-tolerant fungi use differently compared to selenium-sensitive fungi.

Based on their metabolic capacities, microbes may affect plant uptake of selenium by changing the concentration and chemical speciation of Se in the soil. For instance, microbial-mediated Se volatilization and break down or accumulation of organic Se by microbes in the rhizosphere can reduce plant Se availability. Rhizosphere microbes may also induce plant root hair formation and enhance plant sulphate/selenate uptake as a result of increased serine/O-acetylserine concentrations in the rhizosphere (de Souza et al. 1998).

22.7.6 Mycorrhiza and Selenium Bioavailability

An important unresolved question concerns the mechanism of selenite uptake into mycorrhiza and plant roots, whether there is a specific membrane transporter for the selenite ions or if selenite and some other more abundant anion, such as phosphate, may share a common membrane transporter. If the latter should be the case, it must be expected that phosphate will function as a competitive inhibitor of selenite ion transport into the plant roots (Munier-Lamy et al. 2007). The same question could also be raised regarding the mechanism of active transport of selenite ions into other types of organism, e.g. into various groups of planktonic algae or into mammalian cells such as erythrocytes.

22.8 Nano-selenium Biosynthesis and Its Bioavailability

Selenium element is well known for its photoelectric, semiconductor, free radical scavenging, anti-oxidative and anticancer properties (Zhang et al. 2004). Among the various forms of selenium, the red amorphous selenium have the biological activity similar to that of sodium selenite (Wang et al. 2007; Zhang et al. 2007), and it was considered the most potent chemical form to the artificial selenium enrichment. It was generally considered that elemental selenium is to be biologically inert, but researchers proved that nano-Se has similar bioavailability to other selenium forms. Therefore, the red amorphous selenium has attracted much attention (Mishra et al. 2011; Hunter and Kuykendall 2007; Dwivedi et al. 2013; Hunter and Manter 2009). The production of nano-selenium can be achieved through chemical and biological methods. Chemical detoxification of metals is proven to be very expensive and often results in secondary effects on the environment.

In natural environments, the bacteria, fungi, yeasts and plants are known to be capable of converting selenate and selenite to Se^0 (Oremland et al. 1989; Hunter and Manter 2009). Among these, bacteria are preferred for biosynthesis due to their extracellular particle production ability, short generation time, ease of culturing, downstream processing and manipulation (Ramanathan et al. 2013). Microorganisms have been shown to be particularly resistant to SeO_3^{2+} (Kessi et al. 1999; Kessi 2006). This resistance is attributed to the capacity of the organisms to reduce Se oxyanions to their elemental ground state. In recent years, several different bacteria have been reported for the biological synthesis of Se^0 (Fig. 22.1), such

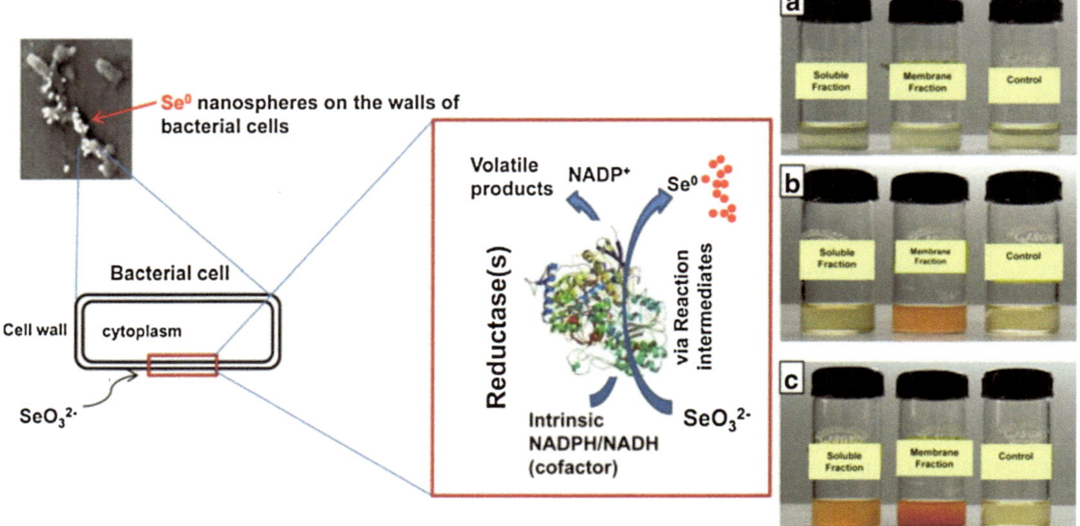

Fig. 22.1 Schematic representation biogenesis of selenium (Se^0) nanospheres. (**a**) Selenite reduction at 0 h. (**b**) Formation of red elemental selenium in membrane fraction after 3–4 h of incubation. (**c**) Prolonged incubation of 12 h resulted in the formation of red elemental selenium in soluble fraction (Dhanjal and Cameotra 2010)

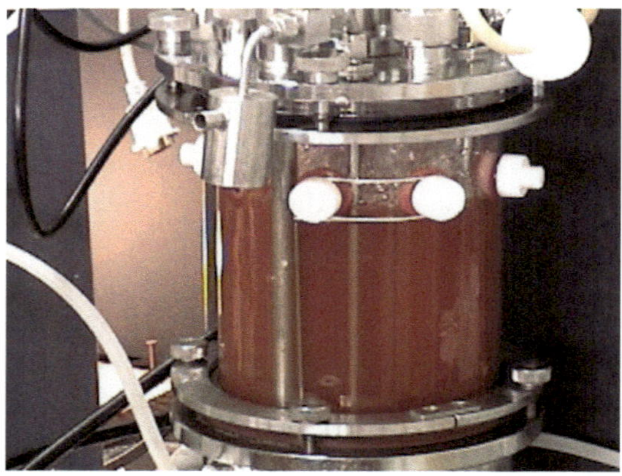

Fig. 22.2 Bioreactor after 72 h anaerobic growth of *P. fluorescens* K27 at 30 °C following 1.0 mM sodium selenite amendment. *Brick-red* coloration is due to elemental selenium in the well-mixed solution (Hapuarachchi et al. 2004)

as *Thauera selenatis* (Bledsoe et al. 1999), *Rhizobium selenitireducens* strain B1 (Hunter and Manter 2009; Euzeby 2008; Hunter et al. 2007), *Escherichia coli* (Avazeri et al. 1997), *Clostridium pasteurianum* (Yanke et al. 1995), *Pseudomonas fluorescens* K27 (Hapuarachchi et al. 2004) (Fig. 22.2), *Bacillus cereus* (Dhanjal and Cameotra 2010) and *Bacillus selenitireducens* (Afkar et al. 2003). Debieux et al. (2011) reported that the production of volatile selenium compounds occurred during the stationary phases (Fig. 22.3). It means that the reduction is independent of strain growth. Moreover, this result suggests that reduction reaction is controlled by the stationary phase of the cultures.

Nano-Se and selenite have large difference in acute toxicity (Zhang et al. 2004). Nano-Se was less toxic than selenite in terms of suppressing growth, liver toxicity and antioxidant status but has similar bioavailability. Wang et al. (2007) concluded that based on animal tests, nano-selenium (nano-Se) is a very effective antioxidant, without high toxicity properties which is typical for other selenium forms. Nano-selenium has at least the same effect on activating glutathione peroxidase and thioredoxin reductase enzymes as selenoproteins have, but according to the LD50, this form is less toxical; it does not trigger acute liver injury and short-term toxications. Furthermore, nano-selenium accumulates less in the treated mice and activates glutathione S-transferase more effectively than

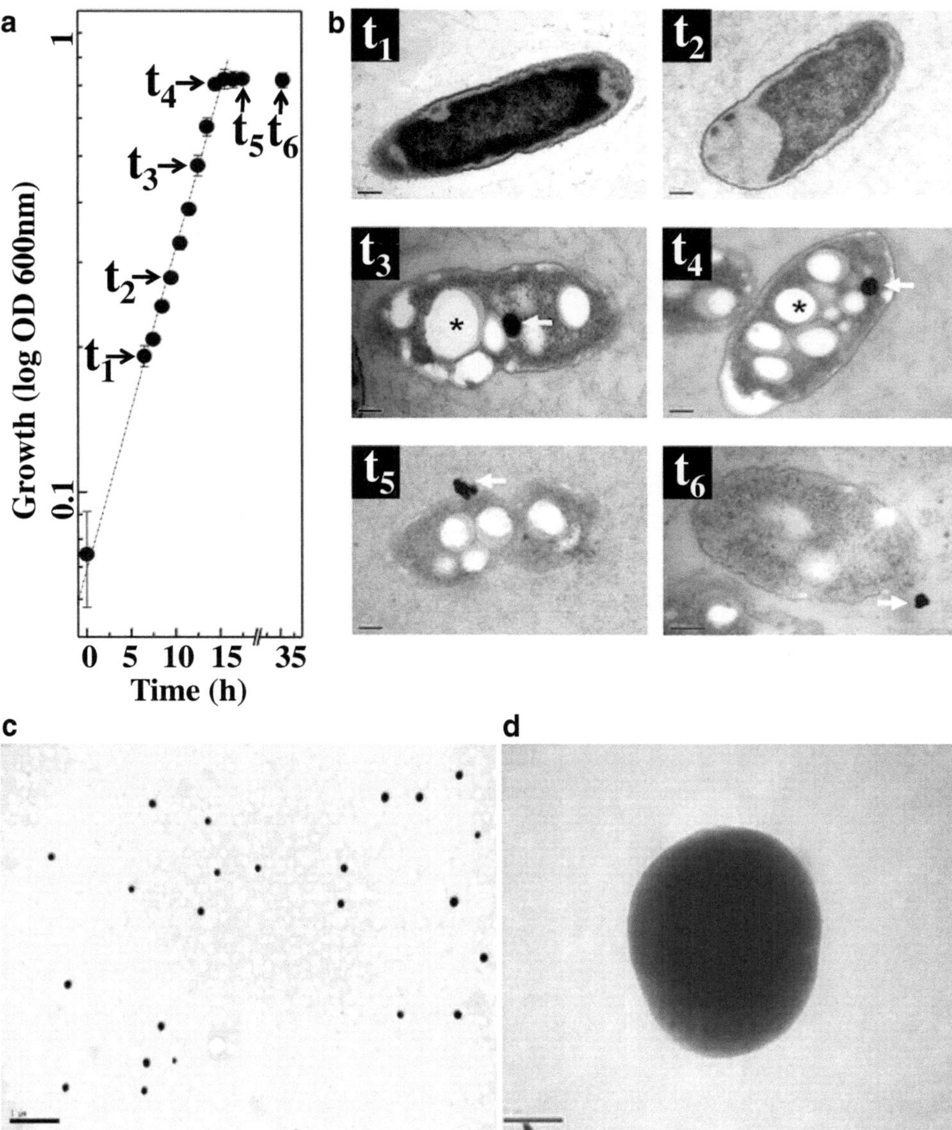

Fig. 22.3 Physiological analysis of Se-nanosphere production. (**a**) Growth curve of *T. selenatis* grown on acetate using selenate (10 mM) as the sole electron acceptor (Error bars are SEM: n + 10 cultures). Time points $t_1 - t_6$ indicate the samples used for EM analysis. (**b**) Transmission electron micrographs of time points from (**a**). Micrographs t_1 and t_2 show mid-exponential phase, t_3 and t_4 show late exponential phase, and t_5 and t_6 show stationary phase. Scale bar, 200 nm. Selenium deposits are indicated by an *asterisk*. (**c, d**) Transmission electron micrographs of purified Se nanospheres. (**c**) Scale bar, 500 nm. (**d**) Scale bar 50 nm (Debieux et al. 2011, *Source*: http://www.pnas.org/cgi/doi/10.1073/pnas.1105959108. Accessed 12 Jan 2015)

selenoproteins, independently from the level of toxication.

Certain lactic acid bacteria and other probiotic bacteria species are able to reduce the selenite in toxical concentration into nano-sized elemental selenium spheres (Patented) (Prokisch et al. 2010), and these bacterial strains produce uniform-sized, high-purity selenium spheres in different size ranges from 50 to 500 nm (Figs. 22.4 and 22.5) (Eszenyi et al. 2011). Advantageously, when used as a food additive, the nanospheres produced according to the processes of the technology need not be fully purified or purified at all because the medium

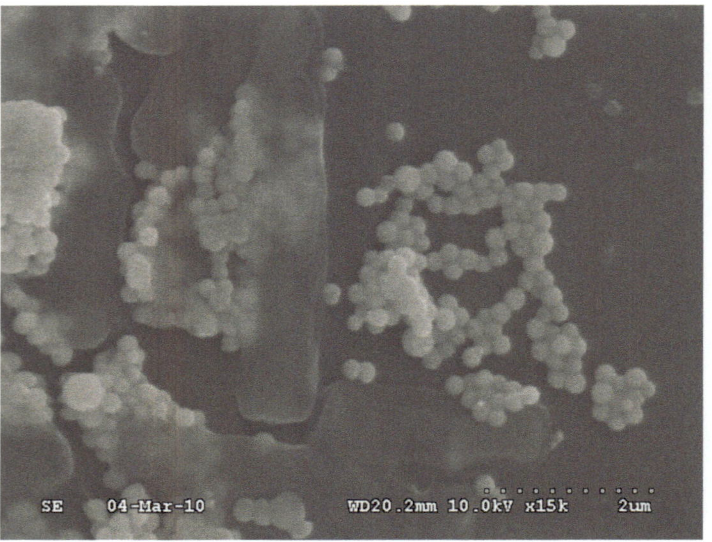

Fig. 22.4 Electron microscopic view of 250 nm-sized selenium nanosphere

Fig. 22.5 Lactomicrosel[R] nano-selenium-rich yoghurt (dried and powdered)

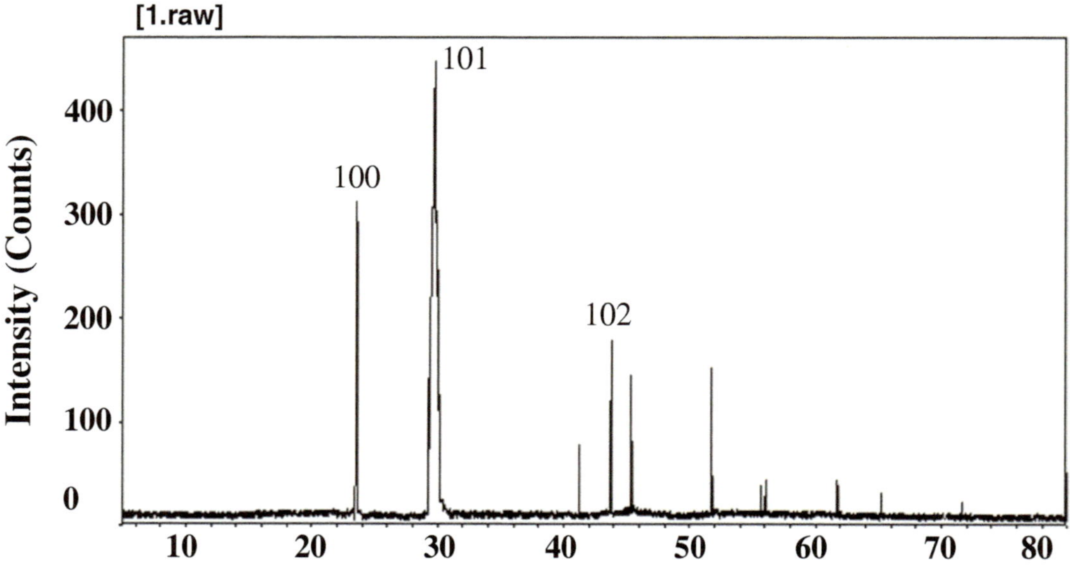

Fig. 22.6 XRD pattern of selenium-containing particles formed by *Rhodopseudomonas palustris* strain N. The characteristic strong diffraction peak located at 29.777u is ascribed to the (101) facets of the face-centred cubic elemental Se^0 structure (Li et al. 2014) (Source: DOI:10.1371/journal.pone.0095955.g005)

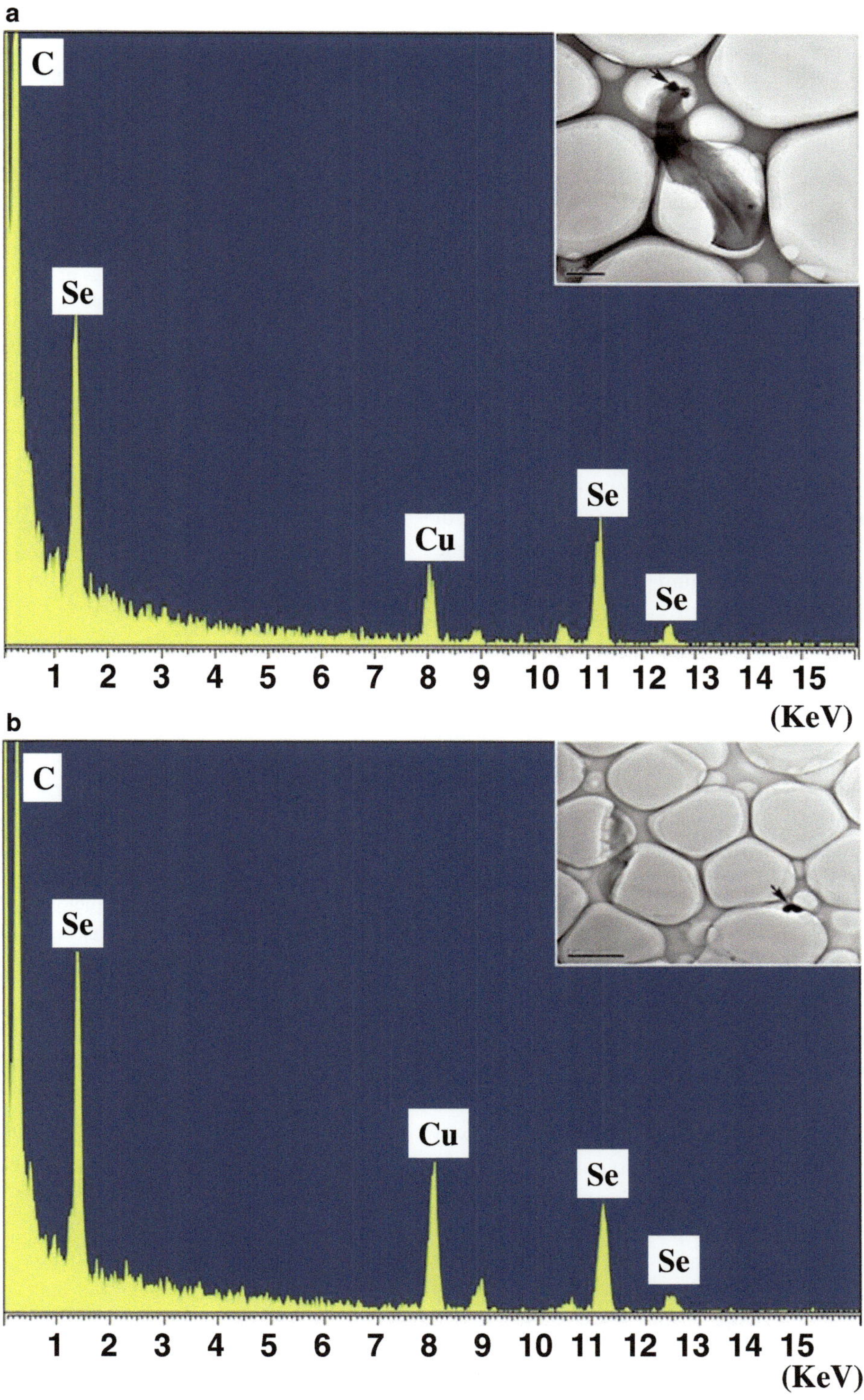

Fig. 22.7 (**a**) Particles on the cell membrane. Energy levels (in kiloelectron volts) are indicated on the x axis. The emission lines for selenium are at 1.37 keV (peak SeLa), 11.22 keV (peak SeKa) and 12.49 keV (peak SeKb). (**b**) Particles in the culture medium. Energy levels (in kiloelectron volts) are indicated on the x axis. The emission lines for selenium are at 1.37 keV (peak SeLa), 11.22 keV (peak SeKa) and 12.49 keV (peak SeKb) (Li et al. 2014) (Source: DOI:10.1371/journal.pone.0095955.g004)

and the bacteria applied in the production process may be fully suitable for human or animal consumption.

The photosynthetic bacteria, *Rhodopseudomonas palustris* strain N, has more digestible bacterial cell wall and are rich in protein, carotenoids, biological cofactors and vitamins (Kobayashi and Kurata 1978). They were already shown to be suitable amendments to health foods for humans and animals. Therefore, the reduction of SeO_3^{2+} to Se^0 by *Rhodopseudomonas palustris* strain N (Li et al. 2014) (Figs. 22.6 and 22.7) provides a potential application to artificially enrich food with selenium for human health. It will provide an eco-friendly and potentially economically viable 'green' synthesis route towards synthesis of red elemental selenium and contribute to the application of selenium for human health (Li et al. 2014).

22.9 Conclusion

Selenium is an important essential trace micronutrient for living systems. There are no deposits that can be mined for selenium alone, only as by-product of copper mining as resource. Direct addition of selenium compounds to food (process fortification) is to be undertaken by the food industry, which is fruitful rather than the wasteful agronomic selenium fortification. New ways to bio-fortify food products are needed, and it is generally observed that there is less wastage if selenium is added late in the production chain rather than early. Bacterial nano-selenium will provide an eco-friendly and potentially economically viable 'green' synthesis route towards synthesis of red elemental selenium with enhanced bioavailability. Selenium is a non-renewable resource. So it should be the concern of all stakeholders that the extracted selenium should be judiciously used and to be stockpiled for use as an essential nutrient by future generations.

References

Afkar E, Lisak J, Saltikov C, Basu P, Oremland RS, Stolz JF (2003) The respiratory arsenate reductase from *Bacillus selenitireducens* strain MLS10. FEMS Microbiol Lett 226:107–112

Aro A, Alfthan G, Varo P (1995) Effects of supplementation of fertilizers on human Se status in Finland. Analyst 120:841–843

Aspila P (2005) History of selenium supplemented fertilization in Finland. In: Proceedings of the twenty years of selenium fertilization, Helsinki, Finland, pp 8–13, 8–9 Sept 2005

Avazeri C, Turner RJ, Pommier J, Weiner JH, Giordano G, Vermeglio A (1997) Tellurite and selenate reductase activity of nitrate reductases from *Escherichia coli*: correlation with tellurite resistance. Microbiology 143:1181–1189

Barceloux DG (1999) Selenium. Clin Toxicol 37:145–172

Bledsoe TL, Cantafio AW, Macy JM (1999) Fermented whey – an inexpensive feed source for a laboratory-scale selenium-bioremediation reactor system inoculated with *Thauera selenatis*. Appl Microbiol Biotechnol 51:682–685

Bradfield G, Somerfield P, Meyn T, Holby M, Babcock D, Bradley D, Segel IH (1970) Regulation of sulfate transport in filamentous fungi. Plant Physiol 46:720–727

Chasteen TG, Bentley R (2003) Biomethylation of selenium and tellurium: microorganisms and plants. Chem Rev 103(1):1–25

Clark LC, Combs GR, Turnbull BW, Slate EH, Chalker DK, Chow J, Davis LS, Glover RA, Graham GF, Gross EG, Krongrad A, Lesher JL, Park HK, Sanders BB Jr, Smith CL, Taylor JR (1996) Effects of selenium supplementation for cancer prevention in patients with carcinoma of the skin. J Am Med Assoc 276:1957–1963

Combs GF (2001) Se in global food systems. Br J Nutr 85:517–547

Cooke TD, Bruland KW (1987) Aquatic chemistry of selenium – evidence of biomethylation. Environ Sci Technol 21(12):1214–1219

Curtin D, Hanson R, Lindley TN, Butler RC (2006) Selenium concentration in wheat grain as influenced by method, rate, and timing of sodium selenate application. N Z J Crop Hortic Sci 34:329–339

Cypionka H (1987) Uptake of sulfate, sulfite and thiosulfate by proton-anion symport in *Desulfovibrio desulfuricans*. Arch Microbiol 148:144–149

de Souza MP, Pilon-Smits EAH, Lytle CM, Hwang S, Tai JC, Honma TSU, YehL TN (1998) Rate-limiting steps in selenium volatilization by *Brassica juncea*. Plant Physiol 117:1487–1494 de Souza MP, Chu D, May Z, Zayed AM, Ruzin SE, Schichnes D, Terry N (1999) Rhizosphere bacteria enhance selenium accumulation and volatilization by Indian mustard. Plant Physiol 119:565–573

Debieux CM, Dridge EJ, Mueller CM, Splatt P, Paszkiewicz K, Knight I, Florance H, Love J, Titball RW, Lewis RJ, Richardson DJ, Butler CS (2011) A bacterial process for selenium nanosphere assembly. Proc Natl Acad Sci U S A 108(33):13480–13485. doi:10.1073/pnas.1105959108

Dhanjal S, Cameotra SS (2010) Aerobic biogenesis of selenium nanospheres by *Bacillus cereus* isolated from coalmine soil. Microb Cell Factories 9:52–66. doi:10.1186/1475-2859-9-52

Dhillon SK, Dhillon KS (2000) Selenium adsorption in soils as influenced by different anions. J Plant Nutr Soil Sci 163:577–582

Di Gregorio S, Lampis S, Malorgio F, Petuzzelli G, Pezzarossa B, Vallini G (2006) *Brassica juncea* can improve selenite and selenate abatement in selenium contaminated soils through the aid of its rhizospheric bacterial population. Plant Soil 285:233–244

Dumont E, Vanhaecke F, Cornelis R (2006) Selenium speciation from food source to metabolites: a critical review. Anal Bioanal Chem 385(7):1304–1323

Dwivedi S, AlKhedhairy AA, Ahamed M, Musarrat J (2013) Biomimetic synthesis of selenium nanospheres by bacterial strain JS-11 and its role as a biosensor for nanotoxicity assessment: a novel Se-bioassay. Plos One 8(3), e57404. doi:10.1371/journal.pone.0057404

El-Bayoumy K, Richie JP, Boyiri T Jr, Komninou D, Prokopczyk B, Trushin N, Kleinman W, Cox J, Pittman B, Colosimo S (2002) Influence of selenium-enriched yeast supplementation on biomarkers of oxidative damage and hormone status in healthy adults: a clinical pilot study. Cancer Epidemiol Biomark Prev 11:1459–1465

Eszenyi P, Sztrik A, Babka B, Prokisch J (2011) Elemental, nano-sized (100–500 nm) selenium production by probiotic lactic acid bacteria. Int J Biosci Biochem Bioinform 1(2):148–152

Euzeby J (2008) Validation list no. 121. List of new names and new combinations previously effectively, but not validly, published. Int J Syst Evol Microbiol 58:1057

FAO (2006) Food and Agriculture Statistics Global Outlook (Produced by the Statistics Division). http://faostat.fao.org/Portals/_Faostat/documents/pdf/world.pdf. Accessed 12 Jan 2015

Frankenberger WT, Engberg RA (eds) (1998) Environmental chemistry of selenium. Marcel Dekker, New York

Gissel-Nielsen G (1998) Effects of selenium supplementation on field crops. In: Frankenberger WT Jr, Engberg RA (eds) Environmental chemistry of selenium. Marcel Dekker, New York, pp 99–128

Graham R, Stangoulis J, Genec Y, Lyons G (2005) Selenium can increase growth and fertility in vascular plants. In: Li CJ (ed) Plant nutrition for food security, human health and environmental protection. Tsinghua University Press, Beijing, pp 208–209

Hapuarachchi S, Swearingen J, Chasteen TG (2004) Determination of elemental and precipitated selenium production by a facultative anaerobe grown under sequential anaerobic/aerobic conditions. Process Biochem 39:1607–1613

Hassoun BS, Palmer IS, Dwivedi C (1995) Selenium detoxification by methylation. Res Commun Mol Pathol Pharmacol 90(1):133–142

Haug A, Graham RD, Christophersen OA, Lyons GH (2007) How to use the world's scarce selenium resources efficiently to increase the selenium concentration in food. Microb Ecol Health Dis 19:209–228

Hopper J, Parker D (1999) Plant availability of selenite and selenate as influenced by the competing ions phosphate and sulphate. Plant Soil 210:199–207

Hunter WJ, Kuykendall LD (2007) Reduction of selenite to elemental red selenium by *Rhizobium* sp. strain B1. Curr Microbiol 55:344–349

Hunter WJ, Manter DK (2009) Reduction of selenite to elemental red selenium by *Pseudomonas* sp. strain CA5. Curr Microbiol 58:493–498

Hunter WJ, Kuykendall LD, Manter DK (2007) *Rhizobium selenireducens* sp. nov.: a selenite reducing α-Proteobacteria isolated from a bioreactor. Curr Microbiol 55:455–460

IBM (2012) Indian minerals yearbook 2011(Part- II) 50th edn. Selenium and tellurium (advance release), Government of India, Ministry of Mines, Indian Bureau of Mines, pp 69:2–5

James LF, Panter KE, Mayland HF, Miller RM, Baker DC (1989) Selenium poisoning in livestock: a review and progress. Selenium in agriculture and the environment. Special publication, 23. Soil Sci Soc Am, pp 123–131

Kessi J (2006) Enzymatic systems proposed to be involved in the dissimilatory reduction of selenite in the purple non-sulfur bacteria *Rhodospirillum rubrum* and *Rhodobacter capsulatus*. Microbiology 152:731–743

Kessi J, Ramuz M, Wehrli E, Spycher M, Bachofen R (1999) Reduction of selenite and detoxification of elemental selenium by phototrophic bacterium *Rhodospirillum rubrum*. Appl Environ Microbiol 65:4734–4740

Kobayashi M, Kurata S (1978) The mass culture and cell utilization of photosynthetic bacteria. Process Biochem 13:27–30

Lechenne B, Reichard U, Zaugg C, Fratti M, Kunert J, Boulat O, Monod M (2007) Sulphite efflux pumps in *Aspergillus fumigatus* and dermatophytes. Microbiology 153:905–913

Li B, Liu N, Li Y, Jing W, Fan J, Li D, Zhang L, Zhang X, Zhang Z, Wang L (2014) Reduction of selenite to red elemental selenium by *Rhodopseudomonas palustris* strain N. PLoS ONE 9(4), e95955. doi:10.1371/journal.pone.0095955

Liu Q, Wang DJ, Jiang XJ, Cao ZH (2004) Effects of the interaction between selenium and phosphorous on the growth and selenium accumulation in rice. Environ Geochem Health 26:325–330

Lyons GH, Lewis J, Lorimer MF, Holloway RE, Brace DM, Graham RD, Stangoulis JCR (2004) High-selenium wheat: agronomic biofortification strategies to improve human nutrition. Food Agric Environ 2:171–178

Meuillet E, Stratton S, Cherukuri DP, Goulet AC, Kagey J, Poterfield B, Nelson MA (2004) Chemoprevention of prostate cancer with selenium: an update on current clinical trials and preclinical findings. J Cell Biochem 91:1740–1752

Mishra RR, Prajapati S, Das J, Dangar TK, Das N, Thatoi H (2011) Reduction of selenite to red elemental selenium by moderately halotolerant *Bacillus megaterium* strains isolated from Bhitarkanika mangrove soil and characterization of reduced product. Chemosphere 84 (9):1231–1237

Munier-Lamy C, Deneux-Mustin S, Mustin C, Merlet D, Berthelin J, Leyval C (2007) Selenium bioavailability and uptake as affected by four different plants in loamy clay soils in particular attention to mycorrhizae inoculated ryegrass. J Environ Radioact 97:148

Myneni SCB, Tokunaga TK, Brown GE (1997) Abiotic Se redox chemistry in the presence of Fe(II, III)-oxides. Science 278:1106–1109

Nakamaru Y, Tagami K, Uchida S (2006) Effect of phosphate addition on the sorption–desorption reaction of selenium in Japanese agricultural soils. Chemosphere 63:109–115

NAS (1980) Recommended dietary allowances, 9th edn. US Food and Nutrition Board, National Academy of Sciences, Washington, DC

Oremland RS, Hollibaugh JT, Maest AS, Presser TS, Miller LG, Culbertson CW (1989) Selenate reduction to elemental selenium by anaerobic bacteria in sediments and culture: biogeochemical significance of a novel, sulfate-independent respiration. Appl Environ Microbiol 55(9):2333–2343

Prokisch, Jozsef (Debrecen, HU), Zommara, Mohsen A (Alexandria, EG) (2010) Process for producing elemental selenium nanospheres, United States patent no. 20100189634

Ramanathan R, Field MR, O'Mullane AP, Smooker PM, Bhargava SK, Bansal V (2013) Aqueous phase synthesis of copper nanoparticles: a link between heavy metal resistance and nanoparticle synthesis ability in bacterial systems. Nanoscale 5:2300–2306. doi:10.1039/C2NR32887A

Rayman MP (2000) The importance of selenium to human health. Lancet 356:233–241

Rayman MP (2004) The use of high-Se yeast to raise Se status: how does it measure up? Br J Nutr 92:557–573

Robbins RJ, Keck AS, Banuelos G, Finley JW (2005) Cultivation conditions and selenium fertilization alter the phenolic profile, glucosinolate, and sulforaphane content of broccoli. J Med Food 8:204–214

Rogerson CT (1957) Kansas mycological notes 1955–1956. Trans Kans Acad Sci 60(4):370–375

Shamberger RJ (1983) Biochemistry of selenium. Plenum Press, New York/London, p 334

Shamberger RJ (1985) The genotoxicity of selenium. Mutat Res 154:29–48

Stolz JF, Basu P, Santini JM, Oremland RS (2006) Arsenic and selenium in microbial metabolism. Annu Rev Microbiol 60:107–130

Vallini G, Di Gregorio S, Lampis S (2005) Rhizosphere-induced selenium precipitation for possible applications in phytoremediation of Se polluted effluents. Z Naturforsch 60:349–356

Wang Z, Gao Y (2001) Biogeochemical cycling of selenium in Chinese environments. Appl Geochem 16:1345–1357

Wang H, Zhang J, Yu H (2007) Elemental selenium at nano size possesses lower toxicity without compromising the fundamental effect on selenoenzymes: comparison with selenomethionine in mice. Free Radic Biol Med 42:1524–1533

Wangeline AL, Reeves FB (2007) Two new *Alternaria* species from selenium rich habitats in Rocky Mountain Front Range. Mycotaxon 99:83–89

Whanger PD (2004) Se and its relationship to cancer: an update. Br J Nutr 91:11–28

Xu J, Hu Q (2004) Effect of foliar application of Se on the antioxidant activity of aqueous and ethanolic extracts of selenium-enriched rice. J Agric Food Chem 52:1759–1763

Xu J, Yang F, Chen L, Hu Y, Hu Q (2003) Effect of selenium on increasing the antioxidant activity of tea leaves harvested during the early spring tea producing season. J Agric Food Chem 51:1081–1084

Yanke LJ, Bryant RD, Laishley EJ (1995) Hydrogenase (I) of *Clostridium pasteurianum* functions as a novel selenite reductase. Anaerobe 1:61–67

Yli-Halla M (2005) Influence of Se fertilization on soil Se status. In: Proceedings of the twenty years of Se fertilization, Helsinki, Finland, 8–9 Sept 2005, pp 25–32

Zayed AM, Lytle CM, Terry N (1998) Accumulation and volatilization of different chemical species of selenium by plants. Planta 206:284–292

Zhang J, Zhang S, Xu J, Chen H (2004) A new method for the synthesis of selenium nanoparticles and the application to construction of H_2O_2 biosensor. Chin Chem Lett 15:1345–1348

Zhang J, Wang X, Xu T (2007) Elemental selenium at nano size (Nano-Se) as a potential chemopreventive agent with reduced risk of selenium toxicity: comparison with Se-methylselenocysteine in mice. Toxicol Sci 101:22–31

Physiological Processes Toward Movement of Micronutrients from Soil to Seeds in Biofortification Perspectives

23

P.S. Basu

Abstract

Out of the 17 elements, zinc and iron are the most important, and their less bioavailability in diets causes deficiency symptoms in human often leading to serious physiological disorder. Therefore, biofortified seeds or grains enriched with these nutrients particularly Zn and iron in staple crops can eradicate malnourishment to large extent. Genetic variability in the micronutrients among crop species has been well documented which could be exploited to improve essential micronutrients in the seeds through conventional breeding or biotechnological interventions. The crop species widely differ in distribution of micronutrients in various parts of the plant and also the diverse routes through which micronutrients move and get accumulated in seeds. The major challenges are to divert more nutrients in the target tissue seeds that are edible part of the crops. The phloem loading and unloading into the target tissue of plants are not clearly known, and the process is considered to be major limiting factors toward biofortification. The other limiting factors are the soil itself in which bioavailability to the plants is sufficiently low in spite of adequate availability of these elements in soil as they remained in bound or fixed in one or another form. The physicochemical properties of soil determine the efficiency of the uptake of nutrients. Soil enriched with organic decomposed matters and plant growth promoting rhizobia facilitate the absorption of nutrients. The major channels of nutrient uptake are through extensive network of root hairs that enter into the root vascular tissue through symplastic or apoplastic pathways. Phloem loading of micronutrients primarily in chelated form is an essential process prior to transport into seeds. The iron chelator nicotianamine (NA) plays a major role in binding copper and zinc in addition to iron and other ions. Unloading of micronutrients present in the phloem sap into the seeds

P.S. Basu (✉)
Division of Basic Sciences, ICAR-Indian Institute of Pulses Research, Kanpur, Uttar Pradesh 208 024, India
e-mail: psbsu59@gmail.com

U. Singh et al. (eds.), *Biofortification of Food Crops*, DOI 10.1007/978-81-322-2716-8_23

takes place through bulk flow created by pressure gradient. The requisite osmotic potential buildup in the source tissue due to sugars, organic acids, and potassium ions is responsible to draw water from sink tissue and in exchange unload the micronutrients dissolved in phloem sap. The movement of micronutrients in the phloem such as zinc and iron takes place along with other major ions like potassium, chloride, and sugars. Foliar application of fertilizer or soil application in soluble forms and bioinoculants of AM fungi and bacteria enhance mobilization of zinc and iron in the plant system. The homeostasis of these elements that is balancing the amount in different tissues and redistribution as per demand is considered to be the important aspects to investigate for enriching nutrient content in the seeds. Several transporter genes have been identified, and several genetically modified crops showed manifold increase in the zinc and iron content. Twelve agriculturally important crops are presently in process of biofortification under the megaproject HarvestPlus. The mobility of the zinc and iron has been elucidated using model plant *Arabidopsis thaliana* and mutant lacking some genes of transporter family. The present review analyzes the mobility of these elements inside the plant, their distribution, limiting factors, and strategies to improve the mineral content in the seeds.

Keywords
Acquisition • Micronutrient • Mineral homeostasis • Physiological processes

23.1 Introduction

Plants and humans require 17 essential minerals; most of these same essential minerals are supplemented through diets or obtained by plants from the soil. The food that we intake supplements most of these essential elements; however, diets that are being consumed from plant and animal resources often lack iron, zinc, copper, calcium, magnesium, iodine, and selenium (White and Broadley 2009). As a result, majority of the population globally suffer from malnutrition or subjected to hidden hunger (Bashir et al. 2013). The importance of these elements has been well recognized in human body in terms of deficiency symptoms caused by inadequate availability of these important elements leading to physiological disorders and loss of vitality. The presence of phytic acid in seeds reduces bioavailability of iron and zinc. Green gram contains less value of phytic acid compared to other pulses; consequently, bioavailability of iron and zinc is more, and often green gram is supplemented as balanced dietary requirement of human beings. The green gram (mung bean) germplasm represents wide genetic variation in micronutrient content in their seeds which could be exploited for biofortification (Nair et al. 2013).

With ever-increasing population and shrinking agricultural land, it is difficult to make the world food sufficient and eradicate malnutrition. Therefore, new strategies are to be made toward development of biofortified food crops rich in mineral contents. The approach of enriching mineral content through biofortification and enhanced bioavailability essentially require a deep understanding of mobility of the minerals from soil to seeds, the limiting factors for the movement of desired elements in the target tissue, and various strategies to overcome these limitations. Research efforts need to be made toward understanding the mechanisms of absorbing minerals by plants from soil and their subsequent distribution throughout the plant, particularly how desired minerals get accumulated in seeds. To maintain proper ion homeostasis, there must be establishment of balanced uptake, utilization, and storage of metals (Waters et al. 2011). Varieties of transporter proteins have been reported to play a major role for mineral transport (Lee et al. 2007; Yang et al. 2009; Pence et al. 1998; Takahashi et al. 2012). The first step of mineral entry into plants is the uptake from rhizosphere followed by xylem loading; thereafter, minerals are transferred from root to shoot tissue and subsequently distributed into various parts like leaves and seed cover and finally loaded to seeds. Conventional breeding or transgenic approaches need to be made to target the genes responsible for mineral movement within plant's internal system and membrane transport proteins playing a crucial role in enriching minerals in the seeds. Greater emphasis needs to be given to understand the underlying mechanisms involved in phloem unloading of metals into the seeds. It is urgently required to develop biofortified crops focused mainly toward modifications of largely consumed crops like cereals, pulses, and oilseeds. Among various metals, perhaps Fe and Zn are the most important minerals to be considered for biofortification as deficiency of these two elements causes serious health problem.

23.2 Physiological Mechanisms of Mineral Transport in Plants

The micronutrients pass through pods prior to transport to seeds. One potential phloem chelator NA (nicotianamine) has been identified that has the ability to bind iron and also other metals like Zn, Cu, Co, Mn, and Ni. The NA is widely present in the phloem sap and is an important chelator for minerals. Once the complexes between NA and minerals formed, the YSL (yellow stripe-like) proteins localized around vascular bundles help to transport the complex. Thus, expression of YSL genes is important in respect to transport of NA-mineral complexes to and from the vascular tissue. Development of pressure gradient between source and sink (seeds) is the driving force toward downloading the phloem sap into the seeds. A positive relationship has been reported with enhanced biofortification in seeds when soil fertilization is increased with nitrogen and zinc. Remobilization of nitrogen into the seeds during leaf senescence mainly in the amino acid forms is usually accompanied by movement of Fe and zinc suggesting cotransport mechanisms of N and minerals into the seeds. There are two parts of the developing seed, i.e., the filial tissue (embryo, endosperm, and aleurone) and maternal tissue (seed coat) where vascular bundle ends. There is no symplastic connection between seed coat and filial tissue; only apoplastic transport of minerals is possible. Hence, this is a kind of barrier for mineral transport to developing embryo and endosperm lacking symplastic connections. Thus, specific transporters belonging to gene families HMA, ZIP, MTP, Nramp, NAS, and YSL may be required to efflux and uptake into the embryonic cells and endosperm. Fe is reported to be stored in vacuoles of the embryo in *Arabidopsis*, while zinc is distributed throughout the embryo. In legumes, iron in large proportion is stored in the form of ferritin. However, seeds of diverse crop species may differ in the distribution of iron in the embryo and seed coat; therefore, efforts need to be made to look for efficient transporter genes that can concentrate more Fe and Zn into the edible parts, embryo and endosperm. Another research of biofortification is likely to reduce the phytates and increase the bioavailability of Fe and zinc as phytates have the binding capacity to iron/zinc, thereby decreasing bioavailability of these two important metals.

Several genes may be involved in translocation of micronutrients into xylem including FRD3, FPN1, HMA2, HMA4, HMA5, and MTP3 (Fig. 23.1). The overexpression of these genes often leads to increased translocation of micronutrients into the shoot. Blair et al. (2011) characterized variations in Fe and Zn concentrations of seed coats of common bean in

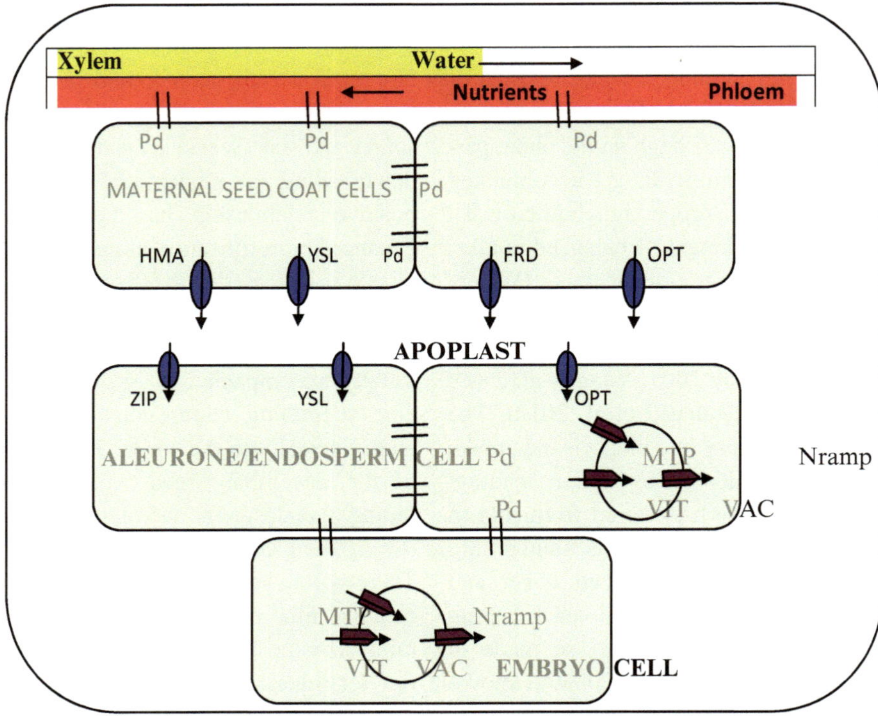

Fig. 23.1 Schematic diagram of location of transport genes of Fe, Zn, and Cu (Reproduced from Waters and Sankaran 2011)

a recombinant inbred population and identified some of the underlying genetic loci responsible for seed coat accumulation of these metals, which could be useful for future biofortification efforts. The constitutive expression of a suite of Zn deficiency-inducible responses through the overexpression of bZI19 and bZIP23 transcription factors can be used to increase Zn accumulation in edible portion of the crop (Assuncao et al. 2010). The Zn accumulation in roots includes transport proteins in the plasma membrane and tonoplast of root cells that facilitate the uptake and sequestration of Zn in the vacuole, together with enzymes involved in the synthesis of compounds that bind Zn^{2+} in the rhizosphere cytoplasm (Fig. 23.2 and Table 23.1).

23.3 Role of Soil and Roots Toward Mineral Acquisition

The availability of nutrients from soil to plant depends upon pH, soil texture, solubility of the ions, and many others. The root hairs are amenable to absorb readily available minerals and water from soils. The root hairs virtually play a major role in increasing the surface area of absorption, and also symbiotic mycorrhizae fungi facilitate in enhancing the area of the root for efficient absorption of water and minerals.

Minerals absorbed by root hairs flow through intracellular route via interconnected plasmodesmata channels, and selected minerals pass through the cell membrane of the cells and reach the xylem called as symplastic movement, while water and solutes also move through extracellular route when passing between the cell wall and plasma membrane and reach to endodermis which is called apoplastic route (Fig. 23.3). The uptake of minerals follows a path in the order of soil > roots > stems > leaves that involves diffusion and through mass flow of water from the soil carrying ions. The minerals, e.g., K^+ and Ca^{2+}, dissolved in the water and often are accompanied by various organic molecules. Minerals from the soil in the ionic forms are absorbed by passive and active transport. Membrane-bound ion pump draws ions from soil

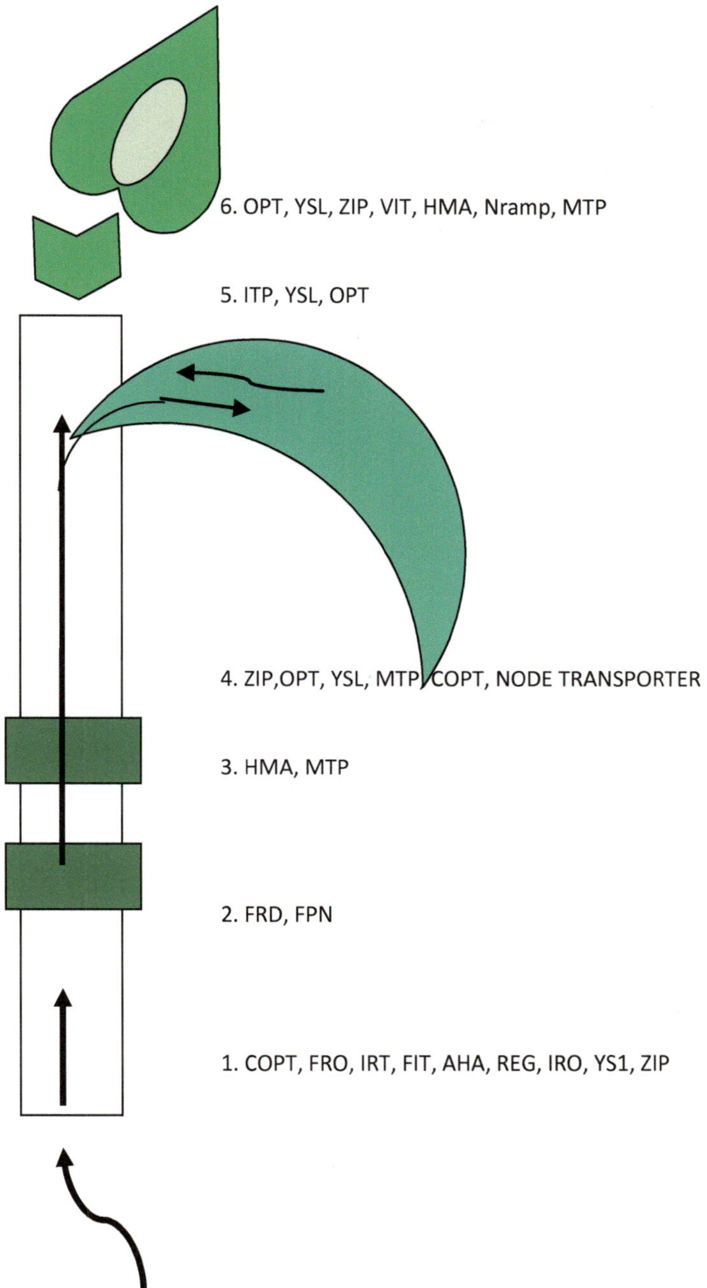

Fig. 23.2 Sequence of Fe and zinc uptake from the rhizosphere, xylem loading followed by root to shoot transfer, distribution to the leaves/seed cover, and loading into the seed. Associated transporter genes are depicted

Table 23.1 Transporter genes and their function in zinc uptake and remobilization

Transporter gene	Function	Crop	Reference
OsHMA9	Efflux of Zn from cells for remobilization	*Oryza sativa*	Lee et al. 2007
OsZIP7a and OsZIP8	Zn uptake from soil in root cells	*O. sativa*	Yang et al. 2009
ZNT1	Zn uptake from soil in root cells	*Thlaspi caerulescens*	Pence et al. 1998
OsHMA2	Root to shoot translocation in rice	*O. sativa*	Takahashi et al. 2012

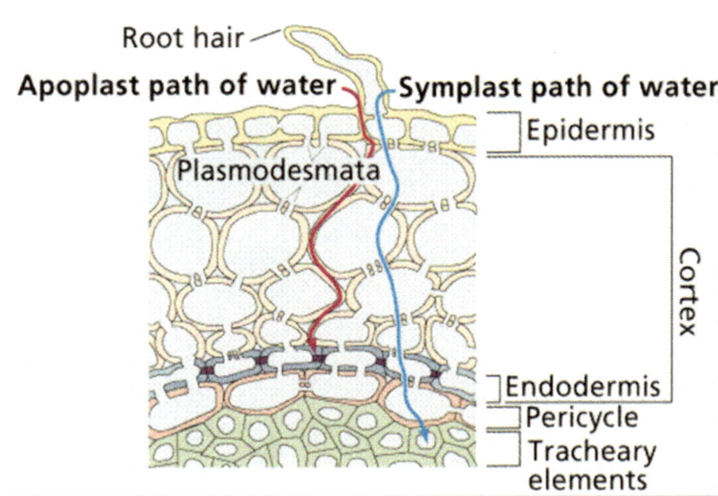

Fig. 23.3 Passage of water and minerals through root

and pumps into the cytoplasms of epidermal cells. Due to higher concentrations of minerals in the root, transport of ions into the root cells has to take place against concentration gradient which is called active transport which is also responsible to develop water potential gradient and osmotic potentials in roots. Subsequent ion diffusion from root hair epidermal cells to xylem takes place through concentration gradient and finally pulled up in the stem by mass flow and transpirational force. While reaching to the leaves through xylem, ions are absorbed by membrane pump of surrounding cells of leaves.

Minerals enter the root into the symplast of epidermal cells by active transport and move into the vascular tissue through the plasmodesmata connecting the cells. They enter the water in the xylem from the cells of the pericycle as well as of parenchyma cells surrounding the xylem through specialized transmembrane channels.

23.4 Pathways of Loading Iron in Seeds and Strategies to Improve Iron in Seeds

To develop biofortified crops with enriched iron content in the seeds, it primarily requires a deep understanding of the mechanisms controlling transport of Fe and how and what form it is stored in the seeds. There are different mechanisms for Fe uptake from soil by plants (Curie and Briat 2003; Schmidt 2003). The first mechanism has been described as the release of proton making Fe^{3+} to solubilize in the soil followed by reduction of Fe^{3+} into Fe^{2+} catalyzed by membrane-bound Fe(III) chelate reductase, and eventually Fe(II) transporter protein carries Fe^{2+} into the cell (Marschner 1995). The second mechanism involves chelation that releases Fe(III)-specific phytosiderophores (PS) and finally uptake of the Fe(III) phytosiderophore complexes via a specific transport system.

Expression and encoding of the genes NRAMP3, NRAMP4, and VIT1 in the roots and shoot tissues provide evidence toward their definite role in vacuolar Fe homeostasis. As cellular iron mostly remains in chelated form, it is interesting to know how proteins access this pool of Fe. The iron (Fe) chelator NA plays a crucial role. The chelator NA binds Fe and the complex (NA-iron) so formed is transported into the sieve tubes of phloem by YSL proteins. Seeds obtain iron from seed coat either through xylem vessels or sieve tubes of the phloem.

Once Fe has entered the symplast, Fe is bound to various chelators such as organic acid citrate which binds Fe^{3+} and nicotianamine (NA) and forms stable complexes with both Fe^{2+} and Fe^{3+} (Hell and Stephan 2003). Fe-chelator complexes also play roles in short- and/or long-distance transport of Fe, facilitating it remaining in solution and preventing it from participating in the generation of hydroxyl radicals. Recent information elucidated the transporters that play a key role in xylem loading and phloem loading/unloading of iron and the involvement of iron chelators in iron homeostasis (van Wuytswinkel et al. 1998). Iron-dependent gene regulation occurs at both the transcriptional and posttranscriptional levels (Kim and Guerinot 2007).

23.5 Storage of Fe in Seeds

Fe is essential for embryo development in seeds (Stacey et al. 2008); therefore, transport of Fe into the embryos and its storage in a stable form is needed before it further remobilizes during germination. Gene *OsYSL2* encodes a PM Fe-NA transporter protein and is expressed in seeds throughout its development (Koike et al. 2004), while citrate efflux transporter FRD3 is expressed in the aleurone layer and in the embryo (Roschzttardtz et al. 2011a, b). Elemental imaging techniques revealed localization of Fe in seeds of diverse crop species like *Arabidopsis*, rice, and pea (Kim et al. 2006). The varied distribution pattern of Fe in seeds of crop species provides an insight on how appropriate strategies can be made to concentrate mineral content in specific edible tissue. As maximum iron concentration was observed in aleurone layers, integument, and scutellum of rice seeds, these are discarded during processing; hence, rice eaters loose substantial amount of iron gained in their daily diet. Therefore, breeding strategies must be focused on increasing endosperm Fe content (Lee et al. 2009) in order to enhance Fe bioavailability (Bashir et al. 2010). In contrast, Fe is mainly stored within the embryo in dicots. Iron can be stored in plastids within ferritins, which assemble as large spherical 24-mer protein complexes able to store up to 4500 Fe atoms in their internal cavities (Briat 1999; Briat et al. 1999). The second mechanism consists in vacuolar sequestration.

Embryos of field pea contain greater proportion of Fe storage pool which is localized in nuclei and plastids. High-iron rice IR68144-2B-2-2-3-2 "Maligaya Special Rice #13" was released by the IRRI Philippines in 2003. Improved bioavailability and level of iron in rice were done through introduction of ferritin (*pfe*) gene from soybean and *Phaseolus* bean and introduction of phytase (*PhyA*) gene.

23.6 Acquisition of Zinc from Soil and Its Transport to Seeds

Zinc (Zn) is an established micronutrient required for normal growth and functioning of plants. Its physiological roles in plants are well documented. Zinc increases proline content in soybean; enhances photosynthetic rate, soluble protein content, and chlorophyll (*a,b*) content; helps stomatal opening and transpiration; and also acts as cofactors of many enzymes and vitamins for various metabolic activities of living system. Zinc deficiency in food crops is widespread with about 30 % of the soil and 50 % of the world's human population being zinc deficient (Kochian 2000). Understanding the functions of zinc in plant is needed to find effective ways to increase the available levels of zinc in soils and to improve crop productivity and to enhance bioavailability levels of zinc in edible portions of food.

Zinc is primarily taken up from soil through plant roots and translocated to other parts where it is needed to perform its specific role(s). Although, mostly, the soil has sufficient amount of total Zn, but bioavailability is low due to several factors. Among these factors, soil microbial communities play a vital role in enhancing bioavailability of Zn to plants. Similarly, application of bioinoculants to soil can promote Zn availability to plants. Bioavailable fraction of zinc in soil is very low due to various soil factors (Alloway 2009). A number of factors including soil texture, pH, soil water content, organic matter, and calcareousness of the soil are known to influence bioavailability of Zn in soil (Alloway 2008). The zinc gets fixed quickly on soil matrix, resulting in poor availability to plants (Zia et al. 1999). Some soils, despite having fair quantity of Zn, cannot support plant growth because of poor bioavailable Zn. Fertilizers containing zinc mineral are good options. Monsant et al. (2011) observed response of nitrogen form and zinc exposure. It can increase bioavailability of Zn by solubilizing fixed Zn and/or by reducing fixation by using organic amendments or solubilizing bioinoculants. Organic amendments improve bioavailability of Zn by increasing microbial biomass and enhance the rate of decomposition of organic matter, but also enhance the bioavailability of indigenous Zn by lowering the soil pH and by releasing chelating agents. Similarly, exogenous application of some potential Zn solubilizing microflora has shown huge capability to improve bioavailable Zn content in soil and its uptake by plant roots (Tariq et al. 2007). Various organic amendments such as compost and farmyard manure can improve soil health, fertility, and crop yields through changes in physicochemical properties of soil (Tejada et al. 2006). Soils having more microbial biomass and microbial activities are supposed to be productive soils as they have good nutrient mobility and availability to plants. Organic amendments improve soil microbial biomass carbon (Cmic) and the Zn content in soil and plant tissues (Saviozzi et al. 1999). The rhizosphere microflora may benefit plants through multifarious mechanisms including fixation of atmospheric nitrogen, mobilization of nutrients, production of phytohormones, alteration of indigenous level of phytohormones, and improvement of plant stress tolerance to salinity, toxicity, drought, metal, and pesticide load and by also acting as a biocontrol agent (Lucy et al. 2004). There are sufficient reports indicating substantial potential of these microbes in improving Zn bioavailable fraction in the rhizosphere of plants and Zn content in plant tissues (Chen et al. 2003; Biari et al. 2008; Yi et al. 1994).

Availability of micronutrients in soil is very much sensitive to soil pH. It has been reported that availability of Zn decreases 100 times with one unit increase in pH (Havlin et al. 2005). Thus, by decreasing the pH of alkaline soil, bioavailable fraction of Zn can be enhanced to an appreciable level. Rhizosphere microflora has been reported to lower the soil pH to a good extent (Wu et al. 2006), which may occur due to secretion of some organic acids and proton extrusion (Fasim et al. 2002).

Due to low persistency/high reactivity of Zn in soil solution, plant-available fraction of Zn in the soil is poor (Alloway 2009); however, bioavailability of Zn could be increased by means of Zn-chelating compounds (Obrador et al. 2003). The chelates of microflora are the metabolites, which form complexes with metal cations like Zn^{2+} (Tarkalson et al. 1998), which reduces their reaction with the soil.

Several studies have revealed that help in mitigation of Zn deficiency in plants through improving mobilization of Zn in soil. Many bacterial and fungal strains have been found capable of solubilizing fixed Zn and consequently increasing its uptake by plants. AM fungus is considered highly effective in improving the availability and absorption of immobile nutrients by higher plants (Liu et al. 2000). AM fungi are well known in improving the availability of phosphorus to plant roots. It has also been reported that mycorrhizal symbiosis is also very effective in improving availability of Zn to plants (Ryan et al. 2007). This bioavailable Zn is taken up by the plant root and accumulates in root or translocated to other plant parts. Thus, concentration of Zn in plant tissues is directly dependent on its availability in soil. There are good reports about increase in Zn uptake by the application of bioinoculants (Liu et al. 2000), which might have occurred through increase in bioavailable Zn in soil. For instance, Swaminathan and Verma (1979) observed a great improvement in bioavailable Zn fraction in soil through fungal (*Glomus macrocarpus*) treatment which subsequently increased the Zn concentration in the leaves of wheat, maize, and potato grown on Zn-deficient soils. Saravanan et al. (2004) found that *Pseudomonas* and *Bacillus* can solubilize various Zn compounds like ZnS, ZnO, and $ZnCO_3$ to a good extent. Bacteria have also shown high mobilization of soil Zn. Tariq et al. (2007) observed almost six times higher bioavailable Zn in inoculated soil compared to uninoculated soil. Whiting et al. (2001) have also documented increase in bioavailable Zn in rhizosphere soil through bacterial inoculation. It has also been widely reported that bacterial inoculation improves plant Zn content (Biari et al. 2008).

For instance, Whiting et al. (2001) observed a twofold more Zn concentration in the shoot of *T. caerulescens* compared to control, while uptake was increased up to fourfold. Similarly, inoculation of corn with *Azotobacter* and *Azospirillum* caused a significant increase in grain Zn content (Biari et al. 2008).

23.7 The Mobility and Distribution of Zinc Within Plant Organs

Due to uneven distribution of zinc within the plant species (Broadley et al. 2012) and also within shoots and leaves, Zn concentrations are often observed greater in root crops and leafy vegetables than in grain, seed, fruit, or tuber crops (White and Broadley 2005). Zn is abundantly localized in the elongation zone of root, in endodermal cells of dicots and pericycle of monocot species, and in older leaves (Monsant et al. 2010). In edible parts of various crops, often zinc is higher and localized in those parts usually not consumed by human and discarded during processing, for example, zinc is high in the husk of cereal seeds and aleurone layers or embryo (Stomph et al. 2011), while in tuber crop like potato, about one-fifth of total zinc remains present in the skin which is not used for eating purposes (Subramanian et al. 2011). White et al. (2009) established relationship between yield and mineral concentrations in potato tubers. These distribution patterns reflect both local and long-distance transport of Zn within the plant (Broadley et al. 2007).

Both soluble and insoluble forms of zinc are present in the plant tissue (Broadley et al. 2007). The soluble form of Zn is complexed with a number of organic compounds and proteins, phosphate salts, and organic Zn phytates. Apoplastic Zn^{2+} binds to negatively charged cell-wall components, organic acids, and phytosiderophores, whereas in cytosol, Zn is complexed by proteins, glutathione, phytochelatins, and NA (Clemens 2010). The vacuoles of root and leaf cells and within xylem contain Zn in the form of Zn^{2+} or Zn-organic acid complexes (Sarret et al. 2009; Terzano et al. 2008), while in phloem sap, Zn is complexed with NA or small proteins (Waters and Sankaran 2011). In general, after uptake of Zn by root cells from soil in the form of Zn^{2+} or Zn-phytosiderophore complexes, it enters plants and is transported either symplastically or apoplastically in regions of the root lacking a Casparian band, to the stele where it enters the xylem (Broadley et al. 2007). In addition, some plasma membrane Ca^{2+} channels are permeable to Zn^{2+} (White et al. 2002). It is believed that most Zn^{2+} influx to the cytoplasm of root cells is mediated by ZRT- and IRT-like proteins (ZIPs), in *Arabidopsis thaliana* principally AtZIP4 and AtIRT1, and that yellow stripe-like (YSL) proteins catalyze the uptake of Zn-phytosiderophore complexes in cereals and grasses (White and Broadley 2009).

Zn^{2+} gets complexed with numerous proteins in the cytosol of root cells that modulate enzymic activities or gene transcription (Broadley et al. 2007). Zinc may also get sequestered in the vacuole as an organic acid complex (Broadley et al. 2007). Release of zinc from vacuole may be triggered by NRAMPs and other transporters AtNRAMP3 and AtNRAMP4 of *A. thaliana* (Roosens et al. 2008). Xylem loading of Zn^{2+} may be mediated by members of the heavy metal P1B-ATPase family, such as AtHMA2 and AtHMA4 in *A. thaliana* (Puig and Penarrubia 2009).

Zinc is transported as Zn^{2+} or as complexed with chelator NA within the xylem, and YSL proteins are likely to load Zn-NA complexes, or orthologs of AtFRD3 load citrate into the xylem to promote Zn transport (Waters and Sankaran 2011). On the other hand, uptake of Zn^{2+} and Zn complexes in the shoot and phloem is facilitated by ZIP and YSL families (Curie et al. 2009). It has been reported that AtYSL1, AtYSL3, and AtOPT3 deliver zinc from vascular tissues to developing seeds in *A. thaliana* (Waters and Sankaran 2011).

The magnitude of mobility of zinc, though it is low in the phloem, usually determines its accumulation in fruits and seeds (Fageria 2009). Foliar application of zinc may improve translocation of Zn in the phloem (Brown 2009).

The transport proteins catalyzing Zn uptake, and the genes encoding proteins helping zinc mobilization from soil and their distribution within the plant are the results of coordinated regulation in response to plant Zn status. Thus, during zinc deficiency, ZIPs, YSLs, ZIF1, FRD3, etc. and enzymes for synthesis of phytosiderophores and NA are upregulated and downregulated when sufficient zinc is present in the plant tissue (Broadley et al. 2007).

The leaf tip application of zinc showed translocation of zinc to wheat grain which was confirmed by using ^{65}Zn suggesting measures to enhance zinc in wheat grains by foliar application of zinc fertilizer. Remobilization of zinc during leaf senescence to the grains is another option to increase the concentration of zinc in the seeds. The flag and lower leaves are the major sources of remobilized micronutrients in wheat while stem in the rice. Zinc concentrations in roots, leaves, and stems can be increased greatly by applying Zn fertilizers. Foliar application of zinc may benefit nonwoody shoot tissues; by contrast, increased production of compounds like NA that chelate Zn^{2+} can increase Zn concentrations in the phloem-fed tissues like fruits, seeds, and tubers. Several biofortified genetically modified crops with higher zinc content in their edible parts have been developed such as cassava roots, brown rice (Johnson et al. 2011), and barley grain (Ramesh et al. 2004).

23.8 Enhancing Remobilization to Increase Mineral Content in Seeds

A substantial amount of nutrients stored in the stems and leaves are translocated to the seeds when plants are subjected to senescence which is known as remobilization. The source-sink synchronization is an essential step to promote remobilization process.

Seeds continue to take up minerals from plants during entire grain-filling process, but remobilization of minerals is usually enhanced if leaves and stems are subjected to senescence during seed filling. This accelerated remobilization is induced by transport of minerals and carbon and nitrogen prestored in the stems and leaves.

The major source of seed carbohydrates and proteins in the seeds is the result of remobilized carbon (C) and nitrogen (N) from leaves and stems. It has been demonstrated that certain elements like Cu, Fe, and Zn are remobilized along with carbon and nitrogen in the developing grains of legumes and wheat. Under nutrient-deficient condition, the remobilization process is upregulated.

Therefore, breeding efforts need to be concentrated to modify remobilization efficieny at genetic level in crops like pulses having primary strategy to support seed filling through remobilization of prestored carbon and nitrogen in leaves and stem.

23.9 Conclusion

There is a requirement of enhanced synthesis of those compounds that form complexes with zinc or else make sequestration within the vacuoles in the perspectives of zinc biofortification. There are wide arrays of genes encoding transporter proteins which facilitate movement of zinc and iron besides other metals like Cu, Co, Ni, Se, Mg, Mn, etc. considered to be vital for eradicating malnutrition. The physiological processes of ion movement from root to shoot and their final accumulation in the desired edible parts like seeds, grains, tubers, etc. are very complex, and distribution of these essential elements is largely uneven. Within the seeds, embryo and endosperm are largely devoid of vascular connection from maternal tissue (seed coat), and only options are apoplastic mobility of minerals to the edible parts (embryo and endosperm). The phloem unloading to the desired tissue like fruits, seeds, tubers, etc. is still unknown and remains a challenge toward development of biofortified seeds. To increase Zn concentrations in edible crops, future research requires an integrated approach of physiology, agronomy, and genetics to enhance zinc transport to the edible parts. It is possible that breeding can increase Zn tolerance in root and leaf crops and increase Zn mobility in the phloem in fruit, seed, and tuber crops.

The understanding of mechanisms of absorbing minerals by plants from soil and their subsequent distribution throughout the plant is an essential step to develop biofortified seeds. Extensive research efforts need to be initiated to modify crops with enriched minerals only partitioned into the desired edible parts. There are a number of rate-limiting steps for mineral transport in plant in various tissues. The major channels of nutrient uptake in plants are through extensive network of root hairs that allow entering of the water and nutrients into the root vascular tissue through symplastic or apoplastic pathways. There are both active and passive modes of translocation of minerals in plants. Plants must balance uptake, utilization, and storage of these metals in order to maintain proper ion homeostasis.

The metals can be transported by a variety of transporter proteins. Essential steps involving absorption of minerals and storing into seeds are root uptake; xylem loading; entry to the shoot from root; distribution to the various plant parts, e.g., leaves, stems, or seed-covering tissues; phloem loading; and loading into the seed. Loading of micronutrients into the phloem is a requisite step for translocation to seeds. The micronutrients pass through pods prior to transport to seeds. Mineral micronutrients are in general in chelated forms during phloem transport.

Development of pressure gradient between source and sink drives movement of phloem sap that occurs by bulk flow. The Fe chelator nicotianamine (NA) has the capacity to bind Fe and zinc in addition to other elements. The improved fertilization practices with N and Zn fertilizers also could be a large step toward biofortification.

References

Alloway B (2008) Zinc in soils and crop nutrition. International Zinc Association and International Fertilizer Industry Association, Brussels/Belgium/Paris, p 130

Alloway BJ (2009) Soil factors associated with zinc deficiency in crops and humans. Environ Geochem Health 31:537–548

Assuncao AGL, Herrero E, Lin YF, Huettel B, Talukdar S, Smaczniak C, Immink RGH, Van Eldik M, Fiers M, Schat H, Aarts MGM (2010) *Arabidopsis thaliana* transcription factors bZIP19 and bZIP23 regulate the adaptation to zinc deficiency. Proc Natl Acad Sci U S A 107:10296–10301

Bashir K, Ishimaru Y, Nishizawa N (2010) Iron uptake and loading into rice grains. Rice 3:122–130

Bashir K, Takahashi R, Nakanishi H, Nishizawa NK (2013) The road to micronutrient biofortification of rice: progress and prospects. Front Plant Sci 4:15. doi:10.3389/fpls.2013.00015

Biari A, Gholami A, Rahmani HA (2008) Growth promotion and enhanced nutrient uptake of maize (*Zea mays* L.) by application of plant growth promoting rhizobacteria in arid region of Iran. J Biol Sci 8:1015–1020

Blair MW, Astudillo C, Rengifo J, Beebe SE, Graham R (2011) QTL analyses for seed iron and zinc concentrations in an intra-genepool population of Andean common beans (*Phaseolus vulgaris* L.). Theor Appl Genet 122:511–521

Briat JF (1999) Plant ferritin and human iron deficiency. Nat Biotechnol 17:621

Briat JF, Lobreaux S, Vansuyt G (1999) Regulation of plant ferritin synthesis: how and why. Cell Mol Life Sci 56:155–166

Broadley MR, White PJ, Hammond JP, Zelko I, Lux A (2007) Zinc in plants. New Phytol 173:677–702

Broadley M, Brown P, Cakmak I, Rengel Z, Zhao F (2012) Function of nutrients: micronutrients. In: Marschner P (ed) Marschner's mineral nutrition of higher plants, 3rd edn. Academic, London, pp 191–248

Brown P (2009) Development of a model system for testing foliar fertilizers adjuvants and growth stimulants. In: Proceedings of the California Department of Food and Agriculture Fertilizer Research and Education Program Conference 2008 Visalia, p 17–23

Chen BD, Li XL, Tao HQ, Christie P, Wong MH (2003) The role of arbuscular mycorrhiza in zinc uptake by red clover growing in a calcareous soil spiked with various quantities of zinc. Chemosphere 50:839–846

Clemens S (2010) Zn – a versatile player in plant cell biology. In: Hell R, Mendel RR (eds) Cell biology of metals and nutrients, Plant cell monographs 17. Springer, Berlin, pp 281–298

Curie C, Briat JF (2003) Iron transport and signaling in plants. Annu Rev Plant Biol 54:183–206

Curie C, Cassin G, Couch D, Divol F, Higuchi K, Le JM, Misson J, Schikora A, Czernic P, Mari S (2009) Metal movement within the plant: contribution of nicotianamine and yellow stripe 1-like transporters. Ann Bot 103(1):1–11. doi:10.1093/aob/mcn207

Fageria NK (ed) (2009) The use of nutrients in crop plants. CRC Press, Boca Raton

Fasim F, Ahmed N, Parsons R, Gadd GM (2002) Solubilization of zinc salts by bacterium isolated by the air environment of tannery. FEMS Microbiol Lett 213:1–6

Havlin J, Tisdale SL, Nelson WL, Beaton JD (2005) Soil fertility and fertilizers: an introduction to nutrient management. Prentice Hall, Inc, Upper Saddle River, 13: 9780130278241. ISBN 10: 0130278246

Hell R, Stephan UW (2003) Iron uptake trafficking and homeostasis in plants. Planta 216:541–551

Johnson AAT, Kyriacou B, Callahan DL, Carruthers L, Stangoulis J, Lombi E, Tester M (2011) Constitutive overexpression of the *OsNAS* gene family reveals single-gene strategies for effective iron- and zinc-biofortification of rice endosperm. PLoS One 6(9): e24476. doi:10.1371/journal.pone.0024476

Kim SA, Guerinot ML (2007) Mining iron: iron uptake and transport in plants. FEBS Lett 581:2273–2280

Kim SA, Punshon T, Lanzirotti A, Li LT, Alonso JM, Ecker JR, Kaplan J, Guerinot ML (2006) Localization of iron in *Arabidopsis* seed requires the vacuolar membrane transporter VIT1. Science 314 (5803):1295–1298

Kochian LV (2000) Molecular physiology of mineral nutrient acquisition transport and utilization. Biochem Mol Biol Plants 20:1204–1249

Koike S, Inoue H, Mizuno D, Takahashi M, Nakanishi H, Mori S, Nishizawa NK (2004) OsYSL2 is a rice metal-nicotianamine transporter that is regulated by iron and expressed in the phloem. Plant J 39(3):415–424

Lee S, Kim YY, Lee Y, An G (2007) Rice P_{1B}-type heavy-metal ATPase OsHMA9 is a metal efflux protein. Plant Physiol 145:831–842

Lee S, Jeon US, Lee SJ, Kim YK, Persson DP, Husted S, Schjorring JK, Kakei Y, Masuda H, Nishizawa HK, An G (2009) Iron fortification of rice seeds through activation of the *nicotianamine synthase* gene. Proc Natl Acad Sci U S A 106(51):22014–22019. doi:10.1073/pnas.0910950106

Liu A, Hamel C, Hamilton RI, Ma BL (2000) Acquisition of Cu, Zn, Mn and Fe by mycorrhizal maize (*Zea mays* L.) grown in soil at different P and micronutrient levels. Mycorrhiza 9:331–336

Lucy M, Reed E, Glick BR (2004) Applications of free living plant growth – promoting rhizobacteria. Antonie Van Leeuwenhoek 86:1–25

Marschner H (ed) (1995) Mineral nutrition of higher plants. Academic, Boston

Monsant AC, Wang Y, Tang C (2010) Nitrate nutrition enhances zinc hyper accumulation in Noccaea caerulescens (Prayon). Plant Soil 336:391–404

Monsant AC, Kappen P, Wang Y, Pigram PJ, Baker AJM, Tang C (2011) In vivo speciation of zinc in Noccaea caerulescens in response to nitrogen form and zinc exposure. Plant Soil 348:167–183

Nair RM, Yang RY, Easdown WJ, Thavarajah D, Thavarajah P, Hughes J, Keatinge JD (2013) Biofortification of mungbean (*Vigna radiata*) as a whole food to enhance human health. J Sci Food Agric 8:1805–1813

Obrador A, Novillo J, Alvarez JM (2003) Mobility and availability to plants of two zinc sources applied to a calcareous soil. Soil Sci Soc Am J 67:564–572

Pence N, Larsen P, Ebbs S, Garvin D, Eide D, Kochian L (1998) Cloning and characterization of a heavy metal transporter (ZNT1) from the Zn/Cd hyperaccumulator *Thlaspi caerulescens*. Plant Physiol 118:356–363

Puig S, Penarrubia L (2009) Placing metal micronutrients in context: transport and distribution in plants. Curr Opin Plant Biol 12:299–306

Ramesh SA, Choimes S, Schachtman DP (2004) Overexpression of an *Arabidopsis* zinc transporter in *Hordeum vulgare* increases short-term zinc uptake after zinc deprivation and seed zinc content. Plant Mol Biol 54:373–385

Roosens NHCJ, Willems G, Saumitou-Laprade P (2008) Using Arabidopsis to explore zinc tolerance and hyperaccumulation. Trends Plant Sci 13:208–215

Roschzttardtz H, Grillet L, Isaure MP, Conejero G, Ortega R, Curie C, Mari S (2011a) Plant cell nucleolus as a hot spot for iron. J Biol Chem 286(32):27863–27866

Roschzttardtz H, Seguela-Arnaud M, Briat JF, Vert G, Curie C (2011b) The FRD3 citrate effluxer promotes iron nutrition between symplastically disconnected tissues throughout Arabidopsis development. Plant Cell 23:2725–2737

Ryan MH, McCully ME, Huang CX (2007) Relative amounts of soluble and insoluble forms of phosphorus and other elements in intraradical hyphae and arbuscules of arbuscular mycorrhizas. Funct Plant Biol 34:457–464

Saravanan VS, Subramoniam SR, Raj SA (2004) Assessing in vitro solubilization potential of different zinc solubilizing bacterial (zsb) isolates. Braz J Microbiol 35:121–125

Sarret G, Willems G, Isaure MP, Marcus MA, Fakra SC, Frérot H, Pairis S, Geoffroy N, Manceau A, Saumitou-Laprade P (2009) Zinc distribution and speciation in *Arabidopsis halleri* × *Arabidopsis lyrata* progenies presenting various zinc accumulation capacities. New Phytol 184:581–595

Saviozzi A, Biasci A, Riffaldi R, Levi–Minzi R (1999) Long–term effects of farmyard manure and sewage sludge on some soil biochemical characteristics. Biol Fertil Soils 30:100–106

Schmidt W (2003) Iron solutions: acquisition strategies and signaling pathways in plants. Trends Plant Sci 8:188–193

Stacey MG, Patel A, McClain WE, Mathieu M, Remley M, Rogers EE, Gassmann W, Blenins DG, Stacey G (2008) The Arabidopsis AtOPT3 protein functions in metal homeostasis and movement of iron to developing seeds. Plant Physiol 146 (2):589–601

Stomph TJ, Choi EY, Stangoulis JCR (2011) Temporal dynamics in wheat grain zinc distribution: is sink limitation the key? Ann Bot 107:927–937

Subramanian NK, White PJ, Broadley MR, Ramsay G (2011) The three-dimensional distribution of minerals in potato tubers. Ann Bot 107:681–691

Swaminathan K, Verma BC (1979) Response of three crop species to vesicular arbuscular mycorrhizal infection on zinc deficient Indian soils. New Phytol 114:1–38

Takahashi R, Ishimaru Y, Shimo H, Ogo Y, Senoura T, Nishizawa NK, Nakanishi H (2012) The OsHMA2 transporter is involved in root-to-shoot translocation of Zn and Cd in rice. Plant Cell Environ 35:1948–1957

Tariq M, Hameed S, Malik KA, Hafeez FY (2007) Plant root associated bacteria for zinc mobilization in rice. Pak J Bot 39:245–253

Tarkalson DD, Jolley VD, Robbins CW, Terry RE (1998) Mycorrhizal colonization and nutrient uptake of dry bean in manure and composted manure treated subsoil and untreated topsoil and subsoil. J Plant Nutr 21:1867–1878

Tejada M, Hernandez MT, Garcia C (2006) Application of two organic amendments on soil restoration: effects on the soil biological properties. J Environ Qual 35:1010–1017

Terzano R, Chami ZA, Vekemans B, Janssens K, Miano T, Ruggiero P (2008) Zinc distribution and speciation within rocket plants (Eruca vesicaria L. Cavalieri) grown on a polluted soil amended with compost as determined by XRF microtomography and micro- XANES. J Agric Food Chem 56:3222–3231

Van Wuytswinkel O, Vansuyt G, Grignon N, Fourcroy P, Briat JF (1998) Iron homeostasis alteration in transgenic tobacco overexpressing ferritin. Plant J 17:93–97

Waters BM, Sankaran RP (2011) Moving micronutrients from the soil to the seeds: genes and physiological processes from a biofortification perspective. Plant Sci 180:562–574. doi:10.1016/j.plantsci.2010.12.003

Waters BM, Renuka P, Sankaran B (2011) Moving micronutrients from the soil to the seeds: genes and physiological processes from a biofortification perspective (Review). Plant Sci 180:562–574

White J, Broadley MR (2005) Biofortifying crops with essential mineral elements. Trends Plant Sci 10:586–593

White PJ, Broadley MR (2009) Biofortification of crops with seven mineral elements often lacking in human diets – iron, zinc, copper, calcium, magnesium, selenium and iodine. New Phytol 182:49–84

White PJ, Bowen HC, Demid-chik V, Nichols C, Davies JM (2002) Genes for calcium- permeable channels in the plasma membrane of plant root cells. Biochim Biophys Acta 1564:299–309

White PJ, Bradshaw JE, Dale MFB, Ramsay G, Hammond JP, Broadley MR (2009) Relationships between yield and mineral concentrations in potato tubers. Hortic Sci 44:6–11

Whiting SN, De Souza M, Terry N (2001) Rhizosphere bacteria mobilize Zn for hyperaccumulator by Thlaspi caerulescens. Environ Sci Technol 35:3144–3150

Wu SC, Cheung KC, Luo Y, Wong M (2006) Effects of inoculation of plant growth promoting rhizobacteria on metal uptake by Brassica juncea. Environ Pollut 140:124–135

Yang X, Huang J, Jiang Y, Zhang HS (2009) Cloning and functional identification of two members of the ZIP (Zrt, Irt-like protein) gene family in rice (*Oryza sativa* L.). Mol Biol Rep 36:281–287

Yi Y, Saleeba J, Guerinot ML (1994) Iron uptake in Arabidopsis thaliana. In: Manthey J, Luster D, Crowley DE (eds) Biochemistry of metal micronutrients in the rhizosphere. CRC Press, Boca Raton, pp 295–307

Zia MS, Aslam M, Baig MB, Ashraf A (1999) Fertility issues and fertilizer management in rice – wheat system: a review. Quart Sci Vision 5:59–73

Part V

Biofortification for Value Addition and Novelty

Antinutrients Restraining Biofortification

24

R.P. Srivastava

Abstract

Biofortification to enhance the bioavailability of micronutrients such as iron, zinc, vitamin A, folic acid (vitamin B_9), and cyanocobalamin (vitamin B_{12}) in staple food is the necessity of people living under malnourished conditions. Almost half of the world population is suffering from deficiency of iron, zinc, vitamin A, folic acid, etc. due to poor quality of food they are consuming. They have either low access or no access for these components of food. HarvestPlus has been working to enhance the availability of these nutrients through improved varieties of food so as to provide them health protection. Though the level of micronutrients in the staple food is enhanced by conventional breeding or biotechnology, the presence of antinutrients in food causes hurdle in absorption and bioavailability of Fe, Zn, Ca, folates, etc. These antinutrients are phytates, polyphenols and tannins, protease and α-amylase inhibitors, saponins, lectins, and lathyrogens. Various processing techniques are adopted to remove or reduce these antinutrients to enhance bioavailabilty of essential nutrients of food. Processing techniques such as soaking, germination, milling (dehusking), cooking, and autoclaving are effective in removal of antinutrients, especially from food grains.

Keywords

Antinutrients • Bioavailability • Inhibitors • Lathyrogens • Phytates • Polyphenols

24.1 Introduction

Biofortification is the breeding of crops to enhance the nutritional value of food. This is done either by conventional breeding or through biotechnology. During the process of

R.P. Srivastava (✉)
Division of Basic Sciences, ICAR-Indian Institute of Pulses Research, Kanpur, Uttar Pradesh 208 024, India
e-mail: rpsrivastava1208@rediffmail.com

U. Singh et al. (eds.), *Biofortification of Food Crops*, DOI 10.1007/978-81-322-2716-8_24

biofortification, the nutrients are enhanced by changing the chemical composition of the food by incorporation of gene responsible for biosynthesis of targeted element or nutrient. It differs from ordinary fortification, in which nutrients are added during processing. The rural and poor mass of society gets the advantage of biofortified food as they are not able to afford the commercially fortified food. It is more advantageous especially in developing world for micronutrient deficiencies such as iron, zinc, vitamin A, and folates. As such, biofortification is seen as an upcoming strategy for dealing with deficiencies of micronutrients. Deficiencies of vitamin A, iron, and zinc affect over one-half of the world's population, especially women and preschool children. The biofortification is feasible without compromising agronomic productivity. The challenge is to get producers and consumers to accept biofortified crops and increase their intake of the target nutrients. The biofortification of staple food crops is a new health approach to control vitamin A, iron, and zinc deficiencies in poor countries.

24.2 The Need for Biofortification

Biofortification is the development of micronutrient or vitamin-rich staple crops, especially iron, zinc, vitamin A, and folates, using the best breeding practices or modern biotechnology. This approach has multiple advantages because staple foods predominate in the diets of the poor, and the poor people do not have access to other food supplements which provide them all essential nutrients. Nutritionally improved varieties are therefore required to be grown continuously to feed these people. Biofortification provides a feasible means of reaching undernourished populations in relatively remote rural areas. Biofortification and commercial fortification therefore are highly complementary. Acceptance of mineral-packed seeds among farmers is very popular because these trace minerals provide resistance against diseases and other environmental stresses. Biofortification provides direct benefit for better human health and enhances the immune system by controlling the deficiency of essential nutrients. This requires a multidisciplinary research approach and ultimate dissemination of the biofortified seeds.

Deficiencies of various micronutrients such as zinc and iron and vitamins like vitamin A and folic acid are common in the developing world and affect billions of people. These can lead to higher incidence of blindness and a weaker immune system. The poor, particularly the rural poor, tend to subsist on a diet of staple crops such as rice, wheat, and maize, which are low in these micronutrients. Most of these rural people cannot afford enough fruits, vegetables, or meat products that are necessary to obtain healthy levels of these nutrients. As such, increasing the micronutrient levels in staple crops can prevent and reduce the micronutrient deficiencies. In one trial in Mozambique, eating sweet potatoes biofortified with beta-carotene reduced the incidence of vitamin A deficiency in children by 24 % (Van Jaarsveld et al. 2005). Golden Rice is an example of a GM crop developed for its nutritional value. The latest version of Golden Rice contains genes from a common soil bacterium Erwinia and maize and contains increased levels of beta-carotene which can be converted by the body into vitamin A. Golden Rice is developed as a potential new way to address vitamin A deficiency (Dawe et al. 2002; Zimmermann and Qaim 2004; Paine et al. 2005). The biofortified foods may also be useful for increasing micronutrient uptake in high-income countries. Researchers at the University of Warwick have been looking for ways to boost the low selenium levels in British grains and have been working to help develop a grain to be used in making bread biofortified with selenium (Wikipedia 2015).

The approach of biofortification may have advantages over other health interventions such as providing fortified foods after processing, or providing supplements. Biofortification is also fairly cost-effective, and the implementation costs of growing biofortified foods are much lower than supplementation which is comparatively expensive. The dietary diversification might be one of the methods to reduce the prevalence of iron deficiency. Thus, biofortification (the use of traditional plant breeding methods or genetic engineering to improve the available iron content of staple food crops) holds promise for the future (Nestel et al. 2006). It would make it possible to deliver iron, zinc, vitamin A, or any other desired nutrient to those most in need. They often depend on subsistence farming and have limited access to fortified foods. More researches are needed to develop successful crops with adequate bioavailable iron, zinc, vitamin A, and folate (vitamin B_9) content.

24.3 Common Problems of Biofortification

Although there is acceptance for biofortified food, some people are against the genetically modified (GM) crops such as Golden Rice. There are difficulties in acceptance of biofortified foods, if they have different characteristics to their unfortified counterparts. For example, vitamin A-enhanced foods are often dark yellow or orange in color – this, for example, is problematic for many in Africa, where white maize is eaten by humans, whereas yellow maize is associated with animal feed (McClafferty and Islam 2008), or where white-fleshed sweet potato is preferred than orange-fleshed counterpart (Nestel et al. 2006). Some qualities may be relatively simple to breed out of biofortified crops according to consumer demand. Due care must be taken to convince the local farmers and consumers that the crop in question is worth growing and consuming. This can be done through improving the cultivation qualities of the plant, for example, making the orange sweet potato mature earlier than white-fleshed sweet potato so that it can be taken to market earlier. It is possible through public health education, elaborating the benefits of eating biofortified foods. While other micronutrients such as zinc or iron can be added to crops without noticeably changing their taste or appearance, some of the consumers do not think that their food has been altered and do not bother much about the external appearance.

Some of the people did disapprove of the biofortification programs, because biofortification is a strategy that aims to concentrate more nutrients in few staple foods. The lack of access to a diverse and balanced diet is the major cause of malnutrition. The use of biofortification as part of a larger strategy involving diversification of foods in the developing world is necessary to solve the problem of deficiency of certain essential nutrients. The biofortification is accepted as a long-term strategy. The diversity in substantially increasing diet will take many decades, and the biofortification will be an effective strategy to reduce the micronutrient malnutrition.

24.4 Recompense of Biofortification

The biofortification strategy seeks to take advantage of the consistent daily consumption of large amounts of food staples by all family members, including women and children who are most at risk for micronutrient malnutrition. As a consequence of the predominance of food staples in the diets of the poor, this strategy implicitly targets low-income households. After the one-time investment is made to develop seeds that fortify themselves, recurrent costs are low and germplasm can be shared internationally. The biofortified crops are highly sustainable, and the nutritionally improved varieties will continue to be grown and consumed year after year.

Biofortification provides a truly feasible means of reaching malnourished populations in relatively remote rural areas, delivering naturally fortified foods to people with limited access. Breeding for higher trace mineral density in seeds will not incur a yield penalty. In fact, biofortification may have important effects for increasing farm productivity in developing countries in an environmentally beneficial way. Recent researches have shown that trace mineral-packed plants have high level of resistance against diseases and other environmental stresses. More seedlings survive and initial growth is more rapid.

24.4.1 Consequences of Micronutrient Malnutrition

More than two billion people worldwide are iron deficient (Stoltzfus 2001). Iron-deficiency anemia is by far the most common micronutrient deficiency in the world. Iron efficiency during childhood and adolescence impairs physical growth, mental development, and learning capacity. In adults, iron-deficiency anemia reduces the capacity to do physical labor. Iron deficiency increases the risk of women dying during delivery or in the postpartum period.

The zinc deficiencies have equally serious consequences for health. For example, meta-analyses of recent randomized controlled trials show that zinc supplementation can reduce morbidity from a number of common childhood infections, especially diarrhea, pneumonia, and possibly malaria, by one-third. In addition, zinc deficiency is an important cause of stunting. Billions of people are also at risk for zinc deficiency. As for anemia, prevalence is highest for South and Southeast Asia and Africa. Because there is no widely accepted method for measuring zinc deficiency, no estimates are available of numbers of people who are zinc deficient.

Globally, approximately three million preschool-age children have visible eye damage owing to vitamin A deficiency. Annually, an estimated 250,000–500,000 preschool children go blind from this deficiency, and about two-thirds of these children die within months of going blind. Even more importantly, the last two decades have brought an awareness that vitamin A is essential for immune function. Estimates of the prevalence of subclinical vitamin A deficiency range between 100 million and 250 million for preschool children.

People with subclinical vitamin A deficiency more often experience anemia, impaired linear growth, and morbidity from common childhood infections such as respiratory and diarrheal diseases, measles, and malaria. Most importantly, a number of randomized controlled trials in developing countries have shown that administration of vitamin A capsules among infants and preschool children reduces mortality rates from all causes by 23 %, and that administration of capsules with vitamin A and beta-carotene among women during childbearing years can reduce maternal mortality related to pregnancy by 40 % and 49 %, respectively. Prevalence of vitamin A deficiencies by region is available only for preschool children. Moreover, similar to iron and zinc, prevalence is highest in South and Southeast Asia and sub-Saharan Africa.

24.4.2 Underlying Causes of Micronutrient Malnutrition

Billions of people in developing countries suffer from an insidious form of hunger known as micronutrient malnutrition. Even mild levels of micronutrient malnutrition may damage cognitive development, lower disease resistance in children, and reduce the likelihood that mothers survive childbirth. The costs of these deficiencies in terms of lives lost and poor quality of life are staggering (Graham et al. 2001). It is important to identify who the malnourished are, where they are located, and what they eat in order to develop an effective strategy to reduce micronutrient malnutrition. And they are mostly women and children, who reside in developing countries where dietary quality is often poor. In addition, they have higher nutritional requirements due to reproduction and rapid growth resulting in malnutrition.

The primary underlying cause of micronutrient malnutrition is poor quality diets, characterized by high intakes of food staples, but low consumption of animal and fish products, fruits, and vegetables, which are rich sources of bioavailable minerals and vitamins. As such, most of the malnourished are those who cannot afford to purchase high-quality, micronutrient-rich foods or who cannot obtain these foods from their own production.

What is perhaps most alarming, however, is the upward trend in non-staple food prices. Cereal prices have fallen by 40 % since the early 1970s. The Green Revolution can rightly take credit for its crucial contribution to this tremendous achievement. Falling cereal prices have not only led to increased food security in terms of energy, but also allowed greater purchases of non-staples by freeing up cash. Unfortunately, the rate of production of non-staple foods (e.g., fruits, vegetables) has not kept pace with demand, so that these micronutrient-rich food sources have become ever more expensive for the poor (Bouis 2003).

24.4.3 Role of Iron, Zinc, and Folate in Human Beings

Iron is integral to the structure and function of red blood cells, and its deficiency can result in anemia. Anemia and iron deficiency during pregnancy can cause preterm birth and low birth weight (Allen 2001). In non-anemic mothers, iron supplementation may offer protection against low birth weight (Palma et al. 2008). Iron is also involved in myelination, neurotransmitter function, various cellular and oxidative processes, energy production, and thyroid hormone metabolism. Iron deficiency has been implicated in neurological and cognitive disorders in the mother; these include major depressive disorder, recognized to have health consequences on both the mother and child (Bodnar and Wisner 2005; Leung and Kaplan 2009).

The high prevalence of iron deficiency in developing countries has on people's well-being and productivity. Physical work capacity is reduced. Iron deficiency in pregnancy contributes to the risk of severe anemia and increased maternal morbidity and mortality (Khan et al. 2006). Iron-deficiency anemia in early pregnancy is associated with a higher risk of preterm delivery (Scholl 2005). Iron supplementation (usually in combination with folic acid) in pregnancy has been reported to reduce the risk of postpartum hemorrhage (Christian et al. 2009a), improve birth weight (Cogswell et al. 2003), and reduce early neonatal (Zeng et al. 2008; Titaley et al. 2010) and childhood mortality (Christian et al. 2009b).

Zinc is integral to DNA synthesis and necessary for the structure and function of regulatory, structural, and enzymatic proteins as well as cell membranes. It is involved in neurological function and proper immune function (Fraker et al. 2000; Huang 1997). Zinc deficiency is also implicated in depressive disorders. Moreover, various studies have implicated zinc deficiency in preterm and low birth weight, although routine supplementation is not recommended unless there is an identified deficiency (Ladipo 2000).

Folate is involved in the metabolism of nucleic acids and amino acids and in neurological functioning. While inadequate folate is implicated in various birth defects and poor pregnancy outcomes, its role in neural tube defects has received the most attention. In various countries, women of childbearing age are advised to take supplements. Folate deficiency is also implicated in depressive disorders. Food fortification policies are in effect in response to the strong evidence of the importance of folic acid intake in the very early stages of pregnancy (Wilson et al. 2003).

The RDA is defined as the average daily dietary intake level that is sufficient to meet the nutrient requirements of nearly all healthy individuals in a particular life-stage and gender group. For pregnant women, the RDA of iron and zinc is 27 and 11 mg/day, respectively, while that of folate as dietary folate equivalents (DFE) is 600 μg/day.

24.5 Role of Antinutrients in Biofortification

Nutritional iron deficiency occurs when the diet supplies inadequate bioavailable iron to meet the body's requirements for growth and pregnancy and to replace iron lost from the gastrointestinal tract and in the urine and through menstruation (in women). The causes have been known for over 50 years. The major factor is poverty. Secondly, the agricultural revolution replaced animal foods rich in bioavailable iron by cereals, legumes, and plant-based diets (Cordain 1999).

The classical iron balance study carried out by Widdowson and McCance (1942) demonstrated that less iron was absorbed from bread with high bran content than from white bread. The inhibitory role of an antinutrient called phytate was suspected, because phytic acid was shown to inhibit iron absorption (McCance et al. 1943). Later on, a second major class of antinutrient known as polyphenols was discovered (Disler et al. 1975; Gillooly et al. 1983). This large body of experimental work provided the basis for predicting dietary bioavailability of iron (Reddy et al. 2000; Hallberg and Hulthen 2000) and designing efficacious strategies for alleviating nutritional iron-deficiency anemia (Hurrell 1999; Hurrell et al. 2010).

Wheat flour has been fortified with iron in America since the 1940s, and the marked decline in the prevalence of iron-deficiency anemia in infants and young children in the United States is generally attributed to the fortification of foods (Miller et al. 1985; Fomon 2001). Iron is also added to many processed foods, such as breakfast cereals, in Western countries (Yip 2002). Despite a relatively clear understanding of the physiology of food iron absorption, iron deficiency is estimated to affect as many as two billion people (Zimmermann and Hurrell 2007). Developed countries may have a mechanism to provide bioavailable micronutrient, but the poor mass of developing countries depend on staple food such as maize, rice, wheat, or grain legumes. Therefore, there is an urgent need to supply biofortified food rich in these micronutrients. Truly speaking, these biofortified foods contain antinutrients like phytates and polyphenols, which are required to be minimized by processing such as soaking, germination, fermentation, cooking, or autoclaving so as to make the nutrient available to human beings.

In developing countries, screening for iron deficiency is usually based solely on hemoglobin (Hb) measurements. Sensitivity is low because the overlap in Hb concentrations between healthy and iron-deficient individuals is considerable, especially if the cutoff values used to identify anemia are not appropriately adjusted for age, gender, pregnancy, ethnicity, and altitude. Specificity is poor, because a multitude of disorders other than iron deficiency can cause anemia. They include other nutritional deficiencies due to vitamins A, B_{12} (cyanocobalamin), and B_9 (folic acid). Ferrous fumarate, a less reactive yet bioavailable form of iron, is recommended or used in many countries in Central and South America. The use of encapsulation technologies may provide a practical solution to fortification with more reactive bioavailable compounds. Condiments such as salt, soy sauce, fish sauce, and curry may be an alternative for the delivery of fortified iron. It will be essential to keep legislators and program managers informed about the necessity of providing iron in a bioavailable form and in sufficient quantities to meet the needs of women, children, and adolescents at highest risk.

Biofortification holds promise for the future. It would make it possible to deliver iron to those most in need. They often depend on subsistence farming and have limited access to fortified foods. A modest improvement in iron stores in non-anemic Filipino women consuming iron-biofortified rice was reported in one recent trial. However, the meals provided an average of only 1.42 mg/d additional iron (Lynch 2011). More research is needed to develop a crop/genotypes with adequate improvements in bioavailable iron content.

24.5.1 Processing to Minimize Antinutrients

It is very necessary now to move from food security to nutrition security and improve the quality of foods both in macro- and micronutrients in order to break the transgenerational effects of malnutrition. Enhancing food and nutrition security through innovative diversified agriculture and dietary practices, prevention and control of infection, promotion of food safety, and fortification of staples with appropriate attention on emerging chronic disorders are essential (Krishnaswamy 2001). The low bioavailability of some key micronutrients from foods, such as Fe and Zn, is substantially enhanced with the right food combinations and with appropriate food processing and preparation techniques. Simple appropriate technology for the preservation of micronutrient-rich foods would need further development and promotion for their year-round availability (Tontisirin et al. 2002). As iron deficiency is the most common micronutrient deficiency in the world, iron biofortification is a preventative strategy that alleviates Fe deficiency by improving the amount of absorbable Fe in crops. High Fe-bioavailability maize contains more bioavailable Fe than the low Fe-bioavailability maize (Tako et al. 2013). Maize shows promise for Fe biofortification; therefore, human trials need to determine the efficacy of consuming the high bioavailable Fe maize to reduce Fe deficiency.

International research efforts, including those funded by HarvestPlus, a challenge program of the Consultative Group on International Agricultural Research (CGIAR), are focusing on conventional plant breeding to biofortify staple crops such as maize, rice, cassava, beans, wheat, sweet potatoes, and pearl millet to increase the concentrations of micronutrients that are commonly deficient in specific population groups of developing countries (La Frano et al. 2014). The bioavailability of micronutrients in unfortified staple crops in developing regions is typically low. Reducing the amounts of antinutrients and food processing generally increases the bioavailability of micronutrients. In general, biofortified foods with relatively higher micronutrient density have higher total absorption rates than non-biofortified varieties. There is a need to breed plants with increased micronutrient concentrations in order to decrease the influence of inhibitors and to offset losses from processing.

24.5.2 Important Antinutrients in Food

The antinutrients have impact on lowering the bioavailability of essential nutrients. There is a need to summarize various antinutrients influencing directly or indirectly the bioavailability of these micronutrients. Most of the cereals and legumes are rich in these micronutrients; therefore, techniques to minimize these antinutrients to enhance the bioavailability of micronutrients are required to be developed or brought to practice. Some of the antinutrients influencing bioavailability of micronutrients are summarized below.

24.5.2.1 Phytates

Phytic acid, myoinositol 1,2,3,4,5,6-hexakis (dihydrogen phosphate), occurs abundantly in most of the food legumes. It is a chelating agent for cations as well as for phosphorus storage in many seeds. Phytate rapidly accumulates in seeds during the ripening period. Excessive phytic acid in the diet can have a negative effect on mineral balance because of the insoluble complexes it forms with essential minerals ($Cu2+$, $Zn2+$, $Fe3+$, and $Ca2+$), which cause poor mineral bioavailability (Zhou and Erdman 1995; Urbano et al. 2000). The ability of phytic acid to complex with proteins and particularly with minerals (Ca, Fe, Zn) has been a subject of investigation for several reasons. The interaction between phytate and proteins leads to decreased solubility of proteins and amino acid digestibility. The calcium ions also interact with proteins and phytate and decrease the solubility of proteins. The reduced solubility of proteins as a result of protein phytate complex adversely affects certain functional properties of protein which are dependent of their hydration and solubility. The ability of phytic acid to bind with minerals, proteins, or starch, directly or indirectly, may alter solubility, functionality, digestibility, and absorption of these nutrients. In addition, monogastric animals have a limited ability to hydrolyse phytates and release phosphate for absorption due to a lack of intestinal phytases (Zhou and Erdman 1995; Greiner and Konietzny 1996a; Urbano et al. 2000). Phytic acid is able to make complex with proteins also, decreasing protein solubility. Therefore, phytates have negative impact on enzyme activity, and there is evidence of its negative effects on key digestive enzymes like lipase, α-amylase, pepsin, trypsin, and chymotrypsin (Thompson 1993;

Table 24.1 Phytic acid (mg g^{-1}) during soaking and germination

Crop	Soaking[a]			Germination[b]		
	Seed	Soaked	%Reduction	Seed	Germinated	%Reduction
Chickpea	–	–	–	4.40	2.59	41.1
Pigeon pea	–	–	–	1.24	0.99	20.2
Lentil	–	–	–	3.09	1.46	52.7
Mung bean	–	–	–	3.12	1.28	59.0
Urd bean	–	–	–	14.56	7.10	51.2
Peas	–	–	–	3.23	0.99	69.3
French bean	–	–	–	11.48	2.59	77.4
Kidney beans	5.8	2.8	51.7	–	–	–
Pinto beans	5.5	2.6	52.7	–	–	–
Northern beans	4.6	1.4	69.6	–	–	–

Source: [a]Iyer et al. (1980); [b]Reddy et al. (1982)

Table 24.2 Phytic acid (mg g^{-1}) during processing

Crop	Seed	Germinated	Fermented	Autoclaved	Roasted
Chickpea	9.2	3.3	5.6	5.9	7.0
Pigeon pea	11.7	4.0	5.4	7.2	7.8
Mung bean	14.8	9.3	10.9	12.2	11.6
Urd bean	13.8	8.2	9.6	10.8	9.6
Soybean	39.9	24.4	27.0	27.9	33.1

Source: Chitra et al. (1996)

Greiner and Konietzny 1996b; Urbano et al. 2000). The binding of phytic acid to these enzymes reduces nutrient digestibility. Phytic acid also binds with starch through phosphate linkages (Lajolo et al. 2004). Efforts made in developing high-iron and zinc-biofortified crops remain unutilized due to reduced bioavailability; therefore, technology to remove phytates from the grains is required to be developed to enhance the bioavailability of these micronutrients (Table 24.1).

Phytate-mineral complexes are well documented. The nutritionally important minerals such as Ca, Mg, Cu, Fe, and others form complexes with phytic acid, resulting in reduced solubility of the metals (Biehl et al. 1995). Low iron availability due to iron-phytate complexes is also of concern. As most of us consume enough minerals in common foods which enable us to meet more than our requirement, and small amounts of these micronutrients might be tied up by phytates. The phytate-associated deficiencies of iron and zinc do occur in some third-world countries where people mostly eat grains.

The bioavailability of iron and zinc can be increased by combination of processing techniques of grain, viz., soaking in water, germination, fermentation, cooking, pressure cooking, etc. As far as food legumes are concerned, they are generally consumed after processing. The purpose behind these processing is to remove or minimize the antinutrients present in the grain. Routine processing such as soaking of grain and discarding the soaking water and then cooking or germination and fermentation reduces phytic acid content of food to an extent of up to 50–70 % (Iyer et al. 1980, 1989; Reddy et al. 1982). Soaking in yogurt, buttermilk, or water combined with lemon juice or vinegar is also much helpful to enhance this breakdown (Table 24.2).

Soaking of seed can eliminate up to 69.6 % of phytate in beans (Iyer et al. 1980), whereas germination can reduce phytate content to an extent of

77.4 % in seeds of various food legumes (Reddy et al. 1982). Highest reduction in phytate content of different food legumes was reported during germination, followed by fermentation, autoclaving, and roasting (Chitra et al. 1996). Germination (24–96 h) of seeds of *Vigna unguiculata* has been found to eliminate 38–95 % of total phytates, whereas germination followed by autoclaving removed 96–100 % of total phytates (Kalpanadevi and Mohan 2013). They also reported that soaking of seeds in water for 12 h followed by cooking or autoclaving removed 68–71 % of phytates. Soaking of seed in distilled water or 2 % sodium carbonate solution or 1 % citric acid solution followed by cooking eliminates 45.1–73.4 % of phytates in soybean (Sharma et al. 2013). Removal of phytates enhances the bioavailability of iron, zinc, and other minerals. The phytate has some health benefits, including anti-inflammatory effects. In laboratory research, phytate has helped normalize cell growth and stopped the proliferation of cancer cells. They also may help prevent cardiovascular disease and lower a food's glycemic load. Some myoinositol phosphates, including IP6 from soybean, have been suggested to have beneficial health effects, such as amelioration of heart disease by controlling hypercholesterolemia and atherosclerosis, prevention of kidney stone formation, and a reduced risk of colon cancer (Greiner et al. 2002).

24.5.2.2 Polyphenols

The polyphenols or tannins are one of the most important antinutrient of food legumes and found in the range of 0.45–20.00 mg g^{-1} grain. Among the food legumes, urd bean, mung bean, kidney beans, faba beans, and pigeon peas contain higher content of tannins. The tannin content in pulses varies with the color of the seed coat or testa. The white- or light-colored seeds generally have low tannins than dark-colored (red, brown, bronze, black) varieties. The polyphenol content of seed changes during seed maturation, and it reduces generally during maturation of seed due to polymerization of polyphenolic compounds to high molecular weight insoluble polymers such as lignins.

Table 24.3 Tannins (mg g^{-1}) in important food legumes

Crop	Seed	Dehusked seed
Chickpea	0.8–2.7	0.16–0.38
Pigeon pea	3.8–17.1	0.22–0.43
Mung bean	4.4–8.0	0.21–0.39
Urd bean	5.4–12.0	0.16–0.33
Lentil	3.5–4.9	0.45–0.59
Field pea	5.0–10.5	0.46–0.58
Kidney beans	10.24	0.73
Cowpea	1.75–5.90	0.28–0.48

Source: Reddy et al. (1985)

The seed coat is having the major amount of polyphenols, and the process of dehulling or dehusking reduces the polyphenols of seed drastically (Table 24.3). Reducing the polyphenolic compounds from the grain helps in enhancing the bioavailability of minerals such as iron and zinc and protein. Indian diet is composed of several ingredients (rice, wheat, maize, sorghum, dehulled beans, leafy and other vegetables, roots and tubers, eggs, fish, meat, sugar, and oil), and about daily intake of 10 % of tannins is derived from food legumes.

The polyphenolic compounds cause decreased secretion of digestive enzymes, toxicity of absorbed tannins, increased excretion of endogenous protein, and formation of tannin complexes with dietary protein and other food components. The greater tendency of tannins to form complexes with proteins rather than carbohydrates and other food polymers is attributed to the strong hydrogen bond affinity of the carboxyl oxygen of the peptide group. One tannin molecule binds two or more carboxyl oxygen of peptide group with possible formation of cross-links between the protein chains. It is also reported that tannin-protein complexes are responsible for growth depression, low protein digestibility, decreased amino acid availability, and decreased fecal nitrogen (Reddy et al. 1985). Tannins also reduce the bioavailability of iron and form tannin-iron complexes. Low tannin varieties have higher bioavailability of micronutrients (Fe, Zn), and processing such as soaking in water or other solutions removes major part of tannins. Other processing such as germination, soaking before cooking, or

Table 24.4 Reduction (%) in tannin content of various food legumes on processing

Crop	Soaking in water[a]	Soaking in sodium carbonate[b]	Germination[a]	Dehulling[c]	Cooking[c]
Chickpea	53	–	59–64	74.5–92.6	77.0
Pigeon pea	48	–	52–59	94.2–98.0	58.4
Mung bean	28	–	46–52	92.6–97.4	71.9
Urd bean	22	–	36–53	96.1–98.2	69.4
Pinto beans	21.2–73.6	90.9	–	–	–
Viva Pink	18.1–40.2	77.9	–	–	–
Cranberry	12.2–33.2	67.2	–	–	–
Peas	–	–	–	90.8–94.5	–
Kidney beans	–	–	–	92.9	–
Cowpea	–	–	–	84.0–91.9	–
Winged beans	–	–	–	–	60.8
Horse gram	–	–	–	–	37.5

Source: [a]Rao and Deosthale (1982); [b]Deshpande and Cheryan (1983); [b, c]Reddy et al. (1985)

autoclaving is a boon to reduce tannins and ultimately higher iron and zinc bioavailability (Table 24.4).

The polyphenols can be removed by processing technique such as dehulling, soaking, cooking, germination, fermentation, etc. However, breeding approach is more effective in breeding varieties with lower tannins. Biofortification is therefore finding valuable importance in lowering down the tannin content of seed. Processing of grain before consumption is equally effective in removing tannins to enhance the bioavailability of iron and zinc. Dehulling eliminates about 74–98 % of tannins in beans (Reddy et al. 1985). Dehulling also helps in improving in vitro protein digestibility and ionizable iron absorption. Soaking of grain in water removes 18–74 % of tannins of legumes, whereas soaking in sodium carbonate solution (2 %) for 12 h causes a reduction of 67.2–90.9 % of tannins. Cooking of grain of legumes reduces 37.5–77.0 % of tannins (Reddy et al. 1985), whereas germination of seed for 24–48 h removes 36–64 % of total tannins of seed (Rao and Deostale 1982). Germination of seeds is very effective in removing tannins (32–76 %), whereas soaking followed by cooking or autoclaving remove 79–84 % of total tannins (Kalpanadevi and Mohan 2013). Soaking of soybean seeds in water or 1 % citric acid solution followed by cooking removed 42.5–60.4 % of tannins (Sharma et al. 2013). It is, therefore, established that soaking, germination, and cooking or autoclaving of seed help in removing major amount of tannins. Removal of tannins from seed ultimately enhances the bioavailability of iron, zinc, and protein. Therefore, biofortified grains must be consumed after conventional processing to enhance the bioavailability of micronutrients (iron and zinc).

24.5.2.3 Lectins

Lectins are proteinaceous compounds commonly found in leguminous plants. In 1908, it was reported that the seeds of edible species of some common legumes such as lentils, navy beans, and garden peas contained phytohemagglutinins. Extracts of many edible crude legume seeds agglutinated red blood cells, although hemagglutinating activity has been detected in over 800 different plant species, of which over 600 are from the family Leguminosae. The lectins have molecular weight ranging from 100,000 to 150,000. Some of the lectins sometimes exhibit toxicity. Most of the lectins contain 4–10 % carbohydrates. Concanavalin A of jack beans and peanut lectins are devoid of carbohydrates.

The purified lectins from beans are toxic and sometime cause death of young rats. The lectins from immature seeds of soybean, cowpea, lima beans, pigeon pea, and rice beans when injected in young rats produced liver damage and death. However, lectins from mung bean have been

Table 24.5 Lectins (HU g^{-1}) during processing

Crop	Seed	Soaked	Germinated	Dehusked	Cooked	Autoclaved
Chickpea	1275	–	131	–	214	7
Pigeon pea	1273	–	–	1289	231	00
Lentil	585	–	–	575	35	00
French bean	2685	2507	–	–	230	00

Source: Srivastava and Vasishtha (2013a, b, 2014; Vasishtha and Srivastava (2011, 2014)

reported to be nontoxic. Traditionally, lectins have been measured by their hemagglutinating activity (Grant 1991). However, when possible, enzyme-linked immunosorbent assay (ELISA)-based immunological methods that recognize specific antibodies are being increasingly used due to their higher specificity (Hajos et al. 1996; Muzquiz et al. 2001). Grant et al. (1983) found that among the 15 legumes they studied, chickpea seeds had the lowest hemagglutinin activities and were nontoxic.

Processing such as soaking, germination, cooking, or autoclaving helps in removing most of the lectins (Kalpanadevi and Mohan 2013; Vasishtha 2011; Srivastava and Vasishtha 2013a, b and 2014; Vasishtha et al. 2012 and 2014; Vasishtha et al. 2012). Autoclaving for 5–30 min removes entire lectins of grain, depending on the type of legume. Detoxification of beans can be achieved by ordinary process of cooking (Table 24.5).

Nutritionally, dietary lectins vary considerably in the nature and extent of their antinutritional effects. Lectins can be toxic; they can interfere with hormone balance and deplete nutrient reserves leading to severe growth depression and a high incidence of deaths. However, lectins could be also useful as it stimulates gut function, limits tumor growth, and ameliorates obesity (Pusztai et al. 2004).

24.5.2.4 Protease Inhibitors

Several substances which have the ability to inhibit the proteolytic activity of certain enzymes are found in food grains. Trypsin inhibitors belong to a broad class of proteins (protease inhibitors) that inhibit proteolytic enzymes. Trypsin and chymotrypsin inhibitors are of much importance. Trypsin inhibitor activity increases as seed maturity progresses. Most of the plant protease inhibitors are destroyed by heat resulting in enhancement of nutritive value of protein. Heating is employed for cooking of grains under moist condition. Moisture content, time, temperature, and pressure are the major factors that influence cooking rates, inactivation of trypsin inhibitor, and subsequent increase in nutritive value. Moist heat has been shown to be effective in destroying trypsin inhibitor activity in food grains especially food legumes. Germination also influences the trypsin inhibitor activity.

Over the last two decades, it was observed that protease inhibitors have been linked to health-promoting properties (Champ 2002) and are considered as natural bioactive substances (Hill 2004). Protease inhibitors may act as anticarcinogenic agents (Clemente et al. 2004). It also reduced the incidence and frequency of colon tumors in dimethylhydrazine-treated rats. However, this effect was not observed with autoclaved Bowman-Birk inhibitors (BBI), suggesting that protease inhibitor activity was necessary for anticarcinogenic activity (Kennedy et al. 2002).

24.5.2.5 α-Amylase Inhibitors

α-Amylases (α-1,4-glucan-4-glucanohydrolases) are endoamylases that catalyze the hydrolysis of α-D-(1,4) glycosidic linkages, which occur in starch and related compounds. They play a major role in the carbohydrate metabolism of animals and humans by providing them glucose as an energy source and as a building block for synthesis of other sugars. Among α-amylase inhibitors (α-AIs) found in plants, legume α-AIs, especially α-AIs from beans, have received considerable attention (Whitaker 1988). Jaffe et al. (1973) screened 95 legume cultivars for α-AI levels and found that lima beans (*Phaseolus lunatus*), mung beans (*Phaseolus aureus*), and horse gram (*Dolichos biflorus* L.) had the highest levels of inhibitory activity. Although little is known about the screening of αAIs in chickpea, Mulimani et al. (1994) determined the α-AI activity of 28 varieties of chickpea and found variations ranging from 11.6 to 51.4 inhibitory units g^{-1}. Singh et al. (1982) observed that the amylase inhibitor activity on pancreatic amylase of chickpea cultivars ranged from 7.8 to 10.5 units g^{-1} (*desi*) and 5.6 to 10.0 units g^{-1} (*kabuli*), with substantial variations among these cultivars. A similar variation but of lower magnitude was observed using salivary amylase. A comparison under similar assay conditions indicated that the amylase inhibitor activity had a stronger influence on pancreatic amylase than salivary amylase for both desi and kabuli cultivars. Jaffe et al. (1973) reported that the partially purified kidney bean inhibited the salivary amylase more than the pancreatic amylase. This shows that amylase inhibitors from different legume seeds may exhibit unequal activity against different enzymes.

α-AIs reduce amylase activity and starch digestion in the gut when given orally to humans (Singh et al. 1982). As a result, they lower postprandial increases in circulating glucose and insulin. These inhibitors may therefore prove to be useful in the treatments of obesity or diabetes mellitus.

24.5.2.6 Saponins

Saponins are secondary plant metabolites present in many of the food grains. It contains a carbohydrate moiety (mono or oligosaccharide) attached to an aglycone, which may be steroidal or triterpenoid in structure. The complexity of the saponin structure depends on the variability of the aglycone structure, the attachment position of the glycosidic moieties, and the nature of these glycosides. Aglycones are generally linked to D-galactose, L-rhamnose, D-glucose, D-xylose, D-mannose, and D-glucuronic acids, some of which may be acetylated (Fenwick et al. 1991). All legumes have triterpene-type saponins.

The chemical composition of soybean (*G. max*) saponins has been extensively investigated. They are triterpene saponins (known as soyasaponins), of which more than ten types have been isolated (Shibuya et al. 2006). Soyasapogenol B was identified as the aglycone in saponins from different legume species (Price et al. 1987, 1988; Ayet et al. 1996). Tava et al. (1993) found 17.7 and 21.9 g kg^{-1} of soyasapogenol B in desi and kabuli types, respectively. Ruiz et al. (1996, 1997) showed the presence of soyasaponin VI, a conjugated form of soyasaponin I, in many legumes and was the only saponin detected in the seeds of chickpea. Soyasaponin VI may have an important physiological role in preventing lipid peroxidation of DNA and proteins by free radical attack.

The saponins are now considered beneficial because of their cholesterol-lowering property and antioxidant and cancer protective effect. The saponins undergo degradation during cooking and pressure cooking. Soaking and germination reduce saponins to some extent. Saponins of food legumes undergo certain changes during soaking, germination, cooking, and autoclaving (Srivastava and Vasishtha 2012 and 2014; Vasishtha and Srivastava 2011 and 2014).

Potential benefits from the consumption of food saponins include a reduced risk of cardiovascular disease and some cancers. Dietary saponins have been repeatedly shown to lower plasma cholesterol in animals. However, their hypocholesterolemic effect in humans is more speculative. They may also have anticarcinogenic properties, as suggested by a recent rodent study in which feeding a saponin-containing diet inhibited the development of preneoplastic lesions in the colon (Koratkar and Rao 1997). Kerem et al. (2005) found that the major saponin in chickpea seeds possesses antifungal properties. In addition, some saponins have been reported to inhibit in vivo human immunodeficiency virus (HIV) infectivity (Thompson 1993).

24.5.2.7 Lathyrogens

Lathyrogens are natural toxins present in seeds of lathyrus. The seeds of lathyrus contain a neurotoxic compound commonly called BOAA (β-N-oxalyl amino alanine) or ODAP (β-N-oxalyl L-α, β-diamino propionic acid). This compound is responsible to cause neurolathyrism in human beings. Neurolathyrism is the condition which involves degeneration of part of the spinal cord, affecting lower limbs and infects disability in locomotion. The central nervous system is affected. Neurological lesions of spinal cord degeneration result in weakness and spastic paralysis of the legs, convulsions, and death in extreme cases (Striepler et al. 1978; Attal et al. 1978). The disease is irreversible, though drugs may give temporary relief from muscular rigidity. Lathyrus is prevalent especially during the period of famines when it is cultivated as an alternative to other crops, and the poor sections of society depend mainly on lathyrus as their staple food for several months. Epidemiological surveys have indicated that young men between the age of 15 and 45 are most severely affected by lathyrism. A survey conducted by ICMR (1964) revealed that a diet consisting of about 25 % lathyrus for a period of 2–6 months produces lathyrism.

The neurotoxin present in lathyrus is BOAA, which exists in alpha and beta isomeric forms. Both forms yield oxalic acid and α, β-diamino propionic acid on hydrolysis. Of the two forms, β-isomer is predominant and accounts for 95–96 % of total BOAA content in different varieties of lathyrus. The BOAA in seed of lathyrus is found in the range of 0.05–0.3 %. Certain varieties are having very low BOAA and considered not harmful from the health point of view if consumed judiciously after proper processing. Soaking of dehusked grain in water for 6–12 h removes almost 1/3rd of neurotoxin (Srivastava and Srivastava 2002). Soaking of dehusked grain in acidic solution of pH 4.0 or alkaline solution of pH 9.2 for 30–60 min at 80–100 °C helps in removing major amount of BOAA (Srivastava and Singh 2013). Heating the dehusked grain at 120–150 °C for 2 h also removes about 40 % of total BOAA (Srivastava and Singh 2013). The BOAA content of less than 0.1 % has been considered to be safer from the neurotoxicity point of view (Siddiq 1995). Firstly, consumption of lathyrus should be advocated after necessary processing so as to minimize the neurotoxin content from the health point of view. Secondly, the genotypes having reasonably low BOAA should be encouraged for human consumption.

24.6 Conclusion

Biofortification enhances the bioavailability of micronutrients such as iron, zinc, vitamin A, folic acid, and cyanocobalamin in staple foods. Almost half of the world population is suffering from these deficiencies as they have either low access or no access for these quality foods. The presence of antinutrients such as phytates, polyphenols and tannins, protease and α-amylase inhibitors, saponins, lectins, and lathyrogens in food causes hurdle in absorption and bioavailability of Fe, Zn, Ca, folates, etc. Various processing techniques like soaking, germination, milling (dehusking), cooking, and autoclaving are effective in removal of these antinutrients.

References

Allen LH (2001) Biological mechanisms that might underline iron's effects on fetal growth and preterm birth. J Nutr 131(2S–2):581S–589S

Attal HC, Kulkarni SW, Choubey BS, Palkal ND, Deotale PG (1978) A field study of lathyrism- some critical aspects. Indian J Med Res 67:608–611

Ayet G, Muzquiz M, Burbano C, Robredo LM, Cuadrado C, Price K (1996) Determinación de saponinas en las principales leguminosas cultivadas en España. (Determination of saponins as the main legumes cultivated in Spain). Food Sci Technol Int 2:96–100

Biehl RR, Baker DH, DeLuca HF (1995) 1-α-hydroxylated cholecalciferol compounds act additively with microbial phytase to improve phosphorus, zinc and manganese utilization in chicks fed soy-based diets. J Nutr 125:2407–2416

Bodnar LM, Wisner KL (2005) Nutrition and depression implications for improving mental health among childbearing aged women. Biol Psychiatry 58 (9):679–685

Bouis HE (2003) Micronutrient fortification of plants through plant breeding: can it improve nutrition in man at low cost? Proc Nutr Soc 62:403–411

Champ MM (2002) Non-nutrient bioactive substances of pulses. Br J Nutr 88:307–319

Chitra U, Singh U, Venkateswararao P (1996) Phytic acid, in vitro protein digestibility, dietary fiber and minerals of pulses as influenced by processing methods. Plant Food Hum Nutr 49:307–316

Christian P, Khatry SK, Leclerq SC, Dali SM (2009a) Effects of prenatal micronutrient supplementation on complications of labor and delivery and puerperal morbidity in rural Nepal. Int J Gynecol Obstet 106 (1):3–7. doi:10.1016/j.ijgo.2009.03.040

Christian P, Stewart CP, LeClerq SC, Wu L, Katz J, West KP Jr, Khatry SK (2009b) Antenatal and postnatal iron supplementation and childhood mortality in rural Nepal a prospective follow-up in a randomized, controlled community trial. Am J Epidemiol 170:1127–1136

Clemente A, Mackenzie DA, Johnson IT, Domoney C (2004) Investigation of legume seed protease inhibitors as potential anticarcinogenic proteins. In: Muzquiz M, Hill GD, Pedrosa MM, Burbano C (eds) Proceedings of the fourth international workshop on antinutritional factors in legume seeds and oilseeds, EAAP Publication No. 110, Wageningen, pp 137–142

Cogswell ME, Parvanta I, Ickes L, Yip R, Brittenham GM (2003) Iron supplementation during pregnancy, anemia, and birth weight a randomized controlled trial. Am J Clin Nutr 78:773–781

Cordain L (1999) Cereal grains humanity's double-edged sword. World Rev Nutr Diet 84:19–73

Dawe D, Robertson R, Unnevehr L (2002) Golden rice: what role could it play in alleviation of vitamin A deficiency? Food Policy 27(5–6):541–560

Deshpande SS, Cheryan M (1983) Changes in phytic acid, tannins and trypsin inhibitory activity on soaking of dry beans (*Phaseolus vulgaris* L.). Nutr Rep Int 27 (2):371–375

Disler PB, Lynch SR, Charlton RW, Torrance JD, Bothwell TH, Walker RB, Mayet F (1975) The effect of tea on iron absorption. Gut 16:193–200

Fenwick GR, Price KR, Tsukamoto C, Okubo K (1991) Saponins. In: D'Mello JPF, Duffus CM, Duffus JH (eds) Toxic substances in crop plants. Royal Society of Chemistry, Cambridge, MA, pp 285–327

Fomon S (2001) Infant feeding in the 20th century formula and beikost. J Nutr 131:S409–S420

Fraker PJ, King LE, Laakko T, Vollmer TL (2000) The dynamic link between the integrity of the immune system and zinc status. J Nutr 130(5S):1399S–1406S

Gillooly M, Bothwell TH, Torrance JD, MacPhail AP, Derman DP, Bezwoda WR, Mills W, Charlton RW, Mayet F (1983) The effects of organic acids, phytates and polyphenols on the absorption of iron from vegetables. Br J Nutr 49:331–342

Graham R, Welch R, Bouis H (2001) Addressing micronutrient malnutrition through the nutritional quality of staple foods: principles, perspectives, and knowledge gaps. Adv Agron 70:77–142

Grant G (1991) Legumes. In: D'Mello JPF, Duffus CM, Duffus JH (eds) Toxic substances in crop plants. Royal Society of Chemistry, Cambridge, MA, pp 49–67

Grant G, More LJ, McKenzie NH, Stewart JC, Pusztai A (1983) A survey of the nutritional and haemagglutination properties of legume seeds generally available in the UK. Br J Nutr 50:207–214

Greiner R, Konietzny U (1996a) Phytate hydrolysis in black beans by endogenous and exogenous enzymes. In: Bardocz S, Muzquiz M and Pusztai A (eds) Effects of antinutrients on the nutritional value of legume diets. Proceedings of the fourth scientific workshop, official publications of the European communities, Luxembourg, pp 19–26

Greiner R, Konietzny U (1996b) Construction of a bioreactor to produce special breakdown products of phytate. J Biotechnol 48:153–159

Greiner R, Larsson AM, Carlsson NG, Muzquiz M, Burbano C, Cuadrado C, Pedrosa MM, Goyoaga C (2002) Pathway of dephosphorylation of myoinositol hexakisphosphate by phytases of legume seeds. J Agric Food Chem 50:6865–6870

Hajos G, Gelencser E, Pusztai A, Bardocz S (1996) Measurements of the antinutrient contents of legume samples. In: Bardocz S, Gelencser E, Pusztai A (eds) Effects of antinutrients on the nutritional value of legume diets, vol I. Official Publications of the European Communities, Luxembourg, pp 130–134

Hallberg L, Hulthen L (2000) Prediction of dietary iron absorption an algorithm for calculating absorption and bioavailability of dietary iron. Am J Clin Nutr 71:1147–1160

Hill GD (2004) Grain legumes and oilseeds: the way ahead. In: Muzquiz M, Hill GD, Pedrosa MM, Burbano C (eds) Proceedings of the fourth international workshop on antinutritional factors in legume seeds and oilseeds. EAAP Publication No. 110, Wageningen pp 353–364

Huang EP (1997) Metal ions and synaptic transmission think zinc. Proc Natl Acad Sci U S A 94 (25):13386–13387

Hurrell R (1999) Iron. In: Hurrell R (ed) The mineral fortification of foods. International Ltd, Leatherhead/ Surrey, pp 54–93

Hurrell R, Ranum P, de Pee S, Biebinger R, Hulthen L, Johnson Q, Lynch S (2010) Revised recommendations for iron fortification of wheat flour and an evaluation of the expected impact of current national wheat flour fortification programs. Food Nutr Bull 31S:7–21

ICMR (1964) Lathyrism -a preventable paralysis. Indian Council of Medical Research, New Delhi

Iyer V, Salunkhe DK, Sathe SK, Rockland LB (1980) Quick cooking beans *(Phaseolus vulgaris L.)* II Phytates, oligosaccharides and anti-enzymes. Plant Foods Hum Nutr 30:45–49

Iyer V, Kadam SS, Salunkhe DK (1989) Cooking. In: Salunkhe DK, Kadam SS (eds) Handbook of world food legumes- nutritional chemistry, processing technology and utilization, vol III. CRC Press, Boca Raton, pp 141–163

Jaffe WG, Moreneo R, Wallis V (1973) Amylase inhibitors in legume seeds. Nutr Rep Int 7:169–174

Kalpanadevi V, Mohan VR (2013) Effect of processing on antinutrients and in vitro protein digestibility of the underutilized legume, *Vigna unguiculata* (L.) Walp subsp. *unguiculata*. LWT: Food Sci Technol 51 (2):455–461. doi:10.1016/j.lwt.2012.09.030

Kennedy AR, Billings PC, Wan XS, Newberne PM (2002) Effects of Bowman Birk inhibitor on rat colon carcinogenesis. Nutr Cancer 43:174–186

Kerem Z, GermanShashoua H, Yarden O (2005) Microwave assisted extraction of bioactive saponins from chickpea (*Cicer arietinum* L.). J Sci Food Agric 85:406–412

Khan KS, Wojdyla D, Say L, Gulmezoglu AM, Van Look PF (2006) WHO analysis of causes of maternal death a systematic review. Lancet 367:1066–1074

Koratkar R, Rao AV (1997) Effect of soya bean saponins on azoxymethane-induced preneoplastic lesions in the colon of mice. Nutr Cancer 27:206–209

Krishnaswamy K (2001) Perspectives on nutrition needs for the new millennium for South Asian regions. Biomed Environ Sci 14:66–74

La Frano MR, de Moura FF, Boy E, Lonnerdal B, Burri BJ (2014) Bioavailability of iron, zinc, and provitamin A carotenoids in biofortified staple crops. Nutr Rev 72 (5):289–307

Ladipo OA (2000) Nutrition in pregnancy: mineral and vitamin supplements. Am J Clin Nutr 72 (1):280S–290S

Lajolo FM, Genovese MI, Pryme IF, Dale M (2004) Beneficial (antiproliferative) effects of different substances. In: Muzquiz M, Hill GD, Pedrosa MM, Burbano C (eds) Proceedings of the fourth international workshop on antinutritional factors in legume seeds and oilseeds, EAAP Pubn #110, Wageningen, pp 123–135

Leung BM, Kaplan BJ (2009) Perinatal depression: prevalence, risks, and the nutrition link- a review of the literature. J Am Diet Assoc 109(9):1566–1575. doi:10.1016/j.jada.2009.06.368

Lynch SR (2011) Why nutritional iron deficiency persists as a worldwide problem. J Nutr. doi:10.3945/jn.110.130609

McCance RA, Edgecombe CN, Widdowson EM (1943) Phytic acid and iron absorption. Lancet 2:126–128

McClafferty B, Islam Y (2008) Fighting the hidden hunger. In TCE, Feb 2008, p 27

Miller V, Swaney S, Deinard A (1985) Impact of the WIC program on the iron status of infants. Paediatrics 75:100–105

Mulimani VH, Rudrappa G, Supriya D (1994) α-Amylase inhibitors in chick pea (*Cicer arietinum* L.). J Sci Food Agric 64:413–415

Muzquiz M, Burbano C, Cuadrado C, Martin M (2001) Analytical methods for determination of compounds with no nutritive value. In: Jacobsen HJ, Muzquiz M, Hassa A (eds) Handbook on common bean related laboratory methods. Galicia, Spain, pp 11–26

Nestel P, Bouis HE, Meenakshi JV, Pfeiffer W (2006) Biofortification of staple food crops. J Nutr 136:1064–1067

Paine JA, Shipton CA, Chaggar S, Howells RM, Kennedy MJ, Vernon G, Wright SY, Hinchliffe E, Adams JL (2005) A new version of golden rice with increased pro-vitamin A content. Nat Biotechnol 23:482–487

Palma S, Perez-Iglesias R, Prieto D, Pardo R, Llorca J, Delgado-Rodriguez M (2008) Iron but not folic acid supplementation reduces the risk of low birth weight in pregnant women without anaemia- a case control study. J Epidemiol Community Health 62(2):120–124

Price KR, Johnson IT, Fenwick RG (1987) The chemistry and biological significance of saponins in foods and feeding stuff CRC. Crit Rev Food Sci Nutr 26:27–131

Price KR, Eagles J, Fenwick GR (1988) Saponin composition of 13 varieties of legume seed using fast atom bombardment mass spectrometry. J Sci Food Agric 42:183–193

Pusztai A, Bardocz S, MartinCabrejas M (2004) The mode of action of ANFs on the gastrointestinal tract and its microflora. In: Muzquiz M, Hill GD, Pedrosa MM, Burbano C (eds) Proceedings of the fourth international workshop on antinutritional factors in legume seeds and oilseeds, EAAP Publication No. 110, Wageningen, pp 87–100

Rao PU, Deostale YG (1982) Tannin content of pulses varietal differences and effects of germination and cooking. J Sci Food Agric 33:1013–1016

Reddy NR, Sathe SK, Salunkhe DK (1982) Phytates in legumes and cereals. In: Chichester CO (ed) Advances in food research, vol 28. Academic Press, Inc, New York, pp 1–4

Reddy NR, Pierson MD, Sathe SK, Salunkhe DK (1985) Dry bean tannins a review of nutritional implications. J Am Oil Chem Soc 62(3):541–549

Reddy MB, Hurrell RF, Cook JD (2000) Estimation of nonheme-iron bioavailability from meal composition. Am J Clin Nutr 71:937–943

Ruiz RG, Price K, Rose M, Rhodes M, Fenwick R (1996) A preliminary study on the effect of germination on saponin content and composition of lentils and chick peas. Z Lebensm Unters Forsch 203(4):366–369. doi:10.1007/BF01231075

Ruiz RG, Price KR, Fenwick RG, Rhodes MJC (1997) Changes in saponin content and composition of chickpeas and lentils during germination and cooking. In: Bardocz S, Muzquiz M, Pusztai A (eds) Effects of antinutrients on the nutritional value of legume diets, proceedings of the fourth scientific workshop, Official Publication of European Communities, Luxembourg, pp 15–18

Scholl TO (2005) Iron status during pregnancy: setting the stage for mother and infant. Am J Clin Nutr 81: S1218–S1222

Sharma S, Goyal R, Barwal S (2013) Domestic processing effects on physicochemical, nutritional and antinutritional attributes in soybean (*Glycine max* L. Merill). Int Food Res J 20:3203–3209

Shibuya M, Hoshino M, Katsube Y, Hayashi H, Kushiro T, Ebizuka Y (2006) Identification of bamyrin and sophoradiol 2,4-hydroxylase by expressed sequence tag mining and functional expression assay. FEBS J 273:948–959

Siddiq EA (1995) Lathyrus genetic resources in Asia. In: Proceedings of the regional workshop on Lathyrus, Raipur, 27–29 Dec, pp 1–2

Singh U, Kherdekar MS, Jambunathan R (1982) Studies on desi and kabuli chickpea (*Cicer arietinum* L.) cultivars. The levels of amylase inhibitors, levels of oligosaccharides and in vitro starch digestibility. J Food Sci 47:510–512

Srivastava RP, Singh J (2013) Soaking and dehusking effect on lathyrus (*Lathyrus sativus*) genotypes. In: Proceedings of the international conference on role of plant biochemistry and biotechnology in food and nutritional security, 11–14 Dec 2013, Sri Venkateswara University, Tirupati

Srivastava RP, Srivastava GK (2002) Nutritional quality of newly developed toxin free lathyrus. Indian J Agric Biochem 15:33–36

Srivastava RP, Vasishtha H (2012) Saponins and lectins of Indian chickpeas (*Cicer arietinum*) and lentils (*Lens culinaris*). Indian J Agric Biochem 25:44–47

Srivastava RP, Vasishtha H (2013a) Dehusking and cooking effect on dietary fibre, soluble protein and lectin of lentils (*Lens culinaris*). Indian J Agric Biochem 26:36–40

Srivastava RP, Vasishtha H (2013b) Dietary fiber, protein and lectin contents of lentils (*Lens culinaris*) on soaking and cooking. Curr Adv Agric Sci 5(2):238–241

Srivastava RP, Vasishtha H (2014) Changes in protein, sapogenols and lectins of different types of chickpeas (*Cicer arietinum*) during soaking and germination. Indian J Agric Biochem 27:116–122

Stoltzfus RJ (2001) Defining iron deficiency anemia in public health terms; a time for reflection. J Nutr 131:565S–567S

Striepler M, Cohn DF, Hirano A, Schujaman E (1978) The central nervous system in a case of neurolathyrism. Neurology 27:1176

Tako E, Hoekenga OA, Kochian LV, Glahn RP (2013) High bioavailability iron maize (*Zea mays* L.) developed through molecular breeding provides more absorbable iron in vitro (Caco-2 model) and in vivo (Gallus gallus). Nutr J 12:3. doi:10.1186/1475-2891-12-3

Tava A, Oleszek W, Jurzysta M, Berardo N, Odoardi M (1993) Alfalfa saponins and sapogenins: isolation and quantification in two different cultivars. Phytochem Anal 4:269–274

Thompson LU (1993) Potential health benefits and problems associated with antinutrients in foods. Food Res Int 26:131–149

Titaley CR, Dibley MJ, Roberts CL, Hall J, Agho K (2010) Iron and folic acid supplements and reduced early neonatal deaths in Indonesia. Bull World Health Organ 88:500–508

Tontisirin K, Nantel G, Bhattacharjee L (2002) Food-based strategies to meet the challenges of micronutrient malnutrition in the developing world. Proc Nutr Soc 61(2):243–250

Urbano G, LopezJurado M, Aranda P, VidalValverde C, Tenorio EJ, Porres E (2000) The role of phytic acid in legumes: antinutrient or beneficial function. J Physiol Biochem 56:283–294

Van Jaarsveld PJ, Faber M, Tanumihardjo SA, Nestel P, Lombard CJ, Benade AJS (2005) β-Carotene-rich orange-fleshed sweet potato improves the vitamin A status of primary school children assessed by the modified-relative-dose-response test. Am J Clin Nutr 81:1080–1087

Vasishtha H (2011) Processing effect on hypocholesterolemic and haemagglutinating characters of important pulses. PhD thesis, CSJM University, Kanpur

Vasishtha H, Srivastava RP (2011) Effect of dehusking and cooking on sapogenols of pigeonpea (*Cajanus cajan*) genotypes. Indian J Agric Biochem 24:60–64

Vasishtha H, Srivastava RP (2014) Effect of soaking and cooking on dietary fiber, protein and lectins of rajmash (*Phaseolus vulgaris*) beans. Indian J Agric Biochem 27:219–222

Vasishtha H, Srivastava RP, Verma P (2012) Milling and cooking effect on dietary fiber, protein and lectins of pigeonpea (*Cajanus cajan*). Curr Adv Agric Sci 4:134–138

Whitaker JR (1988) α-Amylase inhibitors of higher plants and microorganisms. In: Kinsella JE, Soucle WG (eds) Food proteins. Proceedings of the protein. Co-products symposium, American Oil Chemists' Society, Champaign II, pp 354–380

Widdowson EM, McCance RA (1942) Iron exchanges of adults on white and brown bread diets. Lancet 1:588–591

Wikipedia. Biofortification. http://en.wikipedia.org/wiki/biofortification. Accessed 15 Jun 2015

Wilson RD, Davies G, Desilets V, Reid GJ, Summers A, Wyatt P, Young D (2003) The use of folic acid for the prevention of neural tube defects and other congenital anomalies. J Obstet Gynaecol Can 25(11):959–973

Yip R (2002) Prevention and control of iron deficiency policy and strategy issues. J Nutr 132:S802–S805

Zeng L, Cheng Y, Dang S, Yan H, Dibley MJ, Chang S, Kong L (2008) Impact of micronutrient supplementation during pregnancy on birth weight, duration of gestation, and perinatal mortality in rural western China: double blind cluster randomised controlled trial. Br Med J 337:2001–2005

Zhou JR, Erdman JW (1995) Phytic acid in health and disease. Crit Rev Food Sci Nutr 35:495–508

Zimmermann MB, Hurrell RF (2007) Nutritional iron deficiency. Lancet 370:511–520

Zimmermann R, Qaim M (2004) Potential health benefits of golden rice: a Philippine case study. Food Policy 29(2):147–168

Nutritional Enhancers/Promoters in Biofortification

25

Nand Lal Meena, Om Prakash Gupta, and Susheel Kumar Sharma

Abstract

Global population is increasing at an alarming rate which is expected to be around 9.2 billion by 2050 from the present 7.2 billion which urges for higher global crop production to feed world hunger. Cereals and pulses are the two most important crops serving as energy and protein source respectively worldwide. Micronutrients, vitamins, antioxidants, and quality proteins are some of the major areas for crop biofortification in African and Asian countries which is the major caus of annual death of pregnant women and preschool children. Fortification, industrial supplements, dietary variation, and biofortification are the major ways to get rid of these macro- and micronutrient malnutrition. Among them, biofortification is only one of the most viable ways to cater to nutrient-deficient population of the developing countries. Genetic engineering and breeding coupled with agronomic interventions are the only means to take care of this problem. Here, we have summarized the recent progress made in the area of crop biofortification.

Keywords

Antioxidants • Vitamins • Mineral • Vitamins • Essential oils

25.1 Introduction

Using their inbuilt biochemical processes, plants have capabilities to synthesize majority of dietary micronutrients, excluding vitamin D (cholecalciferol, ergocalciferol) and cobalamin (B_{12}) compared to developed nations; crops such as rice, wheat, cassava, and maize are the major sources of energy and proteins in the developing nations which are inadequate to

N.L. Meena (✉)
Division of Basic Sciences, ICAR-Indian Institute of Pulses Research, Kanpur, Uttar Pradesh 208 024, India
e-mail: nd.iari09@gmail.com

O.P. Gupta
ICAR-Indian Institute of Wheat and Barley Research, Karnal, Haryana 132 001, India
e-mail: opguptaiari@gmail.com

S.K. Sharma
ICAR-Research Complex for NEH Region, Manipur Centre, Imphal 795 004, India
e-mail: susheelsharma19@gmail.com

U. Singh et al. (eds.), *Biofortification of Food Crops*, DOI 10.1007/978-81-322-2716-8_25

meet the minimum daily requirements of other essential nutrients. Moreover, the nutrients are not evenly distributed in all parts of the plant (Zhu et al. 2007). For instance, rice leaves are rich in provitamin, while the same is lacking in edible grain. Therefore, attention should be given to enrich the dibble portions of the grains with a specific nutrient or combination of more than one by a process commonly known as biofortification.

Biofortified food crops are characterized by enriched nutrients with enhanced bioavailability of essential micronutrients, which are often preferred by the consumers over food supplementation owing to traditional consumption pattern and trade. Biofortification differs from industrial fortification/supplementation as the latter depends upon external addition of micronutrients, whereas the former relies upon changing biosynthesis or physiological capacity of food plants in order to produce or accumulate desired level of specific essential nutrients. The common approaches practiced for generating biofortified crops involve fertilizer application, conventional breeding, and transgenic technology. Implementation of these above-stated approaches results in micronutrient-dense staple crops, with dense minerals and vitamins. Owing to today's global scenario of increasing nutritional insecurity, biofortification is gaining popularity as it targets the staple food consumed predominantly by the sizeable population inhabiting relatively in the remote areas. Further, developing fortified seeds represents onetime investment which can be consumed year after year and also shared between the communities across the globe (Dwivedi et al. 2012).

Biofortification process pursues various strategies such as optimizing the concentration of antinutrients, mobility of micronutrients within plants, binding capacity/sink strength of seeds, and concentration of metabolites that promote micronutrient absorption (Cakmak et al. 2010). Application of these strategies involves identifying target nutrients suitable for such biofortification. To date, targets for biofortification have been diverted toward essential amino acids (methionine), fatty acids, vitamins A and E, lycopene, flavonoids, Fe, and Zn. In this chapter, we outline the characteristics of the minerals and nutrient elements and evaluate the progress made so far toward enhancement of these targets.

25.2 Antioxidants

Fruits and vegetables supply passable amount of antioxidants such as anthocyanins, carotenoids, vitamins C and E, and polyphenolics such as quercetin (Shehanaz 2013). These antioxidants protect human from various kinds of reactive oxygen species (ROS). Carotenoids along with other similar compounds are generated via general isoprenoid biosynthetic pathways in plants.

25.2.1 Anthocyanins

These compounds are colored (red, purple, or blue) pigments that belong to a water-soluble flavonoid family. It has been seen that tomato berry contains negligible amount of anthocyanins. Significant efforts have moderately enhanced its level mainly in peel parts which is normally destroyed during processing. However, the trait of resulting purple fruit was transferred to several commercial cultivars via backcrossing without compromising growth and yield (Hirschi 2008). Further studies have supported the fact that such transgenic tomato powder has extended the life span of tumorigenic mice. Blackberries and raspberries are among the best sources of dietary anthocyanins, but both are expensive and are consumed in smaller quantities than tomatoes (Crozier et al. 2009). Possibly, these engineered tomatoes could contribute substantially to the antioxidant levels of human diets.

25.2.2 Carotenoids

25.2.2.1 Lycopene

Lycopene is a carotenoid which is poorly absorbed by the human body from fruits because of its fat-soluble nature. Also, its absorption varies across its cis or trans form. Furthermore, ambiguity prevails concerning the enhanced use of lycopene in foods and its subsequent benefits to human body. Lycopene is the major carotenoid pigment found in tomatoes which has received tremendous attention due to its potential cancer chemopreventive property. Its consumption mostly shields from prostate cancer (Kucuk et al. 2002; Miller et al. 2002). The studies in these lines have produced some inconsistent results, which is soberly clarified by problems with the bioavailability of lycopene from different sources (Giovannucci 2002). Consumption of fruits and vegetables rich in lycopene protects breast and colorectal cancers. While many of the studies conducted on a few animal models have assured a protective reaction of lycopene, its mechanism has been resolved to an appreciable extent (Terry et al. 2002a; Cohen 2002). Heber and Lu (2002) have reported that similar to other dietary carotenoids, lycopene might work in various ways to defend cancer.

Various genetic modification (GM) techniques were used to increase the levels of lycopene in tomato plant. Overexpression of phytoene synthase gene, a key enzyme in the carotenoid pathway from the bacteria *Erwinia uredovora* in tomato plants, resulted in 1.8- to 2.2-fold enrichment of lycopene and β-carotene and a twofold to fourfold rise in the total carotenoid levels (Fraser et al. 2002). Furthermore, overexpression of a yeast *S*-adenosyl - methionine decarboxylase has increased lycopene levels in tomato fruit significantly (Mehta et al. 2002). Processed tomato such as pastes mixed with fats enhances lycopene absorption significantly. An artificial formulation made by entrapping lycopene within whey proteins was explained in a recent report by Tucker (2003). This lacto-lycopene has shown a similar bioavailability to lycopene in tomato paste (Richelle et al. 2002).

25.2.3 Vitamin

Biofortification of crops with vitamin A and vitamin C can increase the Fe absorption in the intestine (Garca-Casal et al. 2003). Folate plays a significant role in preventing neuronal disorders in developing fetus. However, it is hard to hold it due to its high water solubility, for example, in rice kernels during boiling (Shrestha et al. 2003). Some of the important biofortified vitamins have been discussed below.

25.2.3.1 Vitamin A

Rice is a staple food crop of the global population. Overdependence on rice has led to vitamin A deficiency (VAD) which directly affects nearly 250,000–500,000 children every year. Vitamin A-based malnutrition in children can be prevented by improved vitamin A nutrition (Wegmuller et al. 2003). Provitamin A-containing rice could significantly decrease VAD (Zimmermann and Qaim 2004). To decrease the rancidity of rice during storage, oil-rich aleurone layer is removed during milling process, and the remaining endosperm lacks vitamin A. Genetic engineering-based enrichment of rice grain with provitamin A has increased its level significantly (Ye et al. 2000). The immature rice endosperm can produce the intermediate compound geranylgeranyl diphosphate (GGPP) which is then used to generate phytoene by expressing the enzyme phytoene synthase (Cunningham and Gantt 1998). The synthesis of β-carotene from phytoene requires three additional enzymes such as phytoene desaturase, β-carotene desaturase, and lycopene β-cyclase. The important concern is the bioavailability of the engineered β-carotene to humans. Although the conversion factor to vitamin A for synthetic β-carotene and β-carotene in fruits is approximately 6:1, however, the conversion factor in case of vegetables like spinach may be as low as 24:1 due to the poor release of the carotenoids during digestion (Institute of Medicine 2000).

25.2.3.2 Vitamin E

Another vital dietary antioxidant that occurs in various isomeric forms in plants is vitamin E (tocopherol), with α- and γ-tocopherol being the most abundant. The α-tocopherol is considered as the most beneficial dietary form; however, it is found less dominantly in many foods than γ-tocopherol. Tocochromanols consist of four tocopherols and four tocotrienols that constitute vitamin E (Cahoon et al. 2003; DellaPenna and Pogson 2006). In the plants, vitamin E content can be increased using nutritional genomics (DellaPenna 2007; Shintani and DellaPenna 1998). Furthermore, the content and types of tocochromanols can be altered by coexpression of several genes related to the tocochromanol biosynthetic pathway. This can increase the vitamin E content of seed oil by tenfold (Henry and Qi 2005). Efforts have been made to extend this technology to soybean, maize, and canola (DellaPenna 2007). An added benefit of this work may upsurge the shelf life of foods (Mehta et al. 2002). It is believed that foods with greater levels of vitamin E are protected from oxidative pressure which could enhance agricultural productivity and improve storage.

25.2.4 Other Antioxidants

Polyphenolics such as flavonols are other important antioxidants which have been enhanced in tomato. In the normal fruit, flavonoids are present in the skin in small amount. Bovy et al. (2002) have increased levels of the kaempferol in a tissue that would normally be lacking this compound in the flesh of the tomato fruit by expression of two transcription factors, LC and C1, from maize.

25.3 Folate

Folate is also known as tetrahydrofolate (THF) which is not synthesized by animals. The recommended dietary allowance for folate ranges from 400 to 600 μg per day for pregnant women (Rosemary and Lisa 2013). Its deficiency leads to megaloblastic anemia (Scott et al. 2000). Other significances are the initiation of hyperhomocysteinemia which is a risk factor for cardiovascular disease (Stanger 2004), misincorporation of uracil in DNA, and chromosomal damage (Lucock et al. 2003). Furthermore, folate deficiency leads to abnormal DNA methylation arrays associated with carcinogenesis (Choi and Friso 2005; Ulrey et al. 2005). Plants are the ultimate source of folate for animals. Cereals such as maize, wheat, and rice contain extremely low levels of folate (USDA National Nutrient Database for Standard Reference. Release 9; http://www.nal.usda.gov/fnic/foodcomp/search/). To reduce the risk of folate deficiency, fortification of cereal foods with synthetic folic acid has been effected in the USA and other countries (Food and Drug Administration 1996). By contrast, the third world countries lack a solid infrastructural platform to enable effective preventive methods in the form of fortification, supplementation, or educational campaigns (Sachs 2005). Therefore, biofortification of staple crops together with conventional public health practices appears as the probable option to meet the folate deficiency especially in developing countries.

Folate consists of three different components such pteridine, para-aminobenzoic acid (PABA), and glutamate. It is synthesized from these precursors within the mitochondria, while these components are synthesized in unlike compartments within the plant cell (DellaPenna 2007). Bioavailability of folate from various cereals and legume foods depends on a number of factors such as the food matrix, the polyglutamyle conjugation, etc. (McNulty and Pentieva 2004).

The polyglutamylation of folate reduce the bioavailability, because dietary folates need to be deglutamylated by the intestinal conjugase before efficient uptake by intestinal (Rebeille et al. 2006). Recently, it has been exposed that the ratio of monoglutamate to polyglutamate in natural folate derivatives has no clear impact on the intestinal absorption (McKillop et al. 2006). This signifies that the amount of the intestinal conjugase is sufficient to remove the polyglutamate tail without affecting the rate of absorption. This proposes that other factors have an impact on bioavailability, such as entrapment of folates in the food matrix, making them inaccessible to the conjugase that is tethered to the intestinal cell membranes. Therefore, one likely strategy to increase bioavailability could be to boost the levels of the plant conjugase activity of gamma-glutamyl hydrolase (GGH), which would be released from the vacuole following maceration and facilitate folate release within the food matrix before digestion. Other strategy is the use of FBPs. One must also consider the fact that the presence of folate binding proteins (FBPs) in the food matrix, although mechanistically unclear, can lead to a decrease in folate absorption (Witthoft et al. 2006).

25.4 Essential Amino Acids

Most crops lack one or more essential amino acids, for example, cereal grains are deficient in lysine and threonine, whereas legumes are in methionine and cysteine (Hartwig et al. 1997; Newell-McGloughlin 2008; O'Quinn et al. 2000; Sautter et al. 2006). As the mainstream population of the world depends on cereals and legumes for their diet, plant biologists have used various approaches to augment anticipated deficient amino acids in these plants (Rapp 2002). For example, expression of storage proteins that comprise high levels of required amino acids has raised lysine content in rice and wheat (Christou and Twyman 2004) and some essential amino acid content in potatoes (Egnin and Prakash 1997). However, attempts to raise sulfur-containing amino acids have not been as helpful (Dinkins et al. 2001).

To address this issue, synthetic proteins have been expressed in cassava to match the amino acid requirements for humans. The inadequate availability of free amino acid pools within the edible portion of plants suggests us to alter the amino acid content of target crops. In higher plants, synthesis of threonine, lysine, and methionine are under complex feedback regulation (Hesse et al. 2004). Therefore, studies now focus on developing feedback-insensitive enzymes for these pathways (Newell-McGloughlin 2008). This has significantly improved the free lysine levels in maize (from 2 % to almost 30 %) (Sofi et al. 2009). Expression of these altered enzymes has also improved lysine content in canola and soybean which has significantly improved the production of tryptophan levels in grains. A fundamental concern with each of these manipulations is to ensure that the total amount and composition of storage proteins are not changed.

25.5 Essential and Very-Long-Chain Fatty Acids

Genetically modified oil seed crops are an abundant, relatively inexpensive source of dietary fatty acids (Abbadi et al. 2004; Anai et al. 2003; Kinney and Knowlton 1998; Liu et al. 2002; Reddy and Thomas 1996; Wallis et al. 2002). Production of these lipids in vegetables could provide an easy mechanism to deliver healthier products without major dietary modifications (Damude and Kinney 2008; Newell-McGloughlin 2008). Plants are sources of the essential fatty acids such as linoleic acid and linolenic acid as well as very-long-chain polyunsaturated fatty acids (VLC-PUFAs) such as arachidonic acid (ARA), eicosapentaenoic acid (EPA), and docosahexaenoic acid (DHA), which are usually found in fish oils.

Given that many of the enzymes tangled in fatty acid biosynthesis and degradation have been characterized, there is an abundance of transgenic approaches to the modification of oil and fat content in plants (Damude and Kinney 2008; Broun et al. 1999). Examples of such modified oils include low- and zero-saturated fat soybean and canola oils, canola oil comprising medium-chain fatty acids, high-stearic acid canola oil, high-oleic acid soybean oil, and canola oil encompassing the polyunsaturated fatty acid linolenic acid (Mensink et al. 2003). Oils abundant in monounsaturated fatty acids provide improved oil stability, flavor, and nutritional qualities. Oleic acid (18:1), a MUFA, can provide more stability and health benefits than the PUFA (18:2 and 18:3). Soybeans have been manipulated to encompass more than 80 % oleic acid against normal 23 % and had a significant decrease in polyunsaturated fatty acids (Kinney and Knowlton 1998). High-oleic acid soybean oil is more resistant to degradation by heat and oxidation requiring little or no post-refining processing.

25.6 Mineral Biofortification

Humans require at least 22 mineral elements for their survivals (Welch and Graham 2004; White and Broadley 2005; Graham et al. 2007) which can be abounding by an appropriate diet. It is estimated that over 60 % of the world's people are iron (Fe) deficient, over 30 % are zinc (Zn) deficient, 30 % are iodine (I) deficient, and 15 % are selenium (Se) deficient (Yang et al. 2007). In addition, calcium (Ca), magnesium (Mg), and copper (Cu) deficiencies are also common in many developed and developing countries (Frossard et al. 2000; Welch and Graham 2002, 2005; Rude and Gruber 2004; Grusak and Cakmak 2005; Thacher et al. 2006).

Fe is an essential component of many enzymes catalyzing redox reactions. More than half of the Fe in the human body is bound to hemoglobin. In the developing countries, Fe deficiency leads to anemia and is estimated that 40–45 % of preschool-age children are anemic. Approximately 50 % of this anemia results from Fe deficiency (Grillet et al. 2014). The significances of early childhood anemia include poor cognitive development (Lozoff 2007). Zn is a cofactor with diverse structural and catalytic functions in about 10 % of all human proteins. In addition, evidence has been accumulating for important regulatory roles of Zn ions in inter- and intracellular signaling (Maret 2013). A major challenge is low bioavailability of Fe and Zn from staple cereals and legumes which have absorption inhibitor such as phytic acid (Olsen and Palmgren 2014). In general, to increase mineral concentrations in edible crops (biofortification), parallel attempts are advocated such as (1) to increase the concentrations of "promoter" substances, such as ascorbate (vitamin C), β-carotene, cysteine-rich polypeptides, and certain organic and amino acids, which excite the absorption of essential mineral by the gut; (2) to reduce the concentrations of antinutrients, such as oxalate, polyphenolics (tannins), or phytate (IP6), which interfere with their absorption; (3) and to increase the mobility of Fe and Zn within the plant as well as the binding capacity/sink strength of seeds.

A recent study reported the absorption of radiolabeled Zn from maize, rice, and barley with low phytic acid mutants in comparison to cultivars with normal seed phytate concentrations. Consistently, reduced phytic acid levels resulted in enhanced Zn absorption (Lonnerdal et al. 2011). In the rice, bioavailability of iron has been targeted by introducing a gene for phytase which resulted in a 130-fold increase in the expression of this enzyme. On the other hand, dietary phytate has been associated with substantial health benefits including strong activity as an anticarcinogenic agent. Several studies reported inhibitory effects of phytate on the growth of different types of tumors. Additional protective effects include the lowering of cholesterol levels (Vucenik and Shamsuddin 2006). Thus, the nutritional benefits of low-phytate grains and seeds have to be weighed against the positive effects of phytate intake (Murgia et al. 2012). The conclusions drawn could well differ between populations depending on the relative importance of micronutrient malnutrition versus cancer incidence.

Transgenic maize plants overexpressing ferritin in combination with a fungal phytase contained up to threefold more bioavailable Fe as determined by the in vitro digestion/Caco-2-system (Drakakaki et al. 2005). Expression of ferritin, an iron-storage protein, in seeds causes a threefold to fourfold increase in iron levels (Goto et al. 2000, 1999; Vasconcelos et al. 2003). Although polishing of rice causes a decrease in mineral levels, ferritin-enhanced rice still has increased iron levels in the transgenic polished rice. Rats fed with a diet containing the transgenic rice demonstrate that the iron in the rice had bioavailability equal to that found in diets containing $FeSO_4$ at equal concentrations (Murray-Kolb et al. 2002).

A potentially promising alternative approach was suggested to indirectly enhance Fe and Zn bioavailability through increasing the concentration of nondigestible carbohydrates, thereby promoting beneficial microbiota that stimulates Fe and Zn absorption by human gut cells (Murgia et al. 2012; Shahzad et al. 2014). The main focus has been on inulin, a polymer consisting of beta-2-1-linked fructose units. Supplemented inulin had enhanced bioavailability of Fe in maize and soybean meals (Yasuda et al. 2006). Variation in concentrations among crop plants and engineering strategies has to be explored more extensively for this prebiotic (Rawat et al. 2013). A combined approach in rice and maize has been developed that involves the expression of iron-storage proteins and fungal phytase (Bashir et al. 2013). This combined approach for mineral biofortification should provide maximal levels of bioavailable iron.

25.7 Conclusions

In conclusion, with the increasing global population and nutritional insecurity mainly in developing nations, research priorities should be redefined and reoriented to a more meaningful approach to solve the problem. Combined approach of genetic engineering, breeding, and agronomic intervention could help figure out above nutritional problem. Future research should be diverted to analyze the interaction of augmented nutrients with other nutrients in vivo. It is also advised that the impact of genetic engineering on the agronomic performance of crops and biotic and abiotic stresses should also be assessed. In addition, the focus should be on studies involving field crop trials and human beings as experimental subjects to analyze the effectiveness of agronomic or genetic biofortification.

References

Abbadi A, Domergue F, Bauer J, Napier JA, Welti R, Zahringer U, Cirpus P, Heinz E (2004) Biosynthesis of very-long-chain polyunsaturated fatty acids in transgenic oilseeds: constraints on their accumulation. Plant Cell 16:2734–2748

Anai T, Koga M, Tanaka H, Kinoshita T, Rahman SM, Takagi Y (2003) Improvement of rice (*Oryza sativa* L.) seed oil quality through introduction of a soybean microsomal omega-3 fatty acid desaturase gene. Plant Cell Rep 21:988–992

Bashir K, Takahashi R, Nakanishi H, Nishizawa NK (2013) The road to micronutrient biofortification of rice: progress and prospects. Front Plant Sci 4:1–7

Bovy A, de Vos R, Kemper M, Schijlen E, Pertejo MA, Muir S, Collins G, Robinson S, Verhoeyen M, Hughes S, Santos BC, van Tunen A (2002) High-flavonol tomatoes resulting from the heterologous expression of the maize transcription factor genes LC and C1. Plant Cell 10:2509–2526

Broun P, Gettner S, Somerville C (1999) Genetic engineering of plant lipids. Annu Rev Nutr 19:197–216

Cahoon EB, Hall SE, Ripp KG, Ganzke TS, Hitz WD, Coughlan SJ (2003) Metabolic redesign of vitamin E biosynthesis in plants for tocotrienol production and increased antioxidant content. Nat Biotechnol 21:1082–1087

Cakmak I, Pfeiffer WH, McClafferty B (2010) Review: biofortification of durum wheat with zinc and iron. Cereal Chem 87:10–20

Choi SW, Friso S (2005) Interactions between folate and aging for carcinogenesis. Clin Chem Lab Med 43:1151–1157

Christou P, Twyman RM (2004) The potential of genetically enhanced plants to address food insecurity. Nutr Res Rev 17:23–42

Cohen LA (2002) A review of animal model studies of tomato carotenoids, lycopene, and cancer chemoprevention. Exp Biol Med 227:864–868

Crozier A, Jaganath IB, Clifford MN (2009) Review article dietary phenolics: chemistry, bioavailability and effects on health. Nat Prod Rep 26:1001–1043

Cunningham FX, Gantt G (1998) Genes and enzymes of carotenoid biosynthesis in plants. Annu Rev Plant Physiol Plant MolBiol 49:557–583

Damude HG, Kinney AJ (2008) Enhancing plant seed oils for human nutrition. Plant Physiol 147:962–968

DellaPenna D (2007) Biofortification of plant-based food: enhancing folate levels by metabolic engineering. Proc Natl Acad Sci U S A 104:3675–3676

DellaPenna D, Pogson BJ (2006) Vitamin synthesis in plants: tocopherols and carotenoids. Annu Rev Plant Biol 57:711–738

Dinkins RD, Reddy MSS, Meurer CA, Yan B, Trick H, Thibaud-Nissen F, Finer JJ, Wayne A, Parrott WA, Collins GB (2001) Increased sulfur amino acids in soybean plants over expressing the maize 15 kDa zein protein. In Vitro Cell Dev Biol Plant 37:742–747

Drakakaki G, Marcel S, Glahn RP, Lund EK, Pariagh S, Fischer R, Christou P, Stoger E (2005) Endosperm-specific co-expression of recombinant soybean ferritin and *Aspergillus* phytase in maize results in significant increases in the levels of bioavailable iron. Plant Mol Biol 59:869–880

Dwivedi SL, Sahrawat KL, Rai KN, Blair MW, Andersson MS, Pfeiffer W (2012) Nutritionally enhanced staple food crops. Plant Breed Rev 36:169–292

Egnin M, Prakash CS (1997) Transgenic sweet potato expressing a synthetic storage protein gene exhibits high level of total protein and essential amino acids. In Vitro Cell Dev Biol Plant 33:52A

Food and Drug Administration (1996) Food standards: amendment of standards of identity for enriched grain products to require addition of folic acid. Final rule, 21 CFR parts 136, 137, 139:8781–8807

Fraser PD, Romer S, Shipton CA, Mills PB, Kiano JW, Misawa N, Drake RG, Schuch W, Bramley PM (2002)

Evaluation of transgenic tomato plants expressing an additional phytoene synthase in a fruit-specific manner. Proc Natl Acad Sci USA 99:1092–1097

Frossard E, Bucher M, Machler F, Mozafar A, Hurrell R (2000) Potential for increasing the content and bioavailability of Fe, Zn and Ca in plants for human nutrition. J Sci Food Agric 80:861–879

Garca-Casal MN, Layrisse M, Pena-Rosas JP, Ramrez J, Leets I, Matus P (2003) Iron absorption from elemental iron-fortified cornflakes in humans. Role of vitamins A and C1-3. Nutr Res 23:451–463

Giovannucci E (2002) Lycopene and prostate cancer risk. Methodological considerations in the epidemiologic literature. Pure Appl Chem 74:1427–1434

Goto F, Yoshihara T, Shigemoto N, Toki S, Takaiwa F (1999) Iron fortification of rice seed by the soybean ferritin gene. Nat Biotechnol 17:282–286

Goto F, Yoshihara T, Saiki H (2000) Iron accumulation and enhanced growth in transgenic lettuce plants expressing the iron-binding protein ferritin. Theor Appl Genet 100:658–664

Graham RD, Welch RM, Saunders DA, Ortiz-Monasterio I, Bouis HE, Bonierbale M, de Haan S, Burgos G, Thiele G, Liria R, Craig A, Meisner CA, Beebe SE, Potts MJ, Kadian M, Hobbs PR, Gupta RK, Twomlow S (2007) Nutritious subsistence food systems. Adv Agron 92:1–74

Grillet L, Mari S, Schmidt W (2014) Iron in seeds–loading pathways and subcellular localization. Front Plant Sci 4:535

Grusak MA, Cakmak I (2005) Methods to improve the crop-delivery of minerals to humans and livestock. In: Broadley MR, White PJ (eds) Plant nutritional genomics. Blackwell, Oxford, pp 265–286

Hartwig EE, Kuo TM, Kenty MM (1997) Seed protein and its relationship to soluble sugars in soybeans. Crop Sci 37:770–773

Heber D, Lu QY (2002) Overview of mechanisms of action of lycopene. Exp Biol Med 227:920–923

Henry EV, Qi Q (2005) Biotechnological production and application of vitamin E: current state and prospects. Appl Microbiol Biotechnol 68:436–444

Hesse H, Kreft O, Maimann S, Zeh M, Hoefgen R (2004) Current understanding of the regulation of methionine biosynthesis in plants. J Exp Bot 55:1799–1808

Hirschi KD (2008) Nutritional improvements in plants: time to bite on biofortified foods. Trends Plant Sci 13:459–514

Institute of Medicine (2000) Carotene and other carotenoids. In: Dietary reference intakes for vitamin C, vitamin E, selenium, and carotenoids. National Academy Press, Washington, DC, pp 325–382

Kinney AJ, Knowlton S (1998) Designer oils: the high oleic acid soybean. In: Roller S, Harlanders S (eds) Genetic modification in the food industry: a strategy for food quality improvement. Blackie Academic and Professional/Thomson Science, London, pp 193–213

Kucuk O, Sarkar FH, Sakr W, Khachik F, Djuric Z, Banerjee M, Pollak MN, Bertram JS, Wood DP (2002) Lycopene in the treatment of prostate cancer. Pure Appl Chem 74:1443–1450

Liu Q, Singh S, Green A (2002) High-oleic and high-stearic cottonseed oils: nutritionally improved cooking oils developed using gene silencing. J Am Coll Nutr 21:205S–211S

Lonnerdal B, Mendoza C, Brown KH, Rutger JN, Raboy V (2011) Zinc absorption from low phytic acid genotypes of maize (*Zea mays* L.), barley (*Hordeum vulgare* L.), and rice (*Oryza sativa* L.) assessed in a suckling rat pup model. J Agric Food Chem 59:4755–4762

Lozoff B (2007) Iron deficiency and child development. Food Nutr Bull 28:S560–S571

Lucock M, Yatesa Z, Glanvillea T, Leemingc R, Simpsona N, Daskalakisd I (2003) A critical role for B-vitamin nutrition in human developmental and evolutionary biology. Nutr Res 23:1463–1475

Maret W (2013) Zinc biochemistry: from a single zinc enzyme to a key element of life. Adv Nutr 4:82–91

McKillop DJ, McNulty H, Scott JM, McPartlin JM, Strain JJ, Bradbury I, Girvan J, Hoey L, McCreedy R, Alexander J, Patterson BK, Hannon-Fletcher M, Pentieva K (2006) The rate of intestinal absorption of natural food folates is not related to the extent of folate conjugations. Am J Clin Nutr 84:167–173

McNulty H, Pentieva K (2004) Folate bioavailability. Proc Nutr Soc 63:529–536

Mehta RA, Cassol T, Li N, Ali N, Handa AK, Mattoo AK (2002) Engineered polyamine accumulation in tomato enhances phytonutrient content, juice quality, and vine life. Nat Biotechnol 20:613–618

Mensink RP, Zock PL, Kester DM, Katan MB (2003) Effects of dietary fatty acids and carbohydrates on the ratio of serum total to HDL cholesterol and on serum lipids and apolipoproteins: a meta-analysis of 60 controlled trials. Am J Clin Nutr 77:1146–1155

Miller EC, Hadley CW, Schwartz SJ, Erdman JW, Boileau TWM, Clinton SK (2002) Lycopene, tomato products, and prostate cancer prevention. Have we established causality? Pure Appl Chem 74:1435–1441

Murgia I, Arosio P, Tarantino D, Soave C (2012) Biofortification for combating hidden hunger for iron. Trends Plant Sci 17:47–55

Murray-Kolb LE, Takaiwa F, Goto F, Yoshihara T, Theil EC, Beard JL (2002) Transgenic rice is a source of iron for iron-depleted rats. J Nutr 132:957–960

Newell-McGloughlin M (2008) Nutritionally improved agricultural crops. Plant Physiol 147:939–953

O'Quinn PR, Nelssen JL, Goodband RD, Knabe DA, Woodworth JC, Tokach MD, Lohrmann TT (2000) Nutritional value of a genetically improved high-lysine, high-oil corn for young pigs. J Anim Sci 78:2144–2149

Olsen LI, Palmgren MG (2014) Many rivers to cross: the journey of zinc from soil to seed. Front Plant Sci 5:30

Rapp W (2002) Development of soybeans with improved amino acid composition. In: 93rd AOCS annual meeting and expo. Montreal/Champaign, Journal of the American Oil Chemists' Society, pp 79–86

Rawat N, Neelam K, Tiwari VK, Dhaliwal HS (2013) Biofortification of cereals to overcome hidden hunger. Plant Breed 132:437–445

Rebeille F, Ravanel S, Jabrin S, Douce R, Storozhenko S, Straeten DVD (2006) Folates in plants: biosynthesis, distribution, an enhancement. Physiol Plant 126:330–342

Reddy AS, Thomas TL (1996) Expression of a cyanobacterial DELTA 6-desaturase gene results in gamma-linolenic acid production in transgenic plants. Nat Biotechnol 14:639–642

Richelle M, Bortlik K, Liardet S, Hager C, Lambelet P, Baur M, Applegate LA, Offord EA (2002) A food-based formulation provides lycopene with the same bioavailability to humans as that from tomato paste. J Nutr 132:404–408

Rosemary AS, Lisa AH (2013) Nutrient Intake values for folate during pregnancy and lactation vary widely around the world. Nutrients 5:3920–3947

Rude RK, Gruber HE (2004) Magnesium deficiency and osteoporosis: animal and human observations. J Nutr Biochem 15:710–716

Sachs JD (2005) Millennium project investing in development: a practical plan to achieve the millennium development goals. First published by Earthscan in the UK and USA

Sautter C, Poletti S, Zhang P, Gruissem W (2006) Biofortification of essential nutritional compounds and trace elements in rice and cassava. Proc Nutr Soc 65:153–159

Scott J, Rebeill F, Fletcher J (2000) Folic acid and folates: the feasibility for nutritional enhancement in plant foods. J Sci Food Agric 80:795–824

Shahzad Z, Rouached H, Rakha A (2014) Combating mineral malnutrition through iron and zinc biofortification of cereals. Compr Rev Food Sci Food Saf 13:329–346

Shehanaz A (2013) The antioxidant effect of certain fruits. Int J Pharm Sci Res 5(12):265–268

Shintani D, DellaPenna D (1998) Elevating the vitamin E content of plants through metabolic engineering. Science 282:2098–2100

Shrestha AK, Arcot J, Paterson JL (2003) Edible coating materials-their properties and use in the fortification of rice with folic acid. Food Res Int 36:921–928

Sofi PA, Wani SA, Rather AG, Wani SH (2009) Quality protein maize (QPM): genetic manipulation for the nutritional fortification of maize. J Plant Breed Crop Sci 6:244–253

Stanger O (2004) The potential role of homocysteine in percutaneous coronary interventions (PCI): review of current evidence and plausibility of action. Cell Mol Biol 50:953–988

Terry P, Jain M, Miller AB, Howe GR, Rohan TE (2002a) Dietary carotenoids and risk of breast cancer. Am J Clin Nutr 76:883–888

Terry P, Jain M, Miller AB, Howe GR, Rohan TE (2002b) Dietary carotenoid intake and colorectal cancer risk. Nutr Cancer 42:167–172

Thacher TD, Fischer PR, Strand MA, Pettifor JM (2006) Nutritional rickets around the world: causes and future directions. Ann Trop Paediatr 26:1–16

Tucker G (2003) Nutritional enhancement of plants. Curr Opin Biotechnol 14:221–225

Ulrey CL, Liu L, Andrews LG, Tollefsbol TO (2005) The impact of metabolism on DNA methylation. Hum Mol Genet 14:R139–R147

Vasconcelos M, Datta K, Oliva N, Khalekuzzaman M, Torrizo L, Krishnan S, Oliveira M, Goto F, Datta SK (2003) Enhanced iron and zinc accumulation in transgenic rice with the ferritin gene. Plant Sci 164:371–378

Vucenik I, Shamsuddin AM (2006) Protection against cancer by dietary IP6 and inositol. Nutr Cancer 55:109–125

Wallis JG, Watts JL, Browse J (2002) Polyunsaturated fatty acid synthesis: what will they think of next? Trends Biochem Sci 27:467–473

Wegmuller R, Zimmermann MB, Hurrell RF (2003) Dual fortification of salt with iodine and encapsulated iron compounds: stability and acceptability testing in Morocco and Cote d'Ivoire. J Food Sci 68:2129–2135

Welch RM, Graham RD (2002) Breeding crops for enhanced micronutrient content. Plant Soil 245:205–214

Welch RM, Graham RD (2004) Breeding for micronutrients in staple food crops from a human nutrition perspective. J Exp Bot 55:353–364

Welch RM, Graham RD (2005) Agriculture: the real nexus for enhancing bioavailable micronutrients in food crops. J Trace Elem Med Biol 18:299–307

White PJ, Broadley MR (2005) Biofortifying crops with essential mineral elements. Trends Plant Sci 10:586–593

Witthoft CM, Arkbage K, Johansson M, Lundin E, Berglund G, Zhang JX, Lennernas H, Dainty JR (2006) Folate absorption from folate-fortified and processed foods using a human ileostomy model. Br J Nutr 95:181–187

Yang XE, Chen WR, Feng Y (2007) Improving human micronutrient nutrition through biofortification in the soil–plant system: China as a case study. Environ Geochem Health 29:413–428

Yasuda K, Roneker KR, Miller DD, Welch RM, Lei XG (2006) Supplemental dietary inulin affects the bioavailability of iron in corn and soybean meal to young pigs. J Nutr 136:3033–3038

Ye X, Al-Babili S, Kloti A, Zhang J, Lucca P, Beyer P, Potroykus I (2000) Engineering the provitamin A (beta-carotene) biosynthetic pathway into (carotenoid-free) rice endosperm. Science 287:303–305

Zhu C, Naqvi S, Gomez-Galera S, Pelacho AM, Capell T, Christou P (2007) Transgenic strategies for the nutritional enhancement of plants. Trends Plant Sci 12:548–555

Zimmermann R, Qaim M (2004) Potential health benefits of golden rice: a Philippine case study. Food Policy 29:147–168

High Beta-Carotene Rice in Asia: Techniques and Implications

26

S.K. Dash, K.K. Sahoo, K. Chattopadhyay, L. Behera, S.K. Pradhan, K.A. Molla, and S.D. Mohapatra

Abstract

Vitamin A deficiency is one of the major problems not only in Asia but also in many countries worldwide. It seriously affects the reproductive and immune systems and mostly leads to blindness in children, maternal mortality, and ultimately death. Its deficiency is widespread among the poor whose dietary source is based mostly on rice or other carbohydrate-rich food, because rice is devoid of vitamin A. So, the concept was perceived to provide the supplementary vitamin A in their staple food, rice. Initially, vitamin-A overproducing Golden Rice developed by inserting only two genes, a plant phytoene synthase (*PSY*) gene from daffodil (*Narcissus pseudonarcissus*) and the carotene desaturase (*CRTI*) gene from the bacterium *Pantoea ananatis*. The vitamin A derived initially in the first phase (Golden Rice 1, GR1) was not sufficient. So, the second generation Golden Rice (GR2) was developed by overexpressing *Zea mays PSY* gene regulated under maize polyubiquitin promoter in place of daffodil *PSY*. The GR2 brought a remarkable improvement by accumulating high quantity of carotenoids (up to 37 μg/g) and β-carotene (31 μg/g), which is enough for vitamin A enrichment. Bioavailability studies showed that around 60 g of rice could be sufficient for 50–60 %

S.K. Dash and K.K. Sahoo contributed equally.

S.K. Dash (✉) • K. Chattopadhyay • L. Behera
S.K. Pradhan • K.A. Molla • S.D. Mohapatra
ICAR-National Rice Research Institute, Cuttack, Odisha 753 006, India
e-mail: skdash139@gmail.com; krishnenducrri@gmail.com; lambodarjamujhadi@gmail.com; pradhancrri@gmail.com; kutubjoy@gmail.com; sdmento73@gmail.com

K.K. Sahoo
Department of Botany, Ravenshaw University, Cuttack, Odisha 753 003, India
e-mail: khirod555@gmail.com

U. Singh et al. (eds.), *Biofortification of Food Crops*, DOI 10.1007/978-81-322-2716-8_26

daily recommended quantity of vitamin A in children which is enough to prevent death and blindness. Effort is under way for introgression of GR2 into the popular *indica* varieties with confined field trials in some countries. However, there are pro- and anti-lobby for this rice because it is a GM crop; hence, it has not reached the farmers and common man yet. In this chapter, an attempt was made to discuss the details of the techniques and implications of high beta-carotene in rice on Asian context.

Keywords
β-carotene • Carotenoid • Golden Rice • GMO • Vitamin A deficiency

26.1 Introduction

Rice is the most important staple food for about 2.4 billion of people in Asia and 19.26 % of the daily protein intake is met from rice (IRRI 2015b). Currently, rice is grown in all six continents inhabited by humans. Rice is grown in about 154 million ha in 114 countries, and world production is about 740 million tons. About 89 % of the world's total rice area is in Asia, 6 % in Africa, 4 % in South and Central America and the Caribbean, 1 % in the USA, and less than 1 % in Europe and Oceania. The rice lands in Asia produce 90 % of the world's rice (Islam et al. 2012). According to IFPRI Global Hunger Index (GHI) 2012, India lags behind in improving its GHI score despite strong economic growth (IFPRI 2012). Rice has a surprising ability to cope up with diverse agroecological situations. This exceptional adaptability is evident by its cultivation in several agroecological zones in diverse soil type under varying climatic situations, although a majority of the rice is grown in warm/cool humid subtropics, warm humid tropics, and warm subhumid tropics. Cutting across the agroecological zones, rice land ecosystems have been categorized into four distinct types occurred in India: irrigated, rainfed lowland, upland, and flood-prone rice.

Rice is mainly a complex carbohydrate used as a source of energy for human body. Moreover, it is a good source of dietary protein, phosphorus, and iron. In comparison to food legumes, rice protein is considered one of the highest quality proteins having high content of methionine with all eight essential amino acids, indispensable building blocks for strong muscles. It is also a good source of other essential nutrients – thiamin, riboflavin, niacin, phosphorus, iron, and potassium. It also contains some amount of calcium. Rice is healthy because of lack of fat, cholesterol, and sodium. Rice contains vitamin B in small quantities, particularly in brown rice (Sarao 2015). However, it does not contain any vitamin A. Biofortified rice offers a sustainable solution to vitamin A deficiency.

26.2 Role of Vitamin A

Vitamin A is a group of unsaturated organic compounds that comprises retinol, retinal, and retinoic acid along with some provitamin A and carotenoids having profuse nutritional values. The vitamin A found in the food could be classified into two categories, viz., preformed vitamin A and provitamin A. Preformed vitamin A is available in animal foods, viz., dairy products (milk, yogurt, cheese, butter, ghee, etc.), meat, fish, and poultry. Provitamin A is found in plant products, viz., fruits and vegetables. Vitamin A is also found in dietary supplements, normally in the form of retinyl acetate or retinyl palmitate, beta-carotene, or a combination of preformed and provitamin A. The most common type of provitamin A is beta-carotene (Evert 2013). Vitamin A is essential for several human body functions including eyesight, reproduction, growth, and immunity. It helps several vital functions such as maintenance of normal visual function, regulation of differentiation of epithelial tissues, embryonic development, fight against diseases (Underwood and Arthur 1996; Ross et al. 2011), cell growth, and skin health including gene transcription as well as protein formation. Vitamin A deficient people get sick more often and there is higher risk of mortality out of infections. Vitamin A deficiency can also cause night blindness and xerophthalmia (Sommer 1982) and is one of the major reasons of blindness in children.

There is growing consciousness regarding vitamin A malnutrition in children as well as in adults. The artificially synthesized vitamin A supplements have been distributed periodically to malnutritious populations of developing countries to prevent vitamin A deficiency (WHO 1997; West et al. 1991; Ribaya-Mercado et al. 2004). This has been considered to be an efficient and more or less safe strategy. However, supplementation programs were difficult to continue for a long period because of high distribution costs. In this context, vitamin A deficiency remains a serious worry throughout the globe. Rice-eating people are especially vulnerable to vitamin A deficiency because rice is devoid of vitamin A or its precursors. Further, consumption of vitamin A-containing diet is mostly low in predominantly rice eaters in the low-income range. In this context, provitamin A carotenoid-rich foods could be considered as a substitute to animal products as the source of vitamin A, because the human body makes vitamin A from these precursor compounds. Results from few recent studies are not very encouraging in relation to higher consumption of carotenoid-rich foods toward vitamin A (ASN 2009). It has been estimated that 40 % of children under the age of five, in the developing world are immunocompromised due to vitamin A deficiency (VAD); as a result they suffered with many common childhood infections which ultimately lead to death. VAD predominantly affects Southeast Asia and Africa. The clinical and population studies suggested that 400 million rice-eating poor people are having impaired vision, epithelial integrity, hemopoiesis, and weak and reduced skeletal growth (Reifen 2002).

Vitamin A deficiency is observed by and large among the poor whose diets are based mainly on rice and other carbohydrate-rich or micronutrient-poor calorie sources. Due to low dietary consumption of iron, vitamin A, iodine, and zinc, there are severe micronutrient deficiencies in the human beings throughout the world. Rice is devoid of β-carotene (provitamin A), which is subsequently converted into vitamin A by human body. Dependence on rice as the major source of calorie consequently culminates in vitamin A deficiency. This affects most seriously to small children and pregnant women. An estimated 250 million kids are vitamin A deficient and it is likely that in vitamin A deficient areas, a major proportion of pregnant women is also found to suffer from vitamin A inadequacy (WHO 2013). VAD is a significant community health concern, with 190 million children under five specifically in Africa and South East Asia. In South East Asia, nearly 4–50 % pre-school going children were afflicted with severe VAD, 85 % belonging to India with xerophthalmia (WHO 2009). Similarly, there is high rate of loss of life due to malnourishment and severe VAD, in neonatal infants and children below 5 years, in India and Bangladesh, to the tune of 33 % of global mortality rate (Akhtar et al 2013). It is matter of concern that momentous increment of VAD has been reported (from 5.9 to 30.3 % during 2001 to 2011) in Indian women (PPDD 2011). Provision of vitamin A to those children could prevent about a third of all kid deaths, which lead to the tune of 2,700,000 children that could be saved from dying (Golden Rice Project 2014) and 670,000 deaths of children below the age of 5 every year (Black et al. 2008). The World Health Organization (WHO) reported an estimation of 250,000–500,000 children with inadequate vitamin A are losing their eyesight each year and an alarming 50 % mortality in a period of 12 months following blindness (WHO 2013). With sufficient supplement of vitamin A, there is chance of reduction of mortality rate to the tune of 30 % in young children due to infections and 40 % in women during or shortly after pregnancy.

One of the approaches for the reduction of vitamin A deficiency could be the consumption of sufficient foods that are inherently high in vitamin A or a form of vitamin A, i.e., beta-carotene, or addition of micronutrients in them, or having supplements. This is available to the body either directly through absorption, from the foods originated from animal source (mainly liver, cheese, and eggs), or indirectly in the form of carotenoids, which are mostly available in fruits and vegetables and subsequently converted to vitamin A. However, it is stocked in the liver inside the body; therefore, it does not require regular consumption. Beta-carotene is an antioxidant similar to other carotenoids. Antioxidants are the molecules that inhibit the oxidation of other molecules. Oxidation is such a reaction which can produce free radicals. In turn, these radicals can cause damage or death of the cells. Eventually, the damage caused by free radicals can cause several chronic illnesses and increases the risk of developing cancer and heart disease. Antioxidants remove free radicals and other intermediates and inhibit other oxidation reactions. They do this by being oxidized themselves (Hamid et al. 2010).

The best option to get vitamin A is to have a well-balanced diet including fruits, vegetables, and low-fat dairy products as well as animal protein. The persons usually having balanced food need to get adequate amount of vitamin A and carotene along with other supplements for their good health. It is recommended to take daily 2,310 IU (0.7 mg per day) of vitamin A for adult women and 3,000 IU (0.9 mg) for adult men (ACS 2012). Rice-biofortified provitamin A could substantially reduce the problems described above. Usually, rice does not have vitamin A or its precursor, beta-carotene. Thousands of rice varieties have been tested for this trait without success. However, a team of scientists from Europe has succeeded in changing this. Through the introgression of two genes from daffodil (*Narcissus pseudonarcissus*) and one gene from a soil bacterium *Pantoea ananatis* (earlier known as *Erwinia uredovora*), Dr. Ingo Potrykus of the Swiss Federal Institute of Technology and Dr. Peter Beyer of the University of Freiburg in Germany with their endless effort have successfully engineered a beta-carotene pathway into Taipei 309 (a *japonica* rice cultivar) and they revealed their golden research product as "Golden Rice" in 1999 (Quijano 2001).

26.3 The Concept of Golden Rice

26.3.1 Carotenoids and Apocarotenoids

Carotenoids are reasonably stable hydrophobic compounds usually of yellow, orange or red in color which accumulate in diverse group of photosynthetically active soft green tissues of plants. The *de novo* synthesis of carotenoids is found in differentiated plastids of various parts of the plant including leaf, root, flower, fruit and seed. It is mostly found in green photosynthetic plastids (chloroplasts) and colored plastids (chromoplasts) and significantly less in dark-grown precursors of the chloroplast (etioplasts), lipid-storing plastids (elaioplasts), starch-containing plastids (amyloplasts), and colorless plastids (leucoplasts) (Cazzonelli and Pogson 2010). Carotenoids are important for photoprotection, photosynthesis, and the manufacture of carotenoid-derived phytohormones (strigolactone and ABA). These are natural isoprenoid pigments that provide distinctive orange, yellow, and some reddish colors as well as several aromas to leaves, fruits, vegetables and flowers in plants. The derivatives of carotenoid play role in relay signaling in plant developmental processes and biotic and abiotic stress stimuli and also mediate response to beneficial and non-beneficial organisms. Since animals are devoid of mechanism to produce carotenoids, they need to consume plant products in daily diets to meet daily health requirements. Animal products are also used as an alternative source of vitamin A in the form of provitamin A. Vitamin A is an essential micronutrient for our body and its importance has already been discussed earlier.

The dietary carotenoids in animals generally break down to produce the precursors for vitamin A biosynthesis and are important for different metabolic processes including antioxidant property, immune-stimulants, yolk nourishment to embryos, photoprotection, eyesight tuning, and macular degeneration (Johnson 2002; Krinsky and Johnson 2005). The nutraceutical industry artificially produces five main carotenoids in mass scale such as lycopene, β-carotene, canthaxanthin, zeaxanthin, and astaxanthin for different food products, health products, cosmetics, and other vitamin supplements (Del Campo et al. 2007; Jackson et al. 2008). Apocarotenoids derived from carotenoid by oxidative cleavage mediated by oxygenases, are also additives used in the food industry. A few species of apocarotenoid such as bixin (annatto), dicarboxylic monomethyl ester apocarotenoid, are usually processed from the *Bixa orellana* plant. The accumulation and sequestration of carotenoid occur inside chloroplasts as well as chromoplasts. All carotenoids are associated with chlorophyll-binding proteins for its sequestration in the chloroplast, whereas a unique mechanism to sequester carotenoids within specific lipoprotein structures has been developed inside chromoplasts (Vishnevetsky et al. 1999). However, this mechanism has been found to be fully active only in leaves but not in the grain of rice plants. Although rice plants have potential to synthesize the β-carotene, very less quantity is produced in the grains.

The fundamental principle for the development of Golden Rice technology was based upon the accumulation of high levels of carotenoids by improvement in its sequestration mechanism, which includes crystallization, oil deposition, membrane proliferation, and protein-lipid sequestration. Moreover, lipid accumulation mechanism was also found to be involved in carotenoid formation by acting as a lipophilic sink (Rabbani et al. 1998). The lipid content of non-carotenogenic starchy rice endosperm is very low and thus the endosperm apparently does not have any lipoprotein structure for carotenoid deposition. So, it was also interesting to see whether Golden rice grain developed necessary precursors for carotenoid biosynthesis as well as for its sequestration.

26.3.2 Carotenoids Biosynthetic Pathway: First Step

The carotenoid biosynthetic pathway in general initiates with condensation of geranylgeranyl pyrophosphate (GGPP) which is the precursor from the upstream methylerythritol phosphate (MEP) pathway to produce the end product cis-phytoene. The GGPP is synthesized by the enzyme GGPP synthase (GGPPS). Upstream to the formation of GGPP, a chain of reactions (Maudinas et al. 1977) leads to the production of isopentenyl diphosphate (IPP) and its isomer, dimethylallyl diphosphate (DMAPP), in the presence of three molecules of isoprenoids. It has been observed that wild-type rice has the biosynthetic capability to produce geranylgeranyl diphosphate (GGPP) through the intermediate product DXP (1-deoxy-D-xylulose-5-phosphate) due to an initial decarboxylation of pyruvate and condensation with glyceraldehyde-3-phosphate (G3P) by the catalytic activity of 1-deoxy-D-xylulose-5-phosphate synthase (DXS). By the early 1990s, Professor Peter Beyer and Ingo Potrykus together put their efforts significantly to overproduce carotene in rice grains. Finally, it was a breakthrough which showed that introgression of only two transgenes, plant phytoene synthase (*PSY*) and bacterial phytoene desaturase (*CRTI*), β-carotene accrues in the rice grains, bringing Golden rice into a reality. (Ye et al. 2000). The first transgene was isolated from daffodil plant encoding a plant *PSY* that utilizes the endogenously synthesized GGPP to produce phytoene. The phytoene is described as a colorless carotene having a triene chromophore (Burkhardt et al. 1997). The second transgene was isolated from soil bacterium (*Erwinia uredovora*) encrypts a bacterial carotene desaturase (CRTI) which conjugates by introducing four double bonds in one step per se. The *CRTI* overexpression leads to yellow or golden color in rice endosperm by transcriptionally activating carotenoid biosynthetic pathway genes through a feedback regulatory mechanism.

26.3.3 Carotenoids Biosynthetic Pathway: Second Step

Rice plant has the potential to biosynthetically manufacture GGPP (geranylgeranyl-diphosphate) through two steps of initial decarboxylation of pyruvate and condensation with glyceraldehyde-3-phosphate due to catalytic action of DXS (1-deoxy-D-xylulose-5-phosphate synthase). The isopreniods like isopentenyl-diphosphate (IPP) and dimethylallyl-diphosphate (DMAPP) are formed through chain reactions from DXP. The enzyme GGPP synthase (GGPPS) converts these isoprinoids to GGPP. The phytoene synthase (PSY), phytoene-desaturase (PDS), z-carotenedesaturase (ZDS) and carotene cis-trans-isomerase (CRTISO) are enzymes which are essentially needed for β-carotene formation in plants. The main function of CRTISO is to convert the specific cis-carotene intermediates into all-trans form of lycopene.

Scientists of Syngenta Company were able to enhance the beta-carotene content in the "Golden Rice" by substituting the daffodil *PSY* with homologous gene from maize plant (Paine et al. 2005). It was important to know whether any of these enzymes are active in the normal wild rice plants or not. PSY is considered to be an important enzyme in carotenoid biosynthesis. Various studies have been performed by engineering the pathway in numerous plants which modify the carotenoid content as desired (Farré et al. 2011; Giuliano et al. 2008). The colored carotenoids are accrued in endosperm of maize seed which is nutritionally enriched. The PSY is controlled by three other paralogous genes, whose expression was evaluated by insertion of *PSY1* gene-specific promoter (Li et al. 2009; Palaisa et al. 2003); *PSY1* expression is anomalously induced in endosperm and thus conditions the characteristic yellow endosperm color in maize. The expression of *PSY1* is found in leaves along with *PSY2*, whereas *PSY3* expression is limited to roots and is induced by different stress conditions (Gallagher et al. 2004; Li et al. 2008a, b). The majority maize *PSY2* and *PSY3*, rice and *Arabidopsis PSYs*, are found to be localized in plastoglobuli. The maize *PSY1* differs from each other only by 1–2 amino acids at positions of 168 and 257. It was necessary to engineer the plant as the seeds of rice plants lack expression of endosperm-specific *PSY* gene. Finally, the bioavailable carotenoids were introduced into the rice seed endosperm to make the carotenoid biosynthetic pathway functional by transforming plants with endosperm-expressed maize *PSY1* and *CRTI* (Ye et al. 2000).

Synthesis of lycopene by *PSY* and *CRTI* genes in transgenic plants provides the substrate for the downstream reaction. Low levels of mRNA transcripts were observed in all the downstream genes of lycopene biosynthetic pathway, which play important role in xanthophyll formation (Schaub et al. 2005). Among all these genes, the expression of *PSY* was found to be the lowest. The mRNA expression for the presence of the carotenoid biosynthetic enzymes encoding phytoene desaturase, ζ-carotene desaturase, carotene cis-trans isomerase, β-lycopene cyclase, and β-carotene hydroxylase (except *PSY* mRNA) in wild-type rice endosperm is still in confliction. The yellow or golden colour endosperm phenotype in Golden Rice plant was predicted because of upregulation of above mentioned downstream pathway carotene desaturase (*CRTI*) rice genes in response to the transgenes. Further, it was confirmed that the formation of β-carotene and xanthophyll in Golden Rice highly depends on constitutive expression of intrinsic rice genes such as carotene cis-trans isomerase, α-/β-lycopene cyclase, and β-carotene hydroxylase. Only PSY needs to be supplemented due to low expression of *PSY* gene, and the need for the *CRTI* transgene in Golden Rice is presumably due to insufficient activity of the phytoene desaturase and/or zeta-carotene desaturase enzyme in endosperm (Schaub et al. 2005). It has been established that CRTI is not rate limiting in contrast to PSY, so it has potential to desaturate large amounts of phytoene which lead to the enhancement of the β-carotene accumulation (Paine et al. 2005). In rice endosperm, rather than CRTI, the tissue-specific expression of the PDS/ZDS system occurs leading to colored carotenoids, showing that the rice endosperm provides the complex requirements for the activity of the plant desaturases.

The carotenoid biosynthetic pathway genes in the endosperm tissue are transcriptionally activated due to the availability of CRTI by a feedback regulatory mechanism. Upregulation of *CRTI* along with *PDS* and *ZDS* produces together a tetra-cis form of lycopene known as prolcopene.

It is subsequently isomerized to the converted tetra-cis form of lycopene to trans-lycopene by lycopene cis-trans isomerase. However, expression of only *CRTI* results in the exclusive formation of all-trans form of lycopene in the case of bacteria. Initially, Peter Bramley showed the advantage of genetically modified tomatoes using a single phytoene desaturase gene (bacterial *CRTI*), rather than introducing multiple carotene desaturases that are normally used by higher plants (Romer et al. 2000). The combined expression of *PSY* and *CRTI* leads to the formation of high lycopene content in genetically modified tomato fruit, a red compound. So far, either lycopene or red color was not observed in any rice transformants. The constitutive expression of *CRTI* transgene in rice plants was used to develop the golden color in rice endosperms (Golden Rice Project 2013). However, provitamin A (α- and β-carotene) is found together with variable amounts of xanthophylls (oxygenated carotenoids), such as zeaxanthin and lutein. So, the endosperm of Golden Rice seems yellow color due to the accumulation of β-carotene (provitamin A) and xanthophylls. The reason why the Golden Rice is golden (yellow) instead of red is due to the activity of intrinsic rice cyclases which has been well described (Schaub et al. 2005). All the important steps involved in carotenoid biosynthesis in relation to Golden Rice (GR) production have been well explained (Fig. 26.1). It was revealed from the high-accumulation carotenoid in rice grain's endosperm that the biosynthetic pathway extends beyond the end point due to the enzymatic action of the two transgenes. The extensions could be due to the activation of other endogenous pathway genes.

26.3.4 Carotenoids Biosynthetic Pathway: Third Step

The enzymes such as lycopene cyclases (LCYs) and α- and β-carotene hydroxylases (HYDs) are produced in wild-type rice endosperm in the subsequent downstream pathway, while PSY and one or both the plant phytoene desaturase (PDS) and ζ-carotene desaturase (ZDS) as well as the ζ-carotene cis-trans isomerase and carotene cis-trans isomerase are not active (Isaacson et al. 2002; Park et al. 2002; Yu et al. 2011; Chen et al. 2010). The α-carotene and β-carotene are produced from all trans-lycopene in two separate pathways due to the activity of α-LYC, β-LYC, and β-LYC enzyme, respectively. Finally, the lutein and zeaxanthin are produced as end product in two separate pathways from α-carotene and β-carotene due to the enzyme action of α- and β-carotene hydroxylases (α- and β-HYD) and β-carotene hydroxylases (β-HYD), respectively (Fig. 26.1). The LYC activity depends on the expression pattern of the respective rice genes downstream to carotenoid biosynthetic pathway in the endosperm for occasional formation of xanthophylls which are catalyzed by the several classes of β-carotene hydroxylases (Tian and DellaPenna 2004). It has been found that dihydroxy-xanthophyll lutein and zeaxanthin possess important roles in the adaptation of plants to highlight stress and maintain photosystem structure.

26.3.5 Golden Rice: First Generation (GR1)

In the first generation Golden Rice (GR1), *Japonica* rice (round grain) was utilized for the transfer of traits followed by *Indica* rice (long grain) varieties (Hoa et al. 2003) with the help of hybridization for the targeted traits into a number of genotypes using breeding methods. GR1 was developed by targeting to overexpress the production of rate-limiting enzymes of the multistep carotene biosynthetic pathways, which actually control the flux for the end products. In GR1, the phytoene synthase (*PSY*) gene from daffodil plant and the carotene desaturase (*CRT1*) gene from the soil bacterium *Pantoea ananatis* (earlier designated as *Erwinia uredovora*) were incorporated to increase the production of carotenoids constitutively in plants using endosperm specific glutelin (Gt1) and constitutive CaMV35S promoter, respectively (Fig. 26.2). Taking this as a proof of concept, many researchers tried directly to develop ways for the enhanced production and collection of carotenoids in the endosperm or seed for the future. Carotenogenic pathway was introduced in various

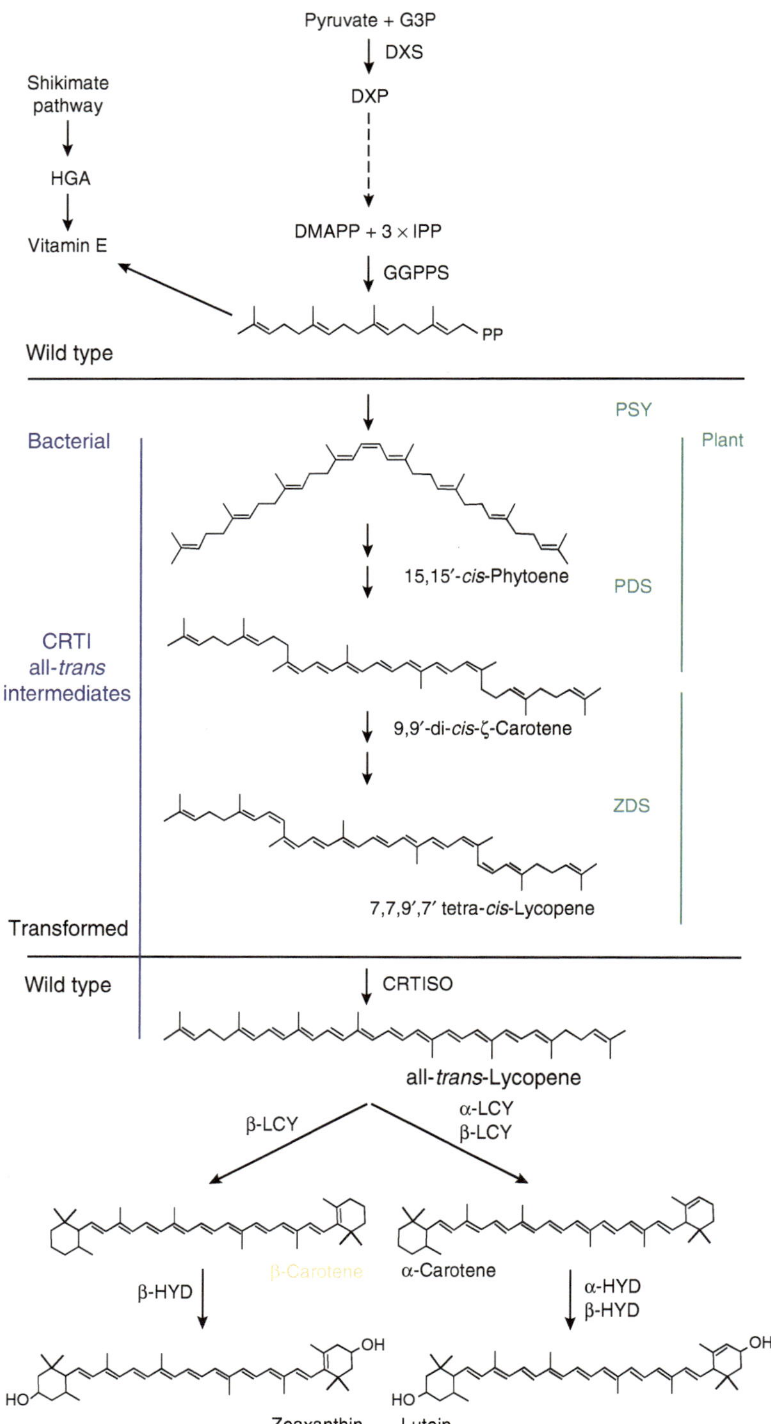

Fig. 26.1 The pathway showing all the important steps involved in carotenoid biosynthesis in relation to Golden Rice (GR) production. Rice plant has the potential to biosynthetically manufacture GGPP (geranylgeranyl diphosphate) through two steps of initial decarboxylation of pyruvate and condensation with glyceraldehyde-3-phosphate due to catalytic action of DXS (1-deoxy-D-xylulose-5-phosphate synthase). The isoprenoids like isopentenyl diphosphate (IPP) and dimethylallyl diphosphate (DMAPP) are formed through a chain reaction from DXP. The enzyme GGPP synthase (GGPPS) converts these isoprenoids to GGPP. The phytoene synthase (PSY), phytoene desaturase (PDS), z-carotene desaturase (ZDS), and carotene cis-trans isomerase (CRTISO) are enzymes which are essentially needed for β-carotene formation in plants. The main function of CRTISO is to convert the specific cis-carotene intermediates into all-trans form of lycopene (Adapted from Al-Babili and Beyer 2005 with permission)

popular *indica* rice cultivars like BR29 (Bangladesh), IR64 (Asian countries), Nang Hong Cho Dao and Mot Bui (Vietnam), and Immyeobaw (Myanmar) using *PSY*, *CRT1* and lycopene cyclase (*LCY*) genes to drive β-carotene accumulation in rice seeds by Prof. Swapan Datta's group at IRRI, Philippine (Datta et al. 2003). The limited quantity of provitamin A

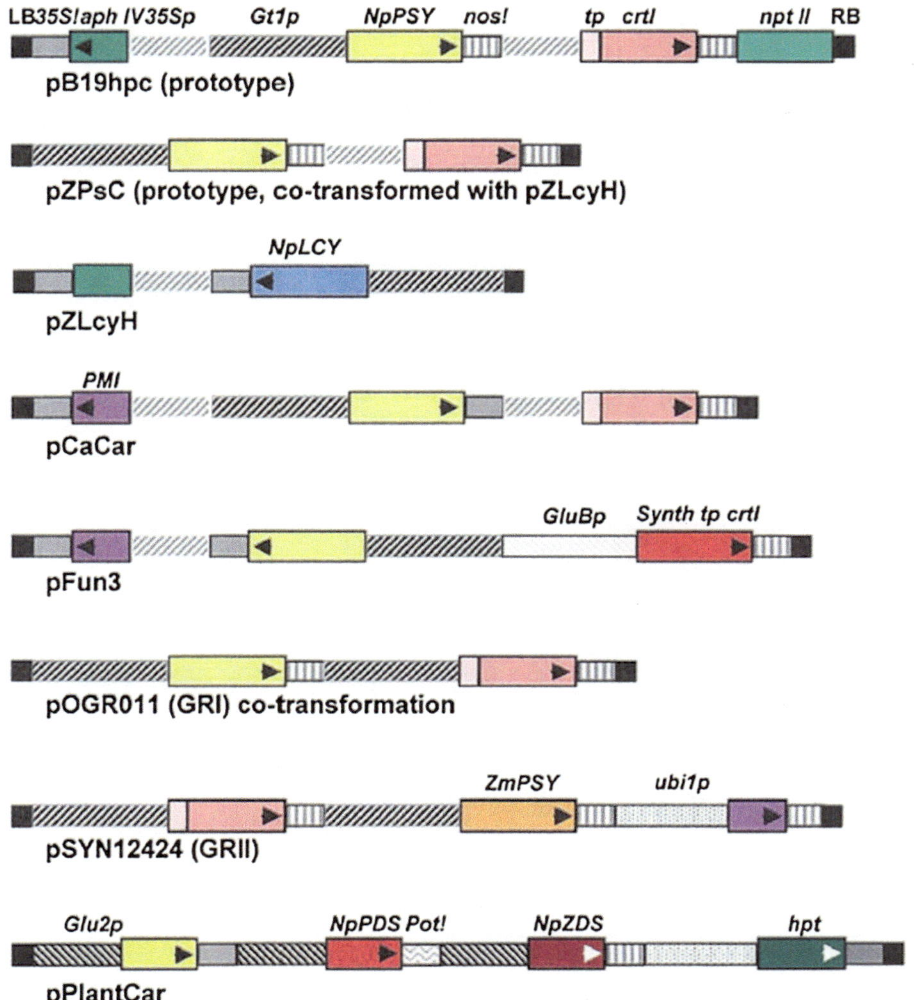

Fig. 26.2 Diagrammatic representation of the DNA constructs used in transformation of Golden Rice (GR) experiments. The gray color represents all the coding sequences regulatory gene (Adapted from Al-Babili and Beyer 2005 with permission) (For detail construct designing etc., please see Al-Babili and Beyer 2005)

(1.6 μg/g) in GR1 seed would not have been sufficient to meet the daily requirement of provitamin A of the target population in the absence of a more diverse diet. The consumption of diverse diets largely varies from place to place and according to daily caloric intake. The level of carotenoids obtained was increased in improved GR1 to an average of 6 μg/g, which is approximately four times greater than the prototype, in field condition. This was probably due to the expression of both the genes (*CRT1* and *PSY*) under the control of endosperm-specific promoter (Gt1) instead of constitutive 35S promoter or possibly because of the availability of high numbers of transgenic events (Fig. 26.4; Mayer 2007; Golden Rice Project 2013).

26.3.6 Golden Rice: Second Generation (GR2)

To meet the high demand of carotenoids and β-carotene, second generation Golden Rice (GR2) was introduced. In 2005, scientists of Syngenta developed a variety of Golden Rice known as "Golden Rice 2", where the new *PSY*

RB Glu tpSSU:CrtI nos Glu Psy nos Ubi1 Pmi nos LB

Fig. 26.3 Gene construct used to develop second generation Golden Rice. (For detail construct designing etc., please see Golden Rice Project 2013) (Reproduced from Golden Rice Project 2013 with due permission)

Fig. 26.4 The yellow or golden color appearance of two different generations of Golden Rice is shown here compared to wild-type rice, as first generation (GR 1) and second generation (GR2) Golden Rice contain very high quantity of β-carotene (Golden Rice Project 2013) (Reproduced from Golden Rice Project 2013 with due permission)

genes from different sources (*Zea mays*) are regulated through maize polyubiquitin promoter and *CRT1* from original Golden Rice (Fig. 26.3; Paine et al. 2005; Welsch et al. 2010). Moreover, a phosphomannose isomerase gene from *E. coli* was also introduced for positive selection of GR2. As a result, GR2 lines were found with an ability to accumulate very high (up to 37 μg/g) carotenoids, of which 31 μg/g was β-carotene which was 23 fold more compared to the original golden rice (Mayer 2007; Fig. 26.4).

26.4 Availability of Vitamin A in Golden Rice

Golden Rice is a genetically modified rice developed to provide a solution to VAD-related disease incidence and mortality in rice-eating populations. It has been shown that the β-carotene contained in Golden Rice is the same as the β-carotene available in other foods in terms of effectiveness in relation to transferring to vitamin A (Grune et al. 2006). It has been shown that the human intestine is indeed capable of extracting β-carotene out of Golden Rice in a highly efficient manner. However, the actual quantity of carotene converted to vitamin A in Golden Rice is not known so far, which could be evaluated through extensive clinical studies in the human being. This is highly important because the bioavailability is the ultimate target of the entire process. When Golden Rice plants were grown in the presence of nonradioactive isotope of hydrogen (deuterium) atom, the incorporation of stably labeled hydrogen atoms in the rice's carotene molecules was observed. This study helps the researchers to think to add the important micronutrients which can be utilized in our body (Tang et al. 2009). The paddy was harvested at maturity and processed and tried in human beings (providing 0.99–1.53 mg of carotene). The result from the above experiment confirmed the absorption of stably labeled β-carotene from the Golden Rice in the gastrointestinal tract. Conversion of β-carotene to vitamin A in subsequent phase was recorded to be at a rate of 3.8:1 in comparison to earlier rate of 10:1–27:1 in the case of colored vegetables, viz., spinach and carrots. This indicated a highly efficient biological conversion rate as compared to other known sources of β-carotene originated from plants. Moreover, it was implied that probably a supply of 50 % of the Recommended Dietary Allowance (RDA) of vitamin A could be possible from even a cup of rice, on daily basis, which is well within the eating habits of most children and their mothers (Tang et al. 2009). Similarly, in another study, Chinese children in the age group of 6–8 years were randomly fed with Golden Rice or spinach, both supplemented with nutrient solution having 23 atom % ^2H_2O or [$^2H_8$] β-carotene in an oil capsule. The result was exciting, because the β-carotene in Golden Rice is as effective as pure β-carotene in oil and better to spinach in supplementing vitamin A to children. 100–150 g cooked GR (50 g dry weight) can contribute 60 % of the Chinese Recommended Nutrient Intake of vitamin A for 6–8-year-old children. This is almost close to 2:1 conversion ratio (Tang et al. 2012; Potrykus 2015). These studies clearly exhibit the potentials of GR for effective solution against VAD.

26.5 Golden Rice at Farmers' Field

With basically 90 % of the rice lands in the globe being in Asia, it is imperative that the trait of high beta-carotene should be transferred to the variety of the choice in respective countries in Asia. There are different types of rice, but out of total traded rice in the globe, *indica* rice has a lion share of 75–80 %, whereas *japonica* rice has 10–12 % and aromatic rice cultivars such as basmati and jasmine with 10 % market share, glutinous rice contributing for the rest (Gulati and Narayanan 2002). In this context, it is important that this trait should be transferred to *indica* rice. Therefore, in the Golden Rice network, the rice breeders have taken burden of transferring the trait with the help of conventional breeding. But for the release of variety, there should be classical breeding, which requires planting of the large number of plants in the field. Successful backcrossing and selection from any of the plant breeding process require proper phenotyping in the open field. It requires a lot of repetitions and conscious selection of phenotypes in response to different environments (Potrykus 2015). The first completely open field trial was done in the USA during 2004 and 2005. In Asia, a confined field trial was done in 2008 and multi-location trial was possible only in 2013 (Dubock 2014). There is no variety released in the globe till date; however efforts are under way for introgression of high β-carotene genes into the popular varieties, viz., IR64, PSB Rc82, and BRRI dhan 29. In India, Department of Biotechnology (DBT) and Indian Council of Agricultural Research (ICAR) have taken positive steps to introduce gene construct of Golden rice for pro-vitamin A into popular Indian varieties, under Indian Network on Golden Rice, funded and coordinated by the DBT. Under this network, Indian Agriculture Research Institute, New Delhi (IARI); Indian Institute of Rice Research, Hyderabad (IIRR); and Tamil Nadu Agriculture University, Coimbatore (TNAU) have initiated work to introgress the genes of Golden Rice lines into popular Indian rice varieties such as Swarna (IARI), MTU 1010 and Improved Sambha Mahsuri (IIRR), ADT43 and ASD 16 (TNAU) (Sahai 2010). International Rice Research Institute and its many research partners are continuously working toward development of an efficient Golden Rice variety in the interest of farmers and ultimately to common man. At present, Golden Rice varieties are in the process of development and evaluation. The first round of Golden Rice multi-location field trials (MLTs) was conducted in different locations of the Philippines during 2012–2013. Similarly, MLTs have also been conducted in Bangladesh and Indonesia (The Daily Star 2012; IRRI 2014). The latest versions of Golden Rice, GR2 event R (GR2-R), were introduced from the donor "Kaybonnet" into two popular *indica* variety IR-64 and PSB Rc 82 (Alfonso et al. 2013). Bangladesh Rice Research Institute is about to go for confined field trial of GR2E-BR29 (The

Daily Star 2015). Marker-assisted backcrossing using SNP markers and line selection was done by IRRI scientists in the Philippines. Selected progenies and introgression lines were tested under confined field trials at IRRI. Another important factor which is of much interest to the farmers is agronomic performance. There was a mixed performance in preliminary results. Even if the expected level of beta-carotene in the grain achieved [5.17–11.20 μg/g] total carotenoid content, with respect to 7.65 μg/g in Kaybonnet (donor) and 0.42 μg/g in IR64 (recipient) at BC3F3 and PSB Rc82 at BC3F1 (Alfonso et al. 2013), average grain yield was unexpectedly less than that of adapted popular cultivars. Golden Rice could only be in the doorstep of farmers and common man if it can successfully accommodate the quality and desired plant type suitable for different countries of Asia, be approved by concerned national regulations, and be proven to be an improved source of vitamin A in community conditions.

26.6 Golden Rice: Social Issues

Golden Rice remains still a distant reality for the poor malnutritious populations of the Asia or even the world. This is because of anti-GMO movements on environmental, biosafety, and health issues. The pros and cons as claimed by different sources have been discussed below.

26.6.1 Supporting Golden Rice

The national academies of sciences have published a large number of reports addressing various aspects, and among them, the National Academies Report 2010 has extensively described the beneficial impacts of GM crops introduced in the USA (NAR 2010). A number of individuals and organizations have supported Golden Rice at large as evident from the editorial in science led by Bruce Albert, past president of National Academies (USA) along with many Nobel laureates (Alberts et al. 2013). They have expressed dissatisfaction that although it is developed and ready for use since the beginning of twenty-first century, still it is unreachable to the needy (Moghissi et al. 2015). Since its inception Golden Rice has passed through extensive tests. The majority of them supports biosafety and is environment friendly (Golden Rice Project 2012).

The genes for β-carotene that have been introgressed into Golden Rice are devoid of any known allergens or toxins (Goodman and Wise 2006; IRRI 2015a). The Joint FAO/WHO Expert Consultation on Biotechnology and Food Safety opined that "Golden Rice" possesses no danger concerning allergenic reactions because of the high digestibility of the transgenic proteins in simulated gastric fluid that further supports the claim of lack of allergenicity. The results of gene expression profiling of thousands of genes showed no unexpected changes or gross perturbances in the expression profile in comparison to the parent material. Again, taste differences could not be detected and the taste was found at par with parental materials with several taste trials. Moreover, it has been shown that Golden Rice diverts only a minute amount of carbon into carotenoids; hence, the compositions remain unchanged at large (Golden Rice Project 2012; Husaini and Tuteja 2013).

IRRI started field trials with the advocacy that there is no anticipated risk to human health from the field trials. The β-carotene in Golden Rice is similar to the β-carotene available in most of the natural foods and supplements (Paine et al. 2005). Rice is essentially a self-pollinating crop; hence the chance of cross-pollination between Golden Rice and other rice varieties is meager (IRRI 2015a).

26.6.2 Opposing Golden Rice

The Golden Rice is one of the major targets of anti-genetically modified organism (GMO) activists. A number of anti-GMO activists including Greenpeace oppose this technology. They believe that allowing Golden Rice would open up the path for the entire GMO applications (Potrykus 2001). They also fear that transgenic would promote the use of monoculture. It is also opposed for being a technological shortcut that ignores the real need of the farmers and their living status (Mayer 2005). The critics also support the existing vitamin supplementation and fortification program and availability from natural sources by growing nontraditional fruits and vegetables, viz., refined palm oil, raw carrots, leafy vegetables, sweet potato, cassava, mango, and papaya in farmer fields. While the classical approaches are successful to a certain extent, the sustainability of the intervention is difficult (Mayer 2005). They are concerned regarding the probable contamination of non-GE rice, if released to the environment. There is still some inherent prediction among the people that GE rice contamination has negative effect on the conventional, traditional, and organic rice farming which would ultimately create bad impact in rural livelihoods. Similar kind of fear of using GE "Golden" Rice contamination may adversely affect other rice varieties or population of rice growing countries, forced to rethink the government prior to grant permission for its cultivation. An ongoing field trials of golden rice in Philippine was vandalized by Green Peace (GP) activists in 2013 (IRRI News). However, like a hitch to GP's anti-GMO campaign, Dr. P. Moore, the former GP president opined that the GP's opposition to GM golden rice costing thousands of lives (Independent.co.uk 2014). Earlier, they were concerned that it could not be effective, because as the vitamin A quantity that can be derived from "Golden Rice" is estimated to be 1.6 ug/g of uncooked rice, an adult has to eat 12 times the normal intake of 300 g (e.g., 3.6 kg) of rice to have the recommended daily nutritional intake, i.e., 500 ug for an adult female. Development of GR2 was a setback for the opponent as it is able to accumulate 31 μg/gm β -carotene in rice seeds. However, Dr. Tang proved in his experiment that 60 g of Golden Rice could provide 60 % of the Chinese recommended daily requirements in recent cultures (Tang et al. 2012).

26.7 Conclusions

It is a fact that science has to work out the solution of malnutrition and vitamin deficiency faced by the poor among rice eaters on earth in general and Asia in particular. About 90 % of the rice lands of the globe are available in Asia. It is obvious that vitamin A deficiency is prevalent among the poor whose source of calories is based mostly on rice or other food that is rich in carbohydrate but poor in micronutrient. These people are economically weak and seldom afford consumption of vegetables, dairy product, or animal products. Golden Rice is supposed to be an alternative solution in the current era, because it can be possible to introduce the requisite amount of vitamin A as an added trait into a traditional staple food. A simple intervention of a new variety could solve the negative impacts of vitamin A deficiency without any additional inputs. Though, the GMO possesses enormous challenge for its acceptance in the public globally due to various biosafety issues but the current Golden Rice come up with a major solution to human health problem considering the biosafety issues. Therefore, the skepticism of "Golden Rice" has to be resolved in an open mind with scientific data so that it could reach the needy and poor children and women and eradicate VAD-mediated blindness and the loss of life and so that it would be a great serve to the humanity and society as well.

References

ACS (2012) American Cancer Society. Vitamin A, retinoids, and provitamin A carotenoids. URL: www.cancer.org/treatment/treatmentsandsideeffects/complementaryandalternativemedicine/herbsvitaminsandminerals/vitamin-a-and-beta-carotene. Accessed 12 June 2015

Akhtar S, Ahmed A, Randhawa MA, Atukorala S, Arlappa N, Ismail T, Ali Z (2013) Prevalence of vitamin A deficiency in South Asia: causes, outcomes, and possible remedies. J Health Popul Nutr 31:413–423

Al-Babili S, Beyer P (2005) Golden rice – five years on the road – five years to go? Trends Plant Sci 10:565–573

Alberts B, Rees M, Sharp P (2013) Standing up for GMOs. Science 341:1320

Alfonso A, Somera J, Espejo E, Miranda R, Avellanoza E, Tabano D, Nazareno E (2013) Confined field test of pro-vitamin a enriched 'golden rice' event GR2-R introgressed lines of IR64 and PSB Rc82. Trans Natl Acad Sci Technol 35:11

ASN (2009) American Society for Nutrition. Researchers determine that golden rice is an effective source of vitamin A. URL: http://ww.nutrition.org/media/publications/ajcnJune809.pdf. Accessed 10 June 2015

Black RE, Allen LH, Bhutta ZA, Caulfield LE, de Onis M, Ezzati M, Mathers C, Rivera J (2008) Maternal and Child Undernutrition Study Group. Maternal and child undernutrition global and regional exposures and health consequences. Lancet 371:243–260

Burkhardt P, Beyer P, Wunn J, Kloti A, Armstrong G, Schledz M, von Lintig J, Potrykus I (1997) Transgenic rice (*Oryza sativa*) endosperm expressing daffodil (*Narcissus pseudonarcissus*) phytoene synthase accumulates phytoene, a key intermediate of provitamin A biosynthesis. Plant J 11:1071–1078

Cazzonelli CI, Pogson BJ (2010) Source to sink regulation of carotenoid biosynthesis in plants. Trends Plant Sci 15:266–274. doi:10.1016/j.t plants

Chen Y, Li FQ, Wurtzel ET (2010) Isolation and characterization of the Z-ISO gene encoding a missing component of carotenoid biosynthesis in plants. Plant Physiol 153:66–79

Datta K, Baisakh N, Oliva N, Torrizo L, Abrigo E, Tan J, Rai M, Rehana S, Al-Babili S, Beyer P, Potrykus I, Datta SK (2003) Bioengineered 'golden' Indica rice cultivars with b-carotene metabolism in the endosperm with hygromycin and mannose selection systems. Plant Biotechnol J 1:81–8

Del Campo JA, García-González M, Guerrero MG (2007) Outdoor cultivation of microalgae for carotenoid production current state and perspectives. Appl Microbiol Biotechnol 74:1163–1174

Dubock A (2014) The politics of golden rice. GM Crops Foods 5(3):210–222

Evert A (2013) Vitamin A. URL: http://www.nlm.nih.gov/medlineplus/ency/article/002400.htm. Accessed 11 June 2015

Farré G, Bai C, Twyman RM, Capell T, Christou P, Zhu C (2011) Nutritious crops producing multiple carotenoids – a metabolic balancing act. Trends Plant Sci 16:532–540

Gallagher CE, Matthews PD, Li F, Wurtzel ET (2004) Gene duplication in the carotenoid biosynthetic pathway preceded evolution of the grasses (*Poaceae*). Plant Physiol 135:1776–1783

Giuliano G, Tavazza R, Diretto G, Beyer P, Taylor MA (2008) Metabolic engineering of carotenoid biosynthesis in plants. Trends Biotechnol 26:139–145

Golden Rice Project (2012) Golden rice humanitarian board. http://www.goldenrice.org/Content2-How/how3_biosafety.php. Accessed 10 June 2015

Golden Rice Project (2013) Golden rice humanitarian board. URL: http://www.goldenrice.org/Content2-How/how1_sci.php. Accessed 10 June 2015

Golden Rice Project (2014) Golden rice humanitarian board. www.goldenrice.org/Content3-Why/why1_vad.php. Accessed 10 June 2015

Goodman RE, Wise J (2006) Bioinformatic analysis of proteins in golden rice 2 to assess potential allergenic cross-reactivity preliminary report. University of Nebraska, Food Allergy Research and Resource Program

Grune T, Lietz G, Palou A, Ross AC, Stahl W, Tang G, David Thurnham D, Yin S, Biesalski HK (2006) β-carotene is an important vitamin A source for humans. J Nutr 140(12):2268S–2285S

Gulati A, Narayanan S (2002) Rice trade liberalization and poverty. Markets and structural studies division discussion paper no 51. International Food Policy Research Institute, Washington, DC

Hamid AA, Aiyelaagbe OO, Usman LA, Ameen OM, Lawal A (2010) Antioxidants: its medicinal and pharmacological applications. Afr J Pure Appl Chem 4(8):142–151, URL: http://www.academicjournals.org/AJPAC

Hoa TTC, Al-Babili S, Schaub P, Potrykus I, Beyer P (2003) Golden indica and japonica rice lines amenable to deregulation. Plant Physiol 133:161–169

Husaini AM, Tuteja N (2013) Imperative for achieving the millennium development goals and sustainability of agriculture in the climate change era. GM Crops Food 4(1):1–9. doi:10.4161/gmcr.22748

IFPRI (2012) Global Hunger Index. The challenge of hunger ensuring sustainable food security under land, water, and energy stresses. IFPRI, Washington, DC, p 72

Independent.co.uk (2014) http://www.independent.co.uk/news/science/former-greenpeace-leading-light-condemns-them-for-opposing-gm-goldenrice-crop-that-could-save-two-9097170.html

IRRI (2014) URL: http://irri.org/golden-rice/faqs/what-is-the-status-of-the-golden-rice-project-coordinated-by-irri. Accessed 12 June 2015

IRRI (2015a) Multi-location field trials of golden rice. http://irri.org/blogs/golden-rice-blog/multi-location-field-trials-of-golden-rice. Accessed 27 Feb 2015

IRRI (2015b) World rice statistics online query facility. URL: http://ricestat.irri.org:8080/wrs2/entrypoint.htm. Accessed 06 Aug 2015

IRRI News (2013) http://irri.org/blogs/golden-rice-blog/golden-rice-field-trial-vandalized

Isaacson T, Ronen G, Zamir D, Hirschber J (2002) Cloning of tangerine from tomato reveals a carotenoid isomerase essential for the production of β-carotene and xanthophylls in plants. Plant Cell 14:333–342

Islam Z, Heong KL, Catling D, Kritani K (2012) Invertebrates in rice production systems: status and trends. FAO Commission on genetic resources for food and agriculture Background study paper, No 62, Food and Agriculture. Organization of the United Nations. Rome

Jackson H, Braun CL, Ernst H (2008) The chemistry of novel xanthophyll carotenoids. Am J Cardiol 101: S50–S57 (Supplement)

Johnson EJ (2002) The role of carotenoids in human health. Nutr Clin Care 5(2):56–65

Krinsky NI, Johnson EJ (2005) Carotenoid actions and their relation to health and disease. Mol Aspects Med 26(6):459–516

Li F, Vallabhaneni R, Yu J, Rocheford T, Wurtzel ET (2008a) The maize phytoene synthase gene family overlapping roles for carotenogenesis in endosperm, photomorphogenesis, and thermal stress- tolerance. Plant Physiol 147:1334–1346

Li F, Vallabhaneni R, Wurtzel ET (2008b) PSY3, a new member of the phytoene synthase gene family conserved in the *Poaceae* and regulator of abiotic stress-induced root carotenogenesis. Plant Physiol 146(3):1333–1345

Li F, Tzfadia O, Wurtzel ET (2009) The phytoene synthase gene family in the grasses sub functionalization provides tissue-specific control of carotenogenesis. Plant Signal Behav 4(3):208–211

Maudinas B, Bucholtz ML, Papastephanou C, Katiyar SS, Briedis AV, Porter JW (1977) The partial purification and properties of a phytoene synthesizing enzyme system. Arch Biochem Biophys 180:354–362

Mayer JE (2005) The golden rice controversy useless science or unfounded criticism? BioScience 55:726–727

Mayer JE (2007) Golden rice, golden crops, golden prospects. Rev Colomb Biotechnol 9:22–34

Moghissi AA, Pei S, Liu Y (2015) Golden rice scientific, regulatory and public information processes of a genetically modified organism. Crit Rev Biotechnol 1–7. doi:10.3109/07388551.2014.993586

NAR (2010) The National Academies Report. Impact of genetically engineered crops on farm sustainability in the United States. The National Academies Press, Washington, DC. URL: www.nap.edu. Accessed 13 Apr 2015

Paine JA, Shipton CA, Chaggar S, Howells RM, Kennedy MJ, Vernon G, Wright SY, Hinchliffe E, Adams JL, Silverstone AL, Drake R (2005) A new version of golden rice with increased provitamin A content. Nat Biotechnol 23:482–487

Palaisa KA, Morgante M, Williams M, Rafalski A (2003) Contrasting effects of selection on sequence diversity and linkage disequilibrium at two phytoene synthase loci. Plant Cell 15:1795–1806

Park H, Kreunen SS, Cuttriss AJ, DellaPenna D, Pogson B (2002) Identification of the carotenoid isomerase provides insight into carotenoid biosynthesis, prolamellar body formation, and photomorphogenesis. Plant Cell 14:321–332

Potrykus I (2001) Golden rice and beyond. Plant Physiol 125:1157–1161

Potrykus I (2015) From the concept of totipotency to biofortified cereals. Annu Rev Plant Biol 66:2.1–2.22

PPDD (2011) Pakistan Planning and Development Division. National nutritional survey 2011. Planning Commission, Planning and Development Division, Government of Pakistan, Islamabad, p 69

Quijano RF (2001) Grains of delusion golden rice seen from the ground. URL: http://www.grain.org/article/entries/10-grains-of-delusion-golden-rice-seen-from-the-ground. Accessed 13 May 2015

Rabbani S, Beyer P, Lintig J, Hugueney P, Kleinig H (1998) Induced beta-carotene synthesis driven by triacylglycerol deposition in the unicellular alga Dunaliella bardawil. Plant Physiol 116:1239–1248

Reifen R (2002) Vitamin A as an anti-inflammatory agent. Proc Nutr Soc 61(3):397–400

Ribaya-Mercado JD, Solomons NW, Medrano Y (2004) Use of the deuterated-retinol-dilution technique to monitor vitamin A status of Nicaraguan schoolchildren 1 year after initiation of the Nicaraguan national program of sugar fortification with vitamin A. Am J Clin Nutr 80:1291–1298

Romer S, Fraser PD, Kiano JW, Shipton CA, Misawa N, Schuch W, Bramley PM (2000) Elevation of the provitamin A content of transgenic tomato plants. Nat Biotechnol 18:666–669

Ross AC, Chen Q, Ma Y (2011) Vitamin A and retinoic acid in the regulation of B-cell development and antibody production. Vitam Horm 86:103–126. doi:10.1016/B978-0-12-386960-9.00005-8

Sahai S (2010) Potential of agricultural genetic engineering for food security in India: research on transgenic food crops. On behalf of Deutsche Gesellschaft für Technische Zusammenarbeit GmbH (GTZ), Eschborn. http://genecampaign.org/wpcontent/uploads/2014/07/Potential_Of_Agricultural_Genetic_Engineering_For_Food_Security_In_India.pdf. Accessed 26 Nov 2015

Sarao C (2015) Does brown rice contain vitamin K? URL: http://www.livestrong.com/article/437748-about-brown-32#rice-vitamin-k

Schaub P, Al-Babili S, Drake R, Beyer P (2005) Why is golden rice golden (yellow) instead of red? Plant Physiol 138:441–450

Sommer A (ed) (1982) Nutritional blindness xerophthalmia and keratomalacia. Oxford University Press, New York

Tang G, Qin J, Dolnikowski GG, Russel RM, Grusak MA (2009) Golden rice is an effective source of vitamin A. Am J Clin Nutr 89:1776–1783

Tang G, Hu Y, Yin S, Wang Y, Dallal GE, Grusak MA, Russell RM (2012) β-carotene in golden rice is as good as β-carotene in oil at providing vitamin A to children. Am J Clin Nutr 96:658–664

The Daily Star (2012) Vitamin-A rich rice get node. URL: http://archive.thedailystar.net/newDesign/news-details.32#php?nid=221236

The Daily Star (2015) Bangladeshi scientists ready for trial of world's first 'Golden rice'. http://www.thedailystar.net/frontpage/worlds-first-goldenrice-ready-32#trial-153676

Tian L, DellaPenna D (2004) Progress in understanding the origin and functions of carotenoid hydroxylases in plants. Arch Biochem Biophys 430:22–29

Underwood BA, Arthur P (1996) The contribution of vitamin A to public health. FASEB J 10:1040–1048

Vishnevetsky M, Ovadis M, Vainstein A (1999) Carotenoid sequestration in plants the role of carotenoid-associated proteins. Trends Plant Sci 4:232–235

Welsch R, Arango J, Bär C, Salazar B, Al-Babili S, Beltrán J, Chavarriaga P, Ceballos H, Tohme J, Beyer P (2010) Provitamin A accumulation in cassava (*Manihot esculenta*) roots driven by a single nucleotide polymorphism in a phytoene synthase gene. Plant Cell 22:3348–3356

West KP Jr, Pokhrel RP, Katz J (1991) Efficacy of vitamin A in reducing preschool child mortality in Nepal. Lancet 338:67–71

WHO (ed) (1997) World Health Organization, Vitamin A supplements a guide to their use in the treatment of vitamin A deficiency and xerophthalmia. WHO/UNICEF/IVACG Task Force, Geneva

WHO (2009) World Health Organization. Global prevalence of vitamin A deficiency in populations at risk 1995–2005: WHO global database on vitamin A deficiency. World Health Organization, Geneva. p 55

WHO (2013) World Health Organization. Micro nutrient deficiencies vitamin A deficiency. URL: http://www.who.int/nutrition/topics/vad/en

Ye X, Al-Babili S, Klöti A, Zhang J, Lucca P, Beyer P, Potrykus I (2000) Engineering the provitamin A (beta-carotene) biosynthetic pathway into (carotenoid-free) rice endosperm. Science 287:303–305

Yu Q, Ghisla S, Hirschberg J, Mann V, Beyer P (2011) Plant carotene cis-trans isomerase CRTISO A new member of the FAD_{red}-dependent flavoproteins catalyzing non-redox reactions. J Biol Chem 286:8666–8676

Biofortified Wheat for Mitigating Malnutrition

27

Hari Ram, Satpal Singh, Neha Gupta, and Balwinder Kumar

Abstract

Zinc, iron, and provitamin A are the critical micronutrients required for structural and functional integrity of biological system. Their deficiency affects billions of people worldwide, by hampering growth and development and destroying the immune systems. Micronutrient-dense wheat (*Triticum aestivum* L.) varieties can be developed by using the existing genetic variability in the germplasm. Even with new screening tools such as gene discovery, marker-assisted selection, and precision phenotyping, selection of high-Zn genotypes would appear an easy task. Wide genetic variation available in primitive and wild relatives, landraces, and synthetic hexaploids is now being intensively exploited under HarvestPlus program to identify quantitative trait loci (QTLs) for enhancing concentration of Zn and its bioavailability in wheat grain. Further regulation of gene expression in identified QTLs in response to biotic and abiotic stresses is also analyzed. The ultimate goal of this program is to improve nutritional status of wheat cultivars. However, enhancement of Zn uptake through utilization of genetic variation would be possible only when the soil environment (i.e., mineral composition) has sufficient zinc pool for absorption. The varieties like BHU 1, BHU 17, and BHU 19 from India and NR 419, NR 420, and NR 421 from Pakistan show 4–10 ppm increase in grain zinc. Agronomic biofortification strategies through application of Zn-containing fertilizer provide an immediate and effective option to increase grain Zn concentration and productivity in wheat, particularly in areas of severe nutrient-deficient soil. Zn fertilizer can be used along

H. Ram (✉) • N. Gupta
Department of Plant Breeding & Genetics, Punjab Agricultural University, Ludhiana, Punjab 141 004, India
e-mail: hari_ram@pau.edu; nehagupta.pau@gmail.com

S. Singh
Department of Agronomy, Punjab Agricultural University, Ludhiana, Punjab 141 004, India
e-mail: satpal.pau@gmail.com

B. Kumar
Department of Animal Genetics & Breeding, Guru Angad Dev Veterinary and Animal Sciences University, Ludhiana, Punjab 141 004, India
e-mail: dr.balwinderkumar@rediffmail.com

U. Singh et al. (eds.), *Biofortification of Food Crops*, DOI 10.1007/978-81-322-2716-8_27

with the pesticide used to control aphids or yellow rust in wheat. Fertilizer application strategies must be practical and economically feasible. However, in developing countries where resource-poor farmers cannot afford fertilizer, breeding for mineral density may remain the sole agricultural intervention to improve the nutritional content of staple crops.

Keywords
Biofortification • Genetic variation • Micronutrients • Wheat • Zinc fertilizers

27.1 Introduction

Zinc (Zn), iron (Fe), and vitamin A deficiencies are a growing public health and socioeconomic issue, particularly in the developing countries (Welch and Graham 2004). Zinc and Fe deficiencies together with vitamin A deficiency have been identified as the top priority global issue to be addressed to achieve a rapid and significant return for humanity and global stability. One of the critical physiological roles of Zn in biological systems is its role in protein synthesis and metabolism. It has been estimated that nearly 2800 human proteins are capable of binding Zn which corresponds to 10 % of human proteome. Zinc is also a critical micronutrient required for structural and functional integrity of biological membranes and for detoxification of highly aggressive free radicals (Cakmak 2008). Any decrease in Zn and Fe concentration of human body will, therefore, result in a number of problems like immune dysfunctions, high susceptibility to infectious diseases, retardation of mental development, adverse pregnancy outcomes, abnormal neurobehavioral development, and stunted growth of children (Black et al. 2008). Moreover, it is reported that deficiency of zinc micronutrient is next to vitamin A being responsible for the mortality of children below the age of 5 years globally (Fig. 27.1).

Rice and wheat constitute two-third of energy source for Indian population and are also the backbone of Indian economy. Wheat alone is responsible for up to 70 % of daily calorie intake of the population living in rural regions and an important source for Zn for human beings living in the developing world (Cakmak 2008). Nevertheless, it is high time that along with food security due attention is also paid to adequate micronutrient nutrition.

Sustainable solutions to malnutrition will only be found by closely linking agriculture to nutrition and health and by formulating interlinked policies (Graham et al. 2007; World Bank 2007). Biofortification (developing food crops that fortify themselves) is the first agricultural tool now being employed to address micronutrient malnutrition worldwide.

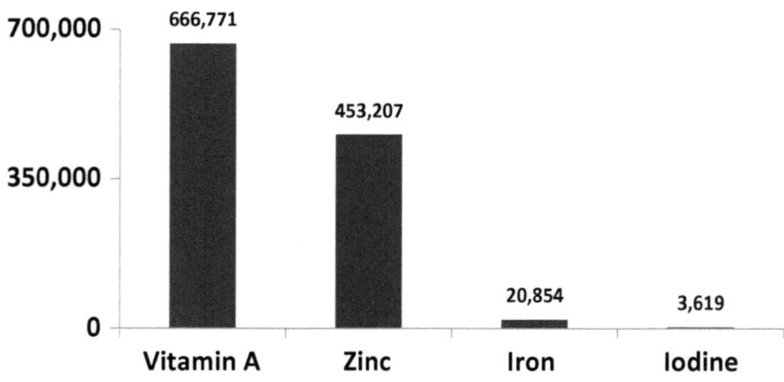

Fig. 27.1 Global mortality of children under 5 years of age in 2004 (Black et al. 2008)

27.2 Reasons for deficiency of Micronutrients in Human Being

The inherited low concentration and low bioavailability of Zn and Fe in cereal grains contributed a lot to Zn and Fe deficiency in people which is widespread mainly in areas where cereal-based foods are dominant in the diets (Cakmak 2008). Cereal grains are inherently low both in concentration and bioavailability of Zn and Fe, particularly when grown on potentially Zn- and Fe-deficient soils (Welch and Graham 2004; Cakmak et al. 2010b). It was reported that about 61 million Indian children possessed symptoms of zinc deficiency, and this deficiency was highest in Orissa (51.3 %), followed by Uttar Pradesh (48.1 %), Gujarat (44.2 %), Madhya Pradesh (38.9 %), and Karnataka (36.2 %) with an average of 43.8 % (Kapil and Jain 2011).

Increasing cropping intensity and accompanying changes in the soil and fertilizer management practices have lowered the macronutrient as well as micronutrient like Zn and Fe status of soils and its availability, especially in the Indo-Gangetic Plains of India. It is reported that 48 % of Indian soils are low in available zinc (Singh 2011). However, in Punjab the deficiency is reduced from 43 % deficient samples in 2002 to 21 % deficient samples in 2010 due to continuous use of zinc sulfate fertilizer by the farmers for rice crop at 25–30 kg/ha (Sadana et al. 2010). The processing of wheat grains substantially reduces the concentration of Zn and other essential elements, which further increases the Zn deficiency in humans (Cakmak 2008; Kutman et al. 2011; Zhang et al. 2010). Nowadays, increasing grain Zn and Fe concentration of cereal grains represents an important challenge to be met by using agricultural tools such as breeding and fertilization.

27.3 How to Correct Micronutrient Deficiency in Human Being

There are two options by which the deficiency of micronutrient can be corrected in human being, i.e., (i) using supplements like liquid, tablets, etc. or (ii) using micronutrient biofortified food. The high-income group can avail the first option, but poor people can't purchase the costly supplements for correction of zinc deficiency. The poor people in India are getting food grains through the public distribution system (PDS) for their consumption. PDS will become strong now, as the "Food Security Bill" has already been passed by the Indian Government. Now about, 50 % of the population will have the right to assess food grains. In these conditions, it will be important to provide the micronutrient-enriched food grains especially zinc to the people to alleviate micronutrient deficiency in the population.

27.3.1 Enriching Cereal Grains with Micronutrient

There are two options to enrich cereal grains with micronutrient especially zinc through genetic biofortification (development of the cereal varieties rich in micronutrients) or through agronomic biofortification (using fertilizers in the crops). Till the new varieties are being developed, the agronomic biofortification can be used to take care of micronutrient deficiency. Different crops can be enriched with zinc through agronomic biofortification. A successful biofortification strategy should meet the following criteria: (i) grain yield capacity of the biofortified genotypes must be maintained or even improved to guarantee farmer acceptance, (ii) the resulted increase in micronutrient levels must have a significant impact on human health, and (iii) the micronutrient levels achieved must be relatively stable across various locations and climatic zones (Welch and Graham 2004). Due to lack of plant-available Zn in soil (Liu 1996) and the limited uptake of Zn by roots at the time of grain filling in the dry season (Liu et al. 2010), the rate of foliar Zn application may be the major factor which could help to determine the size of the Zn pool in vegetative parts of wheat and hence to increase the Zn concentration in grain through nutrient fertilization.

27.3.2 Genetic Biofortification

In genetic biofortification, the varieties of different crops are developed with the help of conventional plant breeding. Among different nutrients, at present, the micronutrients like Fe, Zn, and provitamin A are being focused especially in wheat. First strategy of breeding is to screen the germplasm that is rich in total amount of nutrients, but some other parameters should also be considered, e.g., type of soil mineral and availability of a particular nutrient to plant, antinutritional factors (phytate and tanins) and promoters (inulins), losses in food processing, and bioavailability of nutrient, except that the varieties should not compromise with the ability of biotic and abiotic stresses, potential yield level, and acceptability in general public. The HarvestPlus program has set needed levels for Fe, Zn, and provitamin A carotenoids in target crops after addressing these issues. Once high-yielding biofortified wheat cultivars are developed that meet target nutrient levels, they will be disseminated widely. In genetic biofortification approach, breeders can achieve the mineral target increment by directly breeding for higher micronutrient concentration or breeding for their increased bioavailability. For wheat, bioavailability of Zn and Fe can be assumed 25 and 5 %, respectively.

27.3.2.1 Screening of Genotypes

For rapid and inexpensive screening of wheat for Fe and Zn, the use of X-ray fluorescence spectrometer (XRF) and near-infrared reflectance spectroscopy (NIRS) may be used to assess the organic compounds which are indirectly related with some inorganic constituents of the plant samples (Osborne and Fearn 1986). Analysis through XRF and NIRS is nondestructive and is performed with high speed and accuracy. Choi et al. (2007) used simple, cheap, and semiquantitative analytical methods by adopting spectrophotometer methodologies to screen many field crops.

27.3.2.2 Use of Genetic Variability

Genetic variability for the target traits is one of the most important tools to get the desired results. In case of breeding for Fe and Zn concentration of grain, it becomes very difficult as the concentration of the microelements primarily depends upon the environmental conditions particularly the composition of soil (Fiel et al. 2005). Therefore, it is a hard task to develop a variety especially when the variability of Fe and Zn in soil ranges from extremely low values to high. It is difficult to assess the response of various germplasm as the results are mostly not repetitive. While assessing the germplasm, it is to be ensured whether the soil was added with organic manures like farmyard manure, compost, green manure, vermicompost, or any other type. It has been noted that genotypes with the highest levels of nutrients were low yielding and mostly unadapted (Monasterio and Graham 2000). However, in wheat, there is sufficient genetic variability for Zn, and promising levels were found for Fe which may be used for the development of micronutrient-dense varieties.

In durum wheat, the predominant carotenoid is lutein and zeaxanthin (Leenhardt et al. 2006). Till date, no variety having provitamin A activity has been identified.

27.3.2.3 Genetic Variation in Micronutrient Concentration of Wheat Grain

Germplasm screening for Zn and Fe genes in wheat reveals primitive and wild relatives as genetically rich sources with *Triticum dicoccoides*, *Aegilops tauschii*, *Triticum monococcum*, and *Triticum boeticum* among the most promising sources of high Fe and Zn grain concentration (Cakmak et al. 2000). As we know, breeding of modern wheat cultivars in the past 50 years has been targeted for productivity traits, by selecting for resistance to diseases, short plant height, abiotic and biotic stress tolerance, increased biomass and harvest index, etc. But in spite of the essentiality of grain nutritional composition especially micronutrient density and

Table 27.1 Range of grain concentrations for Fe and Zn in contemporary wheat cultivars

	Genotypes (no.)	Fe (mg/kg dry wt.)	Zn (mg/kg dry wt.)	References
Germplasm	Modern wheat cultivars			
Bread wheat	43	22–34	21–35	Tang et al. (2008)
Bread wheat	384	30–73	27–85	Welch (2001)
Bread wheat	57	34–66	29–46	Ficco et al. (2009)
Bread wheat	14	30–38	26–34	Garvin et al. (2006)
Bread wheat	51	27–42	16–27	Oury et al. (2006)
Bread wheat	34	29–38	8–12	Cakmak et al. (2000)

Table 27.2 Variation for grain Zn concentration in bread wheat, as documented in various studies

		Zn (mg/kg)		
Germplasm	Genotypes (no.)	Mean	Range	References
Bread wheat	150	21.4	13.5–34.5	Zhao et al. (2009)
Bread wheat	20	33.6	32.6–34.8	Joshi et al. (2010)
Bread wheat	1300	30.5	23–52	Velu et al. (2011a)
Bread wheat	600	30.4	16.9–60.8	Velu et al. (2011b)
Bread wheat	40	32.5	29.0–39.5	Velu et al. (2012)
All wheat	>3000		16–142	Monasterio and Graham (2000)

protein quality in wheat, this important trait is often overlooked. This causes a significant loss in genetic variation for Zn in contemporary cultivars which correspond to their low grain Zn concentrations (Table 27.1). Several studies have revealed that there is sufficient genetic variability for Zn concentration that exists in wild relatives, synthetic hexaploid progenitors, and landraces (Cakmak et al. 2000; Ortiz-Monasterio et al. 2007), constituting wider gene pool in developing wheat varieties with high grain Zn (Tables 27.1 and 27.2).

But breeding for high grain Zn concentration in wheat is complicated due to quantitative inheritance of this trait, having low heritability and greater genotype × environment interaction (Trethowan et al. 2005; Trethowan 2007). Evaluation of best high-Zn lines of CIMMYT in a multilocation trial of India's Eastern Gangetic Plains (EGP) also revealed that wheat grain Zn concentrations were highly unstable as the performance of the elite lines varied across locations and years (Joshi et al. 2010). Transgressive segregation for Zn has been encountered in wheat. However, results to date based on limited information of transgressive segregation suggest the presence of complementary genes, particularly in genetically distant sources such as wild relative wheat species.

Further positive correlations between grain Zn and Fe in wheat have been demonstrated in various studies (Morgounov et al. 2007; Peleg et al. 2009; Genc et al. 2009; Zhang et al. 2010; Gomez-Becerra et al. 2010; Velu et al. 2011a, b, 2012), implying that the alleles for Zn and Fe deposition in the grain co-segregate or are pleiotropic and therefore that Zn and Fe can be improved simultaneously. Another study on *Aegilops* species also depicts that Zn and Fe in the flag leaves are positively correlated with grain Zn and Fe (Rawat et al. 2009a, b). Only a few studies have been successful in identification of quantitative trait loci (QTLs) linked to Zn in cereals such as GPC-B1 (250 kb locus) (Distelfeld et al. 2007). Singh et al. (2010) identified two QTL (QFe.pau-2A and QFe.pau-7A) for Fe and a QTL (QZn.pau-7A) for Zn, which they transferred from *Aegilops kotschyi* and *A. peregrina* (both UUSS genome species). In a doubled haploid population, four QTLs for grain Zn concentration and a single QTL for grain Fe concentration were identified (Genc et al. 2009). However, major QTLs that described 92 % of the genetic variation in grain

Table 27.3 First-wave varieties in HarvestPlus program (in which target increment of 4–10 ppm is achieved from original baseline of 25 ppm Zn; Harvest Plus 2014)

Country	Variety name	Zinc increase	Yield (t/ha)	Comments on agronomic properties
India	BHU1	+4–10 ppm	5.0	84 days to heading and 126 days to maturity
	BHU3	+6–8 ppm	4.4	83 days to heading and 125 days to maturity
	BHU5	+4–5 ppm	3.3	86 days to heading and 128 days to maturity
	BHU6	+4–9 ppm	3.4	78 days to heading and 119 days to maturity
	BHU17	+6–10 ppm	4.1	81 days to heading and 122 days to maturity
	BHU18	+6–9 ppm	3.9	87 days to heading and 131 days to maturity
Pakistan	NR 419	+7–9 ppm	4.5	93 days to heading and 130 days to maturity
	NR 420	+7 ppm	3.4	86 days to heading and 128 days to maturity
	NR 421	+14 ppm	3.6	78 days to heading and 119 days to maturity

Zn concentration are located on chromosomes 3D, 4B, 6B, and 7A.

The establishment of HarvestPlus, the Biofortification Challenge Program in 2003 by the Consultative Group on International Agricultural Research (CGIAR), has provided platform to explore genetic diversity for micronutrients existing in global germplasm and to utilize this variation for developing high-zinc genotypes. Till date, breeding effectiveness in developing zinc wheat (from a baseline of 25 ppm to target level of 37 ppm) for developing countries such as India and Pakistan was optimized under this research program. In HarvestPlus Phase I and II, 100–150 promising advanced lines of CIMMYT, based on grain yield and grain zinc, are selected each year for testing in genotype-by-environment (G × E) trials for agronomic attributes and grain zinc at 10–15 sites in India and at five sites in similar agroecologies in Mexico and Pakistan (HarvestPlus South Asia Screening Nursery). The best 40–50 emerging leads are then yield tested in multilocation yield trials (HarvestPlus South Asia Yield Trial) at more than 20 sites in India and Pakistan (Harvest Plus 2014). In India and Pakistan few promising wheat genotypes have been selected on the basis of multilocation performance and zinc data (Table 27.3).

Research efforts are continued in genetic biofortification through classical breeding approaches as well as new approaches such as gene discovery, marker-assisted breeding, etc. in order to identify QTLs associated with grain Zn content, for exploitation of genetic variation and easy selection of micronutrient-rich wheat varieties. Moreover, future researches on characterizing the candidate genes to enhance the promoters (e.g., inulin-type fructans and phytase) and decreasing the inhibitors (antinutritional substances, e.g., phytic acid) for improving Zn bioavailability in wheat grain are in progress (Huynh et al. 2008; Ram et al. 2010).

Due to positive correlation between iron and zinc, it is possible to improve both the nutrients simultaneously (Monasterio and Graham 2000). Oury et al. (2006) showed strong G × E interaction for iron and zinc concentrations, the screening for these elements could be highly unreliable in breeding for increased concentration of micronutrients. If the QTLs are identified and tagged, it will be possible to identify the lines in early generations which may help to develop the varieties with higher concentration of microelements. The incorporation of gene regulating senescence (GpcB1 6BS) in wheat increased not only the level of iron and zinc in grain (Uauy et al. 2006) but also increased their bioavailabilities (Hess et al. 2005). Higher bioavailability of these micronutrients in wheat has also been observed by incorporating natural mutations which reduce the amount of phytate in wheat (Guttieri et al. 2004). Welch and Graham (1999) reported that insoluble fiber and phenolic compounds in bran layer of grain act as antinutritional factors and reduce the bioavailability of iron and zinc. Therefore, bioavailability of different elements and their interaction with other factors is also an important aspect in plant breeding when germplasm is screened.

Compared to breeding approach, agronomic biofortification (e.g., application of Zn and Fe fertilizers) represents a short-term solution to the problem (Cakmak 2008).

27.3.3 Agronomic Biofortification

Shivay et al. (2008) reported that Zn application to soil as zinc sulfate or zinc-enriched/coated urea not only increased yield but also zinc concentration in wheat grain (Table 27.4). However, Cakmak et al (2010a) reported that soil Zn applications have, however, small effectiveness on increasing grain Zn, while foliar Zn applications (0.5 % zinc sulfate) resulted in remarkable increase in wheat grain Zn concentration (Table 27.5). By optimizing the timing and the solute concentration of foliar Zn application, wheat grain Zn concentration could be further increased, not only in whole grain but also in the endosperm. In recent past, a considerable progress has been made on the impact of foliar Zn fertilization in enriching cereal grains with Zn, particularly in wheat (Cakmak 2008; Wissuwa et al. 2008). In wheat, foliar Zn spray, especially at later growth stages (e.g., early milk stage and dough stage), was very effective in increasing Zn concentration of both whole grain and in the endosperm fraction, while soil Zn applications remained less effective (Cakmak et al. 2010b).

Moreover, foliar Zn application (0.5 % zinc sulfate) alone or in combination with soil application (50 kg zinc sulfate/ha) significantly increased grain Zn concentrations from 27.4 mg/kg at nil Zn to 48 and 49 mg/kg (Fig. 27.2) across all of 23 site-years in seven countries (Zou et al. 2012). It was further reported that foliar Zn application did not cause any adverse effect on grain yield, even slightly improved the yield. But in other studies (Karim et al. 2012), increased grain yield was recorded under drought conditions with foliar Zn application, even in a soil containing high DTPA extractable Zn, indicating that foliar-applied Zn met the Zn requirement of wheat plant growth and improved antioxidative defense mechanisms of plants against drought-induced oxidative cell damage.

Table 27.4 Effect of Zn fertilization on Zn concentration in grain and flour (Shivay et al. 2008)

Zn applied (kg/ha)	Wheat grain Zn (mg/kg)
0	38.1
2.6	41.3
5.2	43.8
7.8	47.3

Table 27.5 Effect of Zn fertilization on Zn concentration in grain and flour (Cakmak et al. 2010a)

Treatment	Zinc concentration (mg kg^{-1})	
	Grain	Flour
Soil application rate of $ZnSO_4 \cdot 7H_2O$ (kg ha^{-1})		
0	37.1	9.5
50	38.4	10.1
LSD ($P = 0.05$)	1.0	0.5
Foliar application rate of $ZnSO_4 \cdot 7H_2O$ (%)		
0	26.6	6.4
0.2	37.1	10.3
0.4	42.0	11.3
0.5	46.0	11.3
LSD ($P = 0.05$)	1.5	0.8

In Central Anatolia, a well-known highly Zn-deficient region of Turkey, applying $ZnSO_4$ to soil enhanced both grain yield and grain Zn concentration of durum wheat (Ekiz et al. 1998). An increase in grain Zn concentration by soil Zn application was almost twofold, whereas combined application of Zn through soil and foliar was more effective and resulted in a more than threefold increase in Zn concentration in durum wheat grain. Similar increases in grain concentrations of Zn in wheat following soil Zn application were also seen in Australia (Graham et al. 1992) and India (Shivay et al. 2008) under field conditions. As foliar zinc application at early milk stage cannot increase the grain yield, the farmer will not apply this treatment. But it was reported that application of zinc sulfate heptahydrate (0.5 %) with pesticides (i.e., dimethoate for aphid control and propiconazole for yellow rust control) at heading stage to early milk stage of wheat controlled biotic stress on one hand and improved grain Zn concentration on the other hand (Ram et al. 2013). So foliar Zn can be applied along with pesticides when these are used after anthesis to early milk stage of the

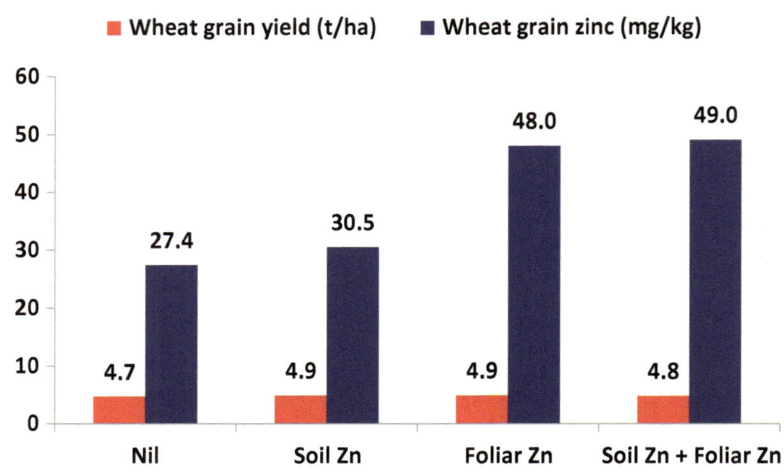

Fig. 27.2 Wheat grain yield and grain zinc content as influenced by zinc fertilization in seven countries (Zou et al. 2012)

crop (Ram et al. 2015). In Central Anatolia (Turkey), the combination of both soil and foliar application of Zn fertilizers to cereal crops seems to be a practical and effective method of maximizing root uptake and Zn accumulation in grain (Yilmaz et al. 1998). Thus adequate fertilization of food crops can partly help in Zn intake by humans. Zn fertilization of crops on Zn-deficient soils helps in attaining both food security and overcoming Zn malnutrition. Zhang et al. (2012) reported that compared with no foliar Zn application, foliar 0.4 % ZnSO4 · 7 H_2O application resulted in best effect on grain Zn, with 58 % increase in whole grain Zn, 76 % increase in wheat flour Zn, and up to 50 % decrease in the molar ratio of phytic acid to Zn in flour.

There is also a need for economic incentive programs to encourage farmers to apply Zn when there is no yield increase by Zn applications. Government policies are required for payment of premium to the farmers for the Zn application. Application of Zn fertilizers to soil and/or foliar seems to be a practical approach to improvement in wheat grain Zn concentration. Cakmak (2010) reported that improving N nutrition of plants may contribute to grain Zn and Fe concentrations by affecting the levels of Zn- or Fe-chelating nitrogenous compounds required for transport of Zn and Fe within plants and/or the abundance of Zn or Fe transporters needed for root uptake and phloem loading of Zn and Fe. Remobilization and retranslocation of Zn and Fe from vegetative tissues into grain through the phloem may also be affected by N nutrition. Zinc and Fe transporter proteins located on root cell membranes have also been identified on the plasma membranes of phloem cells, suggesting that these transporters are possibly involved in phloem transport of Zn and Fe into seeds (Haydon and Cobbett 2007; Curie et al. 2009).

27.4 Conclusions

Micronutrient-rich wheat varieties can be developed by using the existing genetic variability in the wheat germplasm. New screening tools like gene discovery, marker-assisted selection, and precision phenotyping have made selection of high Zn genotypes an easy and practical approach. Wider genetic variability, available in primitive and wild relatives, landraces, and synthetic hexaploids, is now being intensively exploited under HarvestPlus program to identify QTLs for enhancing Zn concentration and its bioavailability in wheat grain. The enrichment of Zn by utilization of genetic variation would be possible only when the soil environment has sufficient zinc pool for absorption. Foliar Zn application significantly increased Zn concentrations alone or along with pesticides in wheat and rice, and this positive impact of foliar Zn application occurred consistently over a wide range of environments and local management practices, including different crop varieties.

There is a need for incentive for farmers to apply Zn when there is no yield advantage by Zn applications. Farmers would also be motivated to spray Zn when they are educated about the knowledge that plants emerging from high-Zn seeds usually have better seedling vigor and field establishment (Cakmak 2008).

References

Black RE, Lindsay HA, Bhutta ZA, Caulfield LE, De Onnis M, Ezzati M, Mathers C, Rivera J (2008) Maternal and child undernutrition: global and regional exposures and health consequences. Lancet 371:243–260

Cakmak I (2008) Enrichment of cereal grains with zinc: agronomic or genetic biofortification? Plant Soil 302:1–17

Cakmak I (2010) Biofortification of cereals with zinc and iron through fertilization strategy. In: 19th world congress of soil science, soil solutions for a changing world, Brisbane, 1–6 Aug 2010

Cakmak I, Ozkan H, Braun HJ, Welch RM, Romheld V (2000) Zinc and iron concentrations in seeds of wild, primitive and modern wheats. Food Nutr Bull 21:401–403

Cakmak I, Kalayci M, Kaya Y, Torun AA, Aydin N, Wang Y, Arisoy Z, Erdem H, Yazici A, Gokmen O, Ozturk L, Horst WJ (2010a) Biofortification and localization of zinc in wheat grain. J Agric Food Chem 58:9092–9102

Cakmak I, Pfeiffer WH, McClafferty B (2010b) Biofortification of durum wheat with zinc and iron. Cereal Chem 87:10–20

Choi EY, Graham RD, Stangoulis JCR (2007) Rapid semiquantitative screening methods for determination of iron and zinc in grains and staple foods. J Food Compos Anal 20(6):496–505

Curie C, Cassin G, Couch D, Divol F, Higuchi K, Jean ML, Misson J, Schikora A, Czernic P, Mari S (2009) Metal movement within the plant: contribution of nicotianamine and yellow stripe 1-like transporters. Ann Bot 103:1–11

Distelfeld A, Cakmak I, Peleg Z, Ozturk L, Yazici AM, Budak H (2007) Multiple QTL-effects of wheat Gpc-B1 locus on grain protein and micronutrient concentrations. Physiol Plant 129:635–643

Ekiz H, Bagci SA, Kiral AS, Eker S, Gultekin I, Alkan A, Cakmak I (1998) Effects of zinc fertilization and irrigation on grain yield and zinc concentration of various cereals grown in zinc-deficient calcareous soil. J Plant Nutr 21:2245–2256

Ficco DBM, Riefolo C, Nicastro G, De Simone V, Di Gesu AM, Beleggia R, Platani C, Cattivelli L, De Vita P (2009) Phytate and mineral elements concentration in a collection of Italian durum wheat cultivars. Field Crop Res 111:235–242

Fiel B, Moser S, Jampatong S, Stamp P (2005) Mineral composition of the grains of tropical maize varieties as affected by pre-anthesis drought and rate of nitrogen fertilization. Crop Sci 45:516–523

Garvin DF, Welch RM, Finlay JW (2006) Historical shifts in the seed mineral micronutrient concentration of U.S. hard red winter wheat germplasm. J Sci Food Agric 86:2213–2220

Genc Y, Verbyla AP, Torun AA, Cakmak I, Willsmore K, Wallwork H, McDonald GK (2009) Quantitative trait loci analysis of zinc efficiency and grain zinc concentration in wheat using whole genome average interval mapping. Plant Soil 314:49–66

Gomez-Becerra HF, Erdem H, Yazici A, Tutus Y, Torun B, Ozturk L, Cakmak I (2010) Grain concentrations of protein and mineral nutrients in a large collection of spelt wheat grown under different environments. J Cereal Sci 52:342–349

Graham RD, Ascher JS, Hynes SC (1992) Selection of zinc-efficient cereal genotypes for soils of low zinc status. Plant Soil 146:241–250

Graham RD, Welch RM, Saunders DA, Monasterio I, Bouis HE, Bonierbale M, De Hann S, Burgos G, Thiele G, Liria R, Meisner CA, Beebe SE, Potts MJ, Kadiajn M, Hobbs PR, Gupta RK, Twomlow S (2007) Nutritious subsistence food systems. Adv Agron 92:1–74

Guttieri M, Bowen D, Dorsh JA, Raboy V, Souza E (2004) Identification and characterization of low phytic acid wheat. Crop Sci 44:418–424

Harvest Plus (2014) Biofortification progress briefs August 2014. www.HarvestPlus.org

Haydon MJ, Cobbett CS (2007) Transporters of ligands for essential metal ions in plants. New Phytol 174:499–506

Hess SY, Thurnham DI, Hurrell RF (2005) Influence of provitamin A carotenoids on iron, zinc and vitamin A status, vol 6, HarvestPlus technical monograph. International Food Policy Research Institute (IFPRI)/International Center for Tropical Agriculture (CIAT), Washington, DC

Huynh B, Palmer L, Mather DE, Wallwork H, Graham RD, Welch RM, Stangoulis JCR (2008) Genotypic variation in wheat grain fructan content revealed by a simplified HPLC method. J Cereal Sci 48:369–378

Joshi AK, Crossa I, Arun B, Chand R, Trethowan R, Vargas M, Ortiz-Monasterio I (2010) Genotype × environment interaction for zinc and iron concentration of wheat grain in eastern Gangetic plains of India. Field Crop Res 116:268–277

Kapil U, Jain K (2011) Magnitude of zinc deficiency amongst under five children in India. Indian J Pediatr 89:1069–1072

Karim MR, Zhang YQ, Zhao RR, Chen XP, Zhang FS, Zou CQ (2012) Alleviation of drought stress in winter wheat by late foliar application of zinc, boron, and manganese. J Plant Nutr Soil Sci 175:142–151

Kutman UB, Yildiz B, Cakmak I (2011) Improved nitrogen status enhances zinc and iron concentrations both in the whole grain and the endosperm fraction of wheat. J Cereal Sci 53:118–125

Leenhardt F, Lyan B, Rock Boussard A, Potus J, Chanliaud E, Remesy C (2006) Genetic variability of carotenoid concentration, and lipoxygenase and peroxidase activities among cultivated wheat species and bread wheat varieties. Eur J Agron 25:170–176

Liu Z (1996) Micronutrients in soils of China. Jiangsu Science and Technology Publishing House, Nanjing

Liu S, Mo X, Lin Z, Xu Y, Ji J, Wen G, Richey J (2010) Crop yield responses to climate change in the Huang-Huai-Hai plain of China. Agric Water Manage 97:1195–1209

Monasterio JI, Graham RD (2000) Breeding for trace minerals in wheat. Food Nutr Bull 21:393–396

Morgounov A, Gómez-Becerra HF, Abugalieva A, Dzhunusova M, Yessimbekova M, Muminjanov H, Zelenskiy Y, Ozturk L, Cakmak I (2007) Iron and zinc grain density in common wheat grown in Central Asia. Euphytica 155:193–203

Ortiz-Monasterio I, Palacios-Rojas N, Meng E, Pixley K, Trethowan R, Pena RJ (2007) Enhancing the mineral and vitamin content of wheat and maize through plant breeding. J Cereal Sci 46:293–307

Osborne B, Fearn T (1986) Near infrared spectroscopy in food analysis. Wiley, New York

Oury FX, Leenhardt F, Remesy C, Chanliaud E, Duperrier B, Balfourier F, Charmet G (2006) Genetic variability and stability of grain magnesium, zinc and iron concentration in bread wheat. Eur J Agron 25:177–185

Peleg Z, Cakmak I, Ozturk L, Yazici A, Jun Y, Budak H, Korol AB, Fahima T, Saranga Y (2009) Quantitative trait loci conferring grain mineral nutrient concentrations in durum wheat × wild emmer wheat RIL population. Theor Appl Genet 119:353–369

Ram S, Verma A, Sharma S (2010) Large variability exits in phytase levels among Indian wheat varieties and synthetic hexaploids. J Cereal Sci 52:486–490

Ram H, Singh S, Singh S, Sohu VS, Dhaliwal SS, Mavi GS, Cakmak I (2013) Use of foliar zinc sulfate fertilizer with pesticides for enriching zinc in wheat grains in India. In: XVII international plant nutrition colloquium held at the Istanbul, 19–22 Aug 2013

Ram H, Sohu VS, Cakmak I, Singh K, Buttar, GS, Sodhi GPS, Gill HS, Bhagat I, Singh P, Dhaliwal SS, Mavi GS (2015) Agronomic fortification of rice and wheat grains with zinc for nutritional security. Curr Sci 109:1171–1176

Rawat N, Tiwari VK, Neelam K, Randhawa GS, Singh K, Chhuneja P, Dhaliwal HS (2009a) Development and characterization of wheat-*Aegilops kotschyi* amphiploids with high grain iron and zinc. Plant Genet Res 7:271–280

Rawat N, Tiwari VK, Singh N, Randhawa GS, Singh K, Chhuneja P, Dhaliwal HS (2009b) Evaluation and utilization of *Aegilops* and wild *Triticum* species for enhancing iron and zinc content in wheat. Genet Res Crop Evol 56:53–64

Sadana US, Manchanda JS, Khurana MPS, Dhaliwal SS, Singh H (2010) The current scenario and efficient management of zinc, iron, and manganese deficiencies. Better Crops 2010:24–26

Shivay YS, Kumar D, Prasad R (2008) Relative efficiency of zinc sulphate and zinc oxide coated urea in rice-wheat cropping system. Commun Soil Sci Plant Anal 39:1154–1167

Singh MV (2011) Assessing extent of zinc deficiency for soil fertility mapping and nutrition security in humans and animals. Indian J Fertil 7:36–43

Singh K, Chhuneja P, Tiwari VK, Rawat N, Neelam K, Aggarwal R, Malik S, Keller B, Dhaliwal HS (2010) Mapping of QTL for grain iron and zinc content in diploid A genome wheat and validation of these loci in U and S genomes. In: Pag conference, San Diego

Tang JW, Zou CQ, He ZH, Shi R, Ortiz-Monasterio I (2008) Mineral element distributions in milling fractions of Chinese wheats. J Cereal Sci 48:821–828

Trethowan RM (2007) Breeding wheat for high iron and zinc at CIMMYT: state of the art, challenges and future prospects. In: Proceedings of the 7th international wheat conference, Mar del Plata

Trethowan RM, Reynolds MP, Sayre KD, Ortiz-Monasterio I (2005) Adapting wheat cultivars to resource conserving farming practices and human nutritional needs. Ann Appl Biol 146:404–413

Uauy C, Distelfeld A, Fahima T, Belchl A, Dubcovsky J (2006) A NAC gene regulation senescence improves grain protein, zinc and iron in wheat. Science 312:1298–1301

Velu G, Ortiz-Monasterio I, Singh RP, Payne T (2011a) Variation for grain micronutrients concentration in wheat core-collection accessions of diverse origin. Asian J Crop Sci 3:43–48

Velu G, Singh RP, Huerta-Espino J, Pena RJ (2011b) Breeding for enhanced zinc and iron concentration in CIMMYT spring wheat germplasm. Czech J Genet Plant Breed 47:S174–S177

Velu G, Singh RP, Huerta-Espino J, Pena-Bautista RJ, Arun B, Mahendru-Singh A, Yaqub Mujahid M, Sohu VS, Mavi GS, Crossa J, Alvarado G, Joshi AK, Pfeiffer WH (2012) Performance of biofortified spring wheat genotypes in target environments for grain zinc and iron concentrations. Field Crop Res 137:261–267

Welch RM (2001) Micronutrients, agriculture and nutrition; linkages for improved health and well being. In: Singh K, Mori S, Welch RM (eds) Perspectives on the micronutrient nutrition of crops. Scientific, Jodhpur, pp 247–289

Welch RM, Graham RD (1999) A new paradigm for world agriculture: meeting human needs. Productive, sustainable, nutritious. Field Crop Res 60:1–10

Welch RM, Graham RD (2004) Breeding for micronutrients in staple food crops from a human nutrition perspective. J Exp Bot 55:353–364

Wissuwa M, Ismail AM, Graham RD (2008) Rice grain zinc concentrations as affected by genotype, native soil-zinc availability, and zinc fertilization. Plant Soil 306:37–48

World Bank (2007) From agriculture to nutrition: pathways, synergies and outcomes. Report 40196-GLB. International Bank for Reconstruction and Development/The World Bank, Washington, DC

Yilmaz A, Ekiz H, Gultekin I, Torun B, Barut H, Karanlik S, Cakmak I (1998) Effect of seed zinc content on grain yield and zinc concentration of wheat grown in zinc-deficient calcareous soils. J Plant Nutr 21:2257–2264

Zhang Y, Song Q, Jan Y, Tang J, Zhao R, Zhang Y, He Z, Zou C, Ortiz-Monasterio I (2010) Mineral element concentrations in grains of Chinese wheat cultivars. Euphytica 174:303–313

Zhang YQ, Sun YX, Ye YL, Karim MR, Xue YF, Yan P, Meng QF, Cui ZL, Cakmak I, Zhang FS, Zou CQ (2012) Zinc biofortification of wheat through fertilizer applications in different locations of China. Field Crop Res 125:1–7

Zhao FJ, Su YH, Dunham SJ, Rakszegi M, Bedo Z, McGrath SP, Shewry PR (2009) Variation in mineral micronutrient concentrations in grain of wheat lines of diverse origin. J Cereal Sci 49:290–295

Zou CQ, Zhang YQ, Rashid A, Ram H, Savasli E, Arisoy RZ, Ortiz-Monasterio I, Simunji S, Wang ZH, Sohu V, Hassan M, Kaya Y, Onder O, Lungu O, Yaqub Mujahid M, Joshi AK, Zelenskiy Y, Zhang FS, Cakmak I (2012) Biofortification of wheat with zinc through zinc fertilization in seven countries. Plant Soil 361:119–130

Specialty Corn for Nutritional Security and Dietary Diversification 28

C.M. Parihar, Bhupender Kumar, S.L. Jat, A.K. Singh, M.L. Jat, V. Chaudhary, and Sain Dass

Abstract

Maize contributes more than half of the coarse cereal production in India. The grain is used for various purposes like feed, food, and several industrial purposes. It has tremendous potential to feed millions of hungry people of the African and Latin American countries as 5 % of world's dietary energy supply comes from maize. In addition, maize has been considered as industrial crop as it is being used as a raw material in many important industries, viz., starch, oil, alcoholic beverages, food sweeteners, pharmaceuticals, cosmetics, textile, paper, film, tire, food processing, packing, biofuel, etc. for developing hundreds of industrial products. Apart from these, quality protein maize (QPM) is being grown to meet the nutritional requirement of underprivileged. Besides these uses, sweet corn (SC) and baby corn (BC) are used for ensuring livelihood and green fodder security in *peri-urban* areas while popcorn (PC) is used as a nutritional alternative snack, etc.

Keywords

Baby corn • Popcorn • Quality protein maize • Sweet corn

C.M. Parihar (✉) • B. Kumar • S.L. Jat • A.K. Singh
ICAR-Indian Institute of Maize Research, New Delhi, 110 012, India
e-mail: pariharcm@gmail.com; bhupender.iari@gmail.com; sliari@gmail.com; aditya_jadon@yahoo.co.in

M.L. Jat
International Maize and Wheat Improvement Center-India, NASC Complex, New Delhi, 110 012, India
e-mail: m.jat@cgiar.org

V. Chaudhary
ICAR-IPNI-CIMMYT Collaborative Project, ICAR-Indian Institute of Maize Research, New Delhi, 110 012, India
e-mail: panwar.vandana21@gmail.com

S. Dass
EX-Project Director, Directorate of Maize Research, Pusa Campus, New Delhi 110 012, India
e-mail: sdass2010@gmail.com

U. Singh et al. (eds.), *Biofortification of Food Crops*, DOI 10.1007/978-81-322-2716-8_28

28.1 Introduction

Maize (*Zea mays* L.) is widely distributed in every part of world as it can be grown in diverse agroecologies including tropics, subtropics, and temperate regions. It grows up to 50 °N and S from the equator to more than 3000 m above mean sea level. The extent of cultivation is categorized from irrigated to semiarid conditions. Corn is the most productive cereal on a worldwide basis and is characterized by a genetic diversity of a very high order that imparts it a very favorable position toward meeting the emerging challenges. Maize is an important staple food in a large number of countries around the world including Latin America, Africa, Asia, etc.

In India maize grain is mainly used for feed (63 %), food (24 %), industrial purposes (12 %), and seed (1 %). Maize, also known as queen of cereals, is cultivated in more than 165 countries on 177 million hectare area, with a production of 875 million tonnes and average productivity of 5 t/ha. The USA has maximum area (35.4 million hectares, mha) and is followed by China (35.0 mha) and Brazil (14.2 mha). India is the fourth maize-growing country in the world with 8.71 mha area. After rice and wheat, maize is the third important crop of the country with current production of 22.26 m tonnes and average productivity of 2556 kg/ha (Yadav et al. 2014). In industry, maize is mainly used for starch, oil and dry milling purposes. As a food crop, maize has a major distinction as compared to other crops with respect to its utility pattern because of its several diverse types, like quality protein maize, sweet corn, popcorn, baby corn, etc. Cultivation of specialty corns can play a very significant role to check the migration of rural peoples toward the urban areas for their livelihood. Further, their cultivation can help to provide the quality nutrition to human beings and green fodder for animal from QPM, baby corn, and sweet corn, respectively.

28.1.1 Quality Protein Maize (QPM)

The nutritional value of quality protein maize has better say than the normal maize. The quality of normal maize protein is poor because of imbalanced amino acid composition and the deficiency of micronutrients. Normal maize protein is deficient in two main essential amino acids, viz., lysine and tryptophan, and possesses excess of leucine content. A major breakthrough came in the 1960s, with the discovery of the enhanced nutritional quality of the maize mutant *opaque 2* (Mertz et al. 1964). The beneficial effects of *o2* mutation resulted in reduced zein synthesis with higher levels of lysine and tryptophan in endosperm protein. But this mutation was also found associated with various deleterious pleiotropic effects, viz., soft chalky endosperm, lesser dry matter accumulation and, thus, lower economic (grain) yield, and dull soft chalky kernel phenotype with more susceptibility to ear rots and stored-grain insect pests. Because of all these reasons, such varieties could not be popularized. But scientists were successful in identifying various endosperm modifier genes that could favorably alter the grain characteristics, thereby overcoming an important obstacle in popularization of high lysine/tryptophan *o2* maize (Vasal 2000, 2001). The resultant germplasm was named as QPM, i.e., quality protein maize. Genetically, this has *opaque 2* gene with hard endosperm *He* gene (which confers kernel vitreousness) and genetic modifiers (many genes with similar, small, supplementary effects). Isolation and characterization of the *opaque2* gene revealed that it encodes a transcriptional factor that regulates the expression of zein genes and a gene encoding a ribosomal-inactivating protein. Quality protein maize is an improved version of maize which contains higher amount of lysine (>2.5 %) and tryptophan (>0.6 %) with lesser amount of leucine and isoleucine in the endosperm than those contained in normal maize (Table 28.1). Such balanced combination of amino acids in the endosperm results into its higher biological value, ensuring more availability of protein to human and animals than normal maize. The biological value of QPM is double than the normal maize, higher than wheat and rice, and matching with milk for true protein digestibility.

India has a large number of people with protein malnutrition. The prices of meat, egg, milk,

Table 28.1 Essential amino acid content in normal maize and QPM

Amino acid	Normal (mg per g N)	QPM (mg per g N)
Lysine	177	256
Isoleucine	206	193
Leucine	827	507
Sulfur amino acid	188	188
Aromatic amino acid	505	502
Threonine	213	199
Tryptophan	35	78
Valine	292	298

Source: Kaul et al. 2014

Table 28.2 Protein quality of maize

Quality measures	Normal	QPM
True protein digestibility	80	92
Biological value (%)	40–47	80
Amount needed for equilibrium	547	230

Source: Kaul et al. 2014

and their products have gone up, and due to that the poor people are able to afford these products. Therefore, high biological value of QPM will reduce food/feed cost and its requirement (Table 28.2 and Fig. 28.1). This will provide solution to malnutrition in human being and will also benefit poultry, livestock, pig, fish sectors, etc. QPM is the cheapest source of protein to the poor masses and can be used as nutritionally superior food for children, pregnant and lactating women, adolescent, and old age population of the country. QPM grain is a biofortified, non-transgenic food that provides improved protein quality to consumers. It looks and tastes like normal maize, but QPM contains a naturally occurring mutant maize gene.

Thus, QPM is solution to food and nutritional security, and several QPM value-added products have also been developed. In India several single-cross QPM hybrids (Table 28.3), viz., HQPM 1, HQPM 4, HQPM 5, HQPM 7, Vivek QPM 9, Shaktiman 1, Shaktiman 2, Shaktiman 3, and Shaktiman 4, have been developed for different agroclimatic conditions.

28.1.2 Baby Corn

Baby corn is a type of maize (*Zea mays* L.) with specialized traits and harvested as young immature unfertilized, tender cobs bearing 2–3 cm silks on the ear. The preferential/desirable size of baby corn is 6–11 cm length and 1.0–1.5 cm diameter with regular row/ovule arrangement. The most preferred color by the consumers/exporters is generally creamish to very light yellow. Baby corn is mainly grown in the peri-urban areas of the country. The dehusked young ears of baby corn is nutritionally rich with minerals, carbohydrates, and vitamins and can be eaten as raw, as vegetable, or in different recipe preparations such as pakora, mix vegetables, pickles, candy, murabba, kheer, halwa, raita, Chinese preparations, etc. The nutritional value of it is comparable to most of popular vegetables (Table 28.4). Nowadays, the popularity of baby corn is increasing, and its cultivation is catching up very fast. If baby corn is cultivated for non-commercial purposes, it can be grown with any type of maize. However, if it is grown for commercial cultivation or export purposes, then prolific varieties with desirable traits should be grown. It is a short-duration income-generating crop that provides avenues for crop diversification and intensification. The by-products such as tassel, young husk, silk, and green stalks generated through baby corn cultivation provide nutritious fodder to livestock. The silage prepared from these by-products of baby corn is nutritious and also provides green fodder during the lean period in the country (Chaudhary et al. 2014). Different vegetables, pulses, and flower crops can be taken as intercrops during the winter season baby corn for enhancing farm income and profitability. Generally, the cultivation practices of baby corn are more or less similar to commercial grain cultivation except these:

1. Preference for early maturing hybrids
2. Dense plant population
3. Higher seed rate
4. Higher doses of nutrients

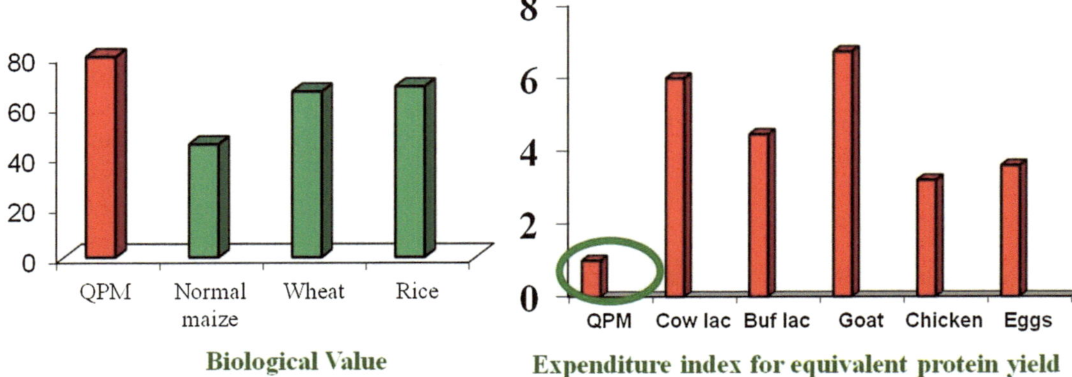

Fig. 28.1 Comparison of biological value of QPM with other cereals and milk, chicken, and eggs (Source: DMR Vision 2030)

Table 28.3 Quality characters of QPM hybrids

Hybrid	Hybrid type	% Protein in grain	% Tryptophan in protein
HQPM 7	Single cross	9.42	0.72
Vivek QPM 9	Single cross	8.46	0.83
HQPM 5	Single cross	9.8	0.76
HQPM 1	Single cross	9.36	0.94
Shaktiman 4	Single cross	9.98	0.93
Shaktiman 3	Single cross	9.63	0.73
Shaktiman 2	Single cross	9.3	1.04
Shaktiman 1	Three way	9.6	1.01

Source: Kaul et al. 2014

5. Detasseling
6. Harvesting of unfertilized cobs

The production cost of baby corn in India is less than many of the other countries of the world; therefore, it has a great potential for earning of foreign exchange through export (Table 28.5). The baby corn cultivation is labor intensive; thus, it has potential for employment generation as well.

28.1.3 Sweet Corn

There has been a significant increase in the demand for fresh market sweet corn in recent years due to the dramatic improvement in the sugar levels of new sweet corn cultivars (varieties). This is primarily due to the introduction of new "high-sugar" cultivars. These higher sugar levels are the result of new genes that improve the corn's sweetness. In sweet corn, the sugary gene prevents or retards the normal conversion of sugar into starch during endosperm development, and the kernel accumulates a water-soluble polysaccharide called "phytoglycogen." As a result, the dry, sugary kernels are wrinkled and glassy. The higher content of water-soluble polysaccharide adds a texture quality factor in addition to sweetness (Table 28.6). Today, the standard sugary corns are being modified with other endosperm genes and gene combinations that control sweetness to develop new cultivars. As a result, growers must consider genetic type while making selections for planting. There are around 13 endosperm mutants, besides sugary, which have been reported in various studies for sweetness in sweet corn. They all act differently with respect to the taste, texture, and

Table 28.4 Nutritive value of baby corn vis-à-vis other common vegetables (per 100 g of edible portion)

Nutrient	Baby corn	Cauliflower	Cabbage	Tomato	French bean	Lady's finger	Radish	Brinjal	Spinach
Moisture content (%)	89.1	90.8	91.9	93.1	91.4	89.6	94.4	92.7	92.1
Carbohydrates (g)	8.2	4	4.6	3.6	4.5	6.4	3.4	4	2.9
Protein (g)	1.9	2.6	1.8	1.9	1.7	1.9	0.7	1.4	2
Calcium (mg)	28	33	18	20	50	66	50	18	73
Phosphorus (mg)	86	57	47	36	28	56	22	47	21
Iron (mg)	0.1	1.5	0.9	1.8	1.7	1.5	0.4	0.9	10.9
Thiamine	0.5	0.04	0.04	0.07	0.08	0.07	0.06	0.04	0.03
Riboflavin	0.08	0.1	0.11	0.01	0.06	0.01	0.02	0.11	0.07
Ascorbic acid	11	56	12	31	11	13	15	12	28

Source: Hooda et al. 2011 (http://cornindia.com/babycorn/)

Table 28.5 Major baby corn-producing, exporting, and importing countries

Particulars	Countries
Producing countries	Thailand, People's Republic of China, India, Taiwan, South Africa
Exporting countries	Thailand, Taiwan, S. Africa, and India
Importing countries	The USA, the United Kingdom, France, Germany, Japan, Hong Kong, Singapore, Australia, Malaysia, Canada, Saudi Arabia, New Zealand, European countries, and India

Source: Dass et al. 2009, 2014

Table 28.6 Sweet corn nutritional value (per 100 g)

Energy	360 kJ	Methionine	0.067 g	Proline	0.292 g
Carbohydrates	19.02 g	Cystine	0.026 g	Serine	0.153 g
Sugars	9–44 g	Phenylalanine	0.150 g	Water	75.96 g
Dietary fiber	2.7 g	Tyrosine	0.123 g	Vitamin A equiv.	9 μg (1 %)
Fat	1.18 g	Valine	0.185 g	Thiamine (Vit. B_1)	0.200 mg (15 %)
Protein	3.2 g	Arginine	0.131 g	Niacin (Vit. B_3)	1.700 mg (11 %)
Tryptophan	0.023 g	Histidine	0.089 g	Folate (Vit. B_9)	46 μg (12 %)
Threonine	0.129 g	Alanine	0.295 g	Vitamin C	6.8 mg (11 %)
Isoleucine	0.129 g	Aspartic acid	0.244 g	Iron	0.52 mg (4 %)
Leucine	0.348 g	Glutamic acid	0.636 g	Magnesium	37 mg (10 %)
Lysine	0.137 g	Glycine	0.127 g	Potassium	270 mg (6 %)

Source: University of Illinois

Table 28.7 Major genes responsible for sweetness

Gene	Type	% sugar
su	Normal	9–16
se	Sugar enhanced	14–22
Sh2	Super sweet	28–44

Source: University of Illinois

sugar content. Major modifier genes for sweetness are shrunken-2 (*sh2*) and sugary enhancer (*se*) (Table 28.7). About 25 % of the kernels are double-mutant endosperm types possessing the enhanced benefits of the modifier. However, the sugary enhancer (*se*) gene along with one of the major modifier genes (e.g., *sh2*) will further modify some of the sugary kernels to about 44 % double-mutant endosperm types.

28.1.4 Popcorn

This type of maize is characterized by a very hard, corneous endosperm containing very small portion of soft starch. Popcorns are generally small-seeded, flint types with less test weight. It can be sold as un-popped on large scale for microwave and conventional uses as well as popped product with added flavor product. The shape of the kernels may be either pointed or round. The recently developed popcorn has thick seed coats in comparison to some primitive semi-popcorn, which has thin pericarps. The pericarp of the popcorn and starch packaging are the two important components for its popping quality. Popcorn is mainly used as snack particularly in metro cities, and its flour can be used to prepare various dishes. The popcorn commercial cultivation can be more beneficial, when it is grown on contract basis between the grower and a processor. Further, before planting there is need to specify the variety/hybrids and area to be planted. Mostly, there are two types of popcorn varieties, viz., butterfly and mushroom, where the mushroom types are considered to be more desirable. The good-quality popcorn variety has maximum volume and minimum percentage of left out un-popped kernel. While growing of popcorn to sell in the open market is risky, it depends on the entrepreneurial capacity or ability of the farmer/grower to sell in open market at reasonable prices. If the producers are also willing to become small-scale processors, then they can also pack and sell it in local market at small towns and cities.

28.2 Other Types of Corns

28.2.1 Waxy Corn

Generally, normal corn contains 75 % amylopectin and 25 % amylose, whereas waxy corn contains almost 100 % amylopectin. Amylopectin consists of branched glucose subunits, whereas amylose consists of unbranched glucose molecules. Waxy corn was found in China in 1908. The waxy trait is controlled by a single recessive gene, the *wx* gene. Waxy corn starch is used by the food industry as a stabilizer/thickener and in the paper industry as an adhesive.

28.2.2 High-Amylose Corn (Amylomaize)

Amylomaize is another special corn which is high in amylase content. The amylose content in amylomaize generally should be more than 50 %. It has an amylase extender mutant (*ae*), which results in an increase of around 60 % of amylase content as compared to normal one. However, there are some modifying factors, which alter the amylase contents along with agronomic performances of the corn. The amylose-extender gene expression is characterized by a tarnished, translucent, sometimes semifull kernel appearance. The starch from high-amylose corn is used in the textile industry, in gum candies (where its tendency to form a gel aids production), and as an adhesive in the manufacture of corrugated cardboard.

28.2.3 High-Oil Corn

High-oil corn contains approximately 7–8% oil which is 2–3% higher over normal corn. In addition to high oil content, the protein quality may be good in high-oil corn because of relatively larger germ size which contains protein of higher quality than the endosperm. The high-oil trait is controlled by many genes and is derived from the Illinois High-Oil selection program.

28.3 Biofortification of Maize

Billions of people in the developing countries suffer from an insidious form of hunger known as *micronutrient malnutrition* because of heavy dependence on maize as their staple foods for their normal diet as maize is more accessible to all the section of the society. So biofortification would provide a low-cost, sustainable strategy for reducing levels of micronutrient malnutrition. The other major limitation of nutritional quality of maize is the deficiency of micronutrients, viz., carotenoid composition and minerals such as iron and zinc. However, accurate assessment of their quantity in maize must be performed to direct effective and efficient breeding efforts. Biofortification of maize with enriched micronutrients has the potential to reduce the malnutrition where maize is a staple crop. Vitamin A being one of the major micronutrient in human nutrition is essential for the normal functioning of the visual system. The deficiency of vitamin A is a major health problem particularly in children and pregnant women. Maize is a staple food for many vitamin A-deficient populations. In typical maize, provitamin A carotenoids include α-carotene, ß-carotene, and ß-cryptoxanthin, but concentrations are low and range from 0 to 1.3, 0.13 to 2.7, and 0.13 to 1.9 nmol/g, respectively. In spite of the fact of the low concentration of provitamin A in maize, promotion of yellow maize in vitamin A-deficient areas should be the top priority in short term. Biofortification of maize with enhanced level of provitamin A, where maize is being used as staple food, would serve as a sustainable solution in the long run. Efforts to biofortify maize with provitamin A carotenoids have been successful. The high β-carotene maize (HBCM) is an agricultural solution to the problem of vitamin A deficiency (VAD) in sub-Saharan Africa. Biofortification of maize by increasing Fe and Zn concentration and/or bioavailability has also a great potential to alleviate their deficiency. Integrated genetic and physiological analyses of Fe and Zn nutrition in maize kernels that influence grain concentration and bioavailability (for identifying loci) are possible. Sufficient genetic variation is available in the maize germplasm for Fe (15–159 ppm) for mid-altitude and 14–134 ppm for lowland maize inbred lines, and Zn concentration ranges from 12–96 ppm for mid-altitude inbred and 24–96 ppm for lowland inbred lines (Dixon et al. 2000). The availability of genetic variation for all these quality traits in maize can be utilized for genetic enhancement of maize with enriched micronutrients, which has the potential to reduce the malnutrition where maize is a staple crop.

28.4 Value Addition in Baby Corn

The main aim of value addition of specialty corn is to enhance the farmer's income and farm profitability especially through baby corn and sweet corn. In the long run, value addition also promotes the installation of small- and medium-scale processing units at local/village level on cooperative basis so that the farmers are able to get safeguard against the price fluctuations of specialty corns. Such types of small- and medium-scale processing units (farmers' cooperative basis) of baby corn and sweet corn already are installed and working well at Atterna and Manouli villages near Sonipat, Haryana, India.

28.4.1 Grading

First steps of baby corn value addition are sorting and grading of raw material with the help of machine or manually as different sizes of baby corn can be used for different purposes. Baby corn of small size is used as salad, while relatively long-size baby corn can be used for making of pickles. However, international market specifications for baby corn on the basis of its size are available (Table 28.8). These above mentioned different grades of the baby corn can be used for following purposes such as short (in packing and processing industry and international marketing), medium (it can be used for local marketing, local consumption), and long (preferably should be used for grinded recipe preparations).

Table 28.8 Specifications for different grades of baby corn

Grade	Length (cm)	Diameter (cm)
Short	4–7	1.0–1.2
Medium	7–10	1.2–1.4
Long	11–13	1.4–1.5

Source: DMR, Hindi Bulletin Baby corn ki kheti and upyog

28.4.2 Packing and Processing

The packing of baby corn varies from unit to unit. It can be either in tin, glasses, or in poly bags. But for longer time of preservation, glass/tin packing is the best one. If it is packed in glass packing then it should have 52 % baby corn and 48 % brine solution. Generally for packing of baby corn farmers are using poly bags.

The baby corn can also be used for various purposes such as raw and salad (it can be eaten as raw and also add a variety in vegetable salad); mixed vegetables (where it can be used for making of various types of vegetables like chilly BC, cheese BC, spinach BC, mushroom BC, Manchurian, etc.) (Fig. 28.2) and recipes (different kinds of recipes like pakora by frying it in oil, chilla, tikki by mixing it with potato, halwa/burfi by frying in butter, etc.) can also be made by using baby corn (Fig. 28.3) and other dishes prepared with grinded BC (halwa/burfi by mixing it with ghee and dry fruits, kheer by mixing it in the milk in place of rice halwa, raita by mixing it with curd, etc.).

Baby corn can be processed for improving its shelf life by canning, jam, murabba, pickles, etc. (Fig. 28.3). The main processing methods used for improving the shelf life of baby corn include canning, dehydration, freezing, and preservation.

Canning is the most common method used for baby corn processing (Fig. 28.**4**). Baby corn can be canned in brine solution, stored for months, and transported to other places. The baby corn ears are generally canned at processing plants.

Fig. 28.2 Use of baby corn in salad and mixed vegetables (Source: DMR, Hindi Bulletin Baby corn ki kheti and upyog)

The following flow diagram explains the details of the processing sequences.

Peeled baby corn → cleaning → boiling → soaking → grading → containing → brine solution → exhaust → lid covering → cooling → quality inspection

Another method, viz., dehydration, can be used to increase shelf life of baby corn for longer period. Baby corn can be cut into ½ cm round pieces and dried in oven [air oven/vacuum oven] or can be solar dried. This dried baby corn can be packed in polythene pack/vacuum pack/tetra pack and can be stored well for longer period. Dehydrated baby corn can be rehydrated by soaking in water and can be used in preparation of food products. Products developed using dried baby corn have been found to be acceptable organoleptically like those prepared from fresh baby corn.

In freezing, like other frozen vegetables, baby corn can be frozen and stored for long period. Frozen baby corn can be used effectively for preparation of food products like soups and vegetables, and these food products are as acceptable as preparation made from fresh baby corn.

In preservation, after filling baby corn in container, brine and water is added in the ratio of 2:98 (brine 2 % and water 98 %); alternatively, a solution of 3 % brine, 2 % sugar, 0.4 % citric acid, and balance water can also be used.

28.5 Current Research Efforts at the Indian Institute of Maize Research

28.5.1 Provitamin A Enrichment Efforts

Vitamin A deficiency is a serious health issue, particularly among children in developing

Fig. 28.3 Use of baby corn in making recipes and processing (Source: DMR, Hindi Bulletin Baby corn ki kheti and upyog)

nations. To overcome vitamin A deficiency, biofortification is a promising strategy in maize-based areas where the supply and availability of animal products, fruits, and vegetables are limited. At IIMR, New Delhi, 40 genotypes (inbred lines having high provitamin A) were screened through column chromatography for total carotenoid and beta-carotene. Of these, four genotypes (IC639247, IC639563, IC632050, and IC639253) had total carotenoid more than 50 μg/g. Further, higher than 3 μg/g beta-carotene content was found in MRCHY4856-3-1-3-1-1, MRCHY5782-2-2-1-2-1, HEYPool-55-2-5, and HEYPool-60-2-3. The correlation studies proved considerable positive correlation between kernel color and total carotenoid. In another study, a separate set of 27 inbred lines were characterized for total carotenoid, beta-carotene, and beta-cryptoxanthin using ultrahigh pressure liquid chromatography (UPLC), as beta-carotene and beta-cryptoxanthin are the two foremost components of provitamin A in maize. The carotenoid content was found in the order: DML26-B (42.36 μg/g) > DML8 (2.08 μg/g) > DML17 (2.07 μg/g), whereas beta-cryptoxanthin content is higher in DML 17 (7.18 μg/g) compared with DML 8 (1.86 μg/g) (DMR 2013–2014)

Fig. 28.4 Baby corn after canning (Source: DMR, Hindi Bulletin Baby corn ki kheti and upyog)

28.5.2 Iron and Zinc Content Enhancement Efforts

As we are aware that micronutrient malnutrition influences substantial fraction of world population, enhancing Fe and Zn content in maize grain might help in alleviating the problem of micronutrient malnutrition or hidden hunger. The research efforts carried out at IIMR, New Delhi, clearly indicate the scope of hope. Forty-five inbred lines were selected for screening of Fe and Zn content in the maize. The inbred lines have ample variation for iron (18–104 ppm) and zinc (16.8–55.7 ppm) content. Among them DML221 and DML136 were superior for both Fe and Zn content in the grain (DMR 2013–2014).

28.6 Conclusion

The demand of baby corn, sweet corn, and popcorn is increasing every year in the world and India as well. As of now in India, there is 63 % shortage of green fodder, and hence, cultivation of baby corn and sweet corn will help for filling this gap without any additional land requirement with additional income in a shortest possible time. The farmers can earn 50–60 thousand rupees per annum per acre with the cultivation of 2–3 crops of baby corn and sweet corn and 100 quintal nutritious fodder/acre/crop. By feeding this fodder, there is an increase of about 15–20 % milk production, and the farmer does not require additional land for fodder as well. Therefore, specialty corn cultivation helps the livestock industry in meeting the regular supply of green fodder for the growing dairy industries to increase the milk production of the country. Due to the increase of urbanization, the change in food habit, and the improved economic status, the specialty corn has gained tremendous importance in peri-urban areas of the country. To check the migration from rural to urban and for enhancing the profitability and livelihood security of the farmer, the suitable hybrids and production technology for baby corn and sweet corn have been developed and popularized. As sufficient genetic variation exists for various quality traits in maize, the same can be used for the development of biofortified normal and specialty corns. Biofortification of corn with enriched micronutrients has the potential to reduce the malnutrition in the areas where maize is a staple crop or used as a food supplement.

References

Chaudhary DP, Jat SL, Kumar R, Kumar A, Kumar B (2014) Fodder quality of maize: it's preservation. In: Chaudhary DP et al (eds) Maize: nutrition dynamics and novel uses. Springer, New Delhi, pp 153–160

Dass S, Yadav VK, Jat ML, Shekar JC, Yadav Y (2009) Baby corn ki kheti and upyog. DMR Technical Bulletin 2009/2, Directorate of Maize Research, Pusa Campus, New Delhi, p 44

Dass S, Jat SL, Parihar CM, Chikkappa GK, Kamboj MC, Singh AK, Chaudhary V (2014) Scientific cultivation of babycorn. In: AK Choudhary et al (eds) Advances in vegetable agronomy pp 260–270. Post Graduate School IARI,NEW DELHI,ISBN 978-93-83168-17-0.

Dixon BM, Kling JG, Menkir A, Dixon A (2000) Genetic variation in total carotene, iron, and zinc contents of maize and cassava genotypes. Food Nutr Bull 21(4), The United Nation University

DMR (2013–2014) Annual report, Directorate of Maize Research, Pusa Campus, New Delhi

Hooda KS, Chikkappa GK, Kumar B, Rajendran RA, Kumar P, Jat SL (2011) DMR Vision 2030. http://cornindia.com/babycorn/. Accessed 4 Apr 2014

Kaul J, Kumar R, Chaudhary DP (2014) Quality protein maize for food, feed and nutritional security. Indian farm 64(4):73–76

Mertz ET, Bates LS, Nelson OE (1964) Mutant gene that changes protein composition and increases lysine content of maize endosperm. Science 145:279–280

Vasal SK (2000) Quality protein maize story. Food Nutr Bull 21:445–450

Vasal SK (2001) Keynote address, quality protein maize development: an exciting experience seventh eastern and southern Africa regional maize conference. CIMMYT, Apdo, Mexico, 11–15 Feb 2001

Yadav OP, Chikkappa GK, Jat SL, Dhillon BS (2014) Overview of maize improvement in India. Indian farm 64(4):5–11

29 Biofortification of Pearl Millet for Nutritional Enrichment

Meenakshi Chaudhary and D. Singh

Abstract

Fuzzy modeling is a methodology that works with partial truths: it can answer questions to which the answers are "yes" and "no" at different times or partly "yes" and "no" at the same time. It can be used to match any type of data, particularly incomplete and imprecise data, and it is able to improve the precision of testing of such data. It can be applied with any type of statistical distribution, and it is, particularly, suitable for uncommon and unexpected non linear relationships. This chapter assesses the use of fuzzy modeling of clinical data.

Keywords

Biofortification • Pearl millet • *Pennisetum glaucum* • Value-added product

29.1 Introduction

Nutrient and micronutrient deficiency is a worldwide health problem in both rural and urban areas around the globe. Around three billion people in the world are deficient in key vitamins, minerals, and micronutrients involving zinc, iron, iodine, etc. (Dahiya et al. 2008), which causes serious disorders like anemia, goiter, dwarfism, hypogonadism, etc. among the masses of meager income group who are mostly ignorant toward health consciousness and are devoid of optimal dietary practices as well. In developing countries of Asian and African origin, micronutrient disorders among the population is rampant which is a serious concern and needs immediate intervention. Hence, to overcome from such disorders, it is necessary to have a balanced diet which is rich in zinc, magnesium, calcium, vitamins, and all other health benefiting minerals. Majority of the population is having their staple food rich in starch but lacks the micronutrients, thus adding more to the existing problem. Further, natural calamities

M. Chaudhary (✉)
Swami Keshwanand Rajasthan Agricultural University, Bikaner, Rajasthan 334 006, India
e-mail: meenu115@yahoo.com

D. Singh
ICAR-Central Institute for Arid Horticulture, Bikaner, Rajasthan 334 006, India
e-mail: dhuren6@gmail.com

U. Singh et al. (eds.), *Biofortification of Food Crops*, DOI 10.1007/978-81-322-2716-8_29

occurring in these regions lead to massive crop loss which can be sustained through growing of minor cereal crops such as pearl millet, jowar, finger millet, proso millet, and foxtail millet among which pearl millet *(Pennisetum glaucum*), also recognized as bajra, is grown on a large scale in tropical semiarid regions of the world primarily in Africa and Asia. It is well adapted to low rainfall (200–600 mm), low soil fertility, and high temperature, thus making it possible to grow in areas where majority of the cereal crops do not do well. Pearl millet forms the staple food among millions of household of poor people dwelling in the hot arid regions of the world as it is having high drought resistance and is generally utilized as a green fodder crop as well as food crop. Moreover its richness in iron, zinc, and antioxidants makes it beneficial for the overall health and well-being of the people. India is soon going to commercially cultivate biofortified pearl millet, enriched with enhanced iron and zinc components. It will help to lessen the hidden hunger underlying the malnourished population especially the poor who are devoid of the minimal micronutrient uptake.

Biofortified pearl millet will help in increasing the iron absorption in people by 5–10 % and grain yield by 5–6 %. Though its volume of production is low, it is a highly strategic crop as it commands security among the poor in dry rural areas (Kelley et al. 1996; HarvestPlus 2009a). Today, 35 million people consume pearl millet as that costs much less than rice and wheat. It has 50–65 ppm iron, about twice more iron than modern wheat varieties and raises the hemoglobin level. Fortification of pearl millet with added zinc helps in improving the person's immunity, neural systems, and reproductive health. As no separate agricultural practices have to be incorporated, hence the agro techniques remain the same as the traditional ones, thus making it farmer friendly (HarvestPlus 2009b, 2010). During the 12th five-year plan, Indian Council of Agricultural Research and the Indian Council of Medical Research (ICMR) propose to introduce biofortified cereals in a larger way to combat the malnutrition. However, the actual intake of measurable iron and zinc in biofortified product in comparison to the non-biofortified commercial varieties in our body needs to be justified.

29.2 Nutritional Value of Pearl Millet

Pearl millet having higher calorific value than wheat is high in protein, iron, calcium, and folic acid and prevents anemia. It reduces the fat level as well as promotes a healthy nervous system. It helps to overcome acidity due to its alkaline nature. It is a rich source of energy (361 Kcal/100 g) and is comparable with commonly consumed cereals such as wheat (346 Kcal/100 g), rice (345 Kcal/100 g), maize (125 Kcal/100 g), and sorghum (349 Kcal/100 g) as per the *Nutritive Value of Indian Foods* (Gopalan et al. 1989). The nutritional details are given in Tables 29.1, 29.2, and 29.3.

29.3 Recent Trends in Research

Two non-GMO biofortified and one traditional pearl millet varieties were compared under abrasive decortication studies to evaluate their potential for increasing iron and zinc content (Hama et al. 2012). The phytate-to-mineral ratios were used to estimate mineral bioavailability. Iron and zinc contents in the biofortified varieties Tabi and GB8735 were two- to threefold higher than in the traditional variety. Iron content reached 7.2 and 6.7 mg per 100 g DM in the biofortified varieties, which corresponds to the target values of biofortification programs. Zinc content was, respectively, 5.6 and 4.1 mg per 100 g DM in the GB8735 and Tabi varieties. Because of the presence of phytate and other chelating factors that were only partially removed during decortication, there was no improvement in iron bioavailability in the biofortified varieties. But whatever extraction rate, phytate-to-zinc ratios ranged between 6 and 18; zinc absorption could be improved by using these biofortified varieties for food processing.

Development of crop cultivars with elevated levels of micronutrients is being increasingly

Table 29.1 Comparative nutritional value of pearl millet and other staple food grains

Constituent	Pearl millet	Wheat	Rice	Sorghum	Maize
Protein (g)	11.6, 11.8[a], 8[b], 11[c]	11.8	6.8	10.4	4.7
Fat (g)	5.0, 4.8[a], 2.4[b], 5.0[c]	1.5	0.5	1.9	0.9
Crude fiber (g)	1.2, 2.3[a], 2.2[c]	1.2	0.2	1.6	1.9
Carbohydrate (g)	67[a]	71[a]	76[a]	70[a]	73[a]
Calcium (mg)	42, 42[a], 39[b], 25[c]	41	10	25	9
Phosphorus (mg)	296	306	160	222	121
Zinc (mg)	3.1, 2.2[b]	2.7	1.4	1.6	1.48[d]
Sodium (mg)	10.9, 5[b]	17.1	–	7.3	51.7
Magnesium (mg)	137, 106[b]	138	90	171	40
Vitamin A (mcg)	132	64	0	47	32
Thiamine (mg)	0.33, 0.38[a], 0.30[b],	0.45	0.06	0.37	0.11
Riboflavin (mg)	0.25,0.21[a].0.15[c]	0.17	0.06	0.13	0.17
Niacin (mg)	2.3, 2.8[a], 3.2[b], 2.0[c]	5.5	1.9	3.1	–
Folic acid (mg)	45.5	36.6	8	20	0.6
Vitamin C (mg)	0	0	0	0	6

[a]Hulse Laing and Pearson (1980) US National Research Council/National Academy of Sciences. 1982. USDA/HNIS. 1984

[b]Source: http://www.wholehealthmd.com/refshelf/Foodsview/1,1523,72,00.html

[c]Gopalan et al. (1989)

[d]Hemalatha et al. (2007a, b)

Table 29.2 Proximate composition (g/100 g) of pearl millet grains

Moisture	Protein	Fat	Ash	Fiber	Reference
–	–	6.1–7.5	–	–	Dalvi (1995)
–	8.1–13.9	3.4–7.4	1.1–2.4	–	Hadimani et al. (1995)
–	12.3	5.3	2.4	–	Bashay (1996)
–	10.2	6.8	–	–	Palande et al. (1996)
7.4–8.4	8.7–10.8	6.0–6.4	–	–	Elkhalifa and Singh (1996)
11.30–11.49	11.3–13.2	7.2–7.8	1.9–2.1	1.64–1.99	Archana (1997)
11.26–11.37	12.8–13.1	6.2–7.4	1.9–2.1	1.74–1.95	Rekha (1997)
–	8.5–15.1	2.7–7.1	1.6–2.4	2.6–4.0	Abdalla et al. (1998)
–	16.9	5.1	1.5	1.6	Malleshi and Kloptenstein (1998)
–	–	6.2	–	–	Banger et al. (1999)
–	11.4	–	–	–	Oshodi et al. (1999)
7.13	12.2	4.5	0.25	0.54	Akubor and Obiegbuna (1999)
8.1–9.26	10.9–13.3	5.8–7.1	1.4–2.0	0.8–1.1	Malik (1999)
8.78	10.36	7.63	2.03	1.26	Poonam (2002)
10.12–13.30	8.83–13.67	4.42–6.90	1.74–2.54	1.30–2.50	Sehgal et al. (2002)

recognized as one of the approaches to provide sustainable solutions to various health problems associated with micronutrient malnutrition, especially in developing countries (Velu et al. 2007; Andrews and Kumar 1996). To assess the prospects of this approach in pearl millet (*Pennisetum glaucum*), a diverse range of genetic materials, consisting of 40 hybrid parents, 30 each of population progenies and improved populations, and 20 germplasm accessions, was analyzed for grain iron and zinc content, deficiencies of which adversely affect human health. Based on the mean performance in two seasons at ICRISAT, Patancheru, India, large variability among the entries was found, both for Fe (30.1–75.7 mg/kg on dry weight basis)

Table 29.3 Mineral composition (mg/100 g) of pearl millet grains

Phosphorus	Calcium	Iron	Zinc	Copper	Manganese	Reference
302	51.4	16.3	2.67	1.21	1.53	Alpana (1989)
–	50.0	18.0	2.7	–	–	Khetarpaul and Chauhan (1991)
290	54.8	11.8	4.09	–	–	Aggarwal (1992)
272–326	27–46	–	–	–	–	Saxena et al. (1992)
–	51.4	16.3	2.67	–	–	Kumar and Chauhan (1993)
–	52.4–58.1	8.5–8.9	3.6–4.4	–	–	Chaudhary (1993)
300.5	–	3.00	–	–	–	Hadimani and Malleshi (1993)
355	79.0	–	–	–	–	Serna-Saldivar et al. (1994)
366	49.0	9.76	1.80	–	–	Sharma (1994)
349.0–376.7	45.7–50.0	8.9–9.7	2.7–2.9	1.11–1.23	1.16–1.20	Archana (1997)
364.0–385.6	44.5–49.7	8.9–9.4	2.7–2.8	1.16–1.23	1.16–1.23	Rekha (1997)
450.0–990.0	10.0–80.0	7.0–18.0	5.3–7.0	1.0–1.8	1.8–2.3	Abdalla et al. (1998)
–	19.9–22.5	8.3–9.9	2.8–3.2	0.9–1.0	–	Malik (1999)
348.40	39.60	8.16	2.85	0.99	1.20	Poonam (2002)

and Zn (24.5–64.8 mg/kg). The highest levels of grain Fe and Zn were observed in well-adapted commercial varieties and their progenies and in the parental lines of hybrids. There were indications of large within-population genetic variability for both Fe and Zn. The correlation between Fe and Zn content was positive and highly significant ($r = 0.84$; $P < 0.01$). These results indicate that there are good prospects of simultaneous selection for both micronutrients and that selection within populations, especially those with predominant IARI germplasm, is likely to provide good opportunities for developing pearl millet varieties and hybrid parents with significantly improved grain Fe and Zn content in pearl millet.

Iron-rich pearl millet is being conventionally bred by the International Center for Tropical Agriculture as part of the HarvestPlus program, coordinated by the International Food Policy Research Institute (IFPRI) and the International Center for Tropical Agriculture, which seeks to develop and disseminate staple food crops rich in micronutrients to improve nutrition and public health. One recently released variety, ICTP8203Fe, commonly known as *Dhanshakti* (meaning prosperity and strength), is now being cultivated by more than 30,000 farmers in Maharashtra (ICRISAT 2013). It is the first biofortified crop cultivar to be officially released and adopted by farmers in India. *Dhanshakti* is an improved version of a pearl millet variety developed from a species found in northern Togo in West Africa. In Benin, another study found that marginally iron-deficient women absorbed twice the amount of iron from biofortified pearl millet than they did from ordinary pearl millet. According to the findings of the study, children under the age of three could meet their full daily iron needs from just 100 g of the iron-rich pearl millet flour. Children under two years old, who might eat less, would still benefit substantially from eating iron-rich pearl millet. A randomized trial on the effects of an iron-biofortified pearl millet intervention on iron status in children (Haas et al. 2013) was conducted on an iron-biofortified pearl millet intervention (vs. control pearl millet), daily for a six-month period in India. Linear and binomial regression models were used to evaluate the effect of the intervention on iron status in children. It was concluded that supplementation with iron-biofortified pearl millet significantly resolved iron deficiency in children within six months.

For effective crop breeding program variation among the wild, landraces and cultivated species form the basic resource to generate improved crop types with desirable traits. Modern crop breeding envisages the identification of valuable alleles in the wild ancestors of crop plants and their reintroduction in the cultivated crops. Potential of landraces for breeding of Fe- and

Zn-rich pearl millet has been demonstrated by several workers. Further, the usage of quantitative trait locus (QTL) for mapping of loci/gene to expedite the marker-assisted selection (MAS) will enhance the estimation of genetic diversity. Kumar et al. (2015) co-localized two QTLs for Fe and Zn in pearl millet using 106 RILs and estimated 19 % and 36 % phenotypic variation for Fe and Zn, respectively. This will help in introgression of genes in locally adapted pearl millet cultivars leading to nutritional efficacy and effectiveness of a biofortified crop.

29.4 Health Benefits of Pearl Millet

Millets are rich storehouse of starch making it a high-energy food (Fig. 29.1). It is also enriched with proteins and fiber. The amino acids in the pearl millet are easily digestible than the ones found in wheat. Due to its rich reservoir of nutrients such as methionine, B complex vitamins, folic acid, lecithin, potassium, magnesium, manganese, and zinc, millets can be effectively used for several purposes. Niacin lowers the cholesterol level while magnesium helps to maintain good heart health, as it reduces blood pressure and checks the risk of heart attacks (Nambiar et al. 2011). Pearl millet being a rich source of phosphorus plays an important part in the structure of body cells. Phosphorus, found in pearl millets, is a significant component of several necessary compounds including adenosine triphosphate (ATP) and a crucial component of nucleic acids. Recent studies have proven that regular consumption of pearl millets help in preventing gallstones in women. They contain insoluble fibers which not only speed up intestinal transit time but also reduce the secretion of bile acids. Pearl millets are known to increase insulin sensitivity and lower the level of triglycerides. Regular intake of millets provides protection against breast cancer in premenopausal women. Apart from that, it has also shown a considerable reduction in the occurrence of wheezing and asthma in children (Huang and Ferraro 1978).

Millets contain an essential phytonutrient, lignin, which is very beneficial for the human body. The lignans in the presence of natural flora are converted to mammalian lignans which fight against hormone-dependent cancer and reduce the risk of cardiac arrests (Kinsella et al. 1990). The risk of type 2 diabetes is reduced if one consumes pearl millets. It has also been reported that pearl millet being a good source of magnesium acts as a cofactor in a number of enzymatic reactions. Pearl millet's attribution toward several health-promoting abilities is enlisted below (Table 29.4).

Fig. 29.1 Greenish grains of pearl millet – nature's capsulated hidden nutrients and view of preparing value-added chappati

Table 29.4 Possible health benefits of pearl millet

Disease/ problem	Possible benefit	Positive factor
Anemia	May help in increasing the Hb	Elevation of iron content (8 mg/100 g)
Constipation	May help in dealing with constipation	Elevation of zinc content (3.1 mg/ 100 g)
Cancer	Anticancer property	Antioxidant property, high flavonoids
Diabetes	Help in dealing with diabetes	Has low glycemic index
Diarrhea	Probiotic treatment	Lactic acid bacteria
NCDs	Inhibits DNA scission, LDL cholesterol, liposome oxidation and proliferation of HT-29 adenocarcinoma cells	Flavonoids, phenolics, omega-3 fatty acids

29.5 Constraints in Utilization of Pearl Millet

In spite of greater availability, low cost, and comparatively good nutritional value, usage of pearl millet in the food industry is negligent. The major constraints which interfere in utilization are its gray color, poor shelf life of flour, rapid development of rancidity, and bitterness in the flour leading to its low acceptability. Pearl millet can be stored for longer periods without qualitative deterioration if the kernels remain intact. Quality of meal deteriorates rapidly if the grain is decorticated and grinded. The lipids of the meals release FFA and peroxides on getting hydrolyzed and oxidized.

29.6 Nutritional Enrichment Through Processing

Pearl millet on getting processed through milling, decortications, germination, malting, blanching, fermentation, and popping leads to improved nutritional quality and better consumer acceptability. Storage stability of ground millet gets improved on dry milling as the major lipid containing portions of grain (i.e., the germ and covering layered) from the endosperm gets removed (Bookwalter et al. 1987). Though decortications of raw grains significantly results in loss of protein, insoluble dietary fiber, fat, ash, lysine, and other amino acids, it increases the apparent protein and dry matter digestibility (Jenkins et al. 1978; Serna-Saldivar et al. 1994). Besides, increased digestibility is noted in consuming debranned pearl millet seeds as a result of reduced levels of phytic acid, amylase inhibitors, and polyphenols (Sharma and Kapoor 1996; Malleshi and Klopfenstein 1998). Higher protein efficiency ratio and bioavailability of minerals as compared to other cereals was observed in processed pearl millet products.

29.6.1 Malting

Malting of grain improves the contents of total protein, fat, ash, certain amino acids, total sugars, and B-group vitamins. It causes significant dry matter loss in grain, and it is attributed to the metabolic activity and separation of vegetative growth. The malt enzyme decreases the viscosity of foods, improving the palatability of products for children. Malting helps to decrease the starch, protein, crude fiber, fat, and polyphenol content and increase soluble protein, free amino acid, and reducing sugar. It produces high amount of reducing sugars and high amylase activity. Malting helps to improve the availability of nutrients and sensory attributes and extends the shelf life. One can add sorghum and millet malts to low dietary bulk and calorie-dense weaning food as supplements. Improvement in the germination of grains on malting helps to degrade or modify the endosperm with a minimal loss in grain weight. Further mobilization of seed reserves and enhancement of a and ß amylase activities, lipids, and protease were observed.

29.6.2 Blanching

Pearl millet on blanching effectively enhances its shelf life and is one of the preliminary processing treatments. The enzyme activity gets slowed down on blanching, thus retaining the nutrient content of the pearl millet flour. The grains are subjected to boiling at 98 °C in container in 1:5 ratio of seeds to boiling water for 30 s and finally dried at 50 °C for 60 min.

29.6.3 Acid Treatment

Acid treatment of any kind helps to lighten the gray color and thus improve its acceptability. Acid solutions like sour milk and tamarind pods when used as a soaking medium markedly reduce the color of the grain. Various acidic solutions have been tried till date among which dilute hydrochloric acid has been found to be more effective in removal of pigments from whole grain before milling as compared to citric acid and acetic acid. The grains are to be soaked in dilute HCl for 15–24 h leading to major removal of the pigments and production of creamy white grains. Soaking of pearl millet in 0.2 N HCl for 6, 12, 18, and 24 h decreased the total phosphorus, phytate phosphorus, and polyphenol content, and protein digestibility in vitro increased. Fat acidity increased by 1.5-fold in acid-soaked grain flour as against sixfold increase in dry heated and untreated grain flour during storage (Kachare et al. 1988).

29.6.4 Dry Heat Treatment

Inactivation of lipase activity is very much necessary before the milling operation as it is responsible for spoilage of pearl millet meal. The lipase activity can be minimized through dry heating of the meal, thereby minimizing lipid decomposition during storage. Dry heat treatment of the pearl millet seeds was done in a hot air oven at 100 ± 2 °C at different intervals of 30, 60, 90, and 120 min followed by cooling to room temperature. Significant reduction in fat acidity and free fatty acids in the processed meal was obtained which was three- to fourfold lower than the meal from the unheated grain. Similarly, lipolytic decomposition of lipids during storage got reduced on heating of grains for 120 min.

29.6.5 Popping

Popped grain is a crunchy, porous, and precooked product. Popping improves taste and flavor and leads to the development of pleasing texture. Popping formulation has very low moisture. This decreased moisture helps in minimizing deterioration during storage which increases the shelf life of products. Popping can be done using common salt as heating medium in an open iron pan, containing sample and salt in the ratio of 1:10 at 240–260 °C for 15–25 s.

Utilization of pearl millet in food products – in India pearl millet is mostly used in the preparation of conventional foods such as porridge, roti, etc. Mostly coarse or finely grounded millet flour is utilized for the preparation of the abovementioned products after separation and removal of bran. To overcome the problem of rancidity in pearl millet flour within a few days of milling, the grains of pearl millet are subjected to processing treatment like blanching, malting, popping, dry heat treatment, etc., before product development, the output being longer shelf life and better nutritional profile.

29.7 Nutrient-Rich Baked Products

Pearl millet flour is not in demand in the baking industry, since its flour lacks gluten and forms dough of poor consistency. Pearl millet biscuit had better amino acid composition; therefore, biscuit prepared from pearl millet and chickpea had better TD and NPU as compared to that prepared from finger millet and chickpea. It appeared from various resources that more than fifty value-added products have been developed in India from pearl millet by cooking and baking.

Table 29.5 Nutritional evaluation of pearl millet-based baked products (g/100 g)

Name of products	Protein	Fat	Crude fiber	Ash
Chocolate cake	12.41	51.51	1.91	0.43
Chocolate cake (eggless)	8.97	23.17	0.41	1.02
Walnut cake	21.73	59.37	2.85	1.06
Chocolate muffin	12.41	51.51	1.91	0.43
Coconut biscuit	11.30	38.28	1.29	1.06
Chocolate ring biscuit	12.81	59.16	0.80	2.30
Sweet salty biscuit	7.71	35.14	0.48	0.68
Sweet paste biscuit	13.18	59.45	0.80	2.36

These products are locally termed as puri, paratha, khichra, matri, kheer, halwa bati, laddu, dhokla, cakes, biscuits, etc. (Dhukwal et al. 2013). Total soluble sugar and reducing sugar were also higher in the product prepared from malted pearl millet than blanched pearl millet. Total minerals, i.e., Ca, P, Fe, and Mn contents, were maximum in products prepared from blanched pearl millet flour than malted pearl millet flour. Dietary fiber components were higher in the products prepared from malted pearl millet flour. Various types of health foods can be prepared from millet. Millet contains a relatively higher proportion of unavailable carbohydrates, and the release of sugar from millet-based diet is slow that adds to its suitability to be used as diabetic food. Six recipes of commonly consumed foods (pearl millet, jowar, ragi, and green gram) were tested in western India for their glycemic index. Recipe based on pearl millet was found to have the lowest glycemic index as compared to other recipes (Mani et al. 1993). Millets, fenugreek seeds, and legumes after processing were used to formulate three nutritious diabetic food products such as dhokla, upma, and laddu which showed that upma had the lowest glycemic index (17.60), followed by laddu (23.52) and dhokla (34.96). RTE mixes like dhokla, chapatti, instant idli, pasta, and biscuit were developed for diabetics after incorporating bleached pearl millet fenugreek or bengal gram seed coat in different proportions. Results showed that minimum glycemic index was observed in pearl millet-based dhokla (37.99) followed by chapatti (48.00), instant idli (52.13), and pasta (54.12) and biscuit (58.09). Products were found to contain high fiber content as compared to control and can be used for diabetics (Table 29.5).

29.8 Conclusion

Breeders apart from working on traditional objectives like disease resistance, yield, drought tolerance, etc., need to emphasize on biofortification based on micronutrient deficiency rates as there is growing demand of its technical feasibility. Crops like pearl millet which are enriched with micronutrients can have biological impact on breeding without compromising on its agronomic traits. Biofortification helps to strengthen the arm for controlling micronutrient deficiencies as evidenced through several cost-benefit analyses studies. Consumer acceptability of the biofortified products needs awareness among the masses, thereby increasing the intake of the target nutrients. All the pearl millet-based products generally have strong flavor and aroma, more nutritional value, and health benefits as compared to similar products developed from major cereals. These value-added products are not available in the market. The cost of pearl millet-based value-added products comes to about 50 % less than that of traditional products. The technology of acceptable pearl millet-based products can be taken up by housewives so as to improve the nutritional status of the family in toto. Efforts can be made to popularize millet-based, low-cost, high-protein, and energy-rich products among population through ongoing nutritional intervention programs.

References

Abdalla AA, El-Tinay AH, Mohamed BE, Abdalla AH (1998) Proximate composition, starch, phytate and mineral contents of 10 pearl millet genotypes. Food Chem 63(2):243–246

Aggarwal AJ (1992) Processing of pearl millet for its more effective utilization. PhD thesis, CCS Haryana Agricultural University, Hisar, India, 183 pp

Alpana (1989) A study on nutritional value and utilization of sprouted pearl millet. MSc thesis, CCS Haryana Agricultural University, Hisar, India, 62 pp

Andrews DJ, Kumar KA (1996) Use of the west African pearl millet landrace Iniadi in cultivar development. Plant Genet Resour Newsletter 105:15–22, ISSN 0048-4334

Archana (1997) Utilization of processed pearl millet for the development of weaning and supplementary foods. MSc thesis, CCS Haryana Agricultural University, Hisar, India, p 121

Banger MU, Bhite BR, Kachare DP, Chavan JK (1999) Role of phenolics and polyphenol oxidizing enzymes in odour generation in pearl millet meal. J Food Sci Technol 36(6):535–537

Bashay AY (1996) Nutritional and chemical evaluation of pearl millet grains [Pennisetum typhoides (Brum, f.) Stapf and Hubbard, Poaceae] grown in the Gizan area of Saudi Arabia. Int J Food Sci Nutr 47(2):165–169

Bookwalter AW, Kirleis AW, Mertz ET (1987) In vitro digestibility of protein in milled Sorghum and other processed cereals with and without soy-fortification. J Food Sci 52:15–57

Chaudhary S (1993) Preparation of nutritional evaluation of some traditional foods from pearl millet. MSc thesis, CCS Haryana Agricultural University, Hisar, India, p 98

Dahiya S, Chaudhary D, Jaiwal R, Dhankher O, Singh R et al (2008) Elemental biofortification of crop plants. In: Jaiswal P, Singh R, Dhankar OP (eds) Plant membrane and vacuolar transporters. CABI International, Wallingford/Cambridge, pp 345–371

Dalvi US (1995) Effect of hot water blanching of seeds on changes in lipids of promising pearl millet cultivars during storage. MSc (Agri.) thesis, Mahatma Phule Krishi Vidyapeeth, Rahuri, India, 112 pp

Dhukwal V, Goyal M, Jain N, Mattar P (2013) Booklet on pearl millet ke paushtik ewam swadisht vyanjan. Nutri Farm Pilot Scheme, pp 1–24

Elkhalifa EA, Singh U (1996) Evaluation of grain quality of newly developed pearl millet cultivars. UK J Agri Sci 4(2):44–55

Gopalan C, Sastri BVR, Balasubramanian SC, Rao BSN, Deosthale YG, Pant KC (eds) (1989) Nutritive value of Indian foods. National Institute of Nutrition, Indian Council of Medical Research, Hyderabad

Haas JD, Finkelstein JL, Udipi SA, Ghugre P, Mehta S (2013) Iron bio fortified pearl millet improves iron status in Indian school children: results of a feeding trial. FASEB J 27:355.2

Hadimani NA, Malleshi NG (1993) Studies on milling, physico-chemical decortication on the nutritional value of sorghum (Sorghum bicolor L. Moench) and fermentation by yeasts and lactobacilli on H C 1 – extractability of minerals from pearl millet. Food Chem 39:347–355

Hadimani NA, Allis SZ, Malleshi NG (1995) Physico-chemical composition and processing characteristics of pearl millet varieties. J Food Sci Technol 32:193–198

Hama F, Icard-Vernière C, Jean-Pierre G, Rochette I, Diawara I, Mouquet-Rivier C (2012) Potential of non-GMO biofortified pearl millet (*Pennisetum glaucum*) for increasing iron and zinc content and their estimated bioavailability during abrasive decortications. Int J Food Sci Technol 47 (8):1660–1668

HarvestPlus (2009a) Breeding crops for better nutrition. HarvestPlus, IFPRI, Washington, DC. http://www.harvestplus.org/sites/default/files/HarvestPlus_General_Brochure_2009.pdf. Accessed 12 Jan 2015

HarvestPlus (2009b) Iron-pearl millet strategy. HarvestPlus, IFPRI, Washington, DC. http://www.harvestplus.org/sites/default/files/HarvestPlus_Pearl_Millet_Strategy.pdf. Accessed 11 Jan 2015

HarvestPlus (2010) Country crop profile: iron pearl millet in India. HarvestPlus, IFPRI, Washington, DC. http://www.harvestplus.org/sites/default/files/Country%20Crop%20Profile_India%20Iron%20Pearl%20Millet.pdf. Accessed 13 Jan 2015

Hemalatha S, Platel K, Srinivasan K (2007a) Influence of heat processing on the bioaccessibility of zinc and iron from cereals and pulses consumed in India. J Trace Elem Med Biol 21:1–7. doi:10.1016/j.jtemb.2006.10.002

Hemalatha S, Platel K, Srinivasan K (2007b) Zinc and iron content and their bioaccessibility in cereals and pulses consumed in India. Food Chem 102:1328–1336

Huang MT, Ferraro T (1978) Phenolics compounds in food and cancer prevention. In: Phenolic compounds in food and their effects on health. II, ACS symposium series, 507:8–34

Hulse JH, Laing EM, Pearson OE (1980) Sorghum and the millets: their composition and nutritive value. J Nutr Dietetics 29:177–180

ICRISAT (2013) Newsletter of ICRISAT, 30 Jul 2013, p 5

Jenkins DJ, Wolever TM, Leeds AR, Gassull MA, Haisman P, Dilawari J, Goff DV, Metz GL, Alberti KG (1978) Dietary fibres, fibre analogues, and glucose tolerance: importance of viscosity. Br Med J 1:1392–1394

Kachare DP, Chavan JK, Kadam SS (1988) Nutritional quality of some improved cultivars of cow pea. Plants Foods Human Nutr 38(2):155–162

Kelley TG, Parthasarathy PR, Weltzien ER, Purohit ML (1996) Adoption of improved cultivars of millet in an arid environment straw yield and quality considerations in western Rajasthan. Exp Agric 32:161–171

Khetarpaul N, Chauhan BM (1991) Effect of pure sequential culture. Academic Press, New York, p 997

Kinsella JE, Lokesh B, Stone RA (1990) Dietary n-3 polyunsaturated fatty acids and amelioration of cardiovascular disease possible mechanisms. Am J Clin Nutr 52:1–28

Kumar A, Chauhan BM (1993) Effects of phytic acid on protein digestibility in pearl millet (Pennisetum glaucum L.). J Cereal Sci 19(1):83–89

Kumar S, Thirunavukkarasu N, Singh G, Sharma R, Kulkarni KS (2015) Biofortification for selecting and developing crop cultivars denser in Iron and Zinc. In: Rakshit A et al (eds) Nutrient use efficiency: from basics to advances. Springer, New Delhi, pp 237–253

Malik M (1999) Functional and nutritional properties of newly developed hybrids and varieties of pearl millet. MSc thesis, CCS Haryana Agricultural University, Hisar, India, 80 pp

Malleshi NG, Klopfenstein CF (1998) Nutrient composition and amino acid contents of malted sorghum, pearl millet and finger millet and their milling fractions. J Food Sci Technol 35(3):247–249

Mani UV, Prabhu BM, Damle SS, Mani I (1993) Glycemic index of some commonly consumed foods in western India. Asia Pac J Clin Nutr 2:111–114

Nambiar VS, Dhaduk JJ, Neha S, Tosha S, Desai R (2011) Potential functional implications of pearl millet (*Pennisetum glaucum*) in health and disease. J Appl Pharm Sci 1(10):62–67

Oshodi AA, Ogungbenle HN, Oladimeji MO (1999) Chemical composition, nutritionally valuable minerals and functional properties of Benny seed, pearl millet and quinoa flours. Int J food Sci Nutr 50:325–331

Palande KB, Kadlag RV, Kakchar DP, Chavan JK (1996) Effect of blanching of pearl millet seeds on nutritional composition and shelf life of its meal. J Food Sci Technol 33:153–155

Poonam (2002) Effect of acid and heat treatment on nutrient composition and shelf life of pearl millet (Pennisetum glaucum) flour. MSc thesis, CCS Haryana Agricultural University, Hisar, India, 106 pp

Rekha (1997) Efficacy of processing techniques in the utilization of pearl millet for value added products. MSc thesis, CCS Haryana Agricultural University, Hisar, India, 125 pp

Saxena AK, Sharma AK, Sehgal KL, Bakshi AK (1992) Proximate composition properties, nutrient composition and dietary fibre content of millets. J Food Sci Technol 30(1):193–198

Sehgal S, Kawatra A, Begum J, Srivastava S, Naikare SM (2002) Annual report of NATP, CRIDA, Hyderabad, India, p 160

Serna-Saldivar SO, Clegg C, Rooney LW (1994) Effects of parboiling and vitro) and HCl – extractability of minerals in pearl millet sprouts. Cereal Chem 70(5):504–506

Serna-Saldivar SO, Clegg C, Rooney LW (1994) Effects of parboiling and decortication on the nutritional value of sorghum and pearl millet. J Cereal Sci 19:83–89

Sharma A (eds) (1994) Fermentative improvement of pearl millet land utilization of the fermented product. PhD thesis, Haryana Agriculture University, Hisar

Sharma A, Kapoor AC (1996) Levels of antinutritional factors in pearl millet as affected by processing treatments and various types of fermentation. Plant Foods Human Nutr 49(3):241–252

Velu G, Rai KN, Muralidharan V, Kulkarni VN, Longvah T, Raveendran TS (2007) Prospects of breeding bio fortified pearl millet with high grain iron and zinc content. Plant Breed 126:182–185. doi:10.1111/j.1439-0523.2007.01322.x

Biofortification for Nutrient-Rich Millets 30

A.C. Shivran

Abstract

There are billions of persons facing the problem of micronutrient deficiency in the world, resulting to undernourishment. This results into severe consequences of health. It is one of the most significant health problems of the humankind. These deficiencies will perpetuate the cycle of poverty especially in rural areas of developing countries. The commercial food fortification alone will not be sufficient to combat the problem of malnutrition; however, biofortification can enhance the nutritional value of the plant-derived foods and feeds and provide a low-cost, sustainable, and long-term means of delivering micronutrients to the poor. The resource-poor people mainly consume a small number of staple crops especially cereals for the vast majority of their nutrition. Millets are important crops in the semiarid tropics of Asia and Africa, with more than 90 % of millet production in developing countries. The crops are favored due to its productivity and short growing season under dry and high temperature conditions. However, millets are highly nutritious and even superior to wheat and rice in certain constituents, so they are now considered as nutria-cereals (nutritious grains). The ingestion of micronutrients in low-income rural families of millet-growing areas is less as compared to recommended diet intake. Any increase in quality of millets might have significant role in combating micronutrient malnutrition for human health over the world. Biofortified millets have a great potential to reduce micronutrient deficiency in the developing countries. The work done on biofortification of millets is still not much. Even after nutrient richness of millets, there is a need to work for more production with quality addition in millets to change the billions of people from nutrient insufficiency to nutrient adequacy.

A.C. Shivran (✉)
S. K. N. College of Agriculture, S. K. N. Agriculture University, Jobner, Rajasthan 303 329, India
e-mail: acs_shivran@rediffmail.com

U. Singh et al. (eds.), *Biofortification of Food Crops*, DOI 10.1007/978-81-322-2716-8_30

Keywords
Biofortification • India • Millets

30.1 Introduction

The WHO, FAO, and many other world organizations reported that there are over two billion people suffering from micronutrient undernourishment throughout the world. The African and South Asian countries are most affected by malnutrition. About 30 % of the developing world's population is suffering from deficiency of minerals and vitamins. It is estimated that India accounts for more than 25 % of the world's undernourished population. The micronutrient deficiency is a great growing public and socioeconomic issue particularly in developing countries (Welch and Graham 2004). A World Bank publication estimated that malnutrition by deficiencies of iron, zinc, and vitamin A can cause economic loss equal to 5–6 % of GNP each year in South Asia due to sickness, poor work performance, and other factors. Improving the nutritional status of children and adults is a highly efficient way to increase economic productivity in agriculture and other sectors (Benoist et al. 2004).

Iron deficiency is the most prevalent nutrient deficiency in the world. It is estimated that more than two billion people or 30 % of the world's population is iron deficit. The average occurrence of iron deficiency among children in 37 African countries has been estimated at 67 % (UNICEF 2004). In India, anemia is reported in more than 70 % of children of 6–35 months and women of 15–49 years of age (Krishnaswamy 2009). Zinc has newly received global awareness, and it was estimated that 30 % of the population of 46 African countries be short of adequacy of zinc uptake in their daily basis diet (Hotz and Brown 2004). Zn and Fe deficiency causes death of about five lakhs children less than 5 years of age annually (Black et al. 2008). Iron deficiency can hamper brain development; decrease immunity, physical growth, mental development, and learning capacity; and augment risk of death and morbidity in children and pregnant women. Iron deficiency anemia reduces the physical working efficiency especially in adults which create a severe setback for the development of a nation. Studies concentrating to identify sources and improve iron content in millet grains would be a significant approach to combat its extensive undernourishment. Zinc deficiency may also results to mortality from a number of diseases. Zinc deficiency in human being results as consequences from reduced nutritional intake, poor absorption, or increased loss. Zinc deficiency affects many systems, including the integumentary, gastrointestinal tract, nervous, immune, skeletal, and reproductive systems. A short of zinc thus has many manifestations, the most common of which are increased rates of diarrhea, pneumonia, and malaria.

Globally, cereal-based diets of poor people are often deficient in indispensable vitamins especially vitamin A. It is estimated that preschool children are more suffering from this deficiency. Again, similar to iron and zinc deficiencies, the prevalence of vitamin A deficiency is more in developing countries. Efforts should be made to increase the concentration of provitamin A carotenoids in staple food crops. Supply of food enriched with vitamins improves the health and longevity of children and adults to a great extent.

Human being and many animals are unable to synthesize certain essential amino acids. Among these, lysine, tryptophan, and methionine are most common as they are present in limited quantity especially in major food crops. Enrichment of food crops in these essential amino acids have both economical and humanitarian interest in developing countries, where these crops are consumed for the majority of the food requirement (Shai and Gad 2008). Other minerals, like calcium, copper, magnesium, manganese, phosphorus, and potassium, are also essential for human health. These elements are important for the development of strong bones, teeth, hair, blood, nerves, and skin and synthesis of vitamins, enzymes, and hormones in human body. These are also critical for the performance of the nerve system, blood circulation, energy production, and muscle retrenchment (MacDowel 2003; O'Dell and Sunde 1997).

Agriculture is the primary source of nutrients that sustain human life. The agriculture crops are the major source of mineral elements, vitamins, and amino acids essentially required for fine human health. Millets are the good resource for micronutrients of malnutrition as compared to other staple food crops like wheat and rice. Fortification and availability of millets to poor inhabitants is not enough to eliminate the mass problem of malnutrition. Moreover, the biofortification approach including breeding for quality varieties as well as agronomic biofortification can play a critical role for ensuring nutritious food to each and every poor. At this point, integrated efforts of agricultural scientists and nutritionists might be impressing to reduce micronutrient deficiency. All the agricultural research centers that comprise the CGIAR have accepted diminution of micronutrient deficiency as a CGIAR goal through increased food production, more stable food supplies, and increased purchasing power of the poor. Minute concentration has been done by human nutritionists to agriculture by itself as a complementary means to solve the problem of quality diet and the contribution that plant scientists might make in this area.

30.2 Food Nutrification

Food nutrification or fortification is addition of minerals and vitamins or improvement in quality of food by addition of supplements either for providing additional nutrition in food or to reduce nutrient deficiency in population.

The World Health Organization (WHO) and the Food and Agricultural Organization of the United Nations (FAO 2014) cleared that the fortification is the way of purposely enhancing the content of a vital micronutrient, i.e., vitamins and minerals (including trace elements), in a food irrespective of whether the nutrients were originally in the food before processing or not, to increase the nutritional quality of the food and to provide a public health benefit with minimal risk to health. Food enrichment is one and the same with fortification, the addition of vitamins and minerals to a food which may lose during food processing (Liyanage and Hettiarachchi 2011).

The four main methods of food fortification named with the procedure that is used in order to fortify the food are:

1. Biofortification, i.e., three strategies have been identified, including conventional breeding of crops, because micronutrient improvement characters exist within genomes that can be used for greatly increasing micronutrient levels in these foods without reducing crop productivity; recombinant DNA technique of genetic engineering and agronomic biofortification through application of micronutrient-containing fertilizers/foliar fertilization or through addition of soil amendments (Carvalho and Vasconcelos 2013).
2. Synthetic biology, i.e., addition of probiotic bacteria to foods.
3. Commercial and industrial fortification, i.e., flour, rice, oils, and common cooking foods.
4. Home fortification, e.g., vitamin D drops.

Biofortification is the process of nutritional quality enhancement of staple food crops through conventional plant breeding and/or use of biotechnology (Bouis 1996). It differs from conventional fortification procedure as the aims of biofortification are to supplement nutrient levels in crops during plant growth and development rather than during processing of the food crops. The biofortification in such a way is also applicable to suppress the malnutrition where the conventional fortification is either not easy or impracticable to employ (Encyclopedia 2014).

The agriculture meets the growing issues of food availability in the world and nutrient supplementation especially in developing countries to a great extent. But the food security was achieved only in terms of quantity not in quality through cereals productivity. Now agriculture must be focused not only on productivity but also on quality to feed and to supply nutrient-rich food to poor for reducing undernourishment. To deal with micronutrient deficiencies, various malnutrition mitigation approaches have been employed, including fortification, supplementation, nutrition education, dietary diversification, and more recently biofortification. Production of staple food crops enriched with bioavailable nutrients especially iron, zinc, provitamin A, and essential amino acids rather than using fortificants or supplements may be referred as biofortification. A method, which results in more transfer of iron and zinc from the soil to the plant parts, is important for biofortification (Zuo and Zhang 2009). It can provide nutrient-rich food to combat malnutrition in children and women of rural poor of developing countries in subcontinent South Asia, Latin America, sub-Saharan Africa, and Caribbean.

Biofortification improves the nutritional quality of the staple foods which provide cheap and sustainable nutrients to people for fighting deficiency especially in women and children. It does not provide food supplementation before consumption, but it is added to crop plants. The micronutrients in biofortified crops are bioavailable when consumed by the people, thus improving the nutritional status of the population.

The food fortification will not be able to cover up micronutrient deficiencies to the great extent alone (Lindsay et al. 2006). The traditional nutrient supplementation or fortification of foods requires continuous investment, whereas quality improvement in staple food crops especially millets through breeding approaches develop a long-term solution without recurrent investment year after year to provide nutrient-fortified food to poor population of rural areas. Breeding of millets for nutrient-rich grains will not reduce crop productivity. Biofortification of sorghum by rising micronutrient content in the seeds provides a sustainable solution to micronutrient deficiency (Pfeiffer and McClafferty 2007). Enhancement of the micronutrient concentration in grains of crops by addition of micronutrient-containing fertilizers is one of the low-priced alternatives to tackle the problem of malnutrition in principally millet-consuming inhabitants of semiarid tropics. Agronomic biofortification of food crops is an important complement to genetic biofortification. Addition of micronutrient-containing fertilizer to the soil results in augmented uptake of zinc in wheat and enhances the bioavailable zinc content in the edible part of the plant (Cakmak 2008). In addition, micronutrient adding up to micronutrient-poor soils can increase crop yields and quality when its adoption is assured by the farmers.

Undernutrition is the most important reason of malnutrition, and it is mainly associated with poor economy. More than 70 % of the world's poor live in rural areas and majority of them are small farmers. The resource-poor families mainly depend on a few staple crops especially millets for their complete nutrition. It decreases the possibility of processed food fortification of micronutrient deficiency removing program for such a group of poor people and emphasizes for the plant-based biofortification to combat the problem of human nutrition (Food Safety Network 2011).

30.3 Millets Uniqueness

30.3.1 Staple Food

Millets are the main staple food crops of millions of rural poor in the regions of developing world. Millets are one of the oldest food crops known to human being. These are first among cereal crops used for domestic requirement. They do not form a taxonomic group, but rather a functional or agronomic one. Millets are unusually hardy and grow well in dry areas as rainfed; thereby consequently there is a progressive increase in the use of these grains as a human food staple, especially in large areas of India and sub-Saharan Africa. They can be grown under poor and marginal soil fertility. The crops are more preferred due to their high productivity and short growing season with dry and high temperature circumstances. However, realizing the nutrient composition of these grains, they are now considered as nutria-cereals.

While millets are indigenous to many parts of the world, they possibly had an evolutionary origin in tropical western Africa, as the maximum number of both wild and cultivated forms of millets exists there. Millets have been important staple food crops in human history, mostly in Asia and Africa, and they have been in cultivation in East Asia for the previous more than 10,000 years.

Millets represent a group of cereal crops. They are divided in two groups, major and minor millets. The major millet includes sorghum and pearl millet. The minor millets are finger millet, proso millet, foxtail millet, kodo millet, little millet, and barnyard millet. Considering the nutrient richness and higher productivity even under extremes of environmental condition, millets have indispensable role in the food security system of the poor world. This will help in eradication of malnutrition even in poor economy of rural population in developing countries. The resource-poor farmers should have to be encouraged for cultivation of millets that are best suited under extreme unfavorable soil, water, and temperature conditions. This will enhance the availability of these staple food crops at farm level with lesser cost to poor families for suppressing micronutrient deficiency.

30.3.2 Nutrient Richness

Millets are the store house of the nutrients. Millets are many times nutritionally superior to the widely promoted rice and wheat in terms of proteins, minerals, and vitamins. They are highly nutritious and nonglutinous. So, they are easy to digest. They are considered to be least allergist. The millet grains contain about 67 % carbohydrate, having major part as nonstarchy polysaccharides which add to many health benefits. Low carbohydrate content makes them most useful for diabetic persons. The millets are better than rice and wheat for their nutritional composition. The foxtail millet and barnyard millet have highest mineral nutrient concentration even among millets. Barnyard millet contains six times of minerals as compared to rice (Table 30.1). The finger millet is the richest source of calcium, potassium, magnesium, and sodium. Almost all millets are richer source of minerals and have manyfold higher concentration compared to rice as shown in Table 30.2. The millets are a good source of most essential amino acids, namely, methionine, lysine, and cystine, and thus offer special benefits to those who are mainly dependent on plant food to complete their dietary requirements. The results of a study indicated that soluble as well as bound portion of millet grains are wealthy sources of phenolic phytochemical compounds with antioxidant, metal chelating, and reducing power (Chandrasekara and Shahidi 2010). The antioxidant content of millets also depends on variety of millets. Thereby, these millets should be an essential component of everyone's diet to increase people's diet diversity and maintaining food nutrition. Total phenolic and flavonoid content of the major millets have been reported by Chandrasekara and Shahidi (2010) for soluble and bound phenolic fractions. It is 411–610 mg/

Table 30.1 Average nutrient composition of various millets (g/100 g)

Crop	Carbohydrate	Protein	Fiber	Fat	Minerals
Rice	78.5	6.8	0.2	1.5	0.7
Wheat	65.7	11.9	1.2	1.6	1.6
Sorghum	72.6	11.3	5.4	1.9	1.6
Pearl millet	69.4	11.6	1.3	4.8	2.4
Finger millet	71.8	7.4	3.8	1.3	2.7
Foxtail millet	60.7	12.3	8.1	4.3	3.4
Proso millet	70.9	12.5	7.2	3.1	1.8
Kodo millet	65.5	8.2	9.2	1.5	2.6
Little millet	67.4	7.8	7.6	4.6	1.5
Barnyard millet	65.7	6.2	9.9	2.1	4.3

Table 30.2 Mineral composition of millets (mg/100 g of edible portion)

Crop	Phosphorus	Potassium	Calcium	Magnesium	Sulfur	Iron	Sodium
Rice	4.1	120	10	36	78	1.2	10.1
Wheat	4.8	281	42	140	130	4.6	16.7
Sorghum	5.6	130	27	138	55	1.6	7.2
Pearl millet	14.1	32	38	122	149	4.8	10.4
Finger millet	17.2	406	346	405	155	2.6	12.0
Foxtail millet	12.9	246	32	124	168	2.8	4.8
Proso millet	11.3	111	14	74	151	1.5	0.8
Kodo millet	30.8	141	29	110	133	0.5	4.8
Little millet	10.2	126	18	61	124	2.3	7.9
Barnyard millet	2.9	–	18	–		6.7	–

100 g in finger millet, 168 mg/100 g in pearl millet, and 140 mg/100 g in proso millet. The nutritive traits in millets have great variability, and it can be exploited for future nutrient thrust in people (Girish et al. 2014).

30.3.3 Resistant to Extreme Ecological Conditions

The millets are well adapted to harsh environmental conditions. Millets are very little water-requiring crops for production as compared to wheat and rice. The amount of water required for successful production of rice and wheat is 1200–1300 mm and 700–800 mm, respectively. This indicates that water needed for production of sorghum (400–450 mm), pearl millet (350–400 mm), and minor millets (300–350 mm) is less than 30 % of rice and 50 % of wheat. Thus, millets are highly tolerant to soil moisture stress condition. Therefore, earlier their cultivation is mostly confined to arid and semiarid regions of tropical and subtropical world. In the coming decades under changing climate with probable low availability of irrigation water and erratic rainfall, millets become more important crops for food availability at affordable prices. Millets can tolerate dry weather and grow luxuriantly even in minimum moisture condition. They have short duration with resistance to pest and diseases and can be grown in rainfed, shallow, and less fertile soils. Millets are also identified for better shelf life. The seeds dried to 10–12 % moisture level can be stored for many years under normal storage conditions at farm level. Millets are going to be important crops under predicted climate change. Climate change warns of low rain, high temperature, reduced water availability, and increased malnutrition. In such a situation, millets can withstand these challenges. These may be the climate-smart crops for feeding the new generations to meet their nutritional requirements.

30.4 Sorghum

Sorghum is the major staple food crop of millions of rural poor in arid and semiarid regions of the world. Sorghum is mainly grown in the USA, India, Mexico, Sudan, Ethiopia, Nigeria, Australia, Brazil, Argentina, China, Egypt, Niger, Mali, Tanzania, and Cameroon. In India, Maharashtra and Karnataka are the main sorghum-producing and sorghum-consuming states. The per capita utilization of the sorghum is 75 kg grain/year in foremost sorghum-growing areas in India (Kumar et al. 2013a). The people of these regions obtain greater part of calories, protein, iron, and zinc intake from this crop. The sorghum improvement work was started earlier in the USA. The crosses between milo and kafir and selection of natural crosses before 1914 by the farmers of the USA were the base for sorghum improvement; after which systematic breeding programs were started. The hybrid vigor of sorghum was utilized extensively for development of sorghum varieties. In India and

Africa, hybrid material has been evolved and suitable hybrid sorghum were developed which were suitable for variable ecological conditions. The sorghum improvement work has prioritized for high productivity, grain quality, and better performance with poor resources availability. Initial efforts for sorghum improvement were started in India after 1960 with germplasm collection from the USA and Africa. The development of high lysine content varieties was started with utilization of Ethiopian high lysine containing genotype IS 11167 and IS 11758. It was possible to develop high-yielding dwarf, photo-insensitive selections combining with high content of protein and lysine. High lysine content genotypes were also developed using mutagens.

Although sorghum is the second cheapest source of energy and micronutrients supplementation after pearl millet, a majority of the population in central India depends on sorghum for their energy and micronutrient requirements (Parthasarathy et al. 2006). Sorghum varieties differ very much in their nutritional quality. Most of the sorghum-producing areas in arid and semiarid regions of India are poor in zinc and iron (Singh 2001). The scientists are working to evolve new varieties of sorghum that are easily digestible and contain high level of vitamins, minerals, and essential amino acids. Their aim is to produce extremely fortified grain with vitamin A, iron, and zinc availability. Protein biofortification of sorghum has also been achieved by both mutation and genetic engineering. Important promising genotypes of sorghum for antioxidant activity, amylase, protein, fat, and phenol and polyphenol content have been identified.

Agronomic biofortification (increasing the grain Fe and Zn status through application of Zn- and Fe-containing fertilizers) in post-rainy sorghum is one of the low-cost options to reduce the problem of hidden hunger in predominantly sorghum-eating populations of semiarid tropics. Mishra et al. (2015) studied that the sorghum cultivars vary in their iron and zinc concentrations of seed. These micronutrients in seed could be increased by selecting suitable variety and application of micronutrient-containing fertilizers. Sorghum cultivar Phule Maulee with soil application of $ZnSO_4$ + $FeSO_4$ each at 50 kg/ha followed by foliar application (0.50 % + 1.0 %) at 45 DAS along with recommended dose of NPK is recommended for producing iron- and zinc-rich post-rainy sorghum. It was observed that sorghum landraces have recorded an increase of 5–12 % for grain iron and 5–8 % for grain zinc content with foliar application of micronutrient fertilizers. This has implications on nutrition as any incremental enhancement in grain micronutrient concentration adds to reduce the micronutrient malnutrition (Kumar et al. 2013b).

Various local products are prepared from sorghum. Lassi prepared from M 35-1, CSV 18R, and C 43 cultivars of sorghum through fermentation technology reduced the phytic content that contribute to anti-nutrition of sorghum and improve protein content. The sorghum pasta prepared with sorghum flour, soya protein concentrate, and channa flour has higher nutrient concentration as compared to native sorghum pasta. Improved sorghum pasta exhibited more than 50 % increase in protein content as compared to native pasta (DSR 2014).

30.5 Pearl Millet

Pearl millet is one of the important staple food crops and is well adapted to low moisture and poor fertile soils. Pearl millet is a multipurpose crop with food, feed, and fodder value. Its grain also has a high feed value for livestock, poultry, and fish. It also provides high-quality green forage in seasons of fodder scarcity. It is reported to be grown in more than 30 countries of Asia and Africa over 70 million ha area of which 50 % is in Asia. India accounts for 15 % of the total world area of pearl millet. Its productivity ranges from 600 to 700 kg/ha in Asia and Africa.

The pearl millet is a very hardy crop as it can be grown very well with low soil moisture condition, high soil pH, and extremely high temperature. In addition, its grains are rich in protein especially gluten-free and micronutrient concentration. It is expected that the pearl millet cultivation will extend in additional regions especially in nontraditional one in the coming near future (Abdalla et al. 1998).

The improvement of pearl millet was started in India in the 1930s of the last century, and it was mainly concentrated on the improvement of yield in local material using simple mass selection and only a few varieties were developed. The import of material from African countries yielded valuable varieties for different Indian conditions. Pearl millet is a highly cross-pollinated crop with crossing rates being more than 85 % which create more possibilities of utilization of genetic variation. The high-quality pearl millet hybrids developed recently in India are KBH 108, GHB 905, Nandi-72, MPMH 17, HHB 234, Kaveri, Bio 70, Bio 448, Pratap, PKV-Raj, CO- 9, HHB 226, PAC 909, RHB 173, RHB 177, HHB 223, RHB 154, GHB 732, GHB 719, PB 180, and RHB 121. The first high-iron-containing pearl millet variety ICTP 8203 Fe-10-2 was evolved through conventional breeding using variation of iron concentration by the International Crops Research Institute for Semi-Arid Tropics (ICRISAT) as part of the HarvestPlus program. This was adopted on commercial scale in 2012 in Maharashtra, India. Another pearl millet variety, Dhanshakti, was also developed in the course of biofortification by the ICRISAT in association with HarvestPlus. The research workers have now to develop more of iron- and zinc-enriched pearl millet varieties to meet the challenges of malnutrition in India (Harvest Plus 2014).

It can withstand drought and simultaneously responds very well to fertilization with improvement in nutrient concentration in grain. Zinc fertilization increases pearl millet yield and improve quality. The zinc application increased lysine and soluble sugar content in the grain of pearl millet cultivars. The results of a study by Zong et al. (2011) in China suggested that foliar zinc application increases yield and also improves grain quality when applied at 1.50–2.25 kg/hm^2 for soils with low zinc content.

In pearl millet, malting decreases antinutritional cause and adds pleasing flavor and taste. Shelf life of pearl millet flour is also improved by the process of malting as it reduces the lipids intensity that impart to bad flavor. Blanching and heat treatment develop the storability and stability of flour. A large number of value-added foods may be prepared from pearl millet processed flour.

30.6 Minor Millets

The minor millets differ from each other in appearance, morphology, maturity, duration, and grain type. According to latest estimates available, India contributes more than 40 % of the total minor millets production in the world. Thereby, India is mostly considered as hub of the minor millets. The dietary consumption of minor millets has been in use from the beginning of ancient civilization. The minor millets have an unusual characteristic of adaptation to drought, high temperature, low soil fertility, and pests. They are free from storage pests; thereby, they can be stored for a longer period of time under ordinary storage conditions. Therefore, the minor millets are also denoted as famine crops. The minor millets include finger millet, foxtail millet, proso millet, kodo millet, little millet, and barnyard millet. The important minor millet-growing countries are India, China, Nigeria, South Africa, Ethiopia, Uganda, Tanzania, and some parts of Europe. In India, finger millet is mainly grown in Tamil Nadu, Karnataka, Maharashtra, and Andhra Pradesh, whereas 50 % of other minor millets are produced from states of Uttarakhand and Madhya Pradesh. The minor millets serve as staple food for millions of people residing in the arid and semiarid areas of the world mainly in Asia and Africa. They have better quality components than rice and wheat. They provide a cheap source of mineral elements, vitamins, and protein to the poor population inhabitating in rural areas of the developing countries. Even after this, much attention has not been paid by the research scientists for the improvement of these crops in terms of productivity and quality. Their productivity potential is still to be realized. Negligible emphasis has been given for quality improvement of minor millets. Nutritional superiority of these crops holds a great potential and importance to challenge the hidden hunger of rural people suffering from malnutrition all over the world.

The minor millets have been the last priority crops as far as their encouragement and development are concerned. Finger millet among minor millets has received a little more attention than the rest. Finger millet occupies an important place after sorghum and pearl millet. Its grains are the richest source of calcium, potassium, magnesium, and sodium. Its straw is also valuable for animal feed. The finger millet embodies higher productivity potential. A number of varieties have been developed which suitable for climate, soil, socioeconomic, and consumer acceptance. In India, varietal improvement work was systematized after establishment of All India Coordinated Small Millets Improvement Project in 1986. In India, technique of mass and pure line selection has been extensively practiced to develop varieties high in productivity and quality. Attempts were made to hybridize indigenous germplasm with exotic germplasm especially from African countries. The work on quality improvement of finger millet has also been done through supplementation of micronutrient fertilization which results in higher nutrient concentration and its uptake in grain.

In India, variety improvement work on other minor millets has been started in the later part of the twentieth century. The cultivated varieties are mostly from the local selection. A number of varieties were developed through pure line selection which accounts maximum varieties evolved for commercial production. The important recently developed varieties of finger millet are VL 324, KMR 301, VL 149, Godavari, PR 202, Divya, Indaf 7, Indaf 9, GPU 67, GPU 66, GPU 28, GPU 45, GPU 48, VR 708, Marua-2, RAU-8, CO-13, HR 374, CO-14, Phule nachani, Birasa, and Hima. The foxtail varieties most commonly grown in India are SiA 3085, PS 4, and Srilakshmi. The improved varieties under most extensively and widespread cultivation are CO 2, RAU 11, and VL 207 for barnyard millet; CO 3, JK 13, JK 48, JK 98, and JK 439 for kodo millet; CO 4, JK 8, GV 2, and Sukshema for little millet; and TNAU 151, TNAU 164, and TNAU 202 for proso millet. In addition to nutrition richness, minor millets offer another advantage of suitability to grow as rainfed with multi-cropping system along with legumes and oilseeds (Pradhan et al. 2010). The grain yield of little millet was significantly increased by 30–40 % under rainfed conditions with nutrient fertilization (Shashidhar et al. 1998; Yargattikar et al. 2004).

30.7 Processing of Millets for Rich Nutrition and Consumer Acceptance

The millets comprise an important staple food in many part of the developing world. In general, the millets are consumed as food in India and developing world, whereas in developed countries these are chiefly used as food products and as a feed for animals. Processing is essential for consumable preparations of acceptable texture, taste, flavor, and quality of millets. Usually millets have hard seed coat and a characteristic flavor which lower consumer acceptability as compared to rice and wheat. The seed of millets is identical to rice in morphology except finger millet. The outer husk and starchy endosperm in finger millet is blinded with soft endosperm. The specialized processing mechanism and machinery are not available for millets. Therefore, these are manually dehusked and decorticated at household level which is a laborious process. Mostly, the traditional methods of processing are adopted in Asian and African countries.

The millet processing involves milling, malting, popping, parboiling, etc. Milling is the primary process of food preparation from grains. It denotes dehusking, debranning, and endosperm sizing for making flour. The milling of millets is done by wheat or rice processing system because of nonavailability of specific system for millets. Millets are malted according to taste and flavor as desired by the consumers. Millets are commercially malted especially sorghum in Africa and finger millet in some parts of India.

Because of nutritious nature of malted millets, they are used as food for children and in milk beverages. The superior malting results to good flavor, adequate enzymes of hydrolyzing, and higher mineral and amino acids content. The parboiling of seed grains of millets with steam harden the endosperm and improve the milling quality of millets. The exposure of millet grain to high temperature results in expansion of volume of a grain, and popped millets are prepared which are extensively eaten as snacks. Popping is traditionally adopted to prepare ready-to-eat snacks from millets. The flour of popped millet, especially when mixed with edible legume, adds to the nutritional amount and quality. The proper processing results to better millet products consumed by people worldwide. These are better in the bioavailable nutrient concentration. Therefore, millet processing adds to nutrition and makes them in a form that is mostly consumed by the people. The millets offer numerous prospects for diversified consumption and addition of value. With proper processing it has potential to prepare numerous different categories of food goods by adopting suitable milling, popping, and other technologies as given in the flowchart Fig. 30.1 to prepare food and food products from millets.

30.8 Cause and Remedies for Millets Productivity

The productivity of millets fluctuates every year. The major reasons that reduce the millet productivity include their production on low fertility soils and soils having low water holding capacity. These are mostly cultivated on marginal and submarginal lands which are having deficiency of nutrients, aeration, water, and extremes of soil pH. The millets are also mostly grown under extreme environmental conditions unfavorable for crop growth and development. In addition, the socioeconomic factors also hinder millet productivity. The research and development activity received lowest priority in the world. So the poor farmers grow millets on hungry soils without adopting new production technologies. The genetic variability among millets is also low which create problem for breeding improvement of crops. The marketing facilities, seed availability, new production technology, processing system, etc. are the other reasons responsible for expansion in cultivation of millets and productivity in grown area.

These reasons of low productivity of millets adversely affect the poor farmer's economy of millet-growing regions. The major remedy of

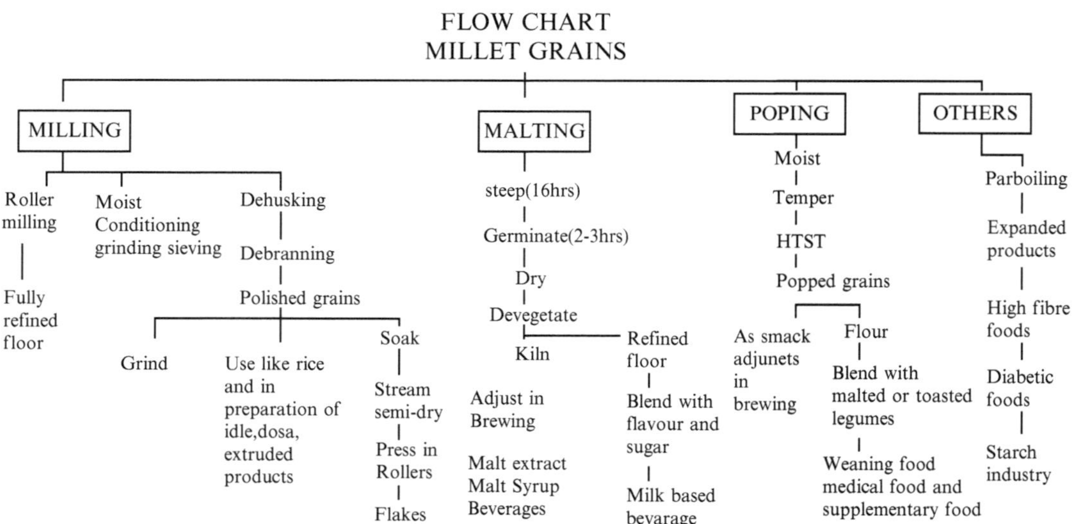

Fig 30.1 The flowchart to prepare food and food products from millets (Source: ICRISAT (2014) URL: http://www.icrisat.org)

poor productivity of millets is the evolution of high-yielding varieties of these crops. The millet seeds of better quality should be produced and their availability to the farmers should be ensured. In order to mitigate low water availability, soil water conservation practices should be developed and popularized among farmers. The development of low-cost agronomic crop production technology is also equally important for enhancement of the level of millet productivity. The use of biofertilizers, plant growth promoting rhizobacteria and chemical fertilizers are important ways to enhance production of quality millets with nutrient-rich grains along with per unit area productivity. Adoption of integrated approaches of weed and pest management with proper marketing facility may be one of the proper solutions for enhanced production of millets. The efforts of research on millets should be streamlined and reach up to the field of poor farmers in millet-growing regions.

30.9 Conclusion

Micronutrient deficiency is the main problem of poor people of developing countries mainly inhabitating in rural areas. The food fortification is the easiest way to remove the hidden hunger of each and every people. However, commercial food fortification/micronutrient supplementation is neither affordable by poor nor it is in access of all the people as most undernourished consume their own produced staple food crops. The biofortification is the cheapest, easy, and approachable way to nourish the poor rural families. The millets offer a better substitute for insufficient nourishment as these crops strangely resist extreme unfavorable soil, water, and temperature conditions. Biofortification of millet crops by plant breeding/genetics and agronomic approaches may be a miracle to remove malnutrition and improve human health of the poor nourished people. It offers a sustainable solution to the enormous global problem of under nourishment and converts it to nutrient sufficiency. Even after nutrient prosperity of millets, there is urgent need to make efforts for more production with quality improvement in millets to change the billions of people from nutrient deficiency to nutrient sufficiency.

References

Abdalla AA, El Tinay AH, Mohamed BE, Abdalla AH (1998) Proximate composition, starch, phytate and mineral contents of 10 pearl millet genotypes. Food Chem 63:243–246

Benoist B, Andersson M, Egli I, Takkouche B, Allen H (eds) (2004) Iodine status worldwide:WHO global database on iodine deficiency. Department of Nutrition for Health and Development, World Health Organization, Geneva

Black RE, Lindsay HA, Bhutta ZA, Caulfield LE, De Onnis M, Ezzati M, Mathers C, Rivera J (2008) Maternal and child under nutrition: global and regional exposures and health consequences. Lancet 371:243–260

Bouis H (1996) Enrichment of food staples through plant breeding: a new strategy for fighting micronutrient malnutrition. Nutr Rev 54(5):131–137

Cakmak I (2008) Enrichment of cereal grains with zinc: agronomic or genetic biofortification? Plant Soil 302:1–17

Carvalho SMP, Vasconcelos MW (2013) Producing more with less: strategies and novel technologies for plant-based food biofortification. Food Res Int 54:961–971

Chandrasekara A, Shahidi F (2010) Content of insoluble bound phenolics in millets and their contribution to antioxidant capacity. J Agric Food Chem 58:6706–6714

DSR (2014) Annual report 2013–2014. Directorate of Sorghum Research, Hyderabad

Encyclopedia (2014) Food fortification. http://encyclopedia.thefreedictionary.com/food+fortification. Accessed 26 July 2014

FAO (2014) Food and Agriculture Organization of the United Nation. http://www.fao.org. Accessed 26 July 2014

Food Safety Network (2011) http://www.uoguelph.ca/foodsafetynetwork/food-fortification. Accessed 30 July 2014

Girish Chandel G, Meena RK, Dubey M, Kumar M (2014) Nutritional properties of minor millets: neglected cereals with potentials to combat malnutrition. Curr Sci 107:1109–1111

Harvest Plus (2014) Iron pearl millet. http://www.harvestplus.org/content/iron-pearl-millet. Accessed 3 Aug 2014

Hotz C, Brown KH (2004) Assessment of the risk of zinc deficiency in populations and options for its control. Food Nutr Bull 25:94–204

ICRISAT (2014) A new generation of pearl millet on the horizon. http://www.icrisat.org/index.htm. Accessed 27 July 2014

Krishnaswamy K (2009) The problem and consequences of the double burden – a brief overview. In: Programme and abstracts. Symposium on nutritional security for India – issues and way forward, Indian National Science Academy, New Delhi, pp 5–6

Kumar A, Reddy BVS, Ramaiah B (2013a) Biofortification for combating micronutrient malnutrition: identification of commercial sorghum cultivars with high grain iron and zinc concentrations. Indian J Dryland Agric Res Dev 28(1):95–100

Kumar AA, Anuradha K, Ramaiah B (2013b) Increasing grain Fe and Zn concentration in sorghum: progress and way forward. J SAT Agric Res 11(12):1–5

Lindsay A, Bruno B, Omar D, Richard H (2006) Guidelines on food fortification with micronutrients. World Health Organization and Food and Agriculture Organization of the United Nations, Geneva. http://www.who.int/nutrition/publications/guide_food_fortification_micronutrients.pdf

Liyanage C, Hettiarachchi M (2011) Food fortification. Ceylon Med J 56:124–127. doi:10.4038/cmj.v56i3.3607. PMID 22164753

MacDowel LR (2003) Minerals in animal and human nutrition, 2nd edn. Elsevier Science B.V. Amsterdam, The Netherlands

Mishra JS, Hariprasanna K, Rao SS, Patil JV (2015) Biofortification of post-rainy sorghum (*Sorghum bicolor*) with zinc and iron through fertilization strategy. Indian J Agric Sci 85(5):721–724

O'Dell BL, Sunde RA (1997) Handbook of nutritionally essential mineral elements. Marcel Dekker, Inc., New York

Parthasarathy RP, Birthal BS, Reddy Belum VS, Rai KN, Ramesh S (2006) Diagnostics of sorghum and pearl millet grains-based nutrition in India. Int Sorghum Millets Newsl 47:93–96

Pfeiffer WH, McClafferty B (2007) HarvestPlus: breeding crops for better nutrition. Crop Sci 47:S88–S105

Pradhan A, Nag SK, Patil SK (2010) Dietry management of finger millet (*Eleusine coracana*) control diabetes. Curr Sci 98:763–765

Shai U, Gad G (2008) Improving the content of essential amino acids in crop plants: goals and opportunities. Plant Physiol 147:954–961

Shashidhar GB, Basavarajappa R, Nadagouda VB (1998) Response of little millet (savi) genotypes to NPK levels. Karnataka J Agric Sci 11(2):343–345

Singh MV (2001) Evaluation of current micronutrient stocks in different agro-ecological zones of India for sustainable crops production. Fert News 42:25–42

UNICEF (2004) Vitamin and mineral deficiency, a global progress report. UNICEF and Micronutrient initiative. http://www.unicef.org

Welch RM, Graham RD (2004) Breeding for micronutrients in staple food crops from a human nutrition perspective. J Exp Bot 55:353–364

Yargattikar AT, Ashvathama VH, Budihal RA (2004) Response of little millet to production technologies under rainfed on farm conditions. Karnataka J Agric Sci 17(3):536–538

Zong X, Wang H, Song Z, Liu D, Zhang A (2011) Foliar Zn fertilization impacts on yield and quality in pearl millet (*Pennisetum glaucum*). Front Agric China 5 (4):552–555

Zuo Y, Zhang F (2009) Iron and zinc biofortification strategies in dicot plants by intercropping with gramineous species. A review. Agron Sustain Dev 29:63–71

Pre- and Post-harvest Management of Physical and Nutritional Quality of Pulses

31

Aditya Pratap, Suhel Mehandi, Vankat Raman Pandey, Nupur Malviya, and P.K. Katiyar

Abstract

Grain legumes are consumed in a variety of ways such as boiled whole grains, cooked *dal* (split grain), powdered and sprouted and in a wide range of snacks. These are extensively used in Asia, Europe and America for gluten-free products, ready-to-eat baked goods, mixes, soups, sauces and other foods along with cereals as one of the ingredients. Nevertheless, major proportion of grain legumes is mostly consumed as natural food products in form of whole grains or dehulled or split grains. Hence, size and shape of seeds, seed coat appearance and its colour, cotyledon colour and grain uniformity are important for trade and commerce. Of late, increasing consumer preferences for healthy and nutritious food have led to greater opportunities for processing, value addition and development of healthy by-products. However, to develop attractive processed foods as well as by-products from grain legumes, their pre- and postharvest management plays an important role. These involve a series of operations which need proper care and maintenance of crop, mechanised production, harvesting and processing and proper storage and disposal of the crop. In recent years technological advancements in processing and value addition provide healthy and nutritious pulse food to the human beings. This chapter describes the food value of grain legumes and their pre- and postharvest management for value addition in enhancing physical and nutritional quality of pulses.

Keywords

Milling • Postharvest • Processing • Value addition

A. Pratap (✉) • S. Mehandi • V.R. Pandey
N. Malviya • P.K. Katiyar
Division of Crop Improvement, ICAR-Indian Institute of Pulses Research, Kanpur, Uttar Pradesh 208 024, India
e-mail: adityapratapgarg@gmail.com; suhelgpb@gmail.com; vankatramanpandey@gmail.com; malviya.nupur@gmail.com; goldikatiyar@yahoo.com

31.1 Introduction

Pulses constitute an important dietary constituent for humans and animals because of their richness

U. Singh et al. (eds.), *Biofortification of Food Crops*, DOI 10.1007/978-81-322-2716-8_31

with proteins (ranging from 15 % to 34 %, depending upon the crop species), as well other essential minerals, vitamins and dietary fibres. The protein content of grain legumes is double the protein content of wheat and three times that of rice (Rodino et al. 2011). They are also high in iron and consequently help in alleviating iron deficiency anaemia. Such a wide range of attributes make grain food legumes an essential ingredient in predominantly vegetarian diets of vast majority of people in India and other countries of Southeast Asia. Grain legumes are also being vastly used in Europe and America for gluten-free products, from ready-to-eat baked goods, mixes, soups, sauces and other foods with pulses as one of the ingredients. The grain legumes are mostly consumed as natural food product in form of whole grains or dehulled or split grains in many parts of the world. Therefore, size and shape of seeds, seed coat appearance, colour, cotyledon colour and uniformity are important for markets. There are several factors including storage, processing and marketing which help to determine quality of grain legumes and minimise the yield losses. Simultaneously, the postharvest technologies help making legume grains more preferable to consumers besides improving their nutrition standards and storability.

Postharvest processing of legumes involves a series of mechanical separations, milling operations and processing for consumption. Accordingly, food legume processing industries comprise three important components, viz. primary processing involving cleaning, drying, storage, packaging, etc.; secondary processing mostly involving dehulling, splitting, sorting and polishing of grains; and tertiary processing for further processing of secondary processed legume grains into useful food products. Tertiary processing mainly aims at value addition and converts the legume grains into ready-to-eat form.

31.2 Nutritional Value of Pulses

Nutritional value of pulses is quite high as compared to most of the cereals. The protein of pulses is generally low in sulphur-rich amino acids. However, it is rich in lysine, an amino acid that is deficient in many cereals. Therefore, when pulses are added in cereal-based vegetarian diets, their nutritive quality comes on a par with animal proteins. Further, pulses are also good sources of the B-group vitamins apart from riboflavin. Although pulses are devoid of vitamin C, a large amount of ascorbic acid is formed during their germination. Sprouted pulses are, therefore, an important food for protection against scurvy. During the sprouting process, vitamins, minerals and protein increase substantially, with a corresponding decrease in calories and carbohydrate content, which leads to an improvement in nutritive value and digestibility of the pulse. The ascorbic acid or vitamin C content rises from negligible levels in the seed to 12 mg/100 g after 18 h of germination. Riboflavin and niacin contents also increase significantly. All these changes are brought about by the enzymes that become active during germination (Singh and Singh 1992). Finally, the digestion of pulses and the absorption of their principal nutrients are practically complete in the gut for people without gastrointestinal disorders. Despite so many advantages, only small quantities of well-cooked pulses should be included in the diets of patients with stomach disorders to avoid any potential risks (Sharma and Garry 1995).

Pulses provide a similar level of energy to grains. All pulses have a low glycaemic index (i.e. the carbohydrate is digested slowly). They are also low in fat (2–6 %), most of it provided in the form of polyunsaturated and monounsaturated fatty acids. Pulses are also a good source of fibre. Evidently, those pulses that are consumed with the seed coat on, such as mung bean, lentils, chickpea, etc., are much higher in dietary fibre than those that have been dehulled before consumption.

31.3 Use of Pulses in Special Diets

High nutrient concentration and good quality proteins make pulses play an important role in predominantly vegetarian diets of the people across countries such as India and the entire

Southeast Asia region. Of late, pulses have emerged as a healthy source of nutrition and therefore become essential ingredients of several special diets.

31.3.1 Diabetic Diet

Pulses have a lower glycaemic index as compared with other foods, especially those having high carbohydrate concentration. As a result, pulse consumption may result in more stable blood glucose levels after meals. Therefore, people with diabetes can manage their blood glucose level by consuming lentils, peas and beans.

31.3.2 Gluten-Free Diets

Pulses are very good for the people with coeliac diseases. If a person with coeliac disease consumes gluten, a protein found in wheat and some other cereal grains, an immune reaction is triggered in the small intestine, causing damage and poor absorption of nutrients. Pulses are free from gluten, and therefore, these are very good nutritional options for people with coeliac diseases.

31.3.3 Vegetarian Diets

Pulses are good sources of protein, vitamins and minerals, especially iron and zinc. These nutrients together make them a preferred nutritional option for vegetarian people around the globe. Further, pulses also contain eight essential amino acids. Consuming pulses with rice provides the full complement of amino acids needed for growth. Due to the high protein content, peas, mung bean and lentils could be supplemented to breakfast cereals in the way as soy has already been used. Mung bean and chickpea sprouts have great nutritional value and are finding a new market in Europe, the Americas and Asia as breakfast cereals.

31.3.4 Weight Management Diet

Most of the pulses are rich in fibres and protein. On contrary, these have less fat and moderate calories. For example, one cup of cooked lentils or dry peas contains about half of the daily fibre recommendation for adults. Due to high fibre content, consumption of lesser amount of pulses helps people feel satisfied and leads to comparatively less consumption of food and therefore low calories. Therefore, it may help with weight management, especially in obese people.

31.4 Preharvest Management

Preharvest management of crops can minimise the losses caused by erratic environmental conditions experienced during the growing season and can ultimately improve the quality of grains. For example, alternating wet and dry periods during later stages of maturity in food legumes can lead to shrivelled and underdeveloped seeds, loss of colour and cracked and/or brittle seed coat. Similarly, to promote mechanical harvesting in stay-green crops like mung bean, chemical desiccation is practised in many areas which may lead to brittle seed coats that chip and spill more easily (Vandenberg 2009). For example, the quality in cowpea grains becomes a serious problem if rainfall occurs soon after maturity. Therefore, preharvest fungicide sprays can have some benefit in the prevention of this. For avoiding grain damage in most of the grain legumes, especially the *Vigna* species, the crops should be harvested well before too much drying. Keeping the drum speed low (250–300 rpm) during combine harvesting also helps in minimising splitting and cracking of grains.

31.5 Postharvest Processing and Management

For good postharvest processing to minimise losses in seed and nutritional quality of pulses, optimum preharvest management, proper threshing and good storage conditions play a vital role. Seeds of poor quality, biotic and abiotic stresses and inappropriate farming practices can cause significant loss of produce even before harvest. While preharvest losses generally occur in the field, postharvest losses which refer to measurable quantitative and qualitative losses during the various stages of the postharvest handling occur at the time of threshing, storage and processing. In countries of the Indian subcontinent, postharvest losses are the major cause of low-realised yield of pulses. Nevertheless, the extent of loss depends considerably depending upon weather conditions, varieties, locations and the processing and storage techniques employed. It has been estimated that total postharvest losses in case of pulses range between 25 % and 50 % (Kurien et al. 1972; Birewar 1984; Jeswani and Baldev 1990) including 1.0–.30 % at the time of harvesting, 1.0–7.0 during handling, 0.5–5.0 % during threshing, 1.0–5.0 % during drying and 15–20 % during milling, while 5–10 % losses may occur due to improper storage of pulse grains.

31.5.1 Harvesting

Harvesting is the most important operation on which the economic gain of the farmer and food and nutritional security of the consumer depend. This process is directly linked with collection of useful parts of the plant and is therefore accomplished when the nutrients have developed fully in the plant and the useful parts reach maturity. Accordingly, the harvesting in pulses should take place at the time when the grain has moisture content in the range 15–20 % (Ali 2004). Higher moisture content in the grains at the time of maturity increases the risks of losses from moulds, diseases and insect attack and also may also sometimes lead to germination before sowing. Unfortunately, until now, not much advancement has taken place for mechanisation of harvesting process in pulses in major pulse-growing countries, and it is still mostly done manually with simple farming implements. In most of the cases, to harvest the pulse crops by hand, either the plants are cut using sickles or the entire plants are uprooted and allowed to predry in the sun.

31.5.2 Threshing

Threshing is the operation of separating grains from other biological mass. This operation is mostly carried out in the field itself or sometimes on the threshing floor. Threshing of food legumes in the Indian subcontinent is generally carried out manually using long wood sticks which leads to more physical damages to seeds including breakage, chipping, removal of seed coats, etc. However, mechanical threshers are now available to separate the seeds from pods although such machines are mostly used for threshing of chickpea, peas and lentil. Crops like pigeon pea are still threshed manually by hand.

The simplest method followed for threshing of pigeon pea crop is to thump the bundles of crop spread over the threshing floor with a stick. This method can be conveniently adopted for several other pulse crops like chickpea, mung bean, urdbean, field pea, rajmash and lentil also. However, for threshing the crops by this method, the crops must have optimum moisture content in their grains so as to minimise the damage. Postharvest losses and mechanical damage may be high in legumes when these are harvested at below 12 % moisture in the grains. Further, reduced drum speeds (280–350 rpm), conveyors, augers and air fronts of threshing machine may be required to reduce the damage to seeds during threshing.

31.5.3 Drying

Drying is generally followed to lower the moisture content of the grain for safe storage and further processing. The moisture content of grains at the time of harvest is generally higher than desired for safe storage of grains (12–14 %). Drying is an integral process of postharvest management of pulses in which the pulse grains are dried until they reach 'safe moisture' level. Safe storage means keeping the products in a way that ensures their safety from physical damage and maintains their food value. Although pulses, if stored properly, can remain in edible condition for several years, they are comparatively more difficult to store than cereals. Pulses, in general, suffer greater damage from insects and microorganisms which results in quantitative losses as well reduction in the quality and nutritive value of grains.

31.5.4 Storage Conditions

A host of storage structures including steel bins, jute bags, earthen pots, mud bins and bamboo baskets or other types of locally available storage structures are used for storing the harvested grain legumes before primary processing. The initial conditions of the grain and storage environment may have an impact on processing quality. The moisture content of each seed lot should be determined before use, using the standard method (Anonymous 1976). During storage of grains of pulses, high temperature, humidity, light, oxygen and moisture content are responsible for changing the seed coat colour. For example, in chickpea, temperature ranging from 33 to 35 °C and 75 % relative humidity for 160 days caused colour darkening of testa (Reyes-Moreno et al. 1987). The combined effect of all these factors can sometimes lead to severe losses in quantity as well as quality of pulses during storage.

Sun-drying is the most common practice for reducing the moisture content of pulses. Alternatively, solarising in solar heater can be used where lethal temperatures are required to kill all insects. For this purpose, solar heaters are available, which are suitable for treating small quantities of grain legumes. However, these heaters are needed to be used correctly to heat the pulse grains at high enough temperatures for a long period for killing all insects, without losing quality of the grains.

31.6 Stored Grain Pests and Their Control

Postharvest damages by insect pests have been an increasingly important constraint to food legume supplies worldwide. Pulses are consumed in a variety of ways, mostly in the form of *dal* or dehusked splits. The *dal* or dehusked splits are more prone to moisture gain and subsequently fungal infestation in absence of seed coat. Likewise, these are also prone to increased insect activities. Grains containing ten insects per 100 g are considered to be unhygienic and are unfit for consumption (Venkatrao et al. 1960).

Pulse beetles or bruchids belonging to *Callosobruchus* spp. (Coleoptera: Bruchidae) are major storage pests of legume crops grown in the tropics and subtropics (Duraimurugan et al. 2014). Among the different species of pulse beetle, *Callosobruchus chinensis* L. is the most destructive in India, and the postharvest seed losses due to the beetle can reach even up to 100 % during severe periods of infestation (Srinivasan et al. 2010). The genus *Callosobruchus* has many species represented by *C. maculatus*, *C. chinensis*, *C. analis* and *C. phaseoli* which are more common in subtropical regions (Lal and Verma 2007). Additionally, *C. theobromae* (Linnaeus) has also been reported in pods of pigeon pea in India. Other insect pests causing considerable damage to stored pulses are *Trogoderma granarium*, *Rhyzopertha dominica*, *Tribolium castaneum*, *Corcyra cephalonica*, *Latheticus oryzae*, *Lasioderma serricorne*, *Stegobium paniceum*, *Oryzaephilus surinamensis*, *Cryptolestes ferrugineus* and few species of mites (Lal and Verma 2007). In rajmash, *Acanthoscelides obtectus* is a serious pest. The rate of deterioration is further enhanced by the action of fungi from genera *Penicillium* and *Aspergillus* in association with these insects, and the situation is more complicated if the moisture content of stored grains is higher than the desired level.

To control bruchid infestation, insecticides or fumigants are commonly used in stored grains. Although these are found quite effective, these pose serious health hazards to farmers and especially consumers, require heavy inputs on costs and cause environmental damage. Continuous use of chemicals may also lead to development of resistance to insecticides, necessitating the application of larger amounts (Boyer et al. 2012). Therefore, breeding of crops for resistance to storage insects having inbuilt resistance, especially against bruchids, is the most effective strategy and has been reviewed by Keneni et al. 2011.

Rodents are also important storage pests of not only pulses but also cereals and other grains. These not only feed on the grains but also contaminate the grains with their urine, faeces and hair. These have very fast multiplication rates and grow multiply near storage places where plenty of food is available. Rodents cause serious damage not only to stored grains but also to packaging material and even to storage buildings, bins and structures while leading to indirect losses caused by spillage, spoilage or contamination. Therefore these are important economic and public health problems worldwide (Brooks and LaVoie 1990).

Temperature management is one of the most promising tools for controlling pests of stored grain (Fields 1992). Studies of the effects of both low and high temperatures on mortality of *C. maculatus* at different life stages gave varying results, depending on the methods (Johnson et al. 2003; Loganathan et al. 2011). Nevertheless, while temperature control offers an effective way to disinfest beans, potential problems include the effect of the treatment on the quality of beans and the cost of heating equipments. Although low temperatures do not harm seeds, high temperature stimulated the germination of mung beans (Purohit et al. 2013). Among other measures neem oil could be combined with resistant cultivars for the management of *C. maculatus* (Lale and Mustapha 2000).

31.6.1 Physical and Chemical Methods

All pulse grains require to be dried to safe moisture level for a safe storage. Moisture level in grains should be preferably below 11–12 % which is generally achieved through sun-drying after the harvest of crop. Further safety of stored grains can be ensured by adopting additional safety measures. For example, adult pulse beetles, being weak and having a short life, cannot move in grain mass and are restricted to top 15 cm layer. Keeping this in view, movement of adult pulse beetle can be prevented by placing a 7–10 cm layer of dry sand at the top of grain mass (Lal and Verma 2007). Of late, several moisture proof, low-cost and thermal absorbance structures are available for storage. These are generally as an improvement over the old traditional structures. Some of such structures are Pusa bin, Pusa cubicle and improved bamboo basket. These structures have been found quite effective for storage of most of the pulses in all weather conditions.

31.6.1.1 Primary Processing

Primary processing involves cleaning of legume grains and their separation into desired quality classes on the basis of physical properties such as diameter, thickness, density and colour of grains (Rodino et al. 2011). For cleaning, different screens, airflows, separator and destoner mechanisms are used to remove unwanted organic or inorganic matters.

Cleaning

The threshed grains have impurities with mud, chaff, straw, inert matters, stones, etc. The simplest cleaning method in most of the grain legumes involves tossing grains in the air and letting the wind carry off the lightest impurities. However, the heavier impurities are not removed by this method. Now with the availability of cleaning machines and equipments, improved quality of pulses can be ensured. The most followed method of cleaning grains of pulses is the 'air-screen' separator. It uses a combination of air, gravity and screens to separate seeds based on size, shape and density.

After air screening, indent cleaning is generally used for separating grains by diameter into different sizes for which commercial separators are now available. Some machines have inbuilt air-screen separators to separate grains based on size, shape and density. The primary aspiration cleans chaff, straw, dust or diseased grains. The gravity separator cleans the grains in machines that have the same size but differ in specific weight. This is used effectively to clean off damaged, broken, immature or undersized seeds

Table 31.1 General outline of steps involved in primary processing of pulses

Operations	Activities
Storage of uncleaned grains	Storage of grains in specific conditions required for a crop to ensure safety from stored pests and rodents and maintain required moisture level
Precleaning	Removal of coarse and fine materials based on stationary screens/vibrators
Air cleaning	Use of a combination of air, gravity and screens to remove chaff, straw, dust, broken and undersized seeds; separation of grains on the basis of size, shape and density
Indent cleaning	Separation of grains uniform in diameter
Gravity separation	Separation of grains based on specific weight
Destoning	Removal of heavier materials such as stone, glass, metal, soil clods, etc.
Grading	Grading of grains on the basis of shape, size and colour as per market demands, mostly mechanical
Packaging	Packing of cleaned and graded grains as per their shipment requirement such as into bags, containers, bulk containers, etc. for delivery

to ensure that the final product is uniform and of high quality. Destoning machines are used to separate heavier materials like stones, glass and metal (Lal and Verma 2007; Rodino et al. 2014). The general steps followed for primary processing of pulse grains are given in Table 31.1.

31.6.1.2 Secondary Processing

Dehulling or Decortication

The cleaned and graded grains are further subjected to secondary processing which mainly includes dehulling or decortication. The dehulled grains of legumes are consumed as whole or split grains in the form *dal*. The *dal* milling is the main secondary processing in South Asia especially in India where about 75 % of total pulses are milled into *dal*. The moisture content of grains has an important role in dehulling (Ramakrishnaiah and Kurien 1983). Mostly, domestic and small-to-medium scale industries are involved in *dal* milling in India. However, most of these mills do not follow standard process because there is variation in size of grains in different pulse crops, as well as varieties, environments and treatments. The use of *chakki* is common in India, but it produces high breakage of grains (up to 20–45 %) and poor quality of *dal* (Ali 2004). Thus extensive efforts were made to develop small capacity mills to give the better recovery of *dal* at several research centres in India. These machines are usually 1–2 horsepower, low-cost machines with dehulling, splitting and aspiration facilities and require pretreatments with various agents including water and/or oil, salts, chemicals and heat alone or in combinations for losing the pericarp. The commercial dehull or *dal* mills consist of cleaning and grading units and pitting, pretreatment and drying units. Cost of these commercial mills depends upon the capacity and degree of automation introduced for material handling. The process of dehulling begins with cleaned and graded grain product of primary processing and requires various additional and optional steps, which are highly variable in the actual commercial set-up of modern dehulling plants.

For decortication of legume grains, loosening of binding between the seed coat and cotyledons is required. Usually a gum (such as galactomannan) or lignin layer binds the cotyledon to hull (Siegel and Fawcett 1976; Rodino et al. 2014). For *dal* milling, grains are initially 'pitted' in the roller mill before pretreatments to crack the husk for improving the absorption of pretreatment agents. Thus pitting, water addition, heating and cooling steps are designed to reduce the natural binding of the seed coat to the cotyledon. Pretreatment is generally followed to loosen the seed coats. Commercial mills use ~1 % edible oil as pretreatment for loosening the husk of those legume grains that are difficult to mill (Sokhansanj and Patil 2003). For example, use of 0.85 % oil treatment with 90 °C temperature in black gram has been reported for maximum recovery of *dal* (Tiwari et al. 2005). The hot milling treatment is also optimised for obtaining maximum recovery of dehulled grains in pigeon pea (Morrow 1991; Phirke et al. 1996).

Dehusking and splitting are the two important steps of *dal* milling process. These processes involve subjecting the seeds to abrasive/scouring forces for removal of husk and for splitting of cotyledons into two equal halves (Benerjee and Palke 2010). For dehusking, several indigenous methods are used. The vertical stone *chakki* is also used to dehusk and split the grains. For splitting of dehusked and moistened grains, vertical disk burr mill is used or grains are allowed to fall on a hard or cemented surface from sufficient height (Sahay 2003). After milling the split and whole grains are polished to remove fine dust as a means of improving the visual quality of the product. Polish with water is common practice that is accomplished in same facilities by adding small amount of water to the product stream as it passes in a horizontally mounted screw conveyer prior to bagging. Adding a small amount of vegetable oil to the product steam is also used for certain markets.

31.7 Constraints in Milling Process of Pulses

For milling of pulses, mostly small- or medium-scale industries are involved which face several constraints in efficient milling process. The major constraints in the milling process of pulses are:

- Poor capacity utilisation of processing units (70 %) due to non-availability of pulses
- Less scientific and technological support to pulse milling
- Following age-old practices in pulses milling with little improvements in the process
- Low recovery and high cost of milling in the milling units
- Reduction in the capacity utilisation of the milling units during rainy season as the pulses are mostly sun-dried
- No mechanism to grade and standardise pulses

31.8 Grading and Packaging

Deterioration and loss of quality as well quantity in pulses during transport and storage depend upon a number of physical, chemical, biological and human factors. Proper packaging, storage and transportation are the most important elements in reducing losses. Cleaned and processed grains are graded for uniformity on the basis of specific diameter or thickness of grains. Grading based on size and quality of grains adds value to the final product. Consumers have different preferences for different types of pulse grains across the globe and accordingly grading is also done as per the need of the consumers. For example, in the case of lentil, consumers prefer extra large-seeded green lentils in Spain. On contrary small red grain lentils are preferred in Bangladesh or other parts of South Asia. Similarly large-seeded *kabuli* chickpea fetches extra remuneration in the market. Keeping this in view, sizing and grading machines have been developed and both hand- and power-operated graders are available commercially.

The grading of legume grains is also done on the basis of colour for which mostly colour-sorting machines are used commercially. However, these machines have been mostly adapted from other industries and are being utilised in legume grain processing. Of late, computerised image analysis system is being used to grade the grains of legumes on the basis of size, shape and colour. These machines employ the principle of computerised digital imaging to compare the colour of grains by sensors that are adjusted to an acceptable colour range. The colours that lay outside the selected range are rejected.

After grading, grains are packed in polypropylene bags or jute bags as whole grains or spilt *dal*. Packaging for retail supply is generally done in consumer-ready packages of 500 or 1000 g polyethene bags. In India, food legume grains are consumed as whole grains as well as in split *dal* form. The cleaned and graded pulses are packed in 50 or 100 kg jute bags for wholesale and retail traders. Packaging of split *dal* is available in 500 or 1000 g polythene bags in market for consumers. Various kinds of packaging machines are now available and commercial *dal* mills have bagging and packaging units as a part of primary and secondary processing (Ali 2004).

31.9 Value Addition

The value addition processing of pulses can increase the consumption as well as market value of food legumes. In general, value addition is inferred as the process of changing or transforming a product from its original state to a more valuable state that is preferred in the marketplace (Gupta et al. 2011). Therefore, it implies to adding economic value to an agricultural produce by processing it into a final product which is more liked by the end user. With increasing household incomes, changing lifestyles and consumer preferences, international market for value-added products and ready-to-eat foods is fast growing. There are now increased opportunities for value-added products because of increased consumer demands regarding health, nutrition and convenience, efforts by food processors to improve their productivity and technological advances that enable producers to produce what consumers and processors desire (Siebert et al. 1997).

31.9.1 Breakfast Foods

Pulses serve as excellent breakfast cereals as these can be easily processed to remove beany flavours and can be directly consumed or even added to cereals or bakery products. Due to high protein content, peas, green gram and lentils could be added to breakfast cereals. Mung bean and chickpea sprouts have great nutritional value as described earlier and are finding a new market in Europe, the Americas and Asia as breakfast cereals.

31.9.2 Snack Foods

Pulses offer a good option for snack food preparation as they can be cooked in a variety of forms to make them crisp and also allow easily for flavour addition. They have a quality of retaining a good crunch after roasting/deep-frying and are therefore used widely in several snack preparations. Pulse-based snack foods are immensely popular in Southeast Asia. In India, snacks made from peas, lentils and chickpeas are quite popular. Indigenous snack foods based on pulses, viz. namkeen, sweets, chips, etc. are immensely popular. In India, Bangladesh and Pakistan, there is a big market for deep-fried dehulled whole or split pulses. A good amount of chickpea, field pea and mung bean goes to this type of value addition in these countries.

31.9.3 Ready-to-Eat Products

Pulse-based ready-to-eat meals are common in the USA, Australia and Asian countries. Consequently, these meals are a growing segment of the retail market. Flavour retention, appearance and ability to retain texture and colour upon reheating are the important considerations for preparing ready-to-eat foods. Pulses are now being increasingly used in preparation of such foods, and these include pulse-based patties and cold meat replacements, vegetarian meals, prepared salads and soups and bean-based packaged foods (Gupta et al. 2011).

31.10 Conclusions

The production of cereals has almost doubled since 1970. However, in case of pulses, the improvement is comparatively slow although considerable strides have been made in these commodities also. With rising demand of vegetarian food due to ever-increasing population and diversification of food habits, demand of pulses is increasing at a fast pace. Increasing family incomes and changing lifestyles have also increased the protein demands of the people, and this has ultimately led to increased demand of protein-rich foods including pulses. However, in the face of already short supply of pulses, postharvest losses of food legumes are a matter of great concern. Preharvest losses due to erratic weather, traditional farm practices and poor crop management and postharvest losses due to improper storage, processing, packaging and preservation need immediate attention to ensure the food security both at macro- and micro-levels. Pre- and postharvest losses of farm produce not only affect the economy of farmers but also affect consumers and traders. Therefore, the important socio-economic and technological issues of postharvest processing, value addition and preservation need to be considered on top priority to reduce pre- and postharvest losses of pulses and to make them more readily available to the consumers.

References

Ali N (2004) Post-harvest technology and value addition in pulses. In: Ali M, Singh BB, Kumar S, Dhar V (eds) Pulses in a new perspective. Indian Society of Pulses Research and Development, IIPR, Kanpur, pp 530–543

Anonymous (1976) Regies internationales pour les essais de semences: Proceedings of the International Seed Testing Association. Seed Sci Technol 4(4):609–743

Benerjee G, Palke LM (2010) Economics of pulses production and processing of India. Occasional paper 51. Department of Economic Analysis and Research, National Bank for Agriculture and Rural Development, Mumbai, pp 80

Birewar BR (ed) (1984) Post-harvest technology of pulses, Pulse production – constraints and opportunities. Oxford/IBH Publishing Co., New Delhi, pp 425–438

Boyer S, Zhang H, Lemperiere G (2012) A review of control methods and resistance mechanisms in stored product insects. Bull Entomol Res 102:213–229. doi: http://dx.doi.org/10.1017/S0007485311000654

Brooks JE, Lavoie GK (1990) Rodent control will reduce post-harvest food losses: grain storage and handling. Agribusiness Worldwide 11–17

Duraimurugan P, Pratap A, Singh SK, Gupta S (2014) Evaluation of screening methods for bruchid beetle (*Callosobruchus chinensis*) resistance in greengram (*Vigna radiata*) and blackgram (*Vigna mungo*) genotypes and influence of seed physical characteristics on its infestation. Vegetos 27:60–67

Fields PG (1992) The control of stored-product insects and mites with extreme temperatures. J Stored Products Res 28:89–118. doi:http://dx.doi.org/10.1016/0022-474X(92)90018-L

Gupta M, Tiwari BK, Norton T (2011) Value addition and international trade. In: Pratap A, Kumar J (eds) Biology and breeding of food legumes. CABI, Oxfordshire, pp 395–404

Jeswani LM, Baldev B (eds) (1990) Advances in pulse production technology. Indian Council of Agricultural Research, New Delhi

Johnson JA, Wang S, Tang J (2003) Thermal Death Kinetics of Fifth-instar *Plodia interpunctella* (Lepidoptera: Pyralidae). J. Econ. Entomol. 96(2): 519–524

Keneni G, Bekele E, Getu E, Imtiaz M, Damte T, Mulatu B, Dagne K (2011) Breeding food legumes for resistance to storage insect pests: potential and limitations. Sustainability 3:1399–1415. doi:http://dx.doi.org/10.3390/su3091399

Kurien PP, Desikachar HSR, Parpia HAB (1972) Processing and utilization of legumes. In: Proceedings of the symposium on tropical agricultural research series No 6, Sept 1972, pp 225

Lal RR, Verma P (eds) (2007) Post harvest management of pulses. Indian Institute of Pulses Research, Kanpur

Lale NES, Mustapha A (2000) Potential of combining neem (*Azadirachta indica* A. Juss) seed oil with varietal resistance for the management of the cowpea bruchid, *Callosobruchus maculatus* (F.). J Stored Prod Res 36:215–222. doi:http://dx.doi.org/10.1016/S0022-474X(99)00035-1

Loganathan M, Jayas DS, Fields PG, White NDG (2011) Low and high temperatures for the control of cowpea beetle, *Callosobruchus maculatus* (F.) (Coleoptera: Bruchidae) in chickpeas. J Stored Prod Res 47:244–248. doi:http://dx.doi.org/10.1016/j.jspr.2011.03.005

Morrow B (1991) The rebirth of legumes. Food Technol 45:96–121

Phirke PS, Bhole NG, Adhaoo SH (1996) Response surface modelling and optimization for dehulling of pigeonpea with different pre-treatments and conditions. J Food Sci Technol 33:47–52

Purohit P, Jayas DS, Yadav BK, Chelladurai V, Fields PG, White NDG (2013) Microwaves to control *Callosobruchus maculatus* in stored mung bean (*Vigna radiata*). J Stored Prod Res 53:19–22. doi: http://dx.doi.org/10.1016/j.jspr.2013.01.00

Ramakrishnaiah N, Kurien PP (1983) Variabilities in the dehulling characteristics of pigeon pea (*Cajanus cajan* L.) cultivars. J Food Sci Technol 20:287–291

Reyes-Moreno C, Okamura-Esparza J, Armienta-Rodelo-E, Gomez-Garza RM, Sahay KM, Bisht BS (1987) Optimization of grades of carborundum for best recovery of dhal in milling of different pulses. Agric Eng Today 9:11–22

Rodino AP, Kumar J, Fuente MDL, Ron AMD, Santalla M (2011) Post-harvest technology. In: Pratap A,

Kumar J (eds) Biology and breeding of food legumes. CABI, Oxfordshire, pp 385–394

Sahay KM (2003) Postharvest technology and value addition in chickpea. In: Ali M, Kumar S, Singh NB (eds) Chickpea research in India. Indian Institute of Pulses Research, Kanpur, pp 315–325

Sharma A, Garry P (1995) Food grains, pulses, oilseeds and cotton in India. In: The potential impact of unilateral trade liberalization (Mimeo). NCAER, New Delhi

Siebert JW, Jones R, Sporleder TL (1997) The VEST model: an alternative approach to value added. Agribusiness 13(6)

Siegel A, Fawcett B (1976) Food legume processing and utilization (with special emphasis on application in developing countries). International Development Research Centre, Ottawa, p 88

Singh U, Singh B (1992) Tropical grain legumes as important human foods. Econ Bot 46:310–321

Sokhansanj S, Patil RT (2003) Dehulling and splitting pulses. In: Chakraverty A, Mujumdar AS, Raghavan GSV, Ramaswamy HS (eds) Handbook of post harvest technology. Marcel Dekker, New York

Srinivasan T, Duraimurugan P, Singh SK, Chattopadhyay C (2010) Bruchids infestation in pulses and its management. Ind Farm 60:13–16

Tiwari BK, Jaganmohan R, Vasan BS (2005) Effect of heat processing on milling of black gram and its end product quality. J Food Eng. doi:10.1016/j.jfoodeng.2005.10.003

Vandenberg A (2009) Postharvest processing and value addition. In: Erskine W, Muehlbauer FJ, Sarker A, Sharma B (eds) The lentil: botany, production and uses. CAB International, Cambridge, pp 391–407

Venkatrao S, Nuggehalli RN, Pingale SV, Swaminathan M, Subrahmanyam V (1960) Effect of insect infestation and stored field bean and black gram. Food Sci 9:79

Enriched Potato for Mitigating Hidden Hunger 32

Rajiv and Prashant Govindrao Kawar

Abstract

Potato (*Solanum tuberosum* L.) is the most important non-grain food crop in the world, ranking third in terms of total production after rice and wheat and is one of main commercial crops grown in the country. Being a short duration crop, it produces more quantity of dry matter, edible energy, and edible protein in lesser duration of time than cereals like rice and wheat. Hence, potato may prove to be a useful tool to achieve the nutritional security of the nation. It can also be fitted suitably into different cropping systems. It is a highly nutritious, easily digestible, wholesome food containing carbohydrates, proteins, minerals, vitamins, and high-quality dietary fiber. It is a very popular food source for human health with potato protein "*patatin*" having high biological value than proteins of cereals and even better than that of milk and can be substituted for meat and milk products with improving taste, lowering energy intake, and reducing food cost. Boiled or baked potatoes are virtually fat-free and palatable and contain about 60–80 % of the fatty acid which is composed of unsaturated fatty acids that increases the nutritive value of the potato fat. Consuming potatoes just once or twice each day lowers high blood pressure almost the same as oats without resulting in increase in weight. Potato protein has 18–20 essential amino acids in varying quantities along with various important minerals and trace elements. In addition, potato is used for various value-added products, viz., chips, French fries, cubes, granules, and canned products with high resistant starch and nutritional value. Its nutritional value can be greatly enhanced with diverse agricultural

Rajiv
Department of Vegetable Science, Chandra Shekhar Azad University of Agriculture & Technology, Kanpur, Uttar Pradesh 208 024, India
e-mail: rajiv.agro69@gmail.com

P.G. Kawar (✉)
Division of Crop Improvement, ICAR-Central Potato Research Institute, Shimla, Himachal Pradesh 171 001, India
e-mail: prashant.kawar@icar.org.in; pkawar@gmail.com

U. Singh et al. (eds.), *Biofortification of Food Crops*, DOI 10.1007/978-81-322-2716-8_32

practices including the modern tools of biotechnology. This paper gives an insight on the role of potato – *food for future* – in the food security of developing nations.

Keywords
Mitigating hunger • Nutritional value • Potato • Processed products

32.1 Introduction

United Nations Food and Agriculture Organization (FAO) records state that an estimated 868 million people are undernourished worldwide (FAO-WFP-IFAD 2012). Undernourishment in combination with vitamin and mineral deficiencies is responsible for the deaths of more than 2.5 million children per year (FAO-WFP-IFAD 2012). With a population that is expected to reach nine billion by 2050 (http://esa.un.org/unpf/index.htm), these problems are only expected to get worse. Potato (*Solanum tuberosum* L.) popularly known as "the king of vegetables" is the most important non-grain food crop in the world, ranking third in terms of total production with over 365 million tonnes (MT) per year, after rice and wheat. It is grown in around 125 countries spread across both temperate and tropical regions and at elevations from sea level to 4000 MT. One third of potato production takes place in developing countries, and over one billion people have potato as their staple diet. It has steadily expanded globally, with 35 % increase in overall production since 1960. The increase in production is still higher in developing countries of Asia and Africa (Fig. 32.1) indicating its budding importance as a staple food source.

Indian vegetable basket is incomplete without potato. Because the dry matter, edible energy, and edible protein content of potato makes it nutritionally superior vegetable as well as staple food not only in our country but also throughout the world. Now, it becomes an essential part of breakfast, lunch, and dinner worldwide. Being a short duration crop, it produces more quantity of dry matter, edible energy, and edible protein in lesser duration of time than cereals like rice and wheat. Hence, potato may prove to be a useful tool to achieve the nutritional security of the nation. It has been observed that during present trend of diversification from cereals to horticultural crops, shifting from wheat/barley cultivation to potato cultivation returns more to the farmers. The Portuguese traders or British missionaries introduced it in India during the early seventeenth century (Singh 2013b).

Potato is one of main commercial crops grown in the country. In India, potato production alone contributes about 4 and 4.5 times higher than that of rice and wheat, respectively, in agricultural GDP from unit area of cultivable land (Singh et al. 2011). The country has achieved a tremendous growth in potato production during the last four to five decades. The annual compound growth rate of potato is higher than other major food crops in respect of area, production, and productivity. Due to the bumper crop and lack of postharvest management, glut situations raised in the market for the surplus yield every year, which ultimately results in drastic price decline. In India, there is a great scope for cultivation of potato suitable for processing. Further, there is a rising demand for quality processed potato products from the country particularly in Middle East. The countries like Japan, Singapore, Korea, Malaysia, and China also have a great demand for processed potato products as well as fresh potato for processing purpose. Thus, the potato processing has opened a new dimension for development of agro-based industries in the country. Indian potato produce is preferred worldwide for its taste and meets the international quality standards in terms of disease freeness, shape, size, skin color, flesh, and dry matter

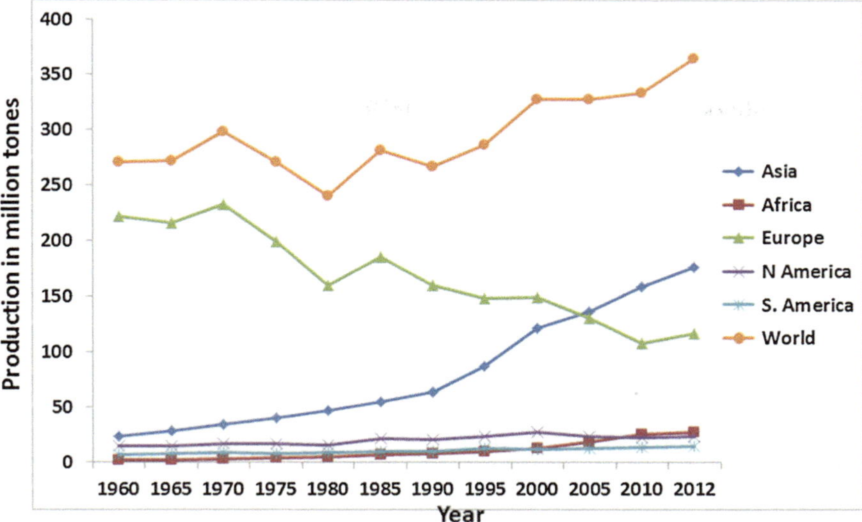

Fig. 32.1 Trends in the world potato production from 1961 to 2012

content. The government of India has set up four Agri Export Zones (AEZs) in Punjab, West Bengal, Uttar Pradesh, and Madhya Pradesh for significant development in this direction.

The demand for potato is expected to grow at 3.80 % annual compounded growth rate. At this rate, total consumption of potato by the year 2030 would be 67.23 million tonnes considering 2007 data of FAO, i.e., 28.51 million tonnes as base. It indicates an excess production of 2.16 million tonnes by 2030 (Singh et al. 2011). Physiologically potato is capable of producing 120 t/ha. Therefore, we have to exploit the potential of this wonderful crop to the maximum extent through variety development program.

Potato was introduced in India as a temperate crop and now it is being grown as a subtropical crop. It is a unique crop and can be harvested early or late depending upon market prices and requirement of field for subsequent crop. Therefore, it can be fitted suitably into different cropping systems. Potato produces almost 2–3 times more dry matter and edible energy/unit area and time than cereal crops like wheat (*Triticum aestivum*) and rice (*Oryza sativa*) and thus augments the productivity of any potato-based cropping system. Potato is an annual, herbaceous, dicotyledonous, and vegetatively propagated plant. It can also be propagated through botanical seed known as true potato seed (TPS). The potato tuber is a modified stem developed underground on a specialized structure called stolon. It contains all the characteristics of a normal stem like dormant bud (eye) and scaly leaf (eyebrow). In potato plants, high night temperatures are much more deleterious to the formation of yield than day temperature. Potato can give good yield even at day temperatures of 30–35 °C provided nights are below 18 °C. However, if night temperature go beyond 22 °C, there is very little tuberization even when the day temperature is 25–27 °C. So heat tolerance in potatoes is concerned more with the minimum night temperature than the maximum day temperature (Burton 1996).

32.2 Importance of Potato

As per FAO, more than one billion people over the world consume potato. It is a high-quality vegetable cum food crop and used in preparing more than 100 types of recipes in India. The popular Indian recipes like *samosas* and *aloo parathas* are prepared from potato. The protein of potato has high biological value than proteins of cereals and even better than that of milk. Hence, potato can be a supplement of meat and milk products for improving their taste, lowering energy intake, and reducing food cost. In a nutritional point of view, potato is a wholesome food and deserves to be promoted as a potential high-quality vegetable cum food crop in the country. Keeping in view the potential of potato in the food security of developing nations, FAO has declared it as the "food for future" (Singh et al. 2011). As being one of the principal cash crops, it gives handsome returns to the growers/farmers due to its wide market demand nationally and internationally for different kinds of utilization. Further, it has been reported by the International Food Policy Research Institute (IFPRI) and International Potato Center (CIP) that India is likely to have highest growth rate of potato production and productivity during 1993–2020. During the same period, demand for potato is expected to rise by 40 % worldwide. This indicates a clear opportunity to capture the huge domestic and international market of potato by producing quality potato and its products.

32.3 Uses of Potato

Potato is utilized in a variety of ways. Potato is utilized as major vegetable throughout the world and in preparation of a number of recipes either by using potato alone or by combining it with other vegetables, pulses, cereals, etc. Its medium-sized tubers are used normally as a seed. It is also utilized in a variety of ways such as dehydrated potato products like chips, flakes, granules, flour, starch, potato powder, and potato biscuits. It is also used to prepare frozen foods like potato patties, puffs, wedges, pancake, dehydrated mashed potatoes, etc.

32.4 Status of the Major Vegetable Crops in India

Potato has the highest (28 %) share among all the vegetable crops in total vegetable production of India during the year 2012–2013 (Fig. 32.2). The details of information about area and production growth trends, production share, and leading producing states of the major vegetable crops in

Production share of poato of total vegetable production in India (%)

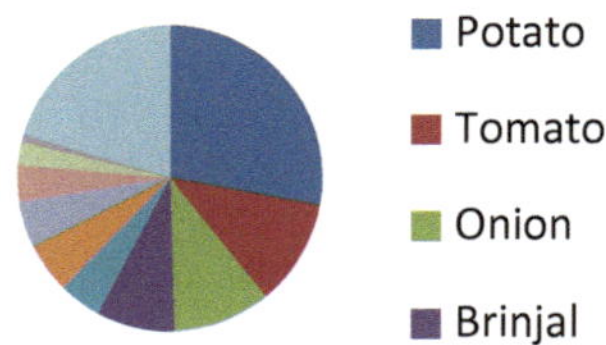

Fig. 32.2 Production share of potato in total vegetable production of India

India are furnished in Tables 32.1, 32.2, 32.3, and 32.4 (NHB 2014).

32.5 Potato in the World

Potato is grown in more than 100 countries in the world with a production of around 365,152 thousand tonnes during the year 2012–2013. China ranks first while India and Russia ranks second and third, respectively, in terms of production while India ranks fourth in potato area (Fig. 32.3). China, India, the USA, Ukraine, Germany, and Poland shared more than 61 % of total global production (Table 32.5). The productivity of potato in India is low (22.8 MT/ha) as compared to that of Netherlands (45.2 MT/ha), Germany (44.8 MT/ha), and the USA (41.8 MT/ha). This is due to nearly 85 % of the crop grown during winters having short photoperiod with 10–11 h sunshine and shorter crop duration (about 90–100 days) in India. The mornings usually have fog, which further reduces the sunshine hours. On the other hand, in Europe, the potato crop is grown in summer having long photoperiod of up to 14 h with long crop duration of 140–180 days. The country-wise potato production during 2012–2013 is furnished in Table 32.5.

32.6 Potato in India

Compared to 1949–1950, when the total production was 1543 thousand tonnes from an area of 239 thousand hectare, we now (2012–2013)

Table 32.1 All India area and production growth trends for vegetable crops

Year	Area (x'000 ha)	Production (x'000 MT)	Productivity (MT/ha)
2005–2006	7213	111,399	15.4
2006–2007	7581	114,993	15.2
2007–2008	7848	128,449	16.4
2008–2009	7981	129,077	16.2
2009–2010	7985	133,738	16.7
2010–2011	8495	146,554	17.3
2011–2012	8989	156,325	17.4
2012–2013	9205	162,187	17.6

Table 32.2 Production share of major vegetable crops in India (2012–2013)

S. no.	Crop	Production share (%)
1.	Potato	28.0
2.	Tomato	11.2
3.	Onion	10.4
4.	Brinjal (aubergine)	8.3
5.	Tapioca	4.5
6.	Cabbage	5.3
7.	Cauliflower	4.9
8.	Okra	3.9
9.	Peas	2.5
10.	Sweet potato	0.7
11	Others	20.5

produce around 45,344 thousand tonnes of potato from about 1992 thousand hectare. As compared to the area, production, and productivity scenario of potato in India during 1949–1950, the increase over this period is 733 %, 2838 %, and 246 %, respectively, as furnished in Table 32.6 (Singh et al. 2011; NHB 2014). Potato production has shown an exponential growth from 15.6 million tonnes (1990–1993) to 43.4 million tonnes (2011–2013) with less increase in area and productivity (Fig. 32.4), and hence focused efforts are needed to increase the potato productivity with respect to its potential.

In India, potato is cultivated in almost all states and under very diverse agroclimatic conditions. Out of the total area under potato, 86 % area is in Indo-Gangetic plains, 6 % in the hills, and remaining 8 % in southeastern, central, and peninsular India (Pandey et al. 2012). The states of Uttar Pradesh, West Bengal, and Bihar accounted for more than 70 % share in total production with the rank of first, second, and third, respectively, in relation to potato production. The highest potato productivity of 30.8 MT is in Gujarat followed by West Bengal (30.0 MT). The state-wise area, production, and productivity of potato during 2012–2013 are furnished in Table 32.7 (NHB 2014).

32.7 Agroecological Zones and Varietal Needs

India has diverse soil types and agroclimatic conditions. Successful potato cultivation requires night temperatures of 15–20 °C with sunny days. Indian subtropical plains offer optimum conditions for potato cultivation, where 85–90 % of potatoes are grown during short winter days from October to February. The hills account for 6 % of the total potato production where the crop is grown during long summer days from April to September/October. The plateau region of southeastern, central, and peninsular India constitutes about 6 % area where potato is grown mainly as rainfed or irrigated winter crop. Based on the diverse soil, climate, and other agronomic features, the potato-growing areas in India can be divided into eight agroecological zones as furnished in Table 32.8 (Pandey and Sarkar 2005; Kumar et al. 2011).

32.8 Quality Attributes of Potato

Appearance, color, size, shape, and defects decide the quality for fresh potato. Total solids or dry matter is highly correlated with texture. Based on dry matter and texture, potatoes can be used for different purposes. A mealy texture is associated with high solids and a waxy texture with low solids. Mealy textured varieties are

Table 32.3 State-wise area, production, and productivity of vegetables (2012–2013)

S. no.	State	Area (x000 ha)	Production (x000 MT)	Productivity (MT/ha)
1.	West Bengal	1347.96	25,466.81	18.9
2.	Uttar Pradesh	912.66	19,571.56	21.4
3.	Bihar	861.80	16,325.70	18.9
4.	Madhya Pradesh	612.80	12,574.00	20.5
5.	Andhra Pradesh	686.10	12,104.70	17.6
6.	Gujarat	537.60	10,520.70	19.6
7.	Odisha	688.10	9464.00	13.8
8.	Maharashtra	474.00	8008.00	16.9
9.	Tamil Nadu	277.80	7897.90	28.4
10.	Karnataka	436.60	7841.90	18.0
	Total	9205.20	162,186.60	17.6

Table 32.4 Leading vegetable-producing states (2012–2013)

	Production		Area	
S. no.	State	% share	State	% share
1.	West Bengal	15.7	West Bengal	14.6
2.	Uttar Pradesh	12.1	Uttar Pradesh	9.9
3.	Bihar	10.1	Bihar	9.4
4.	Madhya Pradesh	7.8	Madhya Pradesh	6.7
5.	Andhra Pradesh	7.5	Andhra Pradesh	7.5
6.	Gujarat	6.5	Gujarat	5.8
7.	Odisha	5.8	Odisha	7.5
8.	Maharashtra	4.9	Maharashtra	5.1
9.	Tamil Nadu	4.9	Chhattisgarh	4.1
10.	Karnataka	4.8	Karnataka	4.7
11.	Others	20.0	Others	24.6

usually considered best for baking or French fries, while waxy texture is more often used for boiling. In India, mostly white-, yellow-, or red-skinned varieties with shallow or medium eyes are the choice of the consumers. Interest now seems to be shifting toward yellow-fleshed varieties. More yellow flesh color is indicative of the higher level of vitamin A. Yellow-fleshed varieties exhibit less darkening after cooking than some red-skinned varieties. The processing purpose varieties should possess high dry matter and low reducing sugars. Specific characteristics of potato varieties for different purposes are listed in Table 32.9 (Kumar 2013).

32.9 Potato for Health

Potato is a very popular food source for human health. Food ranking system qualified potatoes as a good source of vitamin B6 (involved in more than 100 reactions), vitamin C, copper, potassium, manganese, and dietary fiber (Fig. 32.5). Besides that, potatoes also contain a variety of phytonutrients that have antioxidant activity. Among these important health-promoting compounds are carotenoids, flavonoids, and caffeic acid, as well as unique tuber storage proteins, such as patatin, which exhibit activity against free radicals.

32.10 Nutritional Value of Potato

Potato does better than rice and wheat as edible energy source. Consuming potatoes just once or twice each day lowers high blood pressure almost the same as oats without resulting any increase in weight. Potato is a highly nutritious, easily digestible, wholesome food containing

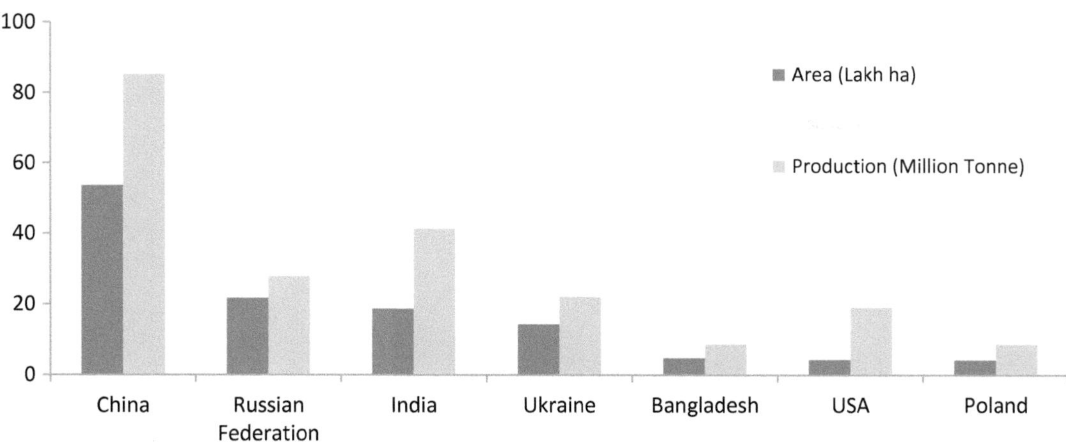

Fig. 32.3 International ranking of India in potato production

Table 32.5 Major potato-producing countries in the world (2012–2013)

Country	Area (x000 ha)	Production (x000 MT)	Productivity (MT/ha)	Production share (%)
China	5429.0	85,860.0	15.8	23.5
India	1992.2	45,343.6	22.8	12.4
Russian Federation	2197.2	29,532.5	13.4	8.1
Ukraine	1444.1	23,250.2	16.1	6.4
USA	458.4	19,165.9	41.8	5.2
Germany	238.3	10,665.6	44.8	2.9
Poland	373.0	9091.9	24.4	2.5
Bangladesh	430.4	8205.5	19.1	2.2
Belarus	332.3	6910.9	20.8	1.9
Netherlands	149.8	6765.6	45.2	1.9
Others	6249.6	120,360.6	19.3	33.0
World + (Total)	19,294.3	365,152.4	18.9	–

Table 32.6 All India area, production, and productivity of potato

Year	Area (x000 ha)	% of total veg. area	Production (x000 MT)	% of total vegetable production	Productivity (MT/ha)
1949–50	239.0	–	1543.0	–	6.6
2005–06	1569.2	21.9	29,174.6	26.4	18.6
2006–07	1743.0	23.0	28,600.0	24.9	16.4
2007–08	1795.0	22.9	34,658.0	27.0	19.3
2008–09	1828.0	22.9	34,391.0	26.6	18.8
2009–10	1835.3	23.0	36,577.3	27.3	19.9
2010–11	1863.0	21.9	42,339.0	28.9	22.7
2011–12	1907.0	21.2	41,482.8	26.5	21.8
2012–13	1992.2	21.6	45,343.6	28.0	22.8

carbohydrates, proteins, minerals, vitamins, and high-quality dietary fiber (Burton 1989; Bandana 2013; Caprara 2012). In fact, potatoes are second only to soybean for the amount of protein produced per ha, having the major storage protein patatin, the most nutritionally balanced plant protein known till date (Liedl et al. 1987). A single 150 g tuber provides up to 45 % of recommended daily allowance (RDA) for vitamin C, 10 % vitamin B6, 8 % niacin, and 6 % folate with

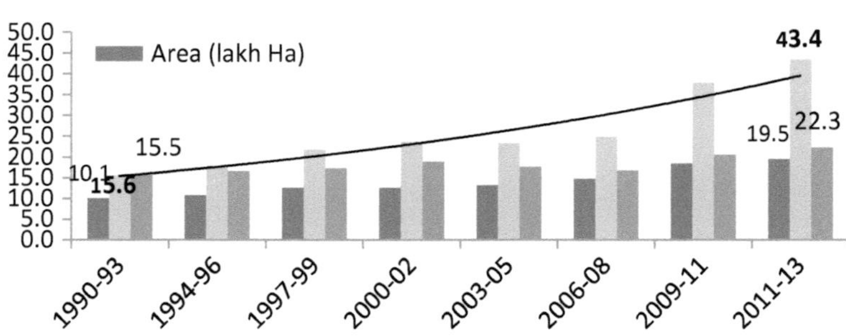

Fig. 32.4 Exponential growth of potato production in India

Table 32.7 State-wise area, production, and productivity of potato (2012–2013)

State	Area (x000 ha)	Production (x000 MT)	Productivity (MT/ha)	Production share (%)
Uttar Pradesh	603.76	14,430.28	23.9	32
West Bengal	386.61	11,591.30	30.0	26
Bihar	322.46	6640.55	20.6	15
Gujarat	81.27	2499.73	30.8	5
Madhya Pradesh	108.87	2299.00	21.1	5
Punjab	85.25	2,132.31	25.0	5
Assam	99.77	975.27	9.8	2
Karnataka	44.40	698.30	15.7	2
Haryana	29.47	676.02	22.9	1
Jharkhand	47.21	659.61	14.0	1
Others	183.10	2741.20	15.0	6
Total	1992.20	45,343.60	22.8	–

Table 32.8 Agroecological zones of potato in India and their varietal requirements

S. no.	Region	State	Varietal requirements
1.	Northwestern plains	Punjab, Haryana, part of Jammu and Kashmir, and Rajasthan	Short day adapted, early bulking, heat tolerance and moderate resistant to late blight, tolerance to frost
2.	West-central plains	Western and central UP, Uttaranchal, MP, Chhattisgarh, and Rajasthan	Short day adapted, early bulking, moderate resistant to late blight, tolerance to frost
3.	Northeastern plains	Eastern UP, Bihar, Jharkhand, WB, Assam, and Odisha	Short day adapted, early bulking, moderate resistant to late blight; red-skinned tubers are preferred in some areas
4.	Plateau region	Karnataka, Maharashtra, and Gujarat	Early bulking, ability to tuberize under high temperatures, resistance to bacterial wilt, mites and potato tuber moth
5.	Northwestern hills	Jammu and Kashmir, HP, and Uttaranchal	Long day adapted, highly resistant to late blight
6.	Northeastern hills	Assam, Meghalaya, Arunachal Pradesh, Manipur, Nagaland, Mizoram, and Tripura	Long day adapted, highly resistant to late blight and bacterial wilt
7.	North Bengal and Sikkim hills	North Bengal and Sikkim	Medium maturity, resistance to late blight and immunity to wart. Red-skinned potatoes are preferred
8.	Southern hills	Nilgiris in Tamil Nadu	Long day adapted, early bulking, resistant to late blight and cyst nematode

Table 32.9 Specific characteristics of potato varieties for different purposes

	Purposes			
	Table potatoes		Processing	
Characters	Boiled	Baking	French fries	Chips
Tuber shape	Long oval/round	Long oval/round	Long oval (>3 in.)	Round (2.5–3.3 in.)
Skin color	White/yellow/red	White/yellow/red	White/yellow	White/yellow
Eye depth	Shallow/medium	Shallow/medium	Shallow	Shallow
Flesh color	White/yellow	White/yellow	White/yellow	White/yellow
Texture	Waxy	Mealy	Mealy	Mealy
Uniformity	High	High	High	High
Defects	Minimum	Minimum	Minimum	Minimum
Dry matter (%)	18–20	>20	>20	>20
Reducing sugars[a]	–	–	<200 mg	<100 mg
Phenols	Less	Less	Less	Less
Glycoalkaloids[a]	<15 mg	<15 mg	<15 mg	<15 mg
Keeping quality	Good	Good	Good	Good
Damage resistance	High	High	High	High

[a]mg/100 g fresh tuber weight

significant amounts of other essential mineral nutrients required, still it lacks in providing many other essential nutrients required for human consumption, and improvement for those constituents is being taken care by using classical breeding and improved genetic engineering techniques. A detailed nutritive value of potato is presented in Table 32.10.

The important nutritional component of potatoes and improvement in them to make potato as "complete food" using advanced biotechnological approaches are described in proceeding section.

32.10.1 Carbohydrate

Carbohydrates are vital to health and their role in nutrition is to provide energy. They are required for brain and are the preferred fuel for the muscles. Starch is the major carbohydrate in potato and sucrose, fructose, and glucose are the main sugars in potato. The main energy-providing nutrient in potatoes is carbohydrate, in the form of starch, and provides primary source of energy for the body and supply at least half of calories for the day. Carbohydrate content of raw potato is approximately 18.2 g per 100 g edible portion (Fig. 32.6), and the digestibility of potato starch is low in raw state but is greatly improved during cooking or processing. Starch, a predominant storage macromolecule in potato tubers, consists two types of high molecular weight D-glucose polymers: amylase (mostly a linear α (1–4) D-glucan polymer with limited branches) and amylopectin-a (much larger molecule with extensive branches resulting from both α (1–4) and α (1–6) linkages in the ratio of 1:3 (Smith et al. 1997; Jansen et al. 2001). The branched structure of amylopectin allows for greater digestibility than linear chain structure of amylase, which leads to higher glycemic response.

Nutritionally resistant starch (or more slowly digested starch) is considered advantageous as it provides similar health benefits to fermentable fiber; similarly resistant starch degradation products are not absorbed in small intestine. Higher-amylose starches have greater retrogradation following processing compared with those having more amylopectins. Higher amylase starches also reduce oil penetration and so are favored in processed food as well (Tarn et al. 2006). Furthermore, potato starches have unique physical and chemical properties, such as the low gelatinization temperature, the high degree of white, and the high degree of polymerization in comparison to starches from other crop

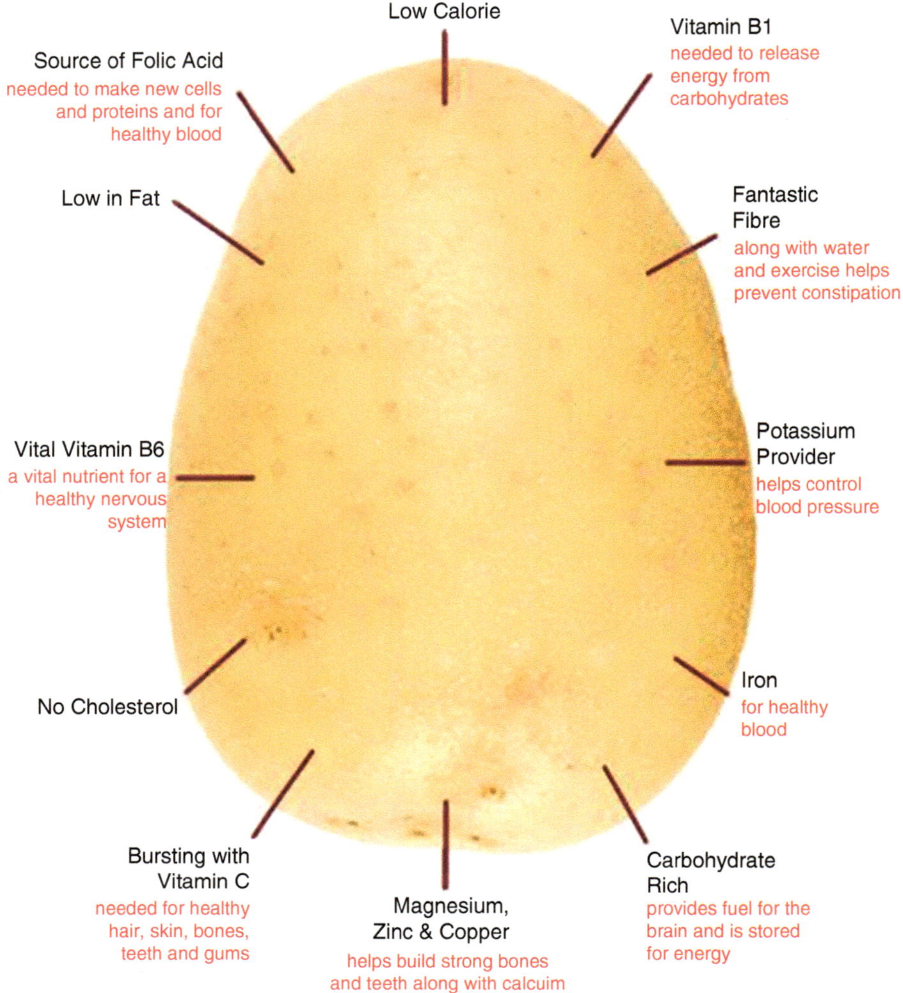

Fig. 32.5 Potato the complete food or bread for life

species, and hence are nutritionally very useful for human consumption (Swinkels 1985). Moreover, the advantage of getting carbohydrates from potatoes is that one can get a considerable amount of certain micronutrients as well.

With the advent of modern biotechnological tools, efforts are underway to improve the health-promoting starch in potato. RNAi-based, *SSIII*, and *SSIII* gene silencing resulted in the increase of amylase content by 2.68–29.05 % with reduction in amylopectin and amylase ratio and phosphorus content of tuber starch by 34–56 % which increased the granular structure quality of potato starch (Hong-hui et al. 2012). High-amylase potato starches were developed by antisense inhibition of *SBE I* and *SBE II* genes increasing the tuber amylose content to 60–89 % as compared to 21–29 % in wild type (Andersson et al. 2006). The average chain length of potato amylose is much greater than that of cereal amylose (Jobling et al. 2002); hence, the new high-amylose potato starches certainly have improved functionality. Overexpression of *SuSy* gene in tubers increased the starch content by 55–85 % than in control tubers with significant increase in tuber dry weight, starch content per plant, and total yield (Fernandez et al. 2009). In another study, transgenic potato plants simultaneously overexpressing endogenous glucose 6-phosphate/phosphate and adenylate

Table 32.10 Nutritional composition of potato

Substance	Range (%)	Mean (%)
Dry matter	13.1–36.8	23.7
Starch	8.0–29.4	17.5
Reducing sugars	0.0–5.0	0.3
Total sugars	0.05–8.0	0.5
Crude fiber	0.17–3.48	0.71
Pectic substance	0.2–1.5	–
Total nitrogen	0.11–0.74	0.32
Crude protein	0.69–4.63	2.00
Lipids	0.02–0.2	0.12
Ash	0.44–1.87	1.1
Ascorbic acid	21.7–68.9[a]	–
Glycoalkaloids	0.2–41[b]	3–10
Phenolic compounds	5–30[b]	–

Source: Schwall et al. 2000
[a]mg/100 g
[b]μg/100 g

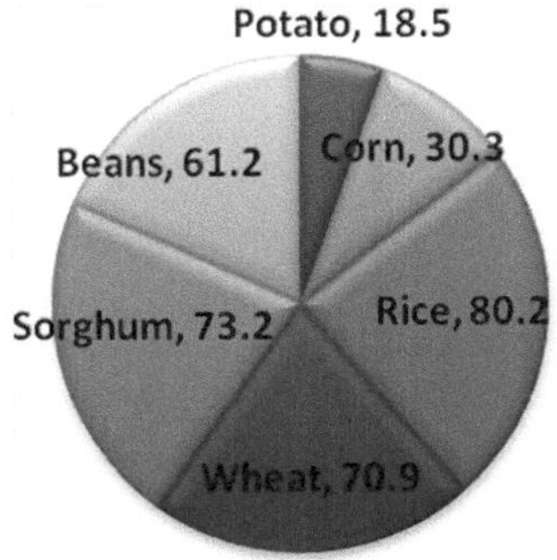

Fig. 32.6 Carbohydrate content of raw potato vis-a-vis other plant foods (g/100 g edible portion)

translocators in tubers increased both starch content and the yield. Using this source and sink approach, the tuber starch content was doubled compared to control (Claidia et al. 2012). Most recently, overexpression of amylosucrase gene from *Neisseria polysaccharea* fused to potato starch-binding domain (SBD) in transgenic potato tubers resulted in starch granules with rough surface, a twofold increase in median granule size, improved freeze-thaw stability, higher-end viscosity, and better enzymatic digestibility. These altered physicochemical properties of potato starch called the "waxy potato starch" have improved paste, clarity, and stability and expected to find applications in the food industry (Xing et al. 2014) for developing nutritionally improved potato products.

32.10.2 Fat

Boiled or baked potatoes are a virtually fat-free food and average fat content of potato is 0.1 %. This little fat present in tuber gives palatability. About 60–80 % of the fatty acid content is composed of unsaturated fatty acids (linoleic acid). This high content of unsaturated fatty acids increases the nutritive value of the fat present in potato. Figure 32.7 depicts the percent fat content of raw potato vis-a-vis other plant foods.

32.10.3 Protein

Protein is considered the most important nutrient for humans and animals and is important constituents of cellular membranes as well as various cytoplasmic structures. Protein malnutrition is essentially caused by poor-quality diets that include a high intake of staple crops with less protein and/or low-quality proteins in terms of amino acid composition. Protein deficiency lowers resistance to disease, delays physical growth and development, and may cause permanent impairment of the brain in infants and young children. Plant proteins contribute about 65 % of the per capita supply of protein on worldwide basis, with cereal grains, tubers, and food legumes as the most important suppliers (Chakraborty et al. 2010). Although potatoes are commonly perceived as a carbohydrate source only, they do contain high-quality proteins, especially the enzymes present in potatoes are the protein nitrogen. On a fresh weight basis, potato tuber protein ranges from 1 % to 1.5 % (Ortiz-Medina 2007). Where potato consumption is high, this vegetable can make a significant contribution to health as a protein source.

Potato proteins comprise 18–20 amino acids in varying quantities in which lysine, aspartic acid, glutamic acid, and valine are present in highest quantities but are deficient in sulfur-

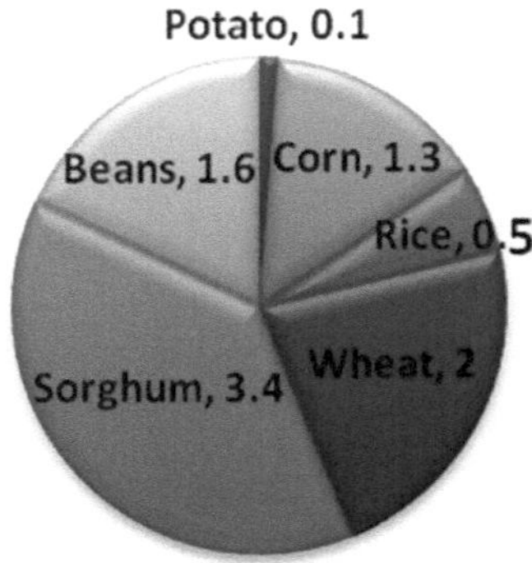

Fig. 32.7 Percent fat content of raw potato vis-a-vis other plant foods

containing amino acids. With its high lysine content, potato can supplement diets which are limiting in lysine. Moreover, major effort is being made to improve the amino acid composition of plant protein because animals, including humans, are incapable of synthesizing 10 of the 21 amino acids required for protein synthesis, and these "essential amino acids" must therefore be obtained from the diet (Chakraborty et al. 2010). Moreover, in comparison with meat, plant proteins are much less expensive to produce. Since the dietary protein is of importance and the fact that plants are its major source, development of strategies to increase protein levels and the concentration of essential amino acids in food crops is of primary importance in a crop improvement program. Since past there have been several attempts through mutant selection and engineering genes encoding key amino acid biosynthesis pathway enzymes to increase free essential amino acids in crop plants (Matthews and Hughes 1993), but with limited success (Falco et al. 1995). Recent advances in biotechnology allowed the use of transgenic approach to increase the content of specific essential amino acids in a target plant. In an attempt to increase the overall protein content of potato, Chakraborty et al. (2010) expressed the *AmA1* gene from *Amaranthus* seed albumin in potato tubers and developed "ProTato" having 48 % increased protein content than the wild/non-transformed potatoes. The paradise nut 2S seed protein has abundant Met residues (16 mol %), and to further increase Met content of this protein in potato, modifications were made in the sequence region between the Cys-6 and Cys-7 codons of *PN2S* cDNA to contain 19, 21, and 23 mol% Met, respectively. All the three modified Met-rich *PN2S* were expressed in transgenic potato revealing that mutated Met-enriched BN2S proteins were accumulated as normal 2S protein in the leaves and tubers of transgenic potato (Tu et al. 1998). In another study, attempts were made to increase the Met content in potato tubers through heterologous overexpression of *Arabidopsis* cystathionine γ-synthase ($CgS_{\Delta 90}$), which is not regulated by Met in potato plants and a storage Met-rich 15-kD zein in Desiree cultivar. There was sixfold increase in free Met content and in the Met content of the zein-containing protein fraction of the transgenic tubers. In addition, in line with higher Met content, the amounts of soluble isoleucine and serine were also increased, and the consecutive overexpression of $CgS_{\Delta 90}$ in *Arabidopsis* caused eightfold to twentyfold elevation of the Met content. However, all the lines with higher Met content $CgC_{\Delta 90}$ expressions were phenotypically abnormal showing severe growth retardation, changes in leaf architecture, and 40–60 % reduction in tuber yield. Moreover, the color of the transgenic tubers was altered due to reduced amounts of anthocyanin pigments (Dancs et al. 2008). Furthermore, efforts are made in the past to produce synthetic proteins in potato. The high essential amino acid encoding DNA (*HEAAE-DNA*) inserted into the chloramphenicol acetyltransferase (*CAT*) coding sequence has shown the accumulation of 0.02–0.35 % of CAT-HEAAE fusion protein of total tuber protein in transgenic potato (Yang et al. 1989). The 284-bp *asp1* gene, under the control of *CaMV 35S* promoter, was normally expressed in transgenic potato which resulted in the accumulation of relatively high levels of ASP1 proteins in transgenic potato and sweet potato (Kim et al. 1992). Robert et al. (2002) changed one amino acid residue of potato gene encoding DHDPS to render the enzyme feedback insensitive, and its introduction back into potato resulted in dramatic increase of lysine content up to 15 % of the total amino acid level as compared to 1 % in the untransformed potato. Gene silencing by RNAi technology also has been tried in potato to increase the essential amino acid content. The threonine synthase (*TS*) involved in threonine synthesis in potato was targeted to divert the cycle and increase the Met content; with reduction of 6 % TS activity, an increase of 30 % methionine levels in transgenic potato was evidenced (Michaela et al. 2001).

32.10.4 Vitamins

Among vitamins, potato contains vitamin C (AsA) in the highest amount, which ranges from 84 to 145 mg per 100 g DW depending on cultivar and soil composition (Camire et al. 2009). Potato provides the almost equal amount of B complex (B1, B2, and B3) vitamins as corn and rice. Vitamin C is water soluble and its main component ascorbic acid helps in strengthening the body's immune system and assists in the iron absorption. Potatoes are a good source of vitamin B6 (folic acid, niacin, pyridoxine, riboflavin, and thiamin), a water-soluble vitamin that plays important roles in carbohydrate and protein metabolism. It helps the body by making amino acids that are later used to manufacture various proteins. A medium serving of boiled potatoes (180 g) contains more than one sixth of the adult daily requirements for vitamins B1, B6, and folate. These B-group vitamins have many functions in the body including being essential components in the metabolism of carbohydrates to provide energy and maintaining a healthy skin and nervous system. Folate is needed for cell growth and development. Different vitamin compositions of potato are presented in Table 32.11, and comparative B complex vitamins in raw potato and other plant foods are shown in Fig. 32.8.

With the advent of transgenic technology, efforts were made to increase vitamin composition of potato. Recently a rat *L-gulone-γ-lactone oxidase* gene was constitutively expressed in potato giving rise to 40 % higher ascorbate accumulation and increased abiotic stress tolerance (Hemavathi et al. 2010). In the previous attempt, overexpression of strawberry *D-galacturonic acid reductase* gene in potato gave rise to twofold increase in tuber ascorbate content, accompanied by increased abiotic stress tolerance with respect to wild-type plants (Hemavathi et al. 2009). Wei et al. (2012) transformed potato with its native cytosolic and chloroplastic targeted *DHAR* cDNAs which led to significant increase in ascorbate content in both tubers and leaves.

Provitamin A is another very important vitamin required for life. Vitamin A deficiency (VAD) is the leading cause of preventable blindness in children and increases the risk of disease and death from severe infections. The carotenoid content of tubers in most potato cultivars ranges between 0.5 and 2.5 μg/g FW. The main carotenoids are the xanthophylls lutein and violaxanthin, which are devoid of provitamin A activity. The main provitamin A carotenoid, β-carotene, is present only in trace amounts up to 0.03 μg/g FW. Although a staple crop, potato is poor in β-carotene (Breithaupt and Bamedi 2002); however, successful efforts were made through metabolic engineering for accumulation of high levels of β-carotene in potato tubers in recent past (Diretto et al. 2006, 2007; Lopez et al. 2008). Two potato cultivars that were transformed with the *crtB* gene (*phytoene synthase*) from *Erwinia uredovora* showed an accumulation of 35 total carotenoids, 11 μg/g DW, and 78 μg/g DW β-carotene in developing tubers of Desiree and Mayan Gold, respectively (Ducreux et al. 2005). Another study with silencing of the gene encoding *Lcy-e* in order to suppress epsilon cyclization of lycopene and direct the flux toward β-β-carotenoid branch indicated a tuber-specific increase in the accumulation of β-carotene (up to 14-fold) and β-β-carotenoids (up to 25-fold) with a decrease in accumulation of lutein (Diretto et al. 2006). When the β-carotene hydroxylation step of the β-β-carotenoid branch was targeted by tuber-specific antisense silencing of the hydroxylase *chy1* and *chy2*, a 38-fold increase in tuber β-carotene content was achieved (Diretto et al. 2007).

Table 32.11 Different vitamin contents of potato

Substance	Quantity (μg/100 g tuber)
Ascorbic acid (AsA, vitamin C)	10,000–25,000
Vitamin B_1 (thiamin)	100
Vitamin B_2 (riboflavin)	70
Vitamin B_6 (pyridoxin)	–
Vitamin B_3 (niacin)	1000
Vitamin B_5 (pantothenic acid)	190–320
Vitamin B_9 (folic acid)	5–33
Vitamin B_7 (biotin)	0.6
Provitamin A (β-carotene)	11–56
Phytonadione	60–80
Vitamin B_{12} (cobalamine)	–
Vitamin D (calciferol)	–

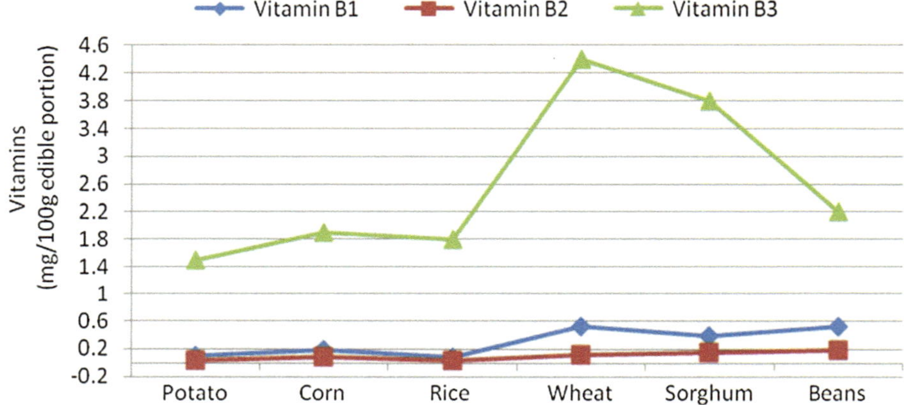

Fig. 32.8 Comparative B complex vitamins in raw potato and other plant foods

32.10.5 Minerals

Humans require various minerals to maintain health and for proper growth (Welch 2002). For example, iron and zinc deficiencies result in decreased immune function and can interfere with growth and development (Zimmermann and Hurrell 2002). It is estimated that iron deficiency affects one third of the world's population and causes 800,000 deaths worldwide each year (Masuda et al. 2012). Plants are essential source of such minerals and trace elements (Welch 2002) and are present in greatest concentrations in raw potato (Buckenhuskes 2005, Table 32.12). Potatoes are rich sources of potassium and have more potassium per serving than any other vegetable or fruit. Diets rich in potassium and low in sodium reduce the risk of hypertension and stroke and help lowering blood pressure. Potato skin is considered a good dietary source of potassium and contains relatively little phosphorous in the form of phytate. Potatoes are a rich source of iron in conjugation with the high vitamin C which helps in its absorption. Moreover, rich in trace element like zinc along with being a good source of phosphorus, magnesium, potassium, and iron makes it a nutritionally important source of these in human diet (Fig. 32.9).

Because plants cannot synthesize these minerals, they must be acquired from soil. As a result, engineering of plant mineral content is quite different from modifications of compounds like vitamins that the plant itself synthesizes. Research to improve the mineral composition of

Table 32.12 Mineral composition of tuber ash

Element	Quantity (mg/100 g DW)
Potassium (K)	1400–2500
Phosphorous (P)	120–600
Chlorin (Cl)	45–800
Sulfur (S)	40–400
Magnesium (Mg)	45–220
Calcium (Ca)	27.1–109.3
Silicon (Si)	5–89
Iron (Fe)	2.9–15.7
Aluminum (Al)	0.2–35
Manganese (Mn)	0.5–8
Zinc (Zn)	1.2–2.9
Copper (Cu)	0.06–2.8

crop plants has mostly focused on iron content, and biofortification has been attempted for micronutrients in potato as well (White and Broadley 2009). Although less research in this regard in potato is carried out worldwide, few attempts were made in recent past, such as *Arabidopsis sCAX* (*cationic exchanger 1*) and H +/Ca2+ transporter genes that were overexpressed in potato displayed up to threefold more calcium content compared to wild type without significant alteration in growth and development (Park et al. 2005). In other works, a chimeric, N-terminus, truncated *Arabidopsis* cation transporter (*CAX2B*) that contains a domain from *CAX1* for increased substrate specificity was overexpressed in potato to improve calcium accumulation (Kim et al. 2006). The transgenic plants had 50–65 % improved tuber calcium content relative to wild type, with stable inheritance and no deleterious effects on plant growth or development.

32.10.6 Antioxidants and Other Health-Promoting Compounds

Diets rich in antioxidant flavonoids and carotenoids have been associated with a lower incidence of atherosclerotic heart disease, certain cancers, macular degeneration, and severity of cataracts (Kruezer 2001). Phenolic acid and polyphenols play an important role in plant health, cooking properties, and human health (Friedman 1997). Potatoes due to its high consumption are considered as the third largest source of phenolic compounds in the human diet after oranges and apples.

Chlorogenic acid is the predominant phenolic compound in potatoes, as other *solanaceous* species such as eggplant (Prohens et al. 2007). Cultivated potato tuber skin contains 2000–5000 pg per g FW phenolic acids and 200–300 ~ tg of flavonoids, whereas the tuber flesh contains lower concentrations ranging from 100 to 600 pg of phenolic acids and 0 to 30 ~ tg of flavonoids (Lewis et al. 1998). Purple- and red-skinned tubers contained twice the concentration of phenolic acids and flavonoids as white-skinned tubers. In potato, flavonoids in order of abundance are catechin, epicatechin, eriodictyol, kaempferol, and naringenin, and the predominant phenolic acids are chlorogenic acid, protocatechuic acid, vanillic acid, and p-coumaric acid (Lewis et al. 1998).

Carotenoids are a class of flavonoids with red, orange, and yellow pigments that are widely distributed in flowers, fruits, and vegetables. Carotenoids and their derivative xanthophylls are diverse lipid-soluble pigments; in potato, xanthophylls are the most abundant carotenoids (Brown 2008). Two of these pigments, present in low concentration in potato (β-carotene and lutein), have an important role to play in eye health. Carotenoid content of potato ranges from 57 to 750 μg/150 g FW, and there are more carotenoids in colored potatoes than in white-fleshed potatoes (Buckenhuskes 2005). A promising genetic tool for carotenoid biofortification is the novel orange (*Or*) mutation isolated from cauliflower (*Brassica oleracea* var. botrytis) (Zhou et al. 2008). The mutant orange phenotype is due to the accumulation of carotenoids caused by the differentiation of proplastids into chromoplasts. The mutant phenotype was also confined in potato tubers carrying the *Or* transgene (Lopez et al. 2008).

Phytochemicals are secondary products of plant metabolism, many of which are implemented in human health as antioxidants, and their amount and composition vary among the potato cultivars (Brown 2008). Anthocyanins

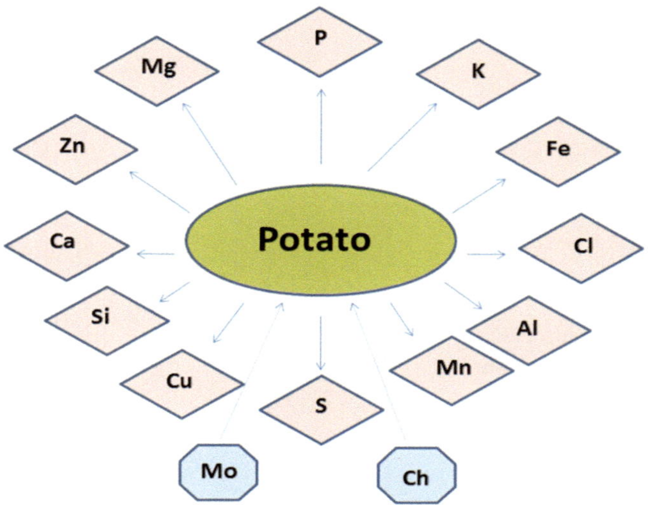

Fig. 32.9 Mineral composition of potato

are potential source of natural coloring in potato; improvement of potato antioxidant capacity through improving tuber anthocyanin pigmentation has also been a focus of research nowadays. A candidate gene for locus D was identified and constitutively expressed in white-skinned, light red-colored cultivars and two white-skinned diploid clones that lack dominant allele of the D locus. Transgenic tubers had uniform skin pigmentation, accompanied by pigmentation of tuber flesh, peals, and foliage (Jung et al. 2005). Transgenic potato plants overexpressing the anthocyanin biosynthetic gene *dihydroflavonol 4-reductase* (DER) showed an increase in tuber anthocyanin content (Stobiecki et al. 2002). Jung et al. (2005) introduced an anthocyanin biosynthetic gene *flavonoid 3′, 5′-hydroxylase* (*F3′5′H*) into the red-skinned potato Desiree and generated transgenic plants with purple-colored tubers and stems. Overexpression of endogenic *3GT* driven by *GBSS1* promoter in a red-skinned cultivar Desiree resulted in increase in anthocyanin content from 0.8 to 1.6 μg/mg FW as compared to 0.4 μg/mg FW in the control (Wei et al. 2012). In another study aimed to enhance anthocyanin synthesis in potato tubers, the enzyme (*3GT*) was overexpressed in the light red-skinned cultivar Desiree with threefold improved anthocyanin content in the skin of the transgenic tubers with respect to wild-type controls (Wei et al. 2012).

32.10.7 Dietary Fiber

Potato is a good source of fiber. It contributes to the feeling of fullness, which may help with weight management and supports healthy digestive functions. Dietary fiber has been shown to have numerous health benefits, including improving lowering the risk of heart disease, diabetes, and obesity. A 180 g portion of boiled potatoes provides about 3 g of fiber, which equates to more than 10 % of the daily recommended intake of fiber. A small amount of the starch in potatoes resists digestion known as "resistant starch," which acts in the body in a similar way to fiber and aids in the control of blood glucose and blood lipid levels. Resistant starch is considered the third type of dietary fiber, as it can deliver some of the benefits of insoluble fiber and some of the benefits of soluble fiber. Fiber is the only nutrient significantly lost when the skin is removed. The amount of resistant starch in potato foods depends upon the time and temperature used during processing or cooking and amount of water used for cooking; highest dietary fiber content is reported in the potato chips among the different forms of cooked or processed potato foods (Fig. 32.10). Among the other popular foods such as bakery products, beans, and parboiled rice, the highest amount of resistant starch was found in potato chips (Fig. 32.11).

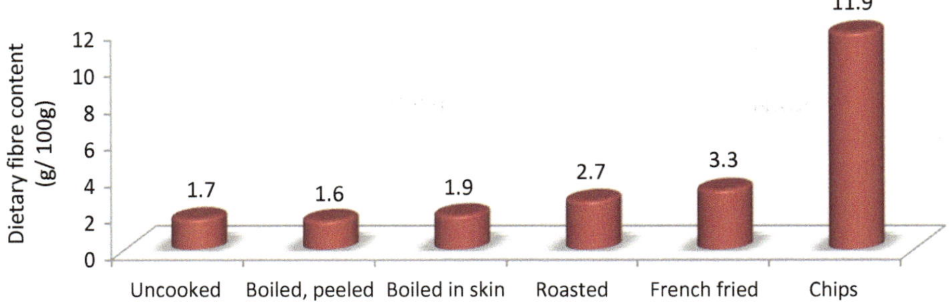

Fig. 32.10 Dietary fiber content of potatoes when cooked by different methods

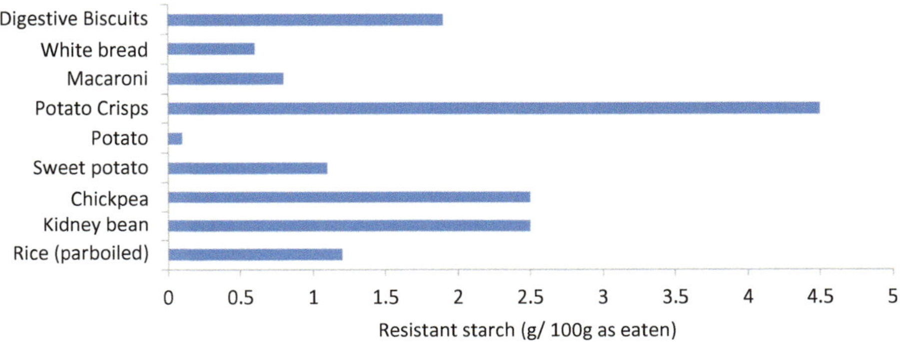

Fig. 32.11 Resistant starch content of different food materials

32.10.8 Toxicant and Allergen Compounds

Glycoalkaloids are found throughout the *Solanaceae* family. They are found in every plant organs (roots, tubers, stolons, stems, foliage, flowers, and fruits) with fresh weight concentrations in potato plants ranging from 10 mg/kg (fresh weight) in tubers to 5000 mg/kg (fresh weight) in the flowers (Smith et al. 1996). In potatoes, they have been a particular concern, due to their toxicity to humans (Friedman and McDonald 1997); although less than 15 mg/kg (fresh weight) is supposed to be safe in human consumption, still the further reduction is necessary to make it completely safe. The use of wild germplasm in potato breeding is extensive and the main source of transmission of unusual SGAs. Elimination of solanidine glycosylation would also decrease toxicity of edible tuber. Antisense DNA constructs of *SGT1* coding for solanidine galactosyl transferase involved in α-solanine biosynthesis (McCue et al. 2005), *SGT2* coding for solanidine glucosyltransferase involved in α-chaconine biosynthesis (McCue et al. 2006), or *SGT3* coding for sterol rhamnosyl transferase, the last step in the triose formation of α-chaconine and α-solanine (McCue et al. 2007), reduced the corresponding glycoalkaloids in transgenic potato plants. Arnqvist et al. (2003) overexpressed soybean (*Glycine max*) type 1 sterol methyl transferase (*GmSTM1*) in potato (cv. Desiree) in an attempt to reduce glycoalkaloids. The transgenic potato showed decreased glycoalkaloid levels in leaves and tubers down to 41 % and 63 % of wild-type levels, respectively.

Allergies to potatoes appear to be relatively uncommon. Patatin (Solt 1) is the primary storage protein (Shewry 2003) and the major allergen in potatoes. Patatin may be cross-reactive for persons with allergy to latex, and children with atopic dermatitis appear to have increased sensitivity to this potato protein (Schmidt et al. 2002). Boiling of potatoes reduces or nullifies the allergic reaction (Lee et al. 2006). Similarly, potato polyphenol oxidases (PPO) are the enzymes responsible for enzymatic browning reaction observed in impacted, damaged, or sliced tubers. These oxidative deterioration reactions alter the organoleptic properties of food and greatly affect potato tuber quality. Silencing of the *PPO* gene in transgenic potato reduced the enzymatic browning and enhanced the shelf life of potato (Llorente et al. 2011).

32.10.9 Potato as Bioreactor

Potatoes are also used as very good bioreactors to produce several health-promoting compounds, enzymes, and vaccines (edible vaccines). Transgenic potato plants producing human lactic β-casein might also be significant for nourishment. Human β-casein produced by plants might be used in the future for the production of human milk proteins such as lactoferrin and lysozyme or for preparation of baby food with increased nutritional value and preventive effects against gastric and intestinal dysfunctions in children (Chong et al. 1997). In the same manner, fructans, another important health compound, are linear or branched polymers of repeating fructose residues connected by β (2–1) and/or β (2–6) fructosyl-fructose linkage, optionally including one terminal glycosyl unit. Inulin, the best characterized fructan, contains predominantly linear molecules with β (2–1) linkage. It is generally believed that inulin biosyntheses in plant occurs through two vacuolar enzymes 1-SST (sucrose:sucrose 1-fructosyltransferase) and 1-FFT (fructan:fructan 1-fructosyltransferase). Ectopic expression of Jerusalem artichoke 1-sst and 1-fft genes in potato accumulated inulin of DP3 to DP8 at 1,8 mg/g similar to the inulin profile found in globe artichoke (Hellwege et al. 2000).

32.11 Nutritional Value of Processed Potato Products

In addition to raw consumption, potatoes can be processed into several products like chips, French fries, cubes, granules, and canned products. For preparing a good quality of these products, potatoes should have high dry matter (more than 20 %) and low reducing sugar content (less than 100 mg/100 g fresh weight). High reducing sugar results in dark-colored finished product due to "Maillard reaction" which is not acceptable in the market. The dark products have bitter taste with the formation of acrylamide, a compound having carcinogenic properties (Marwaha et al. 2008). The processed potato products are classified below:

32.11.1 Fried Products

Potato chips, frozen French fries, and other frozen fries.

32.11.2 Dehydrated Products

Dehydrated chips, flakes, granules, flour, starch, potato custard powder, and soup or gravy thickener.

32.11.3 Non-fried Products

Potato jam, potato murabba, potato candy, potato biscuits, and potato cakes.

32.11.4 Canned Products

Some important varieties, viz., Kufri Chipsona-1, Kufri Chipsona-2, Kufri Chipsona-3, Kufri Jyoti, Kufri Lauvkar, Kufri Surya, Kufri Himsona, and Kufri Frysona, have been developed for processing purposes. Kufri Chipsona-1 and Kufri Chipsona-2 are India's first potato processing varieties developed in 1998. Both the varieties have high tuber dry matter, produce light color chips, possess resistance to late blight, and give good tuber yield. Kufri Chipsona-1 is very popular with the farmers and industry. Kufri Chipsona-3 is resistant to late blight, gives higher yield, and has good storability. All the three Chipsona varieties produce above 21 % tuber dry matter and contain less than 0.1 % reducing sugars on fresh tuber weight basis. These are the most important characters determining the quality of the processed product. Kufri Jyoti and Kufri Lauvkar produced potatoes of acceptable quality for processing in specific areas. In India, potato chips are very popular and constitute nearly 85 % of the total salty snack food business of about 25,000 million rupees every year (Pandey et al. 2008). Demand for processed potato products is increasing and potato products are liked well worldwide. The nutritional values of different processed forms of potatoes are furnished in Table 32.13 (Bandana 2013).

32.12 Value-Added Products

The demands of processed foods are increasing in urban area due to liking of people particularly youngsters for such products. The increase in a number of fast-food outlets is also contributing toward this. Some important value-added potato products are described as given below (Singh 2013a).

32.12.1 Potato Chips

After introduction of liberalization policy of the government of India, the potato chips production by the organized as well as unorganized sector has increased rapidly. The demand of potato chips is increasing because of its increasing popularity as a convenient fast food especially in urban areas. Chip color, crispiness, and taste determine the quality of potato chips. Light or light golden yellow color is preferred while brown or black chips are considered undesirable. The potatoes should be round to round oval in shape with 40–60 mm size and possess a dry matter content of more than 20 % with reducing sugar level of less than 100 mg per 100 g tuber fresh weight. The steps involved in chip making are peeling, trimming, slicing, blanching, and frying. Potato chips provide more crude protein, dietary fibers, and calcium than do the raw potatoes.

32.12.2 French Fries

The quantity of French fries consumed is much less compared to potato chips in India. However, the demand for French fries is increasing gradually. Potatoes are peeled, trimmed, and cut into sticks. The length of sticks is normally 5–7 cm and cross section is generally 10 mm. The French fry sticks are blanched, par-fried, and frozen. Finish frying is done for about 2 min at 180 °C and served hot. French fries can also be prepared from fresh potatoes at home by frying them in deep fat in two stages.

32.12.3 Potato Flour

It may be used in combination with other cereal and pulse flours for preparation of several products. Some products like biscuits, cake, *bhujia*, etc. can be made. 1 kg potatoes will yield about 200 g flour.

32.12.4 Potato Starch

Potato starch has large granule size (5–100 um), low lipid content, higher water binding capacity, and high solubility. It has low protein content which helps in avoiding foam building and color formation. Potato starch is superior to cereal starches in many ways and may be utilized in several industries. It may be used as thickener in sauces, gravy, puddings, soups, etc. and as softening agents in cakes, breads, biscuits, and cookies in the food industry. one kilogram of potatoes will yield 70–100 g of starch.

Table 32.13 Nutritional value of processed form of potatoes (per 100 g)

Form of potato	Total carbohydrates (g)	Dietary fiber (g)	Protein (g)	Fat (g)	Energy (kcal)
Raw	18.5	1.5	2.1	0.1	80
Chips	49.7	11.9	5.8	37.9	551
Frozen French fries	29.0	3.2	3.0	18.9	291
Flakes	14.5	0.3	1.9	3.2	93
Potato flour	79.9	1.6	8.0	0.8	351
Granules	14.4	0.2	2.0	3.6	96

32.12.5 Dehydrated Potato Products

Drying results in the lowering of moisture content leading to lesser chances of microbial growth and product has a longer shelf life. Solar dehydration is quite common in several parts of the country in the form of 2–3 mm thick slices and *papads.*

32.12.6 Dried Potato Slices

Dried potato slices are prepared at home by housewives and also by small-scale manufacturers. The housewives simply hand peel the potatoes, cut them into slices, and then boil the slices in water for 5–8 min. The slices are dried in the sun until they become brittle. Dried potato slices are consumed after frying and salting.

32.13 Agronomical Practices for Biofortification in Potatoes

Potatoes can be enriched with the following agronomical practices.

32.13.1 Integrated/Balanced Fertilization Management

Potato demands high level of soil nutrients due to relative poorly developed and shallow root system in relation to yield. Compared with cereal crops, potato produces much more dry matter per unit area and time. This high rate of dry matter production results in large amounts of nutrients removed per unit time, which generally most of the soils are not able to supply. A healthy crop of potato removes about 120–140 kg N, 25–30 kg P_2O_5, and 170–230 kg K_2O/ha (Dua 2013). Potato crop has strict requirement for a balanced fertilization management, without which growth and development of the crop are poor and quality of tubers are diminished. Nitrogen, phosphorus, and potassium have a great influence on tuber quality. Application of P at optimum rates increases tuber starch and vitamin C content but higher levels adversely affect protein content. High quality (nutritional value) can only be sustained through the application of optimal NPK doses in balanced proportion.

The crucial importance of potassium in quality formation starts from its role in promoting synthesis of photosynthates and their transport to the tubers and to enhance their conversion into starch, protein, and vitamins. Potassium maintains the balance of electric charges in chloroplasts, which is required for ATP formation. Hence, K improves the transfer of radiation energy into primary chemical energy in the form of ATP (photophosphorylation). This energy is required for all synthetic process in plant metabolism, resulting in production of carbohydrates, proteins, and lipids. Potassium is involved in the activation of the enzyme starch synthase, which is responsible of the synthesis of starch. Potassium is the most efficient cation stimulating the activity of this enzyme that catalyzes the incorporation of glucose into long-chain starch molecules. Although potassium activates enzymes involved in starch formation, K can reduce starch content through increased water content in the tubers. High concentrations of K usually lead to an increase in organic acid concentration, also having a beneficial effect on ascorbic acid levels. Nitrogen, phosphorus, and potassium needs of the crop vary with the agroclimatic region, variety, crop sequence, and soil type. The optimum dose for nitrogen, phosphorus, and potassium varies from 180 to 240 kg, 60 to 100 kg, and 100 to 150 kg/ha, respectively (Dua 2013).

In potato, sulfur is required in many metabolic activities, and its importance is being recognized in view of its role in improving crop quality. Calcium and magnesium are the next important secondary nutrients.

Each micronutrient has specific role as far as potato tuber quality is concerned. Translocation of photosynthates from leaves to tubers and their subsequent conversion to starch and the photosynthetic rate have been shown to increase in the presence of zinc and manganese. Zinc, iron, boron, and molybdenum have been reported to increase the tuber number of medium and large grade. Zinc, copper, manganese, boron, and molybdenum have been shown to increase ascorbic acid content of tubers (Dua 2013).

Application of a micronutrient carrier to the soil, foliar application of micronutrients, and treating mother seed tubers with micronutrient compounds are three main approaches for micronutrient management in potato. The critical limits and optimum doses of different micronutrient applications are furnished in Table 32.14 (Dua 2013).

Well-decomposed farmyard manure (FYM), compost, green manuring, and vermicompost can be used in conjunction with inorganic fertilizers. These approaches contribute significantly to increased crop yields, nutritional value and quality of tubers, and soil fertility.

32.13.2 Organic Cultivation

Potato being a heavy feeder of nutrients requires high amounts of nitrogen, phosphorus, and potassium. Chemical fertilizers are the main source of nutrient to potato crop. However, continuous use of chemical fertilizers has resulted in nutritional imbalance, depletion of soil organic matter, contamination of food and water, and adverse effect on biodiversity as well as on human health. Organic potato production depends mainly on crop rotation, minimum tillage practices, manuring through organic materials (viz., crop residues, animal excreta, nitrogen-fixing legumes, green manure, off-farm organic residues), and management of insects and diseases through nonpolluting substances. These different approaches can be used alone or in combination. Research evidences showed that potato crop responds well to organic manure application (Mondal et al. 2005). Supplying of nutrients through organic sources can be opted for avoiding the hazardous effects of chemical fertilizers and maintaining sustainability. Organic nutrient sources like FYM, compost, vermicompost, green manure, cakes, biofertilizers like *Azotobacter* and phosphobacteria, etc. may play a major role in supplementing the crop nutrients through their direct addition, improvement in soil condition, and improvement in tuber quality (nutritional value) of potato.

32.13.3 Proper Management of Diseases and Insects

Incidence of diseases and insect pest adversely affects tuber quality of potatoes. Therefore, effective management of diseases and insects is helpful to keep the plants and tubers healthy, which ultimately improves the tuber quality. Potato crop faces many fungal, bacterial, and viral diseases. Different management practices are recommended for checking the diseases in different agroclimatic zones. All possible practices should be adopted for quality tuber production. Aphids, whiteflies, leafhoppers, tuber moth, and white grubs are major pest of potato, which needs effective and safe management for good tuber quality.

Table 32.14 Critical limits and doses of micronutrient application in potato

Micronutrient	Source	Critical limits in soil (ppm)	Soil application (kg/ha)	Spray application (g/100 l water)	Tuber soaking treatment (g/100 l water)
Zinc	Zinc sulfate	0.75	25	200	50
Iron	Ferrous sulfate	6.60	50	300	75
Manganese	Manganese sulfate	2.00	25	200	50
Copper	Copper sulfate	0.32	25	200	50
Molybdenum	Ammonium molybdate	0.20	2	100	20
Boron	Sodium borate	0.50	2	100	20

32.13.4 Selection of Suitable Variety/Timely Operations

The sowing of recommended varieties for the region is helpful to maintain tuber quality (nutritional value) of potatoes. Timely operations (viz., planting, earthing up, irrigation, roguing, plant protection measures, dehaulming, harvesting) in potato crop contribute to increased crop yields as well as nutritional value of tubers.

32.14 Misconceptions About the Nutritive Value of Potato

Most of the people in India have either less knowledge or wrong notions about the nutritive value of potato. Common misconception about potato is that its consumption causes fattening and worsens the diabetes. But in reality with less than 0.1 % fat, less sugars (except the high sugars due to untreated cold storages), and very low calorie, it does not cause either obesity or diabetes. Due to this misconception, the per capita consumption of potato in India is only 18.58 kg/year. On the other hand, the per capita consumption in Ukraine is 140 kg/year and more than 100 kg in most developed nations (Singh et al. 2011). Most people eat potatoes in the form of greasy, French fries or potato chips. Preparation and serving potatoes with high-fat ingredients raise the caloric value of the dish, and such treatment can make a potential contributor to fattening.

32.15 Conclusion

Potato is a high-quality vegetable cum food crop and used in preparing more than 100 types of recipes in India. The protein of potato has high biological value than proteins of cereals and even better than that of milk. Hence, potato can be a supplement of meat and milk products for improving their taste, lowering energy intake, and reducing food cost. It is highly nutritious and easily digestible and contains carbohydrate, proteins, minerals, vitamins, and high-quality dietary fiber which can be used as wholesome food. Potatoes are a good source of vitamin B6 (involved in more than 100 reactions), vitamin C, copper, potassium, and manganese; therefore, it deserves to be promoted as a potential high-quality vegetable cum food source for human health. Besides that, potatoes also contain a variety of phytonutrients that have antioxidant activity. Among these important health-promoting compounds are carotenoids, flavonoids, and caffeic acid, as well as unique tuber storage proteins, such as patatin, which exhibit activity against free radicals. The demands of processed foods are increasing in urban area due to liking of people particularly youngsters for such products. The increase in a number of fast-food outlets is also contributing toward this. In addition to raw consumption, potatoes can be processed into chips, French fries, cubes, granules, starch, and flour which are the most popular potato products in the world. The more emphasis needs to be given on the quality of the processed potato products because of its increasing demand. As being one of the principal cash crops, it gives handsome returns to the growers/farmers due to its wide market demand nationally and internationally for different kinds of utilization.

References

Andersson M, Melander M, Pojmark P, Larsson H, Bulow L, Hofvander P (2006) Targeted gene suppression by RNA interference: an efficient method for production of high amylose potato lines. J Biotechnol 7:210–218

Arnqvist L, Dutta PC, Jonsson L, Sitbon F (2003) Reduction of cholesterol and glycoalkaloid levels in transgenic potato plants by overexpression of a type 1 sterol methyltransferase cDNA. Plant Physiol 131 (4):1792–1799. doi:10.1104/pp.102.018788

Bandana (2013) Nutritional value of potatoes. A compendium of summer school on advances in quality potato production and post harvest management from 16 July to 05 Aug, Central Potato Research Institute (ICAR), Shimla, pp 154–157

Breithaupt DE, Bamedi A (2002) Carotenoids and carotenoid esters in potatoes (*Solanum tuberosum* L.) new insights into an ancient vegetable. J Agric Food Chem 50:7171–7181

Brown CR (2008) Breeding for phytonutrient enhancement of potato. Am J Potato Res 85:298–307

Buckenhuskes HJ (2005) Chapter 1: Nutritionally relevant aspects of potatoes and potato constituents. In: Haverkort AJ, Struik PC (eds) Potato in progress science meets practice. Wageningen, Academic, pp 17–45

Burton WG (1989) The nutritive value of potato, cooked or processed in relation to human requirements. In: The potato, 3rd edn. Longman Singapore Publishers, Singapore, pp 356–390

Burton WG (1996) Yield and content of dry matter: the underlying physiological processes in potato. Longman Scientific & Technical Essex, Harlow, pp 84–155

Camire ME, Kubow S, Dnnelly DJ (2009) Potatoes and human health. Crit Rev Food Sci Nutr 49:823–840

Caprara C (2012) Potential health promoting properties of potato-derived proteins, peptides and phenolic compounds. In: Potatoes production, consumption and health benefits. Nova, New York, pp 173–193

Chakraborty S, Chakraborty N, Agrawal L, Ghosh S, Narula K, Shekhar S, Naik PS, Pandey PC, Chakrborti SK, Datta A (2010) Next-generation protein-rich potato expressing the seed protein gene AmA1 is a result of proteome rebalancing in transgenic tuber. PANS 107(41):17533–17538

Chong DK, Roberts W, Arakawa T, Illes K, Bagi G, Slattery CW, Langridge WH (1997) Expression of the human milk protein beta-casein in transgenic potato plants. Transgenic Res 6:289–296

Claidia J, Uwe S, Mohmmad R, Ulf-Ingo F, Frank L (2012) Simultaneous boosting of source and sink capacities doubles tuber starch yield of potato plants. Plant Biotechnol J 10:1088–1098

Dancs G, Mihály K, Zsófia B (2008) The effects of enhanced methionine synthesis on amino acid and anthocyanin content of potato tubers. BMC Plant Biol. doi:10.1186/1471-2229-8-65

Diretto G, Tavazza R, Welsch R, Pizzichini D, Mourgues V, Beyer P, Giuliano G (2006) Metabolic engineering of potato tuber carotenoids through tuber-specific silencing of lycopene epsilon cyclase. BMC Plant Biol 6:13. doi:10.1186/1471-2229-6-13

Diretto G, Al-Babili S, Tavazza R, Papacchioli V, Beyer P, Giuliano G (2007) Metabolic engineering of potato carotenoid content through tuber-specific over-expression of a bacterial mini-pathway. PLoS One 2, e350

Dua VK (2013) Nutriment management in potato production. A compendium of summer school on Advances in quality potato production and post harvest management from 16 July to 05 Aug, Central Potato Research Institute (ICAR), Shimla, pp 73–79

Ducreux LJM, Morris WL, Hedley PE, Shepherd T, Davies HV, Millam S, Taylor MA (2005) Metabolic engineering of high carotenoid potato tubers containing enhanced levels of beta-carotene and lutein. J Exp Bot 56:81–89

Falco SC, Guida T, Locke M, Mauvais J, Sanders C, Ward RT, Webber P (1995) Transgenic canola and soybean seeds with increased lysine. Biotechnology 13:577–582

FAO-WFP-IFAD (2012) The state of food insecurity in the world (2012): economic growth is necessary but not sufficient to accelerate reduction of hunger and malnutrition. FAO, Rome

Fernandez EB, Muñoz FJ, Montero M, Etxeberria E, Sesma MT, Ovecka M (2009) Enhancing sucrose synthase activity in transgenic potato (*Solanum tuberosum* L.) tubers results in increased levels of starch, ADPglucose and UDPglucose and total yield. Plant Cell Physiol 50:1651–1662

Friedman M (1997) Chemistry, biochemistry, and dietary role of potato polyphenols. A review. J Agric Food Chem 45:1523–1540

Friedman M, McDonald GM (1997) Potato glycoalkaloids chemistry, analysis, safety, and plant physiology. Crit Rev Plant Sci 16:55–132

Hellwege EM, Czapla S, Jahnke A, Willmitzer L, Heyer AG (2000) Transgenic potato (*Solanum tuberosum*) tubers synthesize the full spectrum of inulin molecules naturally occurring in globe artichoke (*Cynara scolymus*) roots. Proc Natl Acad Sci USA 97:8699–8704

Hemavathi UC, Young K, Akula N, Kim HS, Heung JJ, Oh OM (2009) Over-expression of strawberry D-galacturonic acid reductase in potato leads to accumulation of vitamin C with enhanced abiotic stress tolerance. Plant Sci 177:659–667

Hemavathi CP, Upadhyaya KE, Young N, Akula KE, Young SC, Chun DHK, Park SW (2010) Enhanced ascorbic acid accumulation in transgenic potato confers tolerance to various abiotic stresses. Biotechnol Lett 32:321–330

Hong-hui D, Yang T, Cong-yu MA, Dan F, Zhang N, Huai-jun SI, Wang D (2012) Effects of RNAi silencing of SSIII gene on phosphorus content and characteristics of starch in potato tubers. J Integr Agric 11(12):1985–1992

Jansen G, Flamme W, Schuler K, Vandrey M (2001) Tuber and starch quality of wild and cultivated potato species and cultivars. Potato Res 44:137–146

Jobling SA, Westcott RJ, Tayal A, Jeffcoat R, Schwall GP (2002) Production of a freeze-thaw-stable potato starch by antisense inhibition of three starch synthase genes. Nat Biotechnol 20:295–299

Jung C, Griffiths H, De Jong D, Cheng S, Bodis M, De Jong W (2005) The potato P locus codes for flavonoid 30,50-hydroxylase. Theor Appl Genet 110:269–275

Kim JH, Cetiner S, Jaynes JM (1992) Enhancing the nutritional quality of crop plants design, construction and expression of an artificial plant storage protein gene. In: Bhatnagar D, Cleveland TE (eds) Molecular approaches to improving food quality and safety. AVI Book, New York, pp 1–36

Kim CK, Han JS, Lee HS, Oh JY, Shigaki T, Park SH, Hirschi K (2006) Expression of an Arabidopsis CAX2 variant in potato tubers increases calcium levels with no accumulation of manganese. Plant Cell Rep 25:1226–1232

Kruezer H (2001) Incorporating lutein into foods and beverages. Suppl Food Prod Des 11(4):5

Kumar V (2013) Indian table & processing potato varieties for different agro-climatic regions. A compendium of summer school on "advances in quality potato production and post harvest management" from 16 July to 05 Aug, Central Potato Research Institute (ICAR), Shimla, pp 12–21

Kumar V, Luthra SK, Gopal J, Singh BP (2011) Indian potato varieties and their salient features. CPRI technical bulletin no 78 (Revised). Central Potato Research Institute, Shimla, p 56

Lee SK, Ye YM, Yoon SH, Lee BO, Kim SH, Park HS (2006) Evaluation of the sensitization rates and identification of IgE-binding components in wild and genetically modified potatoes in patients with allergic disorders. Clin Mol Allergy 4:10

Lewis CE, Walkel JRL, Lancaster JE, Sutton KH (1998) Determination of anthocyanins, flavonoids and phenolic acids in potatoes. I: coloured cultivars of *Solanum tuberosum* L. J Sci Food Agric 77:45–57

Liedl BE, Kosier T, Sesbarought SL (1987) HPLC isolation and nutritional value of a major potato tuber protein. Am Potato J 64:1177–1182

Llorente B, Vanina R, Guillermo D, Alonso HNT, Mirtha MF, Fernando FB (2011) Improvement of aroma in transgenic potato as a consequence of impairing tuber browning. PLoS One 5(11), e14030

Lopez AB, Eck J, Conlin BJ, Paolillo DJ, O'Neill J, Li L (2008) Effect of the cauliflower Or transgene on carotenoid accumulation and chromoplast formation in transgenic potato tubers. J Exp Bot 59:213–223. doi:10.1093/jxb/erm299

Marwaha RS, Kumar D, Pandey SK, Singh SV, Kumar P (2008) Status of dry matter and reducing sugar content of potatoes grown in major potato growing states in India. CPRI technical bulletin no 94. Central Potato Research Institute, Shimla, p 18

Masuda H, Ishimaru Y, Aung MS, Kobayashi T, Kakei Y, Takahashi M, Higuchi K, Nakanishi H, Nishizawa NK (2012) Iron biofortification in rice by the introduction of multiple genes involved in iron nutrition. Sci Rep 2:543

Matthews BF, Hughes CA (1993) Nutritional improvement of the aspartate family of amino acids in edible crop plants. Amino Acids 4:21–34

McCue KF, Louise VTS, Paul V, Allen M, Malendia M, David RR, Dennis LC, Davies HV, William RB (2005) Metabolic compensation of steroidal glycoalkaloid biosynthesis in transgenic potato tubers: using reverse genetics to confirm the in vivo enzyme function of a steroidal alkaloid galactosyltransferase. Plant Sci 168:267–273

McCue KF, Allen PV, Shepherd LVT, Blake A, Whitworth J, Maccree MM, Rockhold DR, Stewart D, Davies HV, Belknap WR (2006) The primary in vivo steroidal alkaloid glucosyltransferase from potato. Phytochemistry 67:1590–1597

McCue KF, Allen PV, Shepherd LVT, Blake A, Maccree MM, Rockhold DR, Novy RG, Stewart D, Davies HV, Belknap WR (2007) Potato glycosterol rhamnosyltransferase, the terminal step in triose side-chain biosynthesis. Phytochemistry 68:327–334

Michaela Z, Anna PC, Oliver K, Ute R, Katrin B, Lothar W, Rainer H, Holger H (2001) Antisense inhibition of threonine synthase leads to high methionine content in transgenic potato plants. Plant Physiol 121 (3):792–802

Mondal SS, Acharya D, Ghosh A, Bug A (2005) Effect of integrated nutrient management on the growth, productivity and quality of potato in Indo-Gangetic plains of West Bengal. Potato J 32(1–2):75–78

NHB (2014) Indian Horticulture Database 2013. National Horticulture Board, Ministry of Agriculture, Government of India, Gurgaon

Ortiz-Medina E (2007) Potato tuber protein and its manipulation by chimeral disassembly using specific tissue explantation for somatic embryogenesis. Ph.D. dissertation 152, McGill University, Montreal

Pandey SK, Sarkar D (2005) Potato in India: emerging trends and challenges in the new millennium. Potato J 32(3–4):93–104

Pandey SK, Singh SV, Gaur PC, Marwaha RS, Kumar D, Kumar P, Singh BP (2008) Chipsona varieties: a success story. CPRI technical bulletin no 89, Central Potato Research Institute, Shimla, p 40

Pandey NK, Singh DK, Pandey KK, Sharma S, Singh BP (2012) Agro- techniques for production of quality potato seed. CPRI extension bulletin no 43, Central Potato Research Institute, Shimla, p 14

Park S, Cheng NH, Pittman JK, Yoo KS, Park J, Smith RH, Hirschi KD (2005) Increased calcium levels and prolonged shelf life in tomatoes expressing Arabidopsis H+/Ca_2+ transporters. Plant Physiol 139:1194–1206

Prohens J, Rodrıguez BA, Raigon MD, Nuez F (2007) Total phenolic concentration and browning in a collection of different varietal types and hybrids of eggplant: implications for breeding for higher nutritional quality and reduced browning. J Am Soc Hort Sci 132:638–646

Robert S, van der Meer IM, Raoul B, Andries JK (2002) Increased production of nutriments by genetically engineered crops. J Am Coll Nutr 21:199–204

Schmidt MH, Raulf-Heimsoth M, Posch A (2002) Evaluation of patatin as a major cross-reactive allergen in latex-induced potato allergy. Ann Allergy Asthma Immunol 89:613–618

Schwall GP, Safford R, Westcott RJ, Jeffcoat B, Tayal A, Shi YC, Gidley MJ, Jobling SA (2000) Production of very high amylose potato starch by inhibition of SBE A and B. Nat Biotechnol 18:551–554

Shewry PR (2003) Tuber storage proteins. Ann Bot (Lond) 91:755–769

Singh B (2013) Quality processing parameters and value added products of potato. A compendium of summer school on advances in quality potato production and post harvest management from 16 July to 05 Aug, Central Potato Research Institute (ICAR), Shimla, pp 149–153

Singh BP (2013) Potato scenario- past, present and future. A compendium of summer school on advances in quality potato production and post harvest management from 16 July to 05 Aug, Central Potato Research Institute (ICAR), Shimla, pp 1–7

Singh BP, Chakrabarti SK, Rana RK (2011) Vision 2030. Central Potato Research Institute (ICAR), Shimla, p 32

Smith DB, Roddick JG, Jones JL (1996) Potato glycoalkaloids some unanswered questions. Trends Food Sci Technol 7:126–131

Smith AM, Denyer K, Martin C (1997) The synthesis of the starch granule. Annu Rev Plant Physiol Plant Mol Biol 48:67–87

Stobiecki M, Matysiak-Kata I, Frański R, Skała J, Szopa J (2002) Monitoring changes in anthocyanin and steroid alkaloid glycoside content in lines of transgenic potato plants using liquid chromatography/mass spectrometry. Phytochemistry 62:959–969

Swinkels JJM (1985) Composition and properties of commercial native starches. Starch/Staerke 37:1

Tarn TR, Tai GCC, Liu Q (2006) Chapter 5: Quality improvement. In: Gopal J, Khurana SMP (eds) Handbook of potato production, improvement, and postharvest management. Haworth Press, New York, pp 147–178

Tu HN, Godfrey LW, Sun SSM (1998) Expression of the Brazil nut methionine rich protein and mutants with increased methionine in transgenic potato. Plant Mol Biol 37:829–838

Wei Q, Wang QY, Feng ZH, Wang B, Zhang YF, Yang Q (2012) Increased accumulation of anthocyanins in transgenic potato tubers by overexpressing the 3GT gene. Plant Biotechnol Rep 6:69–75. doi:10.1007/s11816-011-0201-4

Welch RM (2002) The impact of mineral nutrients in food crops on global human health. Plant and Soil 247:83–90

White PJ, Broadley MR (2009) Biofortification of crops with seven mineral elements often lacking in human diets-iron, zinc, copper, calcium, magnesium, selenium and iodine. New Phytol 18:49–84

Xing FH, Farhad NF, Jean-Paul V, Qin J, Richard GF, Luisa MT (2014) Expression of an amylosucrase gene in potato results in larger starch granules with novel properties. Planta 240:409–421. doi:10.1007/s00425-014-2095-1

Yang MS, Espinoza NO, Nagplala PG, Dodds JH, White FF, Schnorr KL, Jaynes JM (1989) Expression of a synthetic gene for improved protein quality in transformed potato plants. Plant Sci 64:99–111

Zhou X, Van Eck J, Li L (2008) Use of the cauliflower Or gene for improving crop nutritional quality. Biotechnol Annu Rev 14:171–190

Zimmermann MB, Hurrell RF (2002) Improving iron, zinc and vitamin A nutrition through plant biotechnology. Curr Opin Biotechnol 13:142–145

Part VI

Biofortification with Futuristic Challenges and Approaches

33 Nutri-farms for Mitigating Malnutrition in India

Purushottam, S.K. Singh, and Riyaj Uddeen

Abstract

The "green revolution" in India had helped to alleviate the crisis of food insufficiency by introducing high-yielding varieties of wheat and rice. Now the time has come to address the issues concerning food quality. Millions of people do not find adequate nutritious foods in their diets, especially foods that are affluent in vitamins and minerals. According to the 2011 census, the farm families comprise the majority of India's population. A high proportion of marginal and landless farm families undergo malnutrition because of too little income and affordability to purchase quality foods. For example, the pulses supplying dietary protein are too expensive for the poor. Here comes the role of nutri-farms. It ensures biofortified food crops that are enriched in critical micronutrients. These foods essentially provide the critical nutrients such as iron and zinc besides supplementing protein and essential vitamins. In the paper, the concept of nutri-farms is discussed in detail so as to mitigate malnutrition in India.

Keywords

Biofortification • Implementation • Malnutrition • Nutri-farms • Self-help groups

33.1 Introduction

In the era of "green revolution", the problem of food scarcity was amicably solved with introduction of high-yielding varieties of wheat and rice in India. Solving food security quantitatively was one issue (Fig. 33.1). Now we are fighting with the issues of nutritional quality of available food. Because of economic disparity and affordability, the poor are not able to feed themselves with the food of appropriate quality, i.e. rich in vitamins and minerals. Here comes the issue of hidden hunger or malnutrition

Purushottam (✉) • S.K. Singh • R. Uddeen
Division of Social Sciences, ICAR-Indian Institute of Pulses Research, Kanpur, Uttar Pradesh 208 024, India
e-mail: purushottam1995@yahoo.com; sushilsinghiipr@yahoo.co.in; riaz_uddin15@yahoo.com

U. Singh et al. (eds.), *Biofortification of Food Crops*, DOI 10.1007/978-81-322-2716-8_33

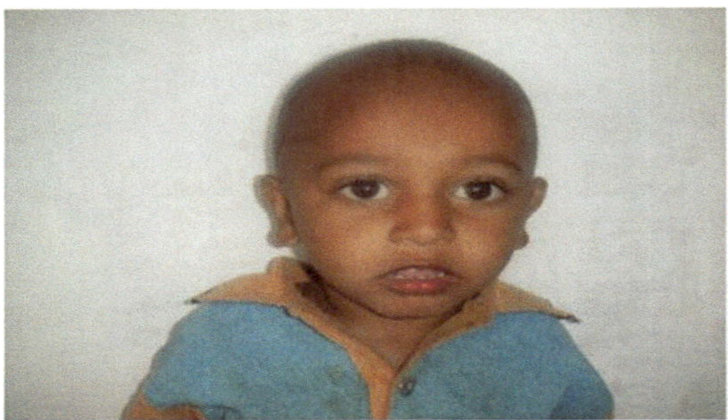

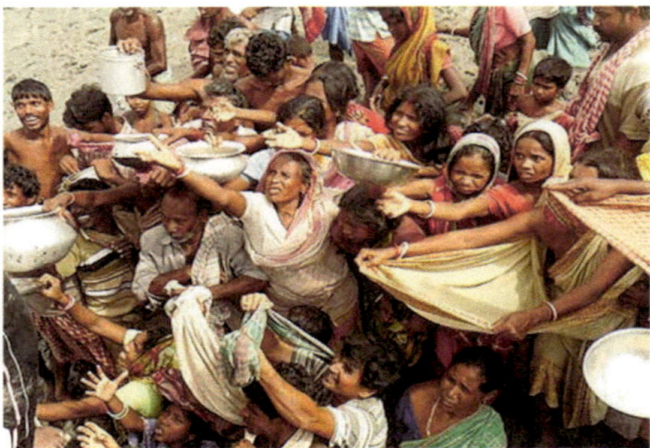

Fig. 33.1 Food is the need of everybody

experienced by a sizable portion of people living on Indian earth. This focuses the issues concerning dietary quality and making the food diet rich in micronutrients and vitamins.

India has about 6.4 lakh villages where 65 % of the population is residing and a high proportion of families of marginal and landless categories still suffer from food (quality)-related malnutrition because of many reasons. The underprivileged sections of the society – the Dalits, SC/STs and other BPL (below poverty line) households whose livelihood is agriculture – are primarily hit hard by the inaccessibility of food. Many of the concentrated quality raw food such as pulses are beyond the reach of the poor. Therefore, the rising demand for food substance and somewhat slower supply response in numerous commodities' need for a second green revolution are being recognized more than ever before. The possibility also lies in biofortification of food or food products ensuring supplementation of critical micronutrients. These foods essentially provide iron and zinc besides supplementing protein and essential vitamins.

33.2 Some Facts and Figures

Many of the facts underlying the need for such a supplementation include the following:

(i) Every third malnourished child in the world today resides in India.

(ii) With more than 200 million starving people, India has the largest number of hungry in the world.

(iii) In the Global Hunger Index, India's position is 66th out of 68 countries.

(iv) In the catalogue, all Indian states are at severe level of hunger. Twelve Indian states fall in the frightening group.

(v) Madhya Pradesh is India's most malnourished state.

(vi) Almost 42 % of the country's children are malnourished, more than the levels in sub-Saharan Africa. Indian Prime Minister Dr. Manmohan Singh had called this as a "national shame".

(vii) India's malnutrition is rampant and is the major cause of concern. This is apparent from the following databases:

- The National Family Health Survey (2005–06) – NFHS-3 – says that malnourished children below 5 are above 40 % and those having low birth weight were 21 %.
- Union Planning Commission (2012) – 217 million are undernourished.
- Global Hunger Index (IFPRI, 2013) – 65th place amid 79 countries.
- UNDP Human Development Report (2013) – 136th place amongst 187 countries.
- UNDP Gender Inequality Index (2013) – 132nd place amongst 148 countries.
- The Nutrition Barometer (Save The Children, 2012) – very modest position.

33.3 Malnutrition Is the Way of Life

Malnutrition, frequently called the "hidden hunger", can lead on to life-threatening illnesses. It is either caused by a lack of protein (protein-energy malnutrition) or micronutrients such as iodine, vitamin A and iron. Malnutrition weakens immune systems; exacerbates the effect of childhood diseases such as measles, malaria, pneumonia and diarrhoea; and can permanently damage long-term physical and cognitive development. One of the major causes for malnutrition in India is gender disparity (Sharma 2013) amongst other issues (Fig. 33.2).

Malnutrition is the way of living both in rural and urban households. The persons are malnourished or suffer from undernutrition if their diet does not offer them with sufficient calories and protein for their protection and growth or they cannot completely utilize the food they eat due to illness. People are also malnourished or continue from overnutrition if they eat too many calories. Malnutrition can also be defined as the deficient, excessive or imbalanced consumption of nutrients. According to the World Health Organization (WHO), malnutrition is the gravest single threat to global public health (Fig. 33.3). The Rome Declaration on Nutrition enshrines the right of everybody to have a way into safe, sufficient and nutritious food and commits governments to prevent malnutrition in all its forms, with hunger, micronutrient deficiency and fatness. The FAO Director General da Silva (2015) viewed "We have the knowledge, expertise and resources required to conquer all forms of malnutrition. Governments should lead the way. But the thrust to improve global nutrition must be a combined effort, linking civil society organizations and the private sector".

Moreover, in the 2013 State of Food Insecurity in the World (SOFI) report, the FAO highlighted that 842 million inhabitants are "undernourished" globally, though it was noted that in some countries, the occurrence of undernutrition – as manifest in the proportion of children affected by stunting – is significantly higher compared to that of undernourishment defined by dietary intake.

The malnutrition problem affects the human being in several ways (Figs. 33.4 and 33.5). These are both physical and psychological types. Efforts are being made to overcome the malnutrition through diverse research and development approaches. As established, the deficiency of various micronutrients with vitamin A, zinc and iron is widespread in the developing world and influences billions of people. These can lead to, amid other symptoms, an elevated incidence of blindness, a weaker immune system, stunted growth and impaired cognitive growth (Islam 2007). The poor, mostly from rural, tend to survive on a diet of staple crops such as rice, wheat and maize, which are short in these micronutrients, and the majority cannot afford or rarely cultivate enough fruits, vegetables or meat products that are necessary to

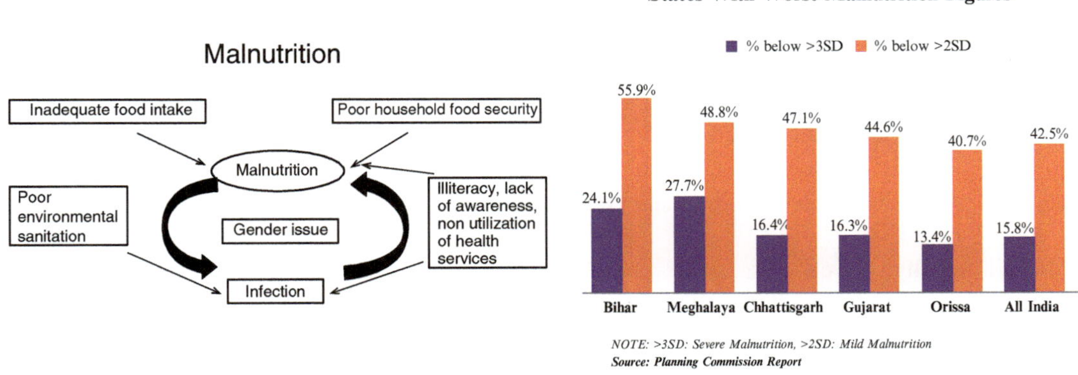

Fig. 33.2 Role of gender issues in malnutrition (*left*) and the Indian states with worst malnutrition figures (*right*)

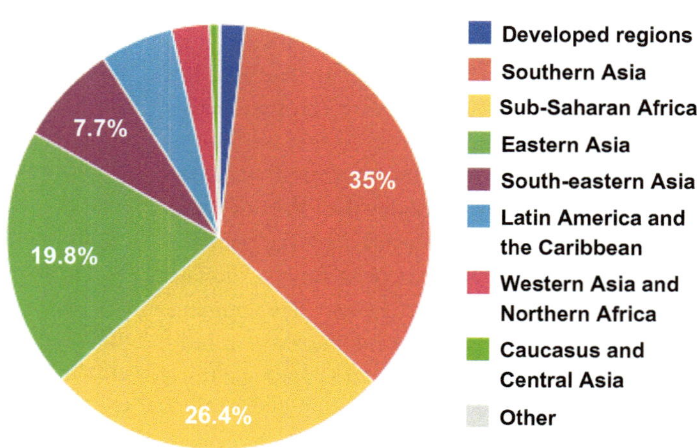

Fig. 33.3 Hunger of the world

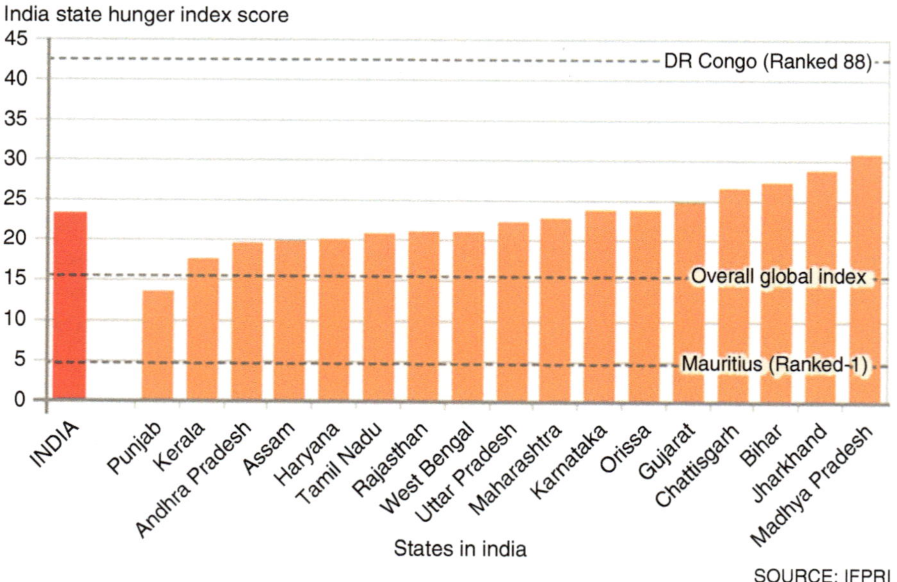

Fig. 33.4 World hunger index depicting India's share

obtain healthy levels of these nutrients (McClafferty and Islam 2008) (Fig. 33.5). As such, intensifying the micronutrient levels in staple crops can assist, prevent and decrease the micronutrient deficiencies – in one trial in Mozambique, eating sweet potatoes biofortified with beta-carotene abridged the incidence of vitamin A deficiency in children by 24 % (Pray et al. 2007). This move may have advantages above other health interventions such as

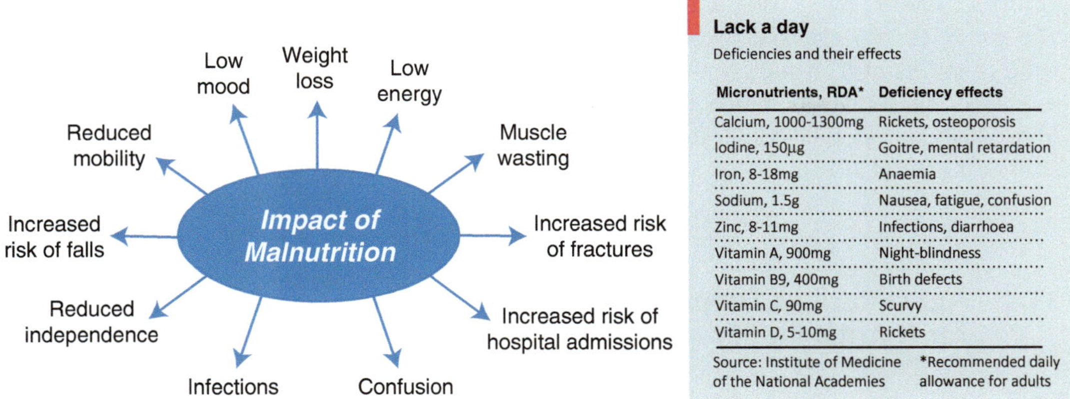

Lack a day

Deficiencies and their effects

Micronutrients, RDA*	Deficiency effects
Calcium, 1000-1300mg	Rickets, osteoporosis
Iodine, 150µg	Goitre, mental retardation
Iron, 8-18mg	Anaemia
Sodium, 1.5g	Nausea, fatigue, confusion
Zinc, 8-11mg	Infections, diarrhoea
Vitamin A, 900mg	Night-blindness
Vitamin B9, 400mg	Birth defects
Vitamin C, 90mg	Scurvy
Vitamin D, 5-10mg	Rickets

Source: Institute of Medicine of the National Academies *Recommended daily allowance for adults

Fig. 33.5 Impact of malnutrition is many folds and effect of deficiency is diverse

providing foods fortified after processing or as supplements. Although these approaches have established unbeaten when dealing with the urban poor, they tend to require access to effective markets and healthcare systems which frequently just do not exist in rural areas (McClafferty and Islam 2008). Further (Nestel et al. 2006), it is in contrast to supplementation which is reasonably expensive and demands continued financing over time, which may be jeopardized by unpredictable political interest.

Poverty and lack of purchasing power were identified as the two main factors responsible for the low dietary intake in India. In spite of the fact that the calorie intake has not increased, there is a rise in the overnutrition; this is mainly because of changes in the lifestyle of peoples and consequent reduction in the energy expenditure. Over the past three decades, there has been a substantial boost in the fat intake inside both rural and urban areas in India. In view of the adverse nutrition (obesity) and health (non-communicable diseases) implications of increased fat intake particularly amongst the prosperous group, this has to be reduced through appropriate nutrition education.

There has been a shift in the type of cereals consumed amongst the lowest-income group. With the accessibility of wheat and rice through the public distribution system (PDS), the poorer segments of the population have changed over to rice and wheat as staple cereals. Coarse cereals such as bajra, ragi, maize and jowar, which are well off in micronutrients and minerals, are no longer being consumed in sizeable quantity by the lowest-income group. Further, the pulse crops are the major source as vegetative protein in the Indian diet especially in the lowest-income group. In spite of increased expenditure on pulses, there is a decline in household "consumption" of pulses in all the income groups both in the urban and rural areas from 65 g/day in 1960 to 35 g/day per person at present.

Although the top 20 % in rural India spent in 2011–2012 almost double the amount on food while compared with the bottom 20 %, there was not much variation between the two classes in terms of proportion of cereal vs. noncereal; both the classes were spending more than 70 % on noncereal food items as per NSSO 2013 (Saxena et al. 2014). The data from the Chhattisgarh and Maharashtra states indicates a very significant upgrading in the nutrition situation. In Chhattisgarh state, the AARR of 4.22 % was observed for the underweight children and 5.64 % for the stunting. In Maharashtra, stunting rate declined from 39 % in 2005 in children under 2 years to 23 % in 2012 (Unicef 2013). Further, according to the NSSO data, the percentage of the households who were not able to afford two square meals in a day for its members had come down from 17.3 % in rural India in 1983 to only 1.9 % in 2009–2010. In the urban areas it is even less, only 0.4 % in 2009–2010 as against 12.1 % in 1983 (Deaton and Drèze 2008). The condition has changed over the years in India as the number of hungry people in India as a percentage of the total residents has declined from 26.8 % in 1990–1992 to 17.2 % in 2011–2013 (FAO 2013).

33.4 Agriculture as Supplement for Nutritional Security

Food safety is not only about the amount of food which we eat; it is also about the superiority and diversity of that food as well. Continued focus on improving agricultural productivity is a vital condition to realizing food security goals as well as nutrition safety. But we also need to look at how many alternatives people have and the quality of the food they consume. The consumption pattern of rural and urban varies in India.

Many in the science and food policy establishments in India have started to think that agriculture provides a means to solving India's severe malnutrition dilemma. India's efforts to battle malnutrition will be well served by the intensification of the linkages between agriculture and nutrition (Howarth Bouis) the Director of HarvestPlus.

Agriculture contributes to reduction in malnutrition by improved consumption from increased food production (production for own consumption), increased income from the sale of agricultural commodities (production for income), improvement of women agriculturists and associated gains in children's nutrition and welfare, lower real food prices ensuing from improved food production and microeconomic expansion arising from agriculture growth. Nearly 60 % of the consumers are also the farmers. The prevalent malnutrition existing in India can be conquered only if farm families are also able to have a balanced diet. In Rwanda beans are staple food. In a country where 44 % of the 12 million citizens undergo malnutrition and micronutrient shortage, biofortified foods, like beans, are seen as an explanation to reducing "hidden hunger" – a chronic lack of vitamins with minerals.

Presently, availability of food grain especially in case of rice and wheat exceeds requirement of food grain in the country. But the level of malnutrition is not declining to desired extent and still there is deficiency of iron, zinc and vitamin A. There are several food crop varieties available in the country that are wealthy in nutrients such as iron- and zinc-rich millets (bajra, ragi and small millets), protein-rich maize and β-carotene-, iron-, manganese- and zinc-rich wheat and iron- and zinc-rich rice. The promotion of cultivation of micronutrient-rich cultivars of these crops and development of their effective supply chain could help in reduction of malnutrition.

Acknowledging this grim reality, policymakers are opening to take on a multi-pronged approach to tackle the problem of malnutrition. Several programmes are under implementation to reduce malnutrition through distribution of food and vegetable supplement rich in iron, zinc and vitamin A to targeted groups in the country. But due to higher cost of processed food supplements and low-income level of the target groups, it is challenging to ensure continuous supply of needed nutrients to target groups.

Nutri-farm is intended to provide concurrent notice to the three key problems as follows:

(a) First, we have to assist farm families in conquering undernutrition as a result of calorie deprivation.
(b) Second, protein hunger is very serious due to the insufficient consumption of pulses and milk (in the case of vegetarians) and eggs, fish and meat (in the case of non-vegetarians).
(c) Third, there is extensive hidden hunger caused by the shortage of micronutrients like iron, iodine, zinc, vitamin A, vitamin B12, etc. in the diet. Here the responsibility of nutri-farms comes (Fig. 33.6).

Acknowledging this grim reality, policymakers are trying to adopt a multipronged

Fig. 33.6 Malnutrition is rampant and spreading its tentacles worldwide and a few missions to address it

strategy to tackle the problem of malnutrition. India on the other hand has been a world leader in designing social shield measures against hunger, such as ICDS, Midday Meal programme in schools, PDS, *Antyodaya Anna Yojana*, *Anganwadi*, Indira Gandhi *Matritva* Sahyog Yojana, nutri-cereals, etc. The National Nutrition Policy (1993) highlights food production, food supply, schooling, information, healthcare, rural advance and women and child progress. Some of the other schemes as the National Food Security Mission, National Horticulture Missions, *Rashtriya Krishi Vikas Yojana*, *Mahila Kisan Sashaktikaran Pariyojana*, *Mahatma Gandhi National Rural Employment Guarantee Act*, *Nirmal Bharat Abhiyan* and National Rural Drinking Water Scheme also support the food chain system. The parliament of India, in order to offer subsidy food grains to the marginalized section of society, has enacted the National Food Security Act on 12 September 2013. This step, along with the food security programmes as the Integrated Child Development Services scheme, public distribution system and Midday Meal Scheme, highlights the attempts made by the government to generate a policy for combating hunger. In spite of all such food carry systems, the burden of malnutrition is towering, mainly in the 200 districts identified by the Prime Minister's Nutrition Advisory Council. Thus, we have to deal with addressing the major issues connecting to hunger if we are to realize food and nutritional safety for all.

33.5 Biofortification as a Means to Eliminate Malnutrition

In the context of biofortification, India had experienced the food fortification as Universal Salt Iodization (USI) to address the crisis of micronutrient deficiencies; fortification of staple foods such as wheat flour, rice, oils and dairy products is on an increase. It also has experiences in salt, cereal and oil fortification. The industrially produced fortified flour is available in the open market and supplied through the public distribution system (PDS). The efforts have also been made in Gujarat and Madhya Pradesh for fortification of flour milled by limited small millers or chakki under village situations. The fortification of rice with micronutrients is being experimented at a small level in the rice-eating states of Karnataka, Odisha and Andhra Pradesh. Production and supply of oil fortified with vitamin A and vitamin D through the PDS is being tried in the Gujarat and

Tamil Nadu states. Use of manifold micronutrients or sprinklers has been experimented (Saxena et al. 2014).

Biofortification is the initiative of breeding crops to augment their nutritional value. This can be carried out either through conventional selective breeding or genetic engineering. In fact, biofortified foods are bred to have superior amounts of micronutrients and can help provide essential vitamins and minerals. For example, golden rice contains higher amounts of beta-carotene and iron, with possible benefits for 250 million children with risk of blindness due to vitamin A shortage and 1.4 billion women who undergo anaemia due to iron deficiency. Biofortification differs from ordinary fortification since it focuses on making plant foods more nutritious as the plants are growing, rather than having nutrients added to the foods while they are being processed (Bailey 2007). This is an upgrading on ordinary fortification when it comes to nutrients for the rural poor, who rarely contain entrée to commercially fortified foods (Islam 2007). As such, biofortification is seen as a forthcoming strategy for dealing with deficiencies of micronutrients in the developing world. In the case of iron, WHO anticipated that biofortification could help in curing the two billion citizens anguished from iron deficiency-induced anaemia (De Benoist and Egli 2008). Further, iron (Fe) shortage is the most prevalent nutrient deficiency worldwide. Biofortification of staple food crops, such as the lentil (*Lens culinaris* L.), may be a useful solution. He analysed Fe concentration, Fe bioavailability and phytic acid (PA) concentration of 23 lentil genotypes grown in five different locations in Saskatchewan, Canada. Relative Fe bioavailability, assessed using the in vitro digestion/Caco-2 cell model, had varied considerably amid the genotypes (DellaValle et al. 2013).

The ideologist and geneticist who led the green revolution in India was a supporter of what is called "biohappiness". "It fortifies in a biological substance and not in chemical material, that is why call it biohappiness"; "I am an enthusiast of biofortification. It is the best way to add nutrients like iron, zinc and vitamin A. In the case of biofortification it is a win-win state of affairs", he says. "We found it is not adequate to give calories, it is important to have proteins and micronutrients". Swaminathan says it is as well a way of aggressive silent hunger caused by extreme poverty. Also, Swaminathan was convinced that all developing countries must hold biofortification to improve nutrition across all demographics, but particularly amongst the poor. In a 2012 editorial in *Science* magazine, he writes, "Efforts ought to be made by developing countries to make each family farm a biofortified farm". In fact, the idea of nutri-farms is based on biofortification technology, in which a crop is made rich in a nutrient either through conventional plant breeding or by means of genetic modification. This is where biofortification comes in – the strategy of breeding crucial vitamins and minerals (e.g. vitamin A, zinc and iron) into staple food crops (such as rice or wheat) that people consume daily (Fig. 33.7).

33.6 Nutri-farms

33.6.1 Basis of Nutri-farm

The technology has to be need based, and a lot of under-utilized crops, in addition to our staple crops like wheat and rice, can be brought under it. We cannot ask people to change their food habits suddenly. You can't force them to take jowar and bajra instead of wheat and rice, and hence biofortification would play a key role. The allocation of funds for agricultural research has to be more since it needs a huge boost. There are different technologies for biofortification, and ICAR will certainly look for promising technologies, like genomic research, where micronutrient-providing genes could be inserted in plants. This can help fight iron and beta-carotene deficiencies.

The development of nutri-farms or biofortification projects which can encourage development of novel food items using modern technological tools is a footstep in the accurate direction (Ram Kaundinya) the Chairman of Association of Biotech Led Enterprises – Agriculture Group (ABLE-AG). The nutri-farm scheme will add nutritional dimensions to the farm sector (Swaminathan 2013). The role of

Fig. 33.7 Biohappiness through biofortification: A possibility

agribusinesses and food companies is to respond with innovative proposals to take the biofortification idea forward. In fact the "Private sector research and the seed industry manly focus on those crops and varieties those have immense markets and continual sales. As a consequence, some crops find little research attention" (Planning Commission 2011). Further, the private sector contribution in agricultural R&D has been more. The estimates disclose that the industry funding (mainly private) for agricultural R&D constitutes about 11 % of the entire R&D funding (Pal and Jha 2007).

The full potential of Indian agriculture will be realized if these measures are supplemented with further reforms in the Agricultural Produce Marketing Act by the United States and the Forward Contracts Regulation Act by the Government of India.

"The Nutri Farm Scheme where the government intends to cultivate new crop varieties loaded in micro-nutrients such as iron-rich bajra, protein-rich maize and zinc-rich wheat etc, will definitely have an answer to both under as well as over nutrition problems of people", explained Chief Nutritionist Ishi Khosla of Whole Foods.

33.6.2 Nutri-farm Movement

Nutri-farms ensure biofortified food crops that are enriched in critical micronutrients. These foods essentially provide the critical nutrients such as iron and zinc besides supplementing protein and essential vitamins. The nutri-farm movement considers the following:

Ascertaining the deficiency of nutrients and vitamins (protein, iron, zinc and vitamin A) under the food crops.

(a) Improving the nutritional status of the population through biofortified food crops enriched in critical micronutrients.
(b) Nutri-fortification through specific nutrient/food: There are several food crops/varieties available which are rich in nutrients such as iron and zinc (rice, millets (bajra, ragi and small millets)), rich in protein (maize) and rich in β-carotene, iron, manganese and zinc (wheat).

33.6.2.1 Components of the Nutri-farm Movement

(i) *Establishment of self-help groups (SHGs)*: Required numbers of self-help groups (SHGs) will be identified for commercial production of nutri-rich varieties of rice, maize and finger millets, sweet potato, etc. The SHG will be identified and established by the State Department of Agriculture and KVK of the implementing district as per the existing Agricultural Technology Management Agency (ATMA) pattern.
(ii) *Cluster demonstrations*: The units of cluster demonstration of rice, maize and finger millets, sweet potato, etc. will be taken through the identified SHGs in the districts.

Assistance of Rs. 5000 per ha will be provided to the farmer beneficiary in terms of critical inputs for organization of demonstration of nutri-rich varieties of identified crops.

(a) The area under cluster demonstration of different crops as per the programme of the Government of India has been updated in several states depending upon the availability of seeds of desired variety from different sources.
(b) Depending on availability of certified seeds of different crops from different sources, use of truthfully labelled (TL) seeds may be permitted for demonstration purposes.

(iii) *Cluster demonstration on postharvest and value addition*: Cluster demonstration on postharvest and value addition of identified crops, i.e. rice, maize and finger millets and sweet potato, is to be organized in identified districts through selected SHGs by SAU/ICAR. Rate of assistance at INR 15,000/per cluster demonstration on each crop is also to be provided to SAU.
(iv) *Market-linked support*: An amount of INR 15 million (INR 2.5 million per district) is allocated to the state for development of a marketing system chain of nutri-rich crops through SFAC.
(v) *Marketing of nutri-rich produce*: The produce of the demonstration of nutri-farms will be procured through the district administration for utilization in Midday Meal, Anganwadi and other such programmes.
(vi) *Role of SAUs/KVK:* The scientists of SAUs/KVKs will help in developing training modules for training of rice, finger millet, maize and sweet potato farmers under the scheme. The desired numbers of cluster demonstration on postharvest and value addition of identified crops in each district will be organized in identified SHGs through the Department of Food and Nutrition of SAU. All critical inputs of demonstration on postharvest and value addition of nutri-rich crop produce will be arranged by SAU. Besides this, the SAU will take up training and capacity-building programme of departmental officials and farm entrepreneurs on postharvest management and value addition of the produce of identified nutri-rich crops.

33.6.3 Coverage Under Scheme

The programme is to be implemented in 100 high-malnutrition burden districts of 9 states, namely, Assam, Bihar, Chhattisgarh, Jharkhand, Madhya Pradesh, Odisha, Rajasthan, Uttar Pradesh and Uttarakhand.

33.6.4 Crops Included Under Scheme

The cereal crops, namely, rice, maize, pearl millets, finger millet, wheat and horticulture crops, viz. sweet potato and moringa (drumstick), are identified for production of nutri-rich foods under pilot scheme (Table 33.1).

33.6.5 Effective Implementation of the Programme

For effective implementation of the nutri-farm programme, the following need-based steps are to be practised judiciously:

- Investigation to identify the key nutritional problems in the region. Such a survey should be engendered, since women undergo more from anaemia due to iron deficiency.
- Studying the current cropping and farming systems (i.e. crop-livestock integration) in the region and recognizing and setting up agricultural interventions, like cultivation of biofortified crops and varieties that will assist to tackle the nutritional deficiencies existing amongst women, children and men.
- Developing impact appraisal criteria for assessing the role of agricultural remedies in

Table 33.1 Details of nutri-farm varieties of crops

Sl. no.	Name of crop	Name of variety	Name of micronutrients
1.	Wheat	HI-8663	β-carotene, iron and manganese
		VL-892	Zinc, copper and manganese
2.	Rice	MSE-9	Fe
		Kalanamak	Fe
		Karjat-4	Fe
		Chittmutyalu	Fe and zinc
		Udayagiri	Fe and zinc
		Mitta Triveni	Fe
		Varsha	Fe
		Poornima	Zinc
		ADT-43	Zinc
		Ranbir Basmati	Zinc
		Pant Sugandha-17	Zinc
		Jyoti	Zinc
		Ratna	Zinc
		Type-3	Zinc
		Kesari	Zinc
3.	Maize	HQPM-1	Lysine and tryptophan
		HQPM-4	Lysine and tryptophan
		HQPM-5	Lysine and tryptophan
		HQPM-7	Lysine and tryptophan
		Vivek	Lysine and tryptophan
		QPM-9	Lysine and tryptophan
		Shaktiman-1	Lysine and tryptophan
		Shaktiman-2	Lysine and tryptophan
		Shaktiman-3	Lysine and tryptophan
		Shaktiman-4	Lysine and tryptophan
4.	Pearl millet	ICTP-8203 FE 10–2 (Dhanshakti)	Fe
		86 M 86	Fe
		Ajit	Fe
		Hybrid Pusa-415	Fe and zinc
		Pusa Composite	Fe and zinc
5.	Finger millet	PRM-1	Fe and zinc
		VL-315	Fe and zinc
		VL-324	Fe and zinc

fighting the nutritional maladies existing in the area.

- Initiation of a nutrition literacy programme in the region through Gram Sabhas and Panchayats and creation of a cadre of community hunger fighters that can facilitate to bring convergence and synergy amongst diverse social protection measures against hunger.

For the effective implementation of nutri-farm programme, the Government of India had allocated funds to the state under the pilot scheme. Statewise fund allocation has been given in Table 33.2.

33.6.6 Critical Analysis

Nutri-farm movement is a unique movement so as to bring people above the level of hidden hunger or malnourished. In this context, while launching a "nutri-farm movement", an eminent scientist "Dr. M. S. Swaminathan" observed the

Table 33.2 Statewise allocation of funds for preparation of action plan under pilot scheme on nutri-farms during 2013–2014

Sl. no.	State	No. of identified districts	Cluster demonstrations (280 units at 10 ha/unit/ district)	Market link support for establishment of production and supply chain by SFAC (₹25 lakh/district)	Publicity of the programme (₹1 lakh/ district)	Cluster demonstrations on food processing and value addition by food and nutritional development of ICAR/SAUs (140 unit/district and ₹15,000/unit)	Establishment of centre of excellence (CoE) by ICAR/SAUs	Fund reserve for additional marketing support	Implementation and monitoring (₹2 lakh/ district)	Total
1	Assam	03	4.20	0.75	0.03	0.63	0	0	0.06	5.67
2	Bihar	12	16.80	3.00	0.12	2.25	0.75	0	0.24	24.68
3	Chhattisgarh	03	4.20	0.75	0.03	0.63	0	0	0.06	5.67
4	Jharkhand	01	1.40	0.25	0.01	0.21	0	0	0.02	1.89
5	M.P.	25	35.00	6.25	0.25	5.25	0.75	0	0.50	49.25
6	Odisha	06	8.40	1.50	0.06	1.26	0	0	0.12	11.34
7	Rajasthan	16	22.40	4.00	0.16	3.36	0.75	0	0.32	32.24
8	U.P.	32	44.80	8.00	0.32	6.72	0.75	0	0.64	62.48
9	Uttrakhand	02	2.80	0.50	0.02	0.42	0	0	0.04	3.78
	National	–	–	–	0.00	0.00	0.00	5.00	3.00	3.00
	Total	100	140.00	25.00	1.00	21.00	3.00	5.00	5.00	200.00

followings key points for the programme to succeed:

- For ensuring the accomplishment of the nutri-farm programme, there is need for introduction of nutrition literacy movement in the selected districts. For this purpose it will be helpful to ask for Panchayats to propose one woman and one man to be skilled as community hunger fighters. The function of community hunger fighters is to publicize to the village community the nutritional maladies widespread in the area and the means to tackle them. Thus, the community hunger fighters will allow farm families to conquer nutritional maladies through the appropriate introduction of agricultural remedies in the existing farming system.
- Agricultural interventions will contain the opening of naturally going on biofortified crops and varieties, such as moringa, sweet potato and maize, as well as genetically developed varieties such as iron-rich bajra and zinc loaded jowar.
- In addition, the cultivation and consumption of pulses will be tremendously significant to battle protein hunger. Improving the productivity and profitability of small holdings will assist to fight undernutrition through improved purchasing power.
- The expansion and reach of a farming system for nutrition (FSN) programme through nutri-farms will help out to mainstream nutritional parameters in the plan of farming system programmes.

33.7 Guidelines of the Scheme: Establishment of Nutri-farm and Its Composition

The Programme Management Group (PMG) will be constituted under the Chairmanship of Additional District Magistrate of the district with the following members:

Collector cum District Magistrate – Chairman
Deputy Director of Agriculture Member Secretary
District Agriculture Officer – Member
District Education Officer – Member
Representative of the Department of W&CD – Member
Representative of Food and Public Distribution Department – Member
Representative of FCI/State Food Procuring Agency – Member
Representative of SAU/KVK – Member

The Deputy Director of Agriculture of the district concerned will finalize the strategic action plan and ensure its implementation as per plan without any deviation. The PD, ATMA/DDA, will intimate the list of beneficiaries, village-wise area, GP and the Block to the DA&FP (O). The salient features of the programme include the following:

- The district action plan will be completed within a week by the PD, ATMA.
- The funds for the purpose will be placed with the concerned PD, ATMA, and the programme will be implemented by involvement of SHGs with active participation of the departmental officials.
- The PD, ATMA, will ensure supply of inputs like fertilizers, seed treating chemicals, PP chemicals, micronutrients, etc. to the field well within the reach of the farmers before the crop season by a specific time decided by the PMG.
- Transparency in preparing the list of beneficiaries for input distribution must be ensured.
- Due participation of schedule caste/schedule tribe/women is to be given importance based on their population in the district.
- The selected SHGs will operate in cluster basis and will create awareness amongst the farmers regarding the demonstrations through awareness campaign. The selection of demonstration site and the beneficiary list will be prepared on the basis of land suitability and

the interest of the farmer to conduct the programme.

- The SHG will prepare a list of beneficiaries along with the list of progressive farmers and submit the same to the concerned PD, ATMA, which will be approved by the PMG.
- The PD, ATMA, of the concerned district will nominate one SMS/AAO/AO in each block as the nodal officer who is accountable for successful implementation of the programme.
- The SHG should ensure all the recommended interventions in the project and proper crop stand as per advice of the field staff of the department.
- The indent for seed/planting material is placed centrally with government agencies.
- The DDA/PD, ATMA, will assess the requirement of various inputs and procure pesticides/ weedicides/micronutrients from the list of chemicals and list of suppliers approved by the state-level technical committee at the latest SLTC-approved rate through OAIC. The cost norms of one hectare demonstration are given in the action plan.
- The PD, ATMA, concerned should release 50 % of fund earmarked for mobility and incentives to the SHGs soon after submission of final countersigned beneficiary list (latest by 15 July 2013). Balanced mobility and incentive cost of the SHG should be released soon after submission of crop cutting results by the SHGs, duly countersigned by the nodal officer concerned.
- The PD, ATMA, should ensure submission of weekly cumulative progress report on Saturday to the state nodal officer of the project, i.e. JDA (SP & C), through e-mail/fax (jdaspc.dag@nic.in/0674-2392935 /RKVY CELL rkvy.dag@nic.in/ 0674–2396855).
- The line sowing/line transplanting of the entire demonstration area is mandatory which should be ensured.
- Display board should be placed in each patch.
- The documented consolidated report of the district will be submitted to the DA&FP(O).
- Documentation should be done by professional agencies having expertise in the work, and coverage is to be done at various stages of crop growth. It should include farmer's perception, success stories, etc. It will be submitted to the DA&FP (O) soon after the crop cutting.

33.8 A Case Study for an Indian State: Odisha

This is highlighted in Table 33.3 where the detailed district-wise and crop-wise demonstration programme as per the GOI allocation in Odisha state is indicated. In Table 33.4, component-specific pattern of assistance is also given. The area under cluster demonstration of different crops as per the programme of the Government of India has been revised due to nonavailability of seeds of desired variety from different sources. Due to nonavailability of certified seeds of different crops from different sources, use of TL seeds may be permitted for demonstration purposes.

33.9 Conclusions

Nutri-farm movement is a unique movement so as to bring people above the level of hidden hunger or malnourished. It is basically a movement to impart awareness to farmers at their farms about the role of adequate and complete nutrition in farming. Here, the major emphasis is given to demonstrate diverse improved production technology to promote cultivation of nutri-rich crop varieties so as to diminish micronutrient-related malnutrition. In addition, emphasis is also given to encourage commercial cultivation of specified nutri-rich crop varieties through self-help groups and developing the supply chain of nutri-rich produce to vulnerable sections of population. The success of the programme will depend on effective implementation of the scheme.

Table 33.3 The detailed district-wise and crop-wise demonstration programme as per the GOI allocation in Odisha state

		Rice (Cluster in numbers, area in ha and seed quantity in quintals)			Maize (Cluster in numbers, area in ha and seed quantity in quintals)			Finger millet (Cluster in numbers, area in ha and seed quantity in quintals)			Sweet potato	
Sl. no.	District	Cluster	Area	Seed quantity	Cluster	Area	Seed quantity	Cluster	Area	Seed quantity	Cluster	Area
1	Dhenkanal	100	1000	600	10	100	15	–	–	–	40	400
2	Boudh	100	1000	600	20	200	30	–	–	–	–	–
3	Kalahandi	430	4300	2580	30	300	45	50	500	50	–	–
4	Koraput	250	2500	1500	100	1000	150	100	1000	100	20	200
5	Gajapati	90	900	540	40	400	60	100	1000	100	–	–
6	Malkangiri	100	1000	600	50	500	75	50	500	50	–	–
	Total	1070	10,700	6420	250	2500	375	300	3000	300	60	600

One cluster = 10 ha; quintals = 100 kg

Table 33.4 Component-specific pattern of assistance (per ha)

Input	Crop						
	Rice		Maize		F. millets		Sweet potato
	Quantity	Cost	Quantity	Cost	Quantity	Cost	Cost
Seed/planting material	60 kg	1500	15 kg	1575	10 kg	800	–
Micronutrients (zinc sulphate)	25 kg	1250	25 kg	1250	25 kg	1250	–
Seedling treating chemical	–	100	–	–	–		73
PP chemical	–	250	–	275	–	450	–
Fertilizer	–	1300	–	1300	–	1900	4327
Display board and publicity materials	–	100	–	100	–	100	100
Documentation	–	50	–	50	–	50	50
Mobility to SHG	–	100	–	100	–	100	100
Incentive to SHG	–	250	–	250	–	250	250
Miscellaneous contingency	–	50	–	50	–	50	50
Visit of GOI/state officials (vehicle/POL) at state level	–	50	–	50	–	50	50
Total		5000		5000		5000	5000

References

Bailey Rebecca (2007) 'Biofortifying' one of the world's primary foods. http://www.dartmouth.edu/~news/releases/2007/11/19a.html

da Silva José Graziano (2015) The way forward: combatting global malnutrition, January 24, Blog, Health, Human Rights

De Benoist, McLean, Egli, Cogswell (eds) (2008) Worldwide prevalence of anaemia 1993–2005. WHO Library Cataloguing-in-Publication Data. ISBN 978 92 4 159665 7

Deaton A, Drèze J (2008) Nutrition in India: facts and interpretations. Princeton University Press, Princeton, NJ

DellaValle DM, Thavarajah D, Thavarajah P, Vendenberg A, Glahn RP (2013) Lentil (*Lens culinaris* L) as a candidate crop for iron biofortification: is there genetic potential for iron bioavailability. Field Crop Res 144:119–125

FAO (2013) The state of food insecurity in the world, Rome Government of India (2011) State of the Economy and Prospects Economic Survey, Ministry of Finance, Government of India, New Delhi. http://exim.indiamart.com/economic-survey10-11/pdfs/echap-01.pdf. Accessed 11 Aug 2015

Islam Y (2007) 'Growing Goodness' in developments, issue no. 38, pp 36–37

McClafferty B, Islam Y (2008) Fighting the hidden hunger, in TCE, (February 2008), p 26. The new face of hunger. In: The economist (17 April 2008)

Nestel P, Bouis HE, Meenakshi JV, Pfeiffer W (2006) Biofortification of staple food crops. J Nutr 136 (4):1066

Pal S, Jha D (2007) Public-private partnerships in agricultural R&D: challenges and prospects. In: Ballabh V (ed) Institutional alternatives and governance of agriculture. Academic Foundation, New Delhi

Planning Commission (2011) Draft on faster, sustainable and more inclusive growth – An approach to twelfth five year plan. http://planningcommission.nic.in/plans/planrel/12appdrft/appraoch_12plan.pdf. Accessed 11 July 2015

Pray C, Paarlberg R, Unnevehr L (2007) Patterns of political response to biofortified varieties of crops produced with different breeding techniques and agronomic traits. AgBioForum 10(3):135–143

Saxena NC, Vir S, Mander H (2014) Review on food and nutrition security: India's domestic story and scope to build global partnerships. www.ipekpp.com/admin/upload.../Report_1_15_Review_4510453616.p

Sharma S (2013) Managing Director, Nutri-Health Systems Pvt. Ltd. New Delhi. *www.nutrihealthsystems.com*

Swaminathan MS (2013) Launching a *Nutri-Farm* movement. www.mssrf.org. Accessed 15 May 2015

UNICEF (2013) Improving child nutrition: the achievable imperative for global progress

Web References Cited

A lot in the budget for agriculture- the Hindu www.thehindu.com

Countries vow to combat malnutrition through firm policies and actions 19 *November 2014* ¦ *Rome* - http://www.who.int/mediacentre/news/releases/2014/icn2-nutrition/en/

Guidelines for establishment of *Nutri – Farms* scheme. *nfsm.gov.in/ Guidelines/ NutriFarms .pdf*

India's 'national shame'- hunger, food insecurity and malnutrition. http://thelogicalindian.com/story-feed/

exclusive/indias-national-shame-hunger-food-insecurity-and-malnutrition

Nutri farm scheme *krishi.bih.nic.in/updation50oct13/*

Nutri-farms to fight hidden hunger. www.downtoearth.org.in

Nutritional *Farming* www.space-age.com/nutri-farm-seminar.doc

Transition in food consumption patterns http://wcd.nic.in/research/nti1947/6.%20Consumption %20expenditure.pdf nti1947 Consumption expenditure

Biofortification: Pathway Ahead and Future Challenges

34

S S Singh, K.K. Hazra, C S Praharaj, and Ummed Singh

Abstract

A large share of global population is affected by mineral and vitamin deficiency, particularly in the developing countries. Recent estimates exposed the problem will be more disappointing in the near future. Biofortification is emerging as a potential crop-based approach to deal with the mineral malnutrition problem by enriching the density of bio-available micronutrients and vitamins in food products. In recent years, significant advancement has been made in the fundamental understanding of micronutrient acquisition and translocation in soil-plant system. However, the current knowledge base in this area needs significant advancement to accelerate the pace of biofortification programme. Apart from the conventional breeding techniques, possible transgenic and agronomic approaches have also been identified for increasing the zinc, iron, selenium and iodine concentrations in the edible parts of food crops. Although these approaches are useful to address the mineral malnutrition problems worldwide, the effectiveness of the biofortification programme essentially relies on the farmers' and consumers' acceptance and future policy interventions. Therefore, strategic research and appropriate policy can lead to biofortification's grand success in the near future. In this chapter, we discussed the current knowledge and future prospects of crop biofortification.

S.S. Singh (✉) • C.S. Praharaj • U. Singh
Division of Crop Production, ICAR-Indian Institute of Pulses Research, Kanpur, Uttar Pradesh 208 024, India
e-mail: sssinghpatna@yahoo.co.in; cspraharaj@gmail.com; singhummed@yahoo.co.in

K.K. Hazra
Agricultural and Food Engineering Department, Indian Institute of Technology Kharagpur, Kharagpur, West Bengal 721 302, India
e-mail: kalikrishna123@gmail.com

U. Singh et al. (eds.), *Biofortification of Food Crops*, DOI 10.1007/978-81-322-2716-8_34

Keywords
Agronomic biofortification • Bioavailability • Micronutrient deficiency • Plant breeding • Vitamin A deficiency

34.1 Introduction

Despite significant advances in alleviating the hunger and malnutrition in some countries, global nutritional security is far from the reality, particularly in the developing world (Stein et al. 2008). Worldwide, billions of individuals are suffering from continuous insufficient intake of micronutrients because of the widespread dietary mineral deficiency in commonly eaten staple food crops (White and Broadley 2005; Thavarajah and Thavarajah 2012; Graham et al. 2007). Recent estimates have suggested that nearly 850 million individuals experience the ill effects of some kind of undernourishment in this planet (United Nations Millennium Development Goals Report 2006a, b). Insufficient intake of micronutrient and vitamins is extensive among the world's population who are dependent on vegetarian diet, consisting of few staple crops (Connolly 2008). Traditionally, to address this issue, we basically depend on mineral supplementation, dietary diversification and food fortification interventions, and their positive impact has already been assessed in different parts of the world (Welch and Graham 2004a; White and Broadley 2005). However, fortification and supplementation programmes largely target the urban population. Further, these types of programmes call for huge amount of fund, which is practically difficult to invest for most of the governments of low-income countries (Meenakshi et al. 2010). In addition, this programme does not include the 'food-based approach' which favours consumption of locally available plant and animal species from diverse origin (Shrimpton 2002). In this situation, a crop-based technique has been developed, in which the edible parts (grain, straw, root and tubers) of food crops are enriched with micronutrients through appropriate breeding methods. This approach is named as 'biofortification' (Bouis et al. 2000; Bouis 2002; CIAT/IFPRI 2004). Worldwide, the technique is progressively becoming popular as a potential strategy to minimise the micronutrient-related unhealthiness (Meenakshi et al. 2010). In fact, the poor people are mostly affected by micronutrient deficiency as they cannot afford micronutrient-rich foods like fruits, vegetables, pulses and animal products. Therefore, micronutrient fortification in common staple food crops can improve their nutrition and health status. Hence, as an alternative low-cost approach, 'biofortification' can play a vital role in mitigating the micronutrient malnutrition and ensure quality of life worldwide (Waters and Sankaran 2011).

34.2 Extent of Deficiency and Its Impact

Mineral and vitamin deficiency affect a major share of the world's population, mostly in the developing world (Stein 2010). Among the minerals that are essential for human health, zinc and iron deficiencies are widespread. Besides this, iodine, selenium and cobalt deficiencies are also observed in different parts of the world. Basically, the deficiency is due to low concentration of these minerals in their day-to-day diets (Fageria et al. 2012). Based on United Nations' estimates, almost one billion individuals are suffering from a kind of trace element deficiency. The problem is more common for women and children, especially in the countries like Africa, Asia and South America (Fageria et al. 2012). Globally, vitamin deficiency is another challenging problem, particularly vitamin A deficiency. Vitamin A deficiency is responsible for early deaths of young children. Every year, almost one million young children die under the age of five due to vitamin A deficiency. Recent estimate also documented that almost 3 % mortality of young children may be directly related to vitamin A deficiency, and besides this all night blindness disorder is attributed to vitamin A deficiency. According to the reports of the World Health Organization (WHO), approximately 250 million pre-school children are deficient in vitamin A. Apart from this, substantial proportion of pregnant women is also suffering from vitamin A deficiency especially in the underdeveloped world. Deficiency of iodine is also widespread in humans. Almost 20 % of the developing world's population is suffering from iodine deficiency disorders. Iodine is very important for pregnant women and iodine deficiency during pregnancy is causing nearly 20 million babies to be born mentally impaired. According to WHO, globally 1.6 billion people are in serious risk of iodine deficiency as some regions are identified under low soil native iodine category (Bruno et al. 2008; Hetzel et al. 2004).

Worldwide, deficiency of iron in human increases the risk of disability as well as death (Boccio and Iyengar 2003). The World Health Organization estimated that out of 3.7 billion iron-deficient individuals, 2 billion people are severely affected and considered as anaemic (Yang et al. 2007). Subsequently, it has been documented that 5 % of all types of maternal mortality is directly associated with iron deficiency (Meenakshi et al. 2010), and every year 50,000 young women are dying during pregnancy/or childbirth due to severe iron deficiency. Likewise, deficiency of zinc also leads to serious health problems and almost one third of the world's population are now deficient in zinc (Welch and Graham 2004b; Hotz and Brown 2004; Muller and Krawinkle 2005). Zinc deficiency in humans is designated as the fifth important cause of diseases and deaths in the developing countries (White and Broadley 2009; WHO 2002) (Fig. 34.1).

The role of selenium is also vital and has a unique value in mammalian nutrition (Diwadkar-Navsariwala et al. 2006). The bioavailability of selenium is usually low in soils of China, United Kingdom, East Europe, Africa and Australia. Therefore, crop plants have lower density of selenium in these regions (Smkolji et al. 2005; Pedrero et al. 2006; Genc et al. 2005). Continuous intake of low-selenium content food causes complicated health-related problems like epilepsy, oxidative stress-related conditions, infertility, immune deficiency, etc. (Rayman 2012; Whanger 2004; Zeng and Combs 2008). Hence, the global statistics of the problem of micronutrient deficiency is really upsetting and needs immediate attention.

34.3 Biofortification: Importance and Relevance

Food fortification, supplementation and dietary diversification programmes have effectively addressed the micronutrient malnutrition in several countries (Khush et al. 2012). However, biofortification has some added benefit over other interventions, i.e. it is a low-cost sustainable technique and virtually addresses the economically backward households and subsistence farmers (Nestel et al. 2006). Apart from this, 'investment is only required at the research and development stage, and thereafter they would become entirely sustainable' (Gomez-Galera et al. 2010; Bouis 1996; Graham et al. 2001; Welch and Graham 1999, 2002, 2004a; Welch 2002). In different parts of the world, food crops are usually deficient in micronutrient because of low native soil micronutrient concentration, cultivation of high-yielding varieties that demands higher quantity of micronutrients and in the long run depletes the soil micronutrient level, liming practices in acidic soils (Fageria and Baligar 2008), unfavourable soil reaction in sandy and calcareous soil for micronutrient availability, continuous application of high analysis fertilizer that contains very negligible amount of micronutrient and minimal use of organic amendments (Fageria et al. 2002). Hence, increasing micronutrient density as well as improving the bioavailability of these trace elements in major food crops is a promising strategy to combat the micronutrient malnutrition problem. In this

a

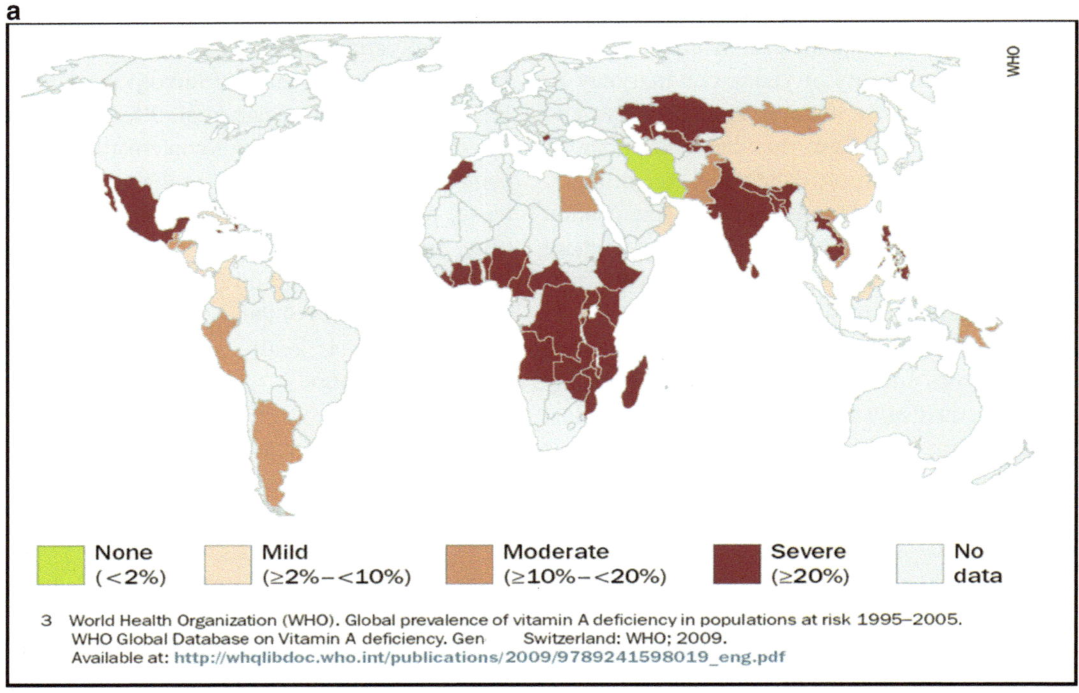

b

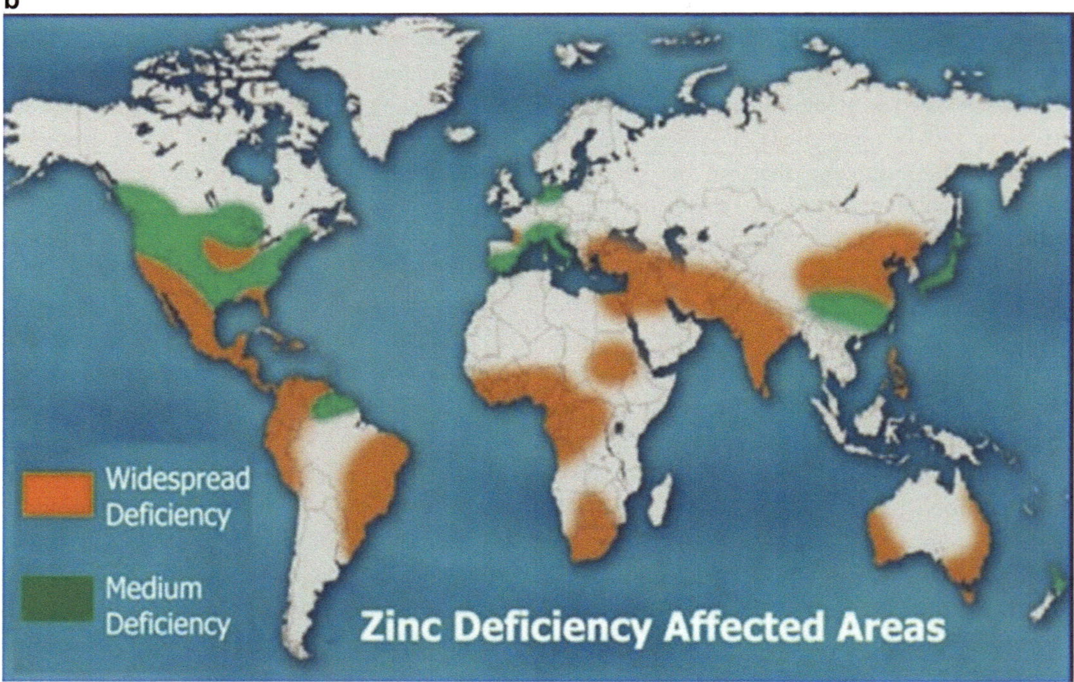

Fig. 34.1 Distribution of global area with variable degree of vitamin A deficiency (**a**) and zinc deficiency (**b**) (Alloway 2004)

programme, the basic target is to improve the levels of important micronutrients of staple food (e.g. rice, potato, etc.), which are widely consumed by a large share of the world's population. Therefore, even very little enhancement in bioavailable micronutrient level in targeted food crops may have considerable impact on alleviating the malnutrition problem, particularly in the developing countries (Sperotto et al. 2012). Additionally, an improved soil-plant system will ensure nutrient cycling and ecologically viable environment (Yang et al. 2007). In the long run, 'biofortification' approach is expected to be more important as a strategy that also has a lower dependency in infrastructure and compliance (Gomez-Galera et al. 2010).

34.4 Research Advances

34.4.1 Understanding the Physiological Basis of Micronutrient Delivery in Soil-Plant System

Understanding the mechanisms of soil-plant nutrient system is essential for successful biofortification method. Lately, a significant progress has been noticed in explaining the plant adaptation mechanisms to micronutrient stress (Ghandilyan et al. 2006) and the complicated transport pathway of micronutrient (iron and zinc) from soil to plant system (Khush et al. 2012). The uptake and translocation procedure of important metals are understood more clearly with physiological and genetic processes.

Research results have been documented that the rice genotype with rich in iron have higher translocation rate of iron from root to the shoot and grain when compared with noniron rich rice genotypes. At the same time, iron-rich rice genotypes have higher accumulation of iron and zinc in endosperm tissues of rice grain (Hao et al. 2005). Scientists have explained this mechanism genetically and confirmed that the expression of specific genes in the phloem cell and grains is basically responsible for the above process. Likewise, the enhanced micronutrient content in transgenic *indica* rice grain is due to expression of the gene 'ferritin' that results in not only higher iron and zinc in whole rice but also in polished rice. Detailed knowledge of physiological and molecular basis of micronutrient translocation and accumulation in plant is crucial for need-based manipulation of the plant system (Vasconcelos et al. 2003). Significant advancement in the understanding of the molecular mechanisms has been observed since the last two decades. The processes of zinc acquisition from soil to plant system have been described by molecular genetics in grain crops like rice, wheat and rye (Yang and Romheld 1999). In the same line, molecular process of detailed transport pathways, distribution pattern and chelating mechanisms of copper, iron and zinc have been explained by Grotz and Guerinot (2006). Based on comparative assessment of phytosiderophore production of different cereals, Romheld and Marschner (1990) confirmed that among the cereals, barley crop has the highest production of phytosiderophore and its associated iron acquisition from soil to plant system. Scientists have identified a metal chelator, nicotianamine (NA), which is responsible for metal transports and homeostasis (Douchkov et al. 2005). Hence, the clear understanding of the pathways of metal transport in plant system enables us to strategically improve the micronutrient mobilisation and transportation to the edible plant parts.

Enhancing bioavailability of micronutrients is also an important challenge under biofortification programme. Scientists proposed that not the absolute density of micronutrients in the edible plant parts but the bioavailability of micronutrient is more important. In fact, phytic acid, which fixes the micronutrients like iron and zinc in the plant system, reduces the bioavailability of micronutrient. Consumption of higher phytic acid content food products is difficult to digest and greatly restricts the absorption of micronutrients. Therefore, attempts have been made to reduce the level of phytic acid in the food crops. Degradation of phytic acid through enhanced activity of phytase enzyme as well as minimising the level of phytin can improve the bioavailability of micronutrients in the human digestive system (Raboy 2002). In this context, the phytate-to-zinc molar ratios or the phytate and calcium-to-zinc ratio are important indicator of micronutrient availability. A greater variability among the species and genotypes has been noticed for phytic acid content and can be used for developing the variety with desired traits (Cakmak et al. 1999). However, as a major plant metabolite the anti-nutrients (e.g. phytic acid) also take part in plant metabolism, stress resistance and pest and pathogen resistance (Yang et al. 2007). The role of phytin in early seedling vigour has been well established, and reduced phytin level greatly hampers the seedling vitality particularly in low fertile soils as it serves energy and is an important source of mineral and phosphorus. Additionally, some anti-nutrients (e.g. polyphenols and phytate) are also important for human health and can reduce the risk of heart disease and diabetes, and also these are potential anticarcinogens (Shamsuddin 1999). Therefore, manipulation of anti-nutrient concentration in edible food crops through breeding approaches should essentially consider the negative consequences on health (Welch and Graham 2004a). Therefore, for successful biofortification, comprehensive knowledge on the physiological process of uptake from the rhizosphere, phloem sap loading, translocation and unloading rates within the reproductive organs, and remobilisation mechanisms are very important (Welch and Graham 2004a; Welch 1986; Sperotto et al. 2012) (Fig. 34.2).

34.4.2 Breeding Approach

Sizeable genotypic variation exists in micronutrient density in different crop species (Yang et al. 2007). Substantial genotypic variation in grain micronutrient concentration of rice, maize and wheat has been documented (Bouis 1996; Graham et al. 2007; Welch and Graham 1999, 2004b). Plant breeder mainly targets to use this existing genotypic diversity for developing micronutrient-enriched crop varieties (Zapata-Caldas et al. 2009). At present, a number of research programmes have been initiated to develop the desired crop varieties with higher micronutrient density (iron and zinc), vitamin A, tryptophan and lysine at the global level. These programmes mainly concentrate on few staple food crops like rice, wheat, cassava, potato, bean, etc. (Pfeiffer and McClafferty 2007). An integrated effort of the Consultative Group on International Agricultural Research (CGIAR) centres on the programme 'HarvestPlus' aiming to breed for increasing concentration and bioavailable zinc and iron in seeds of major stable food crops (Bouis 2002; Pfeiffer and McClafferty 2007). Likewise, in search of potential donors, IRRI (International Rice Research Institute) started a programme in collaboration with the Department of Plant Science, University of Adelaide, Australia. In this programme almost 7000 entries were screened for both iron and zinc contents in the rice grain. Afterwards, a higher variation in iron and zinc level in rice grain has been documented from the study (Khush et al. 2012). In the same line, reports show that the entries of rice, wheat, maize, bean, cassava and yam differ greatly in iron concentration (Frossard et al. 2000; Welch 2001; Genc et al. 2005; Haas et al. 2005; Nestel et al. 2006). The variation in iron content in rice, wheat and maize was noted to be 6–22, 10–160 and 15–360 mg kg^{-1}, respectively (White and Broadley 2005). Therefore, selective breeding might be a potential tool to develop micronutrient-rich staple food crops (Tiwari et al. 2009). An improved breeding line from India has been identified, i.e. IR 68144-3B-2-2-3 (IR72× Zawa Bonday), which is rich in iron level (21 mg kg^{-1} unmilled brown rice), concurrently a good yielder also. Likewise, some emmer wheat accessions were found tolerant to zinc deficiency in soil and maintained a very high level of zinc (up to 139 mg kg^{-1}), iron (up to 88 mg kg^{-1}) and protein (up to 380 g kg^{-1}) in grains (Peleg et al. 2008). Subsequently, latest research also suggests that 'synthetic wheat derived from *Aegilops tauschii* has also a high genetic potential for increasing grain zinc density of cultivated wheat' (Calderini and Ortiz-Monasterio 2003). Recently, some genotypes with low phytic acid content have been identified in rice, maize and barley (Raboy 2000). Hence, from the above points, it is clear that micronutrient concentration in food grain can be increased substantially through adoption of appropriate breeding techniques. As a low-cost and relatively easy strategy, breeding techniques always

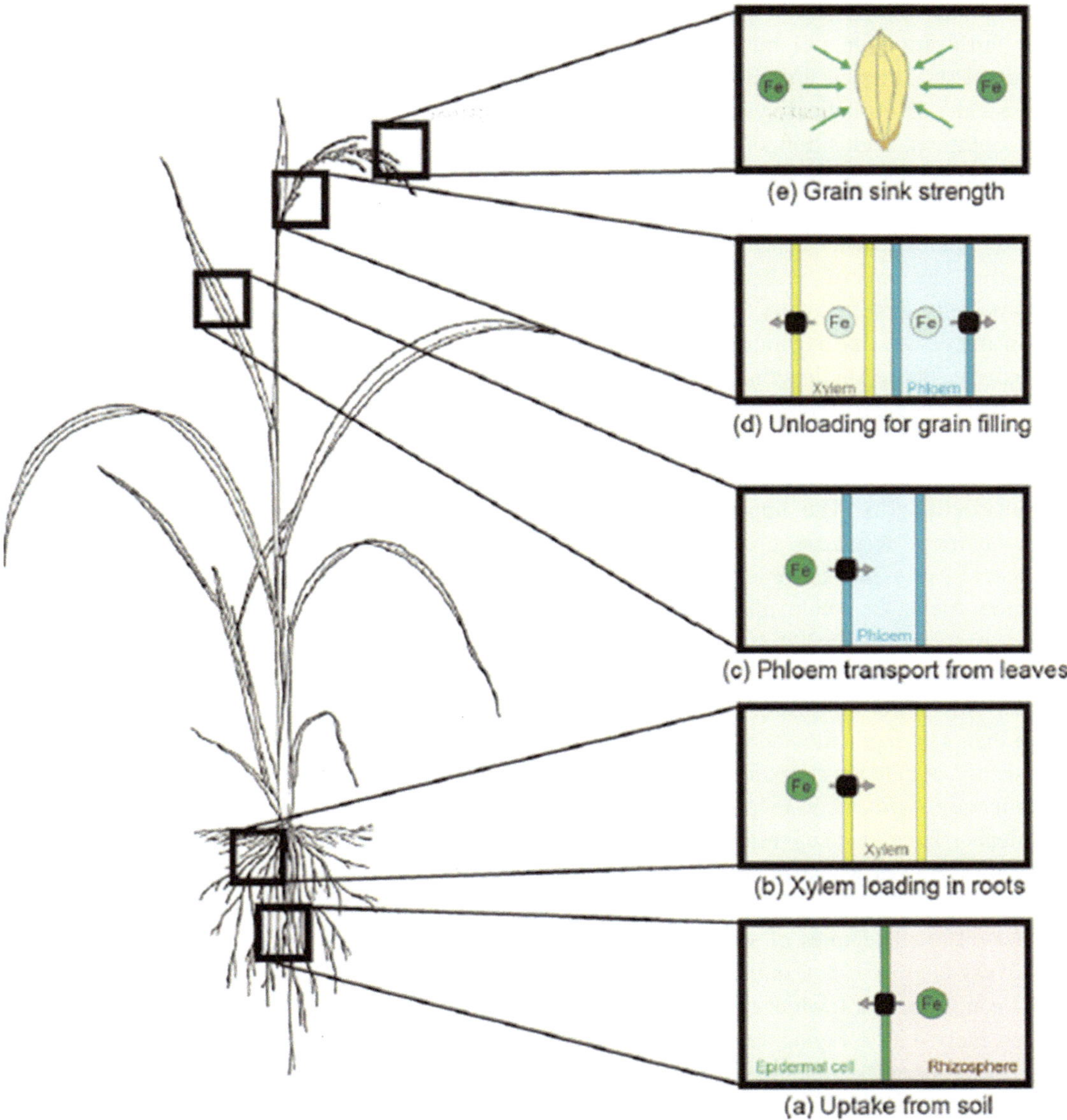

Fig. 34.2 Rate-limiting process of iron transportation from soil to seed (Sperotto et al. 2012)

consider the best biofortification tool to overcome the micronutrient malnutrition problem. However, the availability of micronutrient in soil greatly influences the success of this technique. Meanwhile, continuous cultivation of HYVs of cereal crops aggravated the micronutrient deficiency in soils. Therefore, there is an every need to balance the soil micronutrient pool to get the long-term benefit from these biofortified crops.

34.4.3 Agronomic Approach

Compared to breeding approach, agronomic biofortification represents a short-term solution to the problem. In fact, with introduction of high-yielding exhaustive crop varieties, the soil native micronutrient level has depleted to a large extent (Cakmak 2008). Globally, the extensive deficiency of micronutrient in soil is mainly responsible for low micronutrient density in food crops and associated malnutrition problems. Therefore, crop plants show higher response to micronutrient fertilisation in soil micronutrient-deficient areas. Hence, there is a need to improve the soil chemical and biological system for facilitating the optimum acquisition of micronutrients by the crop plants. With adoption of appropriate fertilisation practice, significant improvement in grain micronutrient concentration has been reported by many researchers. Agronomic approach is therefore very important as short-term and sustainable technique of biofortification (Cakmak 2008). Application of zinc fertilisers in wheat (Hu et al. 2003), rice (Li et al. 2003), pea and cowpea (Fawzi et al. 1993) not only improves the micronutrient density in grains but productivity also. The application method of micronutrient fertilisers greatly influences the efficacy in increasing concentration of micronutrients in edible plant parts. Based on field experimentation, it has been confirmed that the foliar application of zinc spray in wheat at early dough or milk stage crop is more effective than soil application (Cakmak et al. 2010a). Basically, at field situation, foliar or foliar + soil application is found more promising. Research evidences have also suggested that foliar application of zinc in wheat improved the zinc concentration of wheat grain by almost 2–3 folds under different regimes of soil zinc availability (Cakmak et al. 2010b; Cakmak 2008). Likewise, zinc concentration in white rice can be improved extensively with spraying zinc at heading stage. More importantly, application of zinc-based fertilisers reduces the overloading of phosphorus in grain, thereby reducing the phytic acid concentration. Minimising the phytic acid concentration helps to increase the bioavailability of zinc or other micronutrients when consumed (Cakmak et al. 2010b; Erdal et al. 2002). Additionally, micronutrient-enriched seeds have higher germination and vigour and are considered potentially more tolerant to micronutrient stress in soil.

The low availability and mobility of iron in soil also restrict its uptake from soil to plant system. Furthermore, upon application of iron fertiliser in the form of $FeSO_4$, rapidly converts to to Fe (III) form, and become unavailable for plant (Frossard et al. 2000). Meanwhile, iron availability in calcareous soil is one of the challenging problems and effectiveness of iron fertilisers is also noticeably poor. Therefore, rapid conversion of iron to unavailable form and reduced mobility in plant system (phloem) collectively results the lower effectiveness of soil and foliar application of iron-based fertilizers, particularly for cereal crops in calcareous soil (Rengel et al. 1999; Cakmak 2008; Fawzi et al. 1993). In this situation, some researchers emphasised on organic Fe fertilisers to solve the problem. The nutritional quality as well as grain yield of peanut (Xiao et al. 2000), wheat (Hu et al. 2003) and green pea (Zhang et al. 2006) was substantially improved with application of organic Fe fertiliser. Besides this, various cheated compounds (Fe-EDTA, Fe-DTPA, Fe-EDDHA, Fe-citrate and Fe-IDHA, etc.) also can be used to correct the severe Fe chlorosis in crop plants. Interestingly, the transportation mechanism of nitrogen and iron in vegetative part of the plant system is regulated by a common genetic process (Waters et al. 2009), and a positive relationship between grain nitrogen and iron has been confirmed by many researchers. Therefore, integrated or mix foliar application of urea and Fe fertilisers had synergistic effect on grain iron concentration as nitrogen helps to penetrate iron into the leaf tissues (Swietlik and Faust 1984; Rodriguez-Lucena et al. 2010b). Hence, to facilitate iron biofortification of food crops, nitrogen nutrition is also very crucial and needs special attention. Recently, several researchers reported that iron and zinc nutrition can be improved with suitable interspecific root interactions. Intercropping of cereal (*graminaceous* species) with dicot plant, e.g. maize + peanut, chickpea + wheat, guava + sorghum or maize, improves iron and zinc uptake because of interspecific root interactions (Huang et al. 2012; Kamal et al. 2000; Zuo et al. 2004; Inal et al. 2007). Therefore, intercropping with appropriate component crops may be a key approach for improving the micronutrient concentration in the targeted crop by increasing the soil availability of micronutrients. Similarly, agronomic approaches to increase the selenium concentration in food crops are becoming increasingly popular. Foliar applications of selenium fertilisers (Na_2SeO_4 and K_2SeO_4) are very effective and have immediate effect on plant selenium concentration, while some less soluble selenium fertiliser ($BaSeO_4$) and selenite are known to have delayed effect but last for a long time (Broadley et al. 2010). However, agronomic biofortification has several limitations, e.g. a grater variability of micronutrients in soil system, soil chemical reactions that reduce the availability of applied micronutrient fertilisers, non-efficient transportation foliar application that results in micronutrient deficiency in roots, etc. (Yang et al. 2007). Hence, strategic fertilisation strategy needs to be developed at the same time and soil properties should be improved to ensure micronutrient availability. Furthermore, the success of genetic biofortification also largely depends on strategic soil and fertiliser management (Cakmak 2008), and thus agronomic biofortification is basically complementary to breeding biofortification approach.

34.4.4 Biotechnological Approach

Genetic engineering is also a potential tool to enhance the micronutrient content in food grains. Plant genetic makeup may be modified for higher synthesis of vitamins and higher acquisition of minerals and reduce synthesis rate of antinutrient compounds (phytic acid, tannins, etc.) (Bouis et al. 2003; Raboy 2002; Tucker 2003). Presently, significant advances have been achieved in the field of developing transgenic plant with higher micronutrient density. The ferritin (iron storage protein gene) has been successfully transferred from soybean/French bean to rice crop (var. Kitaake) through *Agrobacterium*-mediated transformation (Goto et al. 1999; Lucca et al. 2001) that results in almost threefold increase in rice grain iron concentration over control rice variety (Vasconcelos et al. 2003). Similarly, an attempt has been made to evaluate the expression of two transgenes NAS (AtNAS1) from *Arabidopsis thaliana* and ferritin (Pvferritin) from *Phaseolus vulgaris* in rice crop, which results in a synergistic effect on uptake and storage of iron concentration in grain endosperm (sixfold Fe overcontrol). In the same line, barley grain zinc concentration was improved by incorporating zinc-transporter protein genes from *A. thaliana*. Transgenic approach has been adopted in rice to manipulate the phytosiderophore biosynthetic pathway by introducing the gene for nicotianamine aminotransferase. The transgenic rice crops are resistant to iron-deficient conditions of calcareous soil and produce four times higher yield than the control rice plants (Takahashi et al. 2001). Lactoferrin, a human iron-binding protein present in milk, has been expressed in rice (Nandi et al. 2002) and potato (Chong and Langridge 2000) crops that results in increased availability of iron in plant residues. 'Transgenic approaches to biofortification rely on improving the phytoavailability of mineral elements in the soil, their uptake from the rhizosphere, translocation to the shoot and accumulation in edible tissues. Besides this, transgenic approaches may be used to reduce the concentrations of antinutrients and increase the concentrations of promoter substances' (White and Broadley 2005; Puig et al. 2007).

34.5 The Impact of Biofortification Research and Key Factors

The success of biofortification largely depends on several factors including farmer's acceptance of biofortified varieties and consumption of biofortified food products by the undernourished target population. Hence, cultivation of biofortified varieties and its consumption are the key factors that decide the impact of this programme. Till now, biofortification is in preliminary stage (research phase) in most of the developing countries. Meenakshi et al. (2010) stated that 'The impact of any food-based intervention depends on the dose–response to increased nutrient intakes. Ideally, this would entail determining a biological relationship between enhanced micronutrient intakes and nutritional outcomes'.

Till date, many success stories of biofortification programmes have been documented worldwide. Upon consumption of biofortified cassava, the vitamin A deficiency of northeast Brazil population has been reduced by almost 20 %. Likewise the biofortified maize and sweet potato reduce the occurrence of vitamin A deficiency to an extent of 32 and 64 %, respectively. The biofortification intervention was found highly cost-effective in northeast Brazil and Ethiopia. The consumption of biofortified bean even at very low quantity (45–55 g per day) was found promising and trimmed down the burden of iron deficiency up to 22 and 36 %, respectively (Fig. 34.3).

Under HarvestPlus programme in Mozambique and Uganda during 2007–2009, approximately 24,000 households had been supplied with orange sweet potato rich in provitamin A to reduce the occurrence of vitamin A deficiency. At the end of the project, the adoption rate of provitamin A-rich sweet potato increased up to 61–68 % and almost two thirds of the total women and children are consuming sweet potato for vitamin A (Hotz et al. 2012). Data also reported that the incidence of zinc deficiency reduced by 3–20 % in Latin America with consumption of biofortified beans. Interestingly, the magnitude of reduction of zinc deficiency with biofortified rice and wheat was much higher in Asian countries. Introduction of biofortifed rice in Bangladesh decreased the zinc deficiency by 17– 33 %, while 5–33 % decrease was noticed with high-zinc wheat in Pakistan. Hence, it is apparent that biofortification could have greater impact on mitigating micronutrient malnutrition worldwide.

34.6 Future Prospects

Currently, biofortification intervention is a promising crop-based strategy for eliminating micronutrient malnutrition. Still, substantial exploration hole exists in current biofortification methodology and presently it is a challenging endeavour. The detailed understanding of the mechanisms of mineral translocation from soil to seed are lacking in most of the food crops. Hence, advanced knowledge in the basic understanding of the rate limiting steps of micronutrient acquisition and translocation in soil-plant system should be generated. In addition the safety issues of biofortified crops have to be analysed in detail before making them available to the consumers. Comprehensive knowledge gap also exists in bioavailability of micronutrients in food grain and mineral distribution pattern in plant system. The loss of micronutrient during processing on selective removal of outer tissues is also not analysed for most of the crops and needs to be explored. In the near future, few most recent advances can assume a colossal part in the improvement methodology and can play an enormous role in the enrichment process of plant edible parts. Some of the important strategies would be transferring genes for higher iron and zinc content through molecular cytogenetics, minimising the loss of micronutrients during postharvest processing by uniform allocation of minerals in the grain, manipulation of phytic acid level to enhance the bioavailability, etc. Recently, fertiliser products of nanoscale level are viewed as a potential agro-input for precise micronutrient management even at very low application rate. Consequently, strategic utilisation of these nano-based micronutrient fertilisers may help in biofortification process. Till now, the biofortification technique is confined to some major crops and some crops with local relevance. In the same line, there is a need to explore all the crops that are directly or indirectly associated with micronutrient deficiencies. Thus, an integrated approach of biofortification strategies can improve human health through consumption of micronutrient-fortified food products. However, before we use this tool effectively for mitigating micronutrient malnutrition, some questions are always be there on its scientific feasibility, adoption probability at farmers and consumer level, economic viability and production stability (Nestel et al. 2006). In this background, the success of biofortification programme is directly associated with improved policies including nutrition education, marketing, agricultural policy and finally public

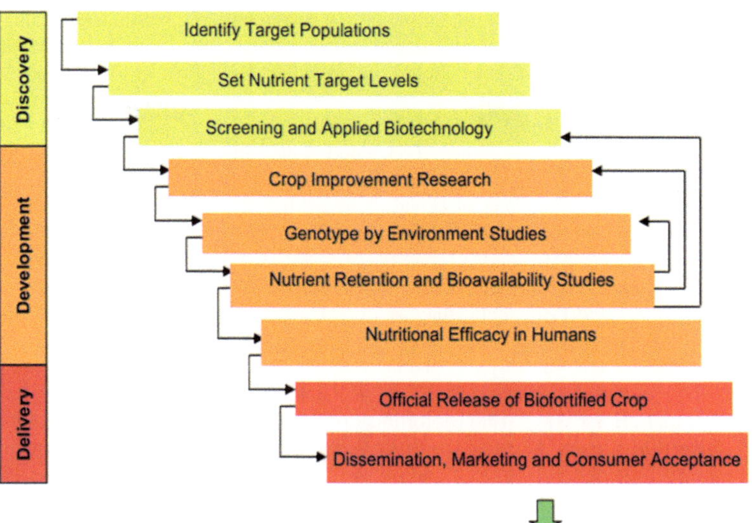

Fig. 34.3 HarvestPlus pathway (Bouis et al. 2011)

awareness. Interdisciplinary research team including human nutrition scientists and crop scientist would essentially work together for developing the final end products with desired nutritional properties. Sometimes, enrichment of micronutrient and vitamins has negative effect on colour and taste of the end product and usually not liked by the consumers. Therefore, biofortified crops will have to have acceptable sensory and cooking qualities for greater adoption. Furthermore, the desired yield level and resistant to biotic and abiotic stresses of these biofortified crop varieties should also be guaranteed. Therefore, in the future more systematic steps towards developing biofortified crops along with suitable agronomic management options are needed to eliminate the micronutrient malnutrition in human and ensure food and nutritional security.

34.7 Conclusion

In the future world, mineral and vitamin deficiency are expected to be more threatening, and biofortification strategy is appearing as a potential tool for addressing the problem. Given a low-cost, easy and crop-based approach, biofortification technique holds a great promise for mitigating the micronutrient malnutrition problem in the developing world. Significant progress has been made in this line, and future strategic research and appropriate policy could lead to biofortification's great success in the coming years.

References

Alloway BJ (2004) Zinc in soils and crop nutrition. IZA Publications. International Zinc Association, Brussels, pp 1–116

Boccio JR, Iyengar V (2003) Iron deficiency: causes, consequences, and strategies to overcome this nutritional problem. Biol Trace Elem Res 94:1–32

Bouis H (1996) Enrichment of food staples through plant breeding: a new strategy for fighting micronutrient malnutrition. Nutr Rev 54:131–137

Bouis HE (2002) Plant breeding: a new tool for fighting micronutrient malnutrition. J Nutr 132:491S–494S

Bouis HE, Graham RD, Welch RM (2000) The Consultative Group on International Agricultural Research (CGIAR) micronutrients project: justification and objectives. Food Nutr Bull 21:374–381

Bouis HE, Chassy BM, Ochanda JO (2003) Genetically modified food crops and their potential contribution to human nutrition and food quality. Trends Food Sci Technol 14:191–209

Bouis HE, Hotz C, McClafferty B, Meenakshi JV, Pfeiffer WH (2011) Biofortification: a new tool to reduce micronutrient malnutrition. Food Nutr Bull 32(Suppl 1):31S–40S

Broadley MR, Alcock J, Alford J et al (2010) Selenium biofortification of high-yielding winter wheat (*Triticum aestivum L.*) by liquid or granular Se fertilisation. Plant Soil 332:5–18

Bruno DB, Erin M, Maria A et al (2008) Iodine deficiency in 2007: global progress since 2003. Food Nutr Bull 29:195–202

Cakmak I (2008) Enrichment of cereal grains with zinc: agronomic or genetic biofortification? Plant Soil 302:1–17

Cakmak I, Kalayci M, Ekiz H, Braun HJ, Kilinc Y, Yilmaz A (1999) Zinc deficiency as a practical problem in plant and human. Field Crop Res 60:175–188

Cakmak I, Kalayci M, Kaya Y, Torun AA, Aydin N, Wang Y, Arisoy Z, Erdem H, Yazici A, Gokmen O, Ozturk L, Horst WJ (2010a) Biofortification and localization of zinc in wheat grain. J Agric Food Chem 58:9092–9102

Cakmak I, Pfeiffer WH, McClafferty B (2010b) Biofortification of durum wheat with zinc and iron. Cereal Chem 87:10–20

Calderini DF, Ortiz-Monasterio I (2003) Are synthetic hexaploids a means of increasing grain element concentrations in wheat? Euphytica 134:169–178

Chong DK, Langridge WH (2000) Expression of full-length bioactive antimicrobial human lactoferrin in potato plants. Transgenic Res 9:71–78

CIAT/IFPRI (2004) Biofortified crops for improved human nutrition. Challenge program proposal. International Center for Tropical Agriculture (CIAT) and IFPRI, Washington, DC

Connolly EL (2008) Raising the bar for biofortification: enhanced levels of bioavailable calcium in carrots. Trends Biotechnol 26:401–403

Diwadkar-Navsariwala V, Prins GS, Swanson SM, Birch LA, Ray VH, Hedayat S, Lantvit DL, Diamond AM (2006) Selenoprotein deficiency accelerate prostate carcinogenesis in a transgenic model. Proc Natl Acad Sci U S A 103:8179–8184

Douchkov D, Gryczka C, Stephan UW, Hell R, Baumlein H (2005) Ectopic expression of nicotianamine synthase genes results in improved iron accumulation and increased nickel tolerance in transgenic tobacco. Plant Cell Environ 28:365–374

Erdal I, Yilmaz A, Taban S, Eker S, Torun B, Cakmak I (2002) Phytic acid and phosphorus concentrations in seeds of wheat cultivars grown with and without zinc fertilization. J Plant Nutr 25:113–127

Fageria NK, Baligar VC (2008) Ameliorating soil acidity of tropical Oxisols by liming for sustainable crop production. Adv Agron 99:345–399

Fageria NK, Baligar VC, Clark RB (2002) Micronutrients in crop production. Adv Agron 77:85–268

Fageria NK, Moraes MF, Ferreira EPB, Knupp AM (2012) Biofortification of trace elements in food crops for human health. Commun Soil Sci Plant Anal 43:556–570

Fawzi AFA, El-Fouly MM, Moubarak ZM (1993) The need of grain legumes for iron, manganese and zinc fertilization under Egyptian soil conditions: effect and uptake of metalosates. J Plant Nutr 16:813–823

Frossard E, Bucher M, Machler F et al (2000) Potential for increasing the content and bioavailability of Fe, Zn and Ca in plants for human nutrition. J Sci Food Agric 80:861–879

Genc Y, Humphries JM, Lyons GH, Graham RD (2005) Exploiting genotypic variation in plant nutrient accumulation to alleviate micronutrient deficiency in populations. J Trace Elem Med Biol 18 (4):319–324

Ghandilyan A, Vreugdenhil D, Aarts MGM (2006) Progress in the genetic understanding of plant iron and zinc nutrition. Physiol Plant 126:407–417

Go´mez-Galera S, Rojas E, Sudhakar D, Zhu C, Pelacho AM, Capell T, Christou P (2010) Critical evaluation of strategies for mineral fortification of staple food crops. Transgenic Res 19:165–180

Goto F, Yoshimura T, Shigemoto N, Toki S, Takaiwa F (1999) Iron fortification of rice seed by the soybean ferritin gene. Nat Biotechnol 17:282–286

Graham RD, Welch RM, Bouis HE (2001) Addressing micronutrient malnutrition through enhancing the nutritional quality of staple foods: principles, perspectives and knowledge gaps. Adv Agron 70:77–142

Graham RD, Welch RM, Saunders DA, Ortiz-Monasterio I, Bouis HE, Bonierbale M, de Haan S, Burgos G, Thiele G, Liria R, Meisner CA, Beebe SE, Potts MJ, Kadian M, Hobbs PR, Gupta RK, Twomlow S (2007) Nutritious subsistence food systems. Adv Agron 92:1–74

Grotz N, Guerinot ML (2006) Molecular aspects of Cu, Fe and Zn homeostasis in plants. Biochemica et Biophys Acta–Mol Cell Res 1763:595–608

Haas JD, Beard JL, Murray-Kolb LE et al (2005) Iron-biofortified rice improves the iron stores of nonanemic Filipino women. J Nutr 135:2823–2830

Hao HL, Feng Y, Huang YY, Tian SK, Lu LL, Yang XE, Wei YZ (2005) Situ analysis of cellular distribution of iron and zinc in rice grain using SRXRF method. High Energy Phys Nucl Phys-Chin Ed 29:55–60

Hetzel BS, Delange F, Dunn JT, Ling J, Venkatesh M, Pandav CS (eds) (2004) Towards the global elimination of brain damage due to iodine deficiency-a global program for human development with a model applicable to a variety of health, social and environmental problems. Oxford University Press, New Delhi, pp 10–12. http://www.iccidd.org/cm_data/hetzel-a-frontpage.pdf

Hotz C, Brown KH (2004) Assessment of the risk of zinc deficiency in populations and options for its control. Food Nutr Bull 25:94–204

Hotz C, Loechl C, de Brauw A, Eozenou P, Gilligan D, Moursi M, Munhaua B, Jaarsveld PV, Carriquiry A, Meenakshi JV (2012) A large-scale intervention to introduce orange sweet potato in rural Mozambique increases vitamin A intakes among children and women. Br J Nutr 108(1):163–176

Hu YX, Qu CG, Yu JN (2003) Zn and Fe fertilizers' effects on wheat's output. Chin Germplasm 2:25–28

Huang Y, Yuan L, Yin X (2012) Biofortification to struggle against iron deficiency. In: Yuan L, Yin X (eds) Phytoremediation and biofortification-two sides of one coin. Springer, Netherlands, pp 59–74

Inal A, Gunes A, Zhang FS et al (2007) Peanut/maize intercropping induced changes in rhizosphere and nutrient concentrations in shoots. Plant Physiol Biochem 45:350–356

Kamal K, Hagagg L, Awad F (2000) Improved Fe and Zn acquisition by guava seedlings grown in calcareous soils intercropped with graminaceous species. J Plant Nutr 23:2071–2080

Khush GS, Lee S, Cho JI, Jeon JS (2012) Biofortification of crops for reducing malnutrition. Plant Biotechnol Rep 6:195–202

Li ZG, Ye ZQ, Fang YY, Yang XE (2003) Effects of Zn supply levels on growth and Zn accumulation and distribution. China Rice Sci 17:61–66

Lucca P, Hurrell R, Potrykus I (2001) Genetic engineering approaches to improve the bioavailability and level of iron in rice grains. Theor Appl Genet 102:392–397

Meenakshi JV, Johnson NL, Manyong VM et al (2010) How cost-effective is biofortification in combating micronutrient malnutrition? An ex ante assessment. World Dev 38(1):64–75. doi:http://dx.doi.org/10.1016/j.worlddev.2009.03.014

Muller O, Krawinkle M (2005) Malnutrition and health in developing countries. Can Med Assoc J 173:279–286

Nandi S, Suzuki YA, Huang J et al (2002) Expression of human lactoferrin in transgenic rice grains for the application in infant formula. Plant Sci 163:713–722

Nestel P, Bouis HE, Meenakshi JV, Pfeiffer W (2006) Biofortification of staple food crops. J Nutr 136:1064–1067

Pedrero Z, Madrid Y, Camara C (2006) Selenium species bioaccessibility in enriched radish (*Raphanus sativus*): a potential dietary source of selenium. J Agric Food Chem 54:2412–2417

Peleg Z, Saranga Y, Yazici A, Fahima T, Ozturk L, Cakmak I (2008) Grain zinc, iron and protein concentrations and zinc-efficiency in wild emmer wheat under contrasting irrigation regimes. Plant Soil 306:57–67

Pfeiffer WH, McClafferty B (2007) HarvestPlus: breeding crops for better nutrition. Crop Sci 47(Suppl 3): S88

Puig S, Andrés-Colás N, García-Molina A et al (2007) Copper and iron homeostasis in arabidopsis: responses to metal deficiencies, interactions and biotechnological applications. Plant Cell Environ 30:271–290

Raboy V (2000) Low-phytic-acid grains. Food Nutr Bull 21:423–427

Raboy V (2002) Progress in breeding low phytate crops. J Nutr 132:503S–505S

Rayman MP (2012) Selenium and human health. Lancet 379:1256–1268

Rengel Z, Batten GD, Crowley DE (1999) Agronomic approaches for improving the micronutrient density inedible portions of field crops. Field Crop Res 60:27–40

Rodriguez-Lucena P, Ropero E, Apaolaza-Hernandez L, Lucena JJ (2010) Iron supply to soybean plants through the foliar application of IDHA/Fe3+: effect of plant nutritional status and adjuvants. J Sci Food Agric 90:2633–2640

Römheld V, Marschner H (1990) Genotypical differences among graminaceous species in release of phytosiderophores and uptake of iron phytosiderophores. Plant Soil 22:147–153

Shamsuddin AM (1999) Metabolism and cellular functions of IP6: a review. Anticancer Res 19:3733–3736

Shrimpton (2002) Nutrition, the millennium development goals and poverty reduction in ECOWAS countries. Paper presented at 7th annual ECOWAS nutrition forum, Banjul, Gambia 2–6 Sept 2002. Centre for International Child Health, London, pp 10. http://www.ich.ucl.acuk/ich/htm/academicunits/cich/pdfs/RogerMGpaper.do/. Accessed 12 Jan 2015

Smkolji P, Pograje L, Hlaston-Ribic C, Stibilj V (2005) Selenium content in selected Slovenian foodstuffs and estimated daily intakes of selenium. Food Chem 90:691–697

Sperotto RA, Ricachenevsky FK, Waldow VDA, Fett JP (2012) Iron biofortification in rice: it's a long way to the top. Plant Sci 190:24–39

Stein AJ (2010) Global impact of human mineral malnutrition. Plant Soil 335:133–154

Stein AJ, Meenakshi JV, Qaim M, Nestel P, Sachdev HPS, Bhutta ZA (2008) Potential impacts of iron biofortification in India. Social Sci Med 66:1797–1808

Swietlik D, Faust M (1984) Foliar nutrition of fruit crops. Hortic Rev 6:287–356

Takahashi M, Nakanishi H, Kawasaki S et al (2001) Enhanced tolerance of rice to low iron availability in alkaline soils using barley nicotianamine aminotransferase genes. Nat Biotechnol 19:466–469

Thavarajah D, Thavarajah P (2012) Evaluation of chickpea (*Cicer arietinum* L.) micronutrient composition: biofortification opportunities to combat global micronutrient malnutrition. Food Res Int 49:99–104

Tiwari VK, Rawat N, Chhuneja P et al (2009) Mapping of quantitative trait loci for grain iron and zinc concentration in diploid A genome wheat. J Hered 100:771–776

Tucker G (2003) Nutritional enhancement of plants. Curr Opin Biotechnol 14:221–225

United Nations (2006a) The Millennium Development Goals report, Department of Economic and Social Affairs

United Nations (2006b) World population prospects: the 2006 revision, Population Division of the Department of Economic and Social Affairs of the United Nations Secretariat, New York. http://www.un.org/esa/

population/publications/wpp2006/English.pdf/. Accessed 12 Jan 2015

Vasconcelos M, Datta K, Khalekuzzaman M, Torrizo L, Krishnan S, Oliveira M, Goto F, Datta SK (2003) Enhanced iron and zinc accumulation with transgenic rice with the ferritin gene. Plant Sci 164:371–378

Waters BM, Sankaran RP (2011) Moving micronutrients from the soil to the seeds: genes and physiological processes from a biofortification perspective. Plant Sci 180:562–574

Waters BM, Uauy C, Dubcovsky J et al (2009) Wheat (*Triticum aestivum*) proteins regulate the translocation of iron, zinc, and nitrogen compounds from vegetative tissues to grain. J Exp Bot 60:4263–4274

Welch RM (1986) Effects of nutrient deficiencies on seed production and quality. Adv Plant Nutr 2:205–247

Welch RM (2002) The impact of mineral nutrients in food crops on global human health. Plant Soil 247:83–90

Welch RM, Graham R (1999) A new paradigm for world agriculture: meeting human needs-productive, sustainable, nutritious. Field Crop Res 60:1–10

Welch RM (2001) Impact of mineral nutrients in plants on human nutrition on a worldwide scale. In: Plant nutrition. Springer, Dordrecht, pp 284–285

Welch RM, Graham R (2002) Breeding crops for enhanced micronutrient content. Plant Soil 245:205–214

Welch RM, Graham R (2004a) Breeding for micronutrients in staple food crops from a human nutrition perspective. J Exp Bot 55:353–364

Welch RM, Graham RD (2004b) Breeding crops for enhanced micronutrient content. Plant Soil 245:205–214

Whanger PD (2004) Selenium and its relationships to cancer: an update. British J Nutr 91:11–18

White PJ, Broadley MR (2005) Biofortifying crops with essential mineral elements. Trends Plant Sci 10:586–593

White PJ, Broadley MR (2009) Biofortification of crops with seven mineral elements often lacking in human diets—iron, zinc, copper, calcium, magnesium, selenium and iodine. New Phytol 182:49–84

WHO (2002) World health report 2002: reducing risks, promoting healthy life. World Health Organization, Geneva

Xiao Y, Li YT, Cao YP (2000) Effects of Fe-fertilizer composition and application methods on the iron chlorosis correction of peanut. Soil Fertil 5:21–28

Yang X, Romheld V (1999) Physiological and genetic aspects of micronutrient uptake by higher plants. In: Jensen A, Gissel-Nielsen G (eds) Plant nutrition-molecular biology and genetics. Springer Netherlands, pp 151–186

Yang XE, Chen WR, Feng Y (2007) Improving human micronutrient nutrition through biofortification in the soil–plant system: China as a case study. Environ Geochem Health 29:413–428

Zapata-Caldas E, Hyman G, Pachón H, Monserrate FA, Varela LV (2009) Identifying candidate sites for crop biofortification in Latin America: case studies in Colombia, Nicaragua and Bolivia. Int J Health Geogr 8:29

Zeng H, Combs GF (2008) Selenium as an anticancer nutrient: roles in cell proliferation and tumor cell invasion. J Nutr Biochem 19:1–7

Zhang J, Wu LH, Kong XJ, Li YS, Zhao YD (2006) Effect of foliar application of iron, zinc mixed fertilizers on the content of iron, zinc, soluble sugar and vitamin C in green pea seeds. Plant Nutr Fert Sci 12:245–249

Zuo YM, Liu YX, Zhang FS, Christie P (2004) A study on the improvement in iron nutrition of peanut by intercropping maize on nitrogen fixation at early stages of growth peanut on calcareous soil. Soil Sci Plant Nutr 50(7):1071–1078